Liberty ship SS *John W. Brown* (courtesy Project Liberty Ship, Baltimore)

A Half Century of Maritime Technology 1943–1993

Written by a group of authorities

Harry Benford, editor
Professor emeritus, Department of Naval Architecture and Marine Engineering, The University of Michigan, Ann Arbor, Michigan

William A. Fox, associate editor
Naval architect, John J. McMullen Associates, Inc., Newport News, Virginia

Published by
THE SOCIETY OF NAVAL ARCHITECTS AND MARINE ENGINEERS
Jersey City, New Jersey

The essays herein reflect the experience, recollections, and opinions of the authors. Neither the Society nor any organization with which authors are, or may have been, associated is responsible for accuracy or expressed opinions.

Library of Congress Catalogue Card Number 93-085493

ISBN Number 0-939773-12-2

A Word from the President

I am sure that as you read this historical volume you will share my great sense of pride in our many accomplishments both as a society and as an industry. Our Society has a long and proud history. We have been at the center of high-quality ship design, construction and operation since our founding. History provides a record of where we have been and what we have done. Now we must focus on our future so that we can continue this tradition of significant contributions and achievements. All of us must create this future by our efforts in both thought and deed. Let's set a course that will keep our Society at the cutting edge of innovation and development in the maritime community, and that will provide an equally impressive basis for the next historical volume.

J. William Kime

Foreword

We celebrate the one hundredth anniversary of the founding of the Society of Naval Architects and Marine Engineers in this Year of Our Lord 1993. The Society of Naval Architects and Marine Engineers was founded in 1893 by a group of professionals who were concerned about the future of United States ship design and construction, and its Navy and merchant marine. The original Articles of Incorporation described the Society's mission as:

> the promotion of practical and scientific knowledge in the arts of shipbuilding and marine engineering and the allied professions, and in furtherance of this object, to hold meetings for social intercourse among its members, and the reading and discussion of professional papers, and to circulate by means of publication the knowledge thus obtained.

In general this mission is still absolutely valid and serves as a testimony to our founders' vision. We have, however, grown within this framework to include interest in (among other things) small craft, offshore engineering, and all manner of marine vehicles regardless of the waters on which they operate.

Keeping up with new developments is important, but periodically a pause to recollect and reflect on what has occurred in the past is an important dimension for us as individuals and as a society. One of the key events marking this Centennial is the publication of this collection of historical essays highlighting the progress and accomplishments achieved over the history of our Society. This is the first comprehensive effort to capture the memories and experience of our members since the *Historical Transactions* published in 1943.

The Executive Committee approved this project in February 1992, as one of the initial group of events and activities chosen to commemorate our first hundred years. A committee was appointed consisting of:

W. C. Cowles, chairman
John Arrison
Harry Benford
Keith Farrell
William A. Fox
William A. Hunley
John A. Kupersmith
Alexander C. Landsburg
Alan C. McClure
David O'Neil
Donald P. Roseman
William duBarry Thomas
Merville Willis

This committee established an initial outline, sought out volunteer authors, and enlisted the services of Harry Benford as editor and William A. Fox as his chief assistant. Everyone concerned worked as a team and so overcame countless obstacles to create this record. Their dedication and commitment is now rewarded with this publication. We have here a set of papers that will form a lasting addition to our libraries.

The Society is indeed indebted to the many authors who took the time to make a contribution to posterity. Their efforts will be appreciated for years to come by all who read this work. The Society is also thankful to Mr. Cowles, Professor Benford, and the entire committee who have worked hard and long to bring this effort to fruition.

Ronald K. Kiss
President 1991–1992

Some Words of Appreciation

Preparation of this volume has been very much a cooperative effort, starting with the initial proposal submitted by Tony Kupersmith and Roy Harrington and endorsed on the occasion of the first meeting of the Centennial Planning Committee on June 14, 1986. Tony was responsible for the preliminary outline that established the general character and content of the volume. The support of the Centennial Planning Committee, initially under the chairmanship of Barry Thomas and later of Monroe Macpherson, is particularly appreciated; even when the viability of the project was seriously questioned, the Committee's support was unfailing.

Each of the members of the subcommittee made significant contributions, first in developing a list of over two hundred potential contributors, then in contacting and verifying authors and overseeing the initial drafts. Of particular note, Dave O'Neil not only organized and secured authors for the Machinery Design chapters, but wrote the chapter on Powering. Alan McClure (now deceased) undertook a similar role for the Offshore section and also contributed a chapter. Keith Farrell organized a committee for the development of the several excellent Canadian contributions. Bill Hunley did a superb job of organizing the Naval and Coast Guard chapters and Don Roseman and Merv Willis contributed significantly to the Commercial Vessel and Construction portions respectively. John Arrison was a major contributor of supporting material for the volume as a whole. Alex Landsburg, in addition to coordinating the work of several authors, served as co-author of the MARAD chapter.

Bill Fox and Harry Benford performed extraordinary roles: Bill in arranging for transcribing all texts onto a common computer base and Harry for the superb and tactful manner in which he carried out the editing role.

Thanks of course go to all sixty-five authors and co-authors and in several cases to a whole team of contributing authors. Many made time from their busy working schedules and others, though retired, in effect went back to work to produce their texts.

Many of the authors, in their enthusiasm for their topic, wrote far more than could be accommodated in a reasonable size volume. Our apologies to them, all of whom were gracious and understanding in accepting the necessary editorial reduction, and thanks to Professor Benford, who carried it out so skillfully. Apologies also to several authors who were unable to complete their texts within our schedule constraints and whose work is therefore not included. We hope these omitted texts as well as the severely edited texts will find publication in an alternative format. Apologies also to those who would have contributed, but whose names, for whatever reason, did not surface in the development of our initial list of potential contributors as well as those who volunteered to write on topics that we ultimately decided were insufficiently pertinent to the general theme of the book.

Special thanks are due Capt. Richards T. Miller, USN (Ret.), who graciously donated his time and experience in obtaining appropriate photographs from the collection of the U.S. Naval Institute, and Merv Willis for obtaining photos from the Philadelphia Maritime Museum. Thanks again to Bill Fox for the donation of photographs from his files, and for his careful proofreading.

Apologies also to those readers who find their favorite subject either not included or not dealt with in a comprehensive manner. Again, it simply was not possible to cover all significant developments, or any topics, in depth. An effort has been made to provide extensive reference sources for those who wish to pursue any particular topic in greater depth.

As authors' original drafts were received, copies were sent to selected individuals for review and comment. Reviewer comments were forwarded to the author for inclusion as he saw fit. Several authors arranged for reviewer comment on their own initiative and have indicated contributing individuals adjacent to their text. In some cases, reviewers offered personal reminiscences of experiences of an anecdotal nature. These have been included in a sidebar format where judged appropriate. The following are noted as reviewers. Again, our apologies to any we might have missed.

Robert G. Allan
David D. Beach
Daniel W. Billingsley
Denny Boyd
Terry E. Brockett
Thomas M. Buermann
Charles B. Cherrix
Richard B. Couch
George H. Curtis III
Lee W. Dana
George A. Dankers, Jr.
Arthur De Fever
Larrie D. Ferreiro
Hugh C. Downer
George R. Duclos
William H. Garzke, Jr.
Jerome L. Goldman
William O. Gray
David A. Groh
Halsey C. Herreshoff
John L. Horton
M. L. Ingwersen
Harry A. Jackson
Robert S. Johnson
Klemme M. Jones
Ronald K. Kiss
Jim Know
Duane H. Laible
Richard Lowery
William Robert McNeilly
Richards T. Miller
Samuel G. Morrison
Albert G. Mumma
Frederick C. Munchmeyer
Hugh F. T. Munroe
John J. Nachtsheim
Don Nicholson
Owen H. Oakley
George H. Palmer, Jr.
E. L. Peterson
Stanley Potter
Robert I. Price
Walter H. Radut
Anil Raj
Harry B. Reed
G. Reynolds
William W. Rogalski
Albert K. Romberg
William D. Roseborough, Jr.
Henry P. Rumble
Kit Ryan
Philip G. Rynn
Mike Saker
Daniel Savitsky
H. S. Shull III
Bruce E. Skinner
Nathan Sonenshein
Irving L. Stern
Richard H. Suehrstedt
Barry F. Tibbitts
Robert J. vom Saal
R. W. Whitwell
Trevor White
Edwin G. Wiggins
Robert E. Williams

Finally, a note of appreciation to Mrs. Connie Jarvis and Miss Barbara Evans of John J. McMullen Associates, Inc., for their processing/transcription work, and to Terry Flanagan, Stephen Brill, and their associates, who assembled the multitude of texts into a handsome volume and carried out a final editorial review.

W. C. Cowles
11 May 1993

Introduction

W. C. Cowles, chairman, Historical Volume Committee

In 1943 the Society celebrated its golden anniversary in the midst of a wartime shipbuilding program unprecedented in history. Nevertheless, in spite of wartime demands on its members, a volume was prepared documenting the developments in maritime technology during the Society's first fifty years. The succeeding half century has seen developments in many areas that could not have been foreseen, some resulting in tremendous advances in the economics of marine transportation, others in the demise of whole segments of the industry and still others in totally new areas.

The Society has deemed it important to take the occasion of its centennial to document in some manner the nature and the cause and effects of many of these developments.

The Society

It is appropriate that we look first at the Society, because it has been the Society through its members, individually and collectively, that has been instrumental in these developments.

Membership. In the wartime period, the Society, along with the industry, experienced a tremendous growth. Indicative of the rate of growth, as of October 1942 the membership stood at 2751. By October of 1943 it had increased to 3204. For the next forty years, membership continued a gradual increase, reflecting both a moderately healthy industry and an increasingly complex technology. Membership peaked at 13,014 in 1983. The succeeding years have seen a gradual decrease to 10,153 in 1993. The decrease reflects the dwindling of the U.S. merchant marine and the decline of merchant shipbuilding to virtually nil in the U.S. and Canada as well as a reduction in naval construction.

Sections. Although membership spread across the continent in 1943, activities centered almost entirely on the Annual Meeting in New York. To provide more opportunity for member participation, the first local Section, Philadelphia, was established in 1941. The New York Metropolitan Section was organized in 1942 and over 500 members attended its first meeting on May 6 of that year. New England and Northern California Sections were organized in 1944 with others following throughout the years to a total of seventeen.

Publications. In addition to publication of the Annual Meeting *Transactions*, augmented later by the Spring Meeting *Transactions*, and Symposia *Proceedings*, the Society began publication of its respected volumes on design and construction with the first edition of *Principles of Naval Architecture* in 1939. *The Shipbuilding Business in the United States* appeared in 1948. The first two-volume edition of *Marine Engineering* came out in 1955 as did *Design and Construction of Steel Merchant Ships*. The first two volumes of Captain Saunders' magnificent text on hydrodynamics appeared in 1957. Updated versions of *Principles of Naval Architecture* and *Marine Engineering* have appeared as warranted by developments in technology. Publishing of books of a somewhat more general nature included such volumes as Professor Harry Benford's *Naval Architecture for Non-Naval Architects* and *Submersible Vehicle Systems Design* by Professor Eugene Allmendinger.

The first issue of the *Members' Bulletin* appeared in 1945 as a four-page newsletter issued three times a year. Over the years this evolved into *Marine Technology*, a well-designed quarterly magazine complete with selected Section papers, news of the Society and Sections and, recently, a generous sprinkling of paid advertisements. *Marine Technology* is now supplemented by the more theoretical *Journal of Ship Research* and the *Journal of Ship Production*. Both appear quarterly.

Finally the Technical and Research Committees continually produce a myriad of Codes, Guides, Bulletins, and Reports, all aimed at improving the tools and methods for designing, building and operating marine vehicles of all types as well as offshore structures and related installations.

Technical and Research Committees. In 1947 the research activities of the Society were placed in the hands of a Technical and Research Committee. Three subcommittees on Hydrodynamics, Ship Structures and Ships' Machinery were established. By 1992 this program had grown to 8 committees and 67 panels operating under the direction of the committees. Approximately 1300 individuals from all branches of the marine and associated industries, including government and regulatory bodies, were involved. Well over a hundred publications resulting from the work of the T&R groups were offered by the Society in 1992.

Education. For several years prior to 1943, the Society had granted two scholarships annually for graduate studies at Massachusetts Institute of Technology. These were discontinued in 1943 owing to war conditions, but resumed in 1946–47 with grants to one student undertaking graduate studies at MIT and one at the University of Michigan. This educational program has continued to grow. In 1991, for example, five graduate scholarships were awarded as well as undergraduate scholarships to students attending MIT, University of Michigan, University of California Berkeley, State of New York Maritime College, Florida Atlantic University, University of New Orleans, Texas A & M (College Station) and Virginia Polytechnic Institute. In addition, annual grants are regularly made to Webb Institute. Through 1991, the total of graduate scholarships was 212. Undergraduate scholarships totaled 512.

In addition to its scholarship program, the Society played a major role in continuing education with its symposiums, each devoted to a specific area of developing technology.

The War Years

The years 1942–1945 and the few years preceding saw the beginning and culmination of the greatest shipbuilding, marine transportation and naval operations program in history.

The following summary appeared in the January 1946 issue of *Marine Engineering and Shipping Review*:

> During the four war years (1942–1945) American shipyards produced the huge total of 4889 steel merchant ships of over 2000 gross tons. These vessels have a deadweight capacity of 52,875,603 tons and measure 36,896,538 gross tons. In addition, American shipyards were called upon to construct a tremendous fleet of combatant and auxiliary vessels for the Navy. Some of these were specially developed designs that had never been built before. A total of 110,965 Navy vessels were turned out during the war years, including 1323 combatant and 109,642 auxiliary types representing 8,300,000 tons. Several thousand small vessels were also built for the Army, the Coast Guard and the Defense Plant Corporation for private interests. This new merchant fleet is twice as large in number and almost three times greater in deadweight tonnage than the largest prewar national fleet, that of the British Empire, which before the war totaled 2553 units of 22,133,400 deadweight tons, including only vessels of 2000 gross tons each.

Although the above quote refers to "shipyards," it should be noted that many of the shipyards were newly built especially for the construction of wartime tonnage. In addition, smaller yards and yacht and boat building firms adapted their facilities for the production of wartime ships. Also, the immense contributions of the nation's manufacturing infrastructure were vital to the shipbuilding and boat building effort. For example, twenty different firms, including such non-marine organizations as Iron Fireman Manufacturing Company and Worthington Pump and Engine Company, built Liberty ship engines [1].

One example of these new yards was a shipyard located on the banks of the Illinois River just above Seneca, Illinois. Here in what had formerly been a farmer's field was a huge assembly line engaged in building LSTs. The building ways were continuous, ending with a launching ramp at the water's edge. Construction began well inland and, as it progressed, the hull would periodically be pushed toward the river and a new keel laid [2].

Established firms undertook construction of vessels far divergent from their peacetime efforts. For example, Manitowoc Shipbuilding Company near the shores of Lake Michigan delivered twenty-eight submarines to the Navy. Delivery was accomplished by loading the boats on specially constructed drydocks for movement down the Illinois–Mississippi waterway to New Orleans. Michigan-based Defoe Shipbuilding Company, a builder of luxury yachts, Coast Guard cutters and similar vessels, produced, among other ships for the Navy, seventeen destroyer escorts. In this Defoe employed a unique roll-over assembly method whereby initial assembly was accomplished with the hull inverted to facilitate downhand welding; then by means of giant wheel-like assemblies affixed to it, the hull was rolled over for completion.

Absolutely critical to the success of the wartime effort was the introduction of welding as the prime technique for fabrication. Prior to the wartime effort, only a few vessels had been of welded construction. Although the technique was new with a minimum of historical precedent, the results were overwhelmingly satisfactory. There were of course several spectacular failures of welded vessels, but this percentage was small.

Supplementing new construction, a wide variety of existing vessels were converted or adapted for wartime service. One of these was the remarkable conversion of two Great Lakes sidewheel passenger vessels to aircraft carriers. These two ships, the *Wolverine*, ex-*Seeandbee*, and *Sable*, ex-*Greater Buffalo*, operating on Lake Michigan served as the platforms on which over 12,000 naval aviators (among them Ensign George Bush) qualified [3].

Maritime Commission

Any consideration of maritime progress during this period almost has to begin with a discussion of the U.S. Maritime Commission and its successor organizations. Established by the Merchant Marine Act of 1936, the Commission had by 1942 already developed designs for several classes of standard cargo ships. It had also contracted for the construction of the *America.* Moreover, in response to wartime demands, it had underwritten the expansion of the U.S. shipbuilding industry to the extent that 1942 saw the production of 750 oceangoing ships totaling 8,000,000 deadweight tons, an increase of 635 percent over the previous year. These included almost 600 Liberty ships.

The years following the war saw the role of the Maritime Commission changing from the production of ships to the disposal of ships and the encouragement of a viable peacetime U.S. merchant marine. In 1950, the Maritime Commission was replaced by the Federal Maritime Commission (FMC) responsible for regulatory and judicial functions, and the Maritime Administration (MARAD), responsible for administrative activities.

The postwar years saw the construction of the highly successful Mariner Class ships and the building of the SS *United States* as well as a number of experimental vessels, including the nuclear powered *Savannah.* The Merchant Marine Act of 1970 resulted in increased R&D programs aimed at improving the competitive position of the U.S. merchant fleet. In spite of some impressive research developments, the commercial fleets remained uncompetitive on a worldwide basis, and with decreasing construction and operating subsidies, continued a decade's long decline until by 1992 construction of oceangoing merchant vessels in the U.S. had virtually ceased.

Rules and Regulations

In 1943, the basic design manual was the American Bureau of Shipping's *Rules for Building and Classing Steel Vessels.* Based largely on experience supplemented with basic engineering concepts, the "Rules" established the basic scantlings and proportions for the design of oceangoing merchant ships. These ABS Rules were augmented by Coast Guard regulations relating to general seaworthiness and safety and compliance with international regulations. The U.S. Public Health Service supervised aspects relating to food preparation and service and potable water requirements.

The years since the war have seen structural design standards become orders of magnitude more sophisticated as new design tools evolved, particularly with the advent of computer-aided design. Regulations relating to seaworthiness have become more stringent, driven to some extent by the International Maritime Organization (IMO), and perhaps even more by perceived demands of the public for ever greater safety. Environmental concerns have resulted in a plethora of regulations covering construction and operation.

Not only have rules and regulations become more detailed and comprehensive, but the scope has broadened to include a myriad of vessels, trades, and structures that were not covered or didn't exist in 1943. These include vessels constructed for the transport of liquefied gases, specialty petroleum and chemical products, sail training vessels, and sailing cruise vessels, as well as oil and gas drilling, production and processing vessels.

Education and Research

In 1943 there were three institutions of higher learning that offered degrees in naval architecture and marine engineering, namely Massachusetts Institute of Technology, the University of Michigan, and Webb Institute of Naval Architecture. In addition to these institutions, a few colleges and universities offered selected courses in the field. The Westlawn School of Yacht Design offered a program in this specific field and several states had academies for the training of ships' officers.

The university programs covered the basics in design of hulls and machinery. In addition to basic engineering courses in strength of materials, geometry, calculus, etc., the specialty courses placed much emphasis on the development of drafting skills, including the fairing of ships' lines. While the slide rule was the symbol of every budding engineer, exercises in hydrostatic, strength and launching calculations were carried out on hand-cranked mechanical calculators. Programs at Kings Point and the state maritime academies were concerned with the elements of navigation, ship handling, and the operation of marine machinery. The past fifty years have seen the slide rule and mechanical calculator, as well as the drafting table, replaced by computers capable of swiftly handling the once tedious calculations and drawings as well as the much more complex finite element analyses and similar calculations. Students at the maritime academies now graduate with a comprehensive knowledge of hydrostatics and hull strength. They also acquire experience in handling diesel and gas turbine machinery as well as conventional steam plants. In addition—again with the assistance of computer-based simulators—they acquire ship handling and marine engineering skills that could have been obtained in earlier times only by years of experience, in some cases at significant risk.

Merchant Vessels

Passenger Ships. The period saw a brief rejuvenation of passenger ships, based initially on the conversion of war-built tonnage to liner services. This was followed by the construction of numerous moderate-size vessels for the Western Hemisphere, African and Mediterranean trades. The construction of oceangoing passenger ships ended with the luxurious *Constitution* and *Independence* to serve the Mediterranean, and finally climaxed with the technically superb *United States* for transatlantic service.

The pre-war coastwise passenger trade never revived. Fire safety regulations adapted after the tragic *Morro Castle* fire tended to eliminate those vessels that survived the war, and the developing interstate highway system as well as air travel essentially eliminated any likelihood of profitability for expensive new construction. While the passenger liner trades virtually disappeared after the mid-1950s, the developing vacation cruise industry initiated the conversion and subsequently the construction of a number of vessels, not, however, in American yards or of American registry.

The later years have seen an increasing number of coastwise, yacht style cruising vessels, a proliferation of dinner boats, and most recently casino gambling craft. There has also been some resumption of interest in ferry services, most tending toward high-speed passenger-only vessels. Although these vessels are considered small, their passenger capacity, ranging from a few dozen to 150 or more passengers for the coastwise and river cruise ships to several hundred for some of the larger dinner and casino vessels, are actually comparable to a vast number of vessels that made up the prewar coastwise, lake and river fleets.

Cargo Vessels. The greatest revolution in maritime technology almost certainly has to be the replacement of break-bulk handling of general cargoes with containerization. Compare the unloading and loading of a conventional ship of eight or ten thousand tons that once required weeks with the eight or ten hours required to turn around a containership of several times greater capacity. The resulting illusion that waterborne trade has declined because we no longer see our waterfronts lined with ships working cargo has to be confronted with the fact that remote container terminals, with their acres of container marshalling facilities, are handling far more tonnage than the more numerous ships of the beginning of our era.

A similar revolution has occurred in the handling of bulk cargoes, where again the traditional three-island general cargo ship loaded with coal, grain or some other bulk cargo has been replaced with the specialized ship designed for efficient loading and discharge. On the Great Lakes, with historically the world's most efficient bulk handling technology, the replacement of run-of-the-mine ore with processed taconite has permitted the use of self-unloading vessels for the handling of this commodity. At the same time, the number of vessels required to handle the ore trade has been reduced from nearly four hundred to about sixty, due to the enlargement of the locks and deepening of the channels and ports.

Tankers. For a few years following the war, U.S. yards supplied tankers, indeed "super" tankers of ever-increasing size, to accommodate the world's rapidly increasing demand for petroleum products. This period was of course short-lived as the shipyards of Europe, and particularly of Japan, were restored and American technology was transferred to the lower-labor-cost overseas yards.

American designers and the U.S. based international oil companies continued to participate in the development of technology, particularly in the exponential increase in tanker size, to provide more efficient transportation. Americans also participated actively in the development of specialty tankers for the delivery of refined products, liquefied gases, molten sulphur, asphalt and other specialties.

Rivers. It is somewhat paradoxical that developments in transportation on the western rivers have been more evolutionary than revolutionary, in part because river operators have always tended to be more pragmatic and less tradition influenced than their deep-water cousins. Consequently, by the early forties, the concepts of "push towing," screw propulsion with tunnel sterns and propeller nozzles, steering and flanking rudders, and diesel power were already well established. The river yards also had already attained extremely efficient fabrication procedures. Accordingly, while the past fifty years have seen increases in the size and power of the towboats, and refinements in hull form of both towboats and barges, basic concepts have remained essentially unchanged.

Naval Vessels

The war saw revolutionary developments in naval tactics, first on the shores of Africa, then in the Mediterranean and on the beaches of Normandy, and finally across the Pacific: whole new concepts of amphibious operations. Then the aircraft carrier centered task force replaced the ancient line-of-battle concept.

Developments in amphibious warfare produced a plethora of new vessel types, beginning with the unlikely concept of the LST, a major vessel intended to deliberately beach itself, unload its cargo and retract. The LSTs were augmented and supported by a whole alphabet of smaller landing craft—LCTs, LCMs, LCVPs, LVTs, etc.—as well as specially designed combat transports and cargo ships such as the APAs and AKAs.

As it became apparent that the aircraft carrier was the new capital ship, the nation's shipyards turned out a fleet of the larger, faster *Essex*-class carriers. The battleship's role changed to carrier defense and shore bombardment. Cruisers and destroyers had their armament altered to serve as screening and air defense vessels for the task force.

An entirely new fleet of convoy escort vessels was constructed: destroyer escorts, some steam powered and others with diesel machinery, and frigates powered with four-cylinder compound engines adapted from a British design.

The shape of ships following the war was influenced by many factors. These included developments in electronics and communications, the phased replacement of guns by missiles, and the pressure for improved habitability in accommodations. In propulsion systems we saw the adaptation of nuclear power for selected carriers and cruisers, and gas turbine or combined gas turbine and diesel for the balance of combatants.

Nuclear propulsion also brought about the realization of the first true submarines: ships that submerged upon departure and didn't surface again until their mission was complete.

Yachts and Pleasure Craft

The war saw the virtual end of "yachts" of the era of "If you have to ask, you can't afford one." Almost all of these vessels, already endangered by the income tax and the Depression, were either commandeered, or sold to or volunteered to the Navy to serve as coastal patrol vessels. Numerous less-prestigious vessels, sail as well as power, also saw coastal service. All were modified and armed for naval services. While a few of the smaller ones did return to postwar civilian use, many were scrapped, or in the case of several of the larger vessels, sold for commercial service as freight or cruise vessels, primarily in the Caribbean area.

Many of the boat and yacht builders returned to the construction of pleasure vessels after the war, but most of these enterprises were short-lived, succumbing largely to changes in technology and materials. The development of aluminum and, much more extensively, fiberglass reinforced plastic for hull construction allowed countless "backyard" firms to undertake the multiple production of hulls favored by the public for durability and minimum maintenance. Although many of the earlier enterprises failed for lack of appreciation of the more technical aspects of the process as well as lack of business acumen, the ultimate result is visible to anyone driving along a lakeshore or harborside where such vessels are moored by the hundreds.

Particularly interesting in this field, if America's Cup classes can be considered either yachts or pleasure craft, in a historical context, are the tremendous resources devoted to, and advances made in, the design of these sophisticated craft. Design development costs now run into the millions of dollars for each boat, with no prospects of a financial return. The ultimate results, however, may well be the application of some of the derived concepts and methods to the design of commercial vessels.

Service Vessels

This is an almost limitless category, so we shall limit our discussion to developments in a few of the more common or more interesting types:

Tugs. Tugs have not changed greatly; indeed as their role in the scheme of things remains essentially unchanged, developments in tug design have been more evolutionary than dramatic. Diesel power has totally replaced steam and steel hulls are accepted, even by the die-hards who long felt that only wood had the resilience to withstand the rigors of towing service. A few tugs have been equipped with Kort type propeller nozzles—some with steering nozzles—thus eliminating rudders, and a few with tractor type vertical axis propellers. Tugs specially designed with elevated or telescoping pilothouses have been developed for use with the increasingly common integrated tug–barge arrangement.

Fireboats. Like tugs, fireboats are now virtually all diesel powered. The big fireboats of the pre-war era have nearly all been replaced with much smaller vessels, more maneuverable but with comparable pumping capacity. To an extent, disappearance of the multiple pile-supported finger piers that once lined our harbors has reduced the need for fireboats. The quays with adjacent open marshalling areas used by container vessels represent a much reduced fire hazard.

Research Vessels. Oceanographic and survey ships played a vital role during the war, proceeding in advance of the fleet to provide hydrographic information necessary for operations throughout the island area of the Pacific, for which reliable charts were not available.

A wide variety of research vessels has evolved since the war. Prominent of course are the hydrographic and survey ships operated by NOAA. The primary purpose of these is the gathering of information for the preparation of navigation charts. Somewhat comparable are the oceanographic research vessels deployed for the study of deeper waters. Fisheries research vessels are employed in the study of these vital food resources. The petroleum industry uses seismic research vessels in the search for undersea petroleum resources. The Navy also operates a significant fleet, some of the vessels especially equipped for the detection of submarines and others for obtaining oceanographic and hydrographic data.

All of these vessels have a good deal in common. Relatively small, they must accommodate scientific personnel as well as their operating crew. Laboratories must be provided for the analysis of data. The vessels are required to be highly maneuverable as well as capable of maintaining a steady course at slow speeds. They must have navigation facilities capable of determining their position within a few meters at all times. Finally, they must be extremely quiet in operation and comfortable in a range of sea states.

Offshore

Scarcely even a concept before the war, offshore technology has become a major part of the maritime field.

Exploration and Production. The drilling of both exploratory and production wells from barges or bottom-supported platforms evolved in the Gulf of Mexico area shortly after the war. These operations soon moved into deep waters, and drillships and semisubmersibles were developed to support the complete "oil patch." In addition to the familiar derrick with its platforms and winches, these offshore rigs provide quarters complete with commissary, recreation and sleeping areas, mud tanks and mud pumps, shale shakers, and laboratory facilities, all powered with the vessel's own diesel power plant. A helicopter platform permits embarkation of personnel and supplies by air. Cranes are provided not only for handling drill pipe and riser sections, but for the transfer of supplies and personnel from supply boats.

Supply and Anchor Handling. Supply of offshore rigs led to the development of a unique class of ship designed especially for support of offshore operations. With quarters and pilothouse well forward, the vessels have a long clear afterdeck suitable for carrying drill pipe and other supplies and equipments. Below decks are special tanks for the transport of cement, barite and bentonite, the basis for the "mud" essential to drilling operations. In many cases, these vessels are also provided with special winches for handling the drill vessel anchors and also to provide towing service to augment the drill vessel's own propulsion motors.

With the decline of offshore operations in the late 1970s and early 1980s, many of these vessels became surplus and found new life converted to sight-seeing cruisers, whale watching or dinner boats, or research vessels. In short, their hull form and general arrangement proved to be remarkably versatile.

Crew Boats. Supplementing the supply boats are the smaller, higher-speed crew boats. Drill vessel personnel are typically rotated on a two- or three-week schedule. Although many of the personnel transfers are handled by helicopter, where operations are not too far offshore the crew boat is a more cost-effective means of transporting personnel.

Navigation and Communication

Developments in navigation and communication have been as far reaching as in any other area of marine activity. These are the results of developments in electronics and the deployment of satellites. The navigator of the early forties relied on the gyrocompass and celestial observations for navigation and on the radio direction finder and lighthouses and lightships for his piloting guides. The war period witnessed the first applications of loran for navigation and radar for piloting as well as for detection of ships and aircraft. Radio communication was largely by code, visual communication by signal flag or signal light. During the war, voice communication was by radio; TBS, sometimes referred to as "Talk Between Ships," though already in use on the Great Lakes, saw initial deployment on ships of the Navy.

In the years following the war, U.S.-established and manned loran stations provided navigational information to ships in the Western Hemisphere while the British-administered sister system, Decca, was established to cover much of the Eastern Hemisphere. The deployment of satellites provided for the ultimate development of GPS, Global Position System, which allows officers to determine a vessel's position within a few meters anywhere in the world.

Our nuclear submarines are navigated submerged for weeks without surfacing, depending on their inertial navigation systems to determine their location.

Satellites together with radio transmission, either by voice or data transfer, allow ship managers at their home office desk to communicate with their vessels virtually instantly anywhere in the world.

Shipbuilding

Developments in the shipbuilding industry have been wide ranging, driven first by the adoption of welding. The desirability of positioning work to permit downhand welding led to adoption of subassemblies that could be positioned for the convenience of welders. The benefits of subassembly fabrication extended well beyond more efficient welding. It enabled work to be spread out, improved dimensional accuracy and significantly reduced residual stresses in the ship. However, it required increased crane capacity for handling the assemblies. The benefits of subassembly were so apparent that units eventually grew to several hundred tons with a commensurate requirement for equivalent cranes and transfer equipment. One of the outfalls of the subassembly procedure was the development of pre-outfitting, wherein hull units were completely outfitted with ladders, piping, cables, etc. before final placement on the ways. Machinery space units could even have much of the machinery pre-installed.

Efforts to reduce the effects of corrosion resulted in the development of high tech coatings that required application under controlled conditions. The preparation of assemblies and application of coatings became a major area of development.

The advent of numerical control of welding and burning operations, and finally the development of computer-aided manufacturing, CAM, further revolutionized the shipbuilding business.

All of these advances required increasingly sophisticated management and control procedures. In effect, shipbuilding tended to become less labor intensive and more management intensive.

Where We're Going

Because many of the developments of the past fifty years were not, and indeed could not have been foreseen, it is unlikely that we can predict future developments with any precision. What we can learn from history is that the future is not particularly predictable. Nevertheless, if we extrapolate current trends, we might gain at least a bit of insight about the near future.

Current negotiations aimed at opening up world trade will result, it is hoped, in a significant expansion of waterborne commerce. This may well be abetted by developments in third world countries as well as the commercial rebuilding of the ex-communist bloc nations.

As the economies of the other developed nations come into closer balance with those of North America, we might expect our own shipping and shipbuilding industries to become more competitive. It would seem that a united will to truly apply America's business and technical talents to our maritime industries could see them once again achieve their rightful place in world affairs, particularly if international trade agreements ultimately result in the elimination of subsidies to favored industries.

It is already apparent that much of the technology will be driven by ecological factors, the growing realization that it is imperative that we preserve our earth's resources and that the two-thirds of the area made up by the oceans are vital to this effort. The same driving force must soon see a surge in shipbuilding to replace older vessels that are not capable of meeting ecological requirements.

As our Navy adjusts from the blue water requirements of the cold war era to the requirements for inshore and anticipated interdictory actions, the mix of vessels can be anticipated to change. This in turn may call for some modification of shipyard facilities to suit. It is conceivable that our naval architects will be asked to come up with entirely new concepts.

Increasingly sophisticated developments in design methods and tools will place greater demands on our educational institutions, not only at the undergraduate and graduate levels but in the area of continuing education. Certainly our own Society, along with our sister societies, will have a major role to play in this area. If indeed there is a resurgence in our shipping industries, the maritime academies should see a healthy demand for graduates trained and perhaps constantly retrained in all aspects of ship handling and management. The anticipated reduction in naval forces in combination with the probability of our nation becoming more frequently involved in a worldwide peacekeeping role may well result in a large cadre of trained maritime and Navy reservists available on call to serve on the ships required to supply military actions anywhere in the world.

As our nation's industry becomes more productive and efficient, in effect, as more can be produced with fewer man-hours, it would seem inevitable that individuals will have more leisure time available. This should bode well for both the commercial cruise industry and the pleasure boat industry. We already see more divergence in the cruise area. On the one hand, huge floating hotels—virtually floating cities—appeal to those seeking a city away from the land with an occasional call at a duty-free port. On the other hand, we see growing interest in exploration type cruises and coastwise or river cruises on yacht-like vessels with a more limited passenger capacity and informal atmosphere. Sail cruise vessels now range from restored windjammers to highly sophisticated sailing machines.

This is just a sampling of what the future might hold. As you peruse the following essays, either in their entirety or simply seeking out those parts that pertain to your particular interests, it may well become apparent that there is indeed a future, and the possibility of a very exciting future, for our industry.

References

1 Sawyer, L. A. and Mitchell, W. H., *The Liberty Ships*, 2nd ed., Lloyd's of London Press, Ltd., 1985.
2 Nelson, W. T., *Fresh Water Submarines, The Manitowoc Story*, Hoeffner Printing, Manitowoc, WI, 1986.
3 Miller, D. E., "Aircraft Carriers on Lake Michigan," *Naval History*, Vol. 2/1/2, U.S. Naval Institute, Winter 1988.

WALTER C. "BUD" COWLES graduated from the University of Michigan in 1942 with a degree in naval architecture. After graduation, he rejoined the staff of the American Ship Building Company where he had been employed as a draftsman the summer before. Except for an eighteen-month spell in the Navy, from which he was discharged with the rating of ETM3, he remained with American Ship Building Company until 1963. He became chief hull draftsman in 1951 and in 1957 was appointed acting naval architect. He left ASB in 1963 and, after interim employment in the offices of Marine Consultants & Designers, joined the Tanker Department of Esso International (now Exxon). While with Esso/Exxon, in addition to routine plan review for tankers building in the U.S. and Canada as well as in Germany, Sweden and the Netherlands, he spent time in Italy in connection with the construction of Esso's LNG ships, and later was a member of an LNG task force formed to monitor worldwide developments in LNG technology. He spent several months on a loan assignment to Exxon Production resources in connection with the construction in Japan of a semisubmersible drilling vessel intended for exploratory drilling in the Gulf of Alaska. Additional assignments included extensive participation in the modification of ships for offshore lightering and in the implementation of Exxon's crude oil washing program. In connection with the latter, he received the Vice Admiral E. L. Cochran award for his paper presented before the New York Metropolitan Section. He also participated in the development of designs for a proposed seismic research vessel. He retired from Exxon in 1984. Bud is a life member of SNAME and the U.S. Naval Institute and a member of the American Society of Naval Engineers. Also, he is a past fellow of the Royal Institution of Naval Architects and past member of North East Coast Institution of Engineers and Shipbuilders. A maritime history buff, Bud is a member of a number of marine historical societies. He has had a lifetime interest in small craft and is the owner of a small fleet of oar, paddle, sail and power boats.

Contents

Part iv Design

Part v Construction

Part vi Ships of Commerce

Part x Fishing Vessels

Part xi The Offshore Industry

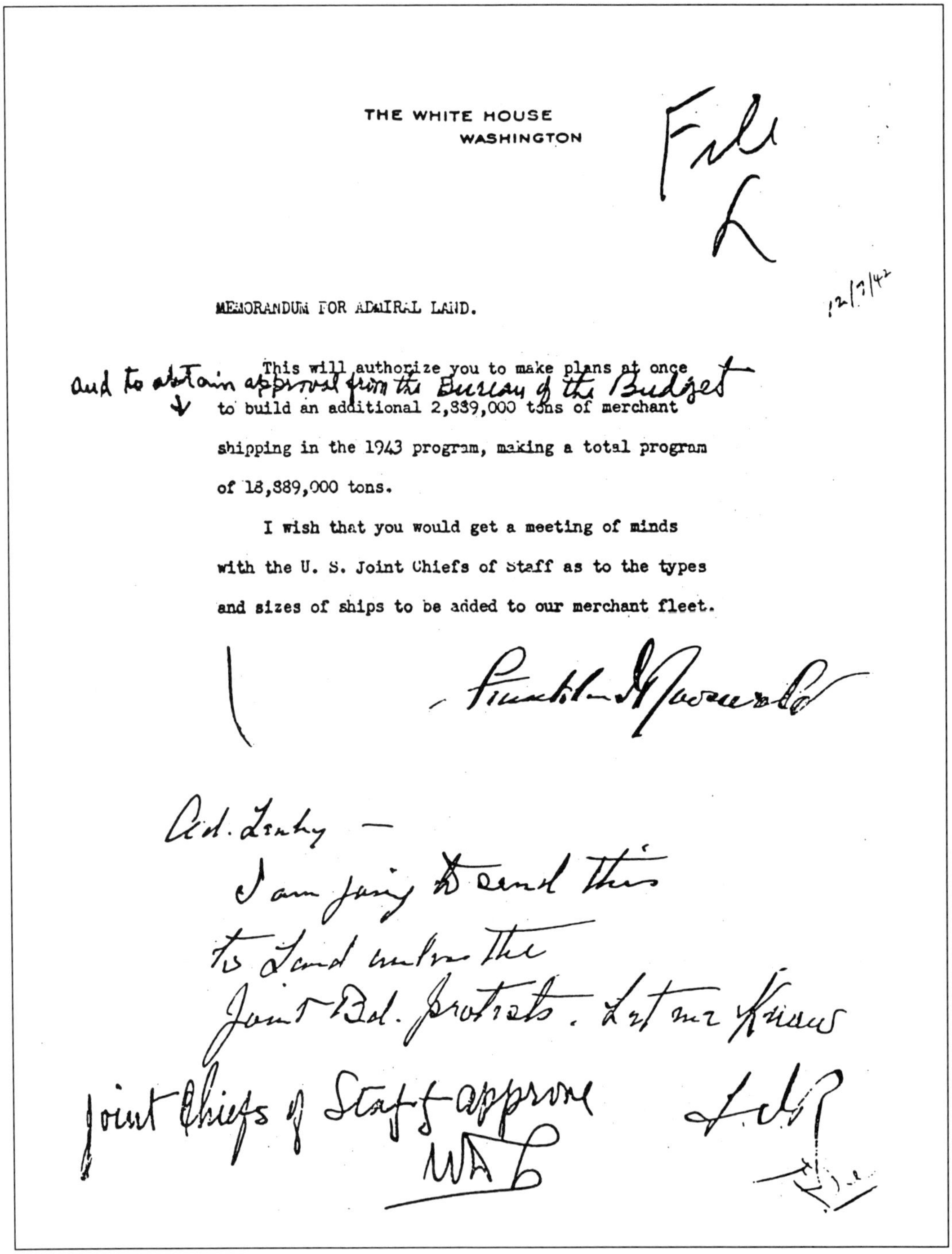

THE WHITE HOUSE
WASHINGTON

File
L

12/7/42

MEMORANDUM FOR ADMIRAL LAND.

This will authorize you to make plans at once and to obtain approval from the Bureau of the Budget to build an additional 2,889,000 tons of merchant shipping in the 1943 program, making a total program of 18,889,000 tons.

I wish that you would get a meeting of minds with the U. S. Joint Chiefs of Staff as to the types and sizes of ships to be added to our merchant fleet.

Franklin D Roosevelt

Ad. Leahy —
I am going to send this to Land unless the Joint Bd. protests. Let me know
F.D.R.

Joint Chiefs of Staff approve
WAL

Order of the century: Dramatic memorandum from President Roosevelt to Admiral Emory Land, December 7, 1942.

Part I

Incentives and Direction

The Maritime Administration and Antecedents

Charles B. Cherrix and Alexander C. Landsburg

From colonial times up until the Civil War, the nation's investment capital was largely focused on maritime enterprises. When the Civil War ended in 1865, the merchant shipping resources of both the North and the South had contracted substantially because of war losses and ships "fleeing the flag." Peace saw investment capital placed in more promising opportunities spurred by the nation's westward expansion.

In 1870, representatives of American maritime interests and some farsighted members of Congress began a long sporadic dialogue on what steps should be taken to revitalize the U.S. merchant marine. But 46 years would elapse before a promotional maritime program was generated by a dire shipping shortage that developed in August 1914, when World War I erupted.

Shipping Act of 1916

Just before the start of World War I, 90 percent of our nation's foreign trade was being carried in foreign-flag ships. Our foreign trade was thus severely disrupted when many of the foreign lines were pulled out to serve their own nations' wartime needs. Our exports quickly piled up in U.S. ports. Those cargoes that did move went at exorbitant freight rates, from $1 per ton in early 1914 to over $20 per ton in the summer of 1917. The Shipping Act of 1916 established a shipping board authorized to buy foreign ships and to start an unprecedented shipbuilding program. Little in the way of merchant ship construction was accomplished, however, until after the declaration of war in April 1917. An Emergency Fleet Corporation was established under the Shipping Board to expedite the construction.

One major war effort involved the completion of commercially contracted vessels requisitioned by the government. New designs were developed for steel construction. Additional designs using wood, concrete, and composite materials further bolstered production capacity. Although development of ships for mobilization did not concentrate on a few designs, an attempt was made to standardize their components. The original construction program shown in Table 1 was ambitious. This table also shows the construction program active in 1921 after the war emergency had passed and many unfinished contracts had been suspended. Table 2 illustrates the ships and their construction materials actively underway in 1921.

Hog Island Construction

Because existing yards were fully booked with naval and commercial vessel orders, the Emergency Fleet Corporation built three new yards to meet the ambitious construction goals. These yards used radically new methods of construction. Bridge and structural steel plants throughout the country produced standard assemblies, which were then assembled at the

Table 1 World War I construction program [1]

	Original Program		Active in 1921	
Class	Number	Deadweight Tons	Number	Deadweight Tons
Requisitioned steel	419	2,963,406	384	2,687,266
Contract steel	1,741	11,914,670	1,309	8,927,695
Contract wood	1,017	3,052,200	589	1,885,250
Contract composite	50	175,000	18	63,000
Contract concrete	43	302,000	12	73,500
Total	3,270	18,407,276	2,312	13,636,711

Table 2 World War I program by type construction active in 1921 [1]

	Steel	Wood	Composite	Concrete	No.	Total Deadweight, tons
Cargo	1,429	304	18	4	1,755	10,777,434
Tanker	138	1	—	8	147	1,427,730
Refrigerator	19	—	—	—	19	161,400
Transport	22	—	—	—	22	179,775
Passenger and cargo	25	—	—	—	25	308,972
Barges	6	28	—	—	34	93,200
Tugs (ocean)	46	13	—	—	59	A
Tugs (harbor)	8	56	—	—	64	A
Hulls converted to barges	—	56	—	—	56	206,000
Finished hulls	—	115	—	—	115	447,700
Hulls converted to sailers	—	8	—	—	8	30,500
Barges converted to schooners	—	2	—	—	2	4,000
Harbor tugs	—	6	—	—	6	A
Total	1,693	589	18	12	2,312	13,636,711

Note: "A"—No tonnage given on tugs.

shipyards. Two hundred-eighty ships in all were built using the "fabricated" concept of construction. The most famous of the yards was the American International Shipbuilding Corporation Plant at Hog Island PA. The general arrangement and principal characteristics of the Hog Island standard ship are shown in Fig. 1 and Table 3, respectively. Features of the design differ little from other ships of the day.

Before the war there were fewer than sixty yards in the United States capable of building oceangoing merchant ships. These yards had a total of 235 shipways. By the signing of the Armistice on November 11, 1918, only 18 months after declaring war, there were 1020 shipways with shipyard labor going from 50,000 experienced workers to 385,000. The program's major success was perhaps not the relatively small number (113) of ships actually delivered before the Armistice, but the demoralizing effect the shipbuilding industrial buildup had on the enemy. Eventually 2312 ships were produced from this program by 1920.

Merchant Marine Act of 1920

Efforts to dispose of the huge surplus of ships led to the Merchant Marine Act of 1920. That law authorized chartering ships to private operators until they could develop sufficient business to undertake vessel purchases. The Board was also empowered to operate ships for government account. Significantly, this law enunciated, for the first time, that it is the policy of the United States to have a strong U.S.-flag merchant marine for national defense purposes as well as for

Table 3 Mobilization cargo ship characteristics (adapted from Reference 2)

Design	Hog Islander	Liberty EC2-S-C1	Victory VC2-S-AP2	Victory VC2-S-AP3	Bland C3-S-OX1	Mariner C4-S-1a
Length overall (ft)	401.0	441.5	455.3	455.3	478.2	563.7
Length between perpendiculars (ft)	390.0	416.0	436.5	436.5	450.0	528.5
Beam (ft)	54.0	57.0	62.0	62.0	66.0	76.0
Depth (ft)	32.0	37.3	38.0	38.0	41.5	44.5
Design draft (ft)	24.4	27.7	28.6	28.6	28.5	29.8
Deadweight (lt)	7,825	10,865	10,900	10,861	10,516	13,409
Displacement (lt)	—	14,245	15,200	15,200	15,910	21,093
Speed (kts)	11.5	11	15.5	16.5	18.5	20.0
Scantling draft (ft)	—	27.7	28.6	28.6	—	31.6
Deadweight (lt)	—	10,865	10,900	10,861	—	14,885
Bale cubic (ft^3)	372,300	499,573	453,210	453,210	561,290	736,723
Shaft horsepower (normal)	2,500	2,500	6,000	8,500	12,500	17,500
Number crew	41	54	43	43	52	58
Number passengers	—	0	28	28	12	12
Gross tonnage	5,735	7,191	7,612	7,612	8,918	9,216
Net tonnage	3,445	4,309	4,553	4,553	5,267	5,366
Stowage factor (ft^3/lt)	47.6	46.0	41.6	41.7	53.4	54.9

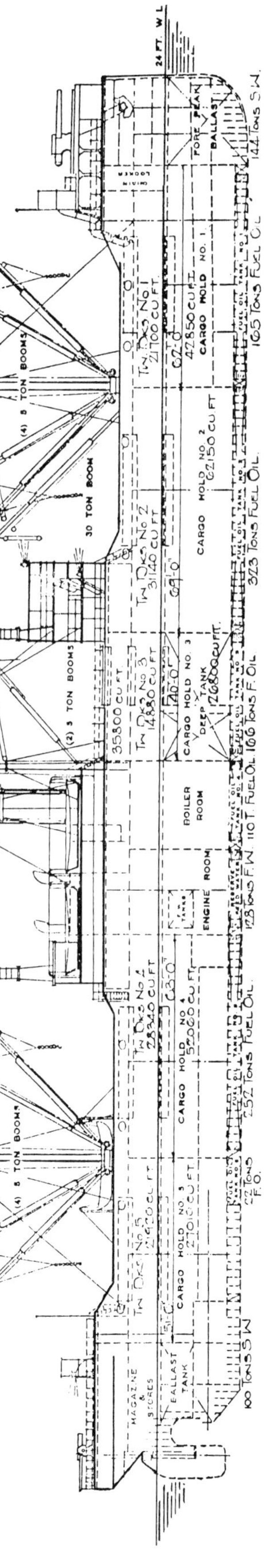

Fig. 1 World War I Hog Island fabricated design [3] (characteristics are shown in Table 3).

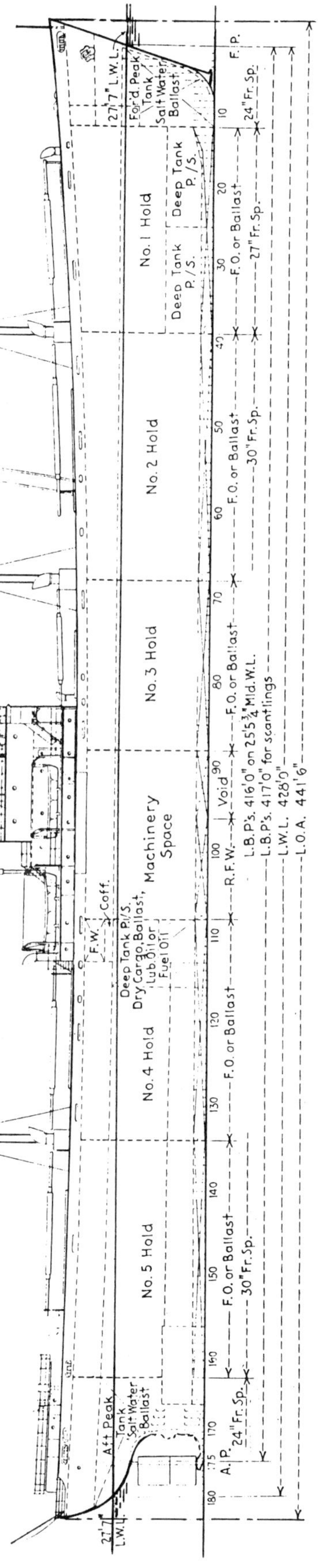

Fig. 2 World War II Liberty EC2-S-C1 [4] (characteristics are shown in Table 3).

moving the nation's peacetime cargo. The Board put over 1500 ships in operation, on more than 200 trade routes, by June 1921. But a year later a worldwide shipping recession reduced the number of active ships to 360.

The 1920 Act contained two other important provisions. One broadened the ban from U.S. domestic trade of foreign-owned or foreign-built vessels to include intercoastal trade and non-contiguous trades. This section, known as the Jones Act, remains in effect today and is the basis for the existence of a U.S. built and flagged domestic fleet. Another section of the Act empowered a regulatory agency, now the Federal Maritime Commission, to retaliate against unfair foreign-flag practices or foreign regulations that discriminate against the operations of U.S.-flag carriers.

Merchant Marine Act of 1928

The Merchant Marine Act of 1928 stimulated an enormous growth in the U.S.-flag passenger fleet by providing liberal payments for mail carriage and shipbuilding loans. During the nine-year period the law was in effect, 64 large passenger ships were built and 61 existing ones were modernized. These ships competed effectively against their foreign counterparts on 45 trade routes. Many of the ships were used as troop transports in World War II.

Merchant Marine Act of 1936

The enactment of the Merchant Marine Act of 1936 followed five years of Congressional hearings on forming a new maritime policy and a program to improve the merchant marine. Over 90 percent of the ships then in the fleet were twenty or more years of age. President Franklin D. Roosevelt, while enunciating the need to modernize the fleet, left it up to Congress to develop the means. The Act's major underlying thrust was that a healthy U.S. merchant marine is essential to our national defense. The 1936 law established the U.S. Maritime Commission (MarCom) and a program goal of building fifty ships per year for ten years.

Among other things, the 1936 Act:

- Provided direct shipbuilding (construction differential subsidy or CDS) and ship operating subsidies (operating differential subsidy or ODS) for ships built for and operated on essential U.S. foreign trade routes.
- Authorized the government to build ships and to charter them to U.S.-flag operators for service on essential foreign trade routes if the operators lacked the funds to purchase vessels.
- Provided for government-guaranteed loans and mortgages to finance shipbuilding projects (Title XI).
- Required subsidized lines to establish funds to replace over-age ships on a timely basis.
- Authorized a training program for officer candidates and crews.
- And, finally, authorized MarCom to develop a series of advanced designs of highly efficient ships that could increase our competitiveness on the world's trade routes.

In addition to agency-sponsored research and development, the Act also required MarCom and Navy design approval and provided for inclusion of national defense features to be paid for by MarCom.

An extensive paper by Dillon et al [5] presents the history of the ship designs and associated innovations since the Merchant Marine Act of 1936. A few of the highlights are described here along with designs that came after the 1976 paper.

The first ship to be built under the 1936 Act was the passenger liner *America*. M.C. Hull No. 1 was awarded to the Newport News Shipbuilding and Dry Dock Company on October 21, 1936. This ship represented a quantum leap in safety for passenger ships with respect to subdivision, damage stability, and fire proof construction. The lessons learned from the *Morro Castle* disaster, were applied by using effective fire-zoning with enclosed stair towers, and fire proof construction throughout. Her prismatic coefficient of 0.60 and speed-length ratio of only 106.8 signified exceptionally fine lines. Operating in the North Atlantic, the ship demonstrated the superior ability of fine-ended hulls to maintain speed in rough weather.

Having an eye to a balanced cargo fleet capable of speedy and economical conversion to military auxiliaries, the MarCom undertook design and construction of what was then considered small, intermediate, and large general-purpose break-bulk cargo ships. These were designated initially as C1A, C1B, C2 and C3. These were notable for their up-to-date design and relatively high speeds (14, 15.5, and 16.5 knots respectively). Geared steam turbines powered most of the ships.

In many respects the original C-class ships established, for the first time in general-purpose dry cargo vessels, standards of safety exceeding those prevailing not only at the time of their introduction but even today's average practice worldwide. Among other things, they all met a one-compartment damage stability standard.

The C3s were the largest, fastest, and beamiest of the group. The first awards went to Federal Shipbuilding for steamers and to Sun Shipbuilding for diesels. Purposely designed with excess transverse stability, their hulls adapted to numerous conversions both military and commercial, including escort carriers and combination passenger-cargo ships. In line with the postwar trend of ever-increasing volume of trade and corresponding increase in size and speed of cargo ships, the C3s stood out for many years as the most competitive and hence most valuable additions to the U.S. merchant marine in the general-purpose dry-cargo ship category.

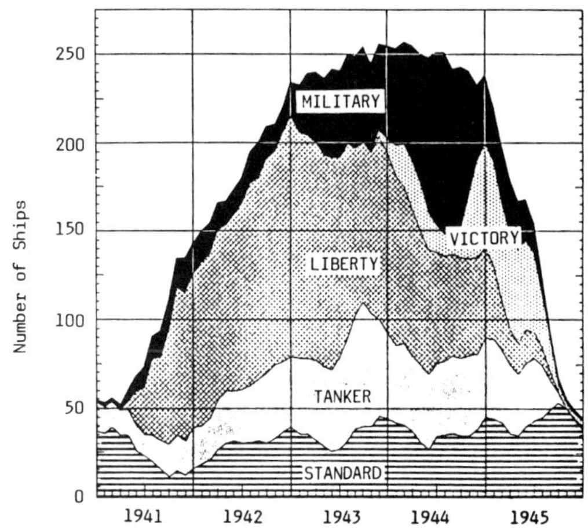

Fig. 3 Types of ships in construction on the ways in Class 1 yards [6].

World War II Liberty and Victory Ships

The shipbuilding program evolving from the 1936 Act and the impetus of World War II enabled the expansion of the shipbuilding industry from 114 building ways on January 1, 1941, to 267 ways at peak production after we had entered the war. Even before then, construction of Liberty ships for Great Britain had begun under the Lend Lease Program (sixty were built). The Japanese attack on Pearl Harbor in December 1941 turned our own emergency program into a war program.

> When the emergency program was decided upon, there was not time to do more than choose among existing designs and then to make some modifications to meet special needs [4].

The British Empire Liberty design was chosen on the assumption that the ships would be strictly for the emergency and any postwar use would be incidental. Most of the changes to the British design were aimed at quicker production. Straight camber replaced the normal curved camber. Oil fuel instead of coal was engineered into the design along with watertube boilers and a single midship house. A contrarudder was used, increasing the speed and maneuverability while costing forty percent less. Habitability was improved, but many standard items on U.S. ships were not included, such as the emergency generator, the spare bower anchor, and some anchor chain. Simpler round bar davits were used and the number of gauges for steel plate dropped from 75 to 27 because of steel shortages. Figure 3 indicates the extent and success of the Liberty program.

While the Liberty ship mass production program met the need for quick mobilization, the later Victory ship design followed from the need for higher speed to outrun torpedoes and the realization that they would be needed after the war for immediate commercial service. Lane [4] chronicles the decision making process and technical developments that resulted in the massive production of Libertys and Victorys.

Between 1939 and 1946, U.S. yards produced nearly 5600 merchant ships totaling more than 56 million deadweight tons. While many factors contributed to defeating the Axis powers, this "bridge of ships" was credited with playing a pivotal role in transporting the troops and war material that turned the tide of victory.

Faced with a chronic shortage of shipbuilding steel throughout the war, while at the same time reinforcing rods remained in ample supply, MarCom turned with some misgivings to a concrete shipbuilding program, which yielded appreciable amounts of usable tonnage for miscellaneous military and civilian support roles. Barges accounted for most of the production. They variously served as floating oil storage tanks at advanced South Pacific bases and mobile storehouses and machine shops. Twenty-four C1-S-D1 ships, powered with 1300 ihp triple-expansion steam engines helped the economy by importing Cuban sugar.

C1-M-AV1's and Other Ships for the War Effort

As General MacArthur began his island-hopping campaign in the Pacific, he foresaw the need for two hundred small seagoing cargo ships capable of lifting 4000 tons deadweight on 18-ft draft. MarCom responded with its preliminary design, refined by George G. Sharp, of the first C-type direct-connected diesel ships with machinery aft. The designers were well aware that single-reduction gearing would have improved propulsive efficiency, but time pressure interfered with such procurement. Moreover, model tests showed the hull form to be less than optimal. Nevertheless, despite poor overall hydrodynamic efficiency, these ships (C1-M-AV1 class) proved useful in war service and popular postwar, especially among foreign buyers.

Several other kinds of ship were designed and built for the war effort, including P1 and P2 troopships and T2 tankers. These are described in References 4 and 5.

Merchant Ship Sales Act of 1946

Under the Merchant Ship Sales Act of 1946 more than 1100 of the government owned ships were sold to foreign operators to replenish their war-ravaged fleets. U.S.-flag operators bought 850 of these ships and operated another 1500 under General Agency Agreements carrying Marshall Aid cargoes to help rebuild a war-torn world. The remaining 1400 vessels were transferred to National Defense Reserve Fleet anchorages to be ready for subsequent emergency needs.

Post World War II Years

The U.S.-flag fleet carried over sixty percent of the nation's waterborne foreign commerce in the immediate post-war years. By 1950, though, the share had dropped to 43 percent. Competition from lower-cost foreign carriers reduced the share to eleven percent only ten years later. Since 1970, U.S. carriage of our waterborne foreign commerce has hovered between four and five percent. Close to ninety percent of the trade consists of liquid or dry bulk cargoes, with almost all of these cargoes moving in foreign ships because of lower costs of operation.

Virtually all of the foreign trade bulk cargoes that are moved under U.S.-flag are government preference cargoes, such as Food for Peace agricultural exports, and imports of crude oil for the nation's strategic petroleum reserve. These account for only three percent of the total bulk cargo import and export tonnage.

Between 1973 and 1983 the U.S. liner fleet, by then principally composed of efficient containerships, carried 25 to 30 percent of the high-value liner tonnage that moved in our waterborne foreign trade. During the next six years, this percentage also tailed off, dropping below 20 percent. It is significant to note that this drop did not represent an actual decline in the carrying capacity of the U.S.-flag fleet, but reflects a tremendous growth in the volume of our overseas trade, along with a much more competitive international shipping environment.

The *Schuyler Otis Bland*

Much progress has been made on the technical side since the initiation of the 1936 Act. Motivated by the onset of the cold war in 1948, Congress appropriated funds for the design and construction of a prototype high-speed cargo ship capable of mass production in case of emergency and also to be internationally competitive. The ship was named *Schuyler Otis Bland* in honor of the chairman of the House Merchant Marine and Fisheries Committee. In development, the *Bland* became somewhat of an experimental ship with innovations, some of which proved remarkably viable. Among other features, the *Bland* pioneered mechanically operated hinged steel weather deck hatch covers. The Ebel burtoning arrangement, named after its designer, Francis G. Ebel, a career MarCom employee, made a major improvement in flexibility and safety of burtoning systems, which had not changed much since Phoenician days.

American President Lines operated the ship initially. After the first year, they reported the cargo gear as the most successful feature on the ship. Thereafter, that company fitted all of its new construction, and retrofitted its Mariners, with Ebel gear.

The Liner *United States*

The passenger liner *United States* (Fig. 4) was the flagship and pride of the U.S. merchant marine. This unique superliner exemplified the ultimate in American passenger ship design. The creation of Gibbs & Cox, Inc., and built at Newport News, this proud ship in 1952 captured the legendary North Atlantic blue ribband by crossing in both directions at an average speed of 35 knots. She held this record until 1992, when a small racing yacht, *Destriero*, broke the record with an average speed of 53 knots. Because the *Destriero* was not a passenger ship, the courts awarded possession of the blue ribband on a fifty–fifty basis.

The *United States* was designed and built as a superliner but with the ability to be quickly converted to a troop ship. She easily qualified on short notice for conversion to the most productive troopship afloat. In addition to speed, superlatives extend to safety from every aspect. She was, for example, the only liner capable of withstanding flooding of any four compartments.

Exceptional light ship weight made possible her speed on a ship length limited by Panama Canal locks. Over 2000 tons of aluminum alloy, the largest amount on any vessel to that date, reduced topside weight such

Fig. 4 S.S. *United States* (courtesy of the Maritime Administration).

that stability requirements for four-compartment flooding are achieved on only a 101.5-ft beam.

The top speed of the *United States* was kept confidential for many years, as was her trial speed of 38.32 knots. A paper by Kane [7] describes the exceptional hull form and propulsion characteristics that make this vessel a record breaker. The quadruple screw power plant was the same as that used on U.S. Navy aircraft carriers.

Changes in Administrative Structure

In 1950 the U.S. Maritime Commission became the Maritime Administration (MARAD). Under President Kennedy in 1961, the Maritime Administration's regulatory functions were split off to a new agency, the Federal Maritime Commission. MARAD retained the promotional functions. Vice Admiral E.L. Cochrane was the first Administrator of the Maritime Administration. He had just completed a two-year term as Dean of Engineering at MIT following his tour of duty as Chief of the Navy's Bureau of Ships. In the course of his MIT activity, he participated in an antisubmarine warfare study, which convinced him of the urgent need for 20-knot cargo ships in case of emergency.

Mariner Class

The revolutionary Mariner design is well chronicled in several noteworthy papers including those by Russo et al [8] and Sullivan et al [9]. In 1948, before the outbreak of the Korean War, and with President Truman's backing, Cochrane obtained authorization from Congress to build up to fifty such ships, with funding for 35 provided. Thereupon, he transmitted MARAD's preliminary design, redesignated C4-S-1a, to Bethlehem's Central Technical Department for contract plan development. During this period Bethlehem redesigned and model tested the final lines. Upon awards in 1951 (in groups of five ships in each of seven yards), Bethlehem became the lead organization in developing working plans furnished to all yards.

As noted later, the basic Mariner design accommodated a wide variety of commercial and military conversions. Noteworthy among its special features are the following:

- First 20-knot general-purpose dry cargo ship.
- Extra-large hatches, 40-ft by 30-ft clear opening.
- Mechanically operated hinged steel hatch covers on every deck.
- Shelter deck hatch covers with minimum coamings.

Fig. 5 *Keystone Mariner* (courtesy of the Maritime Administration).

- 14-ft clear deck height for upper and lower 'tween decks.
- Decks strengthened for 60-ton tanks.
- 10-ton double-ganged burtoning gear at five hatches.
- Two 60-ton heavy-lift booms.
- Seven holds.
- Double-skin protection in way of machinery space.
- Two 600-kW turbogenerator sets with space in engine room for two more.
- Two 12,000 gallons per day saltwater evaporators for self-sufficiency in freshwater.
- Degaussing equipment.
- First 1936 Act cargo ships to use ac current extensively throughout the ships with ac/dc generators installed to supply dc current for each pair of cargo winches.
- 4 percent of midship area bulbous bow.
- Clearwater stern with large semi-balanced streamline rudder mounted on rudder horn.
- One-compartment subdivision and damaged stability with minimum freeboard for shelter decker, that is, subdivision draft equal to minimum freeboard draft, thereby entailing no sacrifice of deadweight as for C1s, C2s, and C3s.
- Complete cargo hold dehumidification.

The Mariner-class ships established many of the machinery practices used during the remaining 1950s and early 1960s. The Mariner power plant development was undertaken concurrently with that on the *Bland*. Owing to the urgency of building 35 ships for Korean War military support, certain changes were made. The selection of machinery components was primarily based on an evaluation of overall cost, availability of material, ease of construction and installation, interchangeability, reliability, and repair experience.

The Mariners were the turning point in U.S. cargo ship propulsion plant trends, both in power output and in higher operating efficiencies. During the ensuing years a series of ships with Mariner hulls was constructed, all having the same propulsion output of 19,250 shp maximum ABS rating. The steam cycles improved, however, and in some ships the engine room lengths were reduced to 55 and 60 feet with the machinery space wing tanks eliminated.

Pacific Far East Lines (PFEL) purchased Mariner hulls 33–35 inclusive, the last of the initial series, before completion, and made some changes to quarters, cargo gear, and tankage to suit special requirements. Most of the other Mariners saw six months' service with the Military Sealift Transportation Service and then went ignominiously into the National Defense Reserve Fleet. Nevertheless, when news broke of their phenomenal profitability in PFEL service, a stampede followed to buy the laid-up ships.

Starting in 1959 and continuing through the 1960s, various operators availed themselves of the proven Mariner hull design to build 18 more ships, modified

Fig. 6 N.S. *Savannah* (courtesy of the Maritime Administration).

internally and changed with respect to cargo handling gear and quarters to suit specific trade requirements.

The *Savannah*

President Eisenhower supported legislation for building the nuclear powered ship *Savannah* to demonstrate the peaceful application of nuclear power and to establish nuclear-powered ship operating requirements in merchant service. The contract for construction was awarded in 1957 to the New York Shipbuilding Corp. The George G. Sharp Co. did the design. The *Savannah* (Fig. 6) fulfilled her mission as the world's first nuclear-powered merchant ship. Although a marvelous engineering success, after having logged over 100,000 miles in perfect safety, the Government decommissioned her in 1971 because of heavy operating expense. This was expected. The *Savannah*, was a hybrid cargo-passenger ship, burdened with high crew costs and extraordinary safety measures. (The latter included $500 million U.S. Government indemnification under the Price–Anderson Act.) In short, the ship was simply not the type to demonstrate economic viability, nor was this ever intended.

Oceanographic, Hydrographic, Fisheries Research, and Other Vessels

In 1959 the Coast and Geodetic Survey, now National Oceanographic and Atmospheric Administration (NOAA), authorized MARAD to design and supervise the building of a new hydrographic survey vessel. The *Surveyor* was built at the National Steel and Shipbuilding Company in San Diego, and delivered to NOAA in 1960. Two coastal survey vessels were designed by J. J. Henry, Inc. and delivered to NOAA in 1963. In 1966 through 1968 MARAD designed and supervised the building of two oceanographic survey ships, the *Oceanographer* and *Discoverer*; five coastal survey ships, and two wire-draggers. The *Researcher* (now *Malcolm Baldrige*) was delivered in 1970.

In 1968 MARAD and NOAA pioneered the concept of using the popular off-shore supply boat hull for a keel-up research vessel. The contracting concept was later adapted by other government agencies. The *Farrell* is a survey-buoy tender used in the mouth of rivers and estuaries and is the mother ship for a 57-ft tender. In 1980 this design-and-build concept was again used to produce the *Chapman*, a fisheries research vessel. Over this time MARAD designed and built a total of 13 research vessels and supervised the construction of two others.

Merchant Marine Act of 1970

The Merchant Marine Act of 1970 called for the construction of thirty ships per year over a ten-year period. The vessels were to be of advanced designs and built under a declining scale of construction subsidies. The 1970 law made dry bulk carriers and tankers eligible for the first time for both building and operating subsidies. Eighty-three new ships were produced under the law before 1977.

The containership, barge carrying ships, and roll-on/roll-off ships also came of age in the 1960s and 1970s. These are well described in Dillon et al [5]. The following descriptions cover the more recent ship designs that were not chronicled in that paper.

In 1976 CCT, Inc. contracted with Seatrain Shipbuilding Corp. (the old Brooklyn Naval Shipyard) for two integrated tug-barge RO/RO vessels. Seatrain, in turn subcontracted to Marinette Marine Corp. for the two tug power units. Before the project was completed, Seatrain forfeited and only partially completed the RO/RO barge units. Both tugs were completed on schedule. One of the barge units was completed at Norfolk Shipbuilding and Drydock Corp. and the other unit was scrapped. These vessels were integrated by way of the patented Autobar (pin) system invented by Fletcher. The combined unit's overall length was 605 ft, and the combination was designated IB5-MT-121a.

In 1977, American Heavy Lift Shipping Co. contracted with Peterson Builders of Sturgeon Bay WI for two heavy lift RO/RO ships, appropriately named the *John Henry* and *Paul Bunyon* (C1-MT-123a). These unique ships had bow ramps and two 216 tonne capacity split topping heavy lift kingpost and derricks located on the port side. They were delivered in 1978 and 1979.

Three 300-ft breakbulk/container cargo ships were awarded to Equitable Shipyards, Inc., by American Atlantic Shipping Co. in 1977 for delivery in 1979 and 1980. The *America, Amazon,* and *Antillia* (designated C1-M-122a) had deadweights of about 1800 long tons. They were fully certified and carried a crew of only eleven.

Levingston Shipbuilding, Co. of Orange TX built three 34,500 DWT dry bulk carriers for Levingston-Falcon I Corp. These were named *Pride of Texas, Star of Texas,* and *Spirit of Texas* (C5-M-129a). They were delivered in 1981 and 1982.

Another type of integrated tug-barge vessel built under MARAD subsidy was the cat-tug. Three integrated vessels were built for Swanee River at Avondale Shipyards, Inc. The oil-chemical barges were 632 ft long, and when integrated with the catamaran tug were over 678 ft long. The *Oxy Trader, Oxy Producer,* and *Oxy Grower* (IB6-MT-130a) were delivered in 1981.

Three container-RO/RO ships were built for Waterman Steamship Co. by Pennsylvania Shipbuilding Co. of Chester PA. These large C7-S-133a class ships had over 107,000 sq ft of RO/RO deck area and a container capability for 1548 twenty ft equivalent units (TEUs). Soon after delivery in 1983 the owner converted them to prepositioned ships and chartered them to the U.S. Navy where they remain today.

Liquefied Natural Gas Tankers

The Maritime Administration aided in the construction of sixteen LNG tankers employing three different containment systems: spherical tanks, waffle membrane, and Conch membrane. The *El Paso Southern, El Paso Arzew,* and *El Paso Howard Boyd* (LG9-S-94a) were delivered in 1979 by Newport News using the waffled membrane (Technigaz) system. Each can carry 126,600 cubic meters of liquefied gas.

The first three of ten LNG tankers using spherical tanks (Moss) were the *LNG Aquarius, LNG Aries* and *LNG Capricorn* (LG9-S-102a). Each of these vessels carries 125,000 cubic meters of liquefied gas. The *Lake Charles* and *Louisiana* were delivered to Lachmar Corp. two years after the *Capricorn.* The remaining five were the *LNG Gemini, LNG Leo, LNG Libra, LNG Taurus,* and *LNG Virgo.* The prismatic self-supporting containment system (Conch) used in the *El Paso Columbia, El Paso Savannah,* and *Cove Point* (LG9-S-107a) suffered insulation failure late in the construction period and were never completed as LNG tankers. Two were later converted to oil-bulk carriers (OB9-S-F107b) and the *El Paso Columbia* was scrapped after running aground while under tow.

Last CDS Ships

The last three containerships to be built with CDS were the *President Lincoln, President Washington,* and *President Monroe,* designated C9-M-132b. A major change under contract revised the power plant from steam to slow speed diesel. Their 43,200 shp provides a service speed of over 24 knots and a range of over 20,500 miles. These 863-ft panamax containerships have the deck house forward to protect the 2492 TEUs carried.

The *Falcon Leader* and *Falcon Champion* (T6-M-136a), built at Bath Iron Works Corp. and delivered in 1983 and 1984, were the last tankers built under the CDS program. They were built for long term charter to the Military Sealift Command. They have a service speed of 16.5 knots on 12,000 shp, and a range of 24,300 miles.

The last ship to be awarded Construction Differential Subsidy in the United States was a small trailer carrier for Blue Lines, Inc. The motor vessel *P'ti Bleu II* (N1-MT-138a) has 1900 sq ft of RO/RO deck area and is capable of handling two 40-ft trailers for over-the-beach cargo handling over the bow ramp.

Section 615 Build Abroad

On July 29, 1981, the Omnibus Budget reconciliation Act of 1981 was enacted. In this landmark legislation, the Merchant Marine Act of 1936 was amended by adding a new Section 615, which, in part, provided authorization for the Secretary of Transportation to permit an operator receiving or applying for operating differential subsidy (ODS) to place orders to construct, reconstruct, or acquire its vessels over 5000 DWT in a foreign shipyard under certain conditions.

This build-abroad policy was designed to increase the competitiveness of the U.S.-flag merchant fleet without direct expenditures of American tax dollars. It did nothing, of course, to assist U.S. shipyards.

The authority provided by Section 615 expired at the end of FY 1982, and no provision for foreign building under Section 615 now exists. Before its expiration MARAD approved new construction of 33 ships and reconstruction and reflagging of 17 ships. Of these, 39 have been delivered, with the remaining authorization in dispute.

To distinguish foreign buildings and conversions from domestic, MARAD added "F" in the design designation, such as C9-M-F141a or C8-S-F81e. The 141a was the designation for the twelve Jumbo containerships built at Daewoo, South Korea, for United States Lines and now owned by Sealand. The 81e was a LASH design converted to a full cellular containership for American President Lines. The shift to diesel propulsion for U.S.-flag ships is dramatically demonstrated by looking at the section 615 foreign built vessels:

Design	No.	Type
C6-M-F139b	2	bulkers
C9-M-F141a	12	containerships
C8-M-F144a	2	dry bulkers
C8-M-F144b	3	container-bulk
C7-M-F145a	3	containerships
C6-M-F146a	2	containerships
C6-M-F147a	4	containerships
C9-M-F148a	2	containerships
C9-M-F150a	5	container-bulk
C9-M-F151a	4	containerships

Since 1982 the Administration, Congress, and industry have been struggling with development of a new maritime program that will have the right balance and incentives to ensure a strong, healthy merchant marine.

MARAD Research & Development

The basic direction for MARAD R&D was set in 1960 by the Maritime Research Advisory Committee comprising an outstanding group of individuals organized by the National Academy of Sciences/National Research Council. Their report [10] was well received, but MARAD had organizational and budgetary difficulties that slowed implementation.

Before 1970 MARAD R&D funds were assigned primarily to a few large development projects such as the *Savannah*, the *Denison* hydrofoil boat, and the joint MARAD/Navy surface effects ship (SES) programs.

A significant feature of the 1970 Act was a commitment to increase research and development efforts. The resulting R&D program has been carried out by the Maritime Administration over the succeeding years in concert with a matching effort by the maritime industry.

In 1969, in anticipation of the Merchant Marine Act of 1970, MARAD organized and convened a conference of almost 200 leaders of the U.S. maritime community at Woods Hole, Massachusetts, to define a comprehensive maritime R&D program. Principal executives and professionals representing steamship companies, shipbuilders, research firms, labor organizations, ship designers, government agencies, and universities joined in the effort. The conference recommended an R&D program to complement the President's new maritime program (see Reference 11). The program was to strike a balance among various maritime research needs, making it clear that the Government's role should be to sponsor projects of higher risk and longer duration than those that normally could be undertaken by industry alone.

The resulting plan and its organization has served as the basis of the MARAD R&D program for the ensuing period, with modifications dictated by international and national events, shifting economic situations, U.S. budget constraints, and MARAD organizational changes. The plan's three functional requirements, derived from national needs, are embodied in the organizational divisions, namely Ship Systems Design and Construction, Ship Systems Operation and Support Research, and Maritime Science and Technology Research.

Program Emphasis

The MARAD R&D program has placed emphasis on system development activities in two areas: (1) Advanced Ship Systems Development and (2) Advanced Ship Systems Operations. MARAD R&D is oriented toward cost sharing as well as active participation in planning by industry. In the National Shipbuilding Research Program, which emphasizes development of new processes and methods to improve productivity, the shipyards have virtually played the role of laboratories. Improvements are developed, tested, and disseminated to other shipyards in a cooperative program cost-shared by the shipyards.

High technology activities like the Nuclear Ship Program and the Advanced Communications/Navigation program have experienced somewhat smaller cost

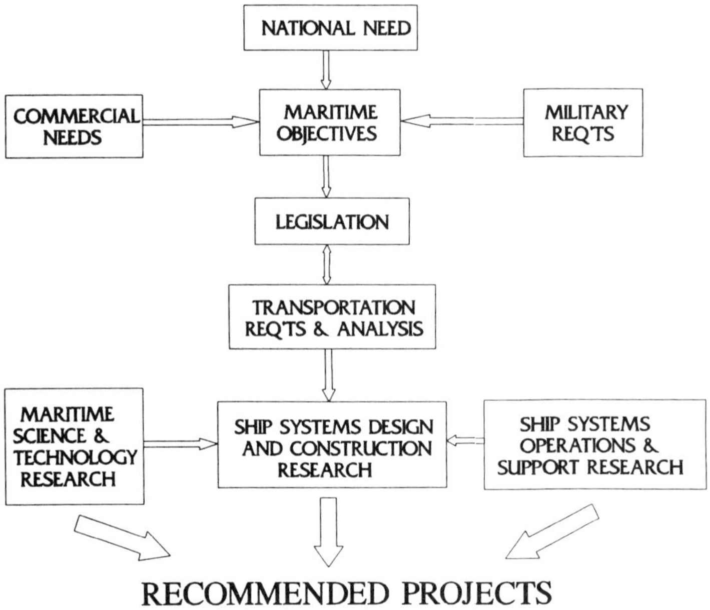

Fig. 7 Functional requirements for Maritime Research and Development (adapted from Reference 12).

sharing by industry owing to the greater risks involved in implementation.

One factor not generally recognized has been the effect of the MARAD R&D program in creating a technology-oriented industry. For example, shipping lines now maintain complex and effective computer control centers, and have adopted satellite communications to conduct the management of their fleets and control of cargo movement. U.S. shipyards have introduced highly efficient integrated shipbuilding procedures that have the potential of matching the productivity of Japanese shipyards.

MARAD R&D Funding

The 1970 Act increased the R&D budget from $6–7 million to the $20–25 million level in the 1970–76 period, with equal emphasis on near term and longer term results. Stringent national budgets, beginning in the early 1980s, as well as the depression in the maritime industries, brought about reductions of funding to the one million dollar range by the 1990s. In spite of the greatly diminished funds available, substantial accomplishments can be cited. Reference 12 describes R&D developments during the 1970–76 period. Some of the highlights of the MARAD R&D accomplishments are briefly described in the following paragraphs.

- Implementation of advanced technologies on the Mariner-class ship.
- Development and proof of the feasibility of nuclear propulsion through the *Savannah* development program and many follow on efforts.
- Pioneering of the hydrofoil craft concept with co-development of the *Denison* (Fig. 8) with the U.S. Navy [13].
- Design and construction of the Computer-Aided Operations Research Facility (CAORF) in 1976

Fig. 8 *Denison* hydrofoil (courtesy of the Maritime Administration).

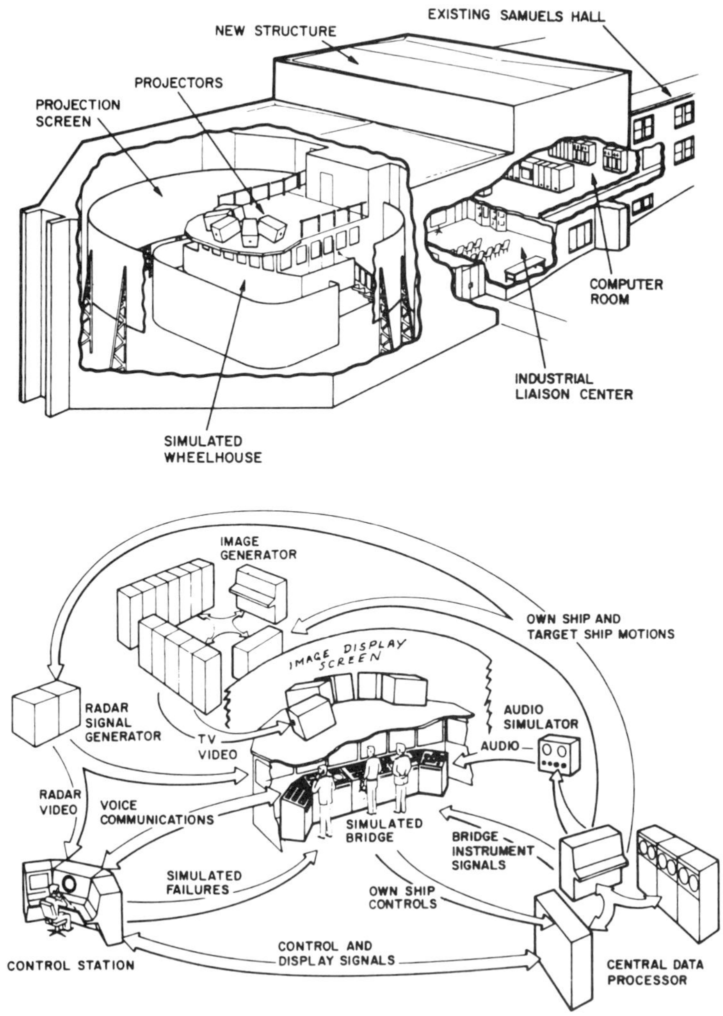

Fig. 9 Computer Aided Operations Research Facility (CAORF) [14].

(Fig. 9), a state-of-the-art, full-bridge, ship handling simulation research facility [14].

- Development and introduction of numerous computer aids into the industry.
- Formation of the National Shipbuilding Production Program (NSRP) through the Society of Naval Architects and Marine Engineers. The initial formation in 1972 started with the introduction of the computer software system, AUTOKON, into U.S. shipyards. Subsequent developments focused in many areas resulting in a vast program that has made great improvements in the productivity of U.S. shipyards.
- Promotion and establishment of other cooperatives in industry for co-development of projects of common interest; for example, the Cargo Handling Cooperative Program, active since 1983, gathers ship operators together to accomplish research projects in common.
- Studies of the navigation of the Great Lakes in winter have improved operations and enabled extension of the shipping season.
- The use of satellite communications in the maritime area has been pioneered and evolved into mature technologies through various efforts at MARAD.
- The improvements in efficiencies through contra-rotating propulsion was well proven out in the late 1970s [15] and a main propulsion planetary gear and contra-rotating propeller system developed and tested in the mid 1980s [16]. The system was tested at 20,000 SHP with the system designed for plants up to 60,000 SHP.
- Marine gas turbine propulsion systems were developed and assisted with evolution into marine use.
- Development of a highly skewed propeller (see Fig. 10) provided methods to significantly reduce or avoid vibration problems [17].

Fig. 10 Highly skewed propeller [12].

- Support of many initiatives in the development of supporting technologies for naval architecture, marine engineering, ship construction, and ship operations.

References

1 Bollenbeck, W. J., "Shipbuilding for War and Peace Times—Work of the United States Shipping Board Emergency Fleet Corporation," *Philadelphia's War History*, Oct. 1921, p. 20.

2 Landsburg, A. C., "Merchant Ships for Wartime Mobilization Prototype Design and Construction for Readiness," *Naval Engineers Journal*, April 1977, pp. 95–112.

3 *Register of Ships Owned by United States Shipping Board*, Government Printing Office, Washington, D.C., 1 August 1920.

4 Lane, F. C., Coll, B. D., Fisher, G. J., and Tyler, D. B., *Ships to Victory—A History of Shipbuilding Under the U.S. Maritime Commission in World War II*, John Hopkins Press, Baltimore, 1951.

5 Dillon, E. S., Hoffmann, L. C., and Roseman, D. P., "Forty Years of Ship Designs Under the Merchant Marine Act, 1936–1976," *Transactions*, SNAME, 1976, pp. 169–219.

6 "Victory Ships," *Marine Engineering and Shipping Review*, Vol. 49, No. 4, April 1944, p. 162.

7 Kane, J. R., "The Speed of the S.S. *United States*," *Marine Technology*, April 1978, pp. 119–143.

8 Russo, V. L. and Sullivan, E. K., "Design of the Mariner-Type Ship," *Transactions*, SNAME, Vol. 61, 1953.

9 Sullivan, F. K. and Scarborough, W. G., "Machinery Design of the S.S. *Schuyler Otis Bland*," *Transactions*, SNAME, Vol. 60, 1952.

10 "Proposed Program for Maritime Administration Research," Maritime Research Advisory Committee, NAS/NRC, 1960.

11 "Report of the Woods Hole Conference on Maritime Research and Development," Maritime Administration, 1970.

12 *Maritime Administration Research and Development Program—An Assessment* 1970–1976, Vols. 1 and 2, Maritime Administration, March 1977.

13 Gross, J. G., "Experience with the Hydrofoil Craft *Denison*," *Transactions*, SNAME Spring Meeting, 1965.

14 Rinehart, V. W., "Maritime Administration's Computer Aided Operations Research Facility," *Proceedings*, SNAME First Ship Technology and Research Symposium (STAR), 26–29 Aug. 1975.

15 Hadler, J. B., "Contrarotating Propeller Propulsion—A State-of-the-Art Report," *Marine Technology*, July 1969, pp. 281–290.

16 Motzer, J. W., "Marine Advanced Planetary Transmission Program Design, Manufacture and Test of System 'A' and 'F'—Final Report," Curtiss-Wright Corp. Report to MARAD, 7 March 1977, pp. *ca.* 350.

17 Valentine, D. T. and Dashnaw, F. J., "Highly Skewed Propeller for San Clemente Class Ore/Bulk/Oil Carrier Design Considerations, Model and Full-Scale Evaluation," *Alpha Proceedings*, SNAME STAR, 1975.

18 *Characteristics and Index of Maritime Administration Ship Designs*, Maritime Administration, Jan. 1991, p. 257.

Charles B. Cherrix currently serves as the Maritime Administration's Chief, Division of Naval Architecture. He graduated from the U.S. Merchant Marine Academy in 1956, after earlier studies at the University of Maryland, and has worked at MARAD in increasingly important positions ever since. His professional achievements have been recognized in his receipt of the Department of Commerce Bronze Medal and the Department of Transportation Silver Medal. He is active in SNAME, ASNE, and various prestigious national and international committees.

Alexander C. Landsburg earned his bachelor's and master's degrees in naval architecture and marine engineering at the University of Michigan. He has been employed by the Maritime Administration since 1968 in positions of increasing responsibility. His professional achievements have been recognized in his receipt of the Department of Commerce Bronze Medal. He is currently Program Manager for Ship Performance and Safety. He is a Life Member and Vice President of SNAME and a recipient of the E. L. Cochrane Award as well as the Distinguished Service Award.

The American Bureau of Shipping

DONALD LIU

By 1893, the year of SNAME's founding, the American Bureau of Shipping (ABS) had already established itself as an integral part of this nation's commercial shipping infrastructure. Following its inception as the American Shipmasters' Association in 1862, the organization had published rules for the construction of wooden vessels in 1870 and for iron vessels in 1877. Rules for steel vessels were published in 1888. These rules were used for "classing" vessels (that is, determining structural and mechanical fitness for the intended service). Classification provided guidance for marine underwriters and others in assessing the fitness of merchant ships. Without such guidance, marine insurance was a high risk enterprise and shipowners were penalized accordingly. Lloyd's Register of Shipping was then already in existence, but at that time American shipowners found communicating with them to be slow and difficult. The advent of ABS was therefore an important forward step for American (and other) shipowners.

The rules have been subjected to continual modifications over the years as dictated by changes in technical knowledge and operating experience, as well as new kinds of ships, some of new extremes of size, or sophistication. Initially, for example, hull scantlings (i.e., basic structural components) were chosen on the rather crude basis of the ship's overall size as approximated by the so-called tonnage numeral (0.0075 × length × beam × depth). In 1900, however, more sophisticated parameters were introduced, and these have been under study and improvement ever since.

The American Bureau of Shipping is an independent, voluntary not-for-profit classification society. Its basic managerial and technical policies are governed by committees with members drawn from all sectors of the marine, offshore, and related industries. Its income is derived from shipowners who pay for the services provided. Classification provides evidence that a vessel is mechanically and structurally fit for the intended service, a fact of interest to many different parties, and without which an owner would have difficulty obtaining insurance.

ABS does more than develop and publish standards for design, construction and survey of ships and other marine structures. It provides review and survey services during the design and building phases, and again after the ship goes into service. It also publishes the *Record*, an annual compilation of ships under its registry. This helpful guide is used by underwriters, chartering agents, and others to obtain important information about each ship's technical characteristics and level of classification. Until 1914, the record and the rules were combined under a single cover. Since that date they have been printed separately.

Although ABS started as an American institution, it soon attracted international attention. This development became most pronounced following World War II, when American war-built ships were sold to many owners overseas. This led ABS to expand its foreign services, opening a network of offices throughout the western world and much of Asia, Africa, and South America. Concurrently, maritime experts from all of the affected regions were brought into the ABS family, serving on technical committees that aided in developing new standards to suit special overseas conditions.

The period between World War II and the present has seen many significant changes in ship technology, and ABS has been quick to adapt to new conditions. It has more than doubled the size of its technical staff since the end of the war. Indeed, with its vigorous research activities, it has often been a pioneer in technical developments. It has been aided in this by the active cooperation of special committees (now numbering fourteen) of top-level experts in special technologies such as welding, propellers, reduction gears, reinforced plastics, underwater systems, and single-point mooring. The following is but a sampling of the important changes in the marine industry of which ABS has played a significant part.

In the early 1960s ABS classed the world's first commercial nuclear ship, the *Savannah*. It classed the newly-developed barge-carriers known as LASH and "Seabee" types. It classed ships of unprecedented size, including tankers of up to 500,000 dwt. This last project was aided by an elaborate computer-aided finite element analysis system developed in conjunction with Chevron Shipping Company and academicians at the University of Arizona. This computer program system, known as DAISY (Displacement Automated Integrated System) continues in use and has been applied to numerous other kinds of ships.

DAISY has also been found useful in analyzing the structure of several forms of offshore drilling units, another field in which ABS has played a pioneering role since the 1940s. By the mid-1980s there were some 575 mobile offshore drilling units (MODUs) classed with ABS. When the International Maritime Organization (IMO) of the United Nations issued its MODU code in 1979, the 1968 ABS rules were used as a basis.

The 1960s also saw ABS issuing rules for such diverse structures as manned submersibles, ice-strengthened ships, and cargo containers.

In the 1970s ABS issued rules for LNG (Liquefied Natural Gas) tankers, ITB (Integrated Tug/Barge) units, aluminum alloy hulls, and small commercial vessels. It developed a comprehensive library of computer programs and, for the first time, issued its rules in foreign languages, namely Spanish, Greek, Italian and German.

In the 1970s ABS developed requirements for verifying the integrity of outer continental shelf platforms. In a widely differing field, it also published a guide for the construction of offshore racing yachts.

Developments during the 1980s include such diverse topics as coal-burning ships, inert gas systems, and vessels for use on rivers and other protected waters. In addition, ABS found itself working with the U.S. Navy in several projects such as developing hull steel specifications. In 1981 ABS reached an agreement with the U.S. Coast Guard to eliminate duplication of efforts, and to act on their behalf, in plan review and inspection of U.S.-flag merchant ships. Other such agreements allocate to ABS responsibility for assigning tonnage certification and stability reviews.

Some unique projects in the 1980s included issuing a guide for preventive maintenance of ships (1985), and guides for lay-up and reactivation of both ships and mobile offshore drilling units (1986).

In 1985, ABS completed development of a family of computer programs known as the EAGLE system. Components of this system are available to the public and can be used for hull girder strength analysis, ship hydrostatic curves (stability, trimming qualities, etc.) and many other calculations carried out in every naval architecture office. During the same decades continuing improvements in equipment allowed offshore drilling units to continue operating during increasingly difficult weather and sea conditions. From a typical 100-day drilling season of 1975, year-round operability was achieved in the case of Arctic units that can submerge and go on drilling during heavy weather.

Current, or recently introduced developments, include new rules for fishing craft, motor yachts, high-speed planing boats, hotel barges, towing vessels, dredges, and SWATH (Small Waterplane Area Twin Hull) craft.

As may be inferred from all of the above, a history of the American Bureau of Shipping is one of an ever-developing institution with a rich history that is difficult to describe adequately in a few pages.

The close relationship that has existed between ABS and SNAME ever since the latter's founding in 1893 merits special emphasis. SNAME's first president, Clement A. Griscom, was chairman of the Committee of Owners of Iron and Steel Vessels of the American Shipmasters' Association. Several presidents of ABS also served in the same position with SNAME: Stevenson Taylor (president of SNAME for two terms: 1909–11 and 1916–17), Walter L. Green (president of SNAME 1957–58), and Robert T. Young (president of SNAME 1977–78).

Members of the ABS technical staff have traditionally helped SNAME in its *raison d'être*, which is the dissemination of technical knowledge through its several publications. Principal among these is the set of annual *Transactions*. Here we find the following significant papers written by members of the ABS technical staff:

In 1954 (Vol. 62) D. B. Bannerman, Jr., proposed what was then an unheard of concept (but which is now fast becoming standard practice): double-skin hulls for tankers.

In 1965 (Vol. 73) J. L. Foley presented guidance for the structural design of oceangoing barges with higher than normal ratios of length to depth. In his paper he made the bold point that classification societies should not discourage new concepts by requiring prior satisfactory service as a condition of classification.

In 1969 (Vol. 77) H. A. Kamel, W. Birchler, and D. Liu (with co-authors) explained the theory and application of finite element analysis applied to ship structures. Their basic approach has since become the most prominent tool in analyzing ship structures.

In 1971 (Vol. 79) R. S. Little (and co-author) reported on measurements of longitudinal bending stresses encountered by ships in service. This work gave naval architects increased confidence as they were called upon to design tankers of unprecedented size.

In 1973 (Vol. 81) R. T. Young spelled out how ABS analyzes innovative designs in which no guidance is available from operating experience. The techniques described included finite element analysis and model test results, and all aided by sophisticated computer-assisted analytical tools.

In 1975 (Vol. 83) S. G. Stiansen (and co-author) explained methods they had developed for applying advanced statistical techniques so as to bring greater confidence to the design of hull structures. A unique aspect of the paper was the recognition of horizontal (or lateral) as well as vertical bending loads in design calculations.

In 1976 (Vol. 84) A. M. T. Elbatouti, H. Y. Jan, and S. G. Stiansen reported on the results of stress measurements made on a 17-foot model of a containership with wide hatch openings. Of particular significance was the close correlation between the model test results and theoretical predictions.

In 1977 (Vol. 85) S. G. Stiansen and Y. N. Chen (with co-author) presented conclusions and recommendations related to comprehensive studies of problems unique to the extreme proportions of Great Lakes bulk carriers. Their work allowed ABS to formulate rules for the structures of ships of increased length.

In 1979 (Vol. 87) S. G. Stiansen, H. Y. Jan, and D. Liu reported the results of stress measurements carried out on a large containership (one of the SL-7 designs). They were able to add that they found excellent agreement between those measurements and stresses predicted by the DAISY computer system. The paper was awarded SNAME's Linnard Prize, as the best contributed to that year's *Transactions.*

In 1981 (Vol. 89) D. Liu (and co-authors) read a paper focusing on how to design machinery foundations so as to combine harmoniously the hull and machinery interaction.

In 1983 (Vol. 91) Y. K. Chen, L. M. Kutt, and C. M. Piaszczyk (with co-author) presented a milestone paper describing advanced procedures for predicting hull stresses subjected to dynamic, as well as static, loads. Their computer-assisted method recognized such complex factors as elastoplastic properties, geometrically nonlinear behavior, and buckling and post-buckling strength.

In 1984 (Vol. 92) H. Y. Jan and Y. N. Chen (with co-authors) explained a rationale for selecting and calibrating a format of reliability-based strength standards for both ships and offshore structures. Their paper included useful recommendations for a checking format, and cited typical values of target reliability.

In 1985 (Vol. 93) L. M. Kutt, C. M. Piaszczyk, Y. K. Chen, and D. Liu read a follow-on paper to that presented in 1983 in which they concentrated on the ultimate longitudinal strength of ships. They described the complexity of applying nonlinear finite element methods to the task at hand.

In 1986 (Vol. 94) Y. K. Chen, J. W. Chiou, and A. K. Thayamballi evaluated methods for predicting ship structure fatigue conditions. Their work was based in part on the SL-7 full scale measurements reported in the 1979 paper. Of particular concern was the effect of fatigue at the vulnerable hatch corners.

In 1987 (Vol. 95) A. K. Thayamballi, Y. K. Chen, and H. H. Chen compared deterministic and reliability-based strength analysis methods for ship structures. Their work, which involved careful study of nineteen oceangoing tankers and bulk carriers, was an important step in easing the transition from deterministic to reliability-based assessments of hull structures.

In addition to the work implied in the above references, ABS engineers have been active in several SNAME technical panels, on many committees, and in the activities of the various local sections. It is doubtful that any other organization has done more to advance SNAME in its stated goals.

Donald Liu is currently Senior Vice President of the American Bureau of Shipping and head of the Technology group. He has been with ABS for 25 years and has held positions in design review, engineering computer program development, and research development with the Bureau.

Dr. Liu is a graduate of the U.S. Merchant Marine Academy, has his Bachelor's and Master of Science degrees in Naval Architecture from MIT, and a Ph.D. in Mechanical Engineering from the University of Arizona.

Regulation

William A. Cleary, Jr.

Introduction

The society's second half century has seen maritime regulatory development unparalleled in history. Before the half-century mark was reached there were only two sets of international maritime regulations. These comprised the agreements of the 1929 Safety of Life at Sea Convention (SOLAS) and the 1930 Load Line Convention. As we shall see in the following paragraphs, those agreements have been complemented and supplemented by many new international initiatives. In addition, the U.S. Congress has unilaterally mandated other regulations, some of which bear on foreign- as well as U.S.-flag ships. SNAME members from all segments of the marine industry have participated on many U.S. and Canadian delegations and advisory groups formulating positions for each of these international forums.

The Congress has on occasion rearranged the administrative structure of its regulatory agencies, which has in some cases caused a different shading of emphasis in interpretation and application of the laws. At the beginning of this half century, regulatory responsibility for merchant shipping lay in the Bureau of Marine Inspection and Navigation (BMIN) of the Department of Commerce. In 1942 BMIN was placed under the U.S. Coast Guard, which was then under the Department of the Treasury (a reflection on its initial primary role in discouraging smugglers). Then, in 1967 the Coast Guard (with its BMIN duties) was transferred to the newly-formed Department of Transportation, where it remains.

The national regulations for ship design and construction have existed in the same approximate location in the Code of Federal Regulations (CFR) for the entire half century, although they have been continually expanded and altered at the request of Congress. The following table is an abbreviated outline of the applicable regulations that have come into force since 1940 (as they now exist in 46 CFR Parts 1–199, subchapters A through V):

	Sub-chapter	Date of Origin
D	Tank Vessels	—
E	Load Lines	—
H	Passenger Vessels	—
I	Cargo and Miscellaneous Vessels	—
O	Bulk Dangerous Cargoes	1980
	Incineration at Sea	1988
R	Nautical Schools	1952
S	Subdivision/Stability	consolidated 1983
T	Small Passenger Vessels	1960
U	Oceanographic Research Vessels	1968
V	Marine Occupational Safety & Health, Commercial Diving Regulations	1978

In addition to their own in-house capabilities, the federal regulatory agencies rely for assistance on the American Bureau of Shipping, the National Cargo Bureau, the National Institute of Electrical Underwriters, the National Fire Protection Association, and other private organizations.

During the past half century regulations have expanded in scope and in the number of safety functions that federal agencies ask to review. There are now well over a dozen safety functions that must be satisfied in order to comply with national or international regulations. In alphabetical order these are:

Safety Function	Reviewed by U.S. Coast Guard or to USCG Satisfaction by
Automation	
Chemical cargoes	
Elec. installations/hazard. areas	
Environmental pollution	
Fire protection	
Freeboard	American Bureau of Shipping
Grain cargoes	National Cargo Bureau
Hazardous materials	
Lifesaving appliances	
Machinery	American Bureau of Shipping
Manning	
Navigation	
Offshore structures	American Bureau of Shipping
Stability	
Hull structure	American Bureau of Shipping
Subdivision	
Vapor collection systems	
Watchkeeping	
Weathertight integrity	American Bureau of Shipping

At least brief mention should be made of the Maritime Administration's role in advancing the cause of ship safety through its design standards. In particular, that agency deserves credit for keeping alive subdivision of merchant ships when other world interests tended to ignore it.

Developments in the 1940s

During the first half of the 1940s the maritime nations were too involved in World War II to give much thought to changing regulations. In 1948, however, the United Kingdom, which had been the host country for the 1929–30 SOLAS and Load Line Conferences, organized a follow-up conference to reconsider the SOLAS convention. At that conference a number of new safety functions were introduced with the intent that all nations would place them in their regulations. These new functions included stability reviews for all ships, damaged stability requirements for passenger ships, and calculations for the carriage of grain in bulk. Of the 46 executive and technical members of the U.S. delegation to the 1948 SOLAS conference, 18 were members of SNAME, including the U.S. Delegate: ADM Joseph F. Farley, USCG; among others were David Arnott, ABS; J. L. Bates, Maritime Commission; H. Gerrish Smith, Shipbuilders Council; George Sharp; and CDR C. P. Murphy, USCG (later RADM).

Attending the same conference was a six-person Canadian delegation which included another SNAME member, Lieut. D. I. Moore, RCN.

Also in 1948 the newly formed United Nations adopted a convention that led, in 1958, to the formation of the Inter-Governmental Maritime Consultative Organization (IMCO), which in 1982 changed its name to the International Maritime Organization, commonly referred to as IMO.

During the latter half of the 1940s our Coast Guard regulations for ship safety remained centered on the original passenger ship regulations; most other ships being grouped in the Cargo and Miscellaneous, or Tankship, subchapters except for the specialized regulations on engineering and electrical installations.

American regulatory practices were spread throughout the world during the post-war period as U.S. war-built ships were sold overseas to replace wartime losses.

Developments in the 1950s

In 1952, the Coast Guard began requiring approved stability information on cargo as well as passenger ships.

In 1956 the *Stockholm/Andrea Doria* collision exposed certain shortcomings in the SOLAS rules. Once again the conference met in London, this time under the United Nation's new organization for marine safety, IMCO. (At the invitation of the United Kingdom, that organization was headquartered in London.) This led to the 1960 SOLAS Convention, which differed from the 1948 Convention because of lessons learned from the *Stockholm/Andrea Doria* collision. Other changes arose from the availability of new lifesaving systems, and from new operating procedures suited to the use of radar. A new part was added on the safety of nuclear ships.

The U.S. delegation to the 1960 SOLAS conference was led by ADM A. C. Richmond, USCG, and again included a wide cross section of members from the U.S. maritime industry. Among these were Robert T. Merrill, John P. Comstock, Vito L. Russo, Owen Oakley, J. B. Robertson, Jr., E. Bentzenberg, Matthew G. Forrest, RADM H. C. Shepheard, USCG (ret), and LCDR R. I. Price, USCG (later VADM). Canada sent a delegation of seven, including three SNAME members: Alan Cumyn, J. H. Kay, and H. O. Buchanan.

In 1958, after the loss of the passenger fishing boat *Montauk*, off Long Island, Congress directed the Coast Guard to develop regulations for all small passenger vessels. This led to the origin of Sub-chapter T, which eventually became the most actively applied portion of the regulations.

Also during this period the United States began pressing for a new Load Line Convention. To help generate interest, the government created a Load Line Committee entirely from U.S. industry, the Coast Guard, and the American Bureau of Shipping. (Almost all of this committee's members were also members of SNAME.)

In 1954, the first attempt was made by the international community to reduce oil pollution resulting from routine operations, such as tank cleaning, in the Convention for the Prevention of Pollution of the Sea by Oil.

Developments in the 1960s

IMCO played host to the 1960 SOLAS conference in London and began to meet regularly thereafter to discuss not only matters of ship safety, but also questions of highseas environmental protection. Since that time SNAME members have been deeply involved in the development of maritime safety matters worldwide. These SNAME members were not only federal employees, but also members of shipping companies, naval architecture firms, shipbuilders, ship repairers, and entities dedicated to environmental protection. With so many organizations involved, and with so many new technological developments, it is not surprising that the process required increasingly frequent meetings, some at IMCO, but others closer to home. In the United States, for example, a two-level advisory group system has now evolved. Preliminary positions are developed by USCG staff, often in consultation with industry. Public meetings are held to discuss and modify positions before the official national positions are sent to IMO. Industry representatives are regularly invited to contribute throughout this process.

In 1961, IMCO formed the Maritime Safety Committee, which in turn began to create special subcommittees for such functional areas as stability, subdivision, and fire protection. A subcommittee on safety of fishing vessels was formed in 1968. By 1974 this produced a U.N. code for fishing vessels, and this became the basis for a 1977 conference on fishing vessels.

In a purely North American development during the mid-1960s, at the request of Great Lakes shipping interests, SNAME itself began research into the strength and seakeeping capabilities of Great Lakes bulk carriers. This work was instrumental in a parallel regulatory joint effort by Canada and the United States. This was centered in the Joint Technical Committee for Great Lakes Load Lines whose membership was formed, as it happened, exclusively of SNAME members from industry, classification societies, universities, and government agencies. Based on this work, in 1974 the two neighboring nations reached a joint accord on new load line regulations, and this was modified some four years later to establish new strength standards for Great Lakes ships up to 1000 feet in length.

During the 1960s there were significant developments in tanker arrangements. The tanker *Manhattan*, built early in the decade, was divided into no fewer than 45 cargo tanks. Subsequent ships (mostly built overseas) featured far fewer (but, of course, far larger) tanks. This development led to structural complications brought on by inertia forces of the liquid cargo that were not always kept under control by swash bulkheads. SNAME, the American Bureau of Shipping, and the Federal Government all supported research that eventually found satisfactory solutions to the problem.

The 1960s were also marked by massive demands for overseas grain transport. Many American ships of World War II vintage carried grain to India and other nations suffering from shortage of food. The need arose to develop and apply special bulk grain stability regulations. Owners were required to supply their officers with information necessary to ensure safety against catastrophic shift of cargoes.

A second load line conference met in early 1966 and completed the international convention that is in effect today. Of the 18-member U.S. delegation, nine were SNAME members. These included David Bannerman, ABS; R.T. Cunningham, AMMI; Captain A. H. McComb, Jr., USCG; J. B. Robertson, Jr., USCG; C. S. Smith, Lake Carriers Association; J. L. Stevens, Jr., Shipbuilders Council; and E. V. Stewart, Jr., API. The Canadian delegation included another SNAME member: H. O. Buchanan.

Another late 1960s international development, which came to regulation in the next decade, concerned bulk chemical carriers. Then, later, came the "drugstore" chemical carriers, capable of carrying more than a dozen different bulk products simultaneously and delivering any amount at any port. This sometimes created unique loading, offloading, and stability problems that came under international discussions leading to an international code for bulk chemical ships.

In the United States during this period, published figures on commercial fishing fatalities led some authorities to recommend special regulations for boats engaged in that industry. Nevertheless, the independently minded American fishermen refused all regulatory assistance and kept on losing lives at a rate far above any other industry. Improvements by regulation had to wait for another two decades, until 1990.

In 1969, another IMCO conference generated a completely new approach to tonnage measurement, one that overcame the shortcomings of the old Moorsom system and its burden of illogical rule-beating clauses that led to lessened seaworthiness. Again, SNAME members were the majority of the delegation, which was headed by RADM Murphy, USCG, and which included LCDR H. J. Kelly, USCG; R. C. Krack, MarAd; C. S. Smith, and E. V. Stewart, Jr.

Developments in the 1970s

Again, throughout the 1970s, SNAME members were active participants in national and international conferences that led to developing necessary regulations in several important maritime sectors. One example was the aforementioned code for ships carrying dangerous chemicals in bulk: CHEMCODE (1971). Another IMCO initiative was the International Convention for the Prevention of Collisions of Ships at Sea: COLREGS (1972).

During this period there was increasing international concern over oil pollution. This led IMCO to sponsor the Marine Pollution Convention of 1973 (MARPOL 73). Once again the U.S. delegation was dominated by SNAME members under the leadership of ADM C. R. Bender and RADM W. M. Benkert. The United States was a strong proponent of these regulations. Congress accordingly directed that the new regulations be enforced in U.S. territorial waters without waiting for international signatory support.

In 1974 an IMCO conference on the SOLAS convention upgraded the 1960 SOLAS regulations, so that we now use SOLAS 74 and its amendments in lieu of the older SOLAS convention. The U.S. delegation to the 1974 SOLAS convention included SNAME members RADM W. M. Benkert, USCG; J. B. Robertson, Jr., USCG; S. F. Sammis, U.S. Cargo Bureau; and E. H. Middleton.

A unique set of problems accompanied the introduction, in the early 1970s, of liquefied natural gas (LNG) carriers. This type of ship had to carry cryogenic liquids safely in ocean storms. This led IMCO, in 1975, to produce a special code for such ships: GASCODE, which recognized both design and operating problems peculiar to LNG carriers.

In 1977 IMCO held a conference at Torremolinos, Spain at which it developed the Torremolinos Convention for the Safety of Fishing Vessels. The U.S. delegation was headed by RADM Benkert and included SNAME members E. H. Middleton, W. A. Cleary, Jr., and George C. Nickum. Mr. Nickum deserves special mention since he served, at the invitation of the U.S. Coast Guard and concurrence of the State Department, as the U.S. Representative to the Subcommittee on Safety of Fishing Vessels for the decade 1967–1977.

Three members of the Canadian delegation at the Torremolinos conference were SNAME members: W. E. Bonn, H. I. Shenker, and a Mr. Laing.

Also during the 1970s, the original few subcommittees of the 1960s were developed into a series of technical subcommittees specializing in each technical safety function, such as fire protection, stability and subdivision, bulk cargoes, etc. There are now more than ten subcommittees serving the two main committees: Maritime Safety and Marine Environmental Protection. These committees and their subcommittees meet at least once a year. During the decade leading up to the SNAME Centennial literally hundreds of SNAME members have served on these committees, subcommittees, and advisory groups. Space does not allow us to name them all, but the foregoing paragraphs have recognized at least a few of the key participants.

During the second half of the 1970s IMCO produced a steady stream of codes and conventions, some of which are mentioned below. Each conference noted as having "convened" actually sat in session, completed its assignment, offered the resulting convention or code to all nations (through IMCO), and then *disbanded* itself.

1973: Convened the Conference on Marine Pollution
1975: Formed the Maritime Satellite Organization (INMARSAT)
1977: Convened the Conference for Safety of Fishing Vessels
1977: Produced the Code for Dynamically Supported Craft
1978: Convened the Conference on Training and Watchkeeping for Seafarers
1978: Convened the Conference for Protocol on Marine Pollution
1979: Convened the Conference on Maritime Search and Rescue
1979: Produced the Code for Offshore Drilling Units

The U.S. delegation to the 1978 Protocol on Marine Pollution included SNAME members RADM Benkert, LCDR W. D. Snider, E. H. Middleton of USCG, and J. J. Nachtsheim of the Maritime Administration.

Developments in the 1980s

The 1980s saw a severe slump in shipbuilding activities, which led to a slowdown in the growth of regulations. Nevertheless, IMO turned out the following codes:

1981: Noise Levels
1981: Safety of Nuclear Merchant Ships
1983: Safety for Special Purpose Ships (training, fish factory, etc.)
1983: International Gas Carriers (superseding existing codes)
1983: International Bulk Chemicals
1983: Safety for Diving Systems
1989: MODU Code, revised

In the United States the Coast Guard consolidated the stability regulations into sub-chapter S, and rewrote the regulations pertaining to tonnage measurement and electrical engineering. The Coast Guard also developed new regulations applied to such diverse vessels as those engaged in commercial diving, sail training operations, and incineration at sea.

SNAME played an active part in the preparatory work for the total revision of SOLAS Chapter III, Survival Systems, which was adopted in 1973. Under the chairmanship of SNAME member Robert Markle, SNAME Panel 0-25 developed text for a number of new requirements and served as a sounding board for international proposals. The new Chapter III embodies a totally new philosophy about lifesaving system design, which gives naval architects and marine engineers greater design flexibility as well as responsibility.

One of the most significant developments toward improving the safety of ships worldwide came in early 1985. The IMO Maritime Safety Committee reported then that there was a compelling need to develop a standard of subdivision and damage stability for dry cargo vessels based on the probabilistic method. Prior to that time, dry cargo ships were not required to meet any damage stability criteria. The development of these rules was a high priority for the Coast Guard and was advanced through IMO in short order through the diligent efforts of many Coast Guard, MarAd, and industry SNAME members, including: RADM J. W. Kime (now ADM), Capt. J. Card, Capt. J. Maxham, LT R. Gilbert, J. Spencer, USCG; R. Sonnenschien, MarAd; R. Tagg, Herbert Engineering; and W. Peters, C. R. Cushing & Co. The rules were approved by IMO in 1988 as amendments to the SOLAS Convention and entered into force for new ships in 1992. In 1989 SNAME President Arthur Haskell wrote a letter to the Coast Guard expressing the Society's support for the new international damage stability standards for dry cargo ships.

The passage of the Commercial Fishing Industry Vessel Safety Act of 1988 was the first successful attempt by the U.S. Congress to enact comprehensive safety legislation for one of the most dangerous industries in the country. Major provisions of the Act include a study addressing safety problems conducted by the National Academy of Engineering, development of safety regulations, development of a plan for the licensing of operators, distribution of educational material to fishermen, and formation of a safety advisory committee. Prior to the Act, the Coast Guard had set up a task group, in 1985, headed by SNAME member CAPT Gordon Piché, USCG, to study the fishing vessel problems and develop a voluntary safety program to improve the awareness level of commercial fishing vessel operators.

Developments in the 1990s

In 1990 one of the most far reaching pieces of marine legislation was passed by the U.S. Congress in response to the *Exxon Valdez* disaster—The Oil Pollution Act of 1990 (OPA 90). Major provisions of OPA 90 include: double hull construction for all tankers entering U.S. waters, certificates of financial responsibility for oil spill liability and compensation, response plans, discharge removal equipment carriage requirements for vessels, manning standards, escorts for certain tank vessels, pilotage, vessel traffic systems, and various reports and studies. To manage the enormous burden of developing the necessary regulations, studies, and reports, the Coast Guard established (under the Office of Marine Safety, Security, and Environmental Protection) the OPA 90 staff headed by SNAME member Norman Lemley. The 75-member OPA 90 staff includes naval architects, economists, lawyers, editors, and the other necessary support positions to manage the monumental undertaking of implementing the provisions of OPA 90.

Conclusion

As indicated at the start, the past fifty-year period has seen a virtual explosion of new and revised maritime regulations. Most of these have resulted from clearly perceived needs and have been hammered out in international conferences. Members of the Society of Naval Architects and Marine Engineers have played important roles in all this, and the Society, itself, has made contributions through the activities of its technical and research committees. As we look ahead to the next half-century we can expect to witness a continuation of enlightened regulatory work advancing the safety and reliability of the many kinds of boats and ships that carry out our nation's commerce and industry.

William A. Cleary, Jr., is a maritime consultant and Adjunct Professor at Florida Institute of Technology. For two decades he served as Chief, Naval Architecture, Office of Marine Safety and Environment, U.S. Coast Guard. He represented the United States on several IMO subcommittees and special topic meetings. He chaired one of the technical committees at the Torremolinos Convention on Safety of Fishing Vessels, co-chaired the USA/Canada Joint Technical Committee for Great Lakes Load Lines, and was an active participant in international safety discussions.

He was an Engineering Duty Officer in the Naval Reserve and is now Captain, USNR (ret). He has been a registered professional engineer for over 30 years and is a Life Member of SNAME and a Fellow of the Royal Institution of Naval Architects and a European Engineer. He has served on SNAME T&R committees since 1967.

David Taylor Research Center

JUSTIN H. MCCARTHY

Introduction

THE DAVID TAYLOR RESEARCH CENTER (DTRC) is the largest and most comprehensive ship research facility in the Western World. Its mission is twofold: (1) to serve as the principal U.S. Navy RDT&E (Research, Development, Test, and Evaluation) center for naval vehicles, experimental aerodynamics, and naval logistics; and (2) to provide RDT&E support to the Maritime Administration and the nation's maritime industry. During the late 1980s DTRC was composed of eight technical departments and a number of support departments with a total of almost 3000 employees located primarily at facilities in Carderock and Annapolis MD, but also at field stations from Alaska to Florida. The technical departments taken together encompass the disciplines required for conducting nearly all aspects of RDT&E on ships, submarines and other vehicles: hydrodynamics, aerodynamics, structural mechanics, materials, machinery, acoustics, electromagnetics, logistics, and systems integration. The principal sponsor of work at DTRC has been the Naval Sea Systems Command (NAVSEA) and its predecessor organizations, most notably, the Bureau of Ships (BuShips).

DTRC was established in 1967 when the David Taylor Model Basin (DTMB) at Carderock was merged with the Marine Engineering Laboratory (MEL) at Annapolis MD. At the time of the merger each laboratory had established long and distinguished records of service to the Navy and maritime industry, dating back to the origins of their predecessor laboratories at the turn of the century: the Experimental Model Basin (EMB) at the Navy Yard on the Anacostia River in Washington D.C. and the Engineering Experiment Station (EES) on the Severn River in Annapolis MD. Both laboratories were conceived during a period of major advances in ship technology and extensive expansion of the U.S. Navy and merchant shipping to support the emergence of America as a world power. Over the years these component laboratories and their successor organizations have been the first stops for evaluation of virtually all new naval surface ship and submarine designs, at both model and full scale.

The Years Up to 1939

In the first two decades of operation both EMB and EES had to overcome similar difficulties in acquiring equipment, establishing experimental procedures and building staffs. By 1920, however, different personalities had developed at each laboratory. EES was recognized as the Navy's center for testing equipment and had evaluated over 1000 items for shipboard use; and metallurgical and chemistry laboratories had been established. EMB tested over 2000 ship models, and it had produced a body of pioneering fundamental research that advanced the state-of-knowledge of naval hydrodynamics and aerodynamics.

The David Taylor Model Basin, known familiarly throughout the world as DTMB or TMB or the "Model Basin," and its successor, David Taylor Research Center, are named for David Watson Taylor who, as a brilliant young naval officer, was the driving force behind authorization by Congress in 1896 of the nation's first ship model basin. David Taylor served as Director of EMB for 15 years, from its opening in 1898 until 1913, and laid both the scientific and engineering foundations for future naval research in hydromechanics and aerodynamics. This period resulted in many important technical papers and in 1911 culminated in publication of Taylor's renowned book, *The Speed and Power of Ships*. This book summarized Taylor's pioneering work on the effects of bulbous bows and parallel middle body on ship wavemaking resistance, and presented the Taylor Standard Series experimental data that continue to be used by naval architects throughout the world for estimating the effects of changes in ship form on wavemaking resistance. Subsequently Taylor and his successor, William McEntee, were successful in building the Navy's first wind tunnel, thereby initiating in 1914 a long tradition of aerodynamics research and development (R&D) at EMB.

Fig. 1 Rear Admiral David Watson Taylor.

Taylor attained the rank of Rear Admiral in 1916, subsequently serving as Chief of the Navy's Bureau of Construction and Repair, and after retirement serving from 1925 to 1927 as President of SNAME.

On May 6, 1936, recognizing the need for new and expanded facilities, the 74th Congress authorized, under Public Law No. 568, the construction of DTMB at Carderock, a facility where the Bureau of Construction and Repair

> shall conduct the work of investigating and determining the most suitable and desirable shapes and forms to be adopted for United States vessels, including aircraft, and the investigation of other problems of ship design . . . experiments may be made at this establishment for private parties, who shall defray the cost thereof

DTMB was dedicated at the advent of World War II on November 4, 1939 in the presence of Admiral Taylor, who died the following year. The construction of the new facilities was directed largely by Captain Harold E. Saunders who had worked out the detailed plans as early as 1933, and served as Technical Director from 1941 to 1946. The main building, faced in gleaming white cement imbedded with quartz, still stands as a fine example of the art deco style of the 1930s. In 1989 it was designated as a building of historical importance by the National Trust for Historic Preservation.

Over the years, those who have worked at DTMB have taken special pride in identifying with an institution bearing the name of David Taylor the man, the father of American ship research. In the words of Emory S. Land, a longtime younger friend and naval colleague, Taylor's "sense of humor was so keen, his interests were so varied, his knowledge so great, and his character so splendid" that he influenced deeply all who came into contact with him or his works.

Soon after the time that Taylor was petitioning Congress for a model basin, Rear Admiral George Wallace Melville was lobbying for another sort of laboratory. Melville was a famous, colorful arctic explorer, author and naval engineer. He was selected in 1887 as the Navy's first Chief of the Bureau of Steam Engineering. The aim of his lobbying was to establish a steam engineering experiment station and testing laboratory for the development of fleet standards for the operation and maintenance of shipboard machinery.

Fig. 2 Main building of David Taylor Model Basin.

Fig. 3 Rear Admiral George Wallace Melville.

Congress authorized the Engineering Experiment Station at Annapolis in 1903. Melville had realized very early the urgent need for an R&D site where naval machinery could undergo examination and trial for reliability before shipboard installation and operation. During his last decade of service he also succeeded in completing the extensive reform of the entire U.S. naval engineering community and in the founding of the Naval Postgraduate School.

Operations began at EES in 1908, following construction and equipment acquisition, under the supervision of Commander Walter F. Worthington, who led the Station in the early years. In 1910, however, the Philadelphia Naval Shipyard took over the testing of main power plant boilers and turbines. The EES engineers carried on in such miscellaneous areas as heat transmission equipment, gasket and packing material, metallurgy, and lubricants. In the 1920s the laboratory carried out important fundamental work in corrosion fatigue and in boiler water treatment. In the 1930s the laboratory was distinguished for its work in welding technology and internal combustion engines. The latter was in support of the design and development of American diesels for submarines. Before 1932 all diesel engine designs were of German origin. By 1939 EES had five laboratories: Mechanical, Chemical, Metals, Welding, and Internal Combustion Engines.

To quote from Carlisle's summary of the history of EMB and EES during the period between the two World Wars:

> both laboratories [EMB and EES] flourished. For, despite the national isolationism and the opposition to naval rearmament, the interwar period was one in which both laboratories grew and matured. Their identities as major facilities became even more firmly established in these two decades. Ship, propeller, and naval aircraft designs would be tested at the Basin, and theoretical scientific work would be kept up [in 1939 approximately 8 percent of the total budget at EMB was identified as "research funding"]; specifications for metals, gaskets, valves, gauges, packing, engine parts, and oils would be set at the Station and products would be evaluated against the specifications. Although most of the categories of work at the two laboratories in 1939 bore a striking resemblance to those in the period 1910–1915, the value, the mission, and the significance of the two laboratories had all changed drastically. By the end of the interwar period, each had established that it was crucial to the modernization of the Navy. American ships, aircraft, and submarines of World War II would go into battle carrying the ideas, the ingenuity, and the hard work of Basin and Station staff.

The most famous and most important fundamental output of EMB between the wars was Karl Schoenherr's 1932 dissertation on the frictional drag of flat plates. Based on careful experimental measurements in the EMB towing tank, married to a theoretical framework derived by Theodore von Karman, the "Schoenherr Line" was adopted by the American Towing Tank Conference in 1946 as the standard for scaling frictional resistance from model to full-scale Reynolds numbers. It served as the baseline for the "Correlation Line" adopted by the International Towing Tank Conference in 1957.

The War Years and Aftermath (1939–1950)

The first two years after dedication of DTMB in November 1939 were devoted to rapid installation and break-in of facilities and equipment. Under the supervision of Captain Saunders, the main deep water towing tank was placed in operation following painstaking

Fig. 4 Wartime workers on DTMB towing carriage.

alignment of carriage rails to within a demanding 0.002-in. tolerance. The original length of this basin was 963 ft. Funding was appropriated in July 1944 to extend the length to 2968 ft and to build a high-speed basin for torpedo experiments next to the deep water basin. The extended basins were opened in August 1946 with regular testing beginning in 1948. The 12-in. Variable Pressure Water Tunnel (VPWT) and 600,000 lb universal materials testing machine at EMB were moved to DTMB before 1941 and a 24-in. VPWT was opened at DTMB in 1941. Additional facilities, opened later, during the war, included an explosives/structures test pond in 1943 and a large circulating water channel in 1944. Two major subsonic wind tunnels were also opened at DTMB in 1944 to supplement two existing wind tunnels that continued to be operated at EMB throughout the war.

Two months after our entry into the war DTMB was operating at two full shifts per day and EMB at one shift per day. The tremendous expansion in facilities could not have been more opportune. Between 1940 and 1945 DTMB employees grew in numbers from 200 to 870. These comprised 164 naval personnel and 706 civilians, who worked eight hours on Mondays, Wednesdays, Fridays, and Saturdays and twelve on Tuesdays and Thursdays. During the same period EES grew from about 400 to 1200 employees. Although there was an enormous influx of new personnel during the war, long-time employees, from the 1930s and earlier, continued to provide important leadership. At EES examples include Bela Ronay as head of the welding group, and at DTMB, Karl Schoenherr and Louis Landweber in hydrodynamics, and James Windenburg in structural mechanics.

The changes brought about by war in Europe were accompanied early on by a change in the "ownership" of DTMB and EES. In June 1940 the Bureau of Engineering and the Bureau of Construction and Repair, to which EES and DTMB, respectively, reported, were merged to form the Bureau of Ships (BuShips). BuShips thus became the common master and sponsor during the war of the hundreds of ship, submarine, and small craft designs evaluated at DTMB, and the thousands of equipment components tested at EES.

At EES the traditional testing and evaluation of new products continued at a vastly quickened pace with added emphasis on diesel engines. At one point as many as 40 large diesel engines were in operation to solve a multitude of development problems. By war's end, owing in part to the successful diesel program at EES under W. F. Joachin, there was more total installed diesel horsepower than steam horsepower in U.S. Navy vehicles. Still another horizon was anticipated with establishment in 1942 of a Gas Turbine Laboratory at EES to investigate a revolutionary new "blue-sky" technology that would replace steam on combatants beginning in the 1970s. An early open-cycle regenerated gas turbine of 3500 horsepower underwent evaluation at EES in 1944 and operated through 1949. In parallel with the crucial evaluation work to support the war effort, EES worked with industry to identify problems, and supplied manufacturers with ideas on how to improve their products, thereby affirming an important role for EES in product development. For a two-year period during the war Robert H. Goddard, the famous rocket pioneer, conducted research at EES on jet-assist take-off (JATO) devices for airplanes.

At DTMB the war focus was almost entirely on practical work involving the test and evaluation of myriad ship, submarine, aircraft, and weapon designs. There was little time for theoretical or experimental research. The projects, rather, dealt with streamlined housings, sound projectors, and hydrophone struts. Shallow water experiments were devised to demonstrate that the aircraft carrier *Essex* could negotiate the Panama Canal. Specifications for the landing craft to be used in the Sicilian invasion were delivered to BuShips after only 42 hours of testing. Parachutes were tested underwater; and the effects of exploding mines on boots were evaluated on cadavers. It was a period of long hours and high *esprit de corps* in support of the war effort.

At war's end, tremendous changes were taking place in public attitudes toward the importance of science and engineering to the nation, particularly with regard to the vital connection between war capability and sustained scientific research. This was vividly demonstrated by the Manhattan Project, which produced the world's first atomic bomb. In the aftermath of the war, national science figures, most notably Vannevar Bush, Chairman of the Wartime Office of Scientific R&D, and James B. Conant, President of Harvard University, effectively articulated the need for a strong national science program in the United States. They also argued for a permanent science-based arms development program. In the words of Vannevar Bush, science was the "endless frontier" to be vigorously promoted for the future health, welfare, and security of the United States. Within this environment, the Office of Naval Research (ONR) was established in 1946 as the first permanent U.S. agency devoted to funding basic research in the physical and social sciences at American universities and industrial and government laboratories. When the National Science Foundation was founded later many of its personnel were recruited from ONR, including its first director.

The desirability of science-oriented naval research did not elude the leaders at DTMB. Captain Saunders and Captain Herbert S. Howard, Commanding Officer of DTMB in 1942-1946, lobbied hard at BuShips and elsewhere to distinguish DTMB from testing-oriented laboratories. To quote Captain Howard: "Unlike most of the laboratories under the Bureau of Ships, which serve primarily for quality control of material, DTMB is responsible for fundamental research and specific development in the fields of structural mechanics, hydrodynamics and aerodynamics."

Captain Saunders, in a 1946 SNAME address concerning the American Towing Tank Conference, stressed the need for new thinking and new technology for ships and submarines, which would be required to perform in an expanded, three-dimensional operating envelope with six degrees of freedom:

> Time was when all we had to worry about was the heeling and rolling and turning of ships, but with the advent of the scientific attack on the hydrodynamics of the torpedo, and of mine-sweeping gear and guided missiles, . . . it became necessary for us to think very definitely in the same sort of general three-dimensional terms that have been standard for airplane work for the past fifty years. These involve roll, pitch, yaw, heave and all the other motions of ships.

At the same SNAME meeting, Commander E. A. Wright, who had served as a Deputy Technical Director of DTMB, presented a paper entitled "A Pattern for Research in Naval Architecture." Warmly received by a score of distinguished discussers, the wide-ranging paper laid out the requirements for basic and systematic research, new experimental facilities, and scientific personnel, including special emphasis on the funding and coordination of research at universities by academics. Specific topics recommended for intensive research included plasticity in steel, strain in ship structures, seakeeping loads on hulls, boundary layer and free-surface flows, propeller-hull interaction, and cavitation physics.

Fig. 5 Captain Harold E. Saunders.

It is thus not surprising that the years 1946–1950 proved at DTMB to be a period of drastic realignment to research-oriented growth as a result of the national shift in priorities. It entailed investigations of both fundamental issues that had been dormant during the war and new issues raised by wartime experiences. In 1949 DTMB had grown to about 1200 personnel, an increase of 330 from war's end. From Germany four wind tunnels were installed at Carderock, one large transonic tunnel, and three blow-down supersonic tunnels. In addition, a detachment was established at Lake Pend Oreille in Idaho for conducting underwater acoustics measurements. The post-war years were also a period of educational incentives, leading to an increase of graduate studies by employees. This time, too, was marked by the arrival of German scientists under "Project Paperclip," a government program put in place to enrich the scientific makeup of U.S. research establishments. Included in this group were three men who were to have a profound influence on the future work and attitudes at DTMB: Georg Weinblum, a ship hydrodynamicist; Herman Lerbs, a propeller theoretician; and Alfred Keil, an underwater explosions expert who came to DTMB later, following some years at the Norfolk Naval Shipyard. Weinblum returned to Germany during the early 1950s to become Director of the Institut für Schiffbau at Hamburg University, and Lerbs returned to become director of the Hamburg Ship Model Basin. Keil became the first civilian technical director of DTMB, serving from 1963 to 1967.

In the last three years of the decade no fewer than six systematic series of model powering experiments were initiated at DTMB. Each constituted an extension of the range of hull form parameters originally studied in the first decade of the century by David Taylor as part of his standard series. Series 56 and Series 61 investigated the effects of size and length/beam ratio on planing boat resistance; Series 57 and Series 60 investigated the effects of block coefficient and longitudinal center of buoyancy on merchant ship powering performance; Series 59 dealt with the effects of bow and stern shape on aircraft carrier resistance; and Series 58 investigated the effect of a full range of form parameters on the resistance of fully submerged bodies of revolution. Work on these series, comprising hundreds of models, extended well into the 1950s.

Additional series work was begun in the 1950s and 1960s, most notably Series 64 for high-speed transom-stern hulls. These series, together with the Taylor Standard Series, provide valuable data used worldwide by naval architects to estimate the powering performance of a wide range of vessels. The Series 58 work, which identified bodies of revolution of optimum shape for minimum submerged resistance, led to the U.S. Navy's first streamlined single-screw submarine, the *Albacore* (AGSS 569), and late in the 1950s to the revolutionary *Skipjack* (SSN 585) which combined a near optimum hull form with nuclear power, and formed the basis for all future submarine designs.

In addition to the extensive series experiments carried out on merchant ship hull forms, between 1945 and 1970 DTMB conducted model powering experiments for more than 150 merchant ship designs under contract to the Maritime Administration and virtually all of America's leading shipbuilders, shipowners and design offices. On average, this amounted to about six evaluations per year, down from approximately twice that number tested at EMB each year in the five years preceding World War II. In 1945, however, with the rapid drawdown of the wartime economy and conversion to peacetime, no fewer than 15 merchant ship designs were contracted for evaluation at DTRC. In the ensuing years tests were conducted on cargo and passenger ships, ore and bulk carriers, RO-RO and containerships, tug/barges, ferries, towboats, yachts and ocean platforms. Special investigations were conducted on the application of contra-rotating propellers to tankers (1964) and containerships (1972); and the first highly skewed propeller for a merchant ship was successfully designed and evaluated at full scale in 1973. By 1970, owing to intense foreign competition, shipbuilding had declined in the United States, with the majority of U.S.-owned merchant ship construction and associated model testing being conducted in Europe and Japan. Although after 1975 very little commercial work was to be conducted at Carderock, over the years many of the technology advances already realized for naval vessels ultimately benefitted the U.S. maritime industry.

At EES, two new areas of development were initiated in 1946 as a result of German submarine technological advances during the war. The first involved adaptation of snorkels to U.S. diesel submarines. Many iterations and test evaluations were conducted to perfect a snorkel that operated properly with two-cycle, as opposed to the German four-cycle, diesel engines. This work was successfully concluded in 1950. The second project evaluated closed-cycle engine systems of the Krieslauf and Walter types, tested in Germany during the war and intended to do away with the need of a snorkel air intake for diesel engines. Following investigation of the German concepts, three American closed or semi-closed cycle designs were evaluated in the late 1940s and early 1950s: the "Alton," "Wolverine" and "Ellis" gas turbine designs, all of which used hydrogen peroxide as oxidant. Rear Admiral W. D. Leggett, Commanding Officer of EES from 1947 to 1951 and later Deputy Chief of BuShips, wrote in 1951 about the importance of EES for the future:

> We have by no means reached a dead end in the improvement of naval power plants. On the contrary, the horizons are broader and farther away than ever before.
>
> We also have greater need than ever before to discriminate among the many possible engineering solutions to our propulsion problems . . . on a laboratory scale.

In 1949 BuShips expanded the EES charter to conduct R&D investigations as well as the traditional tests on engineering equipment and materials. With the important and major exceptions of the snorkel and closed-cycle engine work, this largely translated into determining causes of equipment failures, developing specifications, and testing items for compliance with specifications. With regard to new facilities, a wave mechanics laboratory was established in 1946 for evaluation and reduction of machinery vibration and noise. In 1950, EES acquired the diesel submarine *Turbot* to use as a floating laboratory for noise reduction experiments; and in 1952 a large anechoic noise facility, capable of handling complete diesel or turbine power plants, was completed and became the Navy's principal installation for such work.

In 1954, following the success of nuclear propulsion on the submarine *Nautilus*, work on closed-cycle engine systems was shelved. Henceforth, EES would work no longer on the main propulsion plants of submarines, but would continue to do significant work on submarine auxiliary machinery and electrical and mechanical systems. All of this work fed quickly into the new nuclear submarines.

The Research Years (1950–1970)

By the beginning of the 1950s, DTMB had grown to become one of the "big science" institutions along with the better known NACA and AEC national laboratories. In 1950 four major departments, dubbed laboratories, were in existence at DTMB: Hydromechanics, Aerodynamics, Structural Mechanics, and Applied Physics, the latter being absorbed by Hydromechanics in 1956. With the advent of large computers, the Applied Mathematics Laboratory was established in 1952 with a computing center built around a UNIVAC-A. This was replaced in 1960 by a LARC, and successively updated by other rapidly developing top-of-the-line main-frame digital machines. One of the most important early projects undertaken with the first UNIVAC-A computer was the numerical solution of a mathematical model of submarine nuclear reactor core life, leading to an accurate prediction of refueling requirements. Within a short time computers of every description entered the full range of work at DTMB, from on-line collection, sorting and analysis of experimental data to numerical solution of some of the never-before-attempted, most complex problems of computational fluid and structural mechanics. Computers added an all-important third leg, numerics, to the analytical and experimental legs already available to ship researchers and naval architects at DTMB.

By 1960, DTMB extended far beyond its Carderock site. In addition to the acoustic research detachment established in 1946 at Lake Pend Oreille, a mobile noise measuring barge (MONOB) was acquired to serve as the Navy's principal East Coast acoustics laboratory for ship radiated noise evaluations, initially operating out of Charleston and then Cape Canaveral. An underwater explosions research detachment, with the logistics support barge *UEB 1*, was formed at Portsmouth VA; and administrative control was assumed in 1961 for the high-speed hydromechanics facilities at Langley Field VA.

At Carderock itself two spectacular new hydromechanics facilities were commissioned: the Maneuvering and Seakeeping (MASK) Facility, dedicated in 1962 to honor Harold E. Saunders; this housed under one roof, covering five acres, the world's largest rotating arm basin (260-ft diameter) and a combined maneuvering/seakeeping basin (360-ft by 240-ft); and the 36-in. Variable Pressure Water Tunnel, opened in 1961 and designed for cavitation and noise measurements on large model propellers at speeds up to 50 knots.

The 1960s were marked by experimental and theoretical research at DTMB that led to greatly improved predictions of ship motions in a seaway and the introduction of highly skewed propellers for reduction of propeller-induced hull and shaft vibratory forces. Following the loss of *Thresher* (SSN 593) in 1963, both DTMB and EES became lead laboratories for the Navy's Submarine Safety Program. DTMB determined safe operating envelopes and developed maneuvering simulation software for crew training on all classes of submarines. EES conducted important engineering investigations of welding HY-80 steels and piping and emergency ballast blowdown systems.

With the increasing requirements for ship and submarine silencing associated with the race between the United States and Soviet Union to produce nuclear submarines, in 1963 an Acoustics and Vibration Laboratory was formed at DTMB from related elements of the other laboratories. Over the next two decades this department was to grow at a faster rate than any other, eventually taking on the lead Navy role for measurement and diagnosis of the full-scale radiated noise signatures of ships and submarines. To support this work, in addition to the East Coast measurement facility, acoustic ranges were added at Carr Inlet in Puget Sound WA, and at Santa Cruz Island CA. In order to provide rapid evaluations of the full-scale noise data, a secure Acoustic Data

Fig. 6 Aerial view of DTRC Carderock site.

Analysis Center was created at Carderock. For investigation of flow and flow-induced noise sources on ships and submarines, a unique Anechoic Flow Facility was built at Carderock in the late 1960s. It is a subsonic wind tunnel with a 21- by 21-ft square cross-section anechoic chamber.

The Structural Mechanics Laboratory also prospered, moving into new and expanded spaces at Carderock in 1967. These housed five high-pressure deep-submergence tanks ranging in diameter from 12 ft down to 1.5 ft, with test pressures from 1500 psi up to 25,000 psi, for evaluation of the hulls of underwater vehicles. Also included was a test bed with a hydraulic loading apparatus for stressing large model ship structures under compression or tension loads up to 250,000 lbs. Included in significant structures work undertaken in the 1950s and 1960s were investigations leading to the certification of HY-80 steel for submarine hulls; efforts associated with the strength of aircraft carrier decks; examination of failures of stiffened cylinders of various materials under hydrostatic pressure; and advances in elastic shell theory. Deep submergence trials for the first submarine of a class began in the 1930s and peaked with 44 such trials during the 1950s and 1960s. In 1957 BuShips gave DTMB the lead for contract and in-house research on ship structures that could withstand airblast effects of a nuclear detonation. DTMB was involved in planning and conducting every nuclear test involving ship targets. This work quickly led to an improved deckhouse design for new destroyers. Joint responsibility was given to DTMB and EES in 1965 by BuShips and BuWeps for "hardening" of ships and new weapons against air blasts and underwater explosions. Armor development work began in the 1960s and was highlighted by quick assistance tasks on site during the Vietnam War.

By 1963, the Aerodynamics Laboratory at Carderock had six identifiable areas of principal research: carrier jet approach-path aerodynamics, aircraft weapons launch, aircraft dynamic stability theory, tests of supersonic and hypersonic jet aircraft and missiles, vertical short take-off and landing (VSTOL) planes, and "interface aerodynamics," such as for air cushion vehicles. This last area of work, involving all types of interface craft, would grow to become one of the principal topics of investigation at DTMB during the 1970s.

In contrast to the expansion underway at DTMB during the 1950s, EES underwent some contraction, going from about 1200 employees down to about 1000. During the Eisenhower presidency more and more product testing was turned over to private industry, with the resulting loss to EES of some of its traditional work. One new line of work was initiated in 1959 with completion of the Stray Magnetic Field Facility for diagnosis and reduction of magnetic fields created by diesel engines and electrical machinery aboard minesweepers. In 1958, Harold Nutt, who as a naval officer had served in the fuels division during the war, became technical director of EES, leading the further transition of EES from a test station to a development laboratory. This was reflected in the 1963 change in name to Marine Engineering Laboratory (MEL). During the 1950s, more and more work of an in-house development nature was initiated to supplement the traditional equipment specification and evaluation tasks of the past. Among the in-house contributions were non-flammable fluids for hydraulic catapults on aircraft carriers (two explosions with loss of life had occurred in 1953), numerous improved features for fleet diesel engines, gas turbines, noise and shock reduction mounts for machinery, bearing and gear tooth noise and materials. By 1960 the development of 90/10 and 70/30 copper/nickel alloys for seawater piping was completed, and the new field of direct energy conversion was explored to develop fuel cells and thermoelectric devices. The innovations made in purely in-house development efforts during the 1950s outnumbered what had been achieved in all of the preceding years. EES had become a fully interdisciplinary R&D organization.

During the 1960s, programs were broadened at Annapolis to include the challenging new areas of deep submergence vehicles, automated ship control, and superconducting electric motors. By 1970 machinery and fuel cell projects were underway related to deep-dive submersibles. A 12,000 psi horizontal test tank complex, the largest chamber being 10 ft in diameter and 22 ft long, had been built to accommodate complete submersibles with their operating power systems. Titanium technology advances were also made that found application in an improved pressure hull for the deep-diving submersible *Alvin*.

Up to now the focus of these paragraphs has been on organizational changes, facility additions, and areas of work that were put in place during the years from 1950–1970. Of still greater consequence during these years was the flowering of individual researchers and their pioneering work at DTMB, particularly as related to ship hydrodynamics. Many researchers matured during these years and became well known as a result of their important contributions to the field. Some call this period the Golden Age of Research at DTMB. Although the seeds had already been sown, probably the single most important factor in this flowering of hydromechanics research was the establishment in 1952 of the BuShips Fundamental Hydromechanics Research (FHR) Program, later named the General Hydromechanics Research Program (GHR), administered by DTMB. Between 1955 and 1967 this program supported an average of about 80 man-years of research each year, about 75 percent at DTMB and the remainder mainly at American universities. The impact of this program can be judged from the first six issues of SNAME's *Journal of Ship Research*, published initially in April 1957, where nearly two-thirds of the papers were written by present or former DTMB employees, or by academics with support from the FHR program. The singular success of this program can be attributed to the interaction and collaboration of DTMB personnel with academics, the direct transfer of research from DTMB to design technology at BuShips with active BuShips participation, and the collective identification of research funding priorities by senior personnel at BuShips, DTMB, and ONR.

To give but a few examples of DTMB researchers and their seminal hydromechanics research contributions during the period 1950–1970, the following may be cited:

1952 Herman W. Lerbs: Theory of moderately loaded propellers with arbitrary circulation distribution.

1953 Manley St. Denis and Willard J. Pierson (NY University): Theory of ship motions in confused seas; ship response is the sum of the responses to individual wave components.

1955 Marshall Tulin: Linearized theory of supercavitating foils.

1956 Leonard Pode: Theory of cable towed systems.

1957 William E. Cummins: Theory of time-dependent forces and moments on moving bodies.

1961 Pao C. Pien: Evaluation of propeller lifting surface theory.

1962 Morton Gertler and Alex Goodman: Patenting of planar motion mechanism for vehicle dynamics.

1964 J. Nicholas Newman and Ernest O. Tuck: Slender-body theory of ship motions.

1968 Paul S. Granville: Velocity similarity laws for drag-reducing polymer solutions.

During the same period DTMB became an important breeding ground for and provider of faculty members who would distinguish themselves in the academic community: Landweber to the University of Iowa; Wehausen to the University of California, Berkeley; Abkowitz, Keil, and Newman to MIT; Couch to the University of Michigan; Ogilvie first to the University of Michigan and later to MIT; Breslin to Stevens Institute of Technology; Tuck to the University of Adelaide, Australia; and some years later Hadler to Webb Institute.

A further strengthening of research support was initiated in 1963 with introduction of the ONR-originated Independent Research/Independent Exploratory Development (IR/IED) Program, a source of discretionary funding to be used by naval laboratories to bolster their R&D efforts. At DTMB this helped offset future reductions in hydrodynamics research under the GHR Program, and extended support to all of the other disciplines represented at DTMB and MEL. At its high point in 1967 the IR/IED Program provided about 100 man-years per year of support at DTMB and MEL. Ten years later this program would support only about half that level of effort, and the GHR Program would be down from 80 to 25 man-years per year of effort, divided about equally between DTMB and universities.

Concomitant with the strengthening of research support in the early 1960s was the initiation by Robert McNamara, Secretary of Defense in the Kennedy and Johnson administrations, of steps intended to bolster the morale and improve the organization of military laboratories in general. Several new and often controversial theories of management, involving centralization of control and the "systems approach," had been advanced during the 1950s and were introduced by McNamara to Department of Defense (DoD) planning, budgeting, contracting, and cost benefit analyses. Between 1961 and 1967 no fewer than four high level assessments, followed by several proposed plans, were undertaken to determine how DoD systems commands and laboratories could be better organized. Changes instituted during this period began with the creation in 1963 of civilian technical directorships at naval laboratories (Nutt at MEL and Keil at DTMB), and ended with the merger in March 1967 of MEL and DTMB to form the Navy's first "Center of Excellence" laboratory, the Naval Ship Research and Development Center (NSRDC). The merger led to the formation of two departments at Annapolis: a Materials Department and a Propulsion and Auxiliary Systems Department that conducted work on ship electrical and mechanical systems. Dr. Alan Powell, who served as a department head and then technical director of DTMB for one year following a distinguished academic career in acoustics research, became the technical director of NSRDC, a position he held for the next 18 years.

Other laboratory mergers followed, including the merger with NSRDC in June 1967 of the Mine Defense Laboratory (MDL) located in Panama City FL. In January 1972, following a short and difficult union, MDL was separated from NSRDC and became the Naval Coastal Systems Center. In 1969 the Applied Science Laboratory at the New York Naval Shipyard in Brooklyn was disestablished and 94 of its personnel transferred to NSRDC, to the great benefit of the Materials Department in Annapolis; this group included specialists in high-strength steels, titanium, plastics and synthetic rubber.

In parallel with the general realignment of Navy laboratories during the late 1960s, the lines of command within the Navy's bureau structure were also changed. In 1966 the Bureau of Ships, to which DTMB and MEL had reported since 1940, ceased to exist and was replaced by the Naval Ship Systems Command (NAVSHIPS). The position of Director of Navy Laboratories (DNL) had been created one year earlier within the Naval Material Command (NAVMAT). Thus DTMB and MEL, and soon NSRDC, reported to a director in NAVMAT while NAVSHIPS largely retained control over laboratory budgets and programs. This command structure basically continued in place through 1991, except that the DNL position was moved in 1986 to the newly-created Space and Naval Warfare Systems Command (SPAWAR) following a brief assignment to ONR. In 1975 NAVSHIPS became NAVSEA (Naval Sea Systems Command), which continued to be the principal sponsor of work at Carderock and Annapolis. One year earlier NSRDC had become DTNSRDC (David Taylor Naval Ship Research and Development Center), thereby regaining its long-standing association with its namesake.

The Development Years (1970–1990)

In 1971, fours years after the merger of DTMB and MEL, Center personnel numbered 3122, including 40 military staff; the total center budget was $92 million. Nine years later, at the end of the Carter presidency, following the U.S. naval drawdown of the 1970s, there were 640 fewer Center personnel. By 1989, with the naval buildup during the Reagan administration, the Center staff had risen to 2920 personnel with a total annual budget of $396 million, a significant portion of which was contracted to private industry. During the two decades from 1970–1990 major changes were to take place with regard to the types of work conducted at the Center, the manner in which the work was carried out, and the relationships between the Center and its sponsors.

Within one year of becoming the Chief of Naval Operations in 1970, Admiral Elmo Zumwalt proclaimed the breathtaking new technological goal of a 100-knot surface Navy, to become operational during the early 1980s. The implied high-risk technological challenge, not unlike President Kennedy's goal of a man on the moon in one decade, would entail the development and demonstration of radically

Fig. 7 Aerial view of DTRC Annapolis site.

non-traditional types of naval craft. It would also test both the technical and managerial strengths of NSRDC during the 1970s. Two types of craft were of primary interest: hydrofoil craft and air cushion vehicles, especially surface effect ships (SES), which have rigid side walls and flexible bow and stern skirts surrounding a cavity of pressurized air between the vehicle and the ocean surface. Fortunately, considerable exploratory development work had been conducted at DTMB since the 1950s on hydrofoil craft, and since 1964 on the SES concept by its inventor, Alan Ford, and others in the Aerodynamics Department.

In the second half of the 1960s, DTMB developed and tested the first manned SES, the XR 1; subsequent variants of this craft led in 1974 to the 47-ft XR 5, which incorporated improved features affecting stability, propulsion, and ride control in rough waters. The latter work was sponsored by the Joint SES Project Office (JSESPO) established by NAVMAT at Carderock and jointly supported for three years by the Maritime Administration. The SES 100A and SES 100B, built by Aerojet General and Bell Aerospace, respectively, successfully achieved 90-knot speeds during the mid-1970s and were followed by planning and model testing of an open-ocean 2000-ton SES. This effort was terminated in 1979 for lack of an identified mission for the craft. This was on the eve of prototype construction, and JSESPO was closed shortly thereafter. Subsequent investigations during the 1980s were confined to SES patrol craft and minesweepers for coastal operations. Although Zumwalt's goal of a 100-knot Navy died with the demise of JSESPO, one important by-product of the SES work was the acquisition by NSRDC in 1977 of a new 100-knot carriage for model experiments in the high-speed towing tank at Carderock, and the subsequent closure of the Langley Field high-speed facility.

With the intention of providing a total ship systems, hardware-oriented focus to pull together the discipline-oriented organizational elements of NSRDC, the Systems Development Department was born in 1970. This was placed under William Ellsworth, who had formed the Hydrofoil Development Program Office at DTMB in 1967. In the years to follow, this new department would use both project office and matrix styles of management, playing important coordination roles for a number of high performance vehicles. These included hydrofoil craft, amphibious assault landing craft (AALC) riding on air cushions with full-perimeter flexible skirts, and small waterplane area twin hull (SWATH) ships. The AALC program office was responsible for the design, development, testing and evaluation of two high-speed prototypes, *JEFF(A)* and *JEFF(B)*; the resulting technical data base allowed transition to acquisition of production craft. Lessons learned from hydrofoil trials with *Highpoint* and *Plainview* resulted in improved features that were incorporated into *Flagstaff* (PGH-1) and *Tucumcari* (PGH-2), and culminated in the launching in 1974 of the first of six 235-ton *Pegasus*-class PHMs having speeds in excess of 40 knots.

Although the SWATH concept was conceived by Frederick Creed in Canada during the 1930s, the performance of this novel type of catamaran was not demonstrated until 1971 when a manned 20-ft model was tested by the Naval Ocean Systems Center (NOSC) in the harbor at Marina Del Rey CA. Tests in 4-ft waves indicated the superb seakeeping characteristics of the SWATH concept and led to extensive hydrodynamic, structural, and powering investigations at NSRDC throughout the 1970s and 1980s. With the exception of *Kaimalino*, a 200-ton R&D platform built for NOSC in 1973, the Navy did not find a mission for SWATH vessels until 1986, when a contract was let for *T-AGOS 19*, a 3500-ton open ocean ship for towing sonar arrays. Subsequently, three follow-on ships were built, and the decision was made to construct *T-AGOS 23*, at 5500 tons the world's largest SWATH. Important contributions to SWATH technology made during the intervening years by investigators at DTNSRDC included the placement and sizing of control surfaces for motion control, analysis of cross-structure loads, and optimization of powering performance. Some of the technology derived in this work was used to advantage by Japanese developers of small SWATH passenger ferries and *Kaiyo*, a 3500-ton ocean research SWATH.

Central to the work on high performance craft was an increasing Center involvement in both the design and hardware demonstration phases of vehicle development. Functions that had been carried out previously at the systems commands were being increasingly delegated to the Center, as evidenced for example by the formal designation of the Center in 1973 as NAVSHIPS' propulsor design agent. In general, the Center's design role grew in the 1970s to support NAVSHIPS initiatives related to computer-aided design and manufacturing. Program offices were established, and evaluations were conducted on the actual prototype systems. In order to enhance the Center's ability to evaluate towed systems, two high-speed PG 84 gunboats were acquired by the Center in 1976 and converted to the *Athena I* & *II* research vessels for towed system trials conducted from ports in Florida.

During the 1970s the Center continued its studies related to traditional open-ocean combatants. These included development of new methods to ensure the integrity of controllable pitch propellers, following failure of a research version of this type of propeller on *Barbey* (FF 1088). The extensive hydrodynamic, structural and materials investigations undertaken at the Center led to changes in the structural design of the blade attachment. The lessons learned were immediately applied to the controllable pitch propellers on *Spruance* (DD 963) and *Perry* (FFG 7)-class combatants, and further failures were avoided.

Other surface ship work included evaluation and development of corrosion control coatings, and improved anti-fouling paints to reduce fuel costs up to 15 percent; full-scale trials to demonstrate the ability of ships to carry on combat system operations during exposure to underwater shock, and to provide a basis for refinement of shock hardening criteria for future ships; demonstration of reverse osmosis desalination technology for production of fresh water on gas turbine powered ships; and environmental protection development work on shipboard trash handling, waste disinfection and fuel recovery from bilges.

Following the Middle East oil crisis in 1973–1974, NAVSHIPS (NAVSEA after 1975) sponsored a wide-ranging long-term program at the Center, which identified propulsor, hull and appendage configurations that would provide energy savings for the Navy's surface fleet. Two of the most important findings were the large fuel cost reductions associated with the trailed shaft-and-propeller operating mode, and the use of stern flaps or wedges on transom stern ships.

Another surface ship innovation in the 1980s was the Center's development of rudder roll stabilization, an ingenious and inexpensive scheme soon implemented on some Coast Guard and Navy ships to reduce roll motions in a seaway. The Center also led the development and certification of HSLA-80 and HSLA-100 high yield steels, which eliminated the need for welding preheat and thereby dramatically reduced hull construction costs for the CG 47, DDG 51, CVN and LHD-class ships. Extensive work was also undertaken on the application of light weight composite materials to a multitude of shipboard structures and equipment.

One new technology area, investigated initially in the 1970s but continued through the 1980s, involved homopolar superconducting motors with liquid helium cooled magnets. These were to provide increased shipboard power density, efficiency, and arrangement flexibility. By the end of the 1980s, superconducting motors had been developed and demonstrated at the 400 hp and 3000 hp levels, and the associated

Fig. 8 Large cavitation channel at Memphis.

technology base became available for future fleet applications.

With the Navy's increasing concern about the susceptibility of ships to detection and identification by air and space surveillance systems, in 1987 the Center established an Electromagnetic Signatures Department. This department was charged with the development of model and full-scale measurement capabilities, and the identification of ship features that reduce non-acoustic signatures. In 1988 the Center acquired the Santa Cruz Radar Imaging Facility on Santa Cruz Island CA and a portable radar measurement capability initially used for model imaging in the maneuvering basin at Carderock. Other Center facility acquisitions included construction in 1988–1991 at Memphis TN of the Large Cavitation Channel, the world's largest cavitation facility, with a huge 10 by 10 by 40-ft test section, with test speeds over 35 knots. This quiet facility, specially designed for model propeller noise measurements, will be essential for ship and submarine advanced propulsor developments well into the 21st Century. The Center also acquired the *Hayes* (T-AG 195) for conversion to an acoustic measurement platform, to replace MONOB in the 1990s; and an acoustic measurement facility in southeast Alaska, to be ready for ranging in 1991. At Annapolis a new laboratory was added in 1990 for non-destructive evaluation of advanced welding techniques. Also in 1990, as part of a major internal reorganization of the Center to reflect funding realities, both the Aviation Department and the Computation, Mathematics, and Logistics Departments at Carderock were dissolved with a majority of the personnel and their functions joining the Ship Systems Integration Department.

In 1989, Admiral Carlisle Trost, Chief of Naval Operations, directed that the U.S. Navy's surface combatants of the future employ "integrated electric drive." This would permit flexibility in the choice of new reduced-signature hull-propulsor arrangements, including podded propulsor configurations, and also permit the future use of weapons systems that are electrically driven by pulsed high power energy sources. This new direction resulted in no small measure from the Center's identification, analysis, and vigorous promotion of a number of clusters of surface ship technologies that highlighted integrated electric drive. Leadership for the cluster approach and a mapping of the future surface Navy was provided by the Center's commanding officer from 1987–1991, Captain Clark Graham, and the Center's technical director since 1986, Richard E. Metrey, both of whom were strongly systems-oriented and intent upon DTRC influencing the composition of the Navy in the 21st Century.

With regard to submarine developments, important Center efforts initiated during the 1970s included the design and evaluation of *Trident-* and *Los Angeles-*class depth and course-keeping control systems for safer handling of submarines, and evaluation of novel control surface configurations. This work was enabled by the perfection of an advanced radio-controlled model and computer capability that allowed accurate simulation and evaluation of submarine maneuvering characteristics in the Carderock maneuvering basin. An environmental monitoring and guidance system was also developed and implemented by the Center to reduce the risk of grounding during channel transit to and from the Kings Bay GA TRIDENT base. The Center also contributed directly to the development of reduced-noise propellers through suppression of tip vortex cavitation; quieter auxiliary machinery and improved water-lubricated propeller stern tube bearings.

Several notable structures and material contributions to submarine technology were initiated at the Center in the 1970s and extended well into the next decade. These included the development of analytical techniques and guidelines for submarine structures, which led to the simplification of the structural design of the TRIDENT missile compartment and significant cost savings; the complete development and certification of HY-100 and HY-130 high yield steels; improved welding productivity through automated welding systems having higher weld deposition rates; and development of high-strength titanium alloys for seawater systems.

Fig. 9 *Kokanee* at Lake Pend Oreille.

During the 1980s, a large submarine model, designated *Red Snapper*, was constructed at the Center. It was used to validate the Navy's way of estimating whipping response of submarines to proximity fuzed weapons, and to demonstrate effectiveness of heavyweight torpedo warheads to do structural damage. Of particular merit from the sonar self-noise standpoint was the development and demonstration of quiet sonar domes through the use of new materials and coatings, and the promulgation of improved dome attachment and smoothness specifications. This work was crucially benefitted by using large buoyant free-rise submarine models released from the bottom of Lake Pend Oreille, where dome noise evaluations were undertaken.

The largest, most complex and most demanding experimental and computational effort undertaken at the Center during the 1980s was the SSN 21 (*Seawolf*) submarine development program. Nearly every department at the Center was involved in an important way and, because of the demanding schedule, the staff and facilities were stretched to a degree not experienced since the second World War. Silent operation of the SSN 21 was of primary concern, and the Center became deeply committed to the development of specific quiet submarine components, which included the propulsor, auxiliary equipment, hull, appendages, openings, and bow dome. Much of the propulsor work would not have been successfully concluded without the availability of *Kokanee*, a large unmanned, self-propelled submarine vehicle procured for acoustics experiments at the Center's Lake Pend Oreille detachment. In addition to its significant silencing contributions to the SSN 21 program, the Center developed automatic control algorithms and maneuvering simulations in support of various control surface design investigations. It also developed structural design guidelines and features for increased diving depth, improved survivability to shock during underwater explosions, and improved operations in the Arctic environment. Much of the Center's SSN 21 design support was concluded in 1990, amid congressional debate on how many submarines of the class would be built following the collapse of the communist governments in Eastern Europe and the movement of the United States and the Soviet Union along a path of dialogue, cooperation, and limited bilateral disarmament.

As exemplified by the high-performance craft development efforts during the 1970s and the SSN 21 development program of the 1980s, more and more of the investigations at DTNSRDC dealt with increasingly complex large systems or subsystems of naval vehicles, and total concept development, involving teams of people from more than one discipline. The changes that took place were in distinct contrast to the past where one or two investigators worked largely alone on one aspect of a problem within a single discipline. Additionally, in response to organizational changes at NAVSEA, entailing the establishment of project management offices for each new class of naval vehicle, work at DTNSRDC became more focused on specific vehicles. This was in contrast to previous, more generic investigations of broad groupings of discipline-oriented vehicle problems. With the large buildup of the fleet to a 600-ship Navy during the Reagan administration, this trend was accelerated. Entering the 1980s the technology was not always in place, so the tools to develop new designs had to be devised on a crash basis as part of the vehicle design process itself. One striking example of this was the Center's submarine propulsor design effort for *Seawolf* (SSN 21).

In 1980 the technology base role for NAVSEA was drastically altered with the establishment of the Office of Naval Technology (ONT) to manage the Navy's Exploratory Development Program. ONT subsequently transferred technical management responsibilities for this program from the Navy's systems commands to the laboratories. Although NAVSEA continued to have some technical involvement, by 1985 it was divested of its longstanding central role in this and other technology base programs. This change altered important sponsor-laboratory, or user-developer, relationships that had been developed over many years. By 1989, owing to the great reduction in new ship and submarine designs, and to the loss of much of its technology base role, NAVSEA was providing only 48 percent of the total funding at the Center, down from 67 percent in 1977.

A primary indicator of the great changes that took place throughout the Navy and at DTNSRDC during the 1980s was the steep decline in the amount of support available for fundamental research, exploratory development, and advanced development investigations. Between 1980 and 1989 the total funding at the Center for these categories of work rose only 23 percent from $101 million to $124 million, a large decline in real dollars because of inflation. As a percentage of the total Center budget, funding for these categories of work fell from 60 to 31 percent of the total Center effort. In 1985 the GHR Program, which had supported about 60 man-years of fundamental hydromechanics research at DTMB during much of the 1960s, was terminated and a new program was started at ONR; by 1990 this new program supported only three man-years of hydromechanics research at DTRC. Similarly, the IR/IED Program at DTRC, which had supported 100 man-years of R&D in 1967, supported only 25 man-years of effort in 1990. Although the level of technology base support was severely eroded during this period, the traditions of distinguished fundamental research established in the previous decades were remarkably carried on, and numerous pioneering contributions continued in the face of increasing difficulties.

In closing, it can be fairly stated that throughout its more than four score and ten-year history starting with Taylor and Melville, the Center and its predecessor organizations adapted to all of the changes that occurred, proud of their unmatched capabilities to satisfy the Navy's requirements for future generations of ships, submarines, and small craft, as well as their ability to support the operating fleet.

Looking ahead to the next century it is appropriate to end this history with some remarks made by David W. Taylor in his 1927 presidential address to the Society of Naval Architects and Marine Engineers. In many ways they are as appropriate today as they were then:

> We elders sometimes find it hard to realize, but progress in our profession, as in others, will never cease. One of the peculiar difficulties of our profession insures in a way that progress will continue.
>
> While our young members will see progress in the future, I think they will see it more closely associated than in the past with research, both as to fundamental principles and the myriad details whose sum is becoming more and more important as a factor in the product—the completed ship.
>
> Trial and error were the main dependence of the generation now gone. They have served the present generation well in spite of the steady advance of science and research, but for the next generation science and research should reduce trial and error to a minimum.

Acknowledgments

In preparing this chapter, I have benefitted significantly from Carlisle's history of DTRC and from the historical commentary, thoughtful comments and suggestions of a number of present and former Center personnel: Richard B. Couch, William G. Day, Jr., Thomas T. Huang, Donald L. Martin, Richard E. Metrey, Vincent J. Monacella, William B. Morgan, Alan Powell, Earl R. Quandt, Jr., and Barrick F. Tibbitts. The manuscript was carefully typed by Ramona E. Hatfield and the photographs were assembled by Geraldine R. Yarnall at DTRC.

Bibliography

Allison, D. K., Keppel, B. G., and Nowicke, C. E., *D. W. Taylor*, U.S. Government Printing Office, 1988 (120 pp).

Bush, V., *Science, The Endless Frontier*, Arno Press, New York, 1945.

Carlisle, R., *Where the Fleet Begins: A History of the David Taylor Research Center 1899–1987*, History Associates, Inc., Rockville, MD, Sept. 1987.

Conn, V., and Natelson, R. G., "The David Taylor Model Basin, A Brief History," Naval Historical Foundation Publication, Series II, No. 15, Spring 1971.

DeBaufre, W. L., "The U.S. Naval Engineering Experiment Station," *Scientific American*, Suppl. No. 2091, 29 January 1916.

"Historical Highlights (75th Anniversary)," *The Centerline*, NSRDC, Vol. 5, No. 7, June 1971.

King, E. J., "The United States Naval Engineering Experiment Station, Annapolis, MD," *Journal American Society of Naval Engineers*, August 1913.

Leggett, W. D., "The U.S. Naval Engineering Experiment Station," *Proceedings*, U.S. Naval Institute, May 1951.

Reinertson, J. H., Alsager, L. E., and Morley, "The Submarine Propulsion Plant—Development and Prospects," *Naval Engineers Journal*, May 1963.

Saunders, H. E., "The 1946 Report of the American Towing Tank Conference," *Trans.* SNAME Vol. 54, 1946.

Taylor, D. W., *The Speed and Power of Ships*, U.S. Government Printing Office, 1911.

Vincent, M. da C. "The Naval Ship Research and Development Center: The U.S. Navy's Center of Excellence for Ship Research," NSRDC Report 3039, November 1971.

Wright, E. A., "A Pattern for Research and Naval Architecture," *Trans.* SNAME Vol. 54, 1946.

Appendix

Chronology of laboratory names and sponsoring commands:

Dates	Laboratory Name	Sponsoring Command
EMB & DTMB		
1896–1940	Experimental Model Basin	Bureau of Construction and Repair
1940–1967	David Taylor Model Basin	Bureau of Ships
EES & MEL		
1903–1920	Engineering Experiment Station	Bureau of Steam Engineering
1920–1940	Engineering Experiment Station	Bureau of Engineering
1940–1963	Engineering Experiment Station	Bureau of Ships
1963–1967	Marine Engineering Laboratory	Bureau of Ships
NSRDC, DTNSRDC & DTRC—Merging of DTMB and MEL		
1967–1974	Naval Ship R&D Center	Naval Material Command
1974–1985	David Taylor Naval Ship R&D Center	Naval Material Command
1985–1986	David Taylor Naval Ship R&D Center	Office of Naval Research
1986–1992	David Taylor Research Center (1987)	Space & Naval Warfare Systems Command

Addendum

In January 1992, as part of a general and major reorganization of the Navy's infrastructure, the David Taylor Research Center was merged with the Naval Ship Systems Engineering Station (NAVSSES), Philadelphia to become the "Carderock Division" of a newly-formed "Naval Surface Warfare Center" under NAVSEA. In this reorganization, the Carderock site was designated as headquarters of the division and reverted to its original name, David Taylor Model Basin (DTMB); the Annapolis site became the "Annapolis Detachment" along with other DTRC detachments. In apparent contradiction to the "Naval Surface Warfare" designation for the new Center, DTMB continued to have a charter to conduct both surface ship and submarine research and development and to support the Maritime Administration and the maritime industry.

Justin McCarthy is the head of the Naval Hydromechanics Division at the David Taylor Research Center in Carderock, Maryland, where he has been employed since 1960. In 1986–87 he served as a liaison scientist with the Office of Naval Research in Tokyo. He is a Fellow of SNAME, chairman of the Journal of Ship Research Committee and has served as chairman of the Resistance Committee for the International Towing Tank Conference. He is a graduate of Webb Institute, attended Kings College of Durham University under a SNAME graduate scholarship, and studied mechanics at Johns Hopkins University. He is the author of more than 40 papers and reports and a recipient of the Navy's Meritorious and Superior Civilian Service Awards.

Part II

Form and Function

Hull Forms

JACQUES HADLER

Introduction

THE DEVELOPMENT of a vessel's hull form involves the trade-off between a series of often conflicting requirements. Classically this is depicted by the design spiral leading to the "best" hull form. The development of the hull form is essentially a hydrodynamic problem, which of necessity must interface with the other features of a vessel such as structural requirements, propulsion system, producibility, static stability, and habitability. The recent advent of higher-speed vessels has also brought aerodynamics into consideration. Fortunately, the fundamental science of aerodynamics is similar to that of hydrodynamics.

The hydrodynamic development of the hull form itself encompasses a series of compromises between the desire to minimize the resistance, maximize the performance of the propulsor and its interaction with the hull, minimize the ship motions in the expected environment, and ensure adequate maneuverability and control of the vessel to assure its safe operation. To achieve a "best" hull form the naval architect has available a number of approaches. In the long history of seafaring, the form of vessels has evolved slowly through an evolutionary process of making small changes in a succession of prototypes. This process is still widely used today in conventional ships that have a long history of development and whose operational requirements are similar to those of previous vessels. The rapidly evolving weapon systems for military vessels and the changing modes of commercial transport have frequently required major changes in hull form; thus other means are needed to develop effective designs. The most common approach has been the systematic series, of which many have been developed throughout the world over the past 50 years. More recently, with the advent of the high-speed digital computer, solutions to many classical hydrodynamic problems have become practical. These have assisted substantially in hull form development, particularly for kinds of vessels that had little precedent. Most recently the availability of extremely large and fast computers has boosted the field of numerical hydrodynamics, and this is beginning to have a significant impact on hull form and will only accelerate with time.

A discussion of hull form developments would not be complete without first recognizing the tremendous growth in the number and types of experimental research facilities, as well as the practical application of theoretical hydrodynamics and its recent evolution into numerical hydrodynamics. Before this period there were relatively few experimental model basins. Primarily there were towing tanks, which tested ship models in smooth water for their resistance and self-propelled them for their powering performance. The past 50 years have seen a large growth in the number of experimental research centers throughout the world. They have also seen the addition of seakeeping and maneuvering basins, rotating arms, ice tanks, variable pressure water channels, and many other new types of facilities. These facilities, along with the advances in electronic instrumentation and recording equipment, have increased immeasurably the quantity and types of measurements that can be made. The results of the work from all of these research centers have assisted in the evolution of the hull forms that are prevalent today.

The foundation for theoretical hydrodynamics was well laid before World War II, but was usually not applied because of the difficulty of the computations or the unreality of the simplifying assumptions needed to make the calculations tractable. The advent of the digital computer in the early 1950s made possible the practical application of theoretical hydrodynamics to ship hull form development. Its practical application has continued to grow as theory has been further refined and computational capacity has increased. Among many improvements to hull form, it has contributed to the development of the bulbous bow and the reduction of wave making resistance. It has made possible the development of the strip theory for the prediction of ship motions in a seaway and has contributed to our understanding of ship and submarine responses in maneuvers.

More recently the astronomical growth in computer speed and capacity has made possible the transformation of many of the theoretical hydrodynamic problems into the computationally intensive work now known as numerical hydrodynamics. This field is beginning to have a significant impact on hull form. It has already had a major influence on yacht design through the America's Cup races and is now contributing to submarine design.

I hope in this chapter to highlight some of the changes and improvements that have occurred in hull forms over the past 50 years in various commercial and military vessels as well as some of the wholly new concepts that have come into service during this period.

The material in the ensuing sections will be treated in the usual specific hydrodynamic areas of resistance, propulsion and vibration, seakeeping, and maneuvering and control. The types of vessels that have undergone the most significant changes in each particular hydrodynamic area will be described. In this way I hope that the most beneficial products of the research and development of the past 50 years will stand out.

Resistance

In developing a new hull, an initial concern is a form with low resistance; thus over the years much effort has gone into developing forms with good resistance characteristics. The systematic series has been one of the major methods for developing good hull forms and will constitute one of the major items of this section. Many systematic series have been developed both before and after World War II. Most notable of those before the war was the Taylor series, which is still widely used for both military and commercial ships. Since then there have been numerous systematic series of which the following David Taylor Research Center (DTRC) series are typical examples: Series 58, on bodies of revolution; Series 60, for commercial ships; Series 62, on planing surfaces; Series 64, on high speed displacement ships; and Series 65, on hydrofoil supported hulls, later extended to planing craft. The availability of digital computers and statistical methods has also made possible the synthesis of some of the large bodies of unrelated model test resistance and propulsion data residing in model basins. The initial application was to fishing trawler resistance data by Doust [1] at Britain's National Physical Laboratory (NPL). There are a number of such regression series now available for predicting the resistance and propulsive performance of ships that fall within the data base. The work by Holtrop and Mennen at Wageningen, Holland [2] is probably the most widely used. More recently the combination of hydrodynamic theory and model testing has been used to make significant reductions in hull resistance, or has been instrumental in developing new hull forms for specialized purposes. This section will treat some of the major developments that have occurred during the past 50 years.

Submarines. No vessel has undergone a greater transformation in the period under study than the military submarine. This has been transformed from a surface ship that submerged occasionally to a submerged ship that surfaces infrequently. The advent first of the snorkel mast as a means of providing air to the diesel when running submerged near the surface and, more recently, the advent of non-air-breathing means of propulsion have made this transformation possible.

The World War II submarine was a twin-screw vessel with a ship-shape bow, which was in most instances provided with a deck-mounted gun and a bridge that was intended for surface operation. After World War II it was recognized that these ships could achieve greater submerged speed by removing most of the appendages and fairing the remainder. The *Guppy* conversion program encompassed removing the deck gun, fairing the bridge, removing the wooden deck and the miscellaneous appendages on the deck. This cut the overall resistance in half. Concurrently, the research community was investigating the feasibility of the fully streamlined submerged hull. This culminated in the David Taylor Model Basin (DTMB) Series 58 bodies, which were bodies of revolution with mathematically delineated profiles. The forms developed in this series have served as the prototypes for most of the U.S. Navy submarines in service today and have pointed the direction for the hull forms for most of the military submarines in navies throughout the world. The submerged resistance of these forms is about half that of the *Guppy*-class per unit of volume. Thus just by changing hull form, and reducing and fairing the appendages, the resistance has been reduced to about 25 percent of that of a World War II submarine, with a concomitant increase in submerged speed. The first application of the results of Series 58 was the experimental submarine *Albacore* (SS569), which was the test-bed for establishing the overall performance of the single-screw streamlined modern submarine.

Surface Ships. In the past 50 years merchant ships have experienced numerous changes in hull form to reduce their resistance and improve their propulsive performance. The most notable change in the bow has been the introduction of the bulb in most modern merchant ships, both full and fine formed. The bulb has made significant improvements in the hull resistance by reducing the wave-making at the design speed. In the case of the full-formed vessels (tankers), in the ballast condition the bulb reduces the flow separation around the turn of the bilge in the bow region. This development has evolved from the combined use of hydrodynamic theory, numerical hydrodynamics, and model experiments.

Designing an oceanographic research ship's observation chamber as a new type of bow will reduce the ship's "drag" by 18 percent, according to tests at the University of Michigan. The reduction in bow wave resistance means a corresponding reduction in power and fuel requirements for the 156-ft ship, which will be used by the U.S. Fish and Wildlife Service for research around Hawaii.

The new bow, being installed for the first time for regular use, was invented and announced last year by Prof. Takao Inui of Tokyo University. Adaption of the bow—a bulbous snout that protrudes below the waterline—to the new ship was done in the University of Michigan's Department of Naval Architecture and Marine Engineering, where Prof. Tetsuo Takahei, a colleague of Prof. Inui's, cooperated.

At the time the bow was announced, the Seattle naval architecture firm of W. C. Nickum and Sons was designing the research vessel. Since original plans for the ship called for a two-man underwater observation chamber in the bow, the university was asked to investigate the possibility of making this a bulbous bow. The photo shows Prof. Richard B. Couch, chairman of the University of Michigan Department of Naval Architecture, holding a model of the original bow designed for a two-man observation chamber.

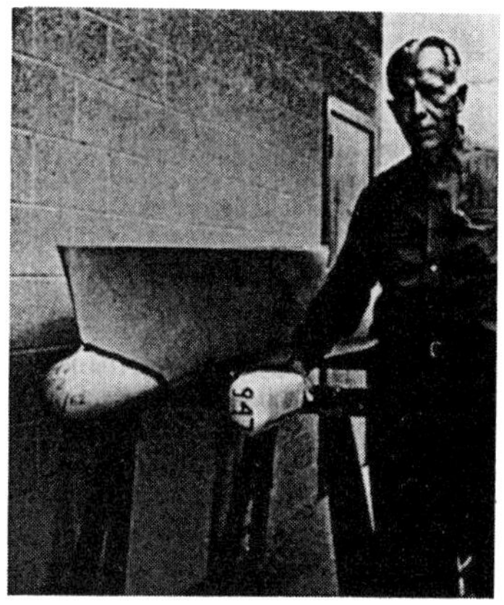

A model of the ship was built and tested in Michigan's 360-ft ship model towing tank, alternating bow types. The researchers reported that the Inui bulbous bow reduced total resistance—and, therefore, power requirements—by more than 18 percent. This means the ship could reach its designed maximum speed with an engine about four-fifths the size that would be required without the new bow, and with a corresponding reduction in fuel requirements. The bulbous bow accomplishes the reduced resistance by virtually eliminating the ship's bow wave, a major resistance force a ship must overcome. In effect, it cancels the normal bow wave set up by the ship's hull by producing another "negative" wave system—exactly out of phase and of the same amplitude as the normal bow wave.

Development of the bulbous bow was hailed as a breakthrough in ship design, in which a reduction of ship resistance of only a few percent has been considered a major accomplishment.

Up to now, the Inui bow has been tested in the open sea only once—last year on a Japanese inter-island ferry, from which it subsequently was removed. Since the bow's development, the University of Michigan has had a long-range research program under way on the application of bulbous bows to a variety of high-speed cargo ships and other vessels.

During this period there has been a tremendous growth in the size and installed power in the major types of merchant ships. This has had an effect on their principal dimensions. The length/beam ratio of the full-formed hulls has tended to decrease, and the block coefficient increase, with the objective of reducing the cost of construction. To minimize the resistance of these fuller forms, systematic model series have been tested. For the finer-formed hulls the trend has been toward an increase in the beam/draft ratio owing to the reduced density of the cargo with the higher center of gravity. Again extensions to systematic series have provided the design resistance data. One example is the extension to the Taylor series by the model basin at Stuttgart (Germany). It should be noted that these trends in hull form have evolved from other requirements and do not denote a major reduction in resistance.

The research that has been done on oceangoing trawlers during this period has significantly reduced their resistance. The most notable effort in this respect was Doust's previously mentioned statistical analysis of a large body of model trawler tests to determine those hull form characteristics that minimized the resistance [1].

As industrial interests have grown in developing the resources of the Arctic much effort has been devoted to reducing the resistance of icebreakers working in ice. The refinement of bow shape and the introduction of air around the hull to reduce friction when in contact with the ice has resulted in significant reductions in power requirements or increased ice breaking capacity.

Combatant ships, particularly the destroyers and frigates, have also experienced some changes. The displacement of destroyers has increased almost three fold. The installed power to displacement ratio has decreased almost 50 percent. The length/beam ratio has tended to decrease and the beam/draft ratio increase. These changes have evolved from military more than from hydrodynamic requirements.

Yacht hull form is being strongly influenced by the recent results derived from the use of numerical hydrodynamics in developing the 12-meter yachts used in the America's Cup competition. The large amount of numerical hydrodynamic research effort devoted to the development of these boats will with time be applied to other vessels with unusual hull forms to ensure that resistance is minimized.

The catamaran, during the past 50 years has grown from a few small novelty pleasure craft into ferries and oceangoing vessels largely used for survey ships where large working platforms are required. Hydrodynamic research, both analytical and experimental, on this hull form has lead to the concept of cambered or asymmetric hulls in order to reduce the interference wave resistance of two hulls in close proximity. The hydrodynamic research on catamarans has also spawned the Small Waterplane Area Twin Hull (SWATH) vessel concept, which holds great potential for relatively small craft that have to operate in ocean environments. The underwater hull form is composed of two parts, a streamlined body of revolution with a long parallel midbody on each side of the vessel and one or two narrow struts connecting the underwater forms to the boxlike superstructure. Numerical hydrodynamics of this form have been developed so that the resistance can be minimized at any desired speed(s).

Advanced Marine Vessels. The past 50 years have seen much research effort devoted to advanced marine vessels. The objective of most of this effort has been to develop the technology to design special vessels that can perform some unique function on the air–water interface much better than a conventional ship. The great preponderance of the effort has been devoted to producing significant reductions in resistance so that higher speeds can be achieved on the water's surface.

Conventional surface ship hull forms have been extended into the higher speed ranges by reducing the displacement/length ratio, increasing significantly the length/beam ratio, and carrying the testing to Froude numbers as high as one (speed/length ratios of 3.0). Several series have been devoted to this region, the most notable of which are the NPL High Speed Series [3], and DTRC Series 64.

The most common vessels that use dynamic lift to reduce resistance at high speed are the planing craft. At high speeds the dynamic lift substantially reduces the wetted surface, thus the frictional resistance, but at the penalty of some induced drag from the lift being created. Much of the hydrodynamic knowledge on dynamic lift has come from the seaplane work done by such noted aeronautical research centers as the National Advisory Committee for Aeronautics (NACA) and the Royal Aircraft Establishment. More recently, ship model basins have published the results of systematic series of typical planing hulls. Most notable in the United States have been DTMB Series 62 and Series 65. I have synthesized the results of these into the design diagram shown in Fig. 1.

The hydrofoil craft is also a dynamic lift vessel that creates lift through foils under the hull, thus minimizing frictional drag, but at the expense of the induced drag from the lift being created by the foils. It, like other dynamic lift vessels, is also intended for high speed waterborne operation. The hydrofoil type has developed into a fully operational vessel used for both military and commercial purposes. There are two essentially different configurations, those with surface piercing foils whose wetted surface varies with speed, and fully submerged foils, which vary the lift by changing the angle of attack. There are also two characteristic planforms for those with fully submerged foils, namely: canard, where the lift is about equally distributed between the forward and after foils, and the airplane, where most of the lift is carried by the forward foils located near the vessel's center of gravity.

The surface-effect ship (SES) is another approach to minimizing frictional resistance so that high speeds can be achieved. In this instance the wetted surface is reduced by using an air cushion to lift all but the bottom of the narrow side walls out of the water. Much of the aero-hydrodynamic research has been accomplished in the development of the U.S. Navy's experimental SES configuration. The technology is now being applied to passenger ferries.

Propulsion and Vibration

The propulsive performance of ships is usually determined by model tests of the particular design under consideration. Thus there is not as large a body of systematic data on the propulsion characteristics of ships as there is on resistance. Some systematic series have included propulsion as well as resistance tests, most notably Series 60 and the British Ship Research Association (BSRA) series for single-screw merchant ships. Several investigators have synthesized specific propulsion factors, such as the effective wake fraction, for use by designers in predicting ship performance. The use of statistical analysis to synthesize large bodies of model test results now provides data on the propulsion factors used in predicting the performance of ships. Again, however, this is limited to ships of the type and dimensions contained in the data base. The most widely known is that of Holtrop and Mennen previously mentioned.

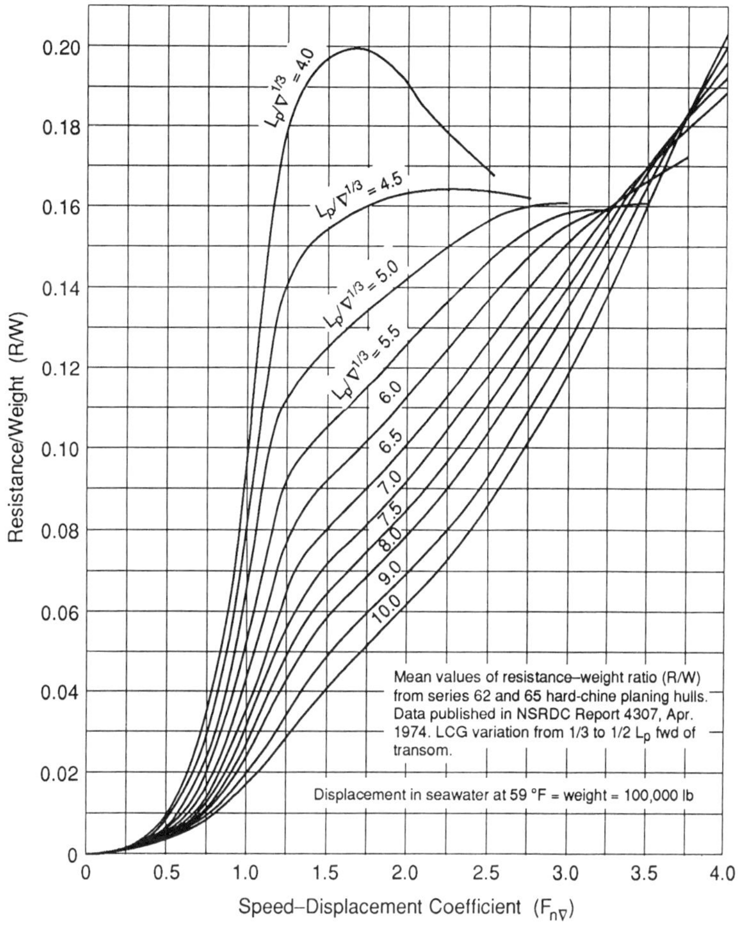

Fig. 1 Resistance/weight contours for efficient planing hulls as a function of speed–displacement coefficient and slenderness ratio.

The tremendous growth in the power transmitted by a single shaft in commercial ships has made propeller induced vibration a major element in designing stern forms. Powers have typically increased from about 15,000 to 18,000 hp in the tankers and cargo ships of the 1950s to as much as 54,000 in the most recent containerships. This has made the wake survey in the plane of the propeller an important experimental tool in the design of a ship and has contributed to the evolution of the stern forms now being used.

Submarines. The submarine of World War II was a twin-screw vessel with long shafts and struts located outside the hull forward of the control surfaces, thus having the low propulsive efficiency characteristics of twin-screw ships as well as adding significantly to appendage resistance. The development of Series 58 pointed in the direction of single-screw propulsion, which would tend to have much higher propulsive efficiency because the propeller is operating in the boundary layer of the hull, thus recovering some of that lost energy. This has improved propulsive efficiency by as much as 25 percent.

Surface Ships. The large growth in power in the single-screw commercial ship over the past 50 years has had a marked influence on the stern form. There the evolutionary change has been to improve the propulsive performance, i.e., maximize the effectiveness of the installed power and improve the flow into the propeller in order to minimize the propeller-induced vibration. Both the transverse section and aperture shapes of the stern form have undergone great change. The single-screw ships of 50 years ago had V-type sections as this type of stern had minimum resistance. Research at DTRC with Series 60 during the 1950s showed that the U-type section usually had higher resistance but better propulsive performance, usually resulting in lower power requirements.

It was also noted that the circumferential velocity variations in the plane of the propeller were somewhat less severe with the U-form than with the V-form stern,

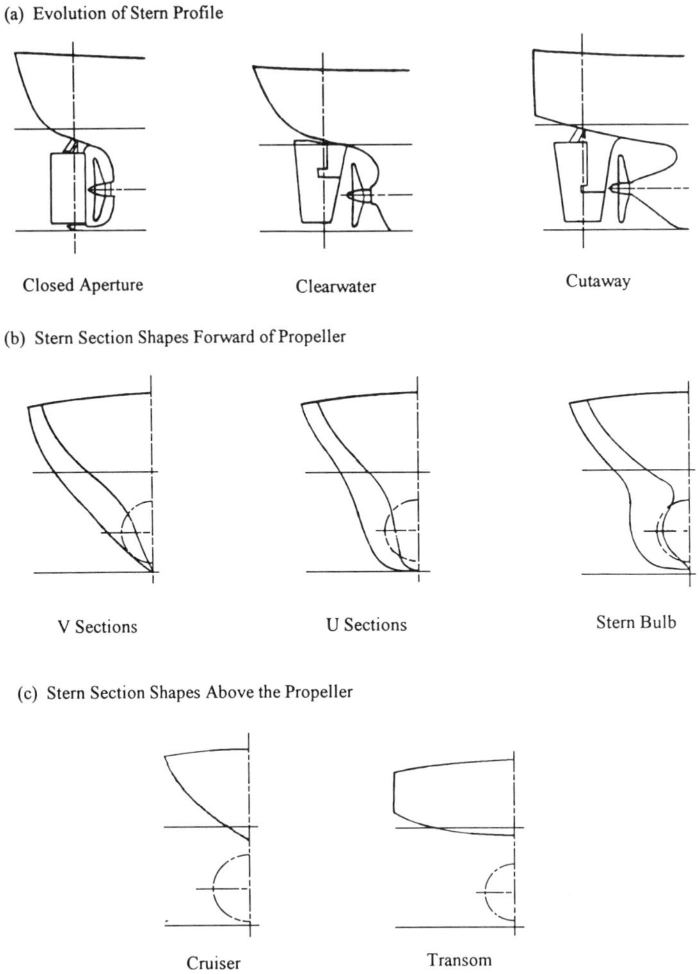

Fig. 2 Single-screw stern forms.

thus reducing the amplitude of propeller-induced vibrations. This trend, in turn, led to the investigation of bulbous sterns with the aim of further reducing the circumferential variations in the wake pattern. This has met with limited success, so the bulbous stern has not been widely adopted. Typical section shapes are shown in Fig. 2*a*.

The need to minimize the propeller pressure-induced vibrations has affected the propeller aperture. The clearwater stern pioneered on the Mariner-class ships has now become standard. As the powers have increased the fore and aft opening of the aperture has become larger, and the skeg below the propeller shaft has been frequently cut away, further reducing the circumferential variations in the wake pattern. The section shape above the propeller has also changed markedly, particularly in the transformation from break-bulk to containerships. Typical stern profiles are shown in Fig. 2*b*.

In containerships, in order to carry as many containers as possible, the main deck aft is almost as wide as the maximum beam. This leads to a transom stern with wide but shallow section over the propeller. This form presents a large surface susceptible to unsteady blade frequency pressure forces from the propeller. This necessitates larger propeller tip clearances on a ship that has a proportionally shallower draft, further exacerbating the propeller-induced vibration problems. Typical section shapes above the propeller are shown in Fig. 2*c*.

To improve the inflow to the propeller, special shapes are given to the stern just forward of the propeller. The contra guide stern, which was used on some ships in the late 1940s and early 1950s, sometimes along with contra guide rudders, attempted to put some rotational flow into the propeller to improve the propulsive efficiency. The idea fell into discard in the era of low fuel oil prices. Recently a more extensive adaptation of that idea has evolved in the asymmetrical stern, which puts a more extensive preswirl into the propeller, measurably improving the propulsive efficiency.

The changes in the propulsion of combatant ships have primarily concerned the reduction of noise. Most of this effort has been devoted to improving propeller performance or to developing special means for suppressing propeller radiated noise. These developments have had little effect upon hull form.

Seakeeping

Surface Ships. The past 50 years have seen a tremendous increase in our ability to predict the seakeeping characteristics of ships. This has come about through both experimental model tests and statistical means for representing the randomness of seaways. Furthering this has been the development of a strip theory to predict the motions of a ship in a seaway. Concurrent with these developments we have evolved the ability to predict slamming and slamming pressures on the hull. Seakeeping basins with the ability to generate random seas have been built in most of the major model basins throughout the world. The statistical representation of the sea by Pierson and St. Denis [4] in the 1950s laid the groundwork for the extensive theoretical hydrodynamic work that has followed and has contributed much to improving the performance of ships in a seaway.

The results of this work have had profound effects upon combatant ships. The most useful finding was the significance of the location of the longitudinal center of buoyancy (LCB) in relation to the longitudinal center of flotation (LCF). This significantly affects the coupling between the pitch and heave motions; thus it is possible to control the longitudinal location of minimum motion on a ship. This is important to many military functions such as in landing aircraft on the deck of a carrier. It can also control the amount of motion at the bow for purposes of minimizing deck wetness or slamming.

This work has also contributed to improving the seakeeping performance of smaller combatants like destroyers and frigates by determining the proper amount of flare at the bow and the effectiveness of bulwarks.

The impact on the hull form of commercial ships has not been as great; nonetheless, the availability of seakeeping tools provides builders and owners with an insight into how their ships will perform at sea. Included here is information on the loss of speed that can be expected. It also helps them to plan the routing and scheduling of their ships.

The availability of these tools has had an important impact on the development of the SWATH concept. Its major feature is the reduction of motions in a seaway owing to the low exciting forces that the seaway can induce because of the small water plane subject to the motion of the sea. The SWATH form is not yet mature and will be going through evolutionary changes for some time to resolve the various control problems, especially in a following sea.

Advanced Marine Vessels. The hull form of the planing boat has seen a number of important developments. The forms of the patrol and pleasure boats of the 1940s were largely spun off from seaplane work of that time. Most were low deadrise, hard chine craft with relatively high length/beam ratios. Many craft of that time were also of round chine with high length/beam ratios; they developed high speed but were not true planing vessels. The offshore racing fraternity evolved the high deadrise, hard chine boats fitted with a series of spray strips as the best form for operating in rough water. This was effective although it was known that smooth water resistance increases with increasing deadrise. This form is now widely used in the pleasure boat industry with much reduced length/beam ratios as compared to their predecessors of the 1940s.

In the past 25 years numerous variations have evolved in hull forms for small pleasure craft. Many are claimed to have special performance characteristics, such as cathedral hulls, which in cross section appear to be two small medium deadrise planing hulls connected by a low flat surface. There are claims that they get lift from the air under the flat surface.

A special two-chine form has been developed for patrol craft. The lower chine is at the boundary of a low deadrise; and there is a high deadrise between it and the upper chine. The patrol craft *CPIC* first incorporated this feature. This configuration maintains the high performance of the low deadrise hull, but with lower vertical accelerations in a seaway.

Maneuvering and Control

The field of maneuverability and control has also witnessed a large growth in the types of experimental research facilities. Most notable among these is the rotating arm used to obtain most of the various derivatives needed in the equations of motion, particularly for submarines, and the planer motion mechanism, which does the same for the surface vessels. The miniaturization of electronics has also made possible the radio controlled model that has been so essential to the development of the high-speed submarine.

More recently the process of systems identification, drawn from the aircraft field, is being used to determine the coefficients for the equations of motion. This is particularly valuable on full-scale trials, thus eliminating some of the scale-effect problems endemic to model testing.

Submarines Control Surfaces. The submarines of World War II had the vertical plane control surfaces behind the propellers where they exploited the added velocity induced by the propeller slip stream. The rudder was on the centerline similar to multiscrew surface ships. The dive planes were located on the bow, where they provided a large moment arm to control the rate of dive or rise. They could also be easily faired into the ship-shaped bow when housed. The advent of the body-of-revolution hull forms with a single propeller required a reevaluation of the location of the control surfaces. The *Albacore* served to evaluate various possible approaches. Stern control surfaces behind and forward of the propeller were evaluated, leading to the conclusion that forward control surfaces provided adequate control with minimum drag. This has now become standard for all military submarines. This control surface arrangement has created special flow problems to the propeller with resulting low frequency noise problems. These are now the subject of intense research effort. The highly streamlined form of the bow makes it difficult to place the dive planes far forward and it was determined that the bridge fairwater was the best location for minimum drag and least interference with the bow sonar. The high speeds of some of the submarines have also presented a problem with snap roll when they go into tight turns.

Surface Ships. The rudders on the higher powered single screw commercial ships have evolved from the simple post type, to the semi-balanced horned, to the fully balanced spade rudders, Fig. 2*a*. This transition is more to increase the propeller clearance in the aperture than to meet steering requirements.

Several special types of rudder have been developed in recent years to improve maneuverability. These include the Schilling rudder, which uses large horizontal plates to reduce flow losses and a special highly effective section with flared trailing edge. The active rudder has a submerged motor and propeller as an integral part of the trailing edge. The shutter rudder consists of three or more rudders, which are mechanically linked. One configuration has a fixed structure ahead of a balanced rudder and behind the propeller. None of these has yet received universal acceptance.

Bow thrusters (and, in some instances, stern thrusters) are now widely accepted in commercial ships to assist in maneuvering alongside piers with minimal use of tugs. They are also being applied to other vessels such as yachts.

The SWATH ships present a special control problem owing to the small waterplane areas, which provide small restoring forces when the equilibrium of the ship is disturbed. This is particularly important in following seas when the encounter frequency is close to the speed of the ship. Special horizontal control surfaces are needed to provide restoring forces and to maintain the desired attitude of the vessel. These are mounted on the inboard side of the main underwater body. Rudder control is also a special problem. If it is mounted behind the strut the disturbed flow from the rudder can enter into the propeller, affecting its performance. If it is mounted behind the propeller where it can gain the effectiveness of the induced velocity from the propeller, it requires elaborate supports, adding to the drag of the vessel.

Conclusions

The past 50 years have witnessed remarkable changes in the tools of ship technology and the hull forms that have evolved from that technology. The growth in the number and types of hydrodynamic research facilities along with the advent of the high-speed digital computer have expanded the tools of ship design beyond the wildest dreams of the naval architect of 50 years ago. The tremendous growth in the size and power installed in ships of all types has resulted in a significant reduction in the cost of transporting the world's commodities and increased the effectiveness of military vessels. The concurrent growth in power available for propulsion and the evolution of hull forms have made these spectacular changes possible. The increase in power has also led to major changes in hull form, such as in submarines, or the evolution of new concepts, such as hydrofoil boats, which have led to dramatic increases in ship speeds. It has been an eventful and exciting time to be a maritime researcher or practicing naval architect.

References

1 Doust, D. J., "Optimized Trawler Forms," *Trans.* North East Coast Institution of Engineers and Shipbuilders, Vol. 79, 1962–63.

2 Holtrop, J., "A Statistical Re-analysis of Resistance and Propulsion Data," *International Shipbuilding Progress*, Vol. 31, 1984.

3 Bailey, D., "The NPL High-Speed Round-Bilge Displacement Hull Series. Resistance, Propulsion, Manoeuvering and Seakeeping Data," RINA Maritime Monograph No. 4, Royal Institution of Naval Architects, 1976.

4 St. Denis, M. and Pierson, W. J., "On the Motions of Ships in Confused Seas," *Trans.* SNAME, Vol. 61, 1953.

JACQUES HADLER spent 31 years at the David Taylor Model Basin where he directed research efforts aimed at improving the hull forms of submarines, surface ships, and advanced marine vehicles. This work covered not only resistance and propulsion, but also seakeeping and maneuverability. He subsequently spent fourteen years at Webb Institute in teaching and research. Now retired from Webb, he continues as a consultant specializing in hull and propeller hydrodynamics and vibrations. He has published over a dozen SNAME papers. He is an Honorary Life Fellow of that society and holds both the Taylor Medal and the Linnard Prize.

Hull Structural Analysis

Roger G. Kline

Introduction

We have made truly impressive advances in ship structural design and analysis since World War II. These advances have been the result of many influences. Some were in response to specific and persistent structural problems. Others were owing to the development of more sophisticated tools of analysis and the increased availability of laboratory test data and full-scale ship structural response data. Other developments followed the introduction of new structural materials and new methods of fabrication and production. Still others came about with the introduction of larger and faster ships, and new types of ships, such as LNG vessels, containerships, vehicle carriers, SWATH, surface-effect ships, and hydrofoil craft.

Historical Background. Over the past 45 years or so, ship structural design has progressed from an evolutionary, to a partially deterministic, to a partially probabilistic process. There are still in use remnants of evolutionary development and considerable use of deterministic design procedures, although the role of probabilistic design and reliability analysis is growing steadily.

When there were fewer kinds of ships, structural arrangements and scantlings evolved on a trial and error basis. Classification societies would establish rules for structural components based on hull dimensions and then adjust those scantlings based on service experience, analyzed with the assistance of basic mechanics of materials principles to compensate for past failures.

As shipbuilding technology progressed, ships increased in size and power, and the number of new kinds of ships, new structural configurations, and new materials came at such a rate that evolutionary methods could no longer keep up with the need to extrapolate from the rules. Thus, the need to employ more deterministic methods for load and response calculations became evident. A prime example of this need was the rapid increase in the size of tankers and bulk carriers that took place, because of economic considerations, in the 1950s, 60s, and 70s. Also during that time, major developments took place in the calculation of wave loadings on ships, and the motions and structural response to those loads.

As tanker size increased, so did problems of transverse and longitudinal strength, which resulted in fatigue fractures and brittle fractures. Brittle fractures are related to the notch toughness of the material, whereas fatigue fractures result from stress concentrations at structural discontinuities or welded details. Both of these problems are aggravated by the heavier scantlings in the larger ships.

Research programs were initiated to collect and analyze statistical data on sea states throughout the world on a seasonal basis. Many other programs were established to collect hull response data on ships of various commercial and military type. It soon became apparent that not only quasi-static loads and responses were important, but that slamming, springing, and other types of dynamic loads and responses contributed to the ultimate structural behavior of the ship.

Also during that time, it became evident that the distribution of cargo throughout the length of the ship was critical in determining the stresses in the structure. Loading manuals became mandatory on some types of ships, and classification society rules were modified to take into account both the still water and the wave induced components of longitudinal hull bending.

Initially, progress was slow in developing accurate methods for predicting seaway induced loads; but progress became rapid after introducing the statistical description of the sea, and then the strip theory of ship motion response.

Now that we have statistically reliable ways to predict hull loading, and a more thorough understanding of the structural response to those loads, ship structural design is truly coming to be a rational process.

Organization and Content. In this chapter we shall review some of the more significant problems in ship structural design that naval architects have had to overcome since World War II. We include here complications such as those brought on by the switch from riveting to welding, noise, vibrations, new materials, and hull flexibility. We shall also look at technologies transferred from other disciplines, along with the advent of the computer and new computational techniques.

The subject of structural design will be broken down into two aspects: synthesis and analysis. (Note that offshore structures are not considered here.) The synthesis aspects are then broken down, first by type of structure, and second by type of loading. The analytical aspects, likewise, are broken down, first by the type of structural phantom, model, or idealization employed, and second by the method of analysis.

Ship Structural Design Problems

Many of the advances that have taken place in the design and analysis of ship structures are the result of persistent structural problems that have required a broad and intensive investigative effort for their solution. Possibly the most significant of these developments was the establishment of prescriptive criteria for longitudinal strength.

Longitudinal Strength Criteria. Because of its dominating importance in ship structural design, there has been greater effort expended in establishing a prescriptive basis for longitudinal strength criteria than for any other aspect of hull design. The midship section is the starting point for defining the structural arrangement and principal scantlings for the entire ship. The section modulus that results from that selection of plate and stiffener scantlings is the universally accepted measure of the hull girder bending strength. The two quantities that determine the required section modulus are the design (i.e., predicted maximum) longitudinal bending moment and the allowable bending stress.

For many years the design bending moment was obtained by imagining the fully loaded ship to be poised on the crest (and again in the hollow) of a standard wave, and determining the midship bending moment that resulted from the distribution of weight and buoyancy along the length of the ship. The $L/20$ trochoidal wave was the standard in commercial ship design for many years, while the Navy preferred the 1.1 times the square root of L trochoidal wave as its standard.

The bending moment that resulted from subjecting the hull design to a standard wave was found to be closely approximated by the product of ship length and displacement divided by a suitable constant. This so-called standard bending moment was then used in conjunction with a design stress to arrive at a required section modulus for the midship section.

The International Load Line Convention of 1930 set a standard for midship section modulus requirements that was used by classification societies until 1968. That provided a table of factors ("f") based on ship's length, which, when multiplied by beam and draft, gave the required section modulus. The value of "f" varied from 2.0 at a length of 120 ft to 22.0 at a length of 600 ft.

Comparing the section modulus given by the load line rules to the standard bending moment mentioned above provided a standard allowable longitudinal bending stress. This standard stress relationship could be closely approximated by a constant times the cube root of the ship's length. It varied from about 8.0 tons psi at 300 ft to 10 tons psi at 600 ft. Clearly, it was applicable only to ships of conventional proportions built of ordinary strength steel and with machinery amidships.

Although such longitudinal strength standards provided a simple and convenient means of assessing a proposed midship section, it eventually became clear that they were far too simplistic. Worse, they were found to be based on erroneous assumptions. With the gathering of statistical data on sea states, and with better understanding of ship motions and sea loads, it became clear that many of the assumptions built into the standard bending moment and stress relationships required better definition.

As more naval architects became aware of the impact of cargo distribution on longitudinal bending moments, two things became clear: first, that the still-water bending loads and the wave-induced bending loads should be treated separately. Second, the load line standard put too much importance on ship draft. This was because the wave-induced moments had little to do with draft; moreover, still-water bending moments were affected by draft only to the extent that the buoyancy curve becomes fuller as draft increases. Further, the collection of sea state data, indicated that the $L/20$ trochoidal wave gave effective wave heights that were too large for the longer ships.

In the late 1960s, the classification societies came to recognize the importance of still-water bending moments, and so based their section modulus requirements on what were essentially allowable still-water stresses. At the same time, the rules required that the bottom section modulus be greater than that for the deck by a margin that varied with ship type and length. At one time, ships with machinery aft were required to have substantially greater section modulus than those with machinery amidships.

Today we can develop a section modulus requirement based on calculated still-water bending moments together with wave-induced bending moments. The latter are derived from ship motion and sea load computer programs that also specify the probability of exceeding the bending moment in question. The combined moments are then applied in conjunction with the allowable longitudinal bending stress to obtain a required section modulus.

The currently used formulation for allowable stress varies with ship length, although its range of variation is much smaller than the standard allowable stress mentioned earlier. It ranges from about 9.9 tons per square inch at 250 ft to about 10.9 at 1400 ft.

Although the reason for the variation with length is not clear, it is sometimes attributed to corrosion effects being more detrimental on smaller ships. Another possible explanation relates to the magnitude of stresses from local loadings. It is generally acknowledged that small ship scantlings, if properly sized for local loads, will be more than adequate for longitudinal strength. Thus as ship size increases, a larger and larger percentage of total stress will be taken up by longitudinal strength requirements. Thus if total allowable stress remains constant, then the percentage of total (and hence the magnitude of primary) stress will increase.

Welding Versus Riveting. Welding was first used for shipbuilding in the 1930s, but it did not become the predominant joining method until World War II. At that time it quickly became an economic necessity. It was a primary factor in the dramatic increase in wartime shipbuilding productivity, and continues to be a contributing factor in that regard.

Welding in place of riveting greatly reduced the number of pieces that went into the structure by eliminating boundary bars, clips, and rivets. It eliminated many labor intensive steps such as punching plates and shapes, joggling plates or frames, and all the reaming, bolting, and driving of rivets on the ship itself. It did away with the need for caulking watertight joints, and made minor shipfitting mistakes less difficult to fix.

Not only were the number of pieces reduced, but the quantity of material was also reduced owing to the elimination of lapped joints and the faying flanges where stiffening members attached to the plates.

Unfortunately, welding brought its own problems. Poor welding methods and poor design practices were major contributors to the problem of brittle fracture. With all natural crack arresters (riveted seams) removed, the sometimes brittle (i.e., notch sensitive) steel plating was susceptible to rapidly propagating fractures that might travel completely across a deck. In unusual cases it did not stop until the ship was in two pieces. Even an accidental arc strike could present enough of a flaw and stress concentration to initiate a brittle fracture.

In time, intensive research efforts overcame the problems described above. These led to the use of tougher steels and better design practices.

Brittle Fracture. One of the most significant problems encountered in ship structural design, and certainly one that has led to a large number of improvements in the design process, is that of brittle fracture. The phenomenon had long been recognized; but it became a major concern during World War II, when many Liberty ships and T-2 tankers suffered catastrophic failures as a result of this problem. Between 1942 and 1952, over 200 ships experienced serious brittle fractures. These included seven Liberty ships and nine T-2 tankers that had broken in two. For the Liberty ships, one of the principal problems was hatch corner design; and for the T-2 tankers, most fractures originated at defects in the bottom shell butt welds.

In 1946, the interagency Ship Structure Committee was organized to initiate, sponsor, and interpret research on improvement in ship structures, but especially to study the brittle fracture problem. The Committee was made up of personnel from the Department of Navy, Military Sea Transportation Service (now MSC), U.S. Coast Guard, Maritime Administration, and American Bureau of Shipping; and an advisory committee was organized under the National Academy of Sciences/National Research Council to administer the research.

The original purpose of the research was to investigate hull fractures in merchant ships. Its activities were later expanded to cover other structural problems such as hatch corner design, ship flexibility, ship response data collection and analysis, computer-aided design, ice strengthening, and probabilistic design procedures.

While investigating brittle fracture, the Committee sponsored many research projects on various metallurgical aspects of steel making, processing, and joining. It investigated ductility exhaustion (the loss of ductility owing to excessive working), crack propagation and arrest, strain rate and temperature effects, and many other aspects of material behavior. The Ship Structure Committee is still active, and is currently investigating such topics as ship structural details in high-strength steel.

Two other organizations need to be mentioned as contributing significantly to understanding ship structural behavior and the advancement of design: one is the International Ship Structures Congress, and the other is the Hull Structure Committee of the SNAME T&R program.

Fatigue. The problem of fatigue fractures in ships has been around for a long time but became more prevalent with the advent of welding. Some of the fractures were directly attributable to improper welding practices, while others were the result of poorly designed welded details. Some of the most common sites for fatigue fractures were at the intersection of longitudinals and deep girders and at bracket toes. Many of these fractures in the vicinity of the waterline were aggravated by tug and docking loads.

There was at one time considerable controversy with regard to whether fatigue cracks were the principal initiators of brittle fractures, or whether they represented only a nuisance that required periodic repair. There is little doubt that if a fatigue crack develops in material that is not sufficiently notch-tough, it is a potential initiation site for a brittle fracture; but if one should develop in a sufficiently notch-tough material, it may extend to a location where stress levels are low enough that the crack ceases to grow, or to where it will be noticed and repaired.

The fatigue problem has been aggravated in recent years with the increase in size of containerships, tankers, and bulk carriers in conjunction with higher-powered propulsion plants and increased use of higher strength steels. The increase in ship size seemed to bring on transverse strength problems that were not anticipated, particularly in tankers and bulk carriers.

Fortunately, as a result of these problems, there have been investigations into the performance of typical structural details. In particular, the Ship Structure Committee has sponsored not only extensive surveys of operating ships, but also the development of detail design procedures based on laboratory testing of configurations representative of most of the details surveyed. These procedures take into account a probabilistic description of the load cycles experienced over the life of the ship together with the *S–N* (stress or strain versus number of cycles to failure) performance of the steel, and provide the maximum allowable stress range for a desired level of reliability.

Such procedures, if properly used, should greatly decrease the incidence of fatigue fractures. Nevertheless, in some of the earlier applications, wherein scantlings were reduced on the basis of the more refined design, the combination of increased stress (and in some cases, the further reduction of thickness resulting from corrosion) resulted in structures with increased fatigue problems and decreased ship life. Corrective design refinements have since been made.

In addition to these advances in the art of fatigue design, the field of fracture mechanics has provided the means for predicting the rate of fatigue crack propagation, and the length of crack having the potential to propagate as an unstable rapid fracture.

Hull Flexibility and Distortion. The problem of ship flexibility or, inversely, ship stiffness has been around for many years. This is not to say that flexibility is of itself a problem. The problem is in determining if too much or if too little flexibility may be responsible for a structural failure.

One source of confusion with regard to the problem is the sometimes close relationship between ship stiffness and ship strength. Originally, classification society scantlings were based principally on ship size or carrying capacity. In those early days ship size and proportions were such that if local strength were satisfied, longitudinal strength was no problem. Then as ships grew in size, it eventually became apparent that in order to provide adequate longitudinal bending strength with local scantlings, sufficient hull depth had to be maintained. For a given thicknesses of bottom, deck and side structure, the deeper ship is not only stiffer but stronger. Unfortunately, stiffness is not always synonymous with strength.

Classification societies still handle the problem in an empirical manner. They limit the applicability of their rules with regard to length/depth (L/D) ratio and midship moment of inertia. These rules put upper limits on flexibility but no lower limit. The L/D ratio may be as small as desired and midship moment of inertia (I) may be as large as desired. If the flexibility limits are exceeded, then scantlings and arrangements will be specially considered.

Several events or circumstances have brought attention to the matter of hull girder flexibility: the increase in ship size together with natural limits on ship draft have tended to promote higher L/D ratios (particularly on Great Lakes bulk carriers and barges); the use of higher strength steels has reduced section modulus requirements without changes in proportions and has thus resulted in more flexible hulls; and finally the installation of higher powered propulsion plants has raised concern over whether ship vibrations would be exacerbated by greater hull flexibility.

The concern over excessive flexibility has centered primarily on the dynamic behavior of the hull girder. On Great Lakes vessels the concern was prompted by the prevalence of a phenomenon that was thought to be almost unique to the Great Lakes—the phenomenon called springing, which is essentially the response of the fundamental mode of vertical hull vibration to periodic wave encounter. The concern about springing led to many full scale, model tank, and computer simulation investigations to help verify the relationship between excitation and response and to determine if the response needed to be reflected in the longitudinal strength requirements for the hull.

Most investigations concluded that springing was the result of encountering waves of the same frequency as the hull fundamental frequency, and because that frequency was generally higher than the sea state's modal frequency, a more flexible hull would have a lower natural frequency and would be subject to excitation from a portion of the sea spectrum having greater energy. As a result of these investigations, the springing bending moment is now calculated and treated as part of the total bending moment (along with the still-water and wave-bending components) when determining the section modulus requirements for Great Lakes ships.

There is little evidence that overall hull flexibility is in itself detrimental to propeller-excited vibratory response. Far more important is the proximity of the blade frequency excitation to one of the main hull natural frequencies. If they coincide, either a stiffer or more flexible hull would improve the situation.

Noise and Vibration. Until 1940 the understanding of ship vibration problems was extremely limited, and with few exceptions the solutions were obtained by trial and error. In merchant ships, structural resonances, balancing of rotating equipment, and low frequency isolation systems were understood and applied, while torsional vibration of reciprocating drive systems was being studied. The average commercial ship had a single screw, less than 15,000 shp and low rpm. Larger ships with higher installed power were likely to be passenger ships with multiple screws. This permitted the propellers to operate in more open water, which was conducive to more even flow into the propeller disc area. Thus serious vibration problems were not noted until the advent of the large shipbuilding program precipitated by World War II.

In 1940 the U.S. Navy was not much further along. Most ships had been built in World War I, or earlier. Battleships generally had reciprocating steam engines and multiple screws of relatively modest power. Destroyers and cruisers were higher speed ships, but they too had multiple screws and moderate powers. Moreover, with better care having been given to stern shapes, good flow of water to the propeller was assured.

World War II marked a series of rapid changes. The Navy put the *North Carolina*-class battleships and *Midway*-class carriers to sea, both of which had serious hull vibrations. Torsional vibration wreaked havoc in the reduction gears of destroyers, and in crankshafts and couplings of diesel drives. Periscope vibrations limited the submerged speed of submarines, and many propeller shafts were lost owing to corrosion fatigue. With the demand for all types of ships, the Navy was forced to obtain engines (from reciprocating steam and diesel to steam turbine) from whatever sources were available. It then tried to match them with propellers and shafts in all types of ships, wooden as well as steel. With this situation, a good many vibration problems were encountered, and vibration engineering in the Navy became a continuous task of problem solving.

During this period shock resistance was identified as a vital consideration in combatant ships. Cast iron fittings were eliminated wherever possible; and shock mountings were attached to all sensitive equipment, exacerbating, in many instances, the vibration problem. Shock and vibration research groups were developed in the various Naval laboratories, while specific research programs were initiated on shock and vibration testing of shipboard equipment. In many cases, industry also participated in specific application and testing, and through research sponsored by various SNAME T&R panels. Parallel efforts were carried out in the development of computers and computer programs necessary for structural and mechanical design and evaluation.

Hull vibration was first studied as an adaptation of uniform beam vibration in terms of mode shapes and their associated natural frequencies. The nonuniformities of mass and stiffness distribution were handled by empirically derived correction factors. Later, as calculation procedures improved, these variations were handled more directly, and the added mass of water vibrating with the ship was also handled more directly as a part of the calculation process.

Computer programs now calculate the real-time variation of hull structural response to a variety of steady and transient excitations from the seaway. They not only deal with nonlinearities in the excitation forces but in the geometric deformation of the structure and in the material properties.

Developments in Ship Structural Design

As already explained, hull structural design generally starts with a proposed midship section, from which is derived a section modulus. Dividing that number into the predicted maximum longitudinal bending moment leads to the predicted maximum bending stress. If that exceeds the design limits, the scantlings are given appropriate increases.

Once longitudinal strength criteria are satisfied, other aspects of ship strength should be investigated to assure that the hull does not fail under the action of torsional loads, racking loads, and local loads of specific types such as slamming, mooring, drydocking, and launching.

Thus the design process consists of alternating exercises in synthesis and analysis until the designer is satisfied that he has proposed a ship with adequate overall strength.

This exercise is carried out for structural components of the ship and for the ship as a whole. As design and analysis tools have developed, the completeness and detail of the whole ship idealization, as well as the description of the loading applied to it, has reached an amazing degree of complexity.

There are many reasons for these advances in the design process, such as a better statistical base for the loads, technology imported from the aerospace and civil engineering fields, and a better knowledge of material behavior; but by far the most important development has been the introduction of the digital computer and the computational power that it provides. Some of the more important advances in the ship structural design process are described in the following.

Design for Longitudinal Strength. As already explained, longitudinal strength analysis has in the past dealt almost exclusively with the beam-like strength of the ship's hull in vertical bending. The load has generally been considered quasi-static.

When evaluating the longitudinal bending strength of a ship's hull in a deterministic manner, it is common to work in terms of a midship bending moment, an allowable longitudinal bending stress, and an associated section modulus of the midship section. The bending moment has traditionally been calculated by imagining the loaded ship poised on a wave profile of standard form. Until very recently the $L/20$ trochoidal wave proposed by William Froude in the late 1800s was used as the standard. This bending moment in conjunction with an allowable stress was then used to determine the required section modulus. The allowable stress was generally in the range of 8.0 to 10.0 tons psi.

For those designing a midship section to classification society rules, those "rule" scantlings were almost guaranteed to produce a section modulus that would satisfy the requirements of the 1930 Load Line Convention. These requirements were in effect until 1968.

In addition to the full scale ship response data, considerable model test data on segmented models were generated to help clarify the relationship between wave-induced loadings and response. Because of all this theoretical development and data accumulation, we can now establish a required midship section modulus on the basis of accurately calculated still-water bending moments and wave-induced bending moments representative of sea states in any locale and for any specified ship life. Not only is it possible to select a proper design bending moment for the mid-length, but also for various points throughout the length. All of this is established on a probabilistic basis.

Besides the significant advances that have been made in the determination of required longitudinal strength, there have been many improvements in the design of the midships structural arrangement. Most of these improvements have come about through the advent of computer-aided design and consist primarily of computer programs that develop rule scantlings or so-called scantlings based on first principles. Other programs concentrate on optimizing the midship section with regard to frame spacing, type of stiffener, and material characteristics. The measure of merit may be based on material weight, cost, or producibility.

Design for Transverse Strength. The term transverse strength refers to the ability of the ship's hull to resist forces tending to change the shape of the cross-section. In a transversely framed ship, the transverse strength is provided by the framing and transverse bulkheads. In a longitudinally framed ship, it is provided by the web frames and transverse bulkheads. Since the transverse structure supports and is supported by various items of longitudinal structure and interacts with it, insufficient transverse strength can be detrimental to longitudinal bending and torsional strength.

Transverse strength is particularly crucial in ships such as tankers or bulk carriers with uneven distribution of weight athwartships. It is also of concern in ships with limited amounts of internal structure. Examples include ferries or roll-on/roll-off ships. Tankers and Great Lakes bulk carriers, in particular, have had more than their share of transverse strength problems. Most of these problems have manifested themselves as fatigue cracks and buckling of the web portion of deep frames.

Early attempts to evaluate transverse strength relied on rigid frame analysis methods and moment distribution applied to the two-dimensional arrangement. Photo-elastic methods have been applied to the analysis of tanker web frames, and plastic design methods have also been used to evaluate web frame strength. Most of these methods ignored or poorly modeled the contribution of the longitudinal structures.

With the advent of the digital computer, it became feasible to conduct three-dimensional frame analyses; and when the finite element method of structural analysis was introduced, it became possible not only to solve the three-dimensional frame problem, but also to obtain a detailed estimate of the stress concentration at bracket toes and cutouts for longitudinals. The analytical capability actually seemed to have advanced beyond the ability to derive applicable loading conditions and allowable stresses.

Some of the loading can be derived from the ship motion and sea loads programs, but for most practical designs the design loading is recommended by the classification society.

At times it is difficult to separate the synthesis from the analysis; and in many of today's ship design computer programs, the two activities are combined and used in conjunction with optimization procedures that can optimize on weight, cost of material, producibility, or any combination thereof.

Tanker web frames and tanker structure in general have been the subject of optimized design studies using computer programs of this type.

Design for Torsional Strength. Until the 1970s, torsional strength was of little concern. More recently, however, it has come to the fore because of specific problems made evident in ships with unusually large hatches. Examples are found in containerships and some bulk carriers. Empirical formulas for torsional moments indicate that such moments are proportional to the ship's length times the cube of the ship's beam. This relationship applies to seaway-induced bending moments; however, cargo-induced torsional moments must also be considered. Obviously, ship beam plays an important part in torsional moments created by the non-symmetric loading of cargo. Thus, large containerships with a low L/B ratio would seem to be prime candidates for hull torsion problems.

Ships with inadequate torsional strength may have problems with hatch cover tightness, cracking around hatch corners and coamings, and possibly shear buckling of side shell and deck plating.

Before the availability of finite element methods and sea load computer programs, designers had to rely on the classical theory of torsion for thin-walled beams, or some adaptation of that theory from the aeronautical field, together with some hydrostatic estimate of the wave-induced torsional moments.

Design for Local Strength. The advances in the design for local loadings have been primarily the result of improved information on their expected magnitude, and in the philosophy with respect to acceptable responses. The theory of ship motions and the study of the hydrodynamic loads corresponding to those motions have provided a much firmer basis for predicting local structural response in a seaway. In addition to the advances in hydrodynamic theory, there has been a wealth of ship response data collected in recent years, which is useful in substantiating the theory as well as providing a basis for empirical design.

The classification societies have made significant advances in analyzing ship response data using probability procedures and have used the results of such analyses to improve their rules. From a practical point of view, most design for local loads is done to classification society requirements, at least for commercial ships and non-combatant naval vessels. Classification society rules increasingly reflect strength-of-material type analyses. Moreover, they will now accept alternative designs based on recognized methods of structural analysis.

Until the late 1950s the American Bureau of Shipping (ABS) rules gave plate and stiffener scantlings in tabular form based on ship length, depth, draft, and frame spacing. More recently plate scantlings have been based on empirical formulas incorporating the same parameters; and stiffeners are designed to section modulus requirements derived from the load supported, the type of end-connection employed, and the level of primary stress expected.

Plate panels subjected to pressure loads if not designed to classification society rules were usually designed for stress at the middle of the long edge and treated as if all edges were built in and all strains elastic, even if the "design" stress were greater than yield. More recently, depending on location, plates may be designed for a certain amount of permanent set, and, if need be, while taking into account some initial distortion. For design against more severe loading, and after a pattern of plastic hinges is assumed, plating may be designed to a specific fraction of the collapse pressure. The basic reason for using plastic design methods is that they may be more closely related to the failure load.

Although plate stiffeners can also be designed on the basis of conventional beam bending moments and shear to a percentage of the yield strength of the steel, plastic design is becoming much more popular; and these stiffeners may frequently be designed to some percentage of the plastic collapse load.

Similar methods of analysis are also being applied to designing web frames and girders. In addition to the plastic design methods, the analysis of web frames, and stanchion and girder systems is now done using moment distribution methods and matrix computation methods to handle redundancies and nonlinearities. The finite element method of analysis is becoming more popular as computing costs come down and programs are easier to use. It has the benefit of rapidly evaluating the effects of stress concentrations in detail design, as well as evaluating the adequacy of overall scantlings under combined loadings.

Design for Dynamic Loading. Relatively simple formulas exist for calculating the natural frequencies of simple single-degree-of-freedom spring/mass systems or of uniformly distributed beam-like systems. But when the mass or elastic properties of the system become non-uniform and complex, the determination of mode shapes and frequencies also becomes more complex.

One of the more common methods of dealing with such non-uniform distributed systems is to divide it by lumping mass and elastic properties into a collection of masses and connecting beams or springs. Mode shapes are assumed and corresponding kinetic and potential energies equated to solve for the corresponding frequencies. The best mode shape is the one that gives the lowest frequency.

This type of solution procedure has been systematized and programmed for computer calculation. The procedure has also been improved to incorporate more realistic representation of the structural characteristics of the system. Because of the manner in which the structural system is represented and analyzed, this method of analysis has become known as the finite element method.

These analytical methods now encompass geometric and material nonlinearities to give steady state and transient response to extreme loads such as the shockwave loadings from underwater explosions and ship collisions.

Developments in Ship Structural Analysis

Since World War II naval architects have made major advances in ship structural analysis. Some of these advances have been in the theory of structural behavior, some in the evaluation of model and full-scale data, some in the modeling of the structure to be analyzed, and some in the area of computational capability.

Structural theory advances have occurred in such areas as structural dynamics, structural stability, plastic design, ultimate strength analysis, hydroelasticity, and fracture mechanics. In general, many of these advances are related to the determination of true structural capability or strength of the ship. The advances in data evaluation are largely in the implementation of probability theory and follow mainly from the dramatic increase in the quantity of ship load and response data collected in recent years. The obvious advances in structural modeling are due largely to the advent of finite element analysis; and the increases in computational capability have come from applying matrix analysis methods and the use of the digital computer.

Ship structural analysis can be divided into two principal activities:

- determination of structural loads, and
- analysis of structural response to those loads.

Hull loads may be environmental, such from waves, wind, ice, and temperature, or they may be cargo (or military supply) loads induced by the motion of the ship; or they may result from docking or launching of the ship. Naval vessels must withstand, in addition, a variety of combat-related loads. And finally, there are dynamic loads induced by the propulsion and auxiliary machinery.

The structural response to these loads are described in terms of bending moments and shears or in terms of yielding, buckling, brittle or fatigue fracture, or some combination of the above. The ultimate concern is that the level of response, in whatever mode, be accommodated by the structure without serious failure for the expected life of the ship.

Loads. *Seaway Loadings.* Traditionally, most seaway loadings were treated as being hydrostatic. The principal ones were based on the standard wave profile as already described in determining the midship bending moment. Various local loadings were estimated on the basis of the head of seawater measured to the design waterline or to the crest of some standard wave.

Occasionally, motion-induced loads were estimated on the basis of a standard roll amplitude and period, a standard pitch amplitude and period, and, at times, a specified heave amplitude and period. These acceleration loads were generally used in the design of foundations for equipment or fittings located well away from the ship's center of motion.

As sea state data became plentiful, and ship motion and sea load programs were developed, the need for standard roll, pitch, and heave was made obsolete, and both short-term and long-term loads and responses could be developed with any specified degree of confidence.

Only recently have attempts been made to predict from basic principles the dynamic loads of slamming and springing. Slam loads are transient impacts on the forward bottom or bow flare structure that result from impact of the wave surface directly on the structure.

Both loadings require the use of sea-state data, together with ship-motions-and-sea-loads computer programs and, in the case of slamming, at least, a significant amount of empirical data. Procedures based on theoretical developments have been proposed for the determination of all probability aspects of slam loads, including their magnitude, frequency of occurrence, and location along the length of the ship.

Cargo Loads. Cargo loads have traditionally been treated on the basis of an average density and the acceleration of gravity. Ship motion and density variations were handled as special situations and covered by a conservative factor of safety. With advances in ship-motion theory, it is now possible to predict with varying degrees of confidence the acceleration loadings at any point in the ship. Container-securing devices are now designed on this basis.

Liquid cargo sloshing loads are also predictable with some confidence, as are the dynamic pressures generated by the movement of bulk cargoes of various types.

Ship-Handling Loads. Launching, drydocking, and berthing loads are more reliably predicted than in the past because of improved analytical capabilities, which treat the dynamic aspects of the loading and the elastic properties of the ship and dock or other structure.

Analysis of Structural Response. All structural analysis problems require a simplified model or idealization of the member, assembly, module, or entire ship whose characteristics are such that mathematical solutions to the load and response can be synthesized. When the available analytical tools are crude, the model is usually simple. As the tools become more sophisticated, the models or idealizations may become more complex.

Phantoms, Models, and Idealization. Not long ago, the most commonly used analytical tools were simple beam equations that accounted for various end conditions and distributions of load. Only bending deflections were considered, although both bending and shear stresses were calculated.

The next step in sophistication was to treat the hull as a prismatic box girder. This allowed a more accurate description of loading, and eventually led to consideration of such phenomena as plate and plate/stiffener buckling, shear lag, torsion of thin-walled members, and grillage behavior. It also allowed for various types of loads and responses to be superimposed so that a more accurate and complete picture of the stress distribution throughout the structure could be obtained.

The box girder analogy led, in turn, to the analysis of transverse strength. For longitudinally framed ships, this meant the analysis of the racking strength of web frames and bulkheads. That type of analysis became particularly important during the 1960s and 1970s, when tankers and bulk carriers were experiencing such rapid increases in size.

The frame analysis used in the evaluation of transverse strength was accomplished by slope-deflection methods, column analogy, and moment distribution. Similar methods were used in the analysis of pillar and girder systems and other statically indeterminant structures.

In the 1950s, following the advent of the electronic digital computer, matrix methods of structural analysis became increasingly popular. The digital computer, together with the matrix analysis methods, made it possible to consider problems far more extensive and complex than could possibly be solved by manual calculation.

In general, those analytical methods were applied to problems in which the structure could be divided into segments of uniform beams and struts; and matrix methods were then used to solve equations relating the displacements and rotations of each end, or node, of each element to the forces and moments applied at the other end, or node.

With the introduction of membrane, plate, and other types of elements, this analytical procedure eventually became known as the finite element method, or finite element analysis procedure.

In addition to providing for the analysis of complex structures, the computer made it possible to analyze a large number of alternative configurations. Thus, structural optimization and the broad area of computer-aided design became an achievable goal and an area of considerable investigation and development.

Analyzing the Ship as a Beam. The realization that ships operating in a seaway respond with beam-like behavior dates back to the mid 19th Century.

Beam-like behavior implies that, in longitudinal hull bending, plane sections remain plane. Further, assuming stress is proportional to strain, bending stresses vary linearly with distance from the neutral axis. Beam theory generally does not compensate for ineffective longitudinally oriented material that results from plate buckling, shear lag, and initial distortion and/or residual stresses.

Since many full scale and model tests verify the beam-like behavior, it has proven to be a useful concept in preliminary design for longitudinal strength and for determining nominal stress levels in the primary structure of the ship. It is also useful in estimating bending deflections of the hull.

For simple cross-sectional geometries, the shear stress, strain, and deflection may also be calculated in accordance with beam theory. For more complex cross-sections, however, shear stresses and strains are more difficult to calculate and are frequently ignored or just roughly approximated.

For the most part, analyzing the primary strength of the hull as a beam has proven to be perfectly adequate as long as members are properly proportioned to avoid instability, and as long as both flanges and web portions of the hull girder are free of large cutouts and other major discontinuities. Generally, this has been accomplished by placing limits on plate panel b/t (breadth/thickness) ratios and slenderness ratios of plate/stiffener combinations.

As could be expected, however, beam theory begins to break down when applied to more local structure, where high stress gradients, large distortions, or complex stress flows at major discontinuities cause warping and buckling, and make a simple analysis inapplicable.

Treating the ship hull as a beam was particularly useful in attacking the problem of overall ship vibration. Success was achieved in basing an estimate of the ship's fundamental frequency of vibration on the characteristics of a free-free beam, and by relating the stiffness of the beam to the moment of inertia of the midship section and the mass per unit length of the beam to the ship's displacement. A coefficient was used to account for the combined effects of non-uniformity of mass and stiffness distribution and the virtual mass of water. The coefficient was empirically derived to reflect ship type and arrangement.

Once the fundamental frequency was estimated, the higher orders of ship vibration were estimated as multiples of the fundamental, also on an empirical basis. Here again, difficulty was encountered when certain aspects of the structural arrangement led to non-beam-like behavior. For instance, for certain types of cargo ships, the local vibration of double-bottom structure in association with cargo and entrained water represented a significant secondary vibrating system coupled to the primary one. This secondary system had to be incorporated into the overall idealization of the ship to allow a reasonably close estimate of vibratory behavior.

Fortunately, the introduction of finite element analysis techniques has provided the means for accomplishing this type of analysis, as well as much more complex modeling of the ship structure for both static and dynamic analyses. A detailed modeling of the structure does not, however, solve all the problems associated with accurately predicting ship vibration characteristics. There are still the problems of the proper representation of the various damping mechanisms, the non-two-dimensional water flow at the ship's ends, and their effects on added mass and damping values, and the neglected effects of operating in a seaway. Another significant problem is that of including the vibratory loads into the total spectrum of seaway loadings for which the ship structure is designed.

In spite of all these problems, much progress has been made in our ability to minimize the tendency of a ship to vibrate and in predicting the characteristics of its vibratory behavior. Currently, finite element analysis codes are being developed that significantly improve the analysis of structural behaviors that include geometric and material nonlinearities.

Analyzing the Ship as a Box Girder. As indicated previously, representing the ship as a thin-walled box girder allows us to understand such phenomena as shear lag, buckling, warping, and racking or transverse distortion of the cross-section.

Much of our longitudinal strength analysis still treats the hull as a beam and considers the remote fiber stress resulting from a given quasi-static bending moment a proof of adequate strength. But, our assumption that plane sections remain plane may be unrealistic if unusual bending moment distributions, or unusual structural arrangements, are considered. The same is true, if the effectiveness of deck houses in longitudinal bending is studied. This is true because in all such cases shear lag comes into play.

By treating the ship as a thinly plated box girder, we can more accurately determine shear stress distribution, hull torsional response, and the response of much of the secondary structure to local loadings.

At one time, stress functions were used to investigate the structural response of plate structures to various idealized loadings. More recently in the aircraft industry, for instance, thin shell structures were idealized in terms of stringers to support tensile and compressive loads combined with shear webs capable of transferring shear loads only. These idealizations, along with developments in isotropic and orthotopic plate theory and grillage analysis, have significantly advanced the naval architect's analytical capabilities.

Structural problems encountered in containerships and others with large or numerous hatches have emphasized the need to analyze torsional strength characteristics. The study of the torsional behavior of open or closed thin-walled prismatic beams has dealt with such topics as free or restrained warping, and the longitudinal direct stresses and secondary shear stresses resulting from this warping action. These theories, together with strain measurements on operating ships, now provide valuable guidance in designing containerships.

Structural problems encountered in bulk carriers and tankers have emphasized the need to analyze the ship's transverse strength characteristics and the transmittal of cargo and buoyancy loads from bottom structure to bulkhead and side structure. The transverse strength problem has been investigated using various frame analysis methods, while the load transfer problems have been investigated using grillage analysis and orthotopic plate theory.

All these problems of non-beam-like behavior have been investigated more thoroughly and more rigorously with the use of finite element analysis procedures. Finite element techniques have made unnecessary and obsolete the more cumbersome and less accurate analyses using shear lag and plate stability concepts.

Conclusion

During the past half century, ship structural analysis has evolved from an evolutionary, to a partially deterministic, to a partially probabilistic process. These changes have been driven in part by changing needs, such as the switch from riveting to welding, and the introduction of more complex ship configurations. Other changes have exploited new analytical techniques, some made possible only by the advent of the electronic computer.

As ever, we owe much of our progress to developments in other branches of engineering; but they owe much to us, as well. For example, our studies of brittle fracture and fatigue have led to understandings that are widely applied in other industries.

Through it all, the Society of Naval Architects and Marine Engineers has played a leading role.

Roger G. Kline earned a degree in naval architecture & marine engineering at the University of Michigan in 1957, and then went on to Berkeley for a master's degree. His professional career has been devoted almost entirely to ship structural analysis. He has developed his expertise in such organizations as MarAd, U.S. Steel Research Center, and R. A. Stearn (where he was Vice President & Technical Director). He is currently employed in the Newport News Shipbuilding Company's Milwaukee office. He is a member of both SNAME and RINA. He is active in technical committees for SNAME and the National Research Council, and is currently serving as Secretary/Treasurer of SNAME's Great Lakes & Great Rivers Section.

Ship Structural Materials

Roger G. Kline

Introduction

In this century, shipbuilding materials have progressed from wood to wrought iron to steel of various chemistries and heat treatments, to aluminum alloys, and to composites of various reinforcements and matrices, including cement. With the exception of wrought iron, these materials, in one form or another, are still in use. Because of developments in joining methods, however, some have been modified so as to make their material characteristics more appropriate to the joining methods used.

Like design and analysis methods, many of the currently used steel and aluminum alloys were developed to solve structural problems such as brittle fracture, ease of fabrication, and corrosion. Composites can be produced with good strength-to-weight ratios and are relatively easy to maintain. The most common incentive for new alloy development has been the desire for greater strength and toughness.

There are few, if any, kinds of ship that benefit from an excess of structural weight. Two of the most important features of a given ship design are its payload and speed, both of which benefit from a reduction of structural weight. Therefore, structural materials with a high strength-to-weight ratio are used whenever affordable. Material cost alone does not determine affordability, nor does yield strength alone determine strength. Cost includes fabrication and maintenance considerations, while notch toughness and fatigue strength are as important to the survivability of the ship as the elastic limit or yield strength of the hull material.

The reason for the large variety of materials in use in shipbuilding today is that different types of ships and different types of ship structure have different requirements with regard to material properties. The desire to increase payload or cargo-capacity of weight-limited ships has led to the application of aluminum and higher-strength steels in ship structures. The desire to minimize wastage and maintenance problems has led to the use of composites, aluminum, corrosion resistant steels, titanium, and copper-nickel alloys. Problems encountered in welding, fatigue, and corrosion are alleviated through the development of new alloys, the introduction of new materials, the improvement in coating systems, and the application of cathodic protection.

Because of its predominance as a ship structural material, the main emphasis will be on steel and its various alloys. Next, aluminum and its alloys will be summarized, along with other non-ferrous metals, and finally composites and wood will be discussed.

Historical Background

Wood was the shipbuilding material of choice well into the 18th century. It wasn't until 1787 that iron was first used as a shipbuilding material, and the first iron merchant ship was built in 1822. Lloyds classed its first iron-hulled ship in 1837 and published rules for iron ship construction in 1854, at the time the *Great Eastern* was being built. By the end of World War I, wood was no longer used in commercial ships, owing to its cost and maintenance difficulty.

The Bessemer process for making steel was invented in 1856; and steel was first used as a shipbuilding material in 1859. Mild steel was first used in merchant ship construction in 1878 and shortly thereafter had largely replaced wrought iron in shipbuilding.

In the early part of the 20th century, higher strength steels were introduced to shipbuilding. In 1907 the *Mauritania* and *Lusitania*, the largest passenger liners of their day, were built using higher strength steel in the upper portions of their hulls. Modern high-strength shipbuilding steels are more than twice as strong as the first mild steels and are much less susceptible to brittle fracture. Stainless steels are often used in chemical tanks to minimize maintenance and cargo contamination.

Aluminum was introduced to shipbuilding in the early 1890s and was used principally in launches, yachts, and torpedo boats until the 1920s, when significant applications of aluminum were made in naval surface combatants worldwide. The first were of Duraluminum (alloy 2017) and were riveted applications in deck house and superstructure. Later, weldable alloys were developed and have seen extensive application in destroyers, cruisers, and aircraft carriers, particularly topsides. Aluminum has also found significant application in the superstructure of large passenger liners, in merchant ship deck houses, and to contain cryogenic cargoes in some LNG ships.

Titanium and cupronickel applications in shipbuilding have been much less extensive than steel and aluminum. Most have been experimental in nature. Titanium has been used in submersibles and cupronickel in shrimp boats and in appendages of larger ships. The principal reason for interest in cupronickel is its resistance to corrosion and marine fouling. It is also known for its ease of fabrication, and its ability to be welded to steel.

Fiberglass reinforced plastic had its start in small pleasure boats because of its appearance and ease of maintenance. It possesses a good strength to weight ratio and is relatively easy to repair. It has also been used in the structure of some non-magnetic mine sweepers.

Although its use has diminished greatly in recent years, wood is still being used as ship and boat-building material. Two developments are primarily responsible for its survival. The most significant advance in wooden shipbuilding is the use of adhesives for joining and laminating. The various marine plywoods and the use of laminated beams, frames, stems, etc., has greatly improved the strength and reliability of wooden ships as well as the ease of construction. Another important advance has been in the use of preservatives to protect against marine borers.

Wood is still used in custom-built pleasure yachts and in small naval craft such as mine sweepers and patrol craft.

Developments in Shipbuilding Steels

The principal developments in steel as a structural material have been in the areas of notch-toughness, strength, and weldability.

During the latter part of, and following, the second world war, research efforts were centered on the brittle fracture problem. Although brittle fracture was not a new problem for steel structures, the dramatic ship failures attributed to it during the war brought it to the forefront of ship structural problems. As a result, the Ship Structures Committee (of the National Academy of Sciences/National Research Council) was organized in 1946 to find a remedy for brittle fracture and other ship structural problems.

Brittle fractures usually result from a combination of circumstances: a brittle material; a notch, stress concentration, or a small crack; low ambient temperature; and frequently a suddenly applied load or stress. Although many brittle fractures occurred under circumstances that were not associated with particularly high stresses, the sudden application of load was found to be more critical than a slowly applied load. High strain rates result in increased yield strengths and more brittle behavior of the steel.

When welding replaced riveting as a method of joining steel plates and shapes, one of the most effective means of arresting a brittle fracture was eliminated; I refer to riveted seams.

In the mid 1960s, higher strength steels were beginning to see significant application in cargo ships, and the American Bureau of Shipping (ABS) began to provide guidance for their properties and scantlings. For all steels, material requirements varied by plate thickness and by location. Gunwale and bilge strake areas required steels with greater notch-toughness than the remainder of the hull. Structural shapes had modest material property requirements, which for ordinary strength steels consisted of maximum limits on phosphorus and sulphur, and a range of tensile strength from 50,000 to 71,000 psi.

Before the advent of welded ships, steel yield strengths were frequently governed by carbon content; but high carbon content resulted in poor welding characteristics and notch-brittleness. Manganese and vanadium were also used to produce higher yield strength; but too much manganese also resulted in poor welding characteristics. Consequently, some of the first specifications for both ordinary and higher strength steels put a maximum value on carbon content and a range for manganese that was at least two and one-half times the carbon content.

Steels for shipbuilding currently can be categorized as

- ordinary strength structural carbon steels,
- high-strength low-alloy steels,
- heat-treated carbon and high-strength low-alloy steels,
- high strength constructional alloy steels, or
- high yield steels.

For many years, the most commonly used shipbuilding steels were the ordinary strength, structural carbon steels, of which the ABS Grades of A, B, D, and E are typical. These grades represent increasing levels of toughness. Their chemistry comprises carbon, manganese, silicon, phosphorus, and sulphur. They have yield strengths usually over 34,000 psi and tensile strengths in the 58,000 to 65,000 psi range. Weldability is assured through restrictions on the carbon and manganese content, and toughness is also provided through chemistry restrictions and the manufacturing process.

Reduced carbon and increased manganese content were found to increase notch toughness, and further improvements were obtained by silicon–aluminum killing and by a normalizing heat treatment, particularly for plates of greater thickness.

Today naval architects have a much wider selection of structural steels from which to choose than they had 20 or 30 years ago. Part of the reason for this wide selection is the development of the high-strength low-alloy steels. Many of these were specifically developed to have greater yield strength and tensile strength, improved toughness at lower temperatures, greater resistance to abrasion and atmospheric corrosion, and improved weldability. Yet these newer steels have workability and forming qualities as good as, or better than, carbon steels.

High-strength low-alloy steels have specified minimum yield points above 40,000 psi and acquire their strength through chemistry and hot-rolling rather than through heat treatment. Typical of the high-strength low-alloy steels are the ABS Grades of AH32, DH32, EH32, and AH36, DH36, and EH36. These steels have yield strengths of 45,500 and 51,000 psi, respectively, and minimum tensile strengths of 68,000 and 71,000 psi. Also in this group are the ASTM Grades of A242, A537, and A588.

High-strength constructional alloy steels, such as ASTM A514 and A517, obtain their strength through a combination of alloying elements in conjunction with a quench and tempering heat treatment. The high yield steels, such as HY-80, HY-100, HY-130, and HY-180 are low carbon, heat treated alloys that have been developed primarily for military applications. The more recently developed high-strength low-alloy (HSLA) steels are more easily fabricated than HY-130 and HY-100 steels, and are rapidly displacing the high yield steels in all but the most critical applications.

The heat-treated carbon and heat-treated high-strength low-alloy steels are normalized or quenched and tempered to obtain more desirable mechanical properties. ASTM A633 and A678 are typical of these steels. They are notable for their strength, which is higher than the high-strength low-alloy steels, but lower than the constructional alloy steels. Their notch toughness is also superior.

Developments in Aluminum Alloys

Some of the most significant early applications of aluminum in shipbuilding were in the deckhouses of large passenger and cargo ships. Its light weight and basic corrosion resistance were the principal attributes that made it an attractive substitute for steel. Other attributes that can be beneficial in the right application are low elastic modulus and notch toughness at low temperatures.

Its low elastic modulus, approximately one-third that of steel, is beneficial for deckhouse and superstructure application and for items of structure subjected to impulsive or impact type loadings. In the upper levels of deckhouse and superstructure, it can sustain high elongation induced by the steel structure below it with only one-third the stress. When properly surface-hardened, it has been used as armor plate and has the advantage of light weight and toughness.

Since the early alloys were not entirely satisfactory for marine use, aluminum was not used in significant amounts until the late 1920s and early 1930s. At that time, Duraluminum was a popular alloy and was used in the deckhouses of several naval combatants.

After the second world war, the 6061 series of aluminum alloys, because of their increased strength, became popular.

Today two types of aluminum alloys are used in ship structural applications: heat-treatable and non-heat-treatable alloys. The heat-treatable alloys, such as 6061-T6, were the first to be used extensively in ship structures but, because of their poor strength and corrosion characteristics when welded, have been largely replaced by non-heat-treatable, aluminum-magnesium alloys such as 5083, 5086, and 5456. These get their increased strength from the alloying elements and cold work, and they possess good weldability. They also have excellent corrosion resistance and are easily cut and formed.

Although both the heat-treatable and non-heat-treatable aluminum alloys have excellent corrosion resistance, when properly applied, great care must be exercised when using them in conjunction with dissimilar metals in a marine environment. Aluminum deckhouses attached to steel hulls, for instance, require special connecting procedures. If mechanical fastenings are used, the faying surfaces must be separated by insulating tape or gasketing materials. An alternative is to use explosively bonded steel and aluminum bi-metallic strips.

Titanium

Titanium and its alloys possess some attractive characteristics for certain marine applications. Their most attractive characteristic is their high strength to weight ratio; but they are also relatively resistant to corrosion, have good fatigue strength, and are non-magnetic. Their principal disadvantage is high cost.

Four alloys suitable for ship structures are

- Ti–0.3Mo–0.8Ni
- Ti–0.6Al–4V
- Ti–3Al–2.5V
- Ti–3Al–8V–6Cr–4Zr–4Mo

Currently the principal applications of titanium in ship structures have been in experimental submersibles and in SSN688 class and SSN-Z1 class nuclear submarines. The Navy is currently evaluating a number of surface ship applications.

ROGER G. KLINE earned a degree in naval architecture & marine engineering at the University of Michigan in 1957, and then went on to Berkeley for a master's degree. His professional career has been devoted almost entirely to ship structural analysis. He has developed his expertise in such organizations as MarAd, U.S. Steel Research Center, and R. A. Stearn (where he was Vice President & Technical Director). He is currently employed in the Newport News Shipbuilding Company's Milwaukee office. He is a member of both SNAME and RINA. He is active in technical committees for SNAME and the National Research Council, and is currently serving as Secretary/Treasurer of SNAME's Great Lakes & Great Rivers Section.

Living and Working Spaces

DANIEL J. WEILER, with contributions by other authorities

DURING the past half century external influences and technological advances have both forced and encouraged radical changes in the design of shipboard accommodations. This has been particularly true in merchant ships largely because economic pressures have led to greatly reduced manning scales. In naval vessels crew numbers have necessarily remained high in recognition of the losses in life that may occur during combat. Although automation has reduced the need for sailors in some duties, added military systems have increased the needs in others.

The design of working areas has been shaped by various influences such as a more widespread appreciation of man–machine systems and automated controls.

Overlying all of the above has been a pronounced raising of shoreside living standards, which has led to parallel rising expectations on board ships. This has been reflected in, and influenced by, improved habitability factors such as sanitary spaces and air conditioning, availability of better materials of construction, and changing tastes in furnishings and decor.

The intent of this chapter is to trace some of the major developments in these various aspects of living and working aboard naval or merchant ships during the past 50 years.

Merchant Ship Accommodations

Of all of the forces leading to change in merchant ship accommodations, none has had greater impact than the reduction of crew complements. Owing to the economic distortions of the 1936 Merchant Marine Act, the typical subsidized post-World War II cargo liner carried a crew of more than 50. Equivalent ships today, under less distorted economic conditions, have reduced those numbers by at least 50 percent. These smaller numbers, coupled with enormous increases in ship productivity, have made it economically feasible for owners to provide much higher habitability standards. Four men to a cabin and centralized sanitary facilities (which were standard in the 1930s) have given way to single-occupancy cabins and private facilities for even the lowest ranks.

The Maritime Administration deserves credit for much of the overall improvement in merchant ship accommodations. The old C-2s, C-3s, and C-4s, for example, introduced special pantries for off-hour meals. The crews were berthed mostly in two-person rooms with hot and cold water. Recreation rooms were introduced. After World War II, the Mariner-class cargo liners established yet higher standards, and set the pattern for U.S.-flag ships throughout the 1960s. Licensed officers were given single-occupancy rooms with private or semi-private sanitary spaces. The unlicensed personnel were in two-person rooms with interconnecting semi-private sanitary spaces. It should be noted, too, that the Maritime Subsidy Board fought hard to reduce crew complements on subsidized ships.

Crew reduction has not been without some unforeseen social and psychological problems. Chief among these is loneliness. To overcome that problem, there has been a trend toward providing attractive social and recreational spaces. Whereas 50 years ago there might be only two social gathering spaces (officers' mess and crew's mess, respectively), today we find not only those, but also swimming pools, gyms, video rooms, lounges, hobby shops, card rooms, and so forth. These public spaces are often arranged in a "village square" to encourage social interaction. Also, in keeping with this the crew quarters are of minimum size so as to encourage the occupants to get out and mix. This is not to say that these cabins have become austere. Berths are larger, storage spaces are more ample, and heating, ventilation, and air conditioning are now under individual control. Comfortable furniture and attractive wall coverings are now the rule.

The United States has particularly strong seagoing labor unions. Improvements in crew productivity have come about only with the full cooperation of these unions. Shipowners and their naval architects have learned to engage union leaders in their discussions throughout all phases of ship design, and progress has

Cargo control room of a modern tanker, ca. 1977.

Deckhouse. Hopeman Brothers, Inc., T-AKX Program, 1984, Quincy Yard, General Dynamics.

Master's dayroom. Hopeman Brothers, Inc., T-AKX Program, 1984, Quincy Yard, General Dynamics.

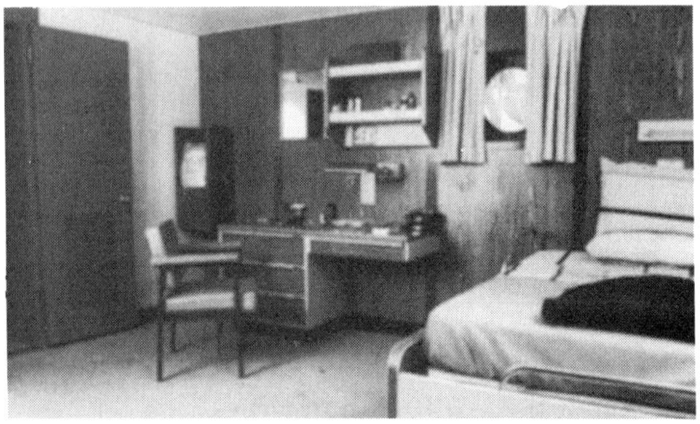

Commercial ship Master's stateroom. (Courtesy Hopeman Bros., Inc.)

Modern interior design. Furniture is designed for 20 years of marine use. When fully extended, drawers will support a static load of 400 lb on the leading edge. (Courtesy Hopeman Bros., Inc.)

Commissioned officers' conference room, USS *Zeus*, ca. 1983. (Courtesy Hopeman Bros., Inc.)

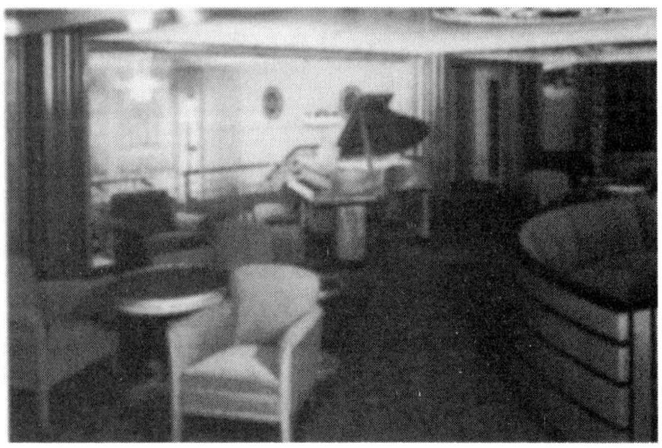

Lounge, SS *Del Norte*, ca. 1952. (Courtesy Hopeman Bros., Inc.)

been most satisfactory when labor and management have recognized that the alternative to fewer jobs may be no jobs.

One advantage that has been found in the provision of single cabins is that a hospital is no longer required because sailors with contagious diseases can now be isolated in their own cabins.

Laundry facilities in ships of the 1940s consisted of a wash room with wash tubs. Clothes were left to dry on hand rails in the upper levels of the engine room. This was something of a nuisance, as they tended to drip—or even drop—on the engineering personnel. Today most ships have completely equipped, automated laundry rooms that include washers, dryers, ironing boards and irons.

On earlier merchant ships, linens were washed on board. By the 1950s, however, it became customary to use commercial laundry services ashore. It is therefore necessary now to have storerooms for clean and dirty linens.

Passenger Ship Accommodations

The chief factors influencing change in passenger accommodations has been (1) the switch in function from transport to cruising, and (2) the virtual elimination of multi-class ships and attendant multiplicity of public spaces, which used to be carefully scaled in size and elaboration of decor commensurate with class.

Fifty years ago, most passengers slept in upper or lower berths, but those are now a thing of the past. Cabins today may have no more clear deck space than 50 years ago, but greater emphasis is now being placed on more attractive public spaces and a much wider range of activities. Nevertheless, some passengers can't get away from shoreside habits, so many cabins are fitted with television, minibars and small refrigerators.

Cruise ship designs are now moving in two opposite directions: small but luxurious mega-yachts, and huge oceangoing resorts. In both cases emphasis is being placed on social interaction and leisure facilities. Promenades have become shopping boulevards (catering to a long-popular pastime), and atria are used to link public spaces and provide a "village green." Inside cabins facing such atria are considered just as desirable as real outside rooms. Stern marinas are provided to allow water sports and entertainment.

Cruise ships are now being marketed as conference centers or business retreats. These functions require suitable meeting rooms, classrooms, and libraries.

Naval Vessel Accommodations

As already mentioned, our naval vessels still carry generous complements to drive the ships, operate the many combat and damage control systems, and maintain the machinery and multitudinous electronic components. In addition, of course, there must be a reasonable level of personnel simply to feed the rest of the crew.

The large crew complements of naval ships preclude the high habitability standards found on merchant ships. Nevertheless, marked improvements have been effected in the past 50 years. Naval designers now pay considerable attention to habitability features, recognizing their importance in keeping the crew at top physical and mental condition—and increasing the probability of re-enlistment of highly trained personnel. Berths are larger; heating, ventilating, and air conditioning are of higher standard; lighting is better; and sanitary facilities are superior both in number and in detail.

Unlike merchant ships, naval vessels are expected to operate for long periods in distant waters without recourse to shore support systems. Self-sufficiency requirements extend into the need for continuing shipboard maintenance and repair. This, too, places extra demands on the numbers of personnel and their accommodations and working spaces.

In 1987–1988 the U.S. Navy conducted the Ship Operational Characteristic Study (SOCS). The team concluded that future reductions in naval shipboard personnel would be difficult to achieve. Nevertheless, the need was seen to move in that direction through various means, such as the use of more reliable equipment, new thinking about operating procedures, and better communications between naval architects and fleet operators.

Introduction of Women

In recent decades we have grown used to the idea of having women help "man" both merchant and noncombatant naval ships. This has caused little impact on merchant ships because, with all hands accommodated in single cabins with private facilities, no rearrangement of accommodations was required. In naval vessels, with their community sleeping quarters and shared sanitary spaces, some rearrangement of spaces has become appropriate. This has been effected without undue difficulty, however, and is working out to the satisfaction of most of those involved. In many

Bridge—seismic research vessel *Shell America.*

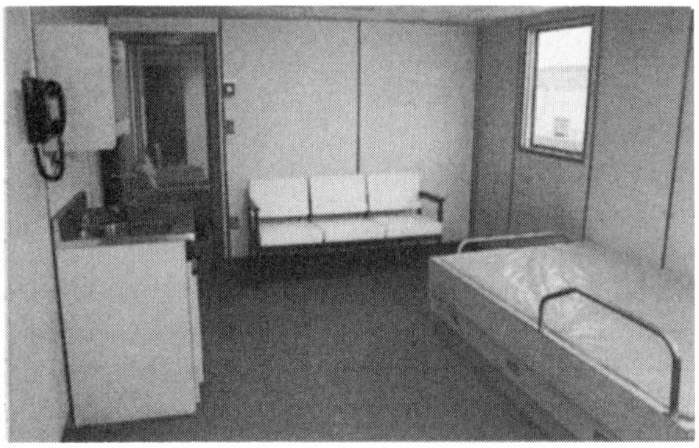

Stateroom—*Shell America.*

Galley—*Shell America.*

A commercial ship's galley. (Courtesy Hopeman Bros., Inc.)

Habitability in the Royal Canadian Navy. Serving line in the HMCS *St. Laurent* galley.

Crew's dining hall—HMCS *St. Laurent.*

Typical tanker wheelhouse and chartroom, arrangement ca. 1971.

Recreation areas and galley—tanker *Burmah Pearl*, ca. 1973.

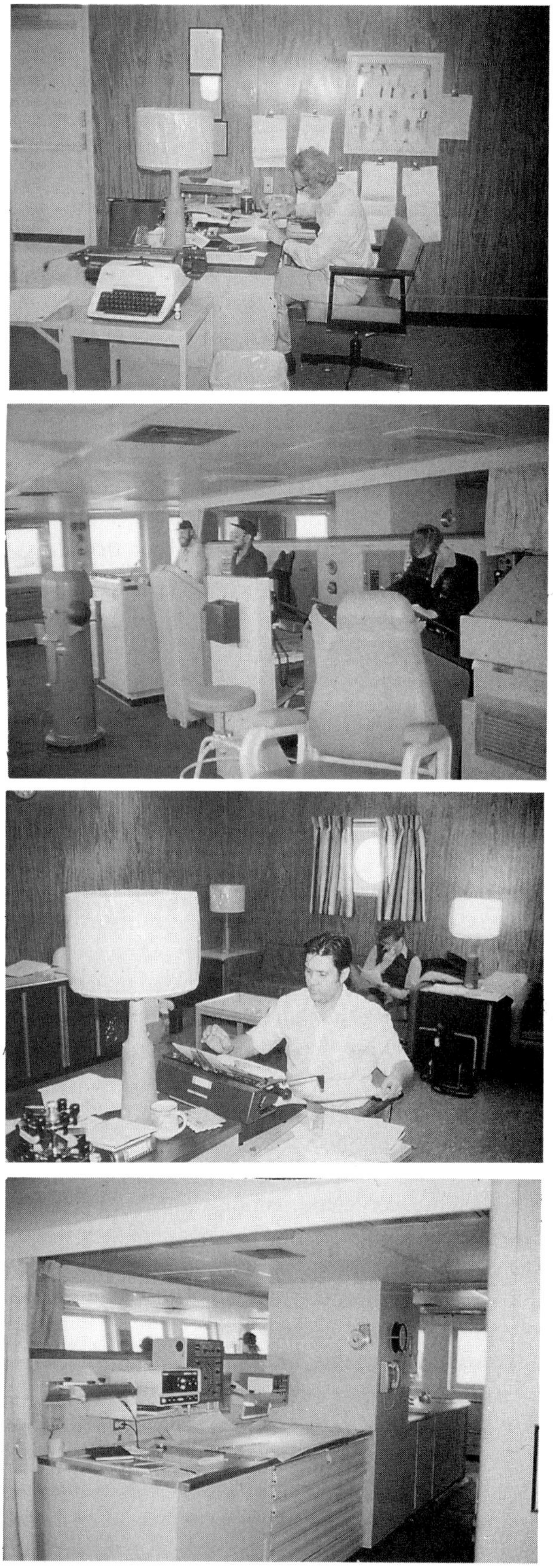

Living and working spaces—LNG vessel *Aquarius*, ca. 1977. (Courtesy Steven Spear.)

naval ships 25 percent of the berths are now allocated to female crew.

Merchant Ship Working Spaces

Perhaps the most radical changes in merchant ship working spaces have come in connection with machinery control systems. Fifty years ago the deck officers communicated their needs to the operating engineers by means of the traditional swing-handle telegraph pointing to full, half, or slow ahead; or full, half, or slow astern; or perhaps "through with engine." Not a very fine-tuned menu of choices. Today, the engine is probably controlled by the deck officers directly from the bridge (if union agreements permit). Engineering officers have duplicate controls, plus complete monitoring displays, in a sound-proofed room perhaps overlooking the machinery space. Their main function today is to maintain the machinery. The controls are on the bridge, and many engine rooms are left unattended overnight.

Navigating bridges have become ever more crammed with sophisticated monitoring and control devices. These are concerned with far more than the main engines. Course recorders, radar, cargo handling, all manner of safety devices, and satellite navigation and communication systems, are included. Meanwhile, however, the need for a radio officer, attendant equipment and radio room have all disappeared, while the need for more elaborate record keeping and other clerical work has increased.

Fifty years ago the captain and chief engineer probably each had an office adjacent to his cabin. Today the offices are more likely to be down on the main deck (for convenience of those coming aboard to carry on business when in port) whereas the officers' quarters remain on a higher deck. Moreover, such offices may be situated next to, or perhaps combined with, one or more conference rooms. These conference rooms are where officers may meet to plan their work. The rooms may also be available for union meetings. Ship's plans, equipment catalogs, and maintenance records may be kept in these rooms.

Considerable change has come about in the food service areas on merchant ships. Perhaps the most significant changes have been involved in finding ways to reduce labor. Pre-cut meats, rather than sides of beef are now arriving on board. That is but one example. Galleys today prepare meals from foods that have been pre-processed, trimmed, and perhaps partially or fully cooked before freezing. Fifty years ago nearly everything was done from scratch. Shipowners recognize that nutritious food, well prepared and served in attractive surroundings is an important part of the social contract that is necessary to crew morale. This means that efforts to economize in the galley must be done with care. Airline type, ready to microwave and eat meals, for example, have been tried but rather quickly abandoned when found less than highly popular. Bake ovens, dough mixers, and potato peelers have given way to microwave ovens, automatic dishwashers, deep fat fryers, convection ovens, infrared heating tables, cafeteria serving lines, ice cube makers, and so forth.

No one in the crew need feel hungry at night. Most messes are equipped with refrigerators, microwave ovens, icemakers, coffee makers, and other such conveniences to go with a goodly stock of pastries, fruit, ice cream, and similar traditional 'tween-meal victuals. Perhaps we should mention here that today's merchant ship is likely to feature an exercise room in which the temptations of the stomach can be offset at least to some degree.

The task of bringing galley supplies on board has been greatly eased. In some instances, one or more cargo containers have been dedicated to that function. Carefully stocked ahead of time, such containers can be lifted aboard and clamped into position where they become the food storage compartments for the subsequent voyage.

Food service on cruise ships remains of highest importance. Here, too, labor-saving steps are being employed, but by no means to the same degree as on cargo ships. The only acceptable compromises are ones that will in no way degrade the quality of what is placed before the passengers.

In cargo, as well as passenger ships government regulations have brought a stiffening in such public health matters as rat proofing and food service sanitation.

Naval Ship Working Spaces

Naval engineers have become increasingly aware of the importance of ergonomics in the design of working spaces. Navigating areas typically include the bridge, bridge deck, chart room, and combat information center (CIC). Into these spaces is packed a vast array of equipment: instrumentation, controls, and displays. These spaces must also accommodate a large number of people who must interact efficiently with one another and with the equipment. Consequently, the selection and arrangement of components within these spaces lend themselves admirably to systematic

Matson's SS *Lurline* (1948). Typical outside single 1st class stateroom made up for day use (Hopeman Bros., Inc.).

Map of carved linoleum in the observation lounge of the Great Laker *Edward B. Greene.* Tiny ships can be moved to indicate location. (W. C. Cowles collection.)

Forward guest lounge on the *Humphrey.* Pantry and bar visible on the left. (W. C. Cowles collection.)

Crew stateroom on the Great Laker *Humphrey.* Two-man rooms, each with private shower. Furniture is metal and division bulkheads, linings, and ceilings are fireproof Marinite. Two lower berths, instead of traditional lower and upper, provide an air of spaciousness. (W. C. Cowles collection.)

Stainless steel galley on the *Humphrey.* Electric ranges and oven in the background with patented "Clean Air" filters to provide ventilation and collect grease. Built-in fire extinguishing systems protect against grease fires. (W. C. Cowles collection.)

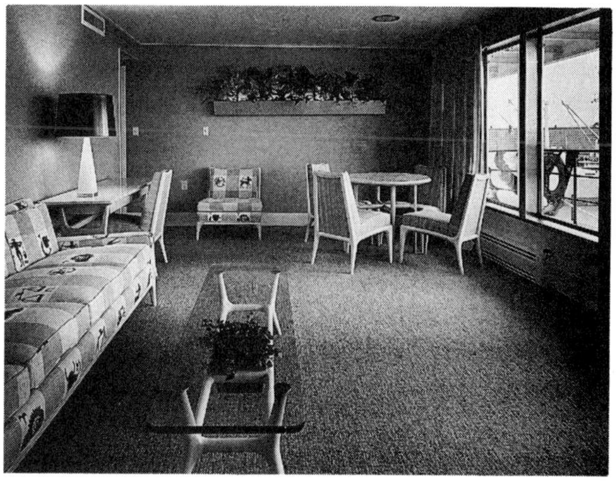

Observation lounge on the *Humphrey.* (W. C. Cowles collection.)

development using well-understood principles of human factors engineering (HFE). Fifty years ago there were far fewer devices involved, and satisfactory arrangements could be conceived without benefit of sophisticated techniques. That is certainly not true today, when military threats may become apparent with only the briefest of warnings and appropriate countermeasures must be taken almost instantaneously. Thus it is that advanced concepts in human engineering have come to play an important role in the design of naval ship working spaces.

The following two distinct steps are typically used to decide what equipment is to go into a working space, as for example a CIC room, and how the items of equipment are to arranged.

1. The first step is called the task analysis (TA). This is a procedure for producing an ordered list of all the tasks that people will do in the given system. These tasks may be broken down into sub-tasks depending on the individual merits of the case, or the degree of sophistication required. Each task on the list may also be qualified by including more detail such as information required, time allotted, environmental conditions, decisions to be made, and so forth. The information may come from sources such as mission scenarios, experience with similar systems, subject matter experts, or analytical development of functional flow diagrams.

2. The second step is called the link analysis (LA). This work aims to optimize the physical layout of all the items in the space so as to achieve the twin goals of maximum efficiency and minimum human errors. Taking the findings of the first step, this second step:

a. Determines who should be linked to whom.
b. Estimates frequency of linkages between different operators or between operators and items of equipment.
c. Weights the linkages to factor in the degree of criticality or importance.
d. Lays out the space starting with the highest link values. Items with lower link values are added in succession and the arrangement is continually adjusted to minimize the linkages.
e. Evaluates the finished arrangement against mission objectives in order to validate, or possibly improve, the design.

Figures 1–3 and Table 1 provide an example of this technique.

In both naval and commercial design, careful attention is now paid to what we call *ergonomics,* the science of designing a work area around human capabilities and limitations. The equipment is therefore designed to accommodate the entire range of the smallest to the biggest attribute, both physical and mental, of the person that may have to use it. Sophisticated items such as monitoring and control equipment, as well as simple things such as tables and chairs, are all made adaptable to the wide range of potential users.

During the past 50 years ships have been transformed. The same can be said for ship design methods. The procedures outlined in the paragraphs above serve as examples.

Materials and Decor

Fifty years ago ship designers were switching from wood to noncombustible (or fire resistant) materials for divisional bulkheads in accommodation spaces. This was triggered by international agreements and Congressional action following the tragic loss of the *Morro Castle* in 1934. While many foreign nations settled for the use of elaborate sprinkler systems, American authorities specified the use of noncombustible materials to the greatest possible extent. Joiner bulkheads, for example, were accordingly built of panels composed of composite materials that included asbestos. Ironically, in more recent years, this material was found to contain threats of its own—namely the danger of lung cancer developing in those who breathed quantities of its dust. Nevertheless, under current regulations, these bulkheads are still legal provided they remain intact. If they are broken, or need to be cut or otherwise changed, then proper asbestos disposal procedures are required. (The real asbestos problems aboard ship are in the insulation and lagging around pipes and valves, not in the joiner bulkheads.) Since the 1970s, other, safer substitutes have been used for such bulkheads.

In developing design criteria, authorities relied on full scale experiments on the steamship *Nantasket,* which led to an understanding of fire containment within controlled spaces by means of noncombustible materials without attempting to fireproof the structural steel. The joiner panel system that evolved required that, in addition to maintaining its structural integrity, confining the fire, and providing a heat barrier, it should perforce also serve as a sound barrier and provide aesthetic appeal.

Doors and frames of metal are used, while window and air port casings may be of fiberglass or metal.

Room furniture, drapes, and so forth, must employ "fire-retardant material in their construction as far as is reasonable and practicable." Still under investigation

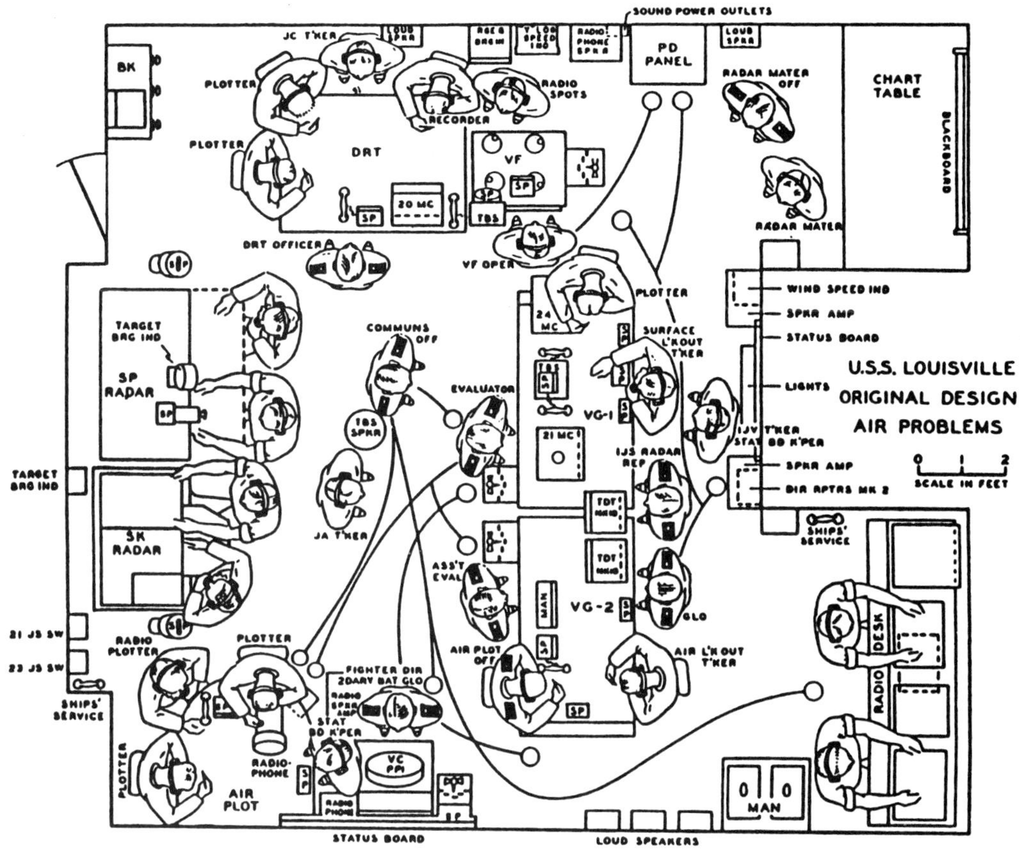

Fig. 1 Original layout of the combat information center aboard the U.S.S. *Louisville*. The men with shoulder boards are officers; all others are enlisted men. Lines show the movements made by certain men during air attacks.

Table 1 Linkages between men and men and between men and machines in the CIC of the U.S.S. *Louisville* during air attacks.

	Men					
	Comm. Off.	Evaluator	Asst. Eval.	GLO	Fighter Dir.	VF Operator
Men						
Communications officer	—	6	—	—	6	—
Evaluator	6	—	5	—	7	—
Assistant evaluator	—	5	—	—	—	—
Gunnery liaison officer	—	—	—	—	—	—
Fighter director	6	7	—	—	—	—
VF radar operator	—	—	—	—	—	—
Machines						
VC radar	—	—	—	—	8	—
Air pilot	1	9	—	—	—	—
Radio desk	1	—	—	—	—	—
PD panel	—	—	—	2	—	1
VG radar no. 1	—	9	—	—	—	—
VG radar no. 2	—	—	8	7	4	—
VF radar	—	—	—	3	—	9
Director repeaters	—	—	—	7	—	—

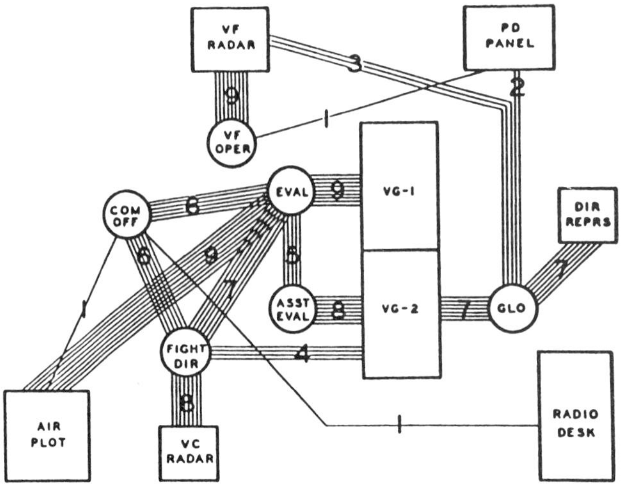

Fig. 2 Schematic representation of CIC using data from Fig. 1 and Table 1.

are other potentially dangerous properties such as toxicity.

Needless to say, all these precautions against the spread of fires have been of interest to naval as well as commercial ship designers.

In the matter of interior design, the past 50 years have seen significant changes. Perhaps of most importance has been an increasing reliance on the advice of professional interior decorators. Their aims have included psychological as well as aesthetic improvements. The result has been a more attractive, warmer living environment (also greatly improved, of course, by the advances in general arrangements discussed earlier).

In passenger ships the switch in primary function from transport to pleasure cruising has been reflected in decor that tends to be more bright and cheerful. Following World War II there was an unfortunate trend toward austerity of interior design, first in keeping with the Bauhaus philosophy, and later with the International style; both movements scorned decoration and natural materials, preferring instead the unadorned plastic box-like look. Fortunately, today's designs reflect the more widespread public desire for luxury. At the same time, the designer has a much greater range of materials from which to choose. In this they now emphasize easy maintenance coupled with less expensive materials. Melamine covered wood cabinet work and vinyl wall coverings have replaced wood, and inexpensive prints have replaced original art work. In bathrooms, what used to be marble is now more likely to be fiberglass. While

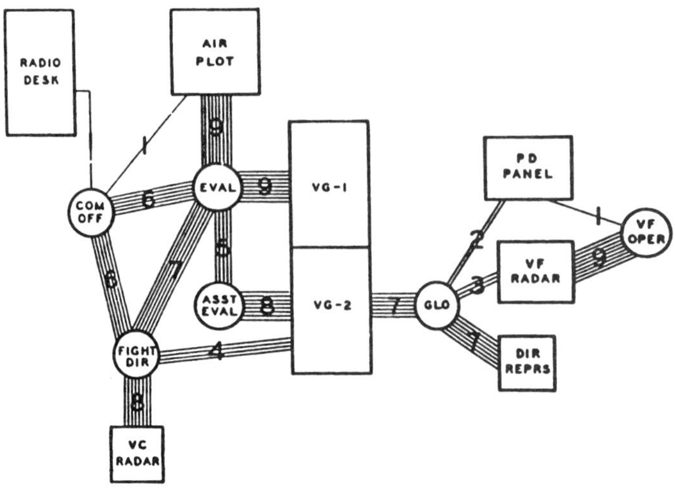

Fig. 3 Schematic optimizing CIC space layout.

formality of ambience may rule in the dining room, bright lights and glitter prevail in the lounge/bar areas. All these developments reflect the decorator's perception of changing public tastes.

In naval craft the changes in fittings and decor have been even more pronounced. During World War II sailors slept in stacked pipe berths, often three-, or even four-high. Today they sleep in relatively comfortable modular berths arranged in segregated groups. Each berth has its own lighting, privacy curtains, and air outlets. The berth front contains its own locker for storing clothes, normally supplemented by wardrobes with hanging space for further storage.

Mechanical and Service Systems

From a human satisfaction perspective, perhaps the greatest advances of the past 50 years have come in the mundane area of sanitary facilities. As mentioned earlier, merchant ship seafarers have graduated from sharing a communal center to private, individual toilet/shower rooms. The sheer numbers of naval personnel have precluded advancing to that degree; their sanitary facilities have nevertheless improved from such primitive fixtures as communal troughs to standard toilets in booths. Particular attention is paid to making the sanitary spaces well lighted and easy to clean.

In all cases, international and national regulations have forced designers to abandon direct overboard discharge of sewage. In some instances holding tanks are used; in others, some form of onboard treatment is employed. In order to reduce the volume of sewage, vacuum suction systems that require only modest amounts of water are now employed. Developments in this entire area are still under way.

The basic design concepts of ventilation and air conditioning have not changed significantly over the past 50 years. Such changes as have occurred result from (a) the evolution of the equipment and materials that make up the system and (b) the tailoring of controls to satisfy individual tastes. Today's systems are more efficient, quieter, lighter in weight, and more reliable than the systems of 50 years ago. Today's trend is toward fully automated systems requiring less human input. The advent of the computer has had a profound impact on both the design and operation of these systems.

Naval ships have experienced a trend toward air conditioning spaces that formerly relied on forced or natural ventilation of ambient air. The design temperature for air conditioned spaces has been reduced from 85 to 80 °F. Some naval ships are now pressurized to eliminate in-leakage of outside air, and use special filters to permit operations in a hostile chemical, biological, or radioactive environment.

The evolution of refrigeration systems has been driven by the ever-increasing need for flexibility. New refrigerants and changing kinds of refrigerated products have both been involved. Self-contained refrigeration units are increasingly employed for such functions as soft drink vending machines and refrigerated salad bars. The extreme example is the self-refrigerated cargo container, which has made obsolete the old reefer cargo ship with its massive refrigeration machinery and heavily insulated holds. Standard temperatures for refrigerated cargo or food stores have been dropped from 0 to –10 or even –20 °F.

Owing to the suspected deleterious impact on the earth's ozone layer, the use of fully chlorinated refrigerants is being phased out. The application of new refrigerants and appropriate equipment is currently still under study.

Lighting

Interior lighting, both lamps and fixtures, has been steadily improved over the past 50 years. Interior systems have progressed from all-incandescent types to virtually exclusive use of fluorescent lighting. This has resulted in greater efficiency and longer lamp life. Continuing improvements are being explored in the design of fixtures.

Lighting of deck areas by means of watertight fixtures has been improved through the use of corrosion-resistant materials and the evolution of high intensity lamps such as mercury, metal halide, and high pressure sodium lamps. These lamps provide not only longer life, but also higher luminous efficiency, so that fewer of them are now required. Exterior use of incandescent lamps is now limited to emergency lights where instant-on is appropriate. Fiber optics provide a new lighting medium that may find application in the future. This can provide non-electrical light transmission into hazardous or inaccessible locations.

Closure

Everything that has been said above has been confined to large cargo, passenger, or naval ships. What about all those miscellaneous types: commercial fishing boats, oceanographic research vessels, tugs, towboats, yachts, and Coast Guard vessels? We need go into no details here, but can simply say that every one of the advances noted above has found its way (to at least some degree) into all sectors of the marine industry.

Predicting future developments is always risky, but we are safe in saying that changes in living and working spaces will be steered primarily by changes in manning practices. In merchant ships extrapolating current trends would point toward ever-smaller crew complements. The ultimate development, which is already technically feasible, might be the ship that is completely unmanned between pilot stations. At the same time, an opposite trend might see our shipowners following the European practice of allowing members of the crew to be accompanied by their families.

In naval vessels future crew reductions, if they occur at all, will probably be far less pronounced than in the merchant marine. Such reductions as may occur will perhaps come as a result not only of economic pressures, but also a humane inclination to place fewer sailors at risk in combat.

In any event, the past 50 years have seen radical and fascinating changes in the design of shipboard living and working space. We have every reason to expect equally interesting changes in the half century ahead.

Acknowledgments

I am grateful to many professional friends who have helped immeasurably in the preparation of this chapter. Chief among these are Amos Baki, Mackie Burcham, Robert Bost, Charles Cherrix, Charles Cushing, Larrie Ferreiro, Edward Hazard, Charles Johnson, Rajiv Khandpur, and Artis Plato.

Daniel J. Weiler is Vice President of JJH, Inc., Washington Office. Before his retirement from the Naval Sea Systems Command in 1985, he directed a design team that made outstanding contributions to the field of shipboard habitability design. This led to his winning the 1984 Silver Medal from the Association of Scientists and Engineers (Naval Sea Systems Command) in recognition of his earned reputation as the world's foremost expert in shipboard habitability design. His work resulted in significant improvements in the health, morale, performance, and retention of naval personnel.

Coatings and Corrosion Protection

David T. Bloodgood

Introduction

The use of coating materials for protection and aesthetics of ships is almost as old as the history of ships themselves. In the past 50 years the development of ship paints and coatings has advanced from a "black art" to a science. There have been major improvements in the binding media (resins), as well as in the pigments and other additives. In the early 1940s, the war years, developments were slowed as most marine paints were confined to shades of gray and applied to ships operated by the government. Nevertheless, during the war newly developed antifouling compositions allowed U.S. naval vessels to remain at sea for greatly lengthened periods, a clear operational benefit that was unavailable to enemy fleets. After the war continual improvements were effected in combatting both corrosion and fouling. Today the industry's major changes are directed toward environmental protection. Parallel developments are being found in surface preparation, equipment, and application techniques.

One item of extreme importance that has played a significant role in the coating and corrosion control process has been the recognition of life cycle costs, and the economic payback to an owner/operator who uses high-performance coating systems.

In this chapter I shall highlight the advances in coatings and corrosion control procedures that have come about during the past 50 years. I shall call particular attention to certain key technical developments that have brought science to what was formerly a "black art."

Developments in Materials

In the area of corrosion control the past half century has seen two major improvements in materials. These comprise zinc-rich coatings and epoxies.

The zinc-rich materials, represented by the inorganic zinc coatings, are made up of silicate vehicles and zinc dust as the major portion of the pigmentation. These compositions are either water based or produced in an organic solvent. Other developments in the zinc products were the reinforced zinc coating and those containing various organic resins such as epoxy, chlorinated rubber, urethane, and others. The zinc-rich coatings came into widespread use in the 1950s and have become standard as part of the coating system above the waterline and in cargo tanks carrying selected materials.

Epoxy coatings came into shipboard use in the 1950s, but did not really become a major part of the corrosion control system until the 1960s. Polymer chemistry has enhanced the sophistication that can be attained in tailoring resins, such as epoxies, enabling the formulation of coatings with a wide variety of capabilities and end uses. Different properties can be achieved by varying the curing agents and modifying with other resin materials. There is no area on a modern day ship where the epoxy coatings have not been used as primers, corrosion control systems, or topcoats where serviceability and coating integrity are required.

More recent developments of "surface tolerant" and low-temperature cure epoxy coatings have added another dimension to these materials. The "surface tolerant" epoxies will provide a coating system that can be applied to a surface with a lesser degree of surface preparation and provide good serviceability and corrosion protection when adverse conditions may preclude providing a "white" to "near-white" abrasive blast in accordance with Steel Structures Paint Council (SSPC) specifications. Rather than limiting application to times when the surface temperatures are above 40 °F it is now possible to apply a coating that will cure at considerably lower temperatures.

Other resin materials with good properties for marine coatings include the vinyls, early favorites, particularly for the underwater hull but also finding favor for exterior topside application. The urethane coatings with their excellent gloss and color retention have had some application in the marine field, but not to any great extent because of certain inherent shortcomings. Chlorinated rubber, long a favorite in Europe, did not develop a large following in the United States until the 1970s and early 1980s. These materials are applied as a single package, with resulting ease of handling, making them a good "shipyard tool." The chlorinated rubber coatings were used on the underwater hull and above-water areas, and also as binder in antifouling paints.

Alkyd coatings, used extensively on both the exterior and interior of marine vessels, have provided inexpensive one-package paints that can be readily applied by crew members with a minimum of surface treatment required. The alkyds, when modified with a vinyl or silicone, provide a somewhat extended exterior service life. Nevertheless, they will not provide the durability, physical damage resistance, and color and gloss retention normally desired. These characteristics are relative, but most shipowner/operators have for most services opted for higher-performance materials such as the acrylics, epoxy acrylics, and vinyl acrylics.

The development of the ablative, self smoothing antifouling coating was another milestone in the development of cost effective operations. One of the major advantages of these paints is the economically extended periods between drydockings that they permit. In the 1970s, when the cost of fuel skyrocketed, these antifouling paints became widely used; and then, after the fuel shortages subsided, the operators continued to use them as their efficacy in operation had been proven.

Corrosion Protection: Coatings

This section consists of a brief review of developments over the past 50 years in corrosion resistant coatings, both exterior and interior, with sub-sections for the specific areas.

Exterior. *Underwater Hull.* The main purpose of the coating materials applied to the underwater hull of a ship, other than the antifouling paint, is to provide corrosion protection for the substrate, which will normally be steel. In the early 1940s the anticorrosive paints were made up of hydrocarbon resins, perhaps in combination with spar/phenolic varnishes pigmented with aluminum or other materials with anticorrosive properties. Vinyl resin primers, formulated with red lead and zinc chromate pigments to inhibit corrosion, were widely used. The U.S. Navy ships and many commercial ships were coated in the underwater area with the vinyl coatings. The commercial ships also used chlorinated rubber materials as well as the hydrocarbon/aluminum pigmented coatings. Since the 1970s we have seen a growing use of epoxy coating systems, until in the 1980s it became the predominate anticorrosive material for underwater hulls.

Boottop. The boottop is the area on the sides of a ship between the light and load waterlines. Being alternately emersed and immersed, it requires special anticorrosive protection. In the early days the boottop priming was the same as that used on the bottom. Now, however, zinc-rich coatings have become standard for boottop application.

Topside. The topside area, the hull above the boottop to the uppermost parts of the ship, will normally receive the same anticorrosive coatings as the boottop. Red lead and zinc chromate pigmented primers had been used for many years on the topside surfaces. These pigments were incorporated into oleoresinous and alkyd resin paints. With the development of the vinyl paints in the mid-40s and the chlorinated rubber coatings, these pigments continued in use as the anticorrosion mechanism. Again, as in the boottop, zinc coatings have now become the basic corrosion control mechanism. The epoxy resin compositions also are used for corrosion control and barrier coats in the topside.

Decks. The weather decks on ships had during the early years been primed with oleoresinous paints and alkyd paints. These were made up of pigments mixed in various varnishes including fish oil. The anticorrosive pigment most often used in deck paints was red lead. With the development of the vinyl resins and chlorinated rubber paints, these materials, incorporating red lead or zinc chromate, were also used. When the zinc-rich coatings came into being they became the standard for corrosion control as with the other parts of the ship exteriors. The epoxy resin coatings have also gained acceptance for decks subject to special service conditions.

Interior Spaces. *Cargo Tanks.* Until the late 1950s cargo tanks were seldom coated. With the refinements to fuels the U.S. Navy developed a coating system with "Saran" as the resin. Because of its extremely low flashpoint, however, it was a safety hazard. Moreover, applying a smooth coat was difficult. Private industry developed an epoxy system derivative that exhibited superior handling, application, and safety characteristics. Both were five-coat systems. These coating systems were used mainly to protect the cargo purity. The first large scale application of tank coating materials was accomplished on the *Yukon*, a tanker owned by what is now the Military Sealift

Command (MSC). The initial applications in the ship came in the late 1950s. The tanks were coated with epoxy materials of different types, inorganic zinc, and other generic types of coatings. These test applications were maintained for a number of years and, as failures occurred, new materials were applied as replacements. The ship was used to carry the standard Navy fuels, such as jet fuel, avgas and mogas. From these tests the Navy developed a cargo/ballast tank coating specification that, with modifications, is still in use today. Coated cargo tanks have now become the rule. Liquid cargoes have become more sophisticated and protection against contamination is critical.

As the cargoes became more sophisticated, the linings applied also became more sophisticated. Many of the cargoes that are today carried in lined tanks were once carried in stainless steel and often in removable tanks of small capacity. Some of the coating materials developed for carrying chemicals required heat or baking for curing. The heating of hold-size tanks is a costly operation and could only be justified when the alternative was the use of stainless steel. Today, however, advances in technology allow many of these special cargoes to be carried in tanks coated with materials that will cure at ambient or reasonable temperatures that can be provided by dehumidification equipment.

Crude oil tanks present special complications. For many years they were left uncoated; but, depending on the makeup of the crudes, extreme pitting was often evidenced, particularly in the tank bottoms and horizontal shelf areas. Now, with the requirement for inert gas systems using stack gas to prevent fires, the overhead structures are also subject to corrosion; and so coating of these areas has become a necessity. Today in crude oil tanker holds the overhead and bottom plating will be coated, as will the topsides of all other horizontal surfaces. One reason for the selective coating is money. Crude carriers usually have exceptionally large holds, so the cost of completely coating them becomes prohibitive.

Ballast Tanks. The history of coating ballast tanks is basically the same as that of cargo tanks. As coatings for cargo tanks were developed, a number of the same coating materials were used in ballast tanks. This was particularly true of the inorganic zinc and epoxy coatings.

Dry Cargo Holds. Dry cargo holds are coated primarily to improve illumination by reflecting light. For corrosion protection, or priming, the steel surfaces, which initially had been weathered or pickled to remove mill scale, were coated with red lead in an oil base, alkyd resin, or oleoresinous compositions. Aluminum pigmented compounds in varnish were also used. Advances in preparation techniques have made it possible to apply coating materials with greater resistance to abrasion associated with cargo handling. The epoxies have found wide acceptance as the total coating system or as the corrosion control barrier coated over with an enamel for light reflectance. Vinyl coatings have been used in cargo holds; but, because of the normally low-volume solids and dangerous flash points, they have faded out of the picture. Aluminum pigmented coatings, such as epoxies, however, are still popular.

Reserve Feed Tanks. Freshwater tanks for boiler feed have in the past been coated with aluminum pigmented varnishes, or cement wash with chemical additives. With the advent of higher surface-preparation requirements for ships, however, those coatings have been replaced with high-performance coatings such as the epoxies.

Potable Water Tanks. Most of the coating systems for freshwater tanks, such as cement wash and aluminum pigmented varnishes, were also used in potable water tanks. Vinyls and asphaltum materials had also been used. As with the feedwater tanks, the epoxy coatings are the most widely used today. Potable water tank coatings require federal agency approval or, for Navy ships, Bureau of Medicine.

Corrosion Protection: Electrolytic

External. *Galvanic Anodes.* Galvanic protection is provided to protect the steel hull from undesirable metal loss in areas where a coating system is damaged or missing. This phenomenon may occur wherever an electrolyte (such as sea water) is available to transport ferrous ions away from the steel. A variation in the composition of the steel itself can cause a galvanic cell to be formed. Sacrificial anodes made of zinc, aluminum, or magnesium have been used for many years to protect portions of a ship's hull against such wastage. Areas of particular concern are around bronze propellers and within sea chests. As the throwing power of the galvanic anodes is limited, they were (aside from a few special naval craft) normally used only in the critical areas listed above. Anodes are still used, but since the 1960s most ships have employed impressed current systems.

Impressed Current Systems. Impressed current systems apply an external source of electric power, using exotic metals such as platinum and platinized titanium, as non-sacrificial anodes to provide protection to the full hull of a ship with a minimum number of anode positions. The anodes are faired into the hull, causing a minimum of drag to the ship's movement. The power supply and the controls are located within the ship, usually in the machinery space. In most cases the electric power supplied to the anodes is automatically controlled, with the output of the system set to provide enough current to protect the hull at locations where there are bare or weak areas in the coating system.

Internal. Galvanic anodes are used to supplement a coating system applied to liquid cargo or ballast tanks. These anodes will be zinc or aluminum and the size will depend on the extent of area to be protected and the length of time protection is desired. The anodes may be installed when the coating system is applied or at a later time when it is felt the life of the coating system in the tank can thus be extended and it is economically expedient to do so.

Galvanic anodes are also used to increase the lives of piping systems. In particular, they may be used to enhance the life of piping interiors at welded joints where any coating system is likely to be the weakest because of the joining techniques or because of the flow characteristics at those areas.

Production Methods

The radical changes in shipbuilding methods that have come about in the past 50 years have triggered equally significant changes in the timing and techniques of surface preparation and coatings. During the 1960s, when large subassemblies were coming into use, major surface preparations were often delayed until the assemblies were complete. They might then be abrasive blasted and spray painted by hand. Now, however, we are likely to see the surfaces of plates and shapes being automatically abrasive blasted while still on the moving production line. This is followed by an automatically sprayed-on preconstruction primer less than 25 microns thick.

Corrosion Resistant and Fouling Resistant Materials

Some liquid cargoes have properties that demand tank coatings superior to those attainable by even the most sophisticated and carefully applied resins, epoxies or inorganic zinc systems. Examples include highly corrosive chemicals. For such cargoes, tanks may be constructed of solid stainless steel or stainless steel-lined structures.

Copper–nickel and copper–nickel clad steels have been used on a trial basis to protect hulls against both fouling and corrosion. Although shown to be successful, the high initial costs of material and labor are discouraging.

Galvanized steel has been used on ships for many years. Galvanizing is accomplished by dipping steel into a bath of molten zinc. Alternatively, a thinner film can be applied by electro-deposition. Galvanizing has been used for the most part to coat sheet metal which can become distorted by the normal surface preparation techniques. As for heavier plating, however, with the advent of inorganic zinc and other zinc-rich coatings it is now more economical on a life-cycle basis to treat the steel with these newer coatings. Nevertheless, there is still a lot of galvanized piping on ships.

Applying paint over a galvanized surface can be difficult. Pretreatment primer was once used extensively as the base coat because it provided good adhesion and, being a vinyl type coating, was not affected detrimentally by the zinc. The pretreatment primer, however, has a dangerously low flashpoint and is a hazard when used in interior spaces. Red lead is also affected by the zinc metal and so the zinc chromate pigmented primers were most often used. Now, however, in both naval and commercial ships, epoxy coatings are most often used on galvanized surfaces. Adhesion is good and anticorrosive properties are not important.

For a number of years flame spray metals, such as zinc and aluminum, have been used to treat surfaces subject to high temperatures. The Navy has in recent years been using this process for many corrosion control applications. A number of government agencies that operate ships are now assessing this procedure and are attempting to establish its cost effectiveness. Flame spray metals, such as copper–nickel alloys, various grades of stainless steel, and other compositions, are being used for repair with a good deal of success.

Antifouling Coatings

Antifouling coatings have in the 50-year period undergone major changes because of government-imposed environmental restrictions, operating economics, and developments by the coating industry. One hundred years ago copper in the oxide form was the most widely used antifouling agent, and that is still true today. Throughout the years many alternative compounds have been studied, including mercury, arsenic and organic arsenicals, lead, DDT, red pepper, enzymes and organotin, to name a few. The biocides have been formulated in resins and compositions such as vinyls, bituminous materials, chlorinated rubber, acrylic compounds, imide resins and others. In the 1970s long life antifouling capability became an economic necessity. Rising fuel costs increased the importance of antifouling measures. At the same time the imperatives of keeping high-investment ships in operation forced owners to prolong periods between drydockings. The ablative, self smoothing antifouling paints were presented to the market with great success. These coatings using organotin copolymers as the sole toxic or, as later developed, a combination biocide of cuprous oxide-organotin, provide a highly effective antifouling capability conducive to cost effective operation. The commercial operators, and later the government-owned and operated ships, used these compounds. In conjunction with a high-performance corrosion control system, the ablative self smoothing antifouling paints have proven economically valuable. They allow a ship to remain in service for five or more years with minimum loss in propulsive efficiency. Moreover, when drydocked, they typically require only a high-pressure water wash, touch-up of damaged areas, and addition of antifouling paint.

Recent tightening of environmental regulations is reducing the number of allowable antifouling paints. In one case it has reduced the amount of any biocide that can be used in a coating by providing a maximum acceptable release rate and limiting the vessel size to which organotin biocides can be applied. By the passage of Public Law 100-333 (June 16, 1988) the maximum release rate of an organotin compound to the water is limited to four micrograms per square centimeter per day; and any coating containing this compound cannot be applied to a vessel with a length of less than 25 meters unless built of aluminum. State regulations in many cases are more stringent and water quality standards approved by a number of states are even more restrictive. The direction for future efforts in the development of antifouling coatings or for use in U.S. waters is uncertain because of the heavy monetary burden involved. Registration of new biocides and antifouling compositions containing new biocides is almost at a standstill at this time.

Habitability and Public Spaces

Shipboard accommodations have through the years changed greatly in character. Areas that 50 years ago would have been painted are today characterized by permanent finishes such as wall coverings, decorative metals and other materials with fire retardant properties. Fire concerns affect details in most of the living areas on board ships. Reduced crew sizes on cargo ships, labor agreements, and maintenance-free materials have reduced the amount of painting required of crew members. The crew accommodations provide for one to two persons per room with almost maintenance-free surfaces. Passageway decks are usually covered with tile. Working spaces, such as offices, galleys and dining areas, are for the most part now provided with tiles and low-maintenance wall covering. Even the navigation and control spaces are treated the same. Passenger spaces are in many cases quite luxurious and finished with coverings and carpeting. Where coatings are used, the modern ship receives high-performance, reduced-maintenance finishes. Long life and minimum maintenance are the key requirements.

Machinery Spaces

The main machinery space in a ship is subject to steam, water, in some cases high temperatures, and petroleum fluids. Before the days when all the steel was abrasive blasted, the coating materials used in machinery spaces were primed with a red lead primer and coated over with an alkyd or oleoresinous paint. As the surface preparation techniques and requirements advanced, the application of higher-grade primers and topcoats proceeded. The alkyd systems were used for many years as they were flexible enough to withstand the conditions found in machinery spaces and provided the colors needed for good light reflectance and ease of maintenance. For better corrosion control, epoxy coating materials are applied along with a topcoat such as an alkyd or an acrylic type. The basic requirements for a topcoat are ease of cleaning, toughness, and some resistance to petroleum products.

The areas below the floor plates or in the machinery space bilges require a coating that will withstand standing water, oil, hydraulic fluids, and other miscellaneous contaminants. The engine room bilges in the 1940s usually experienced high corrosion. Red lead primers were given a topcoat such as an alkyd, modified resin, or varnish base pigmented coating. The crews maintained these areas by continual chipping and painting. Coating systems were upgraded as new ships were constructed, and by the 1960s some epoxy materials were being applied as primers and later as topcoats, often using the same materials as had been specified for tanks coated to carry ballast and petroleum products. As abrasive blasting in a machinery space can cause problems with grit and dust, any repaired structure is now cleaned before painting by chemical compounds abetted by hand or power tools. With the advent of the surface tolerant epoxy coatings, these methods provide a viable preparation for the application of a high-performance coating with good life expectancy and little corrosion for periods of up to five years or even longer.

Preparation and Application

Developments in surface preparation and application equipment have paralleled the advances in the technology of the coatings themselves, and with equal emphasis on environmental protection.

In the 1940s steel surfaces were usually prepared by weathering, hand tool mechanical cleaning, and pickling. There was some sand blasting in shipyards, but it was not until the mid 1950s that blasting became a major method of surface preparation. As was mentioned earlier, surface preparation and coating technology were somewhat synonymous. Sand was the initial abrasive material; but, as its health hazards became known, it was replaced by such materials as the slag from coal burning power plants or copper smelters, principally the former. In today's environment, however, with new restrictions on dust and waste disposal, other abrasive materials are receiving comprehensive technical and economic evaluation. These alternatives include, among others, steel abrasives (where recovery is possible), garnet, and aluminum oxide.

Mechanical closed cycle blasting may be used on plates, shapes, and small parts. There are also available special mechanical devices to blast side shells and exterior decks, and vacu-blast units for cleaning other selected areas. In some cases, however, reliance is still placed on manual blasting. Current trends in this work are driven by increasingly stringent Federal regulations affecting open-air blasting. Among promising dust-reduction developments are slurry blasting, high-pressure water, carbon dioxide pellets, and laser beams.

Paint application equipment has become more sophisticated over the years allowing application of materials with much higher solids content and apparent viscosities.

Fifty years ago paint was applied with brush, air spray, or roller. Spraying was most prevalent, but the equipment was crude and control was difficult. To ensure sufficiency of protective coating, excess paint was likely to be applied. Today, however, several advanced spraying systems are available and uniform paint thickness is rather easily obtained. This has become increasingly important as material costs have continued to increase. Applying a uniform film to irregular shapes is inherently difficult. One solution is the electrostatic spray system, which places electric charges on both the paint supply and the surface itself. This is used primarily in shops where air movement is at a minimum.

Quality Control and Inspection Procedures

Once the proper coatings have been selected, satisfactory results require careful quality control and inspection procedures. There are available special instruments and tests that have been developed for this purpose.

Instruments have been available for many years to obtain dry and wet film thickness readings on steel surfaces. The dry film gauges are of the magnetic type, enabling fairly accurate measurements. These readings can ensure that each coat is at or near the specification. With modern high-performance coatings, good quality control can save a lot of money by making sure the specifications are followed. Film thickness gauges today vary considerably in cost and overall capability. Some have recording capability and can be hooked into a computer for statistical information on a large number of readings. The normal requirement does not need all of this capability and, for the spot checks necessary, less expensive equipment is adequate.

Record keeping is another important task in good quality control. A step by step analysis will be available for future reference with regard to the history of the ship, and this will be important in determining future maintenance requirements and guarantees. Modern hull construction techniques may tend to complicate this work, but it must nevertheless be done.

Developments in Maintenance and Overhaul Procedures

As previously explained, economic pressures now demand longer periods between drydockings. This has led to interim measures such as using mechanical underwater brushing devices to remove at least some of the fouling. Currently, there are even developments in ways to provide anticorrosive protection to bottom shell plating that has sustained modest damage.

Environmental, Health, and Safety Regulations

Environmental protection has been one of the most potent forces for change in marine coatings during the past half century. At the start of the period there were few, if any, government constraints. The paint industry supplied many materials now targeted as being carcinogens. These included lead, chromates, mercury, arsenic, coal tars, and asbestos. As the environmental impact of these materials became better known, the industry reacted quickly, often effecting change long before regulatory action was imposed.

A complication in the matter of environmental regulation arises when local regulations exceed those of the Federal Government. An example of this is found in the case of Californian shipyards, where local constraints against the release of volatile organic compounds impose added costs that do not affect competing yards in other states.

Related to the above is the matter of employees' health and safety. Recent advances here include uniform legal requirements applying to protective clothing and gear for those employed in painting, blasting, and similar occupations. The rules also cover employees who must do their work in the immediate vicinity.

Development of Standards

Through the years standards for surface preparation, paint testing, paint inspection and failure assessment have been promulgated and disseminated to the industry for use as part of a coating specification or to assist quality control. If nothing else, these standards allow shipyard personnel to talk the same language and develop uniform job requirements.

Many qualified organizations have been involved in developing the standards. These include service groups such as ASTM, Steel Structures Painting Council (SSPC), and the National Association of Corrosion Engineers (NACE). The classification societies have also played a role, as have such Federal agencies as the U.S. Coast Guard. SNAME, too, through its Technical and Research committees has developed guides to assist the industry.

Standards properly applied by knowledgeable parties can be an effective tool in ensuring the efficient and economic life of a ship.

Research

Few if any of the major developments described in this chapter would have been possible without the continuing research activities supported by the suppliers of raw materials, the coatings industry, and team efforts such as the National Shipbuilding Research Program. The marine industry would indeed be crippled without the benefits of such research.

Effects on Crew Size and Make-Up

In the last decade or two the availability of better paints and permanent finish materials has been a factor in allowing ships to be operated by smaller crew complements. The old tradition of continual chipping and painting is gone and is unlamented by either management or labor. The economic importance of this change is hard to overestimate.

Reserve Fleet Preservation

The treatment of reserve fleet ships has changed drastically over the years. From minimally trying to preserve the ship for possible future use with compounds made up of pigment and a preservative oil, we have advanced to maintaining ships in a readiness condition able to be placed in service in a minimum of time. The recent Persian Gulf emergency illustrated the readiness of properly preserved idle ships. These were maintained with coating systems designed for use on operating ships, and the underwater hulls were maintained with high-performance coating systems as well as in-place cathodic protection. All aspects of the ships were maintained for rapid operational capability.

Conclusion

During the past half century the dramatic changes we have experienced in marine coating technology have been driven by sheer economic necessity on the one hand, and increased attention to environmental protection on the other. In spite of added costs imposed by the latter influence, the overall improvement in economics has been remarkable. Less painting maintenance is required, whether by crews or by repair yards. Ship productivity has benefited through longer periods between drydockings. Energy needs have gone down and overall reliability has increased. Better resistance to corrosion has extended the economic life of hulls and components and allowed the economic transport of hazardous cargoes.

Remarkable as recent improvements have been, there is still room for improvement and we may have every expectation that the changes of the past are but a prelude to the future.

David T. Bloodgood has been intimately involved in the coatings industry over a period of 40 years. His experience has been as an employee of the Maritime Administration, the U.S. Navy, and the Bethlehem Steel Company's Shipbuilding Division. He is currently the Government Technical Manager for the Devoe Coatings Company. His previous publications include a chapter on marine coatings in SNAME's *Ship Design and Construction* (with co-author).

Part III

Power and Control

Powering

David O'Neil

"Oil squirt cans, dancing connecting rods, sweating coal passers and firemen" characterized the ship's machinery space of 1893. Fifty years of change were implicit in this description in SNAME's 1943 *Historical Transactions.* Ironically, the world's most mass-produced ships were then on the ways, but they, too, were being fitted out with reciprocating steam engines. The development clock, in their case anyway, must have seemed to stand still as the burning of fuel oil was the most visible advancement of their power plant. Technology temporarily regressed in favor of producibility for the building of over 2700 Liberty ships. The majority of ships in the period were typified, however, with a steam plant of two 450 psi watertube boilers and a cross compound turbine/gear set. On the threshold of the 21st century, we, too, now look back for benchmarks. The past 50 years have indeed brought with them not only great advances but also great contrasts in the types of available prime movers.

The early days of Society involvement with power plant development were a direct extension of work that reached much further back, to Newcomen and Watt. The 1893 to 1943 period can be thought of as the period of major mechanical reconfiguration. The current 50-year cycle began with a pursuit of improved steam plant efficiency based on increases in cycle complexity and working fluid conditions. These efforts had the enthalpy of steam as their focus, and this activity grew more intense in the 1960s and early '70s. Then, the governing factors changed, and the contest took on aspects of a race between external and internal combustion schemes. Fossil fueled steam power gradually gave way to large, low-speed, heavy fuel burning diesel engines in oceangoing merchant ships. The development of the large-bore diesel engine had been flourishing in Europe, but was overlooked here, largely because of different economic factors. When the worldwide price of energy soared in the 1970s, there would no longer be much of a contest. The diesel plant required 15 to 40 percent less fuel than the typical steam plant; the "enthalpy chase" was over; the commercial victor would surely reign on the oceans until the end of the millennium.

Coupled with heat from nuclear energy, the steam cycle's primary use is now limited to a few types of warships. Powering for most naval service, however, is now dominated by the same basic type of power plant core that helped put ocean liners out of business in the 1960s (i.e., the aeroderivative gas turbine engine).

While America is certainly well known to be a world leader in many technologies, and continues to have strong inventive fiber, the U.S. of the '90s will have difficulty remaining a major merchant shipbuilding nation and, furthermore, its influence may continue to diminish as a major commercial ships' power plant developer, producer or user. America, however, is likely to remain the dominate force in both naval nuclear power and in naval gas turbine engine production and use.

The Immediate Past History as of 50 Years Ago

The economic depression of the 1930s gave birth to the Merchant Marine Act of 1936. The resulting program produced the C-1, C-2, and C-3 standard cargo ships, which led to standardizing boiler superheater outlet conditions at about 465 psi/750 °F, and led to fuel rates of between 0.55 and 0.60 lb/shp-hr.

Toward the end of the first 50 years of SNAME activity, engineers were thinking about pressurized firing to further reduce boiler space and weight. They had also begun to experiment with the high pressure reheat steam cycle. In 1938 MarAd authorized the last of a series of 8000 hp C-3 ships to be built with a steam reheat cycle. This plant's specific fuel consumption was a respectable 0.525 lb/shp-hr, but high maintenance costs temporarily discouraged further interest.

John F. Nichol, the author of the marine engineering section of the 1943 *Historical Transactions,* concluded his work by noting the growth of the "scientific approach" and solution of problems, assisted by the general advance in engineering education.

The Past 50 Years (1943–1993)

The Period Around World War II. During the pre-World War II period, manufacturers had been preparing for war production. Power plants for new ships were those that had already been fully developed. The primary industrial wartime strategy was to achieve standardization, and the development of increased manufacturing capacity and efficiency.

There was an acute shortage of turbine, gear, and large diesel engine manufacturing capability, so about 3000 ships built during World War II were powered with triple-expansion reciprocating steam engines. Their general efficiency was exemplified by the vast fleet of Liberty ships, with a significantly behind-the-times fuel rate of about 1.2 lb/ihp-hr. More importantly, however, these plants were capable of being operated by relatively inexperienced personnel, and so served the war effort in exemplary fashion. Adapted from a coal-burning British standard tramp design, over 2700 Liberties supplied the Allies with the largest fleet of new vessels ever produced. The American design incorporated sectional header water tube boilers instead of the British design's Scotch boilers. Two Liberty ships are preserved with their power plants still operational as merchant marine War Memorials. These are the *Jeremiah O'Brien* in San Francisco and the *John Brown* in Baltimore.

Later in the war, construction of the 11-knot Liberties was phased out in favor of more than five hundred 15.5- or 16.5-knot Victory ships of the same approximate capacity. They had 450 psi/750 °F steam turbine plants, which had been standardized on earlier C-3s. They were built at two power levels with normal ratings of either 6000 or 8500 shp. Enjoying comparatively increased levels of efficiency, the Victory ships were popular in merchant service for many years after the war. The *Lane Victory* is undergoing restoration as a merchant marine memorial in Long Beach CA.

T-2 tankers (T2-S-A1) were also emergency war-built ships. They were driven by steam turbo-electric plants developing a maximum of 7240 shp. Some of the nearly five hundred built were still operating into the '90s, driving replacement forebodies. A few were converted in the '60s to become some of the first containerships. Limitations of wartime production of reduction gearing, as in the case of the Liberty ships, was the major factor in selecting an electric transmission system. The T-2's two straight-tube boilers' steam output conditions were 450 psi/730 °F, about the same as the pre-war C-3s, and they were equipped with (troublesome) gas air heaters. Forty-four *Mission*-class T2-S-A2's were rated at a higher power level (11,000 shp) and used 600 psi steam at 825 °F. While 72 percent of the 739 oceangoing tankers that were built in the United States between 1940 and 1948 were electrically driven, no electric driven tankers were built in the postwar period (nor were any diesel powered). Perhaps the doubling of the price of fuel oil between 1946 and 1948 was a factor, as it was well known that electric drive losses were about three times those associated with reduction gears (i.e., about 8 versus 2.5 percent).

Just before and after the start of World War II, MarAd sponsored in its C-1, C-2, and C-3 ships a few experimental diesel installations. The intent was to encourage exploitation of the better fuel rates; but it was also a potential way to alleviate the anticipated shortage of steam turbines and gears.

The period's single-reduction geared diesel units were being manufactured by several American companies, such as Nordberg and Busch-Sulzer. The 1700 bhp direct-drive Nordberg and Hamilton diesel engines were installed in 235 C1-M-AV1-class, 5000 dwt cargo ships. Those engines operated at about 180 rpm, which was a bit high for efficient propeller operation.

The direct-drive, British-parented, opposed piston, Doxford diesel engines were built in 3-, 4-, and 5-cylinder versions with 6000 and 7500 bhp continuous ratings. These became one of the more popular American-made, large diesel engines. They were built by the Sun Shipbuilding and Drydock Company of Pennsylvania, and were referred to as the Sun-Doxford engine. Of the engines just mentioned, the Sun-Doxford was the only non-trunk type (i.e., one with an enclosed crank case). The engine design was, in a manner of speaking, configured like a trunk engine on the bottom and a steeple compound engine on the top. The combustion took place between upper and lower pistons in a common cylinder. The upper piston rod was connected to a horizontal beam, which was connected to a pair of side rods (straddling the cylinder), via guides, then to a pair of connecting rods and crank pins 180 degrees out of phase with the crank pins serving the lower piston. While this represented a lot of running gear, the engines ran smoothly and with little vibration.

To reduce maintenance, marine diesels were normally operated below their maximum continuous ratings (MCR), which is about 85 to 90 percent of maximum. The steam turbine ships often operated continuously at their maximum ABS rated power with margin for overload of another 10 percent or so. Significantly, few U.S. operators bought diesel ships following World War II, owing in part to crew and management unfamiliarity with diesel engines.

Again, because of war-time shortages, designers of the fifty 19-knot auxiliary aircraft carriers turned to steam reciprocating plants. They selected twin 4500 shp single-expansion, 5-cylinder, uniflow, steam reciprocating engines built by Nordberg.

Postwar Progress. In 1949 the *Schuyler Otis Bland* was built as a competitive, high-speed, prototype general cargo ship capable of mass production should another major sealift be required. She featured several innovations, some of which proved remarkably viable. Her steam conditions were raised to 860 psi/900 °F, and she had four stages of feed heating.

The passenger liners *Constitution* and *Independence* and the single screw P-2s *Barrett, Geiger,* and *Upshur* were built with steam conditions of 625 psi/850 °F. These (former) commercial passenger ships employed three stages of feed heating and were the first major postwar ships with 440 V ac generators. The (latter) troop ships had four stages of feed heating and regenerative air preheaters, but still retained dc for their cargo winches. Planning for the passenger liners included an alternative with 850 psi/900 °F steam, but the extra cost could not be justified based on existing fuel prices.

In the late 1940s, one of the so-called supertankers was the 26,555 dwt *Esso Zurich.* Among her other innovations were her steam conditions of 850 psi/850 °F.

The Liner *United States.* The zenith in worldwide marine power plant engineering in the 1950s was the passenger liner *United States,* contracted for in April 1949. The concept took advantage of Navy-proven propulsion plants as a means of reducing weight and space requirements, while putting out what still, today, represents the highest level of propulsion power ever used in commercial service. With a maximum ABS rating of 158,000 shp, the quadruple-screw ship further had an impressive defense rating of 240,000 shp. She was equipped with four 925 psi Babcock and Wilcox boilers, four Foster Wheeler boilers (all of B&W design), and four Westinghouse cross-compound geared turbine sets. Her electric plant included six G.E. 1000 kW turbo generator sets. In meeting her 14-day, 32-knot round-trip schedule, the ship normally required between 100,000 and 130,000 shp, depending on weather conditions and length of time out of drydock. During trials at 158,000 shp, the ship developed an exemplary fuel rate of 0.507 lb/shp-hr, with a remarkably flat specific fuel rate curve.

That tremendous power plant operated without any undue difficulties. One of the few exceptions was

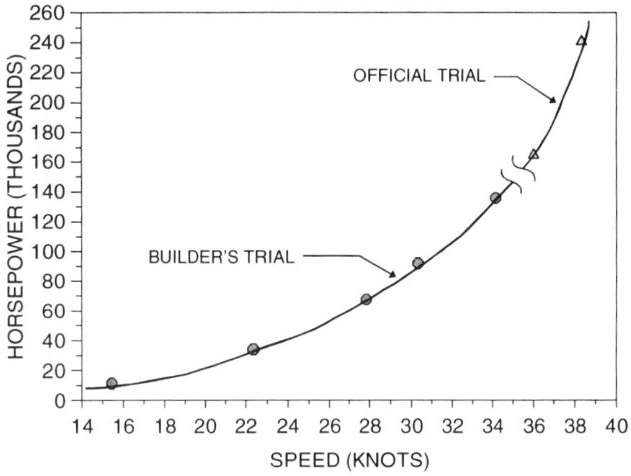

Fig. 1 Trial curves for SS *United States.*

a necessary drop in superheater outlet temperature (1000 to 965 °F) to reduce superheater fouling from vanadium in the fuel oil. It became necessary to obtain a special Venezuelan oil with a total ash content below 400 ppm and a low vanadium content.

Propeller blade erosion was so severe as to require removal and repair of the propellers every year. Redesigned propellers later increased the time between repairs to two years. The blame was generally ascribed to uneven wake distribution. These same adverse conditions appeared, but to a lesser degree, on the *Constitution* and *Independence,* as well as on the *Santa Rosa* and *Santa Paula,* all of which had full propeller shaft bossings.

Postwar Steam Turbine Plant Progress. Late World War II destroyers employed separate cruising and high-pressure turbines. When operating at low powers, the steam flow linked the turbines in series for maximum efficiency. At high power levels, the cruising turbines were bypassed. Aggravated casing sealing problems and windage heating were significant. The solution was a return to a "series–parallel" concept, or "HP–IP combination." In this arrangement, steam enters the HP–IP (high pressure–intermediate pressure) turbine at its center. At high power, steam divides and flows in parallel through both groups of blades, while at cruising power, it flows in series through first one set of blades and then the other. Low-pressure turbine design, although not fundamentally changed, was affected by reducing standard condenser vacuum from 27.5 to 25 inches. This considerably reduced the size of the turbine's last stage. The total number of turbine stages remained essentially constant, thus requiring higher rpm to extract the greater pressure and temperature drop from the steam. The combination of all those factors yielded a significant decrease in turbine weight, although little improvement in efficiency.

These basic advances were accompanied by numerous improvements in such details as interstage seals, rotor and casing materials, moisture removal in lower stages to alleviate blade erosion, and oil and steam seals. Thus, even in a technology as seemingly prosaic as steam turbines, significant advances were made.

DD828—The Experimental *Timmerman*. "The Twentieth-Century *Clermont*," as the chief engineer of the *Timmerman* dubbed her, was fitted with an experimental power plant. The year was 1952. Her hull was essentially a World War II *Gearing*-class destroyer. Special lightweight, high-performance machinery was selected and installed on a premise that if a piece of equipment didn't fail it had too much design margin. Manufacturers were requested to give only a "guarantee of good workmanship, good intentions, and their best technical brains." The developmental features included two side-by-side, advanced steam plants. One was rated with steam conditions of 2000 psi/1050 °F. The other was of a more moderate 875 psi/1050 °F, still greatly exceeding the typical steam conditions of the day. Features also included 400-Hz, 1000-volt electrical equipment; a planetary reduction gear; shaft roller bearings; three different boiler designs (including forced circulation and integral superheat) with automatic combustion controls; one gas-turbine and one diesel-powered emergency generator of novel design. The net result was a plant of some 100,000 shp with a specific weight about half that of a conventional destroyer.

The *Timmerman* project was a "success" in that most of her new equipment failed in one way or another. From her commissioning in September 1952 until June 1953, she was at sea a total of 23 hours, and once attained a speed of 31 knots on two boilers. In the Boston Naval Shipyard she was known as "Building 828"! The machinery plant never achieved full power. The ship's design and performance history were cloaked in secrecy for years. Some innovations, such as higher steam temperatures and pressures, continued in the Navy. Although 400-Hz power was adopted for certain uses on board and on aircraft, it was never again attempted for general ship service power. Planetary reduction gears did not become a standard for U.S. naval vessels; but, in the 1970s they were used in the first reduction stage of many European-built tankers with steam turbines.

Naval Steam Plants of 1200 psi. Our Navy continued to seek higher performance machinery of reduced weight and space. The *Timmerman* had set the stage for the 1200 psi/950 °F steam that became the norm for naval warships, succeeding the 600 psi/850 °F standard of World War II. As often happens with innovations, the transition was not limited to a simple change in steam conditions but was accompanied by a number of ancillary trappings that greatly complicated and confused matters. Among these were automatic combustion controls, integral superheaters, higher fuel oil pressures, and boiler feedwater pumps and deaerating feedwater tanks located in the fireroom.

The *Mitscher*-class were the first new destroyers built after the war. They had a speed of nearly 35 knots on 75,862 shp. They were the first naval ships to employ 1200 psi/950 °F steam. Not surprisingly, they had teething problems. The Navy suffered from a chronic shortage of qualified and experienced petty officers to cope with the newer technology. Operating and maintenance problems continued to plague 1200 psi steam plants throughout the period, giving rise to myriad "get-well" programs. Finally, it became necessary to establish a project managing organization dedicated solely to technical and logistic support of 1200-psi steam plants. The difference between expectations and real world results has seldom been better demonstrated, or on so large a scale.

In a wave of conservatism, owing in part to limits on funding and scarcity of critical chromium, the Navy retained 600 psi/850 °F steam for the carrier *Forrestal* (CVB59) in the early 1950s. Her forerunner, the never-built CV58, was slated to have a 1200 psi plant. Concern for high pressure leaks was being solved with higher quality welding that minimized pipe flanging. A concerted effort was made to construct the *Saratoga* (CVB60) with a 1200 psi plant, among other reasons, to maintain the U.S. Navy's role as the world leader of naval technology, keeping pace with the shore-side electric utilities. Fifty percent of the new central power stations during the early 1950s used 1450 psi and up to 1100 °F steam conditions. Considerable research on thermal shock to austenitic piping, ferritic piping, welding, and valving had been instigated by the Bureau of Ships (BuShips), and results were widely known within the power plant industry.

The plan for an advanced aircraft carrier plant was finally approved. The result, in 1956, was a plant of 20,000 more shp than the 260,000 shp forerunner. A 6.38 percent savings in combined machinery and fuel weight was attained. This was owed largely to the specific fuel consumption savings that were in the 11 to 12 percent range. The *Saratoga* also incorporated steam air preheaters (gas air preheaters were not easily maintained) and a low pressure heater in the cycle to attain some regenerative feed heating from a turbine crossover bleed. *Saratoga*-type fuel oil burners became standard for merchant ship boilers soon after.

Postwar Boiler Progress. The unending quest for reduced weight and space of machinery inevitably led BuShips to pressure-fired boilers, also called supercharged steam generators. In such boilers, combustion air was fed to the furnace at about 65 psi/490 °F. This came from a turbine-driven compressor, powered by furnace exhaust gas. The elevated pressure required less space for combustion and gas flow, and improved heat transfer, reducing overall boiler size and weight by half. The wide-range burners fired downward from the top of the firebox. Natural convection circulation was employed for steam generation.

The first (and only) Navy ships equipped with this type of steam generator were the *Garcia*-class frigates (DE 1040), operating with steam at 1200 psi/950 °F. They experienced the problems typical of other 1200 psi ships mentioned earlier. In addition, the compactness of the boilers made them significantly more difficult to maintain. When problems with the superheaters required a redesigned replacement, they had to be removed rather inelegantly through the bottom of the ship.

Another attempt to reduce boiler size focused on forced internal circulation by pumps, a technique that was first employed in the *Timmerman*, with unpublished results. More definite results, mostly bad, were derived from the installation of "controlled" circulation boilers in the *Mitscher* (DL-2). They were so unreliable that they were replaced with conventional boilers. That experience, along with the *Timmerman*'s, fomented some skepticism, if not hostility, in the fleet with regard to BuShips' innovations.

Commercial ship plants were generally selected with more conservatism, and change was always resisted. In comments to a paper entitled "The Design of Marine Boilers," published in the 1948 SNAME *Transactions*, Vol. 56, Commander L. S. McCready stated, "This introduction of higher ratings and new design features in the merchant marine is usually a slow process. Hard-headed superintendent engineers and port engineers observe that economy of fuel, unless accompanied by economy in maintenance and repairs and a reasonably low turnover in personnel, is no economy."

Postwar Reduction Gearing. Naval use confirmed that reduction gear design had, indeed, been conservative, and pointed the way to a number of approaches to smaller gears without sacrificing reliability. Those developments included such attention to detail as hardened tooth material and techniques to resist pitting; stronger teeth, achieved by coarser pitch, and shot-peening the gear roots against breakage; more accurate gear cutting and shaving; modified tooth profiles to reduce scoring; and improved forging and heat treating.

By the early 1950s, naval reduction gears were being built and operated at two to three times the tooth contact loading of World War II gears. Refinements continued throughout the period, including later emphasis on fine-tooth gears for quieter operation. Nevertheless, the more costly Swiss techniques of hardening and grinding, although proven in many foreign applications, and offering quieter operation, did not take root in the United States for some time.

Propulsion Shafting. After World War II, BuShips began to look for higher strength steels for propulsion shafting to reduce weight. A new alloy with a minimum tensile strength of 120,000 psi was selected, and the allowable stress level was increased over that of conventional steels. In 1954, the *Norfolk* suffered a disastrous shaft fracture at the forward face of a propeller hub, while at nearly full power. The failure was attributed to fretting corrosion. The Navy had been plagued for years by these corrosion fatigue cracks in external shafts in the vicinity of the steel waster rings commonly installed at the ends of bronze bearing sleeves. In most cases, repairs had been made by grinding out or welding the cracks. The *Norfolk* shaft failure, however, led to the replacement of all external shafting made of the new alloy.

Such problems led to several other developments in shaft design. Cold rolling of shafts, in way of the hub and strut bearings, was used to increase corrosion fatigue strength. More conservative estimates of the maximum bending moment were mandated, and shaft designers used lower allowable bending stresses and higher torques. Concurrently, safety factors were slightly reduced. Rubber coverings replaced the steel waster rings. All these factors (and improved cathodic protection) virtually eliminated the corrosion fatigue problem.

Mariners, C4-S. The Mariner-class ships, first contracted for in 1951, established many of the steam machinery practices used during the remaining 1950s and well into the '60s. The Mariner power plant development was undertaken concurrently with that for the *Schuyler Otis Bland*. Owing to the urgency of building 35 ships for Korean War support, certain changes were made. The selection of machinery components was based on an evaluation of overall cost, availability of material, ease of construction, ease of installation, interchangeability, reliability, and repair experience.

The Mariners were designed with a normal output of 17,500 shp. In addition, they met a national defense requirement of 22,000 shp. With superheater outlet conditions of 600 psi/865 °F and with two 600-kW, 440-V ac generators, these ships attained a credible fuel rate of 0.514 lbs/shp-hr. The class sported an uncommonly large power plant that, combined with engine room wing tank requirements, demanded a 70-ft machinery space. The plant operated with few difficulties, and the Mariner-class proved successful despite initial operator disinterest.

The Mariners were a turning point in U.S. cargo ship propulsion plant trends, both in power output and in higher operating efficiencies. During the ensuing years, a series of ships with Mariner hulls was constructed, all having the same propulsion output of 19,250 shp (maximum ABS rating). The steam cycles improved, and in some ships the engine room lengths were reduced to 55 ft and 60 ft with the machinery space wing tanks eliminated. For example, the *Philippine Bear*-class, with a 55-ft engine room, used a 620 psi/905 °F steam cycle with a regenerative air preheater and four stages of feed heating. This gave a trial fuel rate of 0.499 lb/shp-hr. The power plant design was so successful that it served basically unchanged as an American standard for steamships for the next 20 years.

Liberty Ship Modernization and Engine Improvement Program. With more than 1000 nearly new Liberty ships in the National Defense Reserve Fleet, an experimental project ensued in an attempt to recover some of the wartime investment in those ships. MarAd ordered the re-engining of four ships. They were targeted as working test beds with either a steam turbine, a free piston gas turbine, a regenerative gas turbine, or a diesel engine. The new plants were to be capable of driving the old 11-knot Liberties to about 15 knots (with the addition of a longer, more hydrodynamically suitable bow).

In 1956, the *John Sergeant*, the most notable of the converted Liberties, was being fitted with a 6000 shp open-cycle regenerative gas turbine. She also used double-reduction gears and a controllable, reversible-pitch propeller. Low-ash, low-vanadium fuel of the type usually used in the *United States* was further processed and water-washed on board to reduce sodium. A fuel additive was introduced to inhibit vanadium. Her low compression ratio, heavy-duty, industrial-type gas turbine's fuel consumption rate was a respectable 0.52 lb/shp-hr.

In this limited budget program, despite extraordinary planning and execution, problems were encountered with fuel treatment, turbine blade cleaning, and salt in the air intakes. Her controllable-pitch propeller (which was of unproven design) gave problems as well. The regenerated gas turbine was an adaption of an existing General Electric design used for gas pipeline pumping service. It was a two-shaft machine, with a 1450 °F turbine inlet temperature. The combustion turbine drove a 4.9 to 1 ratio compressor, and power was extracted from a second, separate, unconnected ("free") turbine. The exhaust gas was used to preheat the compressed air in a large (automotive-type radiator) regenerator, and then it passed through to the waste heat boiler to generate steam for the ship's service turbo generator. By early 1959, the ship had logged some 7000 hours of almost faultless commercial operation, despite the problem enumerated above. Although U.S.-flag operators were not influenced, some ten or so years later the same sort of engine (advanced in shore-side power applications) reappeared as the G.E. "heavy frame" engine, and was put on U.S.-flag tankers owned by Chevron oil and on several foreign ships.

The Newer Steamships. National Bulk Carriers broke the 100,000 dwt barrier with the Japanese-built, American-designed *Universe Apollo* delivered in 1959. Her power plant was a fairly conventional 600 psi geared steam turbine, but her cargo discharge system was large, being sized for a 30-hr discharge by four 1000 HP steam turbo-driven pumps.

Another 100,000 dwt tanker, the *Manhattan*, came along shortly afterwards, and she was uniquely powered with twin screws. She was designed and built by the Bethlehem Steel Corp. The 940-ft ship remained the largest, highest-powered U.S.-flag tanker for some time. She later tested the Northwest Passage as an icebreaking tanker. Her original design innovations included an angled twin-rudder arrangement. The twin screw plant was rated at 43,000 shp.

In the early 1960s American operators of ocean-going ships began to move toward increased speed and cargo system improvements. Hull lines were improved, and the Mariner-type power plants were refined slightly as reliability became more important. In 1962, the first of U.S. Lines' eleven 21-knot Challenger-class ships was delivered. About the same time,

Fig. 2 Typical GE geared-turbine installation. (Shown is the 9350-shp cross-compound plant aboard ore carrier *George M. Humphrey.*)

Grace Lines' four *Santa Magdalena*-class, with a 20.5-knot speed, became the modern cargo carriers of the day. Both plants, which were of about 18,000 shp maximum, had two oil-fired D-type water tube boilers supplying 600 lb steam to a single set of cross compound steam turbines driving double reduction gears and a single propeller. This design represented a conservative standard for American power plants for the next decade.

Single Boiler Ships. In 1954 and 1955, three Great Lakes freighters owned by the Cleveland Cliffs Steamship Co. were repowered with 5500 shp steam turbine plants, each using a single, oil-fired, Babcock and Wilcox straight tube boiler rated at 515 psi. Each ship was also fitted with a small auxiliary 150 psi boiler with an auxiliary piping connection to the low pressure turbine and to a 185 kW auxiliary diesel generator set for standby power. In 30 years of service, however, none of these vessels had to resort to the jury rig because of a breakdown of the single main boiler.

The *President Van Buren*-class, contracted for in 1964, was designed with a power output of 24,000 shp; these were unique as single boiler ships. Superheater outlet conditions were 870 psi/955 °F. Boilers were equipped with regenerative air preheaters, and the cycle had four stages of feedwater heating. There were two electric generator sets, one being a 1000-kW steam turbine unit, exhausting to the main condenser under all sea and port conditions, and one 800-kW gas turbine driven unit, which acted as a standby and also provided power to a 750-hp take-home motor in case of the boiler's failure. The specific fuel rate under trial conditions was 0.466 lb/shp-hr.

Single boilers were seen to have operational and initial cost advantages, but have been rarely used over many years. At least one major oil company built tankers with one main boiler, but also installed a smaller port (cargo) heating boiler, which could provide steam for emergency take-home use.

American Diesel Engine Drives of the Fifties and Sixties. In the years preceding World War II, most diesels installed in this country were direct drive reversing engines. They were usually adapted from commercial power station or pipeline pumping designs as produced by Nordberg, Fairbanks Morse, Cooper Bessemer, and Enterprise. For smaller powers, adaptations of commercial truck and tractor engines from the Detroit Diesel Division of General Motors, Caterpillar, and Cummins were used; and, since the railroads were changing from steam to diesel power in this period, there was great incentive to adapt the lightweight locomotive engines. These were being mass produced and had advantages in low cost and quick delivery. In those years there was a rapid expansion of demand for engines for small craft in the offshore drilling developments, tugs and towboats, fishing vessels, and other miscellaneous small commercial and military craft. That made it feasible for the large volume locomotive engine manufacturers such as Alco, Fairbanks Morse, the Electro-Motive Division of General Motors, and G.E.'s Erie Division, to market families of engines of various powers and speeds together with controls, transmissions, auxiliaries, and engineering design services.

Since locomotive engines are constrained by clearance for tunnels, bridges, etc., large increases in power are obtained by increasing the number of cylinders, increasing rpm, increasing piston displacement, and supercharging with exhaust turbines. Aboard ship, multiple engines driving combining reduction gear sets were a viable solution.

The range of engines for propulsion varied from about 100 to about 5000 bhp. Single reduction gear sets fitted with controllable reversing propellers or reversing reduction gears were available in a variety of sizes. The installation and capital costs of this arrangement were considerably lower than those of a steam plant.

The trend in post-World War II U.S. submarine design toward greater submerged speed provoked a demand for diesel engines of greatly reduced weight, capable of snorkeling, and resisting high shock. To meet a goal of 4000 hp in one engine room, Fairbanks Morse developed an opposed-piston engine, the first of which was installed in the *Harder*. General Motors developed a 16-cylinder, 1000 hp "pancake" design, which was installed in the *Trigger*-class. Both types of engines immediately demonstrated a greater propensity for trouble than for performance.

Some 160 minesweepers were built, with a special design of Packard nonmagnetic engines of comparably low specific weight; but they were so casualty-prone that they eventually had to be replaced.

On the commercial side, the 37,000 dwt *Falcon Lady* was one of the era's pacesetting diesel-powered tankers in U.S.-flag service. The design was developed by Litton for Falcon Tankers, Inc., for long-term charter to the Military Sealift Command. It had originally been intended to power each of these vessels with a newly developed, uncommon, single Fairbanks-Morse 38A20 opposed piston engine, with single-reduction gearing. Because of delays in development of the new engine, Colt Industries substituted a pair of imported S.E.M.T.-Pielstick, 4-stroke, 16PC2V engines in the first two ships, and later built the Pielstick engine sets in this country, under license. Similar developments were occurring on the Great Lakes, as described in another chapter.

Nuclear Power's Roots and Naval Applications. Nuclear power, which has found considerable use in the U.S. Navy, had its first application in submarines. Those developments are well described in the chapter on submarines and little more need be said here.

Contrary to popular misconception, the nuclear propulsion program office studied many alternatives to pressurized water reactors, and probably knew more about their potentials and pitfalls than any other agency, government or private. Opposition to these alternatives, many of which were little more than crackpot ideas, was based on solid technical fact and engineering analysis, always tempered by the experience of naval engineers who had operated machinery in real ships, and with the overriding considerations of safety. If the resulting technical decisions were conservative, they were well-founded.

The United States' Nuclear Powered Merchant Ship Project. The nuclear ship *Savannah* was conceived in the mid-1950s, in a period when the words "nuclear" or "atomic" in the civilian world elicited a sense of terror. The Navy's submarine *Nautilus* had gone into service, ushering in a reality of undersea warfare that only Jules Verne could have imagined. Warfare then seemed to be the exclusive "benefit" to be derived from atomic fission. Under this cloud, the Eisenhower Administration conceived the "Atoms for Peace" Program, and the *Savannah* was to be one of the cornerstones. The goals for the vessel were two-fold: (1) demonstrate peaceful and commercial potential of nuclear power and (2) use the ship to "open" ports throughout the world to visits from nuclear-powered ships.

The *Savannah* was never intended to operate as a commercial success. As a break bulk, combination passenger/cargo ship, her configuration offered little hope of commercial viability.

The prime contractor for the nuclear plant was Babcock and Wilcox, along with Combustion Engineering and General Electric. They were also the primary suppliers to the Navy's nuclear program. Concerned about problems of technology leakage, the Navy was unenthusiastic about the project. The design called for a pressurized water reactor to be fueled with only slightly enriched (4.5 to 5 percent) uranium oxide fuel pellets. The average primary loop temperature was 508 °F producing saturated steam at 425 psi/454 °F at full power.

The containment vessel was a considerable undertaking in itself. It was a heavy steel pressure vessel surrounded by lead, polyethylene, and concrete for radiation shielding purposes. Additional protection (padding) in the form of redwood timber and steel reinforcing was fitted along its outboard sides. This was to afford protection in the event of a collision. Additionally, provisions were made so that, in the event the *Savannah* were to sink for whatever reason, seawater would enter the containment vessel, thus equalizing the pressure and avoiding an implosion. It would re-seal itself to prevent the discharge of significant quantities of radioactive material. The *Savannah* had elements of political genius but with a poorly communicated mission.

The Navy's Transition to Light Distillate Fuel (DFM). The U.S. Navy made a transition from burning Navy Special Fuel Oil (NSFO—a blend of heavy residual oils with distillate cutter stock) to Diesel Fuel Marine (DFM or F-76, the NATO designation of the fuel) in the late 1960s and early 1970s. During that period, the large majority of our naval combatants were steam powered. The boilers were compact, with higher heat release rates, and operated at much higher temperatures and pressures than their World War II counterparts. While technical advances were made in the design of marine boilers, the standard boiler fuel, NSFO, remained relatively unchanged from its introduction in the early 1940s. The U.S. Navy began to experience numerous problems with the more modern high pressure boilers that were directly attributable to the use of NSFO. The major problems are:

- The continuous fouling of boiler tubes caused significantly higher fuel consumption.
- Additional maintenance was required in the form of frequent soot blowing, water washing boiler firesides, and additional cleaning of fireroom and decks.
- The deposits and fouling of the boiler tubes caused insulating slag to form and corrosion to take place. Substantial repairs and reboilering became more frequent, and boiler reliability dropped.
- Operational readiness, especially for two-boiler ships, dropped significantly owing to excessive boiler maintenance and repair.
- The retention of boiler technicians was lower than for any other shipboard personnel.
- Soot, ash, and other particulates caused fouling and corrosion of topside decks, electronic equipment, and aircraft, adding to the maintenance and repair requirements.
- Air pollution caused from smoke, particulates, and sulphurous emissions was troublesome and noticeable, especially while steaming in port.
- The availability and quality of NSFO and other residual fuels was beginning to diminish as production of gasolines and other distillates increased.

The problems associated with burning NSFO coincided with logistical problems, as well. Fleet oilers were limited to but two grades of fuel: NSFO and JP-5 (for jet aircraft); and that was the limit of their replenishment capability. Naval vessels were therefore burning distillate fuels in all diesel and gas turbine applications. The Navy's 250 makes and models of fuel oil service and transfer pumps also presented difficulties. The Navy wanted to reduce this number to improve its pumps' standardization and support posture. These logistical problems also helped to spark the shift to a standard distillate fuel.

The problems associated with burning heavy fuels were not unique to the U.S. Navy. Many foreign navies shared the same problems. The use of diesel and gas turbine main propulsion engines on foreign combatants was also increasing. As a result, distillate fuels were tested by the British, Canadian, and Dutch navies, and the positive results promptly led to a shift to a standard marine diesel fuel. The U.S. Navy studied the use of diesel oil in marine boilers. It concluded that a shift from NSFO to distillate fuel was justified and necessary. The Navy, with help from the petroleum industry, developed a distillate fuel (DFM or F-76) that could also be used successfully in diesel engines and gas turbines. The shift from NSFO to DFM required the replacement of, or modification to, fuel oil service, transfer, and stripping pumps, burners, sprayer plates, automatic combustion control equipment, settling and storage tank heating coils, and fuel oil heaters for 527 steam-powered vessels. This represented a high initial capital investment for the Navy, but the payback of cleaner combustion, reduced maintenance, (and ultimately improved material condition and operational readiness) more than offset this expense and the additional cost of burning the expensive DFM.

The shift to DFM also paved the way for a transition to more diesel- and gas turbine-powered combatants. Thus, by effectively equalizing the cost of the basic propulsion fuel, the cost advantage of the steam plant was lost, and a logistical advantage for gas turbine power was born. The two Navy fuels (DFM and JP-5) were both compatible with gas turbine engines.

Gas Turbines. As is well known, the gas turbine concept predates the 20th century. Like the diesel engine cycle, the gas turbine's output was a function of the difference between the work in expansion and the work in compression. The power output and efficiency depended on the maximum cycle temperature. Many gas turbines were built in Europe before World War II. After the war they began to find wide acceptance and application in power generation and propulsion. At first, their development was largely spurred by the aircraft industry, since their favorable characteristics of compactness, high power per unit weight, and fuel efficiency offered major advances in aircraft performance.

The so-called "heavy-duty" or frame-type engines were also gaining popularity in scattered applications worldwide. This type of engine resembled the heavy casings of high pressure steam turbines. The bulk of applications for gas turbines of either type was for such uses as pumping stations in gas pipelines. An adaptation of one of these units powered the first U.S. gas turbine-propelled merchant ship, the previously mentioned *John Sergeant.* In the late 1960s Chevron Shipping undertook the construction of a mini-fleet of smaller tankers using the heavy-duty gas turbine engines. By putting the gas turbine and its electric generator on the main deck, and powering an electric motor coupled to the propeller shaft, the traditional engine room was eliminated, simplifying crew operations.

Most naval gas turbines were derived from aircraft engines and underwent different degrees of "marinization." A factor quickly realized was the benefit from aircraft industry levels of funding for extensive development and testing, unavailable to traditional builders of handfuls of engines for ship propulsion.

Early interest on the part of the U.S. Navy was heavily oriented toward the need for lightweight power plants for small craft, and for novel concepts such as hydrofoil craft and air-cushion vehicles. As late as 1966, BuShips experts observed that, "This preoccupation with weight has tended to obscure the fact that the gas-turbine power plant system has many advantages in shipboard applications based on criteria other than weight and volume." By 1965, the Navy had installed many gas turbines for electric and auxiliary power, but only 78 for propulsion, of which all but three were for small craft. One major step was the Navy's combined-diesel-or-gas (CODOG) propulsion plant for patrol motor gunboats (PGM), using diesel engines for cruising and a G.E. LM1500 gas turbine of 14,000 hp for high-speed operations.

The Coast Guard was also a pioneer with the selection of CODAG propulsion in five of its sixteen medium-endurance *Reliance*-class cutters built between 1963 and 1969. CODAG referred to combined diesel and gas turbine running together in parallel. In this application, CODAG proved unfavorable and was abandoned in favor of 2500 hp diesels. The boost from the 1000 hp gas turbine added only a knot or two.

In the meantime, the Coast Guard also pioneered the use of CODOG in American oceangoing propulsion plants. Most notable perhaps were its twelve *Hamilton*-class, high-endurance cutters, first built in 1967. These twin-screw, 378-ft ships used either two 3500 hp relatively efficient diesel engines for cruising or two considerably higher powered 18,000 hp FT-4 gas turbine engines driving two CRP propellers for a designed top speed of 29 knots. The gas turbine's use as a boost engine to provide an occasional increase in speed was a favorable trade-off for the higher specific fuel rate. The CRP propellers were the largest on any American ship up to that time.

Although gas turbine propulsion for merchant ships was not destined to play a major role in the last 50 years, the *ADM. William M. Callaghan* represented an excellent application. She certainly was the ground-breaker that opened the door for the U.S. Navy's near total changeover from steam to gas turbine power for most new combatants. Built for quasi-commercial service, this RO/RO was originally operated for the Military Sea Transportation Service (MSTS). She was initially powered by two Pratt & Whitney FT-4 gas turbines delivering 39,000 shp to two shafts. The *Callaghan* was, in 1967, the world's largest and probably fastest RO/RO ship, setting a transatlantic record for dry cargo vessels. (Her trial speed was 26.5 knots.) Unlike most other high-powered gas turbine ships, the *Callaghan* did not have CRP propellers. Instead, her designers at Sun Shipbuilding selected reversing gearboxes and inflatable Airflex, 900 rpm, intermediate shaft clutches. These gearboxes provided excellent service and maneuverability, with capability of full power, both ahead and astern.

By 1971, the *Callaghan* had completed three years of impressive service. That same year, the port FT-4 engine was removed and replaced by a General Electric LM 2500 engine so that the ship could serve as a Navy test-bed for this more advanced (second generation) engine, which had the benefit of some ten more recent years of development. It was reliable and saved about 20 percent in relative fuel consumption. Its fuel rate was about 0.40 lb/shp-hr. It was so favorable that it was quickly slated for the U.S. Navy's *Spruance*-class (DD-963) destroyers and, soon after, the *Oliver Hazard Perry*-class (FFG-7) frigates. For Operation Desert Shield in 1990, the *Callaghan*, then 23 years old, was broken out of mothballs, and her plant was restarted in a matter of a few hours after reinstalling its cocooned gas generator.

The improved fuel rate and modular maintenance scheme made the change to this engine logical. Since that time, the LM2500 has been the predominant engine for non-nuclear warships. In 1987, there were 235 ships committed to this engine, and 607 were sold worldwide for marine service. As experience with this gas turbine has grown, the rating for shipboard service has increased from 20,000 to 25,000 shp, and the time between overhauls has increased significantly. These engines were installed on the PHM-1, DD-963, FFG-7, and CG-47 class ships, and the DDG-51, AE-36, and AOE-6 classes.

The simple-cycle gas turbine had served the Navy well, but the part-load fuel rate left room for improvement. Operationally, the twin-screw ships could improve overall fuel consumption by running with one shaft powered and the other "free wheeling" or in "drag shaft mode." One combined-cycle improvement that began development in the late 1970s was the Rankine Cycle Energy Recovery (RACER) system. The promise of about 30 percent fuel savings kept it alive for about ten years or so. The intercooled regenerative engine, however, has taken up the interest in fuel rate savings, particularly at part-power conditions.

Experiments in burning other than distillate fuels in gas turbines were tried, but with only limited success, despite elaborate pre-combustion treatment. For example, special additives were tried to inhibit corrosive metals, mainly vanadium. Sodium sulfidation was a major problem for even distillate fuels with unsatisfactory fuel dewatering. Those efforts were most notable for Seatrain's Euroliner-class gas turbine-powered containerships.

In the 1960s, the U.S. and Canadian Navies funded a small regenerative gas turbine project that was unsuccessful. In those early attempts, the pressure drop in the heat exchanger caused a loss in efficiency about equal to the thermal gain. The addition of the intercooler and the high pressure ratio (16:1) of modern gas turbines has made this concept viable. The concept is still being seriously pursued at this writing and is dubbed the Intercooled Regenerative or ICR program.

Developing aero engines is costly, and this level of cost conflicts with the competitive pricing of most civilian businesses. We tend to expect power plants for public utilities and marine service to evolve from prior technology without further development cost. The LM2500 and FT4A development and support costs were mostly borne by the aircraft applications. In the mid-'70s, the Navy funded development of a Pratt & Whitney FT9 engine in the 30,000 hp class, but abandoned that effort largely because of the waste heat recovery system's potential to increase the power output of the "upratable LM 2500" and perhaps the possibilities of developing the ICR engine. The regenerative intercooled gas turbine will probably be successful if it can be developed using proven components from aircraft or industrial gas turbines. The team of Westinghouse and Rolls Royce was awarded initial contracting work in 1992 for the Navy's ICR engine. The more complex cycle and hardware are targeted for significantly greater efficiency in the part power range, enabling a so-called "flatter SFC curve." This is mainly important for the cruising speed range of all gas turbine-powered ships.

The Less Efficiently Powered, Large "Economy of Scale" Ships. The late 1960s and early '70s saw a brief end to a period of unprecedented growth in ship size. Those large ships were designed to take advantage of economy of scale and, in some cases, speed. Large tankers and high-speed containerships broke new ground. The Sea-Land SL-7 containerships and those of the Euroliner-class were notable.

The four Euroliner-class ships were powered by uprated versions of the same gas turbines that were installed on the *Admiral Callaghan* a few years before. Capable of about 35,000 hp per shaft, the twin-engined, twin-screw ships were fitted with controllable-pitch propellers (the most powerful up to that time) instead of reversing reduction gears as on the *Callaghan*. The ships made nearly 30 knots on trials and were initially put into 26-knot transatlantic service. The Euroliners took the Atlantic cargo ship speed record away from the *Admiral Callaghan.* As in that installation, the engines were mounted over the propeller shaft on the afterside of the gear boxes to permit a shorter machinery space. In fact, the engine rooms were only one 40-ft container bay in length. Early service saw relatively low engine life at the uprated power levels. The class's specific fuel rate averaged about 0.49 lb/hp-hr, or about 2200 bbls per day of clean marine diesel fuel (gas oil). When the ships entered service, this was priced at about $1/bbl more than Bunker-C fuel (at about $4/bbl).

The eight SL-7 ships had two basic, side-by-side, separate (and with no cross connection) steam power plants that were ultra-simple, considering the state of the art. The stress was on reliability because the ships were designed to make scheduled ocean crossings at unprecedented speeds of over 30 knots. The SL-7s soon took the cargo ship transatlantic speed record away from the Euroliner-class. With a simple two-heater cycle, each plant's twin geared G.E. turbines were rated at 60,000 shp. Superheater outlet conditions were 875 psi/945 °F from a single, D-type Foster-Wheeler boiler serving each shaft line. They were huge by merchant ship standards, with an evaporation rate of about 388,000 lbs/hr each. The plant's designed specific fuel consumption rate was 0.47 lbs/hp-hr, but the full power daily rate was a remarkable 3900 bbls, perhaps four to five times the average rate for high-speed ships of the 1970s.

The sudden increase in fuel costs of the mid-1970s made both of these high-powered classes grossly uneconomic. The SL-7s were sold to the U.S. Government and converted to military RO/RO vessels with their original power plants intact. The Euroliner ships were first converted to burning blended residual fuel, then totally repowered with medium-speed, 4-stroke engines rated for a 19-knot speed. They were equipped with main engine-driven shaft generators for use at sea, among the first American-owned vessels to successfully use this power take-off (PTO). The units, however, were frequently used in port because of the poor reliability of the ships' single diesel generator set. The shaft generators remained as the prime electrical source even after the ships were converted to diesel. They were readily uncoupled from the propeller shaft but were hard coupled to the engine driven, high-speed pinion shaft.

Electric Drive Developments. In the 1970s the Navy again began to take interest in electrical transmission for propulsion systems. Flexibility in machinery arrangements and in system operation had long been recognized as an attribute of such systems. The ease with which they could be reconfigured with switchgear could enhance survivability. Cross connection at cruise power levels (e.g., one engine powering two shafts), together with speed ratio control, could permit propulsion engines to operate at favorable load and speed conditions (an important factor, since simple cycle gas turbines have poor part-power fuel rates). On the other side of the coin, electric transmission systems have higher first cost, lower transmission efficiency, and greater weight and volume compared with mechanical drives.

Most notable of the electric drive efforts of the 1970s was the development and demonstration of superconductive excitation and liquid metal current collection in a direct current (dc) system at the David Taylor Research Center. With direct conductor liquid cooling, these technologies would provide dc motors and generators, which are smaller, more efficient, and easier to control than their alternating current (ac) counterparts. A 40,000 hp, 180 rpm conventional ac air-cooled motor was estimated to weigh 320,000 lbs with a diameter of 20 ft. Its advanced dc electric motor was projected at 80,000 lbs and 6.5-ft diameter. A 300 kW acyclic dc system, consisting of a superconducting generator and motor, was demonstrated at the Annapolis Laboratory of DTRC. (Subsequently, the system was installed on the *Jupiter II,* a 65-ft crew boat, and demonstrated at sea.) This effort set the stage for acquiring 3000 kW components incorporating this new technology.

The Diesel Engine's Rise to Significance in U.S.-Flag Commercial Ships. The higher cost of fuels in Asia and Europe tended to keep power plants lower in total output. In those parts of the world more attention was paid to the number of barrels of fuel a ship burned per day. Ships were slower and smaller. The diesel engine developed steadily in this scenario before and long after World War II. The development process continued almost without notice in the United States. As European and Asian powering demands grew, steam was a significant contender only in the larger tankers that also had large cargo pumping demands. In fact, some tankers required greater steam generating capacity for cargo discharging than for propulsion. Therefore, on the commercial front, Europe and Asia were inherently prepared to "hunker down" when energy costs suddenly made such an impact in the shipping business. Diesel engine development proceeded in two almost diametrically-opposite directions. For naval applications, engines were small, high-speed, and lightweight. For merchant ships, the trend was toward large, slow-speed engines of higher power, highest efficiency, ability to burn the least costly fuels, and lower rpm to match optimum (lower) propeller speeds.

In merchant ship propulsion, diesels became more and more dominant, even before the oil shortage overshadowed all other considerations in selecting machinery. An idea of the pace of growth in diesel engine size for merchant ships can be gleaned by two reports from *The Motor Ship*. At the end of 1958, this source noted that no marine diesel engine above 15,000 bhp was in service, but cited plans for engines up to 24,000 bhp. By 1968, an article in the same publication listed several "super-large-bore" (over 36-in.) engines producing up to 44,000 bhp with twelve cylinders, at slightly over 100 rpm. Smaller, higher-speed, geared diesels continued to be used in some commercial ships, but the major trend was toward large, slow-speed, direct-coupled engines because of their simplicity, "carefree operation on boiler-fuel," lower maintenance costs, lower fuel and lubrication oil consumption, less noise, and simpler remote control. Downward trends in fuel oil quality, however, soon undermined the diesel's ability to do this without complex fuel treatment systems.

Marine diesel propulsion took a different tack in the United States. On the Great Lakes geared railroad type engines were used in the new 1000-ft bulk carriers that were built after the opening of the new Poe Lock in 1969. These engines were limited to burning relatively expensive distillate fuels. Across the border, on the other hand, Canadian owners preferred the large, slow-speed, direct-drive engines of European design, ignoring the advantages of lower first cost and greater availability of repair parts.

The Great Rise in Fuel Prices—The Embargoes of the 1970s. The 450 percent increase in fuel prices that occurred in the 1970s has had a more pronounced effect on large American commercial marine power plants than perhaps any other single event in the last century. It is no understatement to say that it significantly heralded the end of steam power plants for most conventional applications. But, perhaps more important, it opened the door to our dramatic shift to large imported diesel engines. Simultaneously, the price of fuel also shut the door on further significant use of the gas turbine in commercial service because they required distillate fuels costing more than twice as much as typical bunker grades. Commercial gas turbine ships with ten or more years of remaining life were converted to diesel power.

A few steam ships were also converted to diesel. Most of these conversions were to medium-speed engines because the machinery spaces did not have the height or length to accommodate the more popular low-speed diesels.

Liquefied natural gas (LNG) carriers were being built in increasing numbers and size. The cryogenic cargo lineage began its successful genesis with the aptly named *Methane Pioneer* in 1954. She carried liquefied methane at minus 258 °F. The "boil off" on this pioneering ship was vented to the atmosphere, and so wasted. In the 1970s, two recovery methods were developed. MarAd investigated using a closed Brayton cycle compressor drive engine, burning some of the gas to reliquefy the boil off via refrigeration. The other approach involved power plant dual fuel augmentation schemes. Steam emerged as a leading LNG power plant contender because of its ease in employing dual fueled boilers. The boil-off gas supplemented oil firing in otherwise conventional 850 psi/950 °F steam plants. Energy efficiency took a back seat because a major percentage of the fuel was free. Their power plants required the gas compression/forwarding systems to be double-cased owing to the volatility of the gas and difficulty in its detection. Power plants for LNG carriers are large (50,000 hp) as these ships, with their valuable, perishable cargo, justify speeds of around 20 knots.

Energy efficiency was elevated in importance in military circles as well. The U.S. Navy responded by establishing several R&D efforts. One looked at the availability of distillate fuel worldwide for combustion in its gas turbine ships. (Ship mobility would be compromised if suitable fuel were not available.) As our civilian population grappled with the gasoline shortage and became exposed to "gasohol," the Navy, in conjunction with the Department of Energy, developed, refined, and successfully tested at sea a distillate fuel derived from shale oil. These efforts toward improvement in energy use not only included the mechanical efficiency of different components, but also looked at the energy use and heat balance in total systems, including the hydrodynamic losses. Concepts for more energy-efficient propulsors, rudders, hull lines, and coatings were developed. Underwater propeller polishing and hull cleaning techniques were also investigated. Thus, the oil embargo sparked an exhaustive effort to improve the efficiency of the ship systems that carried over into the decade of the 1980s.

The Two-Stroke Diesel Engine's Rise to Dominance for All Large Commercial Ships. While not an American machinery development, the low-speed diesel's dominance today deserves more exposition. There were seven leading European designers of low-speed, 2-stroke diesel engines in 1960. Each had its own distinctive features. Today the names of Doxford, Gotaverken, Stork, Fiat and MAN have gone into the history books as far as this type of engine is concerned. Today the world market is dominated by what is now MAN–B&W, based in Copenhagen, and Sulzer, based in Winterthur, Switzerland.

The reasons for this remarkable condensation are to be found in the weak shipbuilding market that has persisted since 1973, the intense competition in manufacturing costs, which has led designers to become almost entirely dependent on license income, and the inordinate cost of development and testing of the modern low-speed engine.

Coincidentally, the period immediately following the oil embargo has seen one of the most remarkable periods of engine development and improvement by the two remaining world market leaders. Competition in thin markets has been the spur for this development; but, technically, it has been made possible by application of sophisticated methods of theoretical analysis and detailed design, both of which have been dependent on modern computer software. The application of such technology has greatly reduced design lead times and has eliminated much of the trial-and-error testing on which advanced engine design formerly depended. While equipped with a waste heat recovery system employing a boiler/steam turbine (i.e., combined cycle system), low-speed engine plants could be operated at a system efficiency of over 50 percent. This barrier was broken by a "bare engine" while equipped with turbo-compounding (i.e., multiple stages of turbocharging).

Sulzer Engine Development. Sulzer's RD engine, introduced in 1957, was rated at 1600 hp/cylinder. It was cross-scavenged and pulse-supercharged with under-piston pumping into a divided scavenge trunk to enable operation over the full load range without independent scavenge assistance.

By 1982 Sulzer realized that the limits of development of the cross-scavenged engine had been reached; and in a dramatic move designed, in a remarkably short time, the RTA engine with uniflow scavenge through a single exhaust valve in the head. Its stroke was no less than 2.0 meters. The piston itself was now bore-cooled so that the entire combustion space operated under the most favorable thermal stress conditions. The culmination of this and other developments was the construction of a 12RTA84 engine developing 57,000 hp, with no less than 4750 hp/cylinder.

MAN–B&W Engine Development. The development of the B&W engine over 35 years has been through continuous and systematic improvement of the uniflow engine. In 1963, the production engine was the VT2BF, built with bores up to 840 mm operating at a BMEP of 8.6 kg/cm^2.

B&W's super-long-stroke engine emerged in 1982 with the LMC engine. The results of continuous research and development were evident in many other features of the engine. The general development of the low-speed two-stroke engine added up to 75 percent increase in power per cylinder and 25 percent reduction in specific fuel consumption.

American Slow-Speed Diesel Applications. In 1978, Sea-Land was installing its first main propulsion diesel engines in their D6-class containerships rebuilt in Japan, these being 6-cylinder Sulzer 900 mm bore slow-speed engines. The project gave this leading American containership operator an early leg up on slow-speed diesel engine technology. The high cost of fuel brought about by the oil embargo of 1973 naturally gave primary impetus to the selection of this kind of propulsion power. The selection process for the D9-class, after the large 1975 oil price increases, was done in more depth. While steam propulsion remained a candidate for trade-off studies, the economies of diesels were unchallengeable, as was the initial operating success of the D6 ships.

The D9 ships were planned around a Sulzer turbocharged, 9-cylinder, RND 90M, two-cycle main engine built by the Japanese licensee, Mitsubishi. Because of the potential for further fuel price escalation, a waste heat boiler and a 900 kW steam turbo generator were fitted along with main diesel generator sets. Propeller efficiency was given particular attention as was the question of associated hull induced vibration. All eleven D9 ships, delivered in 1980, represented this country's threshold into the use of diesel engines in its dry cargo fleet. It also marked the end of steam power's dominance of American-owned newbuildings, although built in Japan with Eurasian technology.

In these plants, three other features were significant. The main engines were fitted with slow-speed fuel injector nozzles to permit improved combustion at the reduced power levels that the ships might use over long periods. (Maximum output was thereby limited to about 80 percent power.) A fresh-water central cooling system was installed to serve both the main engine and auxiliary generator engines with multiple pump selectivity to give the required flow arrangement. Finally, the automation system provided a sixteen-hour, unattended capability—one of the first on U.S.-flag ships. Sea-Land followed suit, authorizing three D7-class ships for Jones Act service. These ships, built by Bay Shipbuilding in 1987, were powered by 7-cylinder Mitsui-built L70MC engines.

In 1980, American President Lines acquired two Japanese-built, Sulzer-powered vessels. They found them satisfactory, which led them to select diesels for their three *President Lincoln*-class ships, delivered in 1982. As a subsidized operator, the ships and their engines had to be built in the United States. There were no domestic licensees at the time. Under a cooperative license arrangement, Allis Chalmers built APL's three 12-cylinder RNA 90 engines rated at 43,200 bhp, mostly from parts made abroad. The scheme was successful, and set the stage for APL's entrance into the slow-speed diesel operating arena. The next wave of ships for APL also had slow-speed engines. These were containerships, built by a German licensee in Kiel. These engines had the highest power rating of any built in the world at the time at 64,000 bhp. They were Sulzer 12-cylinder RTA 84 long stroke (840 mm by 1200 mm), low rpm models.

Another major American ship diesel engine powering project was for the Jumbo Econs built for U.S. Lines in 1984 to 1985. These twelve 950-ft containerships were built in South Korea. The engines selected were of Sulzer RLB 90 design but had only seven cylinders.

In 1989 the Matson Navigation Company made the switch from steam to diesel in placing an order for a MAN–B&W slow speed engine to power a containership in their Hawaiian trade.

Additional Marine Engineering Problems of the Final Decade. Contemporary ecological awareness is leading to ways to compact and store or incinerate garbage and sludge as well as to reduce stack emissions. The reduction of stack emissions is being aggressively pursued, at this writing, on the West Coast. Debates about the worth and problems of "cold ironing" (i.e., plant shutdowns and switchovers to shore electrical power) versus stack gas and combustion process alterations are ongoing. Some diesel manufacturers are proposing catalytic converters (as on automobiles) nearly the size of the engine itself. Others are successfully testing fuel emulsification coupled with combustion chamber coatings using stabilized zirconia, a space-age ceramic.

Fuel oil quality continued to degrade during the 1980s, but began to settle down in the '90s. It became a major problem in diesel ships. Shipyards oriented more toward initial cost tended to supply the minimum in fuel processing and clean-up equipment, whereas the operators badly needed comprehensive fuel treatment. Spent fines left over from cat-cracking at the refinery produced abrasives that raised havoc with bore wear. These could ruin an engine in a single voyage. Having 10 to 25 percent on-board spares was useless on a ship with an engine that might have lost 100 percent of its liners and rings at one time. The advanced refinery practice of Visbreaking also affected fuel stability, often making centrifugal purification difficult, if not impossible. By the 1990s, however, most experienced diesel operators managed to solve their fuel quality problems. Steam plants were affected to a much lesser degree, and low excess air control went a long way toward keeping superheater lanes open.

Coal's Brief Reappearance. Coal for ship propulsion reappeared in the *Energy Independence*, delivered in 1983. The ship is in service transporting coal from Hampton Roads to New England Electric Company plants. She was the first coal-burner built in the United States in over 30 years. In interests of minimizing air pollution, the plant could shift to oil when in port. Coal's relatively low price may eventually make it commonplace, but that is unlikely in the first half of the 1990s.

Closure

From the earliest periods of the industrial revolution, the limits of practical know-how, availability of materials, fabrication capabilities, and sources of fuel governed the accomplishments of early inventors, designers, builders, and operators. A broad understanding of basic principles along with some knowledge of past accomplishments and failures, coupled with an imaginative borrowing of technology from other more advanced applications of the day, permitted progress toward hoped for (and usually foreseeable) accomplishments. When arrived at, the accomplishment soon became just another step up the endless ladder.

In many ways, power plant problems have always been much the same regardless of industrial segment. This is mainly because most kinds of power plants require the combustion of fossil fuels. Heat and its physical and chemical associates and by-products are usually at the heart of the larger problems. In the marine environment, using the corrosive sea and the air immediately above it as heat sinks, has always kept engineers busy solving related problems. The advancements and sophistication gained from using improved materials and techniques always push the limits further, in turn creating new problems to be solved in the continuing spiral of advancement.

Whenever new developments are proposed, they must not only hold promise of significant benefit but must also be able to overcome the penalties incurred by the deed of introducing the equipment in the first place. NIH, "not invented here," has proven many times to be a force of overwhelming strength.

Current equipment and methods have the advantage of being familiar, providing a "level of comfort." Their strengths and weaknesses are well understood, and there is usually plenty of data available on performance and maintenance requirements. For the proposed new development, however, such data will likely be insufficient for credibility within the marine community. Sometimes the hurdles are realistic and lend credence to the resistance. Sometimes, however, valuable ideas may be suppressed for years.

A century ago prototype ships permitted new propulsion concepts to be tested and examined in the marine environment. Today, such a test bed is usually unaffordable. The *Timmerman*'s and *Admiral Callaghan*'s examples are no longer followed. Shore testing of new systems, before going to sea, has produced uncertain benefits. Shipboard problems have arisen at sea that were never discovered ashore, and vice versa. Additionally, lab test personnel tend to find problems *ad infinitum*, greatly delaying introduction of developments that might actually be ready for sea. The more sophisticated the shore test, the more expensive it is to set up and run, and the more time it consumes.

Acknowledgments

I appreciate the assistance and guidance of several individuals and the published works of others that greatly contributed to this chapter. In particular, were the considerable contributions made by RADM Lauren S. McCready, along with valuable help from Nicholas Bachko, John Barich, Harry Benford, Robertson H. Bennett, Dr. George Billy, William duBarry Thomas, Chris Evers, Jose Femenia, Robert Giblon, Jerry Gruber, George Palmer, Spencer Reitz, William Van Cott, Charles Wilson, and Jack Woodward.

Two publications, in particular, provided considerable information and were of such value in both content and volume that they have been used and paraphrased extensively. These include ASNE's centenary historical publication edited by RADM Randy King, *Naval Engineering and American Sea Power*, and Dillon, Hoffmann, and Roseman's 1976 SNAME paper "Forty Years of Ship Designs Under the Merchant Marine Act."

Bibliography

Barnes, W. C., "Korea and Vietnam" in *Naval Engineering and American Sea Power*, The Nautical & Aviation Publishing Company of America, Baltimore, 1989.

Bates, J. L., and Wanless, I. J., "Aspects of Large Passenger Liner Design," *Trans.* SNAME, Vol. 54, 1946.

Baumler, R. J., Watanabe, T., and Huzimura, H., "Sea-Land's D9 Containerships: Design, Construction, and Performance" *Trans.* SNAME, Vol. 91, 1983.

Bennett, R. H., Letter to D. A. O'Neil regarding N/S *Savannah*, Nov. 29, 1990.

Bowen, T. L., "Current Marine Applications of Gas Turbine Technology," *Global Gas Turbine News*, Jan. 1992.

Crews, E. R., "The Big Ship," *Invention & Technology*, Spring/Summer 1990.

DeLuce, H. and Budd, W. I. H., "The Design of a Class of 28,000-Ton Tankers," *Trans.* SNAME, Vol. 58, 1950.

Dillon, Hoffmann, and Roseman, "Forty Years of Ship Designs Under the Merchant Marine Act, 1936-1976," *Trans.* SNAME, Vol. 84, 1976.

Ferris, T. E., "Design of American Super-Liners," *Trans.* SNAME, Vol. 39, 1931.

Ford, L. R., "Diesel Versus Steam Drive for Cargo Ships," *Trans.* SNAME, Vol. 41, 1933.

Henry, J. J., "Modern Ore Carriers," *Trans.* SNAME, Vol. 63, 1955.

Jane's Fighting Ships, 1968 to 1969, McGraw-Hill, New York.

Johnson, C. H., "A Marine Gas-Turbine Installation," *Trans.* SNAME, Vol. 54, 1946.

Kane, J. R., "The Speed of the SS *United States*," *Marine Technology*, Vol. 15, No. 2, April 1978.

King, R. W., *Naval Engineering and American Sea Power*, The Nautical & Aviation Publishing Company of America, Baltimore, 1989.

Korakianitis, T. P. and Wilson, D. G., "Low Pressure-Ratio, Regenerative, Brayton-Cycle Engines: The Next Generation of Marine Prime Movers?" The Institute of Marine Engineers, Vol. 99, Paper 1, 1986.

Mac Millan, D. C. and Rohde, E. C., "Improved Steam Propulsion Plant to Reduce Building and Operating Costs," *Trans.* SNAME, Vol. 70, 1962.

McMullen, J. J., "A Gas-Turbine Plant with Reference to a Cargo Vessel," *Trans.* SNAME, Vol. 59, 1951.

Miller, R. T., et al., "ASW Surface Ship of the 80's Study," *Naval Engineers Journal*, Dec. 1972.

Orbeck, F., "Development of the Doxford Engine from 1960," Institute of Marine Engineers, Vol. 102.

Phillips, D. G., "U.S.S. *Timmerman*—An Advanced Design Destroyer," The Institute of Marine Engineers Guild of Benevolence.

Plumb, C. M., "Warship Propulsion System Selection," Marine Management (Holdings) Ltd. for The Institute of Marine Engineers, 1987.

Pluymert, N. J., "Modern Tanker Design," *Trans.* SNAME, Vol. 47, 1939.

Polmar, N., *The Ships and Aircraft of the U.S. Fleet,* 14th ed., Naval Institute Press, Annapolis, MD, 1987.

Pounder, C. C., "Human Problems in Marine Engineering," *Trans.* SNAME, Vol. 67, 1959.

Powell, S. C., "Estimation of Machinery Weights," *Trans.* SNAME, Vol. 66, 1958.
Robinson, H. F., Roeske, J. F., and Thaeler, A. S., "Modern Tankers," *Trans.* SNAME, Vol. 56, 1948.
Robinson, H. F. and Worthen, E. P., "The Ore Carrier S.S. *Venore*" *Trans.* SNAME, Vol. 53, 1945.
Sawyer, W. T., "The Marine Gas-Turbine Plant in 1951," *Trans.* SNAME, Vol. 59, 1951.
Sonenshein, N., "What's New in Navy Petroleum," Address given before American Society of Naval Engineers Flagship Section, March 25, 1969.
Taggart, R., *Marine Propulsion: Principles & Evolution,* Gulf Publishing Company, Houston, 1969.
Thompson, R. V., "To Sink or Swim—Some Thoughts on Marine Transportation and Its Future," Institute of Marine Engineers, Vol. 103, Part 1, Presidential Address, 1991.
Ulrich, G. D., "How To Calculate Utility Costs," *Chemical Engineering,* Feb. 1992.
Wilson, D. G., "Turbomachinery—From Paddle Wheels to Turbojets," *Mechanical Engineering,* Oct. 1982.
Young, R. T., "The Lessons of the Liberties," Address given before The Propeller Club of the United States, New York, May 8, 1974.

DAVID O'NEIL is a graduate of the U.S. Merchant Marine Academy, with additional studies at such institutions as MIT, The Wharton School, and Navy training centers. He has seen shipboard duty in both the Navy and merchant marine, and has worked for the Bureau of Ships, Military Sea Transportation Service, and Pratt & Whitney. He is a member of ABS, SNAME (former vice president), and IME. He holds a SNAME Distinguished Achievement Award. He is currently president of Seaworthy Systems, Inc.

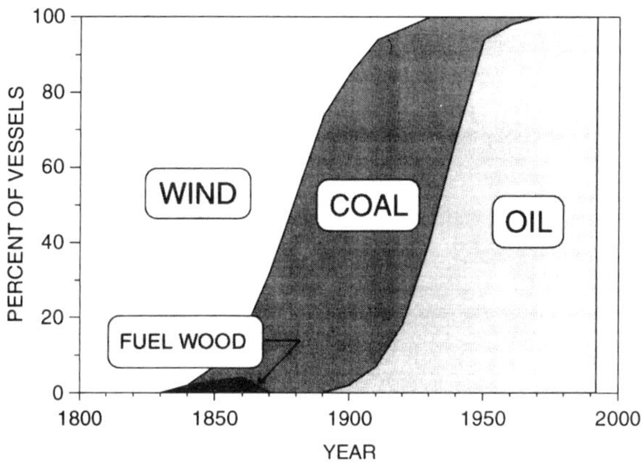

Fig. 3 Adoption of oil from coal and wind.

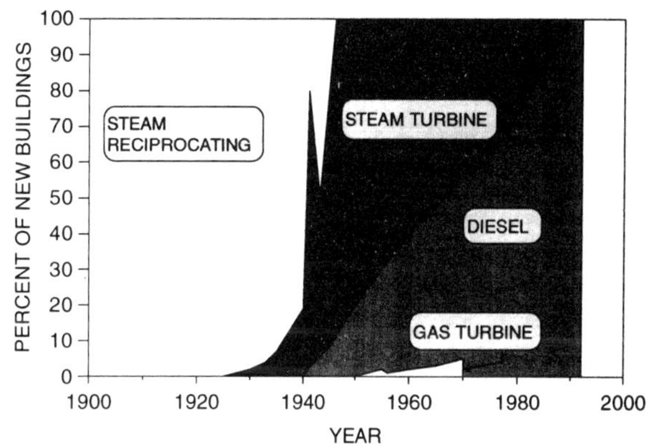

Fig. 4 Evolution of power plant types (large U.S. commercial vessels.)

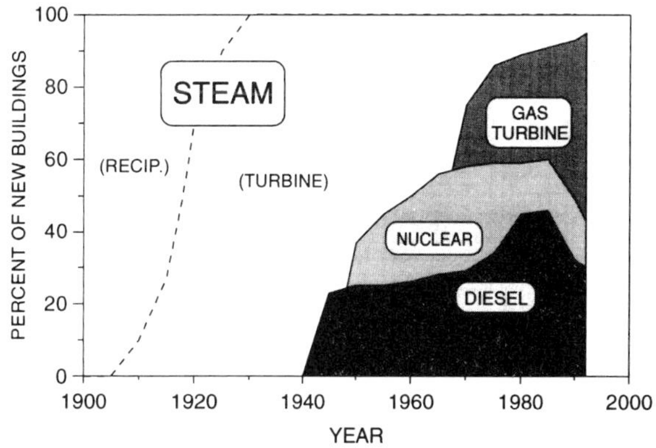

Fig. 5 Evolution of power plant types (large U.S. military vessels).

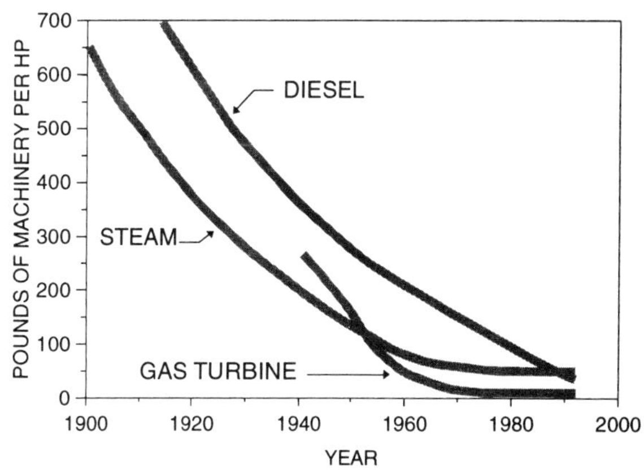

Fig. 6 Machinery weight reduction.

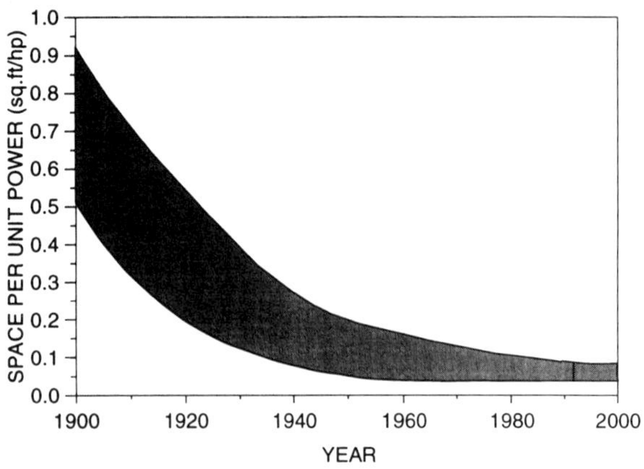

Fig. 7 Machinery space demand trend.

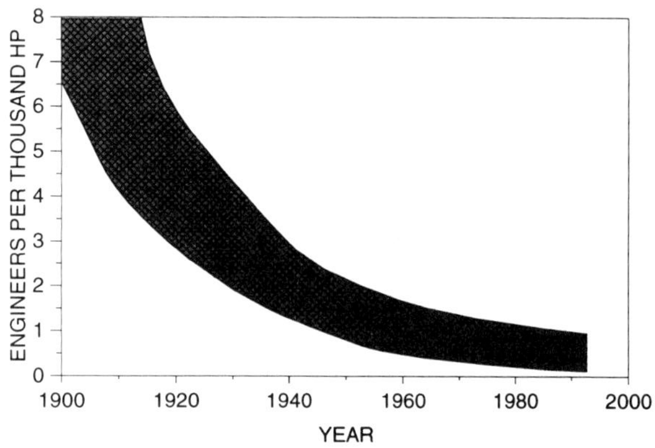

Fig. 8 Variation in engineering complement (commercial).

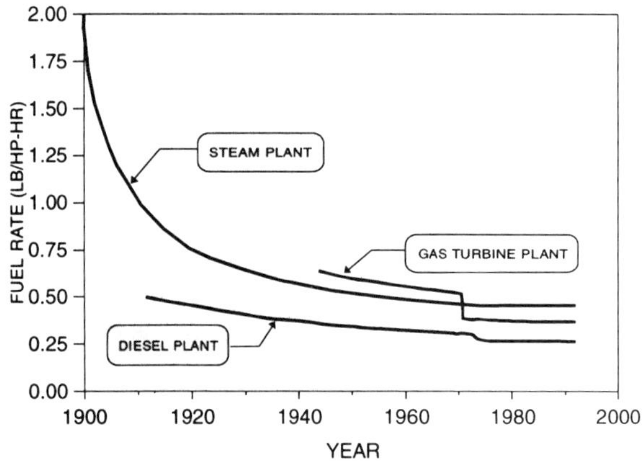

Fig. 9 Full-power fuel oil rate trends.

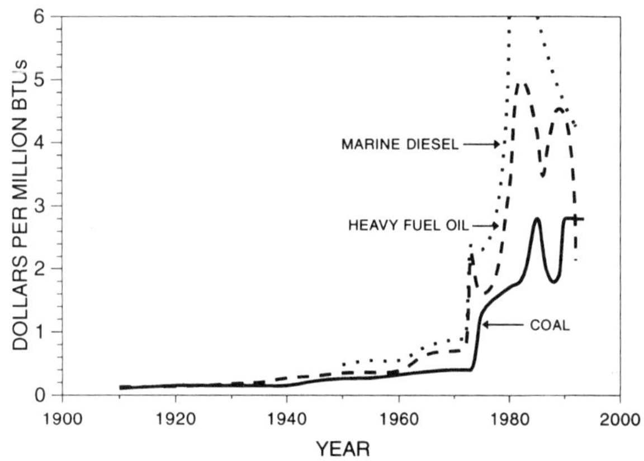

Fig. 10 Prices of marine fuels.

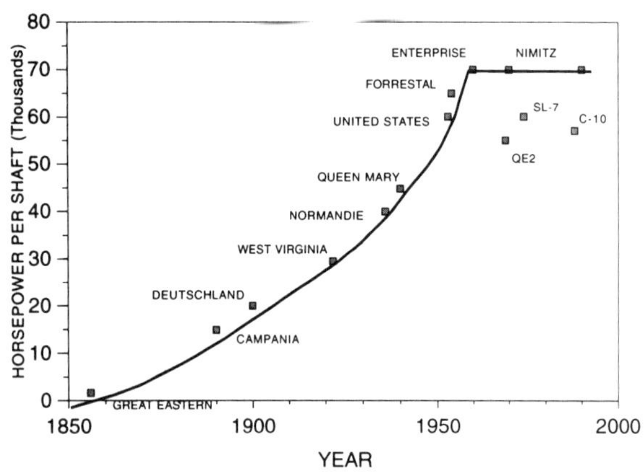

Fig. 11 Increase in power per shaft.

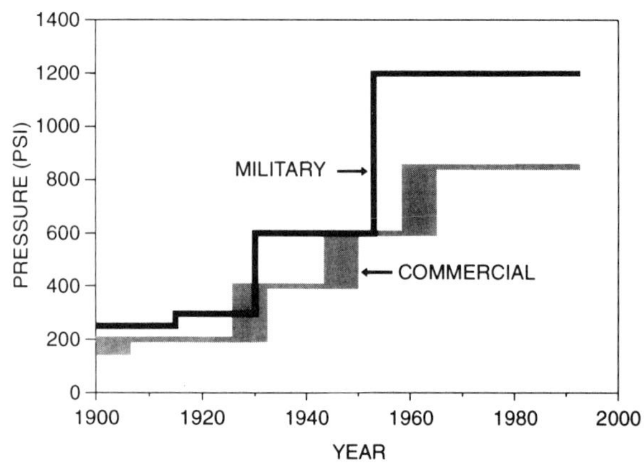

Fig. 12 Superheater outlet pressures.

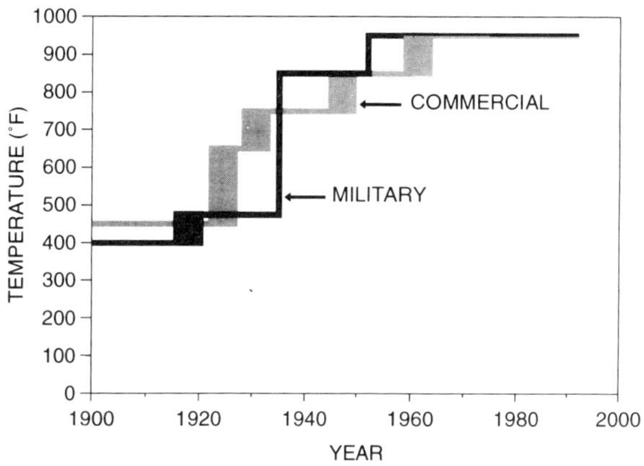

Fig. 13 Superheater outlet temperatures.

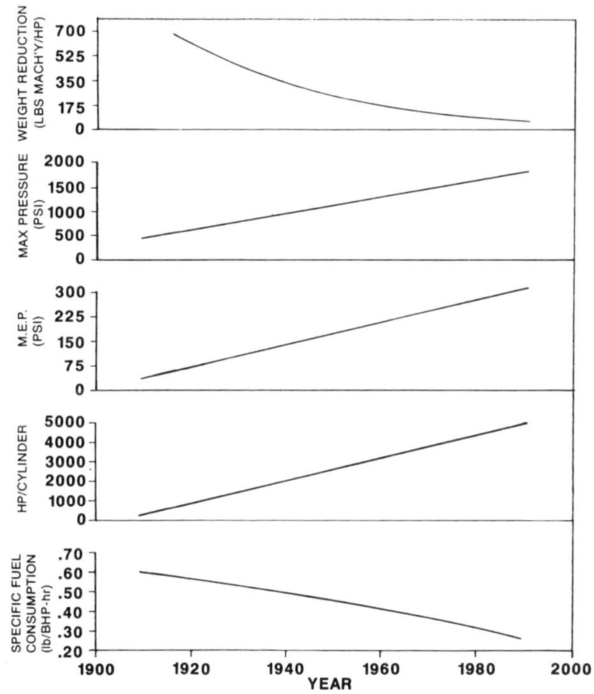

Fig. 14 Performance changes in 2-stroke diesels.

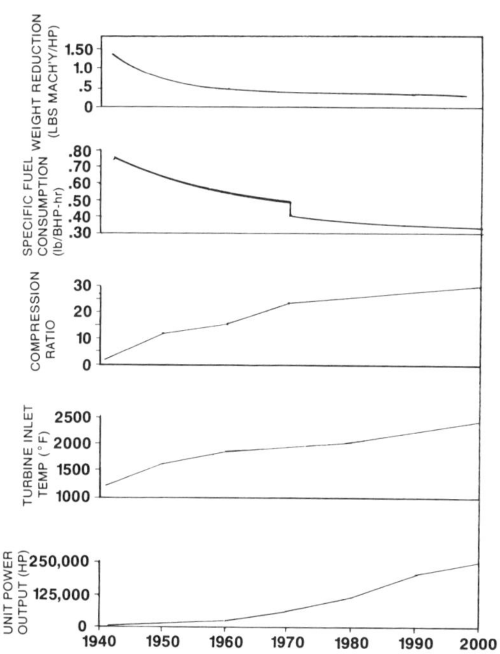

Fig. 15 Performance changes in gas turbines.

Propellers

Donald E. Ridley

Introduction: Terminology

When our Society was founded, the use of the term *propeller* was analogous to our current usage of the term *propulsor*. Barnaby [1] recognized four types of propellers as "embodying some intelligent idea." These were the paddle wheel, the screw propeller, the hydraulic propeller or pump-jet (waterjet), and the screw turbine. The latter was described as a shrouded propeller with the hub configured to provide acceleration and fitted with swirl recovery vanes. Only two, the paddle wheel and screw propeller, were of practical importance, with the latter having supplanted the former for most new construction. Since the screw propeller was dominant over the first century of our Society, as evidenced by the use of the term *propeller* (alternatively *screw*) to mean screw propeller, this chapter will deal almost entirely with it, and will refer to it as simply a *propeller*.

Early History

Three aspects of a momentum theory for propellers had been propounded before 1893. These were by Rankine (1865), Greenhill (1888), and R. E. Froude (1889). While momentum theory can match diameter with rate of revolution, it does not address the required blade section shape. The principal method of design was by extrapolation from the design and known performance of an existing similar propeller/ship combination. This procedure was complicated by the absence of reliable wake estimates and lack of any reliable means for measuring shaft torque. The favored blade section was ogival.

Thornycroft made open water tests of propellers mounted forward of a steam launch (1879–1880). Robert E. Froude extended this work through a limited series of open water propeller tests at Torquay (1886). These results were used by David W. Taylor in a design method that he published in the year of the founding of the Society [2].

Merchant ship propellers were almost always made from cast iron, naval propellers from bronze. Several companies held patents on special alloys used for propellers. Built-up propellers were at first most common because foundry equipment was of limited size. Detachable blades, moreover, lessened the expense of repairing damages and allowed a modest degree of pitch adjustment. Most of the blades were used as cast without benefit of machining [3].

Such was the state of the art when our Society was founded.

The First 50 Years (1893–1943)

David Watson Taylor authored his first publication in the year that the Society was founded, titled *Resistance of Ships and Screw Propulsion* [2]. It was the precursor to his classic treatise, *The Speed and Power of Ships* [4]. In 1935 SNAME established its first award, which it named in Taylor's honor; and he was the first recipient [5–8].

Since the majority of ships used reciprocating steam engines, the measure of power was indicated horsepower (IHP) as determined by engine indicator cards, and propulsive efficiency was defined as the ratio of EHP (effective horsepower) to IHP.

Much time and expense was incurred in conducting progressive trials in an effort to better determine the basis for extrapolating the propeller design from one ship to an appropriate design for a similar ship. Early SNAME *Transactions* carried many papers dealing with these and other problems in propeller design [9–14].

Hewins and Curtis [15] reported results of systematic model tests. Stevens [16] compared the results of earlier publications to achieve some degree of correlation. Troost published the first of several series of systematic model tests, originally known as the "Troost B Series," in 1938 [17]. These were augmented in 1940 [18].

Taylor [14] expounded on the law of comparison between model and full scale propellers in 1906. In 1913 Luders [19] authored the first paper devoted to a systematic attempt to correlate full scale trial results for an experimental boat using propellers designed from model tests with performance predictions based thereon.

Efforts to produce an analytical method of propeller design were summed up by Schoenherr [20] as follows:

> It was not, however, until Grammel (1916) and Foettinger (1918) had introduced an entirely new viewpoint, by taking account of the vortices in the slip stream of a propeller, that the ground was prepared for a satisfactory propeller theory. This was finally developed in the next ten years by Betz, Prandtl, Helmhold, Pistolesi, Bienen, von Karmen, Glauert, Kucharski, and others.

Schoenherr's excellent paper then goes on to put forth a design method based on vortex or circulation theory. This work was paralleled by Lerbs [21]. Goldstein had published an earlier paper [22] that proposed a somewhat more elegant solution, but was cumbersome to apply. No additional work on propeller theory was published in the *Transactions* during this period.

With the advent of steam turbines, torsion meters came into use. Peabody [23,24] reported on tests with thrust measuring devices. Attention was now directed toward consideration of the propeller in its operating environment behind the hull. Some rather crude assumptions were in use, however, such as allowing ten percent for shaft bearing and stuffing box losses. The latter was not rectified until the early thirties [25,26]. These developments brought an awareness that losses in efficiency were being caused by blade inaccuracies.

Both Peabody [24] and Saunders [25] reported loss of propeller performance from improperly formed blade backs. The problem of proper contour of the blade backs was partially solved in the early forties when blade gages were introduced.

During this period model basins started to make significant contributions to propeller design through wake measurements and similar tests. The U.S. Navy completed the first model basin in this country in 1898. This was followed in 1904 by the model basin at the University of Michigan [27]. In due course facilities were constructed at the Stevens Institute of Technology, MIT, the University of California (Berkeley), and others.

Parsons built the first cavitation tunnel in 1897. He followed this with an improved and larger version in 1910. Similar facilities were constructed in Germany and France in the mid-1920s, while the first cavitation tunnel in the United States was built by the Navy in 1928 [28]. New units were placed in service at MIT in 1938 [29], and at Carderock in 1940 [30–32].

Propeller materials evolved slowly during this period, as did regulatory requirements. The American Bureau of Shipping (ABS) introduced its first propeller requirements in the 1922 edition of the *Rules*. Cast iron, semi-steel, cast steel and bronze were listed as acceptable materials, although bronze was recommended. The only other material requirement concerned the studs for securing detachable blades, which were to be made from "Grade 1 steel (Section 40, par. 11) or equally satisfactory material. No test is required unless the Surveyor has reason to question the quality of the material." Except for the addition of a requirement to test the stud material, the *Rules* remained essentially unchanged through 1962.

In the early 1930s the Philadelphia Navy Yard was designated as the Navy's propeller manufacturing yard. From 1940 through 1945 the yard produced more than 5500 propellers, mostly of manganese bronze. They ranged in weight from 75 lb (for LCTs) to 43,000 lb (for battleships).

Solid propellers, and the blades for built-up propellers were molded in loam or green sand until the end of this period, when sand/cement came into use in some foundries. This molding medium is discussed by Towne [33] in his 1943 paper on propeller manufacture, the only paper devoted to this subject that was presented to the Society in its first half century. By the end of this period solid propellers dominated. Langham [34] gives the following distribution of manufacture:

	Percentage by Weight	
Period	Solid Propellers	Built-Up Propellers
1891–1910	44.3	55.7
1911–1930	62.8	37.2
1931–1940	80.0	20.0

Controllable pitch propellers, which came into use in the mid thirties for other than very small propellers, were not yet statistically significant.

By 1943 paddle wheels had all but disappeared from the scene, except for a few shallow draft applications, and the numbers of these were being diminished by the increasing use of nozzles [35]. In 1925 Kort revived and improved on the mid 19th century concept of a propeller in a tube. In 1934 he published a paper on nozzles [36] which led to their application and the general use of the generic term *Kort nozzle*.

Two types of vertical-axis propellers were put forth: Kirsten–Boeing in 1928 [37], and Voith–Schneider in

1931 [38]. The latter became popular in Europe as propulsors for harbor craft. Other forms of propulsors awaited development in the second half of SNAME's first century.

The Second 50 Years (1943–1993)

As the second half of the Society's first century began, the Allies had not quite turned the tide against the Axis countries, but this was soon to come as was the end of the World War II. There is little doubt that the advent of the digital computer within the first decade of this period was the period's most significant event. It made possible the practical application of propeller theories that would otherwise have been far too cumbersome, if not impossible, to use in propeller design. Moreover, when applied to machine tools it led to the production of accurate propeller blade surfaces. It also made possible the use of finite element techniques for blade stress analysis. The increasing role of propeller technology in the affairs of the Society is testified to by the number of symposia conducted on this subject, and now held on a regular basis. The increasing number of papers devoted to both propeller theory and materials provides further evidence.

In 1951 Troost [39] published the third set of data for the "B Series." Cross fairing of the four- and five-blade series was published by van Lammeren et al [40] in 1969. The approach used in the cross-fairing was later modified, and two revised sets of charts were published by Oosterveld et al [41] in 1975. Data for a number of other systematic series have been published, including a wide blade series by Gawn and Burrill [42], then extended by Burrill and Emmerson [43] in 1963.

Application of propeller lifting line theory for steady state conditions became practicable in 1952 based on Lerbs's work [44], and could be used for wake-adapted propellers with arbitrary load distribution. Morgan [45] reported the first application of this theory by means of the digital computer to contra-rotating propellers in 1955, and to single propellers in 1956. The shortfall of lifting line theory is its inability to account for the effect of blade thickness. This can be accounted for with lifting surface theory, first addressed by Ludwieg and Ginzel [46] in 1944, reduced to practice by Kerwin [47] and Pien [48] who independently developed computerized lifting surface programs in 1961. Neither of these programs included the induced velocity field owing to blade thickness. This was accounted for by Kerwin and Leopold [49] in 1963. Brockett [50] presented full boundary conditions in a lifting surface model in 1972. Brown [51] produced a lifting line program for unsteady flow in 1964, while Tsakonas et al [52–54] led in the original development of a rigorous three-dimensional lifting surface theory for the unsteady case during the period of 1968 to 1976.

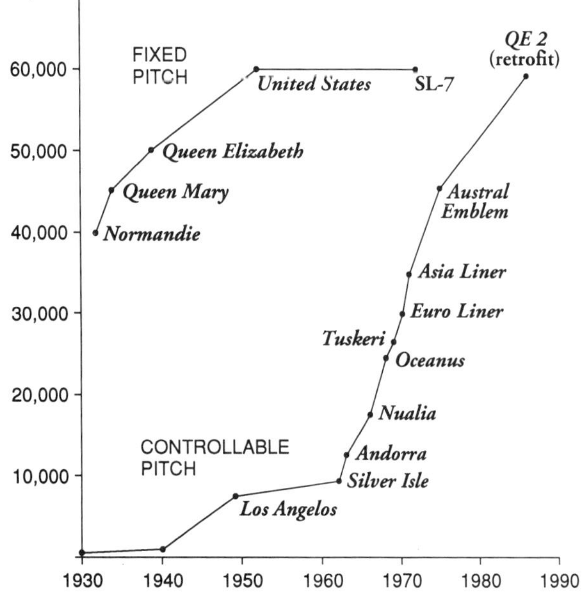

Fig. 1 Propeller horsepower trends for fixed- and controllable-pitch propellers on commercial vessels.

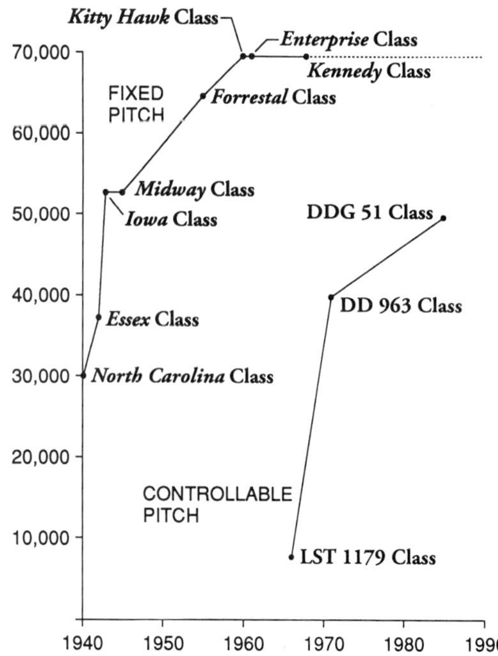

Fig. 2 Propeller horsepower trends for fixed- and controllable-pitch propellers on U.S. naval vessels.

During this time ship size, speed, and thus power density at the propeller were increasing dramatically. The increase in power for commercial ships is shown in Fig. 1, and for U.S. Navy ships in Fig. 2. This led quite naturally to increasing problems with propeller induced hull vibrations. Much of the early work on this subject was done by Breslin [55–57] from 1958 through 1970 and together with Tsakonas and Jacobs [58] in 1962. A method of estimating propeller induced hull vibrations was authored by Vorus [59] in 1974, and in 1975 SNAME issued T&R Bulletin No. 1-34 [60] on this subject.

Blade skew is one means of minimizing the blade/wake generated pressure pulses. When the term *skew* is used without a modifier it means "skew back," that is the blade tip is behind the center line of the blade in reference to the direction of rotation. While the majority of propeller blades were designed with some skew (normally 6 to 10 degrees), serious consideration of the use of high skew started in the late 1960s, in both the United States and in Europe. A comparison of a conventional propeller to one with high skew is shown in Fig. 3. Some early work was performed by Boswell [61,62], and design considerations examined by Cumming et al [63]. Unless compensated for, skew tends to increase the load at the tip of the blade. Conversely, forward skew tends to unload the tip—see Brown and Norton [64]—and to cause leading edge sheet cavitation to roll up, thus resembling a tip vortex. While it was originally believed that no unusual stress conditions would be found [65], experience showed this not to be true, particularly when backing [66]. Adequate methods are now available to permit design of highly skewed propellers with acceptable levels of stress [67]. Application of skewed propeller blades to controllable-pitch propellers avoids the problem of high backing stresses, but limits the amount of skew that can be used owing to the overturning moment generated by centrifugal force. This has been overcome by the use of balanced skew, that is sweeping the leading edge forward at the lower radii and the back at the outer radii, usually in combination with forward rake of the blades. This approach was conceived by Boswell, and developed together with several others at the David Taylor Research Center. The first published work was by Denny et al [68]. A Bird-Johnson Co. controllable-pitch propeller embodying this blade design is shown in Fig. 4.

A lifting-surface program that would predict unsteady propeller forces was essential to successful design, particularly in light of the propeller-induced hull vibrations cited above. A series of widely accepted programs was originally developed at MIT. The first, published in 1978 by Kerwin and Lee [69], was known as PUF-2 (Propeller Unsteady Forces, Version 2). This was followed in 1982 by PUF-3 [70], a program that incorporated the propeller-induced hull pressure work of Breslin. PUF-3A [71] released in 1987 incorporated the analytical work of Wilson at DTRC. The accuracy of any unsteady force prediction is, however, limited by the accuracy of the full scale wake values used. Even though much work has been done in this field, there are still limitations, as spelled out in detail by Blake et al [67].

Improvements in stress analysis capability kept pace with improvements in hydrodynamic analysis and design techniques. Schoenherr [72] codified the then current beam stress model in 1963. This was followed by a shift to finite element calculations, a direct result of the digital computer. At the Propellers '75 Symposium, five papers based on this method were

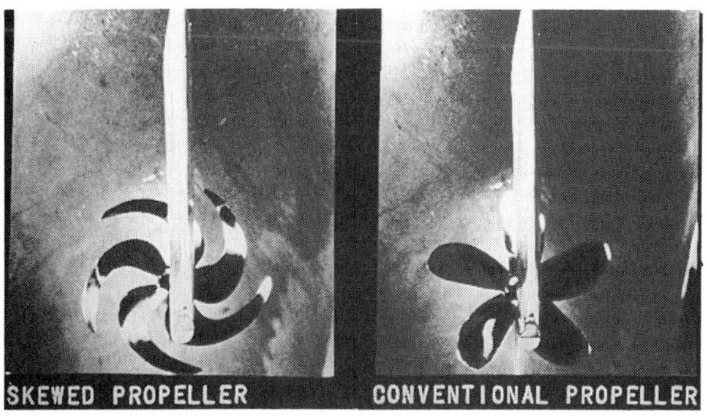

Fig. 3 Comparison of skewed and conventional propellers on the San Clemente-class ore-bulk-oil carrier. (Credit: DTRC.)

Fig. 4 Controllable-pitch propeller on the *Doyle* (FFG 39) at Bath Iron Works. (Credit: B-JCO.)

presented [73–77]. Single papers were presented at Propellers '78 [78], and Propellers '84 [79].

Model testing facilities proliferated, both towing basins and cavitation tunnels. Improvements in the latter tended toward larger test sections, thus permitting larger propeller models, and increased water speed. Some, including several built at Penn State University, were designed for specific types of investigations. As cavitation tunnels were originally used, only open water characteristics were obtained. For wake adapted propellers it became desirable to simulate the wake in some fashion. This was first accomplished by the introduction of wake screens upstream of the propeller to simulate the axial wake. In some facilities it was also possible to cant the tunnel drive shaft to duplicate the shaft rake. As time passed and the power density at the propeller increased, it became necessary to simulate the radial and tangential components as well; and to this end abbreviated hull models, that is full after body with a truncated forebody, were used. This was made necessary by the restrictions of the size of the test sections, and could also lead to a decreased diameter of model propeller. While the use of shortened hull models was an improvement, it did not simulate the wake to the degree of accuracy required for the study of hull pressure fluctuations induced by the propeller. To overcome this, super-size cavitation channels have been constructed, most notably the large tunnel at SSPA in Gothenberg, Sweden, one at the Hamburg Ship Model Tank, and the Large Cavitation Channel for DTRC at Memphis TN. The need for such facilities was addressed by Blake et al [67]. An alternative approach is MARIN's unique variable pressure towing tank. A list of the facilities available, together with principal characteristics is maintained in the International Towing Tank Conference *Catalog of Facilities* [80].

Early in this period the use of coatings to inhibit cavitation erosion of propeller blades was studied. Many forms were tried, ranging from paints to hard surfacing, to resilient coatings such as rubber. Insufficient adhesion defeated most coatings, while potential differences that could lead to corrosion removed hard surfacing from consideration. The deterioration of propeller performance owing to roughening in service became a concern as fuel prices increased [81]. This led to some operators undertaking periodic underwater cleaning of propellers [82].

Materials used for propellers have been the subject of considerable study. In 1963 ABS modified the list of acceptable materials for the first time since the adoption of rules for propellers in 1922. Cast steel and semi-steel were deleted from the list, and nickel-manganese bronze, nickel-aluminum bronze, and manganese-nickel-aluminum bronze added. Some failures had occurred in manganese bronze propellers from a corrosion phenomenon known as dezincification. It had been determined that this could be precluded by proper control of the alloy balance. To this end, the 1963 issue of the *Rules* also invoked a zinc equivalent requirement for the manganese and nickel-manganese bronzes. Dezincification is a selective attack at the grain boundaries, resulting in stress-raising cracks and ultimate fatigue failure. A similar circumstance occurs in either austenitic or martensitic stainless steel; as grain boundary attack from carbon pickup in austenitic (18 percent chrome, 8 percent nickel) varieties, and as oxygen concentration cells in martensitic (13 percent chrome) stainless steel.

The U.S. Navy also started to use nickel-aluminum bronze in the early 1960s. Also at this time some attempts were made to use cast titanium for propeller blades in view of its high strength to weight ratio and successful application in the forged form on the experimental hydrofoil craft *Dennison.* It soon became apparent, however, that in saltwater the cast form did not exhibit the material's normally excellent corrosion resistance. Fatigue failures followed.

It also became apparent that little was known about the corrosion fatigue properties of many of the other materials used for propellers. At the Propellers '75 Symposium, Czyryca and Niederberger [83] presented the results of tests on the three bronze alloys, and May [84] outlined a corrosion fatigue test program for four materials: manganese bronze, nickel aluminum bronze, manganese nickel aluminum bronze, and CA6NM stainless steel. Results for the three bronzes and partial data on CA6NM stainless steel were presented at Propellers '78 by Jaske et al [85]. Stainless steel of both the 13 percent chromium martensitic and 18 percent chromium–8 percent nickel-low carbon austenitic had been in use since the beginning of this period. These were approved by ABS on a case-by-case basis. In 1981 the low carbon austenitic variety, grade CF-3, was added to the acceptable propeller materials in the *Rules.* Nickel aluminum bronze remains the favored material for large propellers. Intermediate sized propellers are usually made from manganese bronze or stainless steel, while small propellers are generally of manganese bronze. Outboard motor propellers are produced mainly in aluminum.

Large electric induction furnaces came into use for melting propeller alloys early in the period. They are preferred for stainless steel, but are used in addition to reverberator furnaces for melting bronze. Air-set or no-bake molding materials were introduced in the 1960s. These were used successfully with bronze from their introduction; but it was not until the 1980s that satisfactory mold treatments had been developed to permit their use with stainless steel without carbon pick-up from the binder materials on the casting surface. Other improvements include various methods of increasing the static head on the metal as it enters the mold. In small propellers this took the form of vacuum casting, while in larger sizes a tilt mold process was used. Tilt molding is particularly applicable to single-blade castings.

Naval architect Stanley Potter of Beaufort, North Carolina, offered the following description of a propeller installation on the oceanographic research vessel Albatross IV, *designed and built for the Bureau of Commercial Fishing, U.S. Department of the Interior, in 1954 and operating out of Woods Hole, Massachusetts.*

The propeller was a Liaan design, controllable and reversible, 3-blade stainless steel, about 60 inches diameter, operating in a steerable Kort nozzle. The nozzle was shipyard-constructed to a Dutch design and proved to be successful in augmenting thrust but, as it was not flared at the exit, was thought not effective backing.

Propulsion and control seemed ideal during trials and it was only on delivery that it was found that flattening pitch to reduce way caused a complete loss of steering control. Tests in the MIT propeller tunnel by Professor Kerwin showed that the flow was choked by the flattened pitch and that the flow was around the outside of the nozzle in a turbulent fashion. The nozzle had a fin but only the small portion above the nozzle was clear of the turbulence.

A hydrofoil ring was fitted to correct the separation at the nose of the nozzle and restore some directed flow, thus providing some improvement for steering effect by the outer surface of the nozzle.

Tracer milling machines came into use for blade machining at the beginning of the period. The Philadelphia Naval Shipyard had several machines that worked both sides of a blade simultaneously. In the early 1970s numerically controlled milling machines of similar configuration were procured by Philadelphia. Roughing out of propeller blades on multi-axis numerically controlled milling machines began in Scandinavia in the mid 1960s. Finishing to final dimension of controllable-pitch propeller blades by numerically controlled machining began in the United States in the early 1970s. Photogrammetric means for lining out propeller castings, and for in situ measurement of propellers on ships in drydock was pioneered at this time. Theodolites coupled to computers now permit accurate inspection measurement obviating the need for gages, and may also be used for in situ inspection.

The phenomenon of a fully cavitating, or supercavitating, propeller was first observed by Parsons during his model experiments on propellers for the *Turbinia.* For many years the propellers of many small high speed craft have operated in this regime. It was not until some 50 years ago, however, that serious study of this subject was undertaken, attendant to interest in high speed craft such as hydrofoils; see for example Venning and Haberman [86]. A good historical overview of this subject is contained in *Principles of Naval Architecture,* Chapter 6, by van Manen and van Oossanen [87].

Controllable and reversible pitch (C&RP) propellers of significant size had been used in Europe since the mid 1930s, and many of smaller size were used by the U.S. Navy on landing craft during World War II. Rupp [88] summarized the state of the art in 1948. Commercial use became important in the 1950s, with the repowering of Great Lakes bulk carriers [89]. C&RP propellers were also used with diesel engines in both merchant and naval auxiliaries [90]. They also found use with gas turbines in surface combatants. Today we find them with diameters as great as eleven meters, and powers up to 45,000 kW. Power trends are shown in Figs. 1 and 2.

A number of special propeller configurations have been examined, all with a view to improving efficiency. Among these are included contrarotating propellers, vane-wheels, ring propellers, fixed vanes, tandem propellers, and overlapping propellers. Contrarotating propellers have been used on torpedoes for many years. In 1960 Morgan [91] published a lifting line method for designing these propellers. Application to large propellers and long shafts, however, creates complications in both the reduction gear and the internal shaft bearings required for this arrangement. Hadler [92] discussed the state of the art in 1969. The most recent application was presented by Nishiyama et al [93] in 1990, including detailed discussion of the shaft bearings, reduction gear and an emergency take-home feature. In 1983 Grimm [94] proposed a free wheeling, contrarotating vane-wheel for energy recovery from the propeller wake. The inner portion of the blades operate as a turbine, while the outer portion produces thrust from the energy absorbed by the turbine portion. The problems that have been encountered with this device are associated with the support bearings for the vane wheel.

A ring that mounts on the tips of the blades, either as a complete ring or individual segments on each blade (fins), has been tried. The purpose is to recover some of the energy normally lost in tip roll off. There is, of course, a friction drag attendant to the ring or fins. In 1970 van Gunsteren [95] proposed that this concept be used in combination with a pre-swirl stator. The use of stator vanes both ahead of and behind the propeller was first discussed by Wagner in 1929 [96]. Nothing came of this, however, until the use of preswirl vanes, or SISUP (Swirl Inducing Stator Upstream of the Propeller) was discussed by Mautner et al in 1988 [97]. This arrangement seems to hold the most promise. Hadler et al [98] tested several alternative propeller arrangements in 1964, including the use of tandem propellers. The object of the tandem arrangement is to divide the energy absorption between two equal size propellers where diameter restrictions would not permit use of the optimum diameter single propeller. It appeared from the test results, however, that little was gained as against the single propeller. Overlapping propellers were proposed by Pien and Strom Tejsen in 1967, and subsequently studied by Kerlen et al [99]. While significant improvement in efficiency was demonstrated, applications of this system have yet to be made.

Several modifications of the after body of full form hulls to provide an improved wake field at the propeller disk have been examined. These include a stern bulb, and cutaway deadwood to increase the aperture above the shaft line, Harvald [100]. These may be symmetrical or not, the latter also providing some pre-swirl. Pre-swirl can also be achieved by using reaction fins—Takekuma et al [101]. A form of ducted propeller design in which the duct is made part of the hull forward of the propeller was discussed by Narita et al [102], and was combined with a rudder ball by Dayuan and Hwaishu [103].

The use of a nozzle surrounding the propeller was made popular by Kort [104] in the early 1930s as a means of thrust augmentation for heavily loaded propellers. Much work has subsequently been done to refine the propeller and nozzle combination (see Morgan and Caster [105] for a comparison of theory and experiment). Kerwin et al [106] provide computer based design methods. Still more recently Hughes and Kinnas [107] examined the combination of nozzles and vanes as an efficient and quiet propulsor unit.

Vertical-axis or cycloidal propellers were discussed by Mueller [108] in 1955. It was noted earlier that the excellent maneuverability inherent in this design had led to many applications in harbor craft in Europe. Similar success has not been experienced in the United States, save for the application of significant size units in a class of ferries [109] in 1975, and most recently for the U.S. Navy's new mine-hunter class.

Water-jets offer an attractive propulsive means for high speed craft where propellers are subject to cavitation erosion. Since the pumps are located inside the hull, they also offer a means of propulsion for craft operating in shallow water where propellers would be subject to grounding damage. Some design considerations were given by Kim [110] in 1966, while Svensson and Bjorheden [111] discussed operating experience with a wide range of power in 1985.

Right-angle or Z-drive units have been in use since the early 1940s. Many of these are portable units mounted over the stern with a hinged retraction feature, thus resembling an oversize outboard arrangement. Wanzer [112] presented a paper in 1960 discussing these units.

Auxiliary maneuvering/propulsion devices came on the scene in the late 1950s. While a few of these have been water-jets, by far the majority have been propellers in transverse tunnels, or retractable right-angle drive units, the latter offering a means of auxiliary propulsion or controlled direction positioning. Perhsson and Mende [113] made the first presentation to a SNAME meeting in 1960 discussing tunnel thrusters. The history of thrusters and the state of the art in their application was given by Norrby and Ridley [114] in 1980.

Conclusion

As we come to the close of the first century of our Society's existence, we can see that tremendous strides have been made in our knowledge of, and our ability to design and apply, propellers. Yet, returning to Barnaby's four basic propeller types mentioned in the Introduction, most of the practical concepts were specified by him, i.e., screw propellers in their many forms (having already supplanted the paddle wheel prior to 1893), screw turbines (e.g., preswirl and contrarotation), and water-jets. Great as progress has been, much remains to be accomplished as we enter the second century, particularly in the area of cavitation/vibration/noise reduction and propulsor/hull interaction.

References

1 Barnaby, S. W., *Marine Propellers*, E. & F. N. Spon, London, 1891.
2 Taylor, D. W., *Resistance of Ships and Screw Propulsion*, MacMillan, New York, 1893.
3 Chase, I. and McKim, M. E., *Screw Propellers and Propulsion*, Wiley, New York, 1895.

4 Taylor, D. W., *The Speed and Power of Ships*, Wiley, New York, 1910.
5 Capps, W. L., "Rear Admiral Taylor's Contributions to Naval and Marine Progress," *Trans.* SNAME, Vol. 38, 1930.
6 Durand, W. F., "Rear Admiral Taylor's Achievements in Aeronautics—An Appreciation," *Trans.* SNAME, Vol. 38, 1930.
7 McFarland, W. M., "Career and Achievements of Admiral Taylor," *Trans.* SNAME, Vol. 38, 1930.
8 Hovgaard, W., "Biographical Memoir of David Watson Taylor, 1864–1940," National Academy of Sciences, Washington, D.C., 1941.
9 McDermott, G. F., "Screw Propellers," *Trans.* SNAME, Vol. 4, 1896.
10 McEntee, W., "The Limit of Propeller Efficiency," *Trans.* SNAME, Vol. 14, 1906.
11 Durand, W. F., "An Experimental Study of the Influence of Surface upon the Performance of Screw Propellers," *Trans.* SNAME, Vol. 5, 1897.
12 Durand, W. F., "Experimental Research on the Performance of Screw Propellers," *Trans.* SNAME, Vol. 13, 1905.
13 Taylor, D. W., "Some Recent Experiments at the U.S. Model Basin," *Trans.* SNAME, Vol. 12, 1904.
14 Taylor, D. W., "Model Basin Gleanings," *Trans.* SNAME, Vol. 14, 1906.
15 Hewins, L. F. and Curtis, A. V., "Some Results of Tests of Model Propellers," *Trans.* SNAME, Vol. 13, 1905.
16 Stevens, E. A., "A Comparison of Durand and Curtis & Hewins Propeller Experiments of 1905," *Trans.* SNAME, Vol. 14, 1906.
17 Troost, L., "Open Water Test Series with Model Propeller Forms," *Trans.* NECIES, 1938.
18 Troost, L., "Open Water Test Series with Model Propeller Forms," *Trans.* NECIES, 1940.
19 Luders, A. E., "Model Experiments and Speed Trials of 60-Foot Motor Cruiser *Kathmar II*," *Trans.* SNAME, Vol. 21, 1913.
20 Schoenherr, K. E., "Recent Developments in Propeller Design," *Trans.* SNAME, Vol. 42, 1934.
21 Lerbs, H. W., "Kerventafeln zur Berechnung starkbelasteter Freifahrtschrauben nach der Tragflugeltheorie," *Werft-Reederei-Hafen*, 1933, No. 3.
22 Goldstein, S., "On the Vortex Theory of Screw Propellers," *Proceedings of the Royal Society*, London, 1929.
23 Peabody, C. H., "Experiments on the Froude," *Trans.* SNAME, Vol. 19, 1911.
24 Peabody, C. H., "Experiments on the Fulton and the Froude," *Trans.* SNAME, Vol. 20, 1912.
25 Saunders, H. E., "Full Scale Trials on a Destroyer," *Trans.* SNAME, Vol. 41, 1933.
26 Kempf, G., "Measurements of the Propulsive and Structural Characteristics of Ships," *Trans.* SNAME, Vol. 40, 1932.
27 Sadler, H. C., "The Experimental Tank at the University of Michigan," *Trans.* SNAME, Vol. 14, 1906.
28 Saunders, H. E., "The Variable Pressure Water Tunnel of the U.S. Experimental Model Basin," *Trans.* SNAME, Vol. 38, 1930.
29 Lewis, F. M., "Propeller Testing Tunnel at the Massachusetts Institute of Technology," *Trans.* SNAME, Vol. 47, 1939.
30 Saunders, H. E., "The David W. Taylor Model Basin—Part 1," *Trans.* SNAME, Vol. 46, 1938.
31 Saunders, H. E., "The David W. Taylor Model Basin—Part 2," *Trans.* SNAME, Vol. 48, 1940.
32 Saunders, H. E., "The David W. Taylor Model Basin—Part 3," *Trans.* SNAME, Vol. 49, 1941.
33 Towne, Jr., S. H., "Propeller Manufacture," *Trans.* SNAME, Vol. 51, 1943.
34 Langham, J. M., "The Manufacture of Marine Propellers with Particular Reference to the Foundry," 21st International Foundry Congress, Florence, Sept. 1954.
35 Dawson, A. J., "Recent Towboat Developments with Particular Regard to Kort Nozzle Propulsion," *Trans.* SNAME, Vol. 50, 1942.
36 Kort, L., "Der neue Dusenschrauben-Antrieb," *Werft-Reederei-Hafen*, Jahrgang 15.
37 Kirsten, F. A., "A New Type of Propeller," *Journal of the Society of Automotive Engineers*, Jan. 1928.
38 Kreitner, H., "Die Hydraulischen Grundlagen des Voith-Schneider Antriebes," *Werft-Reederei-Hafen*, 1931.
39 Troost, L., "Open Water Test Series with Modern Propeller Forms," *Trans.* NECI, 1951.
40 van Lammeren, W. P. A., van Manen, J. D., and Oosterveld, M. W. C., "The Wageningen B-Screw Series," *Trans.* SNAME, Vol. 77, 1969.
41 Oosterveld, M. W. C. and van Oossanen, P., "Further Computer-Analyzed Data of the Wageningen B-Screw Series," *International Shipbuilding Progress*, Vol. 22, 1975.
42 Gawn, R. W. L. and Burrill, L. C., "Effect of Cavitation on the Performance of a Series of 16 in. Model Propellers," *Trans.* The Institute of Naval Architects, 1957.
43 Burrill, L. C. and Emmerson, A., "Propeller Cavitation: Further Tests on 16 in. Propeller Models in the Kings College Cavitation Tunnel," *Trans.* NECIES, Vol. 79, 1962–63.
44 Lerbs, H. W., "Moderately Loaded Propellers with Finite Number of Blades and an Arbitrary Distribution of Circulation," *Trans.* SNAME, Vol. 60, 1952.
45 Morgan, W. B., "The Marine Propeller," *ASME Journal of Fluids Engineering*, March 1975.
46 Ludwieg, H. and Ginzel, T., "On the Theory of Screws with Wide Blades," *Aerodyamische Versuchenstalt, Goetting*, Report 44/A/08, 1944.
47 Kerwin, J. E., "The Solution of Propeller Lifting-Surface Problems by Vortex Lattice Method," MIT Naval Architecture Department Report, Boston, 1961.

48 Pien, P. C., "The Calculation of Marine Propellers Based on Lifting Surface Theory," *Journal of Ship Research*, Vol. 5, No. 2, 1961.
49 Kerwin, J. E. and Leopold, R., "Propeller Incidence Correction Due to Blade Thickness," *Journal of Ship Research*, Vol. 7, No. 2, 1963.
50 Brockett, T., "Propeller Perturbation Problems," Carderock NSRDC Report 3880, 1972.
51 Brown, N. A., "Periodic Propeller Forces in Non-Uniform Flow," Naval Architecture Report 64-7, MIT, Boston, 1964.
52 Tsakonas, S., Jacobs, W. R., and Rank, Jr., P. H., "Unsteady Propeller Lifting–Surface Theory with Finite Number of Chordwise Modes," *Journal of Ship Research*, Vol. 12, No. 1, 1968.
53 Tsakonas, S., Jacobs, W. R., and Ali, M. R., "An 'Exact' Linear Lifting–Surface Theory for a Marine Propeller in Nonuniform Flow," *Journal of Ship Research*, Vol. 17, No. 4, 1973.
54 Tsakonas, S., Jacobs, W. R., and Ali, M. R., "Propeller Blade Pressure Distribution Due to Loading and Thickness Effects," Davidson Laboratory Report 76-1869, Stevens Institute of Technology, Hoboken, 1976.
55 Breslin, J. P., "The Pressure Field Near a Ship Propeller," *Journal of Ship Research*, March 1958.
56 Breslin, J. P., "A Theory for the Vibratory Effects Produced by a Propeller on a Large Plate," *Journal of Ship Research*, Dec. 1959.
57 Breslin, J. P., "Theoretical and Experimental Techniques for Practical Estimation of Propeller-Induced Vibratory Forces," *Trans.* SNAME, Vol. 78, 1970.
58 Tsakonas, S., Breslin, J. P., and Jacobs, W. R., "The Vibratory Force and Moment Produced by a Marine Propeller on a Long Rigid Strip," *Journal of Ship Research*, March 1962.
59 Vorus, W. S., "A Method for Analyzing the Propeller-Induced Vibratory Forces Acting on a Ship Stern," *Trans.* SNAME, Vol. 82, 1974.
60 "Techniques for Estimating Vibratory Forces Generated by Propellers," T&R No. 1-34, SNAME, New York, 1975.
61 Boswell, R. J., "Design, Cavitation Performance and Open-Water Performance of a Series of Research Skewed Propellers," NERDC Report 3339, 1971.
62 Boswell, R. J., "The Effect of Skew on Cavitation of Marine Propellers," ASME Cavitation Forum, New York, 1971.
63 Cumming, R. A., Morgan, W. B., and Boswell, R. J., "Highly Skewed Propellers," *Trans.* SNAME, Vol. 80, 1972.
64 Brown, N. A. and Norton, J. A., "Thruster Design for Acoustic Positioning Systems," *Marine Technology*, April 1975.
65 Boswell, R. J. and Cox, G. G., "Design and Model Evaluation of a Highly Skewed Propeller for a Cargo Ship," *Marine Technology*, Jan. 1974.
66 Luttmer, B. R. I., Hylarides, S., and Keller, J. A., "Effect of Skew on Stresses in Backing Fixed Pitch Propellers," SNAME Propellers '84 Symposium, 1984.
67 Blake, W. A., Meyne, K., Kerwin, J. E., and Frisch, J., "Design of APL C-10 Propeller with Full-Scale Measurements and Observations Under Service Conditions," *Trans.* SNAME, Vol. 98, 1990.
68 Denny, S. B., Themak, H. A., and Nelka, J. J., "Hydrodynamic Design for the Controllable-Pitch Propeller for the Guided Missile Frigate," *Naval Engineers Journal*, April 1975.
69 Kerwin, J. E. and Lee, C.-S., "Prediction of Steady and Unsteady Marine Propeller Performance by Numerical Lifting-Surface Theory," *Trans.* SNAME, Vol. 86, 1978.
70 Breslin, J. P., van Houton, R. J., Kerwin, J. E., and Johnson, C. A., "Theoretical and Experimental Propeller Induced Hull Pressure Arising from Intermittent Blade Cavitation Loading and Thickness," *Trans.* SNAME, Vol. 90, 1982.
71 Kerwin, J. E., Kinnas, S., Wilson, M. B., and McHugh, J., "Experimental and Analytical Techniques for the Study of Unsteady Sheet Cavitation," 16th Symposium on Naval Hydrodynamics, Berkeley, CA, 1986.
72 Schoenherr, K. E., "Formulation of Propeller Blade Strength," *Trans.* SNAME, Vol. 71, 1963.
73 Ma, J. H., Schnobrick, W. C., and Stuber, C. B., "Propeller Stress Calculation Using Curved Finite Elements," SNAME Propellers '75 Symposium, 1975.
74 Genalis, P., "Experimental Verification of Finite Element Analysis of Propeller Strength," SNAME Propellers '75 Symposium, 1975.
75 Westervelt, W. W., "Evaluation of a Finite Difference Helicoidal Shell Analysis by Comparison with Test Results for a Marine Propeller," SNAME Propellers '75 Symposium, 1975.
76 Atkinson, P., "A Practical Stress Analysis Procedure for Marine Propellers Using Curved Finite Elements," SNAME Propellers '75 Symposium, 1975.
77 Tsakonas, S. and Jacobs, W. R., "Steady and Time-Dependent Propeller Blade Loading and Stress Analysis," SNAME Propellers '75 Symposium, 1975.
78 Jourdain, M., Aucher, M., and Armand, J. L., "Strength of Propeller Blades—A Numerical Approach," SNAME Propellers '78 Symposium, 1978.
79 Sunnersjo, S., "Stress Analysis for Highly Skewed Propellers Using the Finite Element Method," SNAME Propellers '84 Symposium, 1984.
80 "ITTC Catalog of Facilities," International Towing Tank Conference.
81 Townsin, R. L., Spencer, D. S., Mosaad, M., and Patience, G., "Rough Propeller Penalties," *Trans.* SNAME, Vol. 93, 1985.

82 Wilkinson, C. P., "Reductions in Fuel Consumption as a Result of In-Water Propeller Polishing," SNAME Propellers '88 Symposium, 1988.
83 Czyryca, E. J. and Niederberger, R. B., "Mechanical, Fatigue and Corrosion Properties of Propeller Bronzes," SNAME Propellers '75 Symposium, 1975.
84 May, T. P., "Corrosion Fatigue Testing of Propeller Alloys," SNAME Propellers '75 Symposium, 1975.
85 Jaske, C. E., Utah, D. A., and Boyd, W. K., "Corrosion Fatigue of Cast Propeller Alloys," SNAME Propellers '78 Symposium, 1978.
86 Venning, E. and Haberman, W. L., "Supercavitating Propeller Performance," *Trans.* SNAME, Vol. 70, 1962.
87 *Principles of Naval Architecture*, E. V. Lewis, Ed., SNAME, 1988.
88 Rupp, L., "Controllable Pitch Propellers," *Trans.* SNAME, Vol. 56, 1948.
89 Kendall, H., "Controllable Pitch Propellers on Great Lakes Bulk Carriers—The Sensibar to the Barker," Paper presented at SNAME Great Lakes/Great Rivers Section, Cleveland, Oct. 1976.
90 Ridley, D. E., Midttun, O. H., "Economic Considerations of Controllable Pitch Propellers with Diesel Main Propulsion," *Naval Engineers Journal*, April 1970.
91 Morgan, W. B., "The Design of Counterrotating Propellers Using Lerb's Theory," *Trans.* SNAME, Vol. 68, 1960.
92 Hadler, J. B., "Contrarotating Propeller Propulsion—A State-of the-Art Report," *Marine Technology*, July 1969.
93 Nishiyama, S., Sacamoto, Y., Ishida, S., Fugino, R., and Oshima, M., "Development of Contra-Rotating Propeller Systems for *Juno*—A 37,000 dwt Class Bulk Carrier," *Trans.* SNAME, Vol. 98, 1990.
94 Grim, O., "Propeller and Vane Wheel, Contrarotating Propeller," 8th West European Graduate Education in Marine Technology, Gothenburg, 1983.
95 van Gunsteren, L. A., "Ring Propellers and Their Combination with a Stator," *Marine Technology*, Oct. 1970.
96 Wagner, R., "Ruckblick und Ausblick auf die Entwicklung des Contra-Propeller," *Jahrbuch des Schiffbautechnischan Gesellschaft*, 1929.
97 Mautner, T. S., Nelson, D. M., and Gillcrist, M. C., "Investigation of the SISUP (Swirl Inducing Stator Upstream of Propeller) Concept for Marine Propulsion," SNAME Propellers '88 Symposium, 1988.
98 Hadler, J. B., Morgan, W. B., and Meyers, K. A., "Advanced Propeller Propulsion for High Powered Single Screw Ships," *Trans.* SNAME, Vol. 72, 1964.
99 Kerlen, H., Esveldt, J., and Wereldsma, R., "Propulsion- , Kavitations- und Vibrationsverhalten von Ueberlappen Propeller für ein Containerschiff," *Jahrbuch der Schiffbautechnischen Gesellschaft*, 1972.
100 Harvald, S. A., *Wake of Merchant Ships*, Danish Technical Press, 1950.
101 Takekuma, K., Tsuda, S., Kawamura, A., and Kawaguchi, N. "Development of Reaction Fin as a Device for Improvement of Propulsive Performance of High Block Coefficient Ships," *Journal, JSNA*, Vol. 150, 1981.
102 Narita, H., Yagi, H., Johnson, H. D., and Breves, L. R., "Development and Full Scale Experience of a Novel Integrated Duct Propeller," *Trans.* SNAME, Vol. 89, 1981.
103 Dayuan, D. and Hwaishu, Y., "Energy Saving Device—A Research of Rudder-Ball and Wake Equalizing Duct," SNAME Propellers '91 Symposium, 1991.
104 Kort, L., "Der neue Dusenschrauben-Antrieb," *Werft-Reederei-Hafen*, Jahrgang 15, 1934.
105 Morgan, W. B. and Caster, E. B., "Comparison of Theory and Experiment on Ducted Propellers," 7th Symposium on Naval Hydrodynamics, Rome, 1968.
106 Kerwin, J. E., Kinnas, S. A., Lee, J.-T., and Shih, W.-Z., "A Surface Panel Method for the Hydrodynamic Analysis of Ducted Propellers," *Trans.* SNAME, Vol. 95, 1987.
107 Hughes, M. J. and Kinnas, S. A., "An Analysis Method for a Ducted Propeller with Pre-Swirl Stator Blades," SNAME Propellers '91 Symposium, 1991.
108 Mueller, H. F., "Recent Developments in the Design and Application of the Vertical Axis Propeller," *Trans.* SNAME, Vol. 63, 1955.
109 Ciechon, E. J. and Hairston, L. N., "Design and Construction of the New Staten Island Ferries," *Marine Technology*, Oct. 1979.
110 Kim, H. C., "Hydrodynamic Aspects of Internal Pump-Jet Propulsion," *Marine Technology*, Jan. 1966.
111 Svensson, R. and Bjorheden, O., "Experience with Jet Propulsion in Power Range 140 to 10,000 kW," SNAME, South East Section, Feb. 1985.
112 Wanzer, A. W., "Right-Angle Drive Applications," SNAME, New York Metropolitan Section, April 1960.
113 Pehrsson, L. and Mende, R. G., "Design, Model Testing and Application of Controllable Pitch Bow Thrusters," SNAME, New York Metropolitan Section, Sept. 1960.
114 Norrby, R. A. and Ridley, D. E., "Notes on Thrusters for Ship Maneuvering and Dynamic Positioning," *Trans.* SNAME, Vol. 88, 1980.

Donald E. Ridley's professional career has been devoted to propellers. He graduated from the NY State Maritime Academy in 1947. After shipboard service he continued his education at Syracuse and Northeastern Universities. He then worked for Gibbs & Cox for a period, followed by 31 years with Bird-Johnson Co., attaining the position of Senior Vice President before retiring in 1988. He remains active as a consultant specializing in propeller design. His many papers have been published by SNAME, ASNE, and ASME. He is an Honorary Vice President of SNAME.

Auxiliary Systems

ROBERT S. BEHR

Introduction

THIS CHAPTER deals with technological developments in shipboard auxiliary machinery during the past 50 years. The topic is too broad to cover in detail. I have accordingly chosen to concentrate on only a few of what seem to me to be the most important developments.

Background information for the years leading up to 1943 is provided in individual sections as appropriate.

Major Trends Over the Past 50 Years

Size and Efficiency. As ship size and main propulsion power increased over the past 50 years, the size and power requirements for auxiliary machinery increased proportionally. Improvements in main propulsion plant design were followed by improvements in auxiliary system and component design. Electrical plants grew in size from a few hundred kilowatts to megawatts. The increase in electrical plant size was a result of both increased auxiliary loads and the introduction of new systems, such as air conditioning, bow and stern thrusters, and large cargo refrigeration and cargo pumping systems. Increased auxiliary power requirements and dramatic increases in energy costs of the 1960s and 1970s led to efforts to improve the efficiency of auxiliaries. This was reflected in improved component efficiency and improved auxiliary system design.

Powering of Auxiliary Machinery. In early steam plants all auxiliaries were steam powered. By the 1940s the transition from steam to electric driven auxiliaries was nearly complete, with a few notable exceptions. Certain critical or high power auxiliaries such as feed pumps, cargo pumps, fire pumps, and forced draft blowers were still often powered by steam turbines. Steam power provided an independence from electrical power for critical auxiliaries and maintained a reasonably sized electric plant. For some critical services, both steam and motor driven units were provided because of concern for the reliability of electric power. This was particularly true on high value merchant or naval ships.

Even the reciprocating steam pump held out where high suction lifts were essential. Examples include the bilge and ballast system, cargo stripping, and fuel transfer. Such pumps were also retained where high dependability and ease of operation were essential, such as the standby boiler feed pump. Over time, however, the use of steam auxiliaries decreased as electrical installations became larger and more reliable.

The advent of diesel propulsion spelled the end of the steam driven auxiliary except for waste heat steam driven electrical generators, and on tankers where a complete auxiliary steam plant was installed for cargo pumping and heating.

Auxiliary Systems for Diesel Power Plants. By the mid-1970s diesel propulsion had virtually replaced steam in new U.S. merchant ships. The diesel plant generally required fewer, smaller, and simpler auxiliary systems. The entire thermodynamic power cycle took place within the engine itself, eliminating feed, condensate, auxiliary exhaust, and vacuum systems. This also reduced the size of the auxiliary steam and main cooling systems. Exceptions to this general simplification were the fuel and lubricating oil filtration and purification systems. Owing to the diesel engine's intolerance for contaminated fuel and lubricating oils, these systems increased in size and complexity.

Modularization and Packaging. The adoption of modular construction techniques by shipbuilders also had an impact on the design and layout of auxiliary systems. Pre-outfitting of machinery spaces resulted in a more orderly machinery layout process. Components serving the same system were grouped to allow interconnecting piping to be installed in ground assembly. Packaged, pre-piped modules manufactured in shops, or supplied by subcontractors, also became more popular. Packaged boilers, purifier modules, and distillers are examples of this technology.

Auxiliary Systems and Mechanical Components

Pumps and Compressors. Although the basic principle of the centrifugal pump has obviously not changed, improvements have been made in overall efficiency, suction characteristics, pump materials, shaft seals, and overall pump design. Flow inducers provide the suction boost necessary to allow higher suction lifts and improved design flexibility. Better materials allow pumping of corrosive and abrasive materials, and extend the life of pump components. Mechanical seals eliminate leaking glands, and decrease maintenance time and costs. Until the 1970s the close coupled pump was prohibited except in the smallest applications. Since then, however, the development of robust close coupled process pumps has simplified pump design for many applications. In addition to simplifying design, these pumps are often available off the shelf because of their extensive use in the chemical processing industry.

In a similar way, positive displacement pumps saw evolution, not revolution. The period saw improvements in efficiency, cost effectiveness, and increased heads and flows. As required flows increased with power plant size, the use of steam reciprocating pumps declined. As the reliability of vacuum pumps increased, they slowly replaced the main condenser's steam air ejector. This trend was further driven by increasing fuel costs and competition from diesel power plants.

The standard reciprocating air compressor is still the workhorse in the marine field. Diesel engines need compressors for high pressure starting air. As compressor capacity increased, water cooling superseded air cooling. Other types of compressors have come to replace the standard reciprocating units in some applications. Oil-less reciprocating compressors provide oil free air for sensitive control systems. Rotary positive displacement type compressors have replaced reciprocating units in some constant use systems. Although centrifugal compressors are used in many industries, marine installations seldom require the capacity to justify them.

Distillers. Although the first submerged tube, flash-type distillers were used by the Navy in the 1920s, the debate over the economics of installing low-pressure units continued into the 1940s. Until then, most ships had high-pressure evaporators and considerable potable water storage capacity. High-pressure evaporators used desuperheated steam to boil potable or seawater at a low positive pressure. The generated vapor was condensed by a separate condenser or added to auxiliary exhaust, where any excess was dumped to the main condenser. Boiling water at high temperatures caused rapid buildup of scale, which had to be removed from the steam coils by chill shocking with cold feed, or by manual cleaning. The process was extremely inefficient since steam made at over 212 °F was often condensed at seawater temperature with no work extracted.

The 1940s and 1950s saw the rapid acceptance of submerged tube, low-pressure distillers. These units were normally driven by extraction steam and considerably reduced the energy expended to make water. In addition, the reduction in the required capacity of potable water tanks increased cargo carrying capacity.

In the 1960s and 1970s the submerged tube distiller was slowly replaced by the flash-type distiller where all heating of the feed took place in heat exchangers before it was flashed to vapor in the distiller shell. This era also saw the introduction of distiller feedwater chemical treatment, which kept feed heater surfaces free of scale. Chemical treatment also greatly reduced the need to acid clean distillers.

With the advent of diesel propulsion and rising fuel costs, the waste heat flash type distiller came to the forefront. These distillers normally use diesel engine jacket water as a heat source for the distillation process. The first diesel waste heat distillers were developed by two Danish firms, Atlas and Nirex, in the late 1950s. Recent developments in distiller systems include the direct use of waste heat from starting air compressors, and greater use of multiple stages for better efficiency. New, corrosion resistant materials have also come into use.

The popularity of waste heat distillers also resulted in an increased use of potable water purification systems including chlorination, bromination and ultraviolet light. The saltwater feed temperatures available from diesel jacket water are generally too low to assure the complete destruction of microorganisms, thereby requiring this secondary purification.

The 1980s also saw increased interest in vapor compression and reverse-osmosis distillers. The all electric vapor compression cycle gained wider acceptance after units were retrofitted on the DD963 class and then made standard on the DDG51 class destroyers. Reverse-osmosis units are used extensively on offshore drilling units and on some ships for use when there is insufficient waste heat available. The Navy is currently investigating the feasibility of reverse-osmosis units.

Auxiliary Boilers and Contaminated Steam Generators. The sensitivity of modern high-pressure boilers to oil contamination led to installing contaminated steam generators on many high-value ships. The contaminated steam generator used live or, more efficiently, extraction steam to generate steam for oil heating services. Drains from these possible sources of oil contamination returned to the steam generator through a contaminated drain inspection and filter tank, cooler, and feed pump, thus totally segregating possible oil sources from the primary steam cycle.

On diesel ships, steam generators using waste heat from the engine exhaust were installed to serve domestic and oil heating requirements. On larger modern installations such waste heat provides steam to power electrical generators. Capacity is controlled by bypassing gas around the heat exchanger, by dumping excess steam to a condenser, or by controlling the feed rate to the generator.

Diesel ships are usually fitted with independently fired boilers to supplement exhaust gas steam generators and provide steam when the engines are secured. On ships requiring large quantities of steam for cargo pumping and heating, these saturated steam installations are often as large as would be required for propulsion by steam turbine. On ships requiring smaller amounts of steam, packaged auxiliary boilers are usually provided. For even smaller requirements, electrically heated systems are sometimes used.

Steering Gears. The earliest powered steering gears consisted of a steam engine connected to a manual drum, or screw type, steering mechanism. By the turn of the century, steam steering had gained sufficient acceptance that the manual steering system had been eliminated in larger ships. The steam engine drove the tiller directly in response to orders from a telemotor. The most widely used steering gear soon became the steam quadrant type.

Although there were early experiments with hydraulic steering gear, the development of the variable displacement hydraulic pump provided the impetus for electro-hydraulic steering gear as it is known today. The first such unit was installed on the battleship *New Mexico* in 1917. By the 1920s, electro-hydraulic steering had become predominant in the merchant fleet. By the early 1940s electro-hydraulic machines had all but replaced steam steering gears on large oceangoing ships. Notable exceptions were the quadrant steam steerers installed on Liberty ships and on a few Great Lakes ships into the 1970s.

Over the years, the electro-hydraulic steering gear evolved into a smaller, more efficient, and more reliable machine. Major changes included the introduction of new materials, improved hydraulic fluids, reversible flow pumps that could develop higher pressures at higher speeds, closed replenishment loops with replenishment pump integral to the main pump, pressure compensated controls, and improved control valves.

The main types of steering gear in use today are all electro-hydraulic. They include the link type, the Rapson slide, the rotary vane, and more recently, the clevis mounted oscillating cylinder.

Along with changes in the steering gear itself, improved pump control and wheel house control produced more functional and reliable steering systems. A typical early control system was the floating lever or hunting gear. The floating lever was a simple differential positioning system in which the steam valve opening for the pump stroke was directly proportional to the difference between the helm order and the actual rudder position. Floating lever control in many forms was used on both steam and electro-hydraulic units. The system is still used today by some foreign manufacturers and on shallow draft vessels and tow boats.

Most U.S. manufacturers abandoned the floating lever several decades ago in favor of differential gear control. The differential gear is a motion storage mechanism, which allows the helmsman to rotate the wheel as rapidly as desired while the steering gear drives the rudder at its rated speed. This feedback controlled system causes the hydraulic system to move the tiller in response to helm orders, and the movement of the tiller to stop the engine as the rudder angle approaches the helm order.

Although mechanical differential control has proved reliable and effective, it is likely to give way to electronic control systems in the near future. A means of electronically controlling pump stroke would eliminate the need for precision differential gearing and vulnerable follow-up linkages. As with any critical safety-related system, change will come slowly and only after extensive reliability testing.

Pilothouse communication with the steering controls in the steering gear room has also seen dramatic improvements. The hydraulic telemotor replaced the early chain, cable and shafting arrangements on larger, more complex systems before the turn of the century. Although the electrical telemotors were first used in the 1920s, hydraulic systems are still in use today after over a hundred years.

The major movement away from hydraulic telemotors was a result of the introduction of the automatic pilot. Although the earlier automatic pilots simply connected to the existing steering system via a clutchable chain drive, later units act directly on the controls in the steering gear room. These systems use synchronous motors to transmit helm orders directly to the steering gear. Modern autopilot systems use microprocessor control to produce an impressive array of features for ship handling.

A series of steering gear related accidents in the mid-1970s resulted in new regulations aimed at increasing reliability of steering systems. The worst of these accidents occurred on June 2, 1972 when the *Sea Witch* lost steering control in New York Harbor and collided with the anchored *Esso Brussels*, which was loaded with crude oil. The collision and subsequent fire claimed 16 lives and caused an estimated $23 million in damage. The second major accident was the grounding of the *Amoco Cadiz*, which resulted in a massive oil spill off the coast of France. Although the initial thrust of the new regulations was aimed at improving the reliability of tank ship control, later legislation addressed the improvement of all ocean-going ships.

The new regulations contained many requirements aimed at increasing the reliability of steering systems, including independent electrical feeders from the emergency switchboard, increased emergency or independent power for steering, pilot house controls and alarms, dual followup, rapid refill of system oil, and rapid means of isolating portions of the system.

Mooring Equipment. Mooring equipment has evolved over the years from steam to dc and ac electric to electrohydraulic drives. Each method provides various advantages and all have been used in recent years. Perhaps surprisingly, the earliest steam winches provided the first crude constant tension features. By simply leaving the steam admission valves cracked, the winch would haul in and pay out within limited tension ranges. With increased ship size and increased cargo handling rates, the need for effective automatic tensioning also increased. Unlike steam winches, electric motors cannot be operated in the stalled condition; other control mechanisms are required. Many of the early developments in winch control occurred on the Great Lakes, where the common practice of docking without tugs requires constant tension winches for stopping a ship alongside a pier. Such winches are also of great help in keeping a ship positioned during the rapid loading or unloading operations. The first electric constant tension winches on the Lakes were on the *Henry Ford II* and the *Benson Ford* in 1924. Early constant tension schemes included worm and planetary gears, and a sliding base arrangement where the entire winch moved on rails against centering springs in its base. Developments in electrical, electronic and hydraulic control technology have made possible the numerous effective designs of modern constant tension mooring winches.

Feedwater and Condensate Systems. In the 1930s and 1940s, propulsive efficiency increased as significant design improvements were made in the steam propulsion system. Increased attention was then directed to feed systems in order to match improved efficiency of the main plant. Watertube boilers of increased power density also required more precise feed level regulation. Variable speed, triplex, dc motor driven reciprocating pumps had become standard for the newer, more efficient power plants. These featured feed level regulating valves, differential pressure control, and recirculating control for low end pressure control. The development of the variable stroke triplex pump, however, offered new possibilities for improving efficiency while simplifying the feed system. Variable stroke pumps could change flow from zero to maximum as opposed to variable electric motor speed, which had limited turn down. The first C-1 cargo ships built by Federal for Lykes Brothers were fitted with one variable stroke triplex motor driven pump for each boiler. Since each pump served only one boiler, there was no need for a feed regulating valve. The level was adjusted directly by pump stroke. Eliminating the feed regulator and corresponding excess feed pressure, reduced required pumping power by five to ten percent.

As required feed pump horsepower increased, however, the high speed turbine driven pump increased in efficiency. By the 1940s, multistage pumps with independent turbine drives were replaced with single and double stage pumps, with turbines mounted on a common shaft. By the 1950s the turbine driven pump had displaced the motor driven triplex plunger pump as the standard.

Early low pressure plants used open feedwater systems. In the open system, wet air pumps took suction from the main condenser and discharged to an open feed and filter tank, which had little effect as a deaerator. These systems were viable only for low pressure, low circulation boilers with heavy wall tubes to allow for the inevitable oxygen induced corrosion. As boiler pressure, circulation, and efficiency increased, condensate and feed systems evolved from these open systems through semi-closed to a closed system with a deaerating feed heater. Ironically, the closed system was first used before the turn of the century but abandoned, apparently as an unnecessary complication.

The direct contact deaerating feedheater places feedwater in direct contact with heating steam, usually auxiliary exhaust supplemented by extraction steam. This is achieved either by spraying condensate through the steam or by allowing it to fall through the steam from multiple trays. The intimate contact of steam and water causes non-condensible gases to separate from the condensate. The gases are then vented from the deaerator, usually through a vent condenser. By the 1940s almost all steam plants were fitted with deaerators. The normal working pressure was 15 psig. As boiler efficiency rose and stack temperatures fell, this pressure was increased to 30–40 psig to prevent cold end corrosion in boiler economizers. The increased feed temperature resulting from this higher pressure raised the metal temperature in the cold end of the economizer above the dew point of the stack gas. This eliminated the formation of moisture and sulfuric acid on the external surface of the tubes. Other changes to condensate, particularly those relative to increased cycle efficiency are discussed in the chapter on powering.

Fuel and Lubricating Oil Systems. The most significant changes in oil systems over the past 50 years have occurred in fuel and lubricating oil filtering and treatment. Improved fuel treatment has been driven by the need for improved fuel efficiency, the introduction of diesel propulsion, and the general deterioration of fuel quality. Improved lube oil treatment is primarily required for diesel propulsion plants.

Steam plant systems have undergone the least change. In many cases a modern fuel and lubricating system is not significantly different from a system 50 years older. The significant changes have been driven by a quest for fuel efficiency and special requirements for fuel gas inerting systems. The major changes are the near universal adoption of high turndown steam atomized fuel burning systems, and the addition of fuel system automation. Automation and control systems, including features such as oxygen trim, are covered in the Automation section.

Although the technology of chemical and physical treatment of fuel has been vastly improved, there is no universal application in steam or diesel power plants. Fuel additives, such as dispersants, surface tension breakers, vanadium inhibitors, and fuel emulsifiers and homogenizers (that introduce water to the fuel) have all seen limited applications. Attempts to wash fuels were tried with varying degrees of success. Fuel washing refers to the removal of seawater contaminants to avoid high temperature corrosion. That occurs when sea salts, primarily sodium, combine with vanadium. Fresh water is added to the fuel to dilute seawater and raise the total water content to a level that can be removed by purifiers. Although Atlantic Richfield experienced some success on steam plants in the 1950s, the system has never been adopted on a large scale.

Early diesel plants burned primarily light diesel oil. By the late 1940s economic pressure to compete with steam propulsion had forced most high powered diesel manufacturers to begin burning residual oil. Diesels are of course more sensitive to fuel quality than is the case with steam plants. The use of residual oil and its reduced quality caused diesel plant fuel treatment systems to grow in complexity. Further reductions in residual oil quality occurred when improved refining methods removed more of the higher fractions from crude oil. One of these methods, catalytic cracking, also introduced abrasive fines to the residual oil. These changes presented significant challenges to the fuel system designer. The first diesel heavy fuel systems resembled their steam plant counterparts; but as fuel quality continued its decline, suitably sophisticated purification systems evolved.

The primary improvements in lubricating oil treatment have also been associated with diesel plants. This is because diesel engines have lower tolerance for lubricating oil contaminants, which are more prevalent in trunk piston engines. Increasing use of fine filters, sophisticated centrifugal purifiers, and coalescing filters has resulted.

The use and sophistication of centrifugal separation has developed dramatically in the period under study. Credit for the invention of the centrifuge is given to Dr. Gustaf DeLaval in 1879. Although his intent was to separate cream from milk, the device was soon used to clean lubricating oil, and was first used aboard ship in about 1910. By the 1940s lube oil purification had become a standard on most turbine driven ships. The advent of diesel power, particularly medium speed diesels, drove many increases in the capabilities, efficiency, and capacity of lube oil purifiers.

The major gains in purifier technology, however, were applied to fuel oil. The first fuel purifier was used on a motor ship in about 1921. The first attempt to centrifuge heavy fuel for a low speed diesel engine was in 1947. In 1955 the medium speed engines on the *Princess of Vancouver* burned centrifuged heavy oil.

Early purifiers were manually cleaned, but the large amount of contaminants in fuel oil prompted the development of the self-cleaning purifier. The first manually controlled self-cleaning purifier was used at sea in 1952, and automatic controls were introduced in 1956. Early fuel systems had a self-cleaning unit set up as a purifier (to remove solids and water) followed in series by a manually cleaned unit set up as a clarifier (to remove only solids). As the reliability of purifiers increased, the use of series operation has decreased. Most modern installations, however, are still capable of either series or parallel operation depending on circumstances. In the late 1970s residual oil quality further deteriorated, and purifiers were developed to separate water from oil exceeding 1.0 specific gravity.

Refrigeration and Air Conditioning. The first successful marine use of refrigeration was as early as 1880. By the turn of the century, refrigerated ships were carrying frozen beef from Argentina and Australia and fruit from the tropics to the temperate zones.

During World War II, modern refrigeration techniques supplied troops with fresh food, protected ammunition against deterioration, and air conditioned command centers and submarines. By 1945, mechanical refrigeration was producing nearly all of the ice for the fishing industry. Today the use of modern freezer trawlers, factory ships, mother ships, and fish transports has made the worldwide distribution of fish a reality and a major industry.

The first commercial maritime application of air conditioning was in public spaces of passenger vessels. The first ship to be completely air conditioned was the *United States* in 1949. The use of shipboard air conditioning continued to spread until by the 1960s complete air conditioning of the quarters was standard.

Most early refrigeration systems used CO_2 or ammonia as a working fluid. In 1930, Dupont developed fluorocarbon refrigerants, better known as Freon. The first commercial compressor to use the new refrigerant was built by York. Freon soon became the industry standard. Environmental concerns about the effect of fluorocarbons on the earth's ozone layer will present the next major challenge to the industry to convert from Freon to more environmentally acceptable working fluids.

Most refrigeration systems are of the direct expansion type, where liquid refrigerant is expanded into a cooling coil and vapors are drawn back to the compressor. Early cargo systems used brine as the cooling medium. In the 1960s the liquid overfeed system was developed for large frozen cargo holds. In this method the outlet of the expansion device is led to an accumulator. Liquid is pumped from the bottom of the accumulator to cooling coils and vapor is returned. Vapor from the top of the accumulator is then returned to the compressors. Refrigeration is also now used extensively to cool and re-liquefy low temperature liquid cargoes. See the chapter on LNG tankers for further details.

Robert S. Behr is currently a Project Engineer with the Maritime Administration. He was formerly with the Bethlehem Steel Corp., where he served as Machinery Procurement Section Chief, Project Engineer, and Chief Engineer. He is the author of several SNAME papers and has served on various SNAME panels. He is a former member of the ABS Engineering Committee.

Tanker Cargo Systems

MAURICE GORDON

Overview

WHILE THE GENERAL PRINCIPLES of tanker cargo handling have remained the same since the days of the *Glückauf*, the methods have changed significantly. Since World War II, the forward pump room and open loading have gone the way of the crow's nest. On the other hand, virtually all modern tankers are built with remote control and monitoring of cargo operations. Radar has migrated from scanning the sea to plumbing the depths of the cargo tanks.

Advances in technology as well as major regulatory changes have resulted in significant changes in the support systems for cargo handling. All tankers over 30,000 dwt are now built with inert gas systems. Crude oil washing systems are installed on most crude oil tankers. Closed loading is required and, in some areas, air quality concerns dictate that tanker and barge owners install vapor recovery systems rather than venting to the atmosphere while loading.

Pumping

Piping Arrangements. The cargo systems on tankers built until about 1970 were usually arranged with three or four segregated systems. This permitted carrying multiple grades of cargo, and to swing from crude to refined product—possibly within a single voyage. Crossover piping was installed for use when a single grade was to be carried or a pump was disabled. Through the early 1960s, these crossovers were double valved to minimize cargo contamination. Austerity by the shipbuilders reduced this to a single valve except where the shipowner specifically required double valves. After the mid-1960s, the number of cargo tanks on new tankers decreased by about 50 percent, simplifying all cargo systems.

Cargo piping arrangements changed in the age of the very large crude carrier (VLCC). Some ships built specifically for crude oil are fitted with sluice valves. These are installed as part of the bulkhead, or on a short stub of pipe through a bulkhead. In general, one valve is installed in the longitudinal bulkhead of each wing tank and multiple valves are installed in the transverse bulkheads. When the valves are opened, cargo from the wing tanks flows into the center tanks and then aft to the last centerline tank which contains a large suction bellmouth for each pump. Stripping piping, generally in the form of a ring main is also installed to permit final stripping and assist in crude oil wash operations. A typical arrangement is shown in Fig. 1.

Cargo Pumps. To this day, the standard arrangement for cargo pumping on tankers uses centrifugal pumps located in a pump room directly forward of the machinery space. From World War II through the 1950s, most tankers were fitted with a second pump room forward of the cargo block. Some of the early 100,000 dwt tankers, most notably the *Manhattan* (1961), had a third pump room amidships.

The aft pump room generally contained three or four large centrifugal cargo pumps and a centrifugal ballast pump. Double suction, single stage pumps were favored because of their relatively low net positive suction head requirements. Through the 1950s and 1960s, most cargo pumps were horizontal; but in the 1970s vertical pumps became common. Over the past 50 years, cargo pump capacities have increased with ship size—from 1134 m^3/hr (5000 gpm) and about 450 kW (600 bhp) to 5670 m^3/hr (25,000 gpm) and 2250 kW (3000 bhp).

On ships handling high viscosity cargo, positive displacement pumps were installed. Most of these were screw pumps but some owners preferred the Waterous cycloid pumps. These can handle suction vacuums of 560 mm (22 in.) Hg to overcome the high friction losses of these cargoes.

In addition, one or more positive displacement pumps were installed for stripping the cargo tanks. These were usually duplex double-acting reciprocating pumps driven by 7–10 bar (100–150 psig) steam. On occasion, screw or cycloid pumps were used instead.

The forward pump room generally contained reciprocating pumps for stripping the forward tanks,

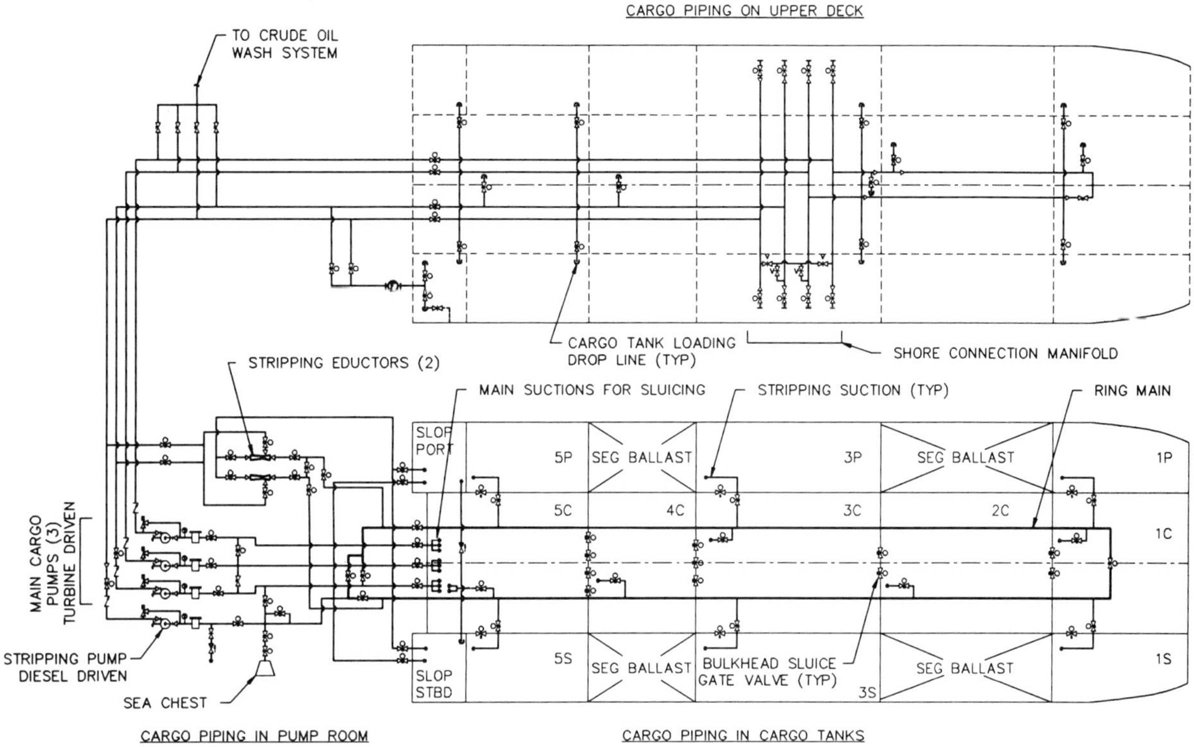

Fig. 1 Typical VLCC cargo piping arrangement. Note the ring main in the cargo tanks, the sluice gates, and the stripping eductors. In this system, each cargo tank has its own loading drop line which also is used as an inert gas purge line.

handling ballast in the fore peak and transferring fuel oil, ballast, or cargo from the forward deep tanks.

Main cargo pumps in pump rooms were generally driven by a single stage geared turbine in the machinery space with a jack shaft through a gastight bulkhead seal. When most tankers were steam propelled, this was the obvious way to use the high boiler capacity while in port.

As the age of the diesel approached, various methods to drive cargo pumps were explored. For VLCCs, where high horsepower was required, steam turbines were still favored, with steam supplied by a large auxiliary boiler. Smaller ships favored increased electrical generating capacity and motor driven pumps.

Another method of cargo handling, which has come increasingly into use, is the vertical deep well pump. These come in a variety of forms, the earliest and most common being the multi-stage vertical turbine pump. In more recent years, submerged, single stage hydraulic driven pumps have become more common. Deep well pumping arrangements are prevalent on chemical and specialty tankers. On these, each cargo tank has its own pump. On crude oil and product tankers, the deep well pumps are often installed in suction casings, which are attached to the cargo piping system. The *Marine Dow-Chem* (1954) made extensive use of deep well pumps for its chemical cargoes. Pumps were arranged one per tank to prevent cargo contamination.

The vertical turbine pump consists of an impeller and bowl assembly at the bottom of the tank. This assembly discharges vertically through a pipe column containing the shaft extension and its bearings to a discharge head at the upper deck. The most common method of driving the vertical turbine pump is a vertical electric motor mounted directly on the pump discharge assembly. Some ships have used a turbine or horizontal motor to drive the pump through a right angle gear.

The submerged hydraulic driven pump (Fig. 2) is a Norwegian development that has become common in chemical tankers. The pump assembly is located at the bottom of the cargo tank. It consists of a hydraulic motor close coupled to a single stage centrifugal pump. The hydraulics and the cargo are separated by a series of seals with an atmospheric cofferdam in the middle. Hydraulic power is developed by multiple power packs in the machinery space and distributed via piping on deck. Cargo pump speed is controlled from the cargo control room by throttling the hydraulic fluid flow. The hydraulic fluid is conveyed to the motor by a concentric pipe arrangement with

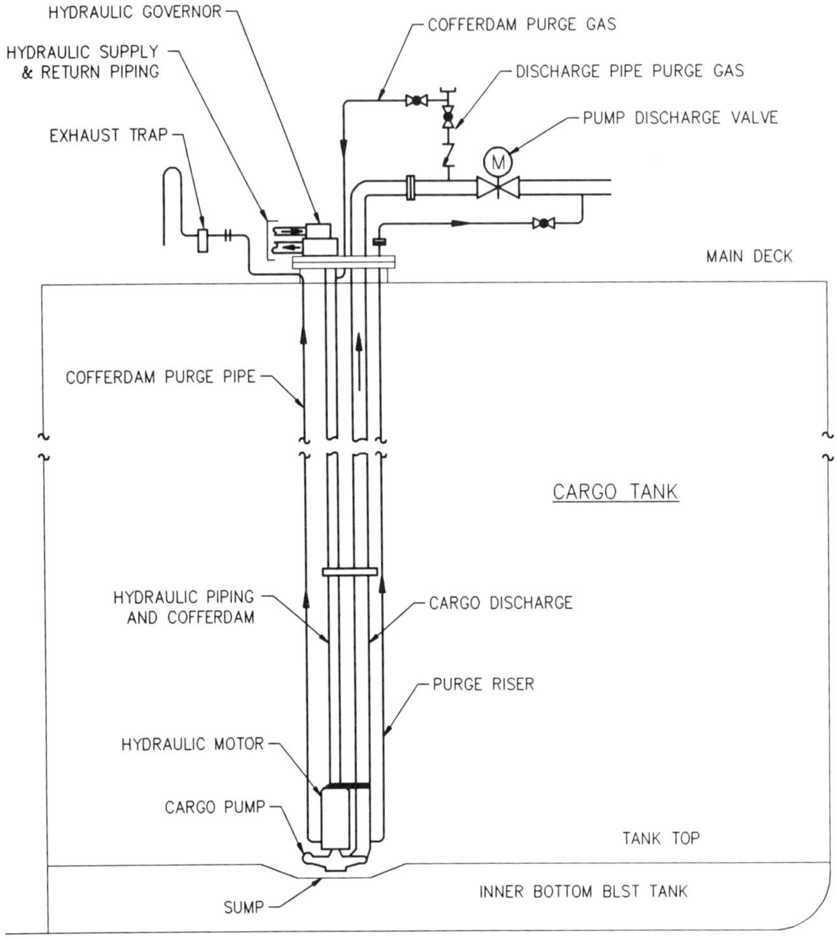

Fig. 2 Submerged hydraulic driven cargo pump when purged, residual cargo can be as little as 38 liters (10 gal).

hydraulic pressure in the center, hydraulic return in the first annular space, and a cofferdam in the outer annular space.

Stripping Pumps. The past 50 years began and ended with the standard stripping pump being the steam powered duplex double acting reciprocating pump; but these are now difficult to obtain because of the decline in overall use of reciprocating pumps. There have, however, been significant developments in cargo tank stripping.

One method that has come of age is the use of eductors to strip cargo tanks. Eductors for this purpose are large and are located in the pump room. The large motive fluid requirements are generally supplied by the main cargo pumps. Eductors are widely used with crude oil washing systems. This arrangement is shown in Fig. 1.

Priming systems have had a large impact on cargo stripping. The first of these is the venturi priming system. Priming valves are arranged to remain closed until discharge flow is reduced owing to loss of suction. These valves open to recirculate liquid to the pump suction to enable the pump to remain primed.

Vacuum priming systems—some quite elaborate—have come into increasing use. The simplest of these consists of a water ring vacuum pump taking suction from the cargo pump casing to maintain or regain prime. Other systems are fully automatic and are based on the level in the main cargo pump suction strainer. As the level falls, the vacuum pump starts and takes suction from the top of the cargo pump casing and from the strainer. When this no longer maintains the level, the system slows the cargo pump turbine to 75–80 percent of full speed. After this, further drops in level cause the system to throttle the cargo pump discharge valve.

One other stripping system worthy of note is that used for the submerged hydraulic pumps. Since many of the cargoes are of high value, and fall under the noxious liquid substances stripping requirements of

MARPOL Annex II, these tanks must be virtually dry when stripping is complete with no more than 150–950 liters (40–250 gal) remaining, this amount depending on cargo classification. Once the pump loses suction, it continues to run, the discharge valve is closed and a small bypass line from the pump casing to the cargo main is opened. Compressed air or nitrogen is introduced upstream of the discharge valve and the liquid is forced down the discharge pipe and up through the bypass. The running pump creates a seal against back flow. With proper sump design, this can result in only 38–76 liters (10–20 gal) remaining in the tank. This arrangement is shown in Fig. 2.

LNG Pumps. Pumping systems on liquid natural gas (LNG) carriers require special cryogenic pumps. These are generally submerged electric motor driven centrifugal pumps. Their most remarkable characteristic is that part of the cargo is pumped through the motor casing between the rotor and the stator. The dielectric properties of the LNG preclude electrical short circuits and grounding problems. The 100 percent gas rich atmosphere in the cargo tanks is outside the explosive range, which removes any concern about problems caused by sparks or mechanical failure.

Cargo Control

Until the mid-1960s, the key to cargo control was muscle power. Valve lineups were done manually via reach rods to in-tank valves. The only concession to automation was remote control of cargo pump speed from pneumatic panels in the pump room access. Monitoring of cargo operations was done by a roving watch mate and seamen. This kind of control led to the likelihood of overfilling a tank with a subsequent oil spill.

Cargo control rooms began to appear on U.S.-flag tankers in the mid-1960s. It is hard to make any general statements about cargo control systems since they seem to be customized to the extent that the owner or shipyard, or both, deem economically attractive. Nevertheless, cargo control rooms contain some or all of the following components:

1. *Cargo and Stripping Pump Control*—Cargo and stripping pump control includes a means to control the pump driver, and suction and discharge pressure indicators. On/off, pump speed or other suitable feedback is provided as appropriate to the type of pump driver.

2. *Valve Control*—Some or all of the valves may be remotely controlled from the cargo control room. The valve controls and indicators are usually laid out on a mimic panel containing a schematic of the cargo piping arrangement. The preferred method of remote valve operation is hydraulic, though some owners prefer electric drive. Hydraulic actuators can be submerged or on deck, while electric actuators are limited to the deck level. In pump rooms, however, because of Coast Guard restrictions on electrical equipment, control is almost exclusively hydraulic. Valve position feedback consists of position indicating lights or meters.

3. *Tank Level Monitoring*—Tank level indication is provided in the cargo control room. Generally, this is combined with overfill alarms. This enables the mate on watch to take early action to avoid oil spills. Tank level information may be fed into loading computers that continuously monitor shear forces and bending moments.

4. *Other Functions*—Additional monitoring and control systems are installed in cargo control rooms. Inert gas system monitoring and tank pressure control are required by the regulatory bodies. Segregated ballast pumps and valves are usually controlled. Oil content of ballast water is monitored. Any specialized systems relating to chemicals and specialty products are usually monitored. For example, tank temperatures for heated cargo may be provided. On LNG ships, the custody transfer computer, which determines the volume and quality of the delivered LNG, is part of cargo control.

The development of the microcomputer and microprocessor based control systems in the 1980s has its application in the cargo control room. While not yet implemented on U.S.-flag tankers, it is possible to combine cargo control, tank level feedback, and hull stress information to enable the chief mate to preprogram cargo tank loading and discharge sequences. These systems are currently available, but this degree of automation has not yet found favor with U.S.-flag operators.

One development that has made cargo control practical in the U.S.-flag fleet is that of electrical components such as transmitters, switches and solenoid valves that are intrinsically safe. These low voltage/low current devices are used in hazardous areas. Zener diode barriers in the safe zones ensure that the excessive electrical energy cannot reach the hazardous area. Intrinsically safe devices and explosion proof light fixtures are the only electrical devices the Coast Guard permits in pump rooms.

Ancillaries

Inert Gas Systems. Probably the most significant change in cargo handling in the past 50 years has been the requirements by International Maritime Organization (IMO) and the Port and Tanker Safety Act of 1978 (PTSA) to install inert gas systems on most tankers. Before that time, cargo tank atmospheres were assumed to be too rich for combustion during the laden voyage. On the ballast voyage during tank washing, cargo tank atmospheres were kept too lean for combustion or, regardless of hydrocarbon levels, all sources of ignition were removed. In the early 1970s, mysterious major explosions occurred on several VLCCs during tank cleaning, sometimes resulting in loss of the ship. Research determined that the high capacity tank cleaning machines in the large tanks created static electric charges, which discharged themselves between the tank cleaning machine nozzle and the tank structure. If the uncontrolled tank atmosphere was in the explosive range, ignition could occur.

As early as the 1950s, cargo tank inerting was investigated as a means to control corrosion in product tanker cargo tanks. Sun Oil Company used these early systems for many years. By 1974, U.S. owners were beginning to install inert gas systems on new VLCCs.

SOLAS 1974 required inert gas systems on all new tankers over 30,000 dwt ordered after mid-1979. Inert gas was also required on most existing tankers by 1983.

The basic requirement for inert gas systems is that they provide inert gas with no more than eight percent oxygen content to the cargo tanks. The production rate during cargo discharge must be 25 percent higher than the pumping rate. The tanks must also be maintained under a positive pressure of no less than 10 mm (4 in.) H_2O at all times during a voyage, including the ballast leg. As the pressure level decays, the system must be used to replenish the gas.

Two major types of gas generating systems have evolved: flue gas and independent generators. In the flue gas system, the inert gas blowers take suction from the boiler uptakes through a scrubber, which removes soot and sulfur compounds and cools the gas. The blowers discharge to the distribution main on deck through a pressure controller and an oxygen content monitor. These systems are well suited to steamships and diesel ships with large boilers to meet cargo pump steam requirements.

Independent generator systems burn diesel oil in closely controlled combustion. The gases are then cooled and scrubbed for distribution. In these systems, the blowers discharge air to the generator. These systems are suitable for diesel ships and for trades in which the oxygen content of the inert gas must be tightly controlled. Some of these generators are capable of controlling oxygen content below one percent without excessive carbon monoxide. Small versions of these systems are used for topping off the tanks at sea on diesel ships with flue gas systems.

Two main methods of distributing inert gas inside the cargo tanks have evolved. Some vessels are fitted with mixing systems where the gas is jetted into the tank through a nozzle at high velocity (30 m/sec [100 ft/sec]). This creates turbulent mixing of the incoming gas with the tank atmosphere. The other method is displacement. In this method, inert gas enters the tank at low velocity through the tank hatch trunk and maintains a constant blanket above the cargo as the tank is emptied. The latter method has a lower first cost, as the cargo piping can be used when purging or gas-freeing an empty tank. (Mixing requires a separate standpipe for each tank.) The blowers can also provide air for gas freeing.

Chemical and specialty tankers use inert gases but there are currently no requirements for general inerting of chemical tankers. Some products degrade in the presence of conventional inert gas while others change when the tank atmosphere is air. Chemical tanker operators do not inert some cargoes while they blanket or pad other tanks with nitrogen gas to prevent problems in transit. Where nitrogen is required, the ships either obtain it from the terminal or they are fitted with liquid nitrogen tanks and vaporizing equipment.

LNG tankers also require an inert gas system. In this case, its use is limited to when tanks are to be warmed up and gas-freed for personnel entry or repairs. The systems are conventional generator systems with tight oxygen control and a desiccant drier after the scrubber. In this case, the inert gas is used as a buffer between the 100 percent gas rich atmosphere during normal operations and the introduction of air. After the cargo tank is warmed above –40 °C (–40 °F), it is purged with inert gas until the methane content is below one percent. At that time, air can be safely introduced.

In the late 1980s, air pollution concerns caused some local governments to limit cargo vapor emissions when loading. This is generally done by connecting a vapor hose between the inert gas main and the terminal. During loading, the vapor from the ship's tanks are returned to shore where they are used or destroyed. During lightering operations, the lightering vessel's vapors are introduced into the lightered vessel's tanks in lieu of inert gas. At this time, requirements for this type of operation vary by locality, but in 1990, the Coast Guard issued rules governing these installations when they are installed.

Tank Cleaning Systems. Until the late 1970s cargo tanks were washed regularly with hot water at the start of the ballast leg of the voyage to prepare them for ballast. On product carriers, additional washing and rinsing were done to prevent cargo contamination.

Early in the period, most of the washing was done using portable rotating jet machines. These were placed on the end of a hose inserted through the tank cleaning openings on deck. The machines were lowered at intervals until the tank was clean. The machines, which had a capacity of about 60 m^3/hr (265 gpm) consisted of two nozzles, which the force of the water rotated around a horizontal axis. A gear train rotated the entire assembly around a vertical axis.

To improve the tank cleaning process, some portable machines were permanently located in cargo tanks. Because there were substantial shadow areas (not reached by direct jet impingement), some cleaning with portable machines was still required. On the whole, this reduced the number of man-hours required to clean a tank.

As the size of tankers increased, it became necessary to increase the capacity of tank washing machines. The new machines had capacities up to 160 m^3/hr (1170 gpm). In the large tanks, these machines create static electric charges and can be used only in controlled atmospheres.

In the mid-1970s the concept of washing tanks with crude oil rather than seawater was developed. Crude oil washing had a number of major advantages. First, it removed wax and sludge buildup by dissolving it into the crude oil so it could be pumped ashore with the cargo. Second, it minimized oil on board prior to ballasting and any subsequent cleaning for clean ballast only required water rinsing to reduce oil content of the ballast to less than 15 ppm for overboard discharge.

Crude oil washing proved so promising that IMO required it in new crude tankers over 20,000 dwt beginning in 1979. It was also required as an alternative to segregated ballast tanks in existing crude oil tankers beginning in 1981. Because of the high capacity machines required, and the potential for static electricity buildup, inert gas is always needed for crude oil wash.

Crude oil washing is done with deck mounted, high-capacity nozzles installed in each cargo tank. These may be supplemented by additional machines mounted inside the tanks (called bottom washers) for cleaning areas under horizontal webs. The IMO specifications required that the machines be arranged so that 85 percent of the vertical surfaces and 90 percent of the horizontal surfaces receive direct impingement. The machines generally have a capacity of 60–150 m^3/hr (440–1100 gpm).

Crude oil washing is performed at the discharge port. It may start with a top wash before the tank is empty. When the tank is empty, however, its bottom must be kept free of standing oil throughout the bottom washing cycle. This high capacity stripping operation is generally done by pump room mounted eductors, Fig. 1.

Even on segregated ballast tankers, operators have found it beneficial to crude oil wash all cargo tanks on a rotating basis: about one third to one half of the tanks at each discharge. The Coast Guard requires that at least 25 percent of the tanks be washed at each discharge.

Most of the deck mounted machines now in use have a single nozzle. The crude oil drives the mechanism and rotates the entire nozzle assembly around a vertical axis. As it rotates in this manner, the nozzle itself is slowly rotated around a horizontal axis. The washing pattern creates a tight helix around the vertical axis. Some of the deck-mounted machines are mechanically programmable to selectively clean certain parts of the tanks. Figure 3 shows a typical deck mounted machine. Bottom washing machines are generally the twin nozzle type like those used in conventional tank cleaning.

One thing that has not changed in the past 50 years is that residual sludge and scale, albeit smaller quantities, must still be "mucked" manually.

Load-on-Top Procedures. In the days before segregated ballast systems, salt water ballast was loaded into some of the cargo tanks at the discharge port to permit maneuvering and minimum seakeeping. During the ballast leg of the voyage, certain cargo tanks were cleaned with hot water to remove oil residue. These tanks were then ballasted with water that would be sufficiently clean to be discharged overboard at the loading port. The tank washing residue and the dirty ballast previously loaded were pumped overboard at sea.

In the early 1960s, environmental concerns caused tanker owners to rethink the way they handled tank washings, and load-on-top became the norm for tanker operations. In that method, rather than pumping tank washings overboard, they are pumped into a slop tank. During the ballast voyage, the water separates itself from the oil. The clean water is decanted from the slop tank and pumped overboard. Starting in 1983, oil content monitors were required in these

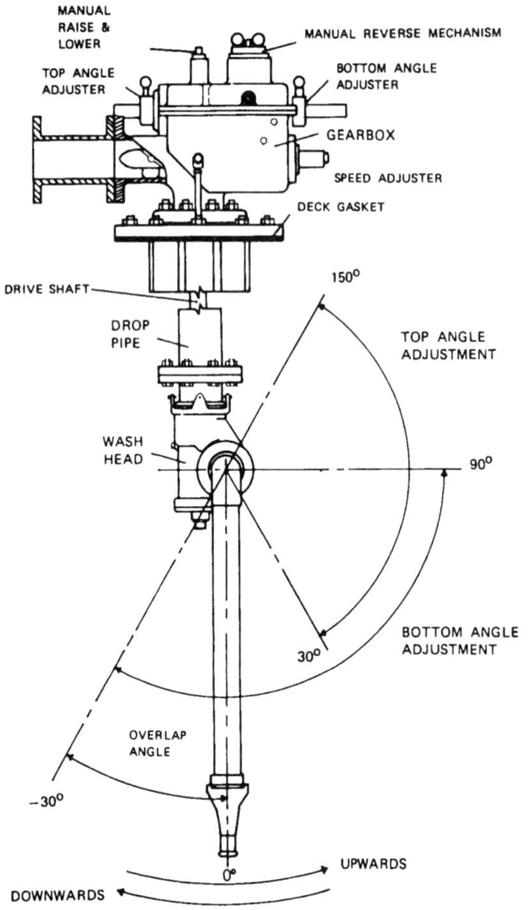

Fig. 3 Deck-mounted, programmable crude oil wash machine (Butterworth Lav-O-Matic). An impeller at the inlet drives the wash head in accordance with the adjustments made at the gearbox. (From W. C. Cowles, *Trans.* SNAME, 1980.)

systems. The next cargo is then loaded on top of the residual oil.

A short ballast leg or heavy weather can have adverse effects on load-on-top operations by impeding proper separation. Some loading ports, such as Valdez, require that all cargo tank ballast and slop tank contents be pumped ashore. In addition, some ships are exempt from some of the requirements on the basis that they trade between specific ports and discharge ballast only to approved facilities at those ports.

Segregated ballast was required on all new tankers over 20,000 dwt contracted after mid-1979. It was also required on existing tankers after 1983 where crude oil washing was not installed. Segregated ballast, of course, minimized the need for clean ballast in cargo tanks.

Tank Gauging Systems. At the time of World War II, virtually all tank gauging and sampling was done through the ullage opening in the hatch cover. Custody transfer measurements were taken using a manual steel tape with a bronze plumb bob. Less formal measurements were made during loading and discharge by the mate or deck hand looking into the tank through the ullage hatch using reflected light from a hand mirror. As late as 1990, many custody transfer measurements were still made this way.

When tanks were pressurized with inert gas, this became inconvenient as pressure had to be dropped to take measurements. Industrial hygiene considerations made it necessary to control sailors' exposure to tank vapors. In the late 1970s, closed manual gauging systems were developed. These consisted of a special tape with an electronic interface sensor. This was connected to a special deck stand. In operation, a valve was opened and the sensor lowered until it indicated it had reached the cargo. The ullage was then read directly from the tape. These are now required by the Coast Guard for certain dangerous chemicals.

Fixed tank gauging systems were available from the 1950s, but came into widespread use only in the 1970s with inert gas and closed loading. These systems rely on a float attached to a steel tape that rides up and down on guide wires or rods in the cargo tank. The floats are read on deck directly from numbers printed on the tape or from a mechanical counter driven by the movement of the tape.

Cargo control from a central location also created a need for remote, continuous reading gauges and level alarms. In some cases, the float/tape gauges have been converted with a transmitter head. Other float gauging systems that use magnetic reed switches have also become common. Some gauging systems use electric resistance to determine liquid level.

Probably the most revolutionary advance in tank gauging since World War II has been the radar gauge. Saab of Sweden saw the potential in using the properties of the vapor/cargo interface to reflect a radar beam and applied the technology to cargo tank gauging. A low power radar transceiver is installed in a deck stand at the top of each cargo tank. The tank ullage is then transmitted to analog or digital gauges in the cargo control room. Radar gauging has been widely used in U.S.-flag new construction.

Another development, which appeared in the 1980s, was pressure gauging of cargo tanks. While this is applicable to all tankers, it is especially appropriate for chemical and product carriers that transport a wide range of cargoes with different specific gravities.

The principle of pressure gauging requires three pressure transducers in each tank connected to a microcomputer in the cargo control room. One transducer is located at the bottom of the tank, one at about the 60 percent level, and one in the deck stand (above the highest liquid level). The upper transducer measures the tank pressure. When loading is started, the bottom transducer measures an approximate level based on a specific gravity provided by the chief mate. After the second sensor is covered, the computer calculates the actual specific gravity and from that point the ullage measurement is extremely accurate. Tonnage information from this program can be loaded directly into the ship's loading computer to obtain a continuous readout of shear force and bending moment.

MAURICE GORDON graduated from Case Institute in 1970 with a degree in mechanical engineering. Following eight years of tanker design experience with Bethlehem Shipbuilding, he spent two years managing maintenance and repairs for Atlantic Richfield and El Paso Marine. Since 1980 he has been with the Exxon Shipping Company, largely as a project engineer for new construction. In 1984 he was given the SNAME Linnard prize as co-author of the best annual meeting paper (on specialty product ships). He is a life member of SNAME, a fellow of the Institute of Marine Engineers, a member of ASME, and a registered Professional Engineer.

Electrical Systems

Thomas H. Bond, John H. Higginbotham, and
Joseph H. Springer

Introduction

Electrical power was first introduced to shipboard use just over a hundred years ago, closely following the introduction of industrial electric lighting in 1878. By 1893 there was an almost universal adoption of electric lighting aboard steamships, consisting primarily of arc lamps and electrical candles driven by a dynamo. The Royal Navy, in recognition of electrical power's vast potential, was experimenting and employing electricity for several purposes in their ships of war. Insulation and methods of distribution were the major problems of the day.

The first 50 years of electrical power use saw great advances in the application, generation, and distribution methods employed for direct current (dc) electricity. Alternating current (ac) was in common use in land installations and many attempts were made to employ ac aboard ships; but there were strong feelings in the marine engineering community that ac was dangerous, expensive to distribute, and too difficult to control for shipboard use. By the middle of this period, sentiment began to change and ac was a subject of interest, but the majority of engineering advances were in dc applications.

In the 50 years from 1943 to 1993, the development of electrical power systems was evolutionary in nature. World War II was instrumental in promoting growth in shipboard electrical power requirements. Developments in the postwar years were concentrated on improvements in plant operations, safety, reliability and maintenance, new power equipment technology, and improvements in plant efficiency.

Growth of Shipboard Power Requirements

As electrical power requirements increased, the discussion in the mid-1940s turned to the topic of using ac generators instead of dc generators for shipboard electrical needs. Most agreed that for larger plants, the ac system offered important advantages [1]. Cost and weight were less than with the dc system, and the simplicity of the squirrel-cage induction motor meant less maintenance than for the dc motor. In addition, squirrel-cage induction motor starters were less complex than their dc counterparts. The biggest disadvantage with an ac system was that adjustable speed could not be obtained with the squirrel-cage motor, as it could for the dc motor. Therefore, for ships having cargo winches, which must operate at various speeds, it was generally agreed at the time that either a dc system should be provided or an ac system with ac-to-dc motor–generator sets to provide dc power to the winches. Later, specialized applications dc systems became more prevalent, largely because of the introduction of the silicon-controlled rectifier in the 1960s. Thus, ac systems were universally adopted on shipboard as speed control of dc motors was made easy.

Along with the adoption of ac generation, shipboard power requirements continued to grow because improved habitability standards required air conditioning, higher illumination levels, and better ventilation. Some of the new types of ship required generous supplies of electricity. Containerships, for example, needed electricity for refrigerated boxes, and some carried electrically-powered cranes. Roll-on/roll-off (RO/RO) ships required extensive ventilation systems to handle exhaust fumes from truck engines. New regulations required shipboard equipment for pollution abatement and fire fighting. To improve cargo operations and ship maneuverability, more powerful cargo gear and electric-motor operated bow thrusters were installed.

To meet this increase in power demand, shipboard generators kept increasing in size until protection against large short circuit currents became a major problem. One solution to reduce the amount of short circuit current for a large capacity power plant was to increase the voltage at which the power was generated. The first application of 4160-volt power systems was on warships [2]. An early application of high voltage distribution in merchant ships was on the Great Lakes self-unloader *Belle River* [3]. The ship's designers were faced with the problem of providing a large amount of power for operating both the cargo unloading system and the bow and stern thrusters. The selection of 4160-volts, 60 Hz, 3-phase power became the first such system in the Great Lakes marine industry. Other naval and merchant vessels have also used 4160-volt systems [4,5].

Improvements in Plant Operation, Safety, Reliability, and Maintenance

Switchboards play an important role in overall safety, reliability, and continuity of service of electrical power systems [6]. Therefore, to improve plant operation, one would look for ways of improving switchboards. In the 1940s switchboards were still being manufactured with fused knife switches to protect electrical circuits from overload and short-circuit currents. These "live front" switchboards with their exposed current-carrying parts were hazardous to engine room personnel. In addition, blown fuses had to be replaced with new fuses before power to that circuit could be restored. Eventually, "dead front" switchboards (no exposed current-carrying parts) were mandated and fused switches were replaced with circuit breakers. Circuit breakers have the advantage that, if they interrupt an overload by "tripping," they can easily be re-closed without replacing parts.

A further requirement of circuit protection is the proper coordination of circuit breakers. This means that when a short circuit occurs, the circuit breaker closest to the short circuit should interrupt the fault current before any breaker "further upstream" operates. Improved circuit breakers today use semiconductor trip units, which shorten interrupting time and facilitate reliable coordination.

When ac generators were first used, the dc power required to produce the generator's magnetic field was applied to the rotor of the ac generator via brushes riding on two slip rings. The dc power was supplied by a dc generator that was either rotated by the ac generator shaft or by a separate ac motor. As time progressed, the dc generator was replaced by rectifiers, which are non-rotating devices that transform ac into dc. At first, shipboard rectifiers were of the mercury type, i.e., thyratrons for smaller currents, or ignitrons for larger currents. With the development of semiconductor rectifiers, which are much smaller and more efficient than the mercury type, it became practical to have "brushless" ac generators. The semiconductor rectifier was attached directly to the rotating shaft, thus eliminating the need for the slip rings. It has decreased the size, complexity, and maintenance required for electric generators.

Many of the ships in the 1940s used only batteries to supply emergency loads. As emergency loads became more numerous and included motor loads for fire pumps and emergency steering, an emergency ac bus supplied by an ac generator was added.

With the advent of propulsion plant automation in the 1960s and the evolution of unattended engine room operation, it became necessary to provide automatic start-up of a standby generator in case of power failure. A diesel-generator set was generally used as the standby generator as it could provide power even when the steam plant was not operating. In case of an overload of the running generator, the standby generator would be automatically started and synchronized with the running generator.

Power Equipment Technology Developments

Power equipment technology since 1943 has played an important role in the development of power conversion equipment, motor speed control, generator voltage regulators, cable insulation materials, and cable installation methods.

The invention of the semiconductor diode and its ability to change ac into dc virtually eliminated the need for conventional dc generators. It also made possible the "brushless" ac generator as mentioned previously. Semiconductor diodes have no moving parts, have efficiencies better than 98 percent, and essentially require no maintenance.

The silicon controlled rectifier was introduced in 1957. This device, called a thyristor, contains a "gate" to control at what part of the ac cycle conduction begins. Therefore, it can convert constant voltage, constant-frequency ac power to adjustable-voltage dc power. As a result, the speed of a dc motor can be varied from zero to its maximum by using thyristors and a constant amplitude ac voltage. This system has replaced the Ward–Leonard system, permitting significant reductions in equipment size and weight, at improved efficiency.

Early in the 1960s thyristors began being used to produce variable frequency from a dc source. This technology provided a high-performance, solid-state ac variable speed drive using the ac induction motor or the ac synchronous motor. Reference 7 explains in detail the operation of this drive and how it can be used for "four quadrant" speed control of ac motors.

This method is far superior to the wound-rotor variable ac motor drive.

Solid-state semiconductors have also been used to improve voltage regulators that control the output voltage of ac generators. These static voltage regulators, in conjunction with brushless ac generators, are quicker in response and more precise in maintaining a set voltage than the earlier mechanical methods, which used relays.

The evolution of shipboard electric cable has been significant. Over the years there has been a tendency for the Navy to lead in the application of new cable designs and materials. Reference 8 gives the historical background and a detailed discussion of the development of both Navy and merchant marine cables. Asbestos (a carcinogen) was eliminated as a component of insulation, as were other materials that give off toxic fumes and large amounts of smoke when exposed to flames. Currently used insulating materials can withstand higher temperatures and are flame retardant. When exposed to direct flames, the new insulations produce a smoke of lower density and toxicity, greatly improving fire fighting methods and effectiveness.

Developments to Improve Operating Efficiency and Reduce Costs

With the energy crisis of 1973 and the soaring price of fuel, it became more important to reduce the amount of fuel required to produce electric power. The propulsion prime mover operated at a lower specific fuel consumption, and often used less expensive fuel than the prime movers of the auxiliary generators. It therefore seemed logical to take mechanical power from the propulsion prime mover to drive an auxiliary generator. A power take-off (PTO) built into the prime mover, or the reduction gears, was proposed. This method presented the problem that the generator frequency changed whenever the PTO speed changed. One method used to correct this was to use the shaft generator only at sea when the propeller shaft was turning at a constant speed. Another method used a controllable-pitch propeller so that the propeller shaft could remain turning at constant speed, even during maneuvering. Still another method used hydraulics and gears to enable constant generator speed during PTO speed changes.

With the advent of the static semiconductor invertor, which can efficiently change dc to ac power, a system was devised whereby the frequency of the shaft ac generator no longer had to remain constant. The ac power from the shaft generator was rectified to dc and the invertor produced the 60 Hz from the dc source. The invertor cannot supply reactive power, so an additional device had to be supplied to provide the reactive power [9].

When the diesel engines began replacing steam turbine propulsion because of its lower specific fuel consumption, efficiencies were further increased by exploiting waste heat recovery. Diesel exhaust gases were passed through a waste heat boiler to provide steam to power a turbo-generator. Reference 10 describes the use of this method with the inclusion of a constant-speed shaft driven ac generator.

Operating efficiency was also improved by using thyristor speed control of dc motors and by using high-efficiency ac motors as has been discussed previously.

References

1 Fox and Coleman, "Alternating Current for Auxiliary Plants of Merchant Vessels," *Trans.* SNAME, Vol. 54, 1946.
2 Thornbury, "Current Electrical Systems and Equipment for Warships," *Trans.* SNAME, Vol. 69, 1961.
3 Bradley et al, "An Electrical Power System for Great Lakes Ore Carrier M/V *Belle River*," SNAME Section Paper, 1978.
4 Kastner et al, "Frigate Electric Cruise Propulsion and Ship's Service Power from a Commercial Distribution Network," *Trans.* SNAME, Vol. 90, 1982.
5 McFadden, "Advanced Integrated Electrical Marine Propulsion Systems," SNAME Symposium, 1984.
6 Dunn, "Marine Switchboards," SNAME Section Paper, 1967.
7 Schoen, "Fundamentals of Solid-State AC Variable-Speed Drives," SNAME Section Paper, 1968.
8 Todd, "Trends in Electric Cable Design for Shipboard Service," SNAME Spring Meeting, 1976.
9 Gott, "Shaft Generator Systems and Electric Energy on Board—Aspects on Methods and Economy," SNAME Symposium, 1982.
10 Corrado, "Economic Power Generation at Sea: Waste Heat Recovery Plant/Constant-Speed Shaft Generator Combination," SNAME Section Paper, 1983.

THOMAS BOND graduated from Webb Institute in 1945 and went on to New York University for a master's degree in electrical engineering. He was a professor of electrical engineering at the SUNY Maritime College for 15 years and at Webb for 20. He retired in 1991. Over his 46-year career he was a frequent consultant to the marine and land power industries. He is an active member of SNAME, IEEE, and ASEE.

JOHN H. HIGGINBOTHAM graduated from Texas A&M in 1956 and went to work for Bell Labs. He entered the marine field in 1962, working for various consulting firms, and is now the Director of Program Management at M. Rosenblatt & Son. He is an active member of SNAME, ASNE, and the Navy League. He has served as Chairman of the Metropolitan Sections of both SNAME and ASNE. He is a recipient of the SNAME Distinguished Service Award and co-author of three previous SNAME publications.

JOSEPH H. SPRINGER earned his bachelor's degree in electrical engineering at CCNY in 1965 and has been engaged in the marine industry ever since. Since 1974 he has been with M. Rosenblatt & Son, and has been the Head Electrical Engineer in the Basic Ship Design Department since 1982. He is a member of SNAME, ASNE, and IEEE.

Automation

Richard B. Ralph

Introduction

As the Society's first 50 years were drawing to a close, automated control system development was in its infancy, with almost no mention of automation cited in the previous 50-year history. As the industry entered the 1940s, the vast majority of ships in service were steam powered, with only feedwater control systems just coming into general use. In addition a few ships were operating with hydraulically powered combustion controls. At this time, the normal steaming watch on a single screw steamship, with separate engine and boiler rooms, included one licensed engineer, one oiler, and one or more firemen and water tenders. Development and application of newer control systems was delayed by the necessity to get ships built and into service quickly to support wartime needs.

Automation of machinery in the marine industry has been driven almost totally by the need to reduce ship operating costs. As new technologies have been developed and proven on stationary power plants, they have been applied to their marine counterparts. Shipowners have accepted these technologies in new control systems that improve efficiency, increase reliability and safety, decrease operating crew size, and reduce maintenance costs. These new technologies have been applied not only to the machinery space, but to the bridge, and to liquid cargo handling systems. The shift away from steam to diesel and gas turbine propulsion has afforded shipowners greater opportunities to automate operations. The main propulsion engine controls are generally integral with the engine, leaving the control system with a primary task of monitoring the plant. This has made it easier to automate a ship for the certification "Periodically Unattended Machinery Space." It should be noted, however, that unattended steam machinery plants were certified for operation overseas as early as 1970, although this certification for U.S.-flag steamships was not granted by the U.S. Coast Guard until 1990.

Each of the decades since 1940 has brought forward new technologies that have allowed a progressive shift from manual controls to electromechanical, pneumatic, electronic and microprocessor-based systems. The impact of each of these technologies on the various operating centers will be described in the following sections.

Propulsion Machinery

Electromechanical Systems. The development of automated control systems began to accelerate in the years immediately following World War II, and was driven by the higher operating pressures and temperatures of marine boilers, and the need to operate them safely and efficiently. The hydraulic combustion controls developed in the late 1930s required high-pressure power supplies and were relatively expensive.

Systems using electric motors and mechanical linkages were developed. These were more reliable and less expensive than the hydraulics. They were used to control forced-draft fan vanes and dampers, and fuel oil valves, by sensing a combination of steam pressure and air flow in the boiler.

Feedwater controls were also receiving new attention. Systems suggested in the early 1940s for controlling the speed of steam driven centrifugal pumps were gradually adopted. With the new combustion controls, these provided the first truly automated marine boilers. These systems proved to be highly reliable and remained the primary control systems in use through the 1950s. Many are still in service today.

This level of automation permitted the boiler room watch to be reduced to a single fireman/water tender.

Pneumatic Systems. In the postwar years, the U.S. Navy recognized the need to automate boilers and began a substantial effort to develop new systems. In the early 1960s, their efforts culminated in the adoption of pneumatic systems in place of electromechanical systems.

The most significant change in the principles of control technology was the concept of assigning mathematical functions to separate instruments and then building up a control system. Previously, each control action had a fixed mechanical relationship to a single sensing instrument. These new principles permitted a single system to control all of a boiler's major functions, and also allowed for remote manual operation.

Also during this period, propulsion turbine throttles had changed from single valve with supplemental hand valves to sequential bar lift mechanisms. Electrical servo motor actuators were added to the bar lift mechanism for remote operation.

These technologies were combined with the use of pneumatics or motors to operate valves remotely. Concurrently, the separate boiler room was eliminated. These improvements resulted in the first reduced machinery space manning certifications where only two watchstanders were required.

Electronic Systems. Solid-state devices with their increased reliability led to much more sophisticated and complex control systems in the late 1960s. This made it possible to operate most machinery from a single location. With this capability, the first one-man watch machinery spaces were approved for operation.

In the case of boilers, solid-state applications also provided for the development of burner management systems, thus relieving the engine room watchstanders from the responsibility of lighting and extinguishing burners in response to steam demand. During the 1970s burner management systems were retrofitted to numerous ships with separate boiler rooms to permit them to be certified for a two-man watch.

Throughout this same period, diesel engines and gas turbines began to find their way into more general marine use, particularly in smaller ships. Electronics afforded the opportunity to automate these machines to an extent to which no ship's machinery had been in the past. As electronic systems became more sophisticated through the end of the 1960s and early 1970s, the first Periodically Unattended Machinery Space was certified for operation on the Norwegian tanker *Thorshammer.*

Microprocessor-Based Systems. The decade of the eighties became the age of the computer. Developments in digital technology developed rapidly, with increasingly smaller devices available to provide more control opportunities.

During this period, the cost of oil and the high efficiency of the trunk piston diesel engine drove more shipowners to accept diesel engines as the preferred method for propulsion. These machines and their auxiliary systems provided new control possibilities for the new computer-based systems.

These systems also caused the first major change in information display technology. Previous manual or automatic systems had used digital and analog gauges and indicator lights to inform the watchstander of plant status. Microprocessing systems employed cathode ray tubes with selectable display screens. The displays showed plant status with a combination of color coding and digital indicators. Owners now had control systems that could reliably provide for unattended machinery space operations. The systems could at the same time provide data to allow for maintenance monitoring of the propulsion equipment and auxiliaries. In addition, the reliability and self-monitoring capabilities of microprocessor control enabled the first U.S.-flag steam ship to be certified by the U.S. Coast Guard for a Periodically Unattended Machinery Space.

As we enter the decade of the 1990s, the development of the microprocessor continues with increasing capability in smaller packages. New systems are being developed that not only will control, but will have the intelligence to recommend corrective action in the event of a machinery casualty, and assist in determining when maintenance action is required.

Bridge Control

Electromechanical Systems. The possibility for practical control of propulsion machinery from the bridge has existed for many years, but its application in commercial use was delayed by cultural differences between deck officers and engineers.

One of the earliest applications of bridge machinery control was on a class of U.S. Coast Guard ships in the mid-1940s. These ships had turbo-electric drive and used servomechanisms to operate propulsion control equipment located in the main machinery space. General acceptance of bridge control did not occur, however, until the development of electronic controls for steam plants and the more general use of diesel propulsion machinery.

Electronic Systems. With the application of electronic control systems in the late 1960s and early 1970s, and the development of burner management systems for boilers, it became possible for shipowners to take advantage of bridge control of propulsion machinery. With burner functions automated, and machinery overload restrictions designed into propulsion controls, traditional arguments against bridge control were weakened.

Pneumatic Systems. As diesel engines in large ships became more widely accepted, so did the use of pneumatics in bridge control. Many engine manufacturers provided a standard pneumatic bridge control panel as part of their engine machinery package. These panels were specifically designed to provide safe operation of the propulsion engines by bridge personnel.

Microprocessor Systems. As bridge control became more widely accepted through the 1970s, and with the development of microprocessor-based systems, significant protection could be provided to allow unarguably safe operation of propulsion machinery from the bridge.

Microprocessors have also had a significant impact on bridge operations. Systems are currently in use or under development that will combine the latest navigation technology with machinery control. These systems include the use of electronic navigation charts and the Global Positioning System to pinpoint a ship's position to within a few meters. Voyage planning is done within the system and the integration of bridge and engine room information permits speed and schedule to be optimized.

Tanker Cargo Systems

Hydraulic Systems. Initially the impetus to automate tanker cargo control systems arose from the construction of VLCCs and ULCCs in the late 1960s and early 1970s. The sheer size of those ships mandated that cargo handling be automated so that crew size did not have to increase. The large valves in these systems also made manual operation impractical. Regulatory body rules for cargo areas (particularly pump rooms) in effect at that time, led to a preference for hydraulic control systems over electrical ones. These systems allowed for centralized operation of cargo operations. Unfortunately, the systems were costly to install and noisy to operate.

Microprocessor Controls. The advent of the microprocessor has had as much effect in cargo controls as it has in propulsion machinery. The noise generated by hydraulic systems is gone, machinery operating reliability has increased, and the cost of control systems has decreased. Further cost benefits are being realized as propulsion system microprocessors are used to perform cargo control functions.

Cargo monitoring and control technology has now reached the point that a tanker can be loaded or unloaded automatically based on the loading plan prepared by the chief mate. The cargo control system monitors tank levels, draft, valve positions, and pump speeds. Using this information, the system calculates shear stress and bending moments to ensure that the maximum levels are not exceeded. It opens and closes valves as required to maintain tank levels and vessel trim.

Summary

In the thousands of years that ships have been in use, the greatest strides in the automation of machinery have occurred only within the past 15 years. The advent of the integrated circuit and the development of the microprocessor-based control system have brought about a level of safety and efficiency of operation that could not have been anticipated as the Society's second 50 years began. The operation of ship's machinery has progressed from essentially manual control to highly sophisticated microprocessor-based control and monitoring. Nor has this transformation been confined to the main propulsion unit alone. It is also found with cargo handling systems, and with auxiliary machinery of all kinds. Overall, these changes have led to definite gains in operating efficiency and at an appreciable saving in crew costs.

References

1 Barnard, O. H. and Vollmer, Jr., H. D., "Progress in the Application of Automation Controls to Naval Boilers," *Trans.* SNAME, Vol. 64, 1956, p. 336.

2 Dirriwachter, J., "The Evolution of Boiler Room and Engine Room Controls Aboard U.S.-Flag Vessels," Pascagoula Section of The American Society of Marine Engineers, March 15, 1977.

3 Gordon, M., "Retrofitting Automation on Steamships for Periodically Unattended Machinery Space—A U.S. Flag Operator's Experience," *Trans.* SNAME, Vol. 97, 1989, p. 245.

4 Dahl, F. T. and Groff, J. C., "Some Recent Developments in High Pressure Feed Systems and Equipment for Low and Medium Powers," *Trans.* SNAME, Vol. 49, 1941, p. 239.

5 *Historical Transactions 1893–1943*, SNAME.

6 Svenning, B., Broman, S., Shaw, D. E., and Butcher, R. O., "The Periodically Unattended Engine Room on the T/T *Thorshammer*," *Trans.* SNAME, Vol. 79, 1971, p. 502.

Richard B. Ralph's educational background includes a Bachelor of Science degree from the U.S. Coast Guard Academy (1964) and two Master's degrees from the University of Michigan: Naval Architecture & Marine Engineering (1969) and Mechanical Engineering (1970). He saw eight years of shipboard engineering service with the Coast Guard in a wide variety of ships and propulsion systems. He then spent seventeen years at Bethlehem Steel's Sparrows Point Shipyard, advancing to the position of Enginering Manager. From there he took a position as Vice President for Tano Marine Systems, a designer and manufacturer of digital, microprocessor-based shipboard control systems. He is now Manager of Marine Equipment for Lake Shore, Inc.

Electronics and Navigation

George Tsirimokos

The Gyrocompass

It can be readily argued that the earliest significant applications of electronics to navigation concerned the need to overcome the inherent errors of the centuries-old magnetic compass. In point of fact, this innovation itself was initially a purely mechanical investigation, and it was only at the end of the last century that electronics were applied to convert what was essentially an intriguing scientific inquiry into practical and usable form.

G. C. Bohnenberger is generally credited with the first recorded construction of a gyroscopic device, ca. 1810; but it was not until 1851 that Foucault first conceived of an inertial reference for ships using the pendulum principle. The following year, he was able to prove, by use of the gyroscope, the earth's rotation, thus fulfilling a proposal posited in 1836 by E. Lang.

It was left to G. M. Hopkins to develop the first electrically driven gyro, ca. 1890. There remained no practical application for several years thereafter. This condition was soon to change. Necessity being the mother of invention, at the turn of the present century three needs demanded solutions: first of all, steel-hulled vessels were becoming common, and the magnetic compass could no longer be relied on for accuracy. The second urgent requirement grew from man's infatuation with polar exploration, where magnetic north can very well be to the south of one's position. And finally the great powers were preparing to conduct underwater warfare, and in steel-hulled boats. In the age of Thomas Edison, the Wright brothers, and Alexander Graham Bell, two further inventors, one on either side of the Atlantic, sought a solution.

Hermann Anschütz and Elmer Sperry built on the properties of gyrostability and precession. If left alone, a gyro's axis will always point to a fixed spot in space. In 1908, Hermann Anschütz, in Germany, invented the world's first north-seeking gyro. That same year, Elmer Sperry, in the United States, introduced the first floating ballistic gyrocompass, which included vertical damping. Horizontal damping was added the following year by M. Schuler at Anschütz, who also developed a compass employing three gyros. In 1916, prompted by the war, England entered the gyrocompass arena in the form of the S. G. Brown compass.

The inherent properties of the gyro, however, do not only accommodate finding true north. The first rate gyro, indicating the velocity of turn, was introduced in 1920, and in 1921 Elmer Sperry gave birth to the famous Iron Mike gyropilot, leading, in 1926, to his invention of the world's first automatic pilot. That same year, Sperry developed the earliest automatic steering device, and in 1930 brought forth the first gyrofin stabilizer.

Other problems remained. The effects on a gyro of earth's rotation vary with latitude, changes in ship speed exert influences, and, most importantly, the gyro must be isolated from the deck of the host vessel. Over the past eight decades these problems have been overcome, and the gyrocompass is now universally accepted as the best heading source for navigation at sea.

Development of Radar

The national origins of radar are disputed. It is clear however that the United States, the United Kingdom, Germany, France, Russia, and Italy were involved in experiments to detect objects using radio waves from the mid-1920s, and that radar as we know it evolved in a series of developmental steps in several countries. Interest in long-range radio propagation using reflections from the ionosphere was behind experimental work of Appleton, Barnett and Watson-Watt in the United Kingdom. Using the cathode ray tube as a timing device they were able to measure the height of the ionosphere by timing the journey of a radio pulse from the ground transmitter, through reflection off the ionosphere, and its return to the ground receiver. Similar experiments were being conducted in the U.S. Naval Research Laboratory.

In 1934 the United Kingdom's vulnerability to air attack from an overwhelming Nazi bomber fleet was a frightening reality. There simply was no defense of any kind available that could prevent the mass destruction of British cities. This was the impetus that drove the radar concept from the laboratory to the first air defense system, which was fully operational at the beginning of the Battle of Britain.

A critical breakthrough occurred in 1940 when the first practical cavity resonator or magnetron was developed in parallel efforts by Randall and Boot at Birmingham University and Hull at the General Electric Research Laboratory in Schenectady, New York. This device, when prodded by an outside stimulus, was able to generate short pulses of high power and at the high frequency necessary for naval applications.

After World War II, the emphasis shifted to the peacetime, commercial applications of radar. In 1948 the first harbor radar system was installed by Sperry in Liverpool, England, to control shipping in the approaches to the port. The first marine radar designed for commercial use was introduced in 1949 by Decca Radar Ltd.

It was not until the late 1970s that any significant advances were made in display technology. Raytheon Marine at that time introduced their so-called bright display. This display improved the brightness of the returns and reduced the need for a restrictive viewing hood in daylight hours. Rather than displaying the radar video information as it was received, and at a timing dictated by the pulse repetition frequency and the antenna turning rate, the bright display stored the video information in a memory and then painted and repainted it independently of those two factors. As a result the increased excitation of the cathode ray tube phosphor made the display much brighter than in previous radars, and hence easier to view in bright sunlight.

The rasterscan display greatly improved the visibility of the display in high ambient lighting but also, being a predominantly digital system, it opened the door to the application of sophisticated graphics display techniques and the use of the first practical signal processing techniques to control and eliminate unwanted clutter from sea returns or precipitation.

The ability to add more features to the basic radar system was hampered, however, by the increasingly complex man–machine interface, often involving eighty discrete controls. Because of this complexity and the difficulty of learning to use the radar system, many of the improved features were not used by the operators. This barrier was broken with the introduction by Sperry of the RASCAR touchscreen radar in 1987. This radar display replaced the many discrete electromechanical controls of the radar with a glass touchscreen overlaid on the display.

Controls were graphically displayed by the software on the display surface and operated by touching the touchscreen. A menu structure organized the operator functions into logical groups only visible to the operator when called up to perform any particular function. As a result, the system was safer and more natural to operate. The time needed to learn to operate the complete system was also dramatically reduced.

Radio Determination Systems

During World War II, various hyperbolic radio navigation systems were developed. All employed the same time-difference theory used in the first war, but differed in certain details, particularly operating frequency. Some of the principal systems are noted below:

System	Operating Frequency	Type
GEE, U.K.	30–80 MHz	Pulsed
Decca, U.K.	70–135 kHz	CW
Loran, U.S.	2 MHz	Pulsed

Loran. By the end of WWII, the Loran (Loran-A) system consisted of some 70 transmitting stations, and 75,000 receivers. We provide here a synopsis of the evolution of Loran:

1945 During the latter part of WW II, experiments were carried out using much longer-range skywaves. Although the stability was poorer, it allowed for longer baselines between stations, thus offsetting poorer measurement accuracy with better geometry. During these skywave-synchronized "SS" loran experiments, phase measurements were employed for the first time in loran.

1946–1948 A proposal by Winslow Palmer of the Sperry Gyroscope Company led to the establishment of three test stations on the U.S. West Coast employing cycle-matching at 180 and 200 kHz.

1949–1951 A single-frequency system was established on the U.S. East Coast by the U.S. Air Force. It was abandoned when the Air Force favored self-contained systems such as inertial navigation.

1951–1955 Naval requirements for precise positioning led to a single-frequency, 100 kHz, multiple-pulse, cycle matching system, Loran-C.

1957 The first operational Loran-C chains were established in the Norwegian Sea and Mediterranean areas.

1971–1974 USCG contracts encouraged the development of low-cost Loran-C receivers for civil use.

1973–1976 Transmitter timing and monitoring equipment were upgraded to solid-state electronics.
1974 U.S. Department of Transportation declared Loran-C as the radio navigation system to serve the coastal confluence zone.
1978–1981 Loran-C was employed in the repeatable mode to achieve precise navigation in rivers and harbors.

Omega

1947 Harvard University scientists proposed a system, Radux, measuring the phase of a 40–50 kHz signal modulated at 200 Hz.
1955 Experiments were extended to VLF, below 14 kHz, resulting in Radux-Omega with the Radux used to resolve the ambiguities of a 10 kHz carrier phase measurement.
1962 The Bureau of Ships established the Omega Implementation Committee to develop an Omega System Plan, which was issued in 1964.
1967 Four Omega stations were operational.
1972–1981 The remaining four Omega stations became operational.

Decca

1991 The U.K. government announced that their Decca station would be upgraded and retained for the foreseeable future. Other European maritime nations had hoped that the U.K. would join in converting all the limited range Decca facilities to Loran. The U.K. decision was unexpected, and is attributed to (1) support for national industry, (2) hesitation to be seen as too cooperative with the rest of Europe, and (3) a feeling that loran itself would soon be made obsolete by GPS (defined below).

Transit Satellites

1957 Johns Hopkins Applied Physics Laboratory (JHU-APL) determined the orbit parameters of Sputnik by measuring the doppler frequency shift.
1958 JHU-APL proposed to the U.S. Navy a navigation system based on satellite doppler frequency shift.
1959 First Transit satellite was launched.
1961 First military receiving equipment was produced.
1964 The system became operational for military use.
1967 The system was released for civil use.

Global Positioning System (GPS)

1966–1973 U.S. Navy conducted the TIMATION program, and the U.S. Air Force conducted its System 62B program, both to develop a satellite-based, world-wide, 3-D position and time capability.
1973 DoD designated the USAF as lead agency in the development of a joint service navigation system to be designated NAVSTAR, Global Positioning System.
1974–1978 The GPS Concept Validation Program was conducted.
1978 Launch of the first GPS satellite.
1986 Destruction of space shuttle *Challenger* added significantly to the already existing 3–4 year delay in implementation.
1989 Launch of the first Block II (operational) satellite.

It can be reasonably agreed that GPS is at long last operational, and that it meets in every respect the expectations and benefits that will accrue to the marine community. It can further be said that GPS represents the ultimate realization of the concept of hyperbolic positioning, which began during World War II.

Communications

Conventional Radio at Sea. Perhaps the most significant application of electronics at sea is in the area of communications. We present a synopsis of the more significant developments that led to the now universal, indeed mandated, installation of communciations equipment aboard oceangoing ships:

1865–1873 British physicist James Maxwell developed the theoretical laws governing the propagation of electromagnetic waves.
1888 Heinrich Hertz demonstrated the existence of electromagnetic waves, verifying the theories of Maxwell.
1895 Guglielmo Marconi produced apparatus that was capable of communication over a distance, using a transmitter. He invented the grounded dipole antenna, resulting in a much greater range of communication.
1899 Marconi demonstrated wireless telegraphy to the U.S. Navy on board the battleships *New York* and *Massachusetts*, exchanging messages at a distance of 36 miles. The *St Paul* was the first U.S. merchant ship to be equipped with wireless telegraphy, and on Nov. 15 sent a message to the Isle of Wight, establishing a distance record for ship-to-shore communications.

1914 As a result of the *Titanic* disaster, the British government convened the first International Safety of Life at Sea (SOLAS) conference, to establish maritime radio communication safety rules (among other measures).

1920 Dr. Lee De Forest invented the three-element vacuum tube, which spawned amplitude modulation techniques leading to radiotelephony, radiofacsimile and television.

1933–1934 Edwin Armstrong invented the "static free" wideband FM broadcasting system, used today in Marine VHF communications.

1934 U.S. Congress passed the Communications Act of 1934 specifying that it is unlawful for subject ships to navigate in the open sea unless equipped with manual Morse code radiotelegraphy equipment.

1938 AT&T in the U.S. launched a transoceanic short wave single-sideband (SSB) service.

1974 SOLAS '74 convention adopted by the International Maritime Organisation (IMO). This required all passenger ships and vessels of 300 or more gross tons to carry certain radio equipment for the purpose of safety. This system is based on manual operations on 500 kHz radiotelegraphy, 2 MHz SSB radiotelephony, and 156 MHz FM radiotelephony.

1988 IMO conference in London amended the 1974 SOLAS convention and defined the requirements for the Global Maritime Distress and Safety System (GMDSS). This defines new communication requirements for all ships of 300 or more gross tons on domestic or international voyages, and will be phased in from 1992 to 1999. The new system will change international distress communications from Morse code and manual operation to automated, or semi-automated communications using satellite and digital technologies.

The most controversial change eliminates the radio officer as a requirement for international voyages. This resulted from the quantum improvement in marine communication systems, which enables anyone to operate them effectively, thus rendering a communications specialist redundant.

Satellite Communications. In 1976, COMSAT, at that time a part of the FCC, inaugurated the era of maritime satellite communications, popularly known as "satcom," with the launch of the first of three "Marisat" satellites. In short order, the second and third Marisats were put into orbit as well, and near-universal satellite coverage of the earth's ocean areas became a reality.

The maritime community thereby acquired a communications system that addressed and remedied the principal problems associated with traditional communications between ship and shore.

Conventional single-sideband required an average of 15 hours between message transmission, receipt, and acknowledgement. This was unacceptable in the highly-competitive spot oil market that characterized the post–Suez-blockage years. Via satcom, the onshore manager can speak directly to his crew, with the assurance they have received and understood his instructions.

Single-sideband and even VHF tansmissions are affected by weather and distance. With satcom, neither condition affects the quality of communications.

Conventional radio broadcasts were not available under all circumstances. There were ocean areas where no suitable coast station service was possible, where weather, distance, and non-compatibility of equipment and frequencies between vessel and coast station precluded contact. In all circumstances, a trained, expert radio operator was needed on the ship to effect communications. During his off-watch periods, there was no way to pass messages from shore. Satcom is universally available, and can be operated by anyone who is capable of using a telephone.

Conventional marine communications, whether HF single-sideband, VHF, or direct printing (SITOR) schemes offered broadcast service only. Only recently has "Digital Selective Calling" offered the privacy that only satcom provided heretofore. There remains no certain method to preclude others tuned to the same frequency from monitoring and intercepting private radio traffic when using conventional marine radio. With satcom, however, only the addressed ship can receive communications.

Satcom's automatic distress signalling features make possible immediate notification of emergencies. Building on satcom's manifest superiority in promoting safety at sea, the international maritime community has now adopted the Global Maritime Distress and Safety System (GMDSS). This encourages the fitting of satellite communications systems, as they better serve to satisfy the new carriage rules.

INMARSAT. Following the Marisat launch in 1976, the advantages of satcom were apparent to all. The United Nations, under whose auspices and authority IMCO (now IMO) functions, convened a "Meeting of Experts" to consider broader management of a truly international maritime satellite communications network.

The deliberations and work of the Experts led to the establishment in 1979 of the International

Maritime Satellite Organization, or INMARSAT, with headquarters in London. In February 1982, INMARSAT assumed management of the network.

INMARSAT has now grown to include some 96 member nations. Each assigns a signatory to the INMARSAT treaty, and each signatory owns a varying percentage of INMARSAT, based on the signatory's stake in the system. Traditionally, the United States's signatory, COMSAT, has maintained the maximum 25 percent ownership, and has accordingly greatly influenced INMARSAT operations.

A more serious problem arose as the satcom SES population approached 10,000 ships in 1989. It was long apparent that the preponderance of satellite messages originated from or were meant for vessels trading in the Atlantic Ocean. This led to saturation of the satellite capacity, and spurred INMARSAT to launch a new series of satellites with greater capacity.

As competition among shipowners for charter business requires ever more efficient fleet operation, only those who are fully prepared to exploit the benefits offered by satcom are likely to survive.

The Automated Radar Plotting Aid (ARPA)

As little as 15 years ago, most of the marine community was unaware of the contributions computer science could make to safer, more efficient ship operations. It was not surprising, therefore, that the idea of manipulating radar returns through a computer to provide collision avoidance was initially greeted with subdued enthusiasm.

Pioneering, but ineffective attempts to harness the computer in this service were made as early as 1971. In the years immediately following, the Marine Digital Company of Boston was able to demonstrate a workable design. Alas, there was little demand from the maritime industry, and the company was forced to fold its tent.

A determined group of people, uniquely experienced in electronics, computer science, navigation, radar, and the marine industry, ventured yet another try, forming Iotron Corporation along Massachusetts's Silicon Valley in 1975, from whence they brought forth the well-regarded and fairly successful Digiplot.

Iotron was given a helping hand by the Congress, which, reacting to public outcry following an oil spill, rushed to judgement, and proclaimed the Port and Harbor Safety Act of 1978. This mandated carriage of this sort of device aboard certain classes of ship. Immediately, the engineering staffs of many major companies were put to work to develop similar capabilities.

Although several non-marine companies sought to enter this field (notably IBM) none was successful. It seemed that, after all, a knowledge of the mariner's needs was more important than pure technical expertise. Major marine electronic companies, however, soon offered their own products; and by 1981 there was a plethora of "collision avoidance" makes and models available.

Each competitor sought to discriminate his product with innovative variations. Iotron used an effective two-color scheme, whereby all radar returns were displayed in green, while all computer-generated data were shown in yellow. Krupp Atlas Elektronik of Bremen, Germany, introduced the roller-ball; while Sperry Marine of Charlottesville, Virginia, patented its Predicted Areas of Danger, or PADs.

Other early entrants into the field were Raytheon, of Manchester NH, which offered a "sidecar" ARPA to complement its highly successful radar line, and Radar Devices Inc. (RDI), San Leandro CA, which offered a sidecar to complement any radar.

The mariners themselves, however, remained unconvinced. They were unprepared to hand over due vigilance to a gadget. Those involved in changing this attitude took pains to point out that "collision avoidance" was an imprecise term for this new technology. The U.S. Coast Guard initially called it an "Electronic Relative Motion Analyzer," or ERMA—a name now lost in time, but nevertheless far more accurate than "Collison Avoidance."

In fact, "ERMA" can be argued to be more descriptive than is the now universally accepted "ARPA," which derives from "Automated Radar Plotting Aid" and which the U.S. Coast Guard adopted following publication of the IMO (at that time IMCO) specifications in 1982. "ARPA" tells us what the gadget does; "ERMA" tells us how it does it.

The Integrated Bridge

The concept of a one-man bridge is no longer just an idea for the future. Scores of ships have already put to sea fitted with specified navigation and communications electronics, and have been certified by legally constituted authorities to operate as envisioned and provided for in relevant regulations.

The concept of the integrated bridge needs to be defined. Apparently, there is a lingering impression that integrated bridge simply means everything being the same color. Well, it can indeed mean that. Not only of one color, but aesthetically designed and ergonomically engineered. Simply stated, the integrated bridge is pleasant to look at and easy to work with. But there's much more to the concept. In general, the integrated bridge enhances the benefits that marine electronic navigation and communications equipment offer. It does this by making these electronics work together more effectively. This aspect of integrated bridge design is a manifestation of the adage that a whole is usually much more than the sum of its parts.

Conventional bridge design provides that a piece of electronics such as a radar needing input from, say, the gyrocompass, be tied to the compass; and if it happens to be a true motion radar, or ARPA, and also needs to have speed information, then the conventional thing to do is to also string a wire to the speed log. Integrated bridge design, on the other hand, can be likened to a well, into which each piece of electronic gear pours its information, and out of which each piece of gear draws whatever information it requires.

As with ARPA, bridge integration relied heavily, indeed primarily, on the application of computer technology. Again, it was necessary to explain computer technology in simplified terms to the digitally innocent shipowners: Computers are machines (hardware) whose life blood is digital data (software). Digital data can be integrated more readily than can analog information, which comes in infinite variety.

The integrated bridge customarily comprises six system elements:

- Shipboard Data Network,
- System Sensors,
- Navigation System,
- Steering Control System,
- Vessel Management Control Station, and
- Satellite Communications System

The Data Network provides a two-way data link between all bridge sensors and systems, and ideally acts as a real-time information data base, as in the Token Ring Network. Some manufacturers offer less effective near-real-time ethernet schemes.

These elements describe the most comprehensive approach to bridge integration today, responsive to industry needs. They provide substantial benefits, including:

- streamlined bridge manning;
- improved efficiency and productivity;
- quicker assimilation and evaluation of data;
- reduced fatigue:
- more timely, accurate information exchange between vessel and office;
- lower support costs;
- enhanced spatial arrangements;
- functional and operational elegance; and
- safety

Bibliography

Elrod, B. D. and Weinberg, A., "Satellite-Aided Concepts Employing the NAVSTAR Global Positioning System," Global Positioning System Papers published in *Navigation*, 1980.

Goodman, G. R. and Oswitt, R. P., "Loran-C Replacement Equipment," *Navigation*, Vol. 23, No. 3, 1976.

Benton, W., "Gyrocompass, Gyroscopes," in *Encyclopaedia Brittanica*, New York, 1958.

Henderson, D. W. and Strada, J. A., "NAVSTAR Field Test Results," Global Positioning System Papers published in *Navigation*, 1980.

Hughes, T. P., *Elmer Sperry: Inventor and Engineer*. Johns Hopkins Press, Baltimore, 1971.

Kayton, M., *Navigation: Land, Sea, Air and Space*. IEEE Press, New York, 1990.

Mixter, G. W. and McLench, D., *Primer of Navigation*. Van Nostrand, Princeton, NJ, 1967.

Mooney, F. W., "Loran-C Expansion in the Coastal Confluence Zone," *Navigation*, Vol. 23, No. 3, 1976.

Pierce, J. A., "Memoirs of John Alvin Pierce: Development of Loran," *Navigation*, Vol. 36, No. 1, 1989.

Pierce, J. A. and Woodward, R. H., "The Development of Long-Range Hyperbolic Navigation in the United States," *Navigation*, Vol. 18, No. 1, 1971.

Olsen D. L. and Stoltz, J. R., "Precision Loran-C Navigation on the St. Marys River," *Navigation*, Vol. 25, No. 3, 1978.

Rawlings, A. L., *The Theory of the Gyroscopic Compass and Its Deviations*, MacMillan, New York, 1944.

Stansell, Jr., T. A., "Transit, the Navy Navigation Satellite System," *Navigation*, Vol. 18, No. 1, 1971.

Truxall, J., *Control Engineers' Handbook*. McGraw-Hill, New York, 1958.

Vass, E. R., "OMEGA Navigation System: Present Status and Plans 1977–1980," *Navigation*, Vol. 25, No. 1, 1978.

Wrigley, W., Hollister, W. M., and Denhard, W. G., *Gyroscopic Theory, Design, and Instrumentation*, MIT Press, Cambridge, 1969.

George Tsirimokos is a member of SNAME and holds degrees from the University of New Hampshire, Boston University, and George Washington University. A retired Colonel from the U.S. Air Force, he acquired expertise in navigation and radar during several thousand hours' flying time. At Sperry Marine he directed the INMARSAT Ship Earth Station programs, and went from there to World Communications Systems, where he is now Director of Sales. His several inventions include the well-known "Starbaud" concept.

Part IV

Design

Education and Research in the United States

RAYMOND A. YAGLE

SUCH DEVELOPMENT—progress and change—as has characterized the U.S. and Canadian marine industries in the past 50 years has largely been accomplished by assimilating the education and research efforts they have managed to support. Thus developments recorded in the various other chapters of this volume need to be supplemented by this description of those institutions, organizations, and programs that have provided the education and research. Education will be addressed first and research second, but it must be acknowledged that these are not separate topics but are inherently coupled and typically linked administratively. The results of research are only beneficial when they become generally known and serve to educate others as well as those who were themselves educated by carrying out the research. This is true whether the subject involved is a sophisticated hydrodynamic problem or the best means for training shipyard workers in piping installation.

Education

Near the end of the 19th Century three distinctly different schools in the United States began offering instruction in naval architecture and marine engineering. The Massachusetts Institute of Technology admitted its first class in 1865 and first taught courses in naval architecture some 20 years later. Webb Institute of Naval Architecture, in Glen Cove, New York, is an academy that has given courses in naval architecture since it was established and endowed in 1889. In 1933 Webb increased the content of its curriculum and was given authority by the University of the State of New York to award a Bachelor of Science degree. The University of Michigan, a large public university dating back to 1817, first offered classes in naval architecture and marine engineering in 1881 and then established the present department in 1904.

Of those three schools, MIT is perhaps best known for its graduate level education. Webb Institute confines itself to undergraduate education, while Michigan maintains a balance between the two. As has been true for the past 50 years, these three schools continue to educate most of the career-oriented naval architects and marine engineers in this country. All three institutions are also actively engaged in research.

The University of California at Berkeley, another large state-supported institution, originally offered only graduate study in naval architecture, but in 1981 was authorized also to grant bachelor's degrees in naval architecture and offshore engineering. About the same year formal classes in naval architecture and marine engineering began at the University of New Orleans, another public institution. Both of these undergraduate programs have been accredited by ABET (Accreditation Board for Engineering and Technology). A similar program initiated in 1982 by the Memorial University of Newfoundland has been accredited by CAB (Canadian Accreditation Board).

Among the maritime academies in the United States the U.S. Merchant Marine Academy, at Kings Point NY, the State University of New York Maritime College, at Fort Schuyler, and the Texas Maritime College, at Galveston, all have ABET-accredited curricula.

Fig. 1 The modified original Pratt residence—now the Stevenson Taylor Hall—on the present campus on Long Island Sound of Webb Institute in Glen Cove, New York. The move from Fordham Heights in 1947 and construction of the Haeberle Laboratory and the Robinson Model Basin buildings were certainly a most significant change for this educational (and research) institution in the last 50 years.

At Kings Point it is in marine engineering systems; at Fort Schuyler it is in electrical engineering, marine engineering, and naval architecture; at Texas it is in marine engineering. The other state maritime academies—Maine, Massachusetts, Northwestern Michigan, and California—also exist primarily to prepare licensed deck and engineering officers, but graduates of these schools have increasingly found engineering careers ashore.

The U.S. Naval Academy, at Annapolis, and the United States Coast Guard Academy, in New London, Connecticut, educate students in various engineering disciplines and these curricula too are ABET accredited. The Coast Guard Academy's major of interest here is designated marine engineering while at Annapolis one can major in marine engineering, naval architecture, or ocean engineering.

Ocean engineering began being recognized as another, separate engineering discipline in the 1960s, but a universally accepted definition still does not exist because the field continues to evolve. In regard to educational programs intended to prepare graduates for practice in this field, ABET-accredited bachelor's degrees are awarded by Florida Atlantic University, Florida Institute of Technology, and Texas A&M University at College Station. Graduates of MIT can actually choose to have their Bachelor of Science degree be in ocean engineering or in naval architecture and marine engineering. Moreover, the academic unit offering the program at MIT was renamed the Department of Ocean Engineering about 1970, suggesting that "ocean," or possibly "marine," may in time become the generic designation to describe a single discipline that includes naval architecture, marine engineering, and ocean engineering—at least as it is defined by those schools with more longstanding association with the maritime industry.

But there are a number of additional institutions that offer curricula in ocean engineering in one form or another, often as an option within the civil or mechanical engineering undergraduate programs. For example, the ocean engineering curriculum at the University of Rhode Island is administered by two such departments plus that of chemical engineering. At Virginia Polytechnical Institute and State University a department has a name that combines aerospace with ocean engineering, while at Stevens Institute of Technology the appropriate department includes civil and ocean engineering. Specific arrangements at many schools have varied widely and are frequently changed, so the ABET accreditation remains the most convenient means of distinguishing among the undergraduate programs. Table 1, from the 1990 ABET Annual Report, includes the names of the degrees and

Table 1 Accreditation Board for Engineering and Technology (ABET) undergraduate naval architecture, marine engineering, and ocean engineering accredited programs as of 1990.

School	Degree	Year First Accredited
Webb Institute	Naval architecture and marine engineering	1936
M.I.T.	Naval architecture and marine engineering	1936
University of Michigan	Naval architecture and marine engineering	1936
U.S. Naval Academy	Naval architecture	1972
U.S. Naval Academy	Marine engineering	1972
U.S. Naval Academy	Ocean engineering	1972
M.I.T.	Ocean engineering	1972
State University of New York Maritime College	Marine engineering	1972
U.S. Coast Guard Academy	Naval architecture and marine engineering[a]	1973
Florida Atlantic University	Ocean engineering	1973
Texas A&M University (Galveston)	Marine engineering	1977
Texas A&M University (College Station)	Ocean engineering	1978
State University of New York Maritime College	Naval architecture[b]	1978
Florida Institute of Technology	Ocean engineering	1980
U.S. Merchant Marine Academy	Marine engineering systems	1982
University of California (Berkeley)	Naval architecture	1982
University of New Orleans	Naval architecture and marine engineering	1985
Memorial University (Newfoundland)	Naval architectural engineering[c]	1986
Virginia Polytechnic Institute and State University	Ocean engineering	1990

[a] From 1990. Formerly, marine engineering only. An accredited ocean engineering Bachelor's degree was granted from 1973 to 1986.

[b] An accredited ocean engineering degree was also granted several years ago.

[c] Accredited by the Canadian Accreditation Board.

the initial year of accreditation for the schools discussed thus far.

No such universally accepted measure is available on the graduate level. Thus, in regard to graduate ocean engineering education, we may add to all of the larger schools listed thus far almost every major university that has graduate instruction in oceanography plus programs in engineering. The Universities of Hawaii, Miami, Washington, Florida, British Columbia, and Alaska deserve mention, and there are probably several others. There is no reason why a graduate student associated with any of the engineering departments in a large, modern, American or Canadian institution should not be able to select among the array of courses normally available to complete a program that could reasonably be called "ocean engineering."

The foregoing paragraphs have dealt with the education of engineers specifically trained for work in the marine industry. Such training equips one to handle the overall design of ships or other marine systems. Nevertheless, much, perhaps most, of the detailed engineering work is carried out by individuals with other educational backgrounds. Probably fewer than a third of SNAME members hold degrees in the traditional maritime-related disciplines. Mechanical engineers are well equipped to analyze gearing problems, for example, while civil engineers or aeronautical engineers are capable of analyzing hull strength problems. These are but two examples. The typical naval architecture office has always made use of engineers and technicians with a wide variety of training and experience. Moreover, such assistance becomes increasingly important when dealing with advanced concepts such as those encountered in designing hydrofoil craft or air cushion vehicles.

What is said above about the need for non-traditional maritime engineering talent is perhaps best illustrated by the offshore oil industry. Petroleum engineers, well acquainted with land-based oil drilling operations, took a leading role in moving their equipment out to sea, with appropriate help from naval architects. Many professional engineering societies (ASCE, ASME, and IEEE, for example) formed separate ocean engineering divisions. SNAME, on the other hand, logically incorporated the evolving field as an inherent portion of its traditional overall interests and activities.

Thus it is appropriate that the advent of ocean engineering as a distinct discipline, and recognition of its incorporation into SNAME, be acknowledged here as being historically significant. Student sections of SNAME at institutions whose maritime-related interests are confined to ocean engineering are a clear manifestation of this. SNAME's Technical and Research program committee structure and publications offer yet further evidence.

Fig. 2 View showing the distinctive location also on Long Island Sound of the State University of New York Maritime College at Ft. Schuyler, New York. Founded more than 50 years later but just about 50 years ago, the United States Merchant Marine Academy at Kings Point, New York, also faces Long Island Sound.

Another important educational development that deserves mention is the attention given in the past decade to ship production. Appropriate consideration is included in the programs offered at many of the institutions already listed, and by others that may owe their interest to their proximity to a shipyard. Some of this latter group, which are too numerous to name here, have found it worthwhile to offer courses in shipbuilding methods and technology. Others have concentrated more on such topics as information control and management, and administration. The National Shipbuilding Research Program, begun in 1973, has been most effective in improving education and training relative to ship production at all levels and in a wide range of pertinent areas.

The concept of a degree in "engineering technology" has existed in one form or another for some years, but the detailed arrangements vary from state to state. About 1969 the National Sea Grant Program (then in the National Science Foundation, now part of NOAA) funded the development of a curriculum designed to produce competent technicians for the ship construction and repair industry. This was sited at what came to be called the Institute of Engineering Technology, affiliated with Mississippi State University. That program in Marine Engineering Technology, and others like it, should also be recognized. So, too, should a number of shipyard training programs for newly employed, newly graduated naval architects. A good example is the Management Training Program at the National Steel and Shipbuilding Company. Another is the Career Development Program at the Electric Boat Division of General Dynamics Corporation. The well known Apprentice School at the Newport News Shipbuilding Company has a long and distinguished history and continues to flourish.

The U.S. Navy sponsors a wide range of educational programs. Of these, some are considerably more academic than those intended primarily to train petty officers and even commissioned officers (Engineering Duty Only, for example). Among the more advanced programs is the XIII-A Course at MIT. This provides opportunities for graduate study in a number of curricula appropriate for naval officers who will later oversee the design and construction of ships. Students of the XIII-A program have intermittently been awarded the degree of Naval Engineer, or the Master's degree, depending on how long they were enrolled. Opportunities for graduate study in a number of engineering disciplines are also provided at the U.S. Naval Postgraduate School at Monterey CA.

The U.S. Coast Guard has also long sponsored graduate level education for some of its officers. Their programs emphasize naval architecture and marine engineering, and have been carried out at various schools, but most notably at Michigan and MIT.

No other single change in the past 50 years has been as influential on education (and research) at all levels in the marine field as the introduction of computer technology. Ship design, shipbuilding, and ship operation have been dramatically revised and enhanced, so it is not unexpected that education for the industry would also need to be. Another chapter will deal with the "CAD/CAM revolution," but this development has brought about many major changes in what is taught, and how, in almost every field at whatever level. Partly as a result of the availability of computers, engineering education has become far more scientific in nature. Gone are the days of hands-on experience in forging steel or testing steam driven pumps. Drafting skills are largely neglected as students learn to visualize three-dimensional forms with the aid of interactive computer graphics. Some observers lament this loss of practical training, but many will admit that today's graduates are truly more capable and creative than were their predecessors. They are, moreover, far better equipped for the demands of our more technically oriented, rapidly changing society.

Some educational problems remain, however. Not all engineering graduates can speak or write well. An equal concern is ensuring that engineers have enough familiarity with the humanities and social sciences to qualify as good citizens. Most engineering students are now required to study classical economics, and practical, design oriented economics also rate attention.

One of the most striking changes that has come about in the past 50 years is the proportion of students who have extended their education past the bachelor's degree level. Related to this is the concurrent development within the major research universities, where candidates for the Ph.D. degree usually earn income as research assistants or as teachers of undergraduate students.

Another improvement of recent decades that should be mentioned here is the expansion of continuing engineering education. This includes the proliferation of short courses (usually one or two weeks) each dealing with some specific topic. At the other end of the scale, we find both government and private industry sending employees back to school for graduate level studies, often leading to advanced degrees. This may take the form of full-time enrollment at an institution, or it may simply involve taking courses

Fig. 3 Two eras are represented in these photographs of the carriages in the towing tank in the West Engineering Building at the University of Michigan: (*top*) perhaps 50 years ago and (*bottom*) probably no more than 10 years ago. Although the Department of Naval Architecture and Marine Engineering at Michigan moved to its own building on the North Campus in 1977 and houses the Computer-Aided Marine Design and an Ocean Engineering Laboratory there, the "tank" and other adjacent new facilities remain in the West Engineering Building but are now referred to as the Ship Hydrodynamics Laboratory.

Fig. 4 One of the carriages for use in the main towing tank at the David Taylor Research Center, one of a number of carriages if those for the high-speed basin and in the maneuvering/seakeeping and the rotating arm basins are included. These along with the cavitation tunnels and circulating water channels and other such equipment at Carderock, Maryland, have made this facility the premier hydrodynamic research center in the world for the past 50 years.

during late afternoons and evenings while still carrying on essentially full-time work.

Finally, one can not overlook the role of correspondence courses. In particular, the Westlawn School of Naval Architecture and the Yacht Design Institute have attained recognition since they have started a number of the better known yacht designers on their careers.

Which individual professors should be cited here for their significant contributions to maritime-related education? Once begun, the list would need to be lengthy: Abkowitz, Adams, Allmendinger, Anderson, Beck, Bhattacharyya, Bragg, Bunch, Chryssostomidis, Cochrane, Compton, Couch, and on through the alphabet to Ward, Webster, Wehausen, Woodward, and so on to Zubaly. It does seem unfair not to mention even a few of the names in between, but it is also true that short shrift must be given to the names of benefactors of the various schools. Names such as Gibbs, Goldman, Henry, Pratt and Rosenblatt are but a few examples. Recognition should also go to senior administrators such as Keil, or Porter, and to department chairmen such as Baier, Dyer, Femenia, Parsons, Troost, and many others. Of course they also made contributions as teachers and researchers, but in fairness to all those whose names are omitted altogether, no one should be mentioned in more than one category.

In conclusion, it is certainly appropriate here to recognize the contribution SNAME has made by its generous scholarship program instituted in 1930. These scholarships have helped hundreds of highly qualified young people in entering the marine industry. SNAME also deserves credit for underwriting the preparation and publication of its several books—one of which is now in its third version. Most are best described as reference books, rather than textbooks, but they are nevertheless of great educational value. Prominent among these are *Principles of Naval Architecture* (first edition, of 1939, edited by Rossell and Chapman, with a second edition edited by Comstock, and a third by Lewis). An important series started with *Design and Construction of Steel Merchant Ships* (Arnott), followed by *Ship Design and Construction* (edited by D'Arcangelo and then Taggart). Two editions of *Marine Engineering* were edited by Seward and

Fig. 5 Two of the current six deep ocean pressure tanks also at the David Taylor Research Center, one about to enclose and test a full-scale submersible. Not just the capacities but the configurations of structural mechanics testing equipment have changed in the past 50 years from the typical load-deflection Universal testing machines once considered adequate.

Harrington. Saunders wrote two volumes of *Hydrodynamics in Ship Design,* with a third volume from Taggart. Korvin-Kroukovsky's *Theory of Seakeeping* and Hughes' *Ship Structural Design* are two examples of several SNAME books dealing with special topics. Two more are expected soon.

Research

What one defines as research is at best a reflection of one's own education and experience, and is thus somewhat time-dependent. Those engineering studies that might have been classified as research 50 years ago might be routine today. An undertaking one practicing engineer would today deem to be research might well be thought of more as development by the person actually conducting it. Consider, for example, the challenge of designing the first tension leg platform for offshore drilling. Owing to the profession's total lack of experience with that configuration, the task could well be described as R&D rather than as simply design. Such matters cannot be resolved here; but it should be apparent that research—whether more basic than applied, or more theoretical than pragmatic—has enabled engineers in the marine industry to practice their profession with increased confidence and with a wider choice of design alternatives.

The design of any marine system can now be carried out more rationally because of the basic research that has led to a more scientific means of describing the environment in which the system is to operate. Research has also led to continued fundamental development of more sophisticated, more capable methods of analysis to predict how the system will respond to the environment. The resistance of a proposed hull form can now be determined accurately, and with enough facility to warrant consideration of a wide range of variations in its geometry. And, in each instance evaluation can be done by criteria reflecting various parameters such as economics, safety, or ease of fabrication. Resistance analysis is often accomplished by using the various standard hull form series, which are the results of exhaustive systematic model testing programs. But the conceptual nature of the approach is the result of years of research. The division of the hull form resistance into components, particularly that portion due to wave-making, has generated extensive study of this particular topic in recent years, and these studies should be classified as research. Thus there exists the paradox that the applied research, which in due course makes engineering design more sound and less of an art, also generates research that leads to a more basic understanding of the phenomenon. Another example is the highly skewed propeller. This was merely a concept initially, but it became the means for obtaining improved efficiency and superior cavitation and noise characteristics only after extensive analytical and concurrent experimental investigation. These and numerous other examples also demonstrate that much of that body of knowledge that is now used by practicing engineers was not known 50 years ago. Research, of whatever level or type, has in fact permitted this to take place.

Perhaps the most efficient means to grasp the extent of the body of pertinent research that has been conducted in the past five decades is to survey the titles of the publications issued during that period. Just as one example, consider those from the David Taylor Research Center. Then there are those that were sponsored by the General Hydromechanics Research Program of the U.S. Navy, or reviewed by the International Ship Structures Congress (ISSC) and the International Towing Tank Conference (ITTC), and other similar sources. The Office of Naval Research, the Maritime Administration, the National Oceanic and Atmospheric Agency, the National Science Foundation, and other government agencies have also supported research of direct benefit to the marine industry. But so have many private industrial corporations. The interagency Ship Structure Committee has been sponsoring research since 1946. Other reports of distinct benefit have come out of the work of committees formed by the National Academy of Sciences–National Research Council. And what of the activities of the American Bureau of Shipping and other regulatory bodies such as the U.S. Coast Guard individually or through the Inter-Governmental Maritime Consultative Organization (now IMO) collectively? But how inclusive would any review be that did not also recognize the extensive research carried out by the other maritime nations of the world?

An easy and representative alternative is to propose that one can grasp at least a hint of the impressive scope of these activities simply by examining the publications of SNAME itself. Among these, the Technical and Research (T&R) Committee reports indicate the problems engineers have felt required study and resolution. Publication of the *Journal of Ship Research* by SNAME was begun in 1957, and it has been widely perceived as a source of scholarly papers dealing with scientific and fundamental principles of engineering. Articles in *Marine Technology* and the much newer *Journal of Ship Production* have not generally been such that they should be classified as research; but, again, they do collectively reflect the changes in practice over the years that often have stemmed from research and development.

The current T&R committee organization itself illustrates the categories of research activities of immediate interest: hull structure, hydrodynamics, offshore, ship design, ship's machinery, ship production, ship technical operations, and small craft. Significantly, the number of categories in the list has doubled within the past decade or two.

Research has made possible marine system configurations not even contemplated 50 years ago. These radical changes have greatly enhanced the functional efficiencies of systems. Some, for example, have been reduced in size or complexity. Alternatively, some have been allowed to grow enormously in size. (Tankers are a compelling example.) On the other hand, where the demands of the service call for complexity (as for example in carrying liquefied gas), research has played a crucial role there as well.

Major improvement has followed from the use of diesel engines in place of steam turbines in the majority of merchant ships. At the same time, gas turbines have become standard in military surface vessels. These developments could never have come about without years of research to improve materials or to advance control theory. And what of nuclear marine power plants, a development foreseen by at most a few visionaries 50 years ago?

A beneficial feature of research is that the results often can be applied in totally unexpected areas and activities. The marine industry has been fortunate in that it has been able to absorb much of the research carried out by those associated with, for example, other transportation modes. The container, which altered general cargo ships in a quantum leap, became operational on highways and railroads before finding its way aboard ship.

Engineers (and those in the "hard" sciences) tend to envision research as pertaining only to physical or natural processes; but research in human behavior and economics, for example, has affected the practice of engineering over the past 50 years every bit as much as that in integrated circuits or combustion.

It is generally true that doctoral theses must contribute new knowledge just to be acceptable. Today it is often at the large research universities that research important to the marine industry is being conducted.

Fig. 6 View of the new wave tank at the Offshore Technology Research Center in the Research Park of Texas A&M University in College Station, Texas, for establishing structural loads in multidirectional wave spectra as well as a variety of other studies. Most of the facilities needed to develop the techniques and equipment for offshore operations associated with oil and gas exploration and production have largely come into being in just the past 50 years.

There are more doctorates being earned in engineering now, by at least an order of magnitude, than 30 or 40 years ago. If one includes just those awarded in areas of direct interest to the marine industry, the number is impressive enough; but, if we also include those whose research has contributed indirectly, this source of research results has probably been more significant than that generated in government laboratories and those of private corporations. It must be acknowledged, however, that most academic research has been funded by the Federal Government.

Looking at modern research laboratories, one is struck by the sophistication (and high price tag) of the instrumentation and other equipment. The towing tank remains a principal research tool specific to the marine industry, but tanks are now more aptly described as hydrodynamic laboratories. Most can replicate ocean wave conditions; some can simulate ice-covered waters; others are fitted for a general range of ocean engineering studies, and all need extensive investments in data processing equipment.

Water tunnels have evolved to about the same degree; and laboratories devoted to structural research and engine improvement (but not exclusively for maritime application) are capable of testing and research undertakings not even imagined a few years ago. But it is the capacity and availability of what are now termed the "super computers" that have allowed researchers to explore with ever greater rigor problems once thought beyond any realistic approach. Dynamic behavior that only recently had to be analyzed with linear theory can now routinely be described with nonlinear theory, and analyses that were two-dimensional can now be replaced with those that are three-dimensional. Ever more extensive and more exact modeling of phenomena are to be anticipated, and more complete understanding will surely follow.

The roll of distinguished maritime researchers is also lengthy. The giants of the past—such as Taylor, Todd, Hovgaard, Davidson, Weinblum, and Lewis—have been followed by others equally capable. Schade's pioneering work in structural analysis has been continued by his colleagues at Berkeley, Paulling and Mansour. Similar sequences could be cited at other institutions, but none can compete with David Taylor in bulk or breadth. Important research has been carried out, for example, by St. Denis and Hadler at the David Taylor Research Center. Some of their contributions can rightfully be termed *classical*. But, additional contributions to hydrodynamic theory have come from Savitsky, Jacobs, Ogilvie, Tuck, Salvesen, Newman, and Mathews, among many others and at many institutions. Improvements in propeller design have followed research by investigators such as Breslin,

Kerwin, Morgan and Vorus. Other researchers, too numerous to mention, have made possible the development of such relatively advanced concepts as hydrofoil craft or air cushion vehicles, and ocean thermal energy conversion systems (OTEC). Similarly, others have found ways to improve marine coatings, to automate ship operations, and to effect advances in composite construction. Benford's pioneering work in practical economics has led the way to more rational ways to make decisions in ship design. But, many others have made significant contributions to the science of ship design: Evans, Mandel, Johnson and Rumble, Roseman, Garzke, Graham, Lamb, and Leopold are well known examples. McClure first envisioned the semisubmersible offshore drilling platform, but others have found varied applications of this configuration. SWATH vessels are finding wider acceptance and application. Lang and others deserve mention for bringing this about.

In all of the above, attention has been concentrated on our American and Canadian researchers, all of whose names are associated with SNAME. There is no intent to slight the admirable achievements of our colleagues overseas. For example, the publications of the Royal Institution of Naval Architects, as well as those of the Society of Naval Architects of Japan, offer proof that maritime research continues to receive active encouragement throughout the world.

Looking into the future, it seems clear that the marine industry cannot prosper without vigorous ongoing programs in education and research. The requisites for such programs are in place, and what is most needed is a continuing flow of financial support. The authors of the publication to succeed this one must be able to look back and remark on how much more development—progress and change—will have characterized the marine industry in that 50-year interim.

Bibliography

Benford, H., "Observations on Education and Research in the Japanese Shipbuilding Industry," Report No. 124, Department of Naval Architecture and Marine Engineering, the University of Michigan, Jan. 1972.

Cantor, J. A., "Multishipyard Cooperative Training: A Report on the State of the Art," *Journal of Ship Production*, Vol. 1, No.3, Aug. 1985.

Femenia, J., and Zubaly, R., "Training of Ship Operators: The Marriage of Theoretical and Practical Education," SNAME Spring Meeting, 1979.

Mansour, A. E., "Implementation of Reliability Methods to Marine Structures," *Trans.* SNAME, Vol. 92, 1984.

Mansour, A. E., "Probabilistic Design Concepts in Ship Structural Safety and Reliability," *Trans.* SNAME, Vol. 80, 1972.

"National Shipbuilding Research Program Bibliography of Publications 1973–1988," U.S. Dept. of Transportation and U.S. Navy, Nov. 1988.

Nicholson, W. M., "Education of Naval Architects and Marine Engineers," 1st Pan-American Congress of Naval Architecture and Maritime Transport, Rio de Janeiro, June 1966.

Ogilvie, T. F., Dyer, J., and Payne, C. N., "Panel Discussion on the Education of Engineers for the Ocean Industry: Future Trends," *Marine Technology*, Vol. 14, No.3, July 1977.

Parsons, M. G., "Marine Engineering, Naval Architecture, and Ocean Engineering in the United States," 8th Pan-American Congress of Naval Engineering, Maritime Transport and Port Engineering, Washington, D.C., Sept. 1983.

Paulling, J. R., and Payer, G. R., "Hull–Deckhouse Interaction by Finite Element Calculations," *Trans.* SNAME, Vol. 76, 1968.

"Research and Education for Maritime Progress," Transportation Research Board, Division of Engineering, National Research Council, National Academy of Sciences, Washington, D.C., Jan. 1973.

St. Denis, M., and Pierson, W. J., "On the Motions of Ships in Confused Seas," *Trans.* SNAME, Vol. 61, 1953.

Salvesen, N., Tuck, E. O., and Faltinsen, O., "Ship Motions and Sea Loads," *Trans.* SNAME, Vol. 78, 1970.

Schade, H. A., "The Effective Breadth of Stiffened Plating Under Bending Loads," *Trans.* SNAME, Vol. 59, 1951.

Stark, R. E., "The Scholarship Program of the Society of Naval Architects and Marine Engineers," *Trans.* SNAME, Vol. 95, 1987.

Weinblum, G., and St. Denis, M., "On the Motions of Ships at Sea," *Trans.* SNAME, Vol. 58, 1950.

Yagle, R. A., "Naval Architectural Education Today," *Trans.* SNAME, Vol. 74, 1966.

Raymond A. Yagle, formally educated as a naval architect, served as a naval officer during World War II and worked in industry for several years thereafter. Most of his career, however, has been spent at the University of Michigan. For some 40 years he has taught courses, first in mechanics, and then in ship strength, ship design, and his concept of ocean engineering. He is well known throughout the industry.

Marine Training and Research in Canada

David G. Parkes

Education

A study of maritime education in Canada is, at first glance, mostly a study of vocational education and training. Beginning in the mid-1730s at the fortress of Louisbourg in Cape Breton, to the present day, Canada has seen the growth and decline of a plethora of marine training facilities. There have been private, provincial, and federal vocational training facilities, but only in the past 15 years has there been any marine education at the university level.

Until recently, there has been little in the way of a single, national maritime training policy, nor has there been a coordinated national training plan.

This is surprising, given Canada's stewardship of the world's longest coast line and her ownership, in 1945, of one of the world's largest merchant fleets. Yet it is also understandable, given the provincial jurisdiction over education, a general lack of any national education policy, and a definite lack of a national maritime policy.

Over the past 300 years Canada's shipbuilding industry developed region by region. Shipyards in the Atlantic provinces built clipper ships for overseas trade; the St. Lawrence and Great Lakes yards produced ferries and cargo ships for inland trade; and West Coast builders developed a vast, economical tug-and-barge system to support our forest industries.

Recognizing the relatively small size of the shipbuilding industry in Canada and its geographical dispersion, the industry has designed and built a remarkable diversity of high-technology, special-purpose ships ranging from icebreakers, ferries, offshore supply vessels, hydrofoil craft, frigates, semisubmersibles, self-unloading bulk carriers, and seismic research vessels. Perhaps what is more remarkable is that these achievements were made by an industry that for many years operated without a source of nationally trained engineers and technologists and also a low level of investment in industry-related research and development. This is probably because, although the Federal Government has always had complete authority to control shipping and navigation, it has been relatively inactive over the years. As McKenzie King, a former Prime Minister of Canada, is sometimes quoted as saying, "Shipping policy in Canada must be constantly studied to insure that no critical decisions are ever taken," and it seems that few have been.

Under the constitution the provinces have retained the right to provide education to Canadians, and they have historically managed to avoid working together to create any common policies or standards.

Canada's original shipbuilders were immigrants, mostly from Europe, who brought their skills with them. Whenever there was a need for additional manpower or skills they either sent "home" for them, or set up local facilities for training young men in their neighborhoods.

Since then, however, many efforts have been directed toward maritime education and training. Most of these efforts were regional in nature and often limited in scope and, although some were private endeavors, a majority were provincially managed. Only in the past 15 years have our major universities offered courses dedicated to naval architecture or marine engineering.

The Schools. Apart from the school at Louisbourg, the earliest recorded schools seem to have been private activities set up in Newfoundland and Nova Scotia. For example, in an advertisement for his private school in 1814, Jeremia McCathyin included navigation as one of the subjects to be offered; and in 1841 a plea went out from the Benevolent Irish Society for extra funds so that it could expand its facilities in St. John's, Newfoundland, to accommodate navigation courses. In the early part of this century, private schools also existed in Halifax, Saint John, N.B. and Montreal.

Various provinces started schools or programs dedicated to marine training at various times. In Newfoundland, in 1928, University College, the forerunner of Memorial University, offered a vocational course in navigation. Interestingly it was not offered on campus but rather by a traveling school. In 1948 navigation was offered at the St. John's Vocational Institute but it was not until 1964 that an act of the provincial legislature created the College of Fisheries.

The Canadian Government apparently opened the Marine Navigation School in Nova Scotia in 1872, and a School for Marine Engineers was opened in 1948. In 1961, the Province of Nova Scotia assumed responsibility for both schools, and in 1974 combined them to form the Nova Scotia Nautical Institute. This was moved into new, purpose-built facilities at Port Hawkesbury, N.S., which subsequently became part of the recently formed provincial Community College system.

In Quebec, a marine school was opened in Rimouski in 1944. Its mission was to train marine officers, both engineers and navigators, which it accomplished for many years, placing its graduates in senior positions in the marine fields in Canada and abroad.

In 1974 education in Quebec was dramatically reorganized. L'Institut Maritime was integrated with the Community College of Rimouski; its location and mission were solidified, and its programs became part of the system of junior college level programs in Quebec. In 1978 l'Institut moved into a converted convent, giving it a campus of 15,000 square meters.

Since the move, the mission of the institute has grown so that it now trains students, at the technologist level, in marine engineering, navigation, naval architecture, and telecommunications and electronics.

In Ontario, the first major school was at George Brown College in Toronto. After a number of successful years the school failed, and its activity was transferred to the Georgian College at Owen Sound. Historically, Owen Sound has been a major source of mariners for the Great Lakes since the early part of the century, when it was one of the busiest ports in central Canada. Training had existed there for years in the form of tutorial schools, which operated only in the winter off season to allow out of work mariners to study for the examinations required by the Department of Transport to upgrade their certificates.

The new school began modestly in a church manse in 1970 when traditional marine upgrading courses moved into the building. Four new buildings were constructed between 1971 and 1989. The school, constantly supported by the Great Lakes shipping industry, continues to train future officers, although reduced demand has led to the curtailment of the training of new navigation officers. Ontario also operates maritime training facilities as part of the Community Colleges of Applied Arts and Technologies at St. Lawrence College in Cornwall and Niagara College in St. Catharines.

In Prince Edward Island, Holland College continues to provide training for mariners. In British Columbia the Pacific Marine Training Institute and Camosun College also provide courses for the industry.

The Canadian Coast Guard College opened in 1965 at a vacant naval base near Sydney, N.S., with the sole purpose of training young men and women to become officers in the Canadian Coast Guard.

The Coast Guard Officer Training Program, a technologist level program, began as a four-year course in either marine engineering or marine navigation. In response to an increased demand for officers in the Coast Guard, the duration of the program was reduced to three years. Changing requirements, however, led to the re-establishment of the four-year program, which is still in place.

In 1981 the College entered new, purpose-built facilities one mile from the old location. The College was immediately expanded to meet a sudden increase in the need for graduates, but that need subsequently declined in parallel with the decline in marine activity in Canada, particularly in offshore oil exploration.

In July 1990 the Department of Transport decided to consolidate, at the College, all the technical marine training taking place at both the college and the Maritime Training Center of the Transport Canada Training Institute in Cornwall, Ontario. Upon completion of the consolidation, the Federal Government's center of excellence in civilian technical marine training will be the Canadian Coast Guard College, which will result in a more effective, efficient and economic use of the Department's resources.

Since the early 1980s, the Coast Guard College has also been involved in international training and has played a large part in the Caribbean Maritime Training Assistance Program, funded by the Canadian Industrial Development Association (CIDA) and administered by the Department. Over 500 Caribbean candidates have been trained at the College in programs ranging from 9 to 26 weeks.

Dr. Edgar Gold, of the Oceans Institute of Canada, in a keynote address presented to the National Conference on Marine Training in Ottawa on October 30, 1989, said "... [Canadian marine training] ranges from studies for certificates of competency under the Canada Shipping Act, upgrading certifications and licenses, and certificates of completion of certain skills courses to limited university-level training. The latter is provided by the Coast Guard College, an establishment with superb facilities which nevertheless is rather under-utilized and basically only provides training for the Canadian Coast Guard personnel. However, the qualifications obtained are of only limited portability and are not designed as stepping stones for broader academic or technical use. The Coast Guard College has all the basic requirements to become a Canadian Maritime University, which should be seriously considered."

The Universities. In 1967 the Department of Industry, Trade and Commerce commissioned Dr. S. Alpay, then Head of the Department of Mechanical Engineering at the University of Waterloo, to produce "a study to identify the requirement for academic education in naval architecture and marine engineering within Canada's Marine Industries." The study was to examine the existing level of education, training, and research and development in Canadian shipbuilding, to estimate the future demand for such education and training, and to prepare an undergraduate curriculum for an engineering degree level naval architecture program.

Dr. Alpay's report became something of a milestone in the history of education for the shipbuilding industry in Canada, but no resulting action took place at that time.

In 1975 Captain K. P. Farrell revived the subject with the members of the Canadian Shipbuilding and Ship Repairing Association (CSSRA), and offered to compile an adequate report for that organization. The resultant work was discussed at a SNAME panel held in April 1976, and was recorded in the proceedings of the Eastern Canadian Section of SNAME dated April 22, 1976. The report included an estimate of the number of professional and technical people required in the maritime engineering field within Canada.

This meeting heard summaries of the work being initiated at the University of British Columbia, the Technical University of Nova Scotia (TUNS), Memorial University of Newfoundland, and the College of Fisheries, Navigation, Marine Engineering and Electronics in Newfoundland; and it gained the support of related government departments.

There had been practically no university level educational programs for professional naval architects in Canada until the CSSRA initiative noted above. The four universities started to offer courses and to set up programs to meet the needs of Canadian marine industries. The formation of these courses and report of their progress was recorded in a SNAME symposium held in Montreal on January 15, 1980. This served to present a coordinated summary of universities in Canada that were offering degree or diploma programs in marine engineering and naval architecture.

As a result of the development of this national approach to education and training, the CSSRA compiled a booklet entitled "Opportunities in Shipbuilding." This was published in 1983 and issued to educational authorities across Canada with the object of instructing young men and women about the activities of the marine industry, and explaining how they might enter the various educational programs. In a large, sparsely populated country like Canada, with its widely dispersed colleges and universities each directed on a regional basis, it is important to put together an integrated picture of the opportunities for maritime training. This booklet was a welcome publication, and its sponsors intend to produce revised editions as appropriate.

Memorial University of Newfoundland (MUN). A fully accredited undergraduate curriculum in naval architecture was introduced at Memorial University of Newfoundland in 1980. The program is composed of 62 courses, each of twelve weeks duration, and six four-month work terms. The Canadian Council of Professional Engineers has accredited the program three times since its start. Graduates have gone on to MIT, Berkeley, Michigan, and Hamburg to obtain Master's degrees.

The Master's degree program is entitled Ocean Engineering and may not be strictly confined to naval architecture or marine engineering. Many graduate students from the other disciplines attend post-graduate courses offered by the naval architecture group.

Memorial University also entered into an agreement with the University of New Brunswick (UNB) in which UNB enrolls students for the first four or five terms, the so-called core program, and then transfers them to MUN for completion of the naval architecture curriculum. Subsequently UNB built on this initiative and now offers a full program in marine engineering. MUN also accepts students from the Engineering Diploma program from the University of Prince Edward Island.

Technical University of Nova Scotia (TUNS). The program at TUNS is a post-graduate program offered in the Mechanical Engineering Department, which provides an opportunity for engineering graduates who have completed non-naval architecture Bachelor's degrees and would like to develop a career in the marine field. With two years of full-time studies, a student can usually achieve his or her specialization requirement in naval architecture and marine engineering and also acquire theoretical foundations and advanced analytical skills qualifying for a Master's degree. Students with various technical backgrounds, such as in physics or electrical, chemical, or mechanical engineering, have completed the program. Students can select from among four areas of specialization, namely ship design and production, ship/marine hydrodynamics, ship/marine structural analysis, and marine engineering systems.

University of British Columbia. At the University of British Columbia, naval architecture is offered as part of the mechanical engineering programs at the Bachelor's, Master's, and Doctorate levels.

University level programs in marine studies also exist at many other universities in Canada. Examples include ocean sciences and technology programs at Memorial University of Newfoundland, and the ocean sciences, marine law and policies, and ocean business programs at Dalhousie University in Halifax.

The Impact of the Federal Government. The Department of National Defense has traditionally trained personnel to meet the Navy's requirement at either the Military Colleges (Royal Military College, College Militaire Royale de St. Jean, and Royal Roads) or at the Fleet Schools in Halifax and Victoria, B.C. In recent years, however, the Navy has contributed greatly to the growth and improvement of marine training by sponsoring specific programs for training marine engineering technicians at St. Lawrence College, Cornwall, and l'Institut Maritime de Quebec, at Rimouski. These are based on nationally accredited, three-year merchant marine programs. Moreover, they have changed their "Commissioned from the Ranks" program so that all candidates are academically upgraded to technologist level before they start their application training, usually at the College of Fisheries in Newfoundland or at Camosun College in British Columbia.

The Navy has recruited large numbers of engineers from various universities as direct entry officers into their Maritime Engineering Classification and has also worked specifically with Memorial University and its cooperative engineering program.

To obtain qualified naval architects, however, the Navy must continue to sponsor engineering officers to post-graduate degrees at either MIT, University College London, or other foreign universities.

The Department of Transportation has influenced marine training in Canada in two ways; first as the operator of the Canadian Coast Guard Fleet, Canada's single largest fleet, which created the Canadian Coast Guard College in 1965, and second as the regulator of marine training in Canada through the Coast Guard Ship Safety Branch, previously known as the Canadian Steamship Inspection Service.

The Department has directly influenced the training of mariners as operators and maintainers of vessels rather than as ship designers or repairers. However, the need of the Coast Guard for technologically sophisticated vessels, particularly for icebreaking, has had a major impact on naval architecture, and indeed on all aspects of shipbuilding in Canada.

Research

Marine research in Canada has taken place in two principal areas, namely research in ice-covered waters, and research for national defense.

The science of icebreaking has been developed over the years in the production of ships to operate in the St Lawrence River and East Coast ports during the winter season, and in the Arctic in the summer. In 1960 a new interest was taken in the transit of ice in order to develop the potential of oil and gas reserves in the Arctic region. Much innovative work was sponsored by the Canadian Coast Guard and various oil companies to investigate the most efficient way of working ships in ice. Between 1970 and 1988, 16 icebreakers were built or converted for the Coast Guard.

New hull forms and different methods of propulsion were developed, leading to a considerable improvement in the science of ship transit through ice. These developments have been reported in SNAME papers presented during the past 20 years.

One unusual development was the discovery that air-cushioned vehicles were particularly effective at breaking ice in a level ice field. Although a full application of this principle has yet to be achieved, the technique has proven effective in preventing flooding from ice jams in shallow rivers around Montreal.

The Transportation Development Center was established in 1970 in Montreal. It is Transport Canada's central research organization. It manages an R&D program aimed at improving the safety, efficiency, and accessibility of the Canadian transport system, while protecting the environment. Projects are planned and implemented in close cooperation with groups from Transport Canada, as well as with other departments and levels of government. Often in cooperation with partners, projects are contracted out to a variety of organizations, mostly in the private sector, including operators, manufacturers, and consultants.

The Center has collected an impressive record of R&D achievements over its relatively short history.

These include the Shipboard Integrated Marine Navigation System, the Shipboard Integrated Navigation Support System (SINSS) for navigation in ice, a bow optimization model used in the modernization of various Coast Guard ships, and a propulsion dynamics model used to evaluate the drive trains for a wide range of ice-going ships.

The National Research Council (NRC) and the Department of National Defense have cooperated in a number of engineering developments that have been adopted in Canada's naval ships.

The National Research Council (NRC) has model tested the hulls of all our naval ships and has participated in full scale evaluation of the results. Of specific interest is the work done in acoustics research which is reported elsewhere in the Canadian Naval chapter. Other important developments have taken place in the operation of manned helicopters from our frigates, and in the design of ships to resist shock and air blast. The Navy's research establishments in Halifax, Suffield, and Equimalt spearheaded these developments.

Another activity largely sponsored by the Canadian Navy, the National Research Council (NRC), the universities, and private consultants is participation in the work of the International Ship and Ocean Structures Congress (ISSC). This congress is organized by the universities of maritime nations to monitor and encourage development in ship structures.

The ISSC took special interest in the Canadian development of structures to withstand ice impact. Largely as a result of our contribution in this area, Canada was invited to participate on the Steering Committee of ISSC. Norman Jeffrey, head of the Institute of Marine Dynamics in St. John's, Newfoundland, is our representative and is planning to sponsor the next ISSC Conference in St. John's in 1994.

David G. Parkes is the Commandant of the Canadian Coast Guard College, in Sydney, Nova Scotia. He is a former ship's officer in the Canadian Coast Guard.

Computers and CAD/CAM

JON MATTHEWS

THE INTRODUCTION and wide acceptance of computer-aided design has produced significant changes in the engineering and manufacturing processes throughout the U.S. and Canadian shipbuilding and ship operating communities. The principal reason for the exponential growth in computer-aided design and computer-aided manufacturing (CAD/CAM) is that manufacturing and service companies are finding it increasingly easy to exploit the speed, power, and versatility of the computer to achieve major gains in productivity. This chapter will describe some of the accomplishments achieved by the marine community and will discuss the technical revolution supported by SNAME and other marine societies as the use of computers continues to change our lives.

Topics to be covered here include the use of computers as an engineering tool for propulsion plant heat balances, ship's hull characteristics data, and high-performance structural design using finite-element analysis. Other topics include the use of the computer as a manufacturing tool, resulting in the migration from optical lofting to computer-generated numerical controlled lofting, plus the current trend toward automated manufacturing methods.

Finally, this chapter will deal with the current worldwide leadership role that the U.S. maritime industry has achieved in the development of high quality intelligent interactive geometric product model databases to be used in the design, manufacture, and operation of current and future ship programs.

Brief History of CAD/CAM in the Marine Industry

The past 20 years have seen the birth and exponential growth of computer aids in the marine industry. As mentioned above, the principal reason is that manufacturing and service companies are finding it increasingly easy to take advantage of the speed, power, and versatility of the computer. This chapter will describe the evolution of computer-aided design (CAD) and computer-aided manufacturing (CAM) technologies, with emphasis on developments by and for the marine industry. (The glossary on page 173 defines some common CAD/CAM terminology.)

The rapid growth in computer supported design applications within the marine community in recent years in support of reduced engineering costs and schedules, has resulted in the commitment to maintain computer skills at the leading edge of practical technology. Development of low-cost computer hardware, coupled with easy-to-use interactive software application programs, and the rapidly growing standard libraries of components in digital form, all reinforce the further expansion of computer applications. This leads to an even greater commitment as new products and tools continually increase the scope of computer usage.

The period between the mid-1960s through the 1970s produced many engineering applications, including the development of both general purpose and customized software tools in the design of marine structures. These include a full range of naval architecture programs, finite-element analyses, propulsion,

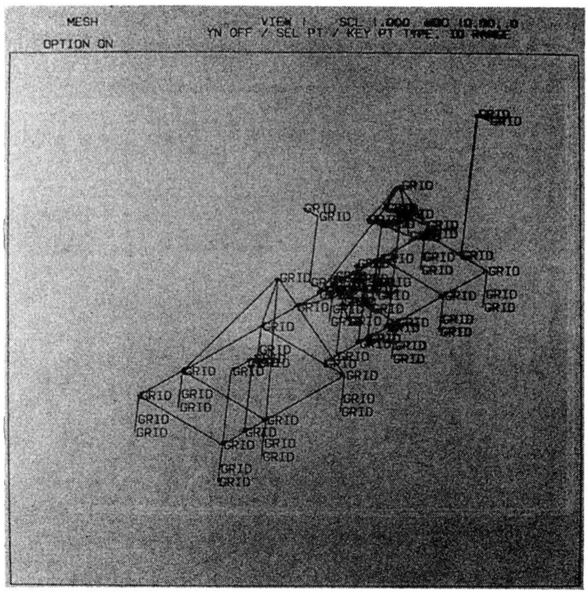

Fig. 1 Finite-element model: Foundation.

shafting, ship motion studies, speed/power estimates, propeller design, shock analysis, and others.

The mid-1970s saw the beginning of real growth in computer applications, as tools, both hardware and software, were beginning to demonstrate their worth as production aids within the ship design and ship building industries. The products offered began to show promise as easily adaptable tools that could perform complex tasks that supported production activities and resulted in reduced engineering time, lower costs, better quality, and less rework in the shipyards.

The clear economic advantages through reduced costs and improved schedules traditionally result in a decision to install computerized facilities.

These early efforts were previously focused on automating the manufacture of structural parts by replacing traditional lofting practices with numerical control (N/C) processes. Through this period (from mid-1970 through 1980), the primary facilities were alpha numeric terminals with "batch" input/output capabilities. This resulted at times in engineers, designers and draftsman becoming expensive keypunch operators who spent a great deal of time waiting for a response from the relatively low powered computers of the day. Nevertheless, the overall throughput was increasing, as a great deal of data was being processed more accurately and faster than by manual techniques.

About the same time, so called "turnkey" interactive-graphics vendors began to show real progress in software development as it affected the marine community, and CAD/CAM terminology became commonplace. It was at this point that the marine industry took a forward-thinking approach relative to CAD/CAM applications, and actively sought solutions for common database generation, use and exchange. In this regard the goal was and remains useable, interchangeable data that can be accessed throughout the design, construction, and maintenance phases of a ship.

The CAD/CAM industry, as a whole, has been growing at impressive rates, doubling sales every one to two years. This growth has been the direct result of the development of powerful, useful tools and the application of those tools to traditional manual techniques.

The tremendous growth in work stations in the past few years has resulted in an even greater push for "friendly" systems that allow for easy interaction with the engineer or designer, and more importantly, offer the engineer or designer more control over what the computer does. In addition, as the quantity and quality of the digital database increase, the significance of "common" data, which can be accessed by any computer, becomes paramount to the future growth of the computer within the marine community.

Data Transfer Standardization

The marine industry has been actively working on database exchange for the past several years, as a computer generated database must often be transmitted to other systems for continuous use. While transmissions of data within similar systems can readily be accomplished, transfer of data between dissimilar systems is more difficult owing to incompatibilities among vendors' proprietary hardware and software. Continued pressure by the user toward exchange standards such as IGES (Initial Graphics Exchange Specification) will serve to enhance this capability over the long term.

The continued, rapid expansion of CAD/CAM usage within the industry has resulted in the desire to rapidly backload manual data into digital databases as this activity supports productivity goals by using the computer as an engineering, production, and maintenance tool. A backloading capability, which by itself offers little productive benefit, is useful to supplement computer- generated data with technical information produced manually, and to enhance the preservation of historical data or external products (usually third-party input). In both instances, the intent is to accelerate the generation of data within the computer and to improve the usefulness of the CAD/CAM database.

The techniques for rapid backloading of manual data into a digital format were developed from the generation of geometric models, analytical results,

Fig. 2 Numerical controlled burning.

drawings, text, and production aids. These techniques progressed from an operation of simply backloading data on a CAD workstation and manually digitizing tables, to automatic drawing scanning techniques and external generation of alphanumeric data. Attached digitizers, owing to their inherent ability to aid positioning details on the drawing, and to recreate them in digital form, help reduce the backloading time by 25 to 50 percent. There is still room for improvement, however, as most text and dimensional data continue to be created, rather than automatically copied.

Optical scanning methods, while still in the early stages of development, will further reduce the backloading time as this technology will essentially automate the copying of manual data into digital form.

The effective exchange of information within the shipbuilding process is one of the most challenging tasks faced by any group of organizations in the world. This is because of

- the complexity of the product,
- the life span of the product, and
- the number of participants in the ship product life cycle.

A ship is among the most complex devices known to man. It accomplishes complex tasks while providing hotel accommodations and complete city services (e.g., fire and police protection, medical and recreational facilities, and waste disposal systems) for its operators. Only a few of each type are built, each differing to some extent from one another. By the standards of most industries, these collections (of eight million or so parts per ship) are all engineering prototypes. In effect, ships are constructed (like buildings) and not manufactured (like other vehicles).

Also, unlike aircraft and most mechanical products, ships are not designed, built, operated, maintained, and modernized by vertically integrated corporate giants. Rather all of these functions are accomplished by a collection of government agencies and private companies. Competitive pressures make it impossible to know in advance who the participants in the process will be. Further, the process itself tends to vary somewhat from ship to ship. For example, many builders have found computer-aided design (CAD) applications a cost-effective way of detecting potentially costly interferences. This allows them to change their drawings before beginning construction. Others still rely on manual techniques.

The automation efforts within each agency or company have required substantial investments in hardware and software (both custom and commercial) and in training, orienting, and adapting work processes to capitalize on computer capabilities. The range and extent of this investment is even more impressive considering the general decline and low profitability of the marine industry. There can be no denying that the marine industry is serious about CAD!

Investment choices made by different agencies and companies have naturally led to the selection of different systems. Even companies with identical computer systems have developed different application techniques. Together with the variations in the process noted above, the marine industry must face the requirement to transfer product information between and among all activities and companies. This transfer must take place at all stages of the product life cycle including design, construction and service life support.

Throughout the design, construction, and operation process, cost effectiveness gains can be achieved through the efficient transfer of digital data. This was recognized in 1985 when SNAME's Ship Production Committee Panel SP-4 identified digital data transfer as the area having the greatest potential cost savings for the industry. These savings are made possible through the following efforts:

- Development of data "one time." These data can be used many times within several organizations. The potential savings possible with the sharing of data through computerized processes are not limited to naval programs.
- Maintenance of developed data. The practice keeps the design process fully up-to-date as changes are made. For instance, a current ship product model would be of significant value to shipyards involved in new construction as well as to repair yards and support entities.

Current and consistent ship product information must be readily available to all potential users. Data must be available in usable form to avoid the expense of regenerating databases. A great deal of work is being accomplished through organizations, such as NIDDESC, that define requests, provide specifications, and provide leadership in this complex area.

The Presentation of Ship Geometry

There are many ways to represent ship design data. In the traditional ship design process, the product model takes the form of drawings and corresponding specifications. Other representations include scale models (e.g., for complex and critical machinery spaces),

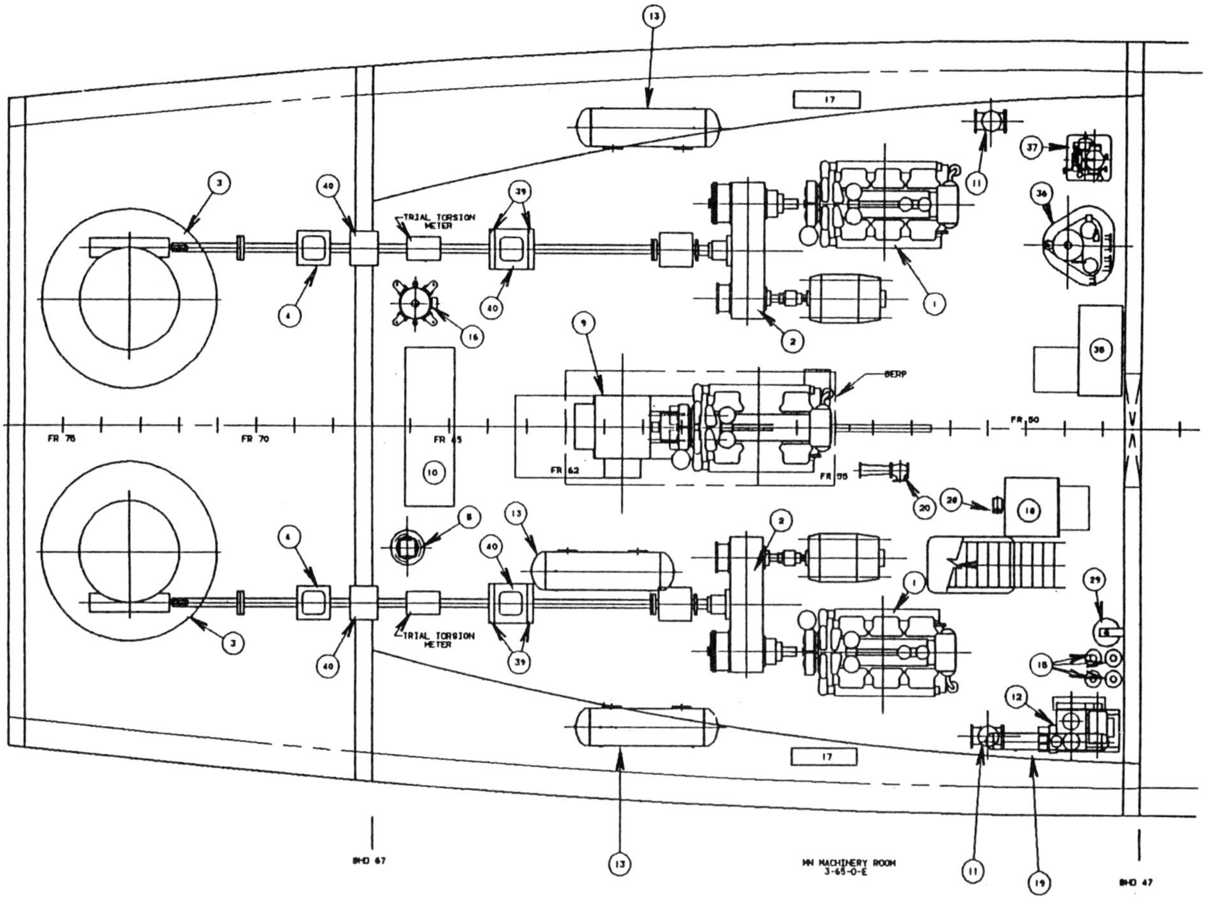

Fig. 3 2-D CAD drawing.

or analysis models (e.g., finite-element models) for special applications. CAD systems likewise employ a number of geometric forms for defining ship geometry. Each of the geometric forms and the visualization techniques associated with them are discussed next.

Engineering drawings represent a formal means of describing design and construction data; the production of engineering drawings is strictly controlled according to developed, accepted standards and conventions. Drawings are two-dimensional representations of three-dimensional objects that are developed in several forms including

- section views, such as elevations, profiles, and transverse sections,
- projections, such as deck plans and outboard profiles,
- isometrics, involving system delineations,
- perspective views,
- schematic representations and one-line diagrams,
- diagrammatic drawings such as damage control plans, and
- plots and graphs—the products of various analyses such as static stability determinations or longitudinal strength calculations.

In all cases, drawings are presented in a 2-D format either on a computer display device or in hard copy format on paper or film. Although many manufacturers use CAD systems to develop 3-D models, they still consider their product to be represented by 2-D engineering drawings rather than by the 3-D solid model in the computer database.

However, much research and development effort is being directed to applying the capabilities of 3-D CAD systems to replace the engineering drawing as the primary form representing a product. In order for this to be completely successful, it is also necessary to provide broad access to this computer-based product model throughout the design, manufacture, and life of the ship.

More sophisticated CAD systems provide for grouping of instanced parts, access to symbols in catalogs, and the association of material properties to

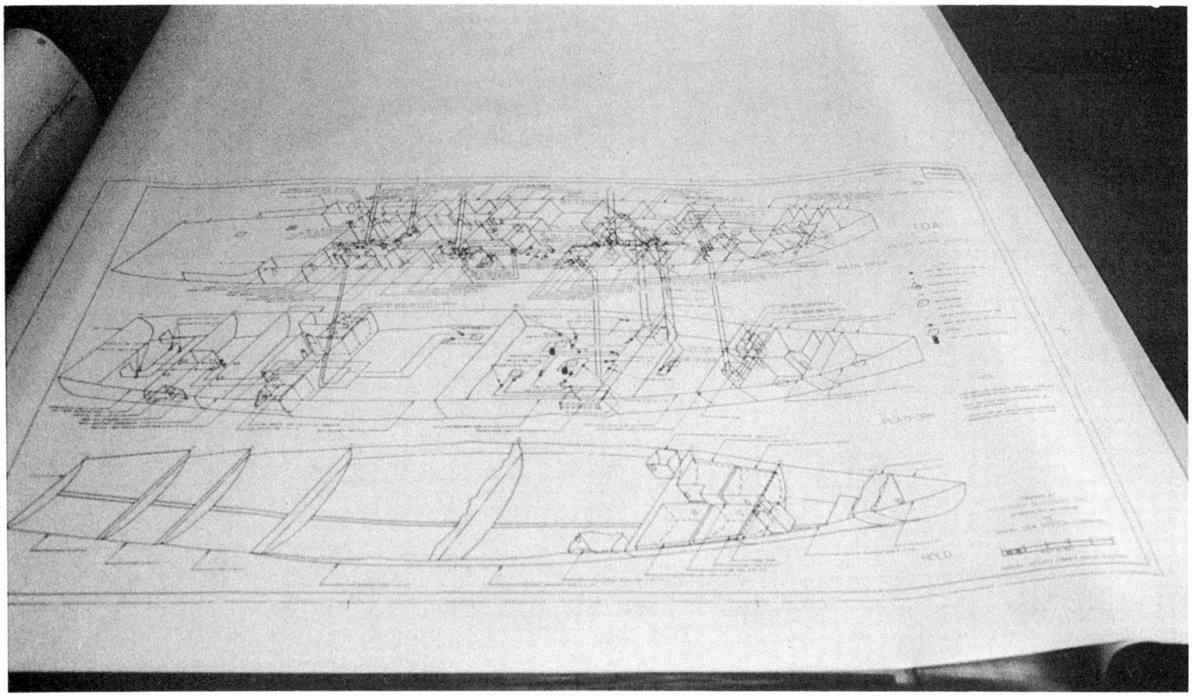

Fig. 4 $2\frac{1}{2}$-D CAD file.

graphic entities such as bills of material. Many of the 2-D systems have utilities to develop auxiliary views using projection techniques familiar to draftsmen. An example would be the ability to draw an isometric projection based on existing orthographic views.

This automatic generation of views is often called 2½-D because it provides the appearance of 3-D in a 2-D system. Some CAD systems provide for converting raster data to the native vector form of the CAD system. These conversions require human intervention and validation of the results. Some systems have utilities to allow overlaying raster and CAD system data so that the raster data can be digitized on the screen by the CAD operator. This type of utility allows the operator to convert only that portion of the file that requires changing and can save time and effort when dealing with raster drawings.

Although there are many variations in 2-D CAD system capabilities, it remains true that the drawing is the product model in all cases. Although it is possible to develop representations of 3-D objects and assemblies, the database remains 2-D. Such data are transferrable between CAD systems using standard file formats, although some loss of information may result. This method has been employed in recent ship design programs using the initial graphics exchange specification (IGES) file format. In contrast to neutral file conversions, direct conversion computer programs attempt to overcome the problem of information loss for a limited number of CAD systems.

Wireframe models define 3-D geometry by incorporating model edges. One way to envision this representation is to imagine that the modeled object is made of a transparent material. All of the edges of the object are visible regardless of the viewing angle. If an object is curved, the curvature would only be visible at the edge of the object. This method of geometric depiction is limited. The wireframe model is composed only of "wires," which can be represented by points, straight lines, arcs/conics, or curves. Even though the wireframe model may appear to define a solid physical entity, there is no implicit definition of surface entities (e.g., surface curvature, areas or area properties) and enclosed surface entities (e.g., thickness, volume or volumetric properties).

Many CAD systems provide the ready ability to convert wireframe geometries into surfaces, or solids, and vice versa. Under these circumstances, surface analysis including calculation of areas, centers, first moments, and moments of inertia may be performed One should remember, however, that these properties are performed on the surface or solid created from the wireframe, not on the wireframe itself.

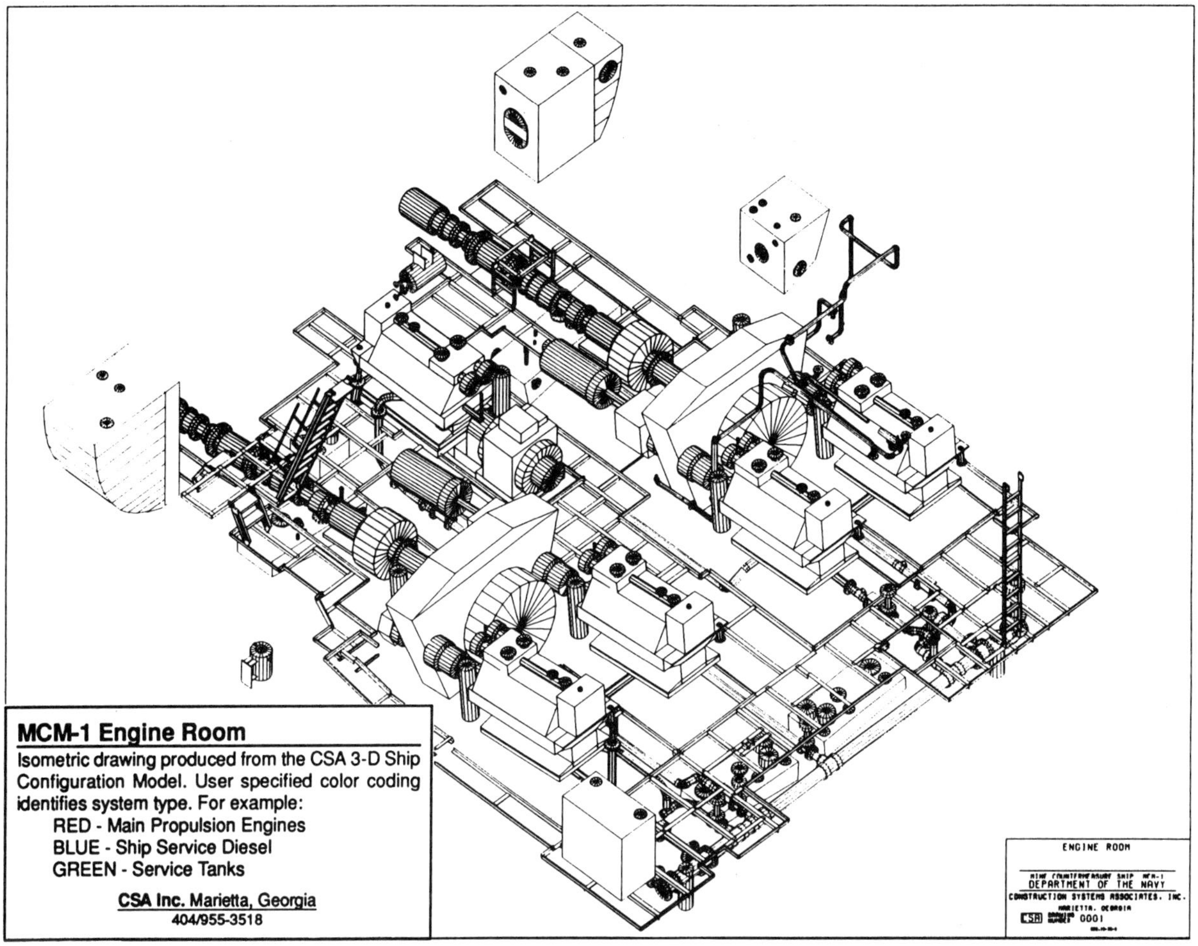

Fig. 5 Wire frame model.

More advanced visualization techniques, such as shading, are not possible with wireframes. You may, however, create the appearance of rendering of a solid model using wireframe geometry. This visualization technique, called hidden line removal, is partially a manual process. The designer designates the viewing angle and the entities in view. The designer can then erase all or part of the lines from the hidden line view. Lines that represent the maximum extent of surfaces (i.e., tangencies) must also be manually created. In comparison, the solid model can always be viewed from any angle and will yield a solid appearance. This is because changes to the solid model geometry are automatically reflected in any generated view.

Surface modeling is a separate consideration from surface visualization. Surface modeling is achieved by a concise mathematical formulation that defines the surface continuously and uniquely. Computer applications make direct use of the surface definitions to perform such functions as surface–surface intersection, surface modification, translation, rotation, scaling, and visualization. Surface visualization techniques allow the operator to "see" the surface. Visualization techniques include grids and lines arranged on the surface to give a feel for the curvature or shape of the surface.

Fig. 6 Solid model.

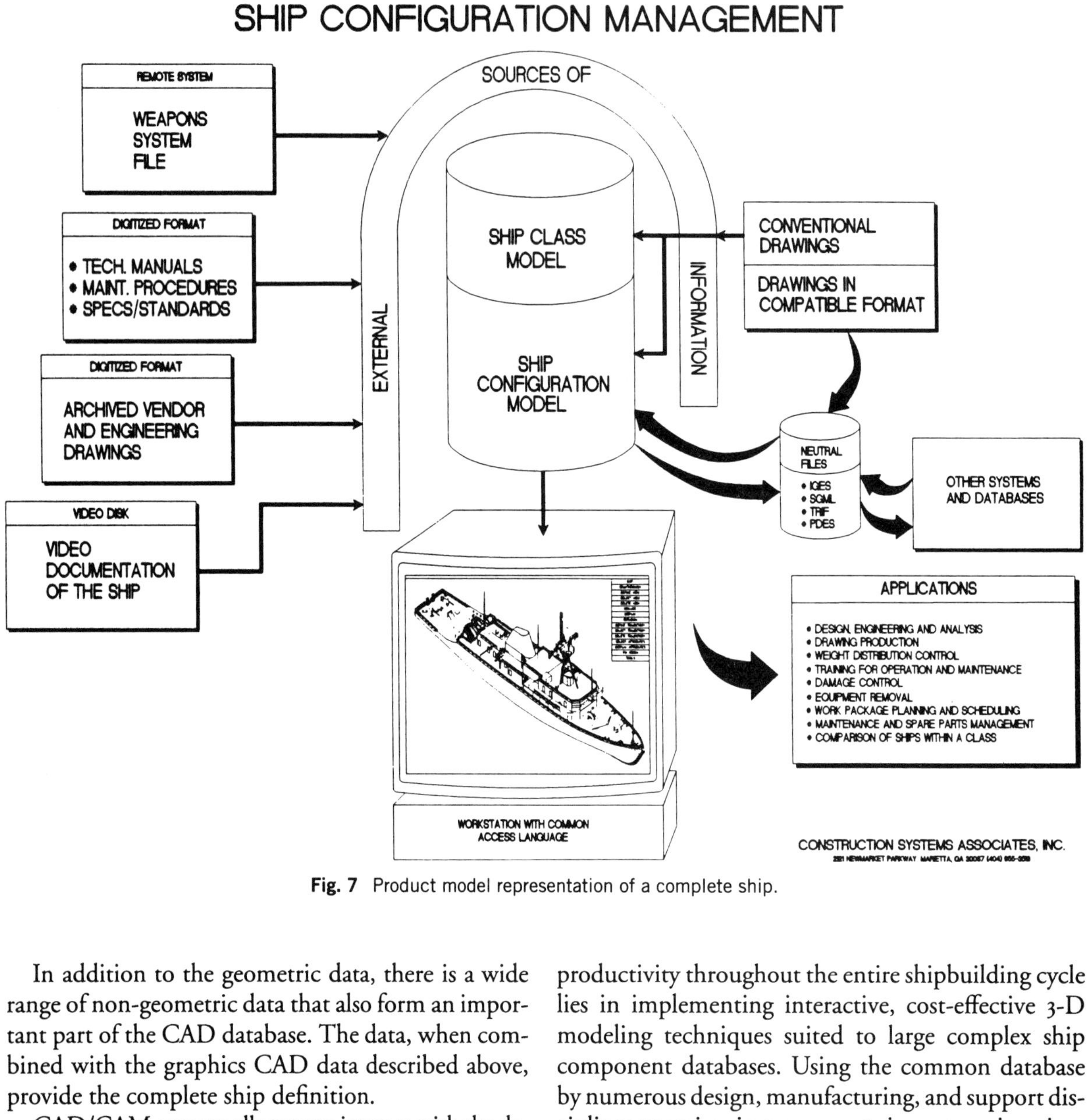

Fig. 7 Product model representation of a complete ship.

In addition to the geometric data, there is a wide range of non-geometric data that also form an important part of the CAD database. The data, when combined with the graphics CAD data described above, provide the complete ship definition.

CAD/CAM systems allow us to interact with the design process so that we approach the optimum configuration through interactive design refinements. Also, we can visualize the product before fabrication, which allows us to evaluate design and production tradeoffs.

Looking Ahead

In looking to the future, the industry must maintain a high quality product at a lower overall cost. This doesn't imply an absolute commitment to any one system; rather it emphasizes the need to find the best operating approach to deal with the work to be accomplished.

It has become clear that the key to maximizing productivity throughout the entire shipbuilding cycle lies in implementing interactive, cost-effective 3-D modeling techniques suited to large complex ship component databases. Using the common database by numerous design, manufacturing, and support disciplines promises improvements in cost-engineering, planning, interference control, and engineering and design. None of these applications, however, eliminates the use of skilled people throughout the design process. Interactive graphics does not make good designers out of bad ones, but instead permits good designers to make better designs.

We believe our industry is an attractive market to spur software and hardware development.

Substantial growth in this complex and rapidly changing environment over the next five years can be anticipated with CAD/CAM playing a major part of the strategy to achieve the productivity gains that will enable us to provide a top quality, cost-effective ship in the 1990s and beyond.

Glossary

CAD/CAM. Computer-Aided Design/Computer-Aided Manufacturing.

Database. Any collection of information.

NIDDESC. Navy Industry Digital Data Exchange Standards Committee—formed in 1986 to develop marine industry standards for CAE/CAD/CAM data to facilitate communication within the industry.

N/C. Numerical Control.

IGES. Initial Graphics Exchange Specification—an ANSI standard for the exchange of CAD system data.

Vector. A format for presenting graphical data in digital form. A vector graphic image contains the end points and directions of line segments that make up the drawing.

Product Model. All the data required to define the geometric, topological, functional, physical and technical aspects of a manufactured or constructed product. Product model data changes in form and quantity throughout the product life cycle from design, construction, maintenance, alteration, and disposal.

Raster. A means of presenting graphical data in digital form. The raster comprises dots on a display device such as a computer terminal, laser printer, or electrostatic plotter. A raster is formed by scanning the source document to form rows and columns of dots.

Jon Matthews has over 25 years of experience in commercial, Navy, and offshore engineering, planning, and management. He is a vice president of JJH, Inc., responsible for the Tidewater Area operations.

Mr. Matthews holds bachelor's and master's degrees in mechanical engineering from Drexel University and the University of Pennsylvania. He has published six papers, including four on computer-aided design. He is a member of several SNAME design panels and chairs the panel on computer-aided design. He is a member of SNAME, ASNE, and ASME.

Consulting Naval Architectural and Marine Engineering Firms

CHARLES R. CUSHING

Introduction: Historic Outline

IN THE 50 YEARS from 1942 to 1992, the profession of naval architecture and marine engineering has moved in parallel with the changes in the maritime industry, each impacting and changing the other. Neither is the same after a half century of transition.

The following narrative does not delve into yacht design, as this facet of the profession is covered elsewhere in this volume. Design work on naval vessels and similar government ships is not dealt with extensively herein due to national security and other factors. However, it should be noted that the total amount of ship and boat design work performed in the United States throughout the period 1943 to 1993 for the U.S. Navy, Coast Guard, Military Sealift Command, and Army vastly exceeds that performed on merchant ships and yachts. Private consulting naval architectural and marine engineering firms have provided the U.S. Government and the U.S. shipbuilding industry with most important, if not vital, support in the design of naval type vessels throughout this 50-year period. This applies particularly to the larger design firms.

The shipping industry expanded rapidly during World War II under the vigorous War Shipping Administration Program. The U.S. Maritime Commission, later to become the Maritime Administration, embarked on an ambitious post-war program to rebuild the merchant marine. In the late 1940s industry absorbed the war-built standard ships, T-2s, C-2s, C-3s, etc., but competition from rebuilt foreign merchant fleets forced technical development on otherwise traditional offshore shipping companies.

During the 1940s, the industry turned to the naval architectural profession to supply the technical support required to prepare conceptual, preliminary and contract designs. Government subsidy approval and shipbuilding bidding procedures also required considerable assistance from independent naval architects and marine engineers. For example, a typical bid package would consist of 20 to 50 drawings and a voluminous set of specifications. The consulting naval architect was often called upon to assist in the bid analysis and yard negotiations. Often, such firms would be responsible for working plan approval, and perhaps provide all or part of the owner's shipbuilding inspection team.

By the 1950s, a number of technical developments made it essential that shipowners reach outside to secure specialized technical assistance. These innovations included

- higher sea speeds,
- larger vessels,
- higher propulsion steam temperatures and pressures,
- metal folding hatch covers,
- bulbous bows,
- higher strength steels,
- advances in welding,
- automation,
- cargo dehumidification,
- improved habitability,
- bow thrusters and stern thrusters,
- cathodic protection,
- more extensive electronic navigation equipment,
- improved hull forms,
- advances in propeller designs,
- oil lubricated stern tube bearings,
- hydraulic and improved electric deck machinery,
- innovative cargo gear rigs and deck cranes, and
- anti-rolling fins and anti-rolling tanks.

The consulting naval architects and marine engineers, together with the equipment manufacturers, shipyards, and owners' staffs, were instrumental in developing and implementing these advances.

By the end of the 1950s and through the 1960s, the revolutionary cargo handling advances embodied in containerships, roll-on/roll-off, and barge carrying ships were a direct result of innovative consulting naval architects or their collaboration with shipowners and their in-house technical staffs.

During the 1940s and 1950s, inland marine transportation flourished, principally on the rivers, with towboats and flotillas becoming larger. Consulting naval architectural involvement was limited. Usually such involvement was restricted to propulsion, improved hull forms and specialized barges, with the main technical thrust coming from the river yards, themselves.

Coastwise trade, unfortunately, was not in a growth mode and did not fare well during this period. The traditional intercoastal (Atlantic–Pacific) domestic trade, except for containers, declined steadily, with Calmar Lines virtually the last carrier, moving primarily steel and bulk cargoes. Sea-Land, Seatrain and Bull Lines (until 1964) were active in the Puerto Rican trades. All of these operators engaged consulting naval architects to carry out ship conversion programs, essentially rebuilding or jumboizing ships built during World War II.

On the Great Lakes two independent design firms were formed immediately after World War II. One was headquartered in Cleveland, the other in Sturgeon Bay WI. The two of them are credited with most of the innovative designs that have come out on the Lakes since that time.

Containerization was an American innovation. The early days of its development in the late 1950s and early 1960s was spearheaded by Sea-Land on the East Coast and Puerto Rico, and by the Matson Navigation Co. for its domestic trade to Hawaii. Sea-Land made extensive use of consulting naval architects, while Matson's own technical staff carried on most of its work; but they did reach out to consultants for special help.

After an early abortive attempt by Grace Lines in the South American trade in the late 1950s, the container concept spread to the U.S.-flag foreign trades in the early 1960s, principally by Sea-Land in the Atlantic and American President Lines in the Pacific. Both companies extensively engaged consulting naval architects to develop the required technology.

The rival barge carrying concepts, LASH and SEABEE, were consultant-driven developments that found applications worldwide. Foreign shipping lines engaged U.S. naval architects to assist them in developing these systems.

With a number of exceptions, the roll-on/roll-off vessels were more extensively developed by the shipyards and owners' staffs.

By the 1970s, the Oil Producing and Exporting Countries (OPEC) had in a short period caused a 20-fold increase in oil prices. Fuel had suddenly become the largest single line item in a vessel's operating costs, especially with the higher-speed liner vessels. Just as suddenly, ship operators became energy conscious. Intense efforts were made to develop hull forms with less resistance, more efficient propellers, hull coatings to reduce boundary layer drag, engines with lower specific fuel consumption, waste heat recovery devices, engine-driven auxiliaries, etc. These efforts came from all branches of industry, academia, and government, but the naval architectural and marine engineering firms were in the forefront.

Rising shipboard labor rates and extensive vacation plans (six months per year) moved labor costs to another prominent position in operating budgets. Hence, automation became important in the period 1960 to 1980. The early dramatic reduction (40 percent) in vessel manning was the result, not of automation, but of work rationalization. A further reduction (of up to 50 percent) was a direct result of extensive use of automatic controls, sensors, data logging, microprocessors and computers. Consulting marine engineers were active in those areas.

Ferry boat design in the United States has been very much a product of consulting naval architects, particularly in Seattle, San Francisco, and New York City.

Besides ferries, many other types of non-naval, government-owned ships are the result of the work of American and Canadian consulting naval architects. Included here is a wide range of craft serving such diverse functions as oceanographic research, fire fighting, oil spill clean-up, coastal patrol, and pilotage. This involvement comes from the tendency of government agencies to follow formal construction-bid-to-specifications procedures.

Before World War II, virtually all trans-Atlantic, trans-Pacific, and South American passenger travel was by ship. It was not until 1957 that the number of passengers crossing the Atlantic by air equalled that by ship. Until that time, American consulting naval architects and marine engineers were almost exclusively responsible for the development of safer, faster and more luxurious passenger ships. Between 1945 and 1965, United States Lines' *United States*, Grace Lines' *Santa Rosa* and *Santa Paula*, Moore-McCormack's *Brazil* and *Argentina*, and American President Lines' passenger ships were all the result of close collaboration between the owners and the independent naval architects.

During this period, a good many cargo-passenger ships were put into service. Some of these were new construction, but others were conversions of war-built cargo ships. During the 1950s there appeared the cargo-passenger vessels of Grace Lines (52 passenger C-2s), Farrell Lines (80 passenger C-3s) and, in the 1960s Grace's *Magdelena*-class, American Export's "Four-Aces," Delta's *Del Mar*-class, APL's cargo-passenger ships, and others. All of these reflected the innovations required to combine into one hull the demanding requirements of a passenger liner and an efficient cargo carrier.

Naval engineering and naval construction is a highly specialized and demanding branch of naval architecture dealing principally with warship design. Most firms specializing in this field have centered in the New York City and Washington D.C. areas, close to the Department of the Navy and other branches of the government requiring naval engineering assistance. Many such firms also maintain branch offices near naval shipyards and commercial yards specializing in naval shipbuilding.

Several U.S. naval architectural and marine engineering firms have been successful in establishing consulting offices overseas. Their work ethic and ingenuity have been sought after throughout the world. The unique, rapid response nature of such consulting services has been found useful to foreign shipowners, especially in the early stages of design. The specialized technology capability, especially in gas carriers, offshore oil rigs, and containerization provided many international opportunities for U.S. naval architects in the period between 1960 and 1980. Other opportunities were found in overseas construction of ships built to U.S.-flag requirements in the period between 1970 to 1990.

In closing this section, one other historic point should be made. In many respects independent naval architects have taken a more prominent role in the United States and Canada than has been true in other parts of the world. Most overseas shipyards maintain large engineering departments, and each yard carries a menu of standard designs and the capability of modifying them to suit individual shipowner's needs. Some American and Canadian yards also employ large design staffs, but many do not and must therefore rely heavily on outside consulting firms.

Organizational Structure

It is difficult to describe the typical naval architecture and marine engineering consulting firm, since they vary so much in size, specialty, corporate structure, and scope of services. The most common feature is that they are privately owned, profit making enterprises. In some cases they may be one-man operations and, in the other extreme, multi-office firms of upwards of a thousand employees.

The one- or two-man firms usually work on only one project at a time. They depend on the services of a hard working secretary, and engage drafting assistance on an as-needed basis. Such firms are usually partnerships or operate on a professional basis.

The ten- to twenty-person firms have a somewhat formal corporate structure. They may have some specialists on the staff, but most of the engineers are generalists and can handle a variety of disciplines. Such firms would have some permanent drafting capability but would hire temporary drafting assistance as required.

The larger naval architectural firms (50 or more persons) would have a more formal structure. Usually they would be privately owned corporations. Such firms would definitely employ at least a few technical specialists.

During the 1940s and 1950s, a large portion of the consulting naval architect's work involved hull design, drafting and fairing lines, carrying out the tedious hand-calculated hydrostatic properties, structural calculations by rule, etc. The marine engineers labored over heat balances, HVAC, and similar calculations, also done by hand. The computer revolution, starting in the 1960s, freed engineers from these dreary, time-consuming calculations and permitted them to do more design work. First, hydrostatic and damage stability calculations were performed by computers. During the 1960s, only the largest consulting firms could afford computers with large capacity. By the late 1960s, however, most consultants were doing powering estimates and propeller calculations on their computers. During the late 1970s and early 1980s, finite element techniques applied to ship and offshore structures could be performed on affordable mini and personal computers. In the 1980s computers had become so efficient and inexpensive that consultants increasingly applied them to computer-aided design (CAD).

Writing voluminous specifications, running as much as 500 to 1000 pages, was another tedious task during the period 1945 to 1970. The development of affordable word processors was a factor in freeing engineers to do more imaginative and rewarding design work.

American tanker companies have traditionally maintained large engineering departments and have been nearly self-sufficient. They have, however, engaged consulting naval architects for specialized technology or to supplement their design staffs. In particular, consulting naval architects were heavily involved in the tanker jumboizing projects of the 1950s and 1960s, and in the gas carrier projects of the 1960s and 1970s.

The offshore oil exploration and production industry originated in the U.S. Gulf. In the 1940s and early 1950s, the relatively shallow water operations required simple bottom rigs, essentially shore-type platforms moved out into the water. As operations moved into deeper water, more innovative rigs evolved, and more naval architectural expertise was called into play. Independent consulting firms, particularly those in the New Orleans and Houston areas, were deeply involved in nearly every aspect of these unique concepts.

In more recent years, although U.S. shipbuilding activity has slowed, independent naval architects and marine engineers have expanded the scope of their clientele by finding work as consultants to overseas shipowners and shipyards.

Range of Services Offered

The traditional role of the naval architect and marine engineer has not changed over the half century 1942–1992. That work, essentially, consists of aiding a prospective owner in the acquisition of a ship or other marine-related object. This work may involve the complete process extending from the initial concept and planning through construction, to testing, implementation in service, and provision of technical services to the operating vessel. On the other hand, it may involve only one or more steps in the process, such as preparation of contract plans, working plan approval, or shipbuilding inspection. Firms engaged as consulting naval architects and marine engineers must be versatile in technical capability. For examples, here are just a few of the principal types of work they may be called upon to perform:

- marine transport planning and economic analyses,
- concept design,
- preliminary design,
- contract design and specifications,
- plan review and approval,
- working drawings (detail design),
- testing/trials/witness same,
- construction inspection,
- contract management (owner's representative), and
- guarantee engineers.

A more complete list, not shown here, contains no fewer than 87 examples of distinct kinds of services that have been provided by consulting naval architects and marine engineers in North America during the second half of this century. In order to accomplish this wide range of work, many engineering disciplines are called into play. Naval architects and marine engineers, in a general sense, may have capability in many of these areas or they may employ specialists in a score of disciplines encompassing such diversities as hydrodynamics, cost estimating, piping systems, and corrosion control.

Appendix: A Compendium of the American and Canadian Independent Naval Architects and Marine Engineers

During the past 50 years the marine industry has been served by many independent design firms. Some have enjoyed only a brief existence, others have become national institutions. Preparing a definitive list of them all is admittedly impractical, but an attempt, however error-prone, seems appropriate simply to give the reader some idea as to the numbers involved. With due apologies to any firms we may have overlooked, below is an attempted compendium (exclusive of purely yacht designers). The asterisks indicate the number of people employed at some time during the years 1943–1993: * for 50 or more, ** for 500 or more, and *** for 1000 or more.

* Advanced Marine Enterprises, Inc.
* John G. Alden, Inc.
Robert Allen, Inc.
* Art Anderson Associates, Inc.
Band, Lavis & Associates
Breit & Garcia, Inc.
Cadcom, Inc.
Camat Int'l Transp. Consultants, Ltd.
*** CDI Marine Co.
Central Design & Drafting, Ltd.
Coast Engineering Co.
Crandall Drydock Engineers, Inc.
* C.R. Cushing & Co., Inc.
DAI, Inc.
Arthur D. Darden, Inc.
Del Breit, Inc.
Design Associates, Inc.
* Designers & Planners, Inc.
* DLI Engineering Corp.
* H.C. Downer & Associates, Inc.
George Drake
Elliott Bay Design Group, Ltd.
* Engineering Visions, Inc.
Friede & Goldman
Fryco
John Gordon German
German Marine, Inc.
* German & Milne
Giannotti & Associates, Inc.
*** Gibbs & Cox, Inc.
Glosten Associates, Inc.
Timothy Graul Marine Design
* Morris Guralnick Associates, Inc.

J.B. Hargrave Naval Architects, Inc.
Peter S. Hatfield, Ltd.
** J.J. Henry Co., Inc.
Herbert Engineering Corp.
C. Raymond Hunt Associates, Inc.
* Hydronautics, Inc.
Jensen Maritime Consultants, Inc.
* JJH, Inc.
Kinetics Group, Inc.
Korkut Engineers, Inc.
James S. Krogen & Co., Inc.
Lindsay Lord
Thomas T. Lunde, Inc.
MacLear & Harris, Inc.
Robert H. Macy
* Marine Consultants & Designers, Inc.
Marine Design Associates, Ltd.
* Marine Systems Corp.
Marine Technical Consultants
Maritech, Inc.
Rudolph F. Matzer & Associates
Alan C. McClure Associates, Inc.
** John J. McMullen Associates, Inc.
George E. Meese
Robert C. Morrell
Navac, Inc.
* George C. Nickum & Son
* Nickum & Spaulding Assoc., Inc.
Guido Perla & Assoc., Inc.
Polar Design Associates, Ltd.
Propulsion Dynamics, Inc.
** Q.E.D. Systems, Inc.
Roland Marine, Inc.
*** M. Rosenblatt & Son, Inc.
Sargent & Herkes, Inc.
* Seaworthy Systems, Inc.
*** George G. Sharp, Inc.
Shipping Research Services, Inc.
SHM International, Inc.
* Phillip F. Spaulding & Associates
* Sparkman & Stephens, Inc.
* The Stanwick Corporation
* R. A. Stearn, Inc.
Robert Taggart
Talbot, Jackson & Associates, Ltd.
Richard R. Taubler, Inc.
* Techmatics, Inc.
Henry M. Tiedemann
* Tracor Hydronautics, Inc.
* John H. Wells, Inc.
* Wheeler Industries, Inc.

Charles R. Cushing is a graduate of the U.S. Merchant Marine Academy, MIT, and SUNY. He has sailed as an officer on U.S. passenger and cargo ships, and was a naval architect for Sea-Land for seven years in its container-pioneering days. For the past 25 years he has been president of the firm C.R. Cushing & Co., naval architects and marine engineers.

Part V

Construction

Shipyards

Howard M. Bunch

Introduction

This chapter discusses the organization and operation of shipyards in the United States starting with the period just before World War II. Origins and driving forces for new technologies and organizational systems within both private and public shipyards are discussed, with emphasis on their effects on cost, quality, and delivery time.

The chapter has three major sections. The overall institutional history of American shipbuilding for the 50-year period 1940–1990 is described in the Introduction. This includes the role of the government in influencing the health and direction of the industry during the period. Then, the role of labor is discussed. This is followed by an overview of U.S. shipyard technology, and changes that have occurred during the 50-year period.

Design and Production Integration focuses on design activities found in the shipyards, and how these activities have been affected by the computer.

Production systems, and their changing characteristics are presented in Shipyard Production Systems. The focus is on how new concepts of production have required significant increases in pre-production planning, and the implementation of concepts of continuous process improvement.

Historical Background. The beginning of the era of modern shipbuilding in the United States occurred just before World War II. In the case of merchant shipbuilding the impetus was passage of the Merchant Marine Act of 1936; with respect to naval shipbuilding, the impetus was the political turmoil that ultimately resulted in the war itself.

The Merchant Marine Act of 1936 was the most comprehensive legislation in American history that dealt with shipping and shipbuilding ([1], p. 37). Among other things it provided a mechanism for the U.S. Government to pay a shipyard the difference between the higher American cost and the lower foreign cost. The Act made shipyard investment reasonably attractive for private capital; the result was a rebuilding of American shipyards—starting modestly at first, but gathering momentum as World War II approached. Activity further accelerated during the war. At the end of hostilities there were 57 major private shipyards in operation [2]. In addition to achieving unusual production levels, the yards were innovative and brought new concepts to the industry, most notably incredibly rapid production rates of standardized designs, the introduction of welded construction, and prefabrication of large subassemblies [2]. Since then, the number of major U.S. shipyards has dropped from 57 to fewer then 20 in the active production base. Employment has declined nearly 90 percent. In 1945, the final year of World War II, over 1000 merchant ships were delivered ([1], p. 26); in 1971 there were 216 ships (over 1000 grt) delivered by American yards ([3], p. 47); in 1991 there were no deliveries by American yards. The situation since 1981 has been that almost all construction by privately owned American yards is for one customer: the U.S. Navy.

Before World War II Navy shipyard activities encompassed a full range of ship production and ship support. The central function revolved around the industrial capacity to repair and, in the case of some yards, to construct naval ships and small craft. Wartime mobilization resulted in a ten-fold increase in employment from mid-1940 to the peak year of 1943. At the end of the conflict significant downsizing occurred, along with transfer of many of the yards' wartime functions to other Navy organizations and private shipyards. Several of the Navy yards were closed. And, by 1973 all new construction activities had been transferred to the private sector. Today, there are eight public-owned yards, six of which have nuclear capability.

Figure 1 shows the aggregate value of shipyard work on order at U.S. shipyards at regular intervals during the period 1940 to 1985. Two factors immediately stand out. The first is the overall decline of the commercial ship production activity as a percentage of the total; the second is the growing dominance of new

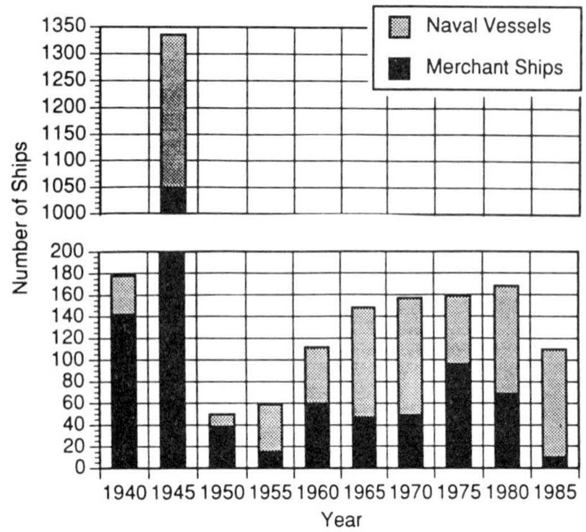

Fig. 1 Industry order book, 1940–1985 (from Shipbuilders Council of America).

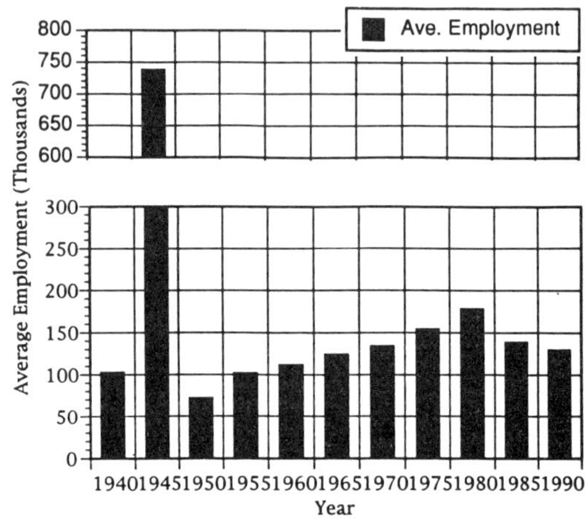

Fig. 2 Average employment in private U.S. shipyards, 1940–1990 (from U.S. Department of Labor).

naval ship construction as a percentage of the total. The U.S. shipbuilding industry faces an uncertain future as the 21st Century approaches; the break-up of the communist system has reduced the need for naval ship construction, and the industry must redirect its market focus back to the commercial sector, where it has virtually no penetration.

Figure 2 shows the shipyard employment figures for the 50-year period 1940–1990. As seen, there has been significant fluctuation in the employment count, with the highest peak occurring during World War II in private yards. One of the historical consistencies of the U.S. shipbuilding industry is the cyclical swing that occurs in employment. Figures 1 and 2 highlight this fact.

Today there are 17 privately owned shipyards [4] in the active industrial base.[1] Their locations are displayed in Fig. 3, and their addresses are given at the bottom of the figure. Of the private shipyards only four were building the Navy's combat ships in 1992: Bath Iron Works, General Dynamics Electric Boat Division, Ingalls Shipbuilding, and Newport News Shipbuilding. Another seven—American Shipbuilding, Avondale, McDermott Shipyards, National Steel and Shipbuilding, Peterson Builders, Marinette Marine, and Intermarine USA—were under contract for auxiliary ship and minesweepers for the Navy [5]. The remainder were involved in commercial and naval ship repair and conversion or small craft construction, or both.

Regardless of the customer or the type of work performed, the U.S. shipbuilding industry has, through the 50-year period under study, made significant contributions to the nation's economy. Total *direct* output of the industry during the period has typically been in the range of $4 to 8 billion (1990 constant dollars) per year. The total annual direct, indirect, and induced impact generated by the industry would thus be in the range of $15 to 30 billion (1990 constant dollars).[2] Input–output studies of the nation's economy indicate that the shipbuilding industry's output affects other industries more than it is affected by their output. Indeed, the industry's impact ratio is the greatest of any of the 50 sectors that exist in the Maritime Administration's input–output tables ([7], pp. 3–12).

Role of Government in the Prosperity and Direction of the Industry. In the United States the health and vitality of the shipbuilding industry has been dominated by the intervention and support of the Government, either as shipowner (in the case of Navy ships) or as a provider of direct and/or indirect subsidy support. The support has come in two forms: the outright purchase of ships, as in the case of the Navy, and in the form of construction subsidy to offset the cost differential between a U.S. procurement and a foreign procurement. The justifying rationale for the construction subsidy has been that survival of U.S. shipyards was essential to the defense posture of the nation, and the subsidy must be provided to assure

[1]Active Shipyard Industrial Base (ASIB) is defined as yards seeking, as well as having, the capability of fulfilling contracts for constructing naval and/or large merchant ships [1].

[2]Estimates are based on the extrapolation of 1967 figures from Ref. 7. An inflation rate of 4.0 percent/annum is assumed.

the yards' survival. The legislation that provided the subsidy was, as mentioned earlier (compare [7], p. 2), the Merchant Marine Act of 1936 (Title V).

In 1938 the Merchant Marine Act of 1936 was broadened to permit federal government insurance of private loans used to finance construction or reconstruction of vessels in U.S. shipyards. (Subsequently the definition of a "vessel" was broadened to include offshore drilling rigs, barges, and tugs.)

In 1951 legislation was passed (Public Law 911, Jan. 6) that provided funds for the government to build thirty-five 13,400 dwt, 20-knot ships, the "Mariners." The next year the construction subsidy option was extended (Long Range Shipping Act of 1952) to all ships operating in U.S. foreign trade that were "suitable for national defense purposes in time of war or national emergency." This legislation also allowed the ship operator to trade in old ships for credit on new construction, and permitted tax breaks by the establishment of a construction reserve funds system.

The next piece of legislation came when Congress passed the Merchant Marine Act of 1970. This law set a goal of building 300 ships for U.S.-flag registry during the decade of the 1970s, with construction subsidies being the primary financing mechanism. The legislation also extended construction subsidy assistance to bulk carriers, which until that time had not been covered. A final element of the Act established the National Shipbuilding Research Program that was to facilitate cooperative research by industry and government.

In the 1980s two major actions by the government had a profound effect on the industry. The first was the decision by the Administration to stop supporting the ship construction subsidy program. The result was the almost total disappearance of commercial ship production in the United States. From 1985 to 1992 fewer than five commercial ships were built and delivered. The second action, which partially offset the negative effects of the elimination of the construction subsidy program, was the decision by the Department of the Navy to expand its fleet to 600 ships. The combined effect of the two programs was that there was a modest decline in the total shipyard employment, and at the end of the decade almost all workers involved in new construction were working on Navy ships.

Role of Labor. In U.S. shipyards between 35 and 60 percent of the direct costs associated with building a ship is attributable to labor, with the average estimated at about 45 percent. In foreign yards the labor percentage is significantly less: 20 to 35 percent ([1], p. 92). The difference is mostly linked with productive efficiency, but not totally. Other factors are differences in cost of material (U.S. material costs are generally higher), and the wage rate. One of the significant shifts in comparative factors between U.S. yards and foreign yards has been that of wages. In the 1940s U.S. shipyards had the highest labor rates in the world. By 1990, however, the U.S. yards had significantly cheaper wage scales than almost all of the European yards, and were at about the same level as those of Japan. Table 1 shows the hourly shipyard wage scales for the major shipbuilding nations for the period 1975–1990, and shows the relative shift in U.S. position.

Data from the 1970 census break down shipyard work force into general worker categories as shown in Table 2. The dominance of the craftsman, as shown in the table, is believed to have persisted. Too, the proportions suggest a flat organizational structure of

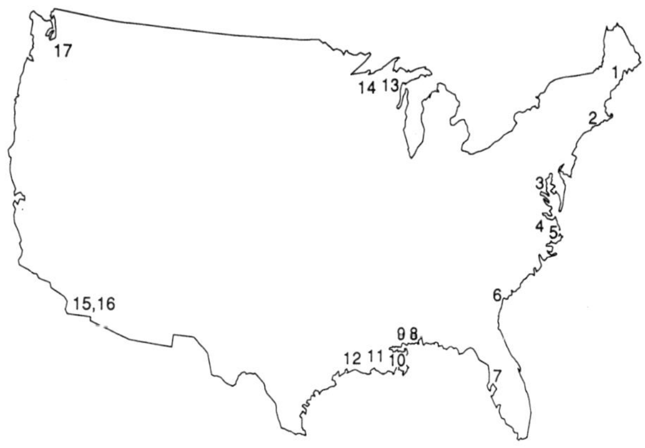

Shipyard	Location
1. Bath Iron Works	Bath, ME
2. General Dynamics–Electric Boat	Groton, CT
3. Bethlehem Steel, Sparrows Point	Baltimore, MD
4. NORSHIPCO	Norfolk, VA
5. Newport News Shipbuilding	Newport News, VA
6. Intermarine	Savannah, GA
7. Tampa Shipyard, Inc.	Tampa, FL
8. Ingalls Shipbuilding	Pascagoula, MS
9. Trinity Marine Group	Gulfport, MS
10. Avondale Industry	New Orleans, LA
11. McDermott International	Amelia, LA
12. Bollinger Machine Shop & Shipyard	Lockport, LA
13. Marinette Marine	Marinette, WI
14. Peterson Builders, Inc.	Sturgon Bay, WI
15. Southwest Marine	San Diego, CA
16. NASSCO	San Diego, CA
17. Todd Shipyards	Seattle, WA

Fig. 3 Location of major private shipyards in the U.S. ([4] and [6], p. 16).

Table 1 Hourly compensation for shipyard workers worldwide (from U.S. Bureau of Labor Statistics, Nov. 1991).

Country	1975	1980	1985	1990
United States	6.85	11.25	14.62	15.66
Canada	6.35	9.24	11.94	16.93
Hong Kong	1.05	1.67	2.32	3.78
Japan	3.92	6.75	8.12	15.04
Korea	0.54	1.65	2.13	6.52
Singapore	1.20	1.95	3.20	5.16
Taiwan	0.57	1.54	2.46	7.50
Belgium	8.75	15.75	10.71	N/A
Denmark	7.04	11.90	8.60	19.37
Finland	5.73	9.15	9.14	23.19
France	5.13	10.23	8.86	16.43
Germany	7.22	14.02	10.91	23.45
Italy	5.82	9.17	8.48	18.87
Norway	N/A	12.50	11.28	21.92
Sweden	8.08	13.22	10.03	20.93
United Kingdom	3.74	7.52	6.45	13.21

Table 2 Worker category by type for all industry and shipbuilding, 1970 (from U.S. Census).

Worker Category	All Industry, %	Shipbuilding, %
Laborers, Service Workers	21.0	6.9
Operatives	23.5	24.8
Craftsmen, Foremen	18.5	‘52.6
Managers, Administrators, Technical Professions	30.9	15.7

shipyards. The craft orientation of the labor force indicates that technological change will be time-consuming, and will be difficult as the crafts tend to resist any change. Attempts to make changes during the 1980s have confirmed this.

With the exception of Avondale Corp., of New Orleans, all shipyards in the active shipbuilding industrial base are unionized. The major unions representing production workers are:

- International Brotherhood of Boilermakers, Iron Shipbuilders, Blacksmiths, Forgers, and Helpers,
- Industrial Union of Marine and Shipbuilding Workers,
- Metal Trades Council, and
- United Steel Workers.

Table 3 portrays the shipyard production workers by skill. About 65 percent of the workers are in ten critical skills: shipfitters, welders, machinists, electricians, pipefitters, flame cutters, riggers, shipwrights, draftsmen, and crane operators.

Training times for selected skills are shown in Table 4. The numbers are even greater in yards where nuclear skills are required. But at a minimum, the cost of replacing a skilled shipyard worker is significant, and is in excess of $50,000 (1990 constant dollars), even for the simpler skill classifications.

Role of Technology and Facilities. The preceding sections have shown that U.S. merchant shipbuilding has generally not been competitive in the world market, and that the domestic market is dependent almost entirely on government policy. Even though the United States is recognized as a leader in the construction of naval combat ships, productivity has continued to be alarmingly low in this sector as well.

Improvements in productivity since World War II have focused on different areas of enhancement in the various shipyards of the world. Japan concentrated on the emplacement of rationalized facilities and processes in the years immediately after the war. The Japanese looked to the United States expertise and guidance, and used concepts developed in the United States during the war. Regarding technology transfer, Mr. Elmer Hann ([10], p. 2), an American who introduced many of the U.S. methods in Japan, indicated the following:

> . . . all-welded construction was introduced into Japan for the first time. . . . We used the American Bureau of Shipping . . . as most of our machinery . . . came from the U.S.A. for the first several years.
>
> Job and material controls were organized into one department. Sequence and scheduling of work as carefully planned and closely monitored along with quality control and inspection which were kept separate from production departments.
>
> . . . [W]e adhered to:
>
> 1. Careful analysis of vessel as to size [of] blocks and shape with refined drawings or sketches of each weldment, together with machinery, piping, etc., to be installed at assembly shop or area.
>
> 2. Coordinated material control.
>
> 3. Allocation of labor and time schedule for each operation.
>
> 4. Installed machinery, piping, and other equipment to a great extent before erection.
>
> 5. Reduced staging to a minimum.
>
> 6. Introduced inorganic-zinc coating in the assembly line.
>
> 7. The key to rapid construction is how to weld without distortion and shape of weldments

Table 3 Shipyard production workers classified by type of skill (from Maritime Administration and University of Michigan data developed in U.S. and Ref 8).

Employee Classification	Percentage of Total Work Force
Nonproduction workers	15.0
Production workers—skill not particular to shipyards (e.g., painter, machinist, carpenter, electrician)	57.0
Production workers—skill unique to shipyards (e.g., pipefitter, rigger, shipfitter, loftsman)	28.0

Table 4 Training time for new shipyard employee to reach journeyman level in selected shipyards [9].

Skill	Training Time, hours
Crane operator	1,000
Shipfitter	8,000
Pipefitter	8,000
Rigger	8,000
Marine draftsman	10,000
Shipwright	8 to 10 years

or modules that defy or resist distortion, especially when such affects the vessel's measurements and locked-in stresses.

Japan's area of interest was exclusively on commercial ship construction. The United States, on the other hand, made no new major investment in assets for the first few years after the World War II; the main focus was on redirection of the facilities that had been active during that war. The Korean War rekindled some investment in facilities, but generally, the ships and processes used to build them were not significantly altered.

In the late 1950s things began to change as a result of market stimulations from legislation passed in 1951 and 1952 ([10], p. 9), U.S. yards began to incorporate European and Japanese practices into their needs. And during the 1960s facility improvements, such as the following, occurred ([3], p. 29):

Avondale Shipyards in New Orleans introduced a modular production system.

Bath Iron Works in Bath ME implemented a facility improvement program that included modifying the ways to accommodate larger ships. A 220-ton crane to serve the ways was installed.

Bethlehem's Sparrows Point MD yard installed a panel line and a large building basin.

Litton Shipyards constructed a highly developed modular production system, unique in the world.

Todd Shipyards, San Pedro CA, installed modern steel-handling facilities, a panel line, and larger cranes.

National Steel and Shipbuilding Co., San Diego CA, installed modern plate handling facilities and more advanced material control techniques.

Newport News Shipbuilding Co. developed an independent shipyard facility adjacent to the existing yard to construct large commercial ships.

Also in the 1960s, and continuing into the 1970s, the government undertook initiatives to improve the shipbuilding infrastructure and the technological assets. The Navy initiated acquisition policies, such as multi-ship contracts, to encourage investment in shipbuilding technology. The Maritime Administration (the former U.S. Maritime Commission) developed standard ship designs to promote multi-ship construction programs. Additionally, the Maritime Administration exploited the National Shipbuilding Research Program, legislated in 1970 ([3], p. 8), to lead active education and technology transfer programs for the shipbuilding industry.

The National Shipbuilding Research Program was directed to establish a collaborative program with the shipbuilding industry to improve productivity. The central theme of this collaboration was the insistence on industry-wide interchange of ideas and expertise for solving common problems. Industry participation was encouraged by the requirement that industry participants assume full responsibility for project identification, direction, and implementation. The government provided seed money for performance of approved projects, but basically took a hands-off approach to the technical directions. Industry was also required to share in the cost. The approach gave industry a sense of ownership, and led to widespread acceptance of the program's output [11].

It is clear that the program has been successful; the program has been producing a continuous stream of technology, innovation, procedures, and other useful information to improve the productivity and reduce the cost of ship repair and shipbuilding—both naval and commercial—in the United States [11].

On a long-term basis, the industry has made significant organizational and technological changes. Even though there have been setbacks, it appears that the gap between U.S. and foreign yards for the production of commercial ships has been reduced as U.S. yards emplace new technologies, including knowledge on how to better organize work. "In 1992, U.S. yards have the potential to build ships in series at world market standards for a cost of about 15 percent more than

Japanese and Korean yards; German costs are comparable to those of U.S. Yards" [12].[3]

Naval Shipyard Facilities. At the end of World War II there were 10 public (or Navy) shipyards in operation. Their mission was ([3], p. 30):

> . . . [to] provide logistic support for assigned ships and service craft; to perform authorized work in connection with construction, conversion, overhaul, repair, alteration, dry docking, and outfitting of ships and craft as assigned; to perform manufacturing, research, development and test work, as assigned; and to provide services and material to other activities and units, as directed by competent authority.

In the 1970s that mission was altered when new construction was terminated at naval shipyards, and assigned to the private yards. At that time, also, two yards—Boston Naval Shipyard and San Francisco Naval Shipyard—were closed. The remaining eight yards continued to operate through the remainder of the 1980s. In 1991 the Navy announced that the Philadelphia Naval Shipyard would be closed by 1996.

Each yard was, and is, capable of rendering a wide range of services to naval ships, yet each, in practice, has a broad sets of specialties. The active yards in 1992, and their assigned missions, are as follows [12]:

Charleston Naval Shipyard—conversion and overhaul of surface ships, less nuclear, and overhaul of all types of submarines.

Long Beach Naval Shipyard—overhaul and repair of all surface ships but nuclear; serves as major fleet repair center.

Mare Island Naval Shipyard—conversion and overhaul of all types of submarines; research center for ocean engineering activities.

Norfolk Naval Shipyard—conversion and overhaul of surface ships, including nuclear; serves as major fleet repair base.

Pearl Harbor Naval Shipyard—conversion and overhaul of all types of surface ships, except aircraft carriers, and overhaul of nuclear attack submarines.

Philadelphia Naval Shipyard—conversion and overhaul of all surface ships, but nuclear; research and production center for ship propellers. Yard is scheduled for closure by 1996.

Portsmouth Naval Shipyard—conversion and overhaul of all types of submarines.

Puget Sound Naval Shipyard—conversion and overhaul of all types of surface ships, and overhaul of nuclear attack submarines.

Figure 4 shows the employment profile since World War II for one of the yards: the Mare Island Naval Shipyard. The peaks associated with the conflicts in Korea and in Viet Nam are typical of that experienced by all of the Naval yards. It is also anticipated that downward trends in the employment figures will continue at Mare Island, and the other yards as well.

Design and Production Integration

Since World War II there has been significant change in the relationships between the design and production functions in ship acquisition. This major section focuses on those changes. First a description of the traditional relationship is presented. This is followed by review of the concept of concurrent engineering, which is the form of the design-production relationships in the advanced shipbuilding systems of today. The section concludes with a discussion of how the introduction and use of the computer has affected both of the functions and their interrelationships.

Traditional Design–Production Relationships. One of the distinguishing characteristics of the ship acquisition system in the United States has been the sharp separation between the design and the

[3]The key phrase in this quotation is "build ships in series." Most observers feel it is unlikely that ships will be built in series unless there is government intervention to support such a concept or unless the shipyard is able to locate a unique niche where it can be the sole supplier.

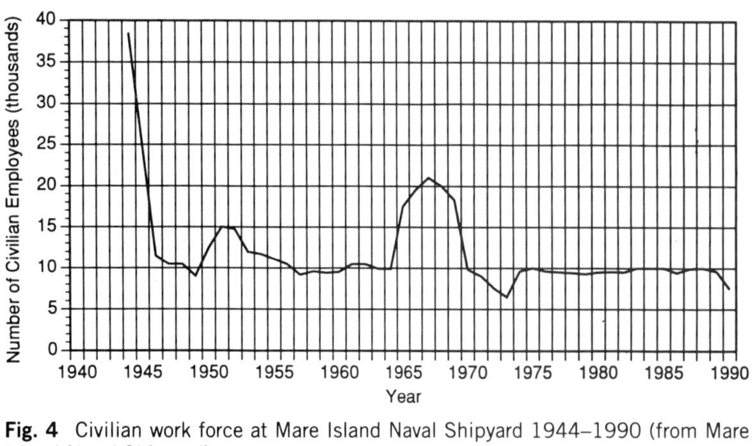

Fig. 4 Civilian work force at Mare Island Naval Shipyard 1944–1990 (from Mare Island Naval Shipyard).

production functions. The owner has traditionally contracted with a naval architecture firm to prepare all of the preliminary design and contract drawings. The drawings have then been presented to several competing shipyards for quotations. From this competition a yard is selected to produce the ship, and a contract is negotiated. In the case of the Navy the procedure is essentially the same, except that preliminary design, and in some cases the contract design, is generally performed (or closely controlled) by the Navy's own design staff. The advantage of this type of acquisition structure has been that the owner could specifically tailor his design to his unique needs and proclivities. The disadvantage has been that the system inhibited the development and emplacement of advanced ship production systems like that employed in the rest of the world. A comparison study [13] of productivity requirements to build a commercial ship in Japan versus the United States shows the Japanese yard required only half the labor hours needed in the American yard. A significant contributor to this difference has been the separation of design and production functions found in the United States.

Concurrent Engineering Activities. Beginning in the 1980s, American industry (including shipbuilders) began to recognize that separating the design and production functions was having a bad effect on our ability to compete in a world market. Too, both design and production dynamics were being radically altered because of the appearance of the computer. The computer has facilitated rapid changes in the technology of design and of production systems, and the requirement to link the two became even more important. Concurrent engineering (also called "simultaneous engineering") is designing a product and designing the process to manufacture that product at the same time. Typical of the reported benefits were improvements of 45 percent in materials cost reduction, 20 percent reduction in rework costs, approximately 50 percent improvement in "manufacturing yield" (productivity improvement), and significant reductions in cycle time [14].

In 1988, the Defense Advanced Research Projects Agency (DARPA) initiated a five-year program called DICE (DARPA Initiative in Concurrent Engineering). The primary mission of the DICE program was to develop a computer-assisted concurrent engineering environment [15]. While there have been no formal programs developed by the shipyards that are explicitly linked with this initiative, the yards, nonetheless, have informally started moving toward the concepts.[4] The most notable example has been the Seawolf submarine design/construction project, which was initiated in the mid-1980s. In this case, the Navy required that the two involved yards (Electric Boat Div. of General Dynamics and Newport News Shipbuilding) produce design details that would be compatible with construction processes in place in both yards. A series of technical papers appeared in the *Journal of Ship Production* over a several-year period that described, in reasonable detail, the concurrent engineering plans and the anticipated benefits that would accrue [16–18].

The Impact of the Computer. The first computer appeared in a shipbuilding operation in 1953: Parsons Engineering and MIT developed a machine for performing hydrostatic calculations. Clyde Shipbuilders in Scotland put into place a similar machine in 1959.

When transistors started to replace vacuum tubes, in 1964, the speed and on-line availability of the machines were dramatically increased. Concurrently, there was a significant increase in the penetration of computers into the ship design and production system. Computer-aided lofting was introduced in Norway in 1965 with the AUTOKON system. During the same general time period the Norwegians and IBM completed the ADAPT system, a computer-aided-manufacturing system for processing steel plate.

In 1973 The Maritime Administration acquired the AUTOKON-71 computer-aided and -manufacturing system for sub-licensing to U.S. shipyards. A contract was given to the Illinois Institute of Technology Research Institute (IITRI) to provide the system support for shipyards that wanted to be licensed users of the system. A consortium, Research and Engineering for Automation and Productivity in Shipbuilding (REAPS), was established to guide the development of supporting software and hardware.

In November 1973, the AUTOKON system was delivered to IITRI and installed on a UNIVAC 1108 EXEC 8 computer. At installation the system consisted of the following modules [19]:

MISC—a program for initialization of the system database and limited utilities.

FAIR—a program to fair offsets; a table of offsets and a set of curves used as reference contours for later use.

DRAW—a program to read curves stored in FAIR, and to produce a data tape for eventual drawing of the curves.

[4] Communication with Mr. Joe W. Meredith on 2 Dec. 1992.

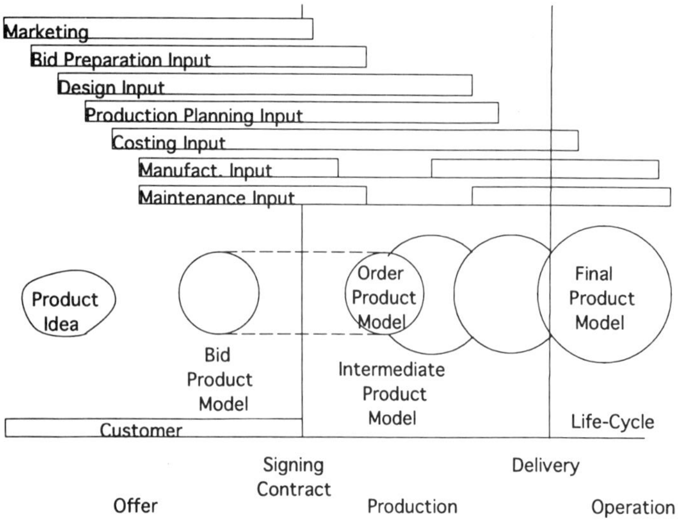

Fig. 5 Evolution of the product model [20].

TRABO—a program to transfer the body plan from the FAIR database to the system database.

LANSKI—a program to fit longitudinal curves on the hull surface and to store the information in the database.

SHELL—a program to produce numerical-control burning tapes for cutting shell plates.

TEMPLATE—a program to produce shell plate templates and frame templates.

ALKON—a parts programming language.

NEST—a program to store nesting formats for parts to be cut from steel plates.

PRODA—a program for the generation of planning and production data.

PRELIKON—a module for preliminary lines design.

DUP—a utility program for manipulation of the database.

Through the interaction of the above-listed programs, a ship could be taken from the initial design stages through to the final production stage.

The REAPS program was the breakthrough for implanting computer systems into American shipyards. Most American medium and large shipyards became affiliated with the initiative, and the program expanded through the rest of the 1970s and early 1980s. During this period (1974–1983) there were annual symposia sponsored by the REAPS organization where presentations were made of computer-aided-design and -manufacturing developments related to shipyard activities. In the early 1980s most shipyards had reached the stage where they had internalized the computer system activities; additionally, the Maritime Administration ceased its financial support of the REAPS program. In 1983 the program was terminated and all of the support assets were transferred to the University of Michigan's Department of Naval Architecture and Marine Engineering for use by students.

The computer has been the facilitating mechanism for visualizing total product modeling of the ship, with a single database being used from conceptualization through engineering and production, to in-service operation. Figure 5 shows the evolution of the product model. Maintaining this model requires a sophisticated computer (both hardware and software) system, and a computer integrated management system (CIMS) that emulates the entire ship acquisition and control process. If trends continue, such a system will be in place by the turn of the century.

Shipyard Production Systems

"Before World War II yards characteristically had limited storage and shop facilities and lift capacities in the five- to ten-ton range. Most of the ship components were assembled on or immediately adjacent to the shipway. Much of the planning was accomplished by highly skilled craftsmen as the ship was erected" ([6], p. 184). Since the beginning of the war there have been major changes in the production system as the industry has attempted to respond to technological advances in design and manufacturing processes. The changes can be roughly divided into those that occurred before 1960, and those that have occurred since.

World War II to 1960. From the beginning of World War II until about 1960 the major changes were linked prefabrication of weldments away from the erection site, and the development of more accurate cutting and welding technology. Sections of the ship were built in shops, and transported to the erection sites as assembled units. Significant improvement in dimensional control was required, and extensive upgrading of pre-production planning and scheduling was necessary. Functional work areas were better defined, and shipyard workers became more specialized, and job-shop oriented.

Important facility changes during this period included fewer erection sites, and more space devoted to storage and shop facilities as work was performed away from the erection site. The crane capacity was

Fig. 6 Photograph of Ingalls Shipbuilding when operating at full capacity (supplied by Jesse W. Brasher, Ingalls).

increased to accommodate the movement of larger weldments in the subassembly and erection stages. More accurate cutting and welding methods were developed. Lofting systems were significantly altered, and improved, as tenth-scale optical systems were introduced ([6], pp. 185–186).

Since 1960. Specialized ship construction has strongly influenced the shipbuilding systems since 1960. Some U.S. shipbuilders began to modify their yards to accommodate a process-type production concept. Work stations were clearly defined and fixed. During this period the Navy adopted a policy of contracting with a single yard for an entire class (or the bulk of a class) of ships. For example, all 30 of the *Spruance*-class destroyers, and 27 of the 46 FF-1052 class frigates were allocated to two yards: Bath Iron Works and Ingalls Shipbuilding ([6], pp. 186–187).

One "greenfield" shipyard was constructed in late 1960s: the West Bank yard of Ingalls Shipbuilding. This yard was specifically organized to build the *Spruance* destroyers, and is an example of a high-capacity in-line process yard. Work stations were fixed and assemblies moved toward the water on a heavy-lift rail system. At each station, subassemblies were combined until at the final station (called the integration area) the ship consisted of three modules. Figure 6 shows the Ingalls yard when it was operating at full capacity building the *Spruance*-class destroyers.

As the decade of the 1990s begins, a central problem facing American yards is the need to adjust their facilities so as better to accommodate commercial construction. In this environment the U.S. yards will need to become world-class competitive, meaning a rationalization of facilities to be consistent with those seen around the world: graving docks (instead of sliding ways), highly sophisticated dimensional control systems using the latest technology (e.g., lasers), computer control of the planning and scheduling operations, and robotic work stations.

References

1 Whitehurst, C. H., *The U.S. Shipbuilding Industry*, Naval Institute Press, 1986.
2 Beazer, W. F., et al, *U.S. Shipbuilding in the 1970s*, Lexington Books, 1972, p. 3.
3 "Report of the Commission on American Shipbuilding," Vol. 2, U.S. Government Printing Office, Washington, D.C., 1973, p. 47.
4 *1991–92 Maritime Services Directory*, Aegis Publications.
5 *Naval Shipbuilding*, Shipbuilders Council of America, 1 Jan. 1992.
6 Storch, R. L., Hammon, C. T., and Bunch, H. M., *Ship Production*, Cornell Maritime Press, 1988.
7 *Economic Impact of the U.S. Merchant Marine and Shipbuilding Industries: An Input–Output Analysis*, GPO Stock No. 003-007-00093-6, U.S. Maritime Administration, 1978.
8 "Analysis of the International Competitiveness of the U.S. Commercial Shipbuilding and Repair Institute, International Trade Commission," p. 19.
9 Ramsey, R., "A Time for Shipbuilding Renaissance," *Naval Engineers Journal*, Sept. 1983, p. 56.
10 Chirillo, L. D. and Chirillo, R. D., "The History of Modern Shipbuilding Methods: The U.S.–Japan Interchange," *Journal of Ship Production*, Feb. 1985, p. 2.
11 Schaffran, R., "The National Shipbuilding Research Program: An Overview and Status Report," *Journal of Ship Production*, Feb. 1990, pp. 35–37.
12 *Shipyard Costs and Capabilities*, Center for Naval Analysis, 1992; cited in *Shipyard Chronicle*, Shipbuilders Council of America, 17 Sept. 1992.
13 Bunch, H. M., "Comparison of the Construction Planning and Manpower Schedules for Building the PD214 General Mobilization Ship in a U.S. Shipyard and in a Japanese Shipyard," *Journal of Ship Production*, Nov. 1987.
14 Allen, C. W., "Simultaneous Engineering: What? Why? How?," Paper No. MM89-493, Society of Manufacturing Engineers, Detroit, 1989.
15 Meredith, J. W. and Blanchard, B. S., "Concurrent Engineering: Total Quality Management in Design," Hampton Road Section, SNAME, 12 Apr. 1990.
16 Brucker, B. R., "Infusing Producibility into Advanced Submarine Design," *Journal of Ship Production*, May 1989.
17 Brucker, B. R. and Merrill, K. J., "Computer Integration of SEAWOLF Class Submarine Life-Cycle Functions," *Journal of Ship Production*, Feb. 1991.
18 Brucker, B. R. and Baseler, R. W., "SEAWOLF Producibility II: Transition from Design to Production," *Journal of Ship Production*, Nov. 1991.
19 Tasks, P. D., "The Updated REAPS AUTOKON System," *Proceedings*, REAPS Technical Symposium, 24–25 June 1975, Palm Beach, FL.
20 Paper VIBI, *Proceedings*, NSRP Symposium, 1991, p. 2.

Howard M. Bunch studied civil engineering at Stanford and then, in 1958, earned an MBA (in transportation) at the University of Texas. He worked with Southwest Research Institute for eleven years, then moved north to accept a position as vice president for a research lab owned by the Northrup Corp. He later joined the University of Michigan as a project manager in the Transportation Research Institute, and later split his appointment to work in the Department of Naval Architecture & Marine Engineering. In the latter association, he revived the dormant educational program in shipbuilding. He is the founding editor of SNAME's *Journal of Ship Production*, and has served on several SNAME committees and the NAS/NRC Committee on Navy Shipbuilding Productivity.

The War Years: Canadian Merchant Ships

MAX REID

IN THE YEARS leading up to World War II, the Canadian merchant marine was really an extension of its British counterpart. Canadian shipyards, which had enjoyed an earlier success in building wooden ships, had little experience in modern shipbuilding. Most of Canada's merchant fleet had been acquired from the United Kingdom. World War II propelled the Canadian merchant marine from 41 to 173 oceangoing ships and her shipyards produced in excess of 400 merchantmen, 500 naval ships, and 3000 special purpose craft, including patrol boats and landing craft.

Canada's merchant marine had reached its peak in the 1890s when it ranked near the top of the world's merchant fleets. These wooden sailing ships were built in small yards mainly in Quebec and the Maritime Provinces. While some ship-rigged vessels and barques were between 300 and 1000 tons, the majority were brigs and brigantines under 200 tons. By 1902, when 6836 vessels were registered in Canada, these wooden hulls were slowly giving way to steel-hulled steamships.

During World War I, the bulk of Canada's shipbuilding was devoted to naval patrol vessels and merchant ship repair; however, it is noted that there were 43 Canadian registered ships, which were armed by the British Admiralty. By 1937 only 20 of 442 vessels registered were built of steel. Canada's disappearing merchant marine and shipyards were about to undergo a major technological and political change.

Although Canada's naval building program started shortly after the outbreak of World War II, the commercial building program did not get underway until after a decision by the British government in 1940. By the middle of that year, Britain faced invasion and her very existence was being threatened by the German U-boat offensive. As a result, a change in Britain's shipbuilding policy was directed. It was decided to concentrate her shipyards on naval construction while asking the United States and Canada to provide merchant ships.

A British technical mission crossed the Atlantic with plans for a 10,000-ton freighter. This design was for a riveted shelterdeck ship with machinery amidships, and it formed the basis for both America's and Canada's wartime merchant shipbuilding programs. In all, some 3000 vessels of the Ocean-, Liberty-, Fort-, and Park-classes were based on this design. All ships shared a common triple expansion steam engine; a boon to running repairs and spare parts in a worldwide emergency.

In the United States, orders were placed for 60 Ocean-class ships with the Todd organization; the order was split between Bath in Portland and Todd-California in Richmond. The huge American Liberty Ship program followed shortly thereafter.

This same mission then went to Canada to arrange the program that catapulted Canada into becoming a major shipbuilder. This program had just two customers: Britain and Canada. The ships built for Britain were named mainly after Canadian forts; and those built for Canada were named after federal, provincial, and municipal parks.

The 244 ships built for Britain included 90 that were ordered by the U.S. Maritime Commission and delivered to the British Ministry of War Transport under the Lend-Lease Agreement; two were sold outright to Britain and 99 Forts/North Sands types, plus seventeen 4700 tonners, were provided under Mutual Aid; the remainder were naval stores ships. The first Fort-class ship delivered, the *Fort St James*, was from the Burrard Yard in North Vancouver on January 15, 1942.

The Park Steamship Company was formed in 1942 as a Canadian Crown corporation with the objective of moving war cargoes. Although owned by the government, these ships were chartered to shipowners and agents in Canada. The Park-class was an improved *North Sands* type and called a Victory Ship. Unlike the U.S. built ships, which were of welded construction, the Canadian Victory Ships were generally riveted. As the war progressed, and riveters were harder to find, a greater amount of welding took place, particularly in the tank tops and superstructure.

The first Park ship delivered was the *Prince Albert Park* from the Davie Yard in Lauzon, Quebec, in June 1942. The Park fleet quickly expanded to 159, including 114 Victory type cargo vessels, 13 tankers of the same basic design, twenty-six 4700-ton Scandinavian type cargo ships, and six 3600-ton tankers of a Canadian design. There were another 51 ships that operated outside the Park Steamship Company, for a total fleet of 210 Canadian registered ships, of which 37 were sold or lost.

While the Fort ships were owned by the British Government, many of the crew members were Canadians, including the naval gunners. Throughout the War, the manning and certification of the crews were almost completely controlled by the Canadian Government.

To provide this shipbuilding capability and to repair both Canadian and Allied ships, some 15 minor yards with a work force of about 3000 exploded to nearly 90 entities employing over 85,000 persons. Nearly 57,000 of these were dedicated to building and repairing merchant ships. As in the 1890s, shipbuilding had become one of Canada's prime industries.

To illustrate how these yards assimilated this technology, the average time to build a Fort/Park 10,000 tonner improved from 307 days in the first order to 163 in the second. As would be expected, the time to build these ships varied from yard to yard. On the first order, this varied from 239 to 426 days, while on the second the range was from 112 to 237 days.

Of the 18 major shipbuilding yards, seven were on the West Coast, five in Quebec, four in Ontario and two in the Maritime Provinces. There were seven engine manufacturers, all in Ontario or Quebec. The (approximate) particulars of the 10,000-ton Victory Ship are noted below:

Deadweight (tons) = 10,000
Total displacement (tons) = 13,770
Gross registered tonnage = 7150
General cargo (tons) = 9000
Fuel oil (tons) = 975
Length overall (feet) = 441.5
Beam (feet) = 57
Depth to upper deck (feet) = 37
Engine, triple expansion (HP) = 2500
Boilers = 3 coal burning Scotch, or 2 oil burning water tube
Speed/endurance = 12 knots/6000 miles
Crew = 32 plus 7–15 naval gunners

In addition to this large shipbuilding and repairing task, the war, with its unrestricted U-boat warfare plus the Luftwaffe's ability to strike shipping, required a major program to provide defensive armament to these ships. In Canada, as in the rest of the Commonwealth, these ships were identified as Defensively Equipped Merchant Ships (DEMS) and were generally more heavily gun-armed than a naval frigate. The typical armament of a Fort or Park ship included the following:

4-in. gun,
3-in. gun,
six 20 mm Oerlikons,
2 twin 50 cal. machine guns,
20 rail anti-aircraft rocket projectors,
6 parachute and cable anti-aircraft devices,
minesweeping paravanes,
anti-torpedo nets (covering about 80 percent of ship),
smoke floats for smoke screens,
degaussing coils and generators,
plastic armor plating around the bridge and AA gun tubs, and
small arms.

This armament would be manned by the naval gunners under a junior non-commissioned officer. When required in action, the naval crew would be supplemented by about 15 of the civilian crew, most of whom would have undergone a short naval gunnery course. The Royal Canadian Navy had to provide some 2000 naval gunners and training for hundreds of merchant seamen. In addition there was the support of four dedicated gunnery ranges, training schools, torpedo net depots, plus ammunition and armament repair depots.

One major change evolved from this conflict: the Canadian merchant marine underwent a Canadianization process. Canada now provided her own merchant marine training facilities and certification process. However, it never really was to establish itself as a thriving and competitive national merchant marine because of two events. With the end of the war, Park Steamships was dissolved with the selling of 130 ships and the placement of 45 on short term charters. The other event involved a union/management problem, which contributed to a noncompetitive shipping situation.

World War II caused the loss of 62 Canadian merchant ships (37 Canadian registered) and 1064 of her merchant seamen. On the other hand, it did force Canadian shipbuilders to move from wooden hulls, and building to someone else's designs, to an industry capable of its own design and building of the most sophisticated vessels. The Canadian merchant marine in 1990 had nearly 100 ships over 1000 tons, but only twelve of these were in foreign trade.

Bibliography

Kennedy, J. de N., *History of the Department of Munitions and Supply*, King's Printer, Ottawa, 1950, p. 247.

Mitchell, W. H. and Sawyer, L. A., "The Oceans, the Forts and the Parks," *Journal of Commerce*, Liverpool, 1966.

Parker, J. P., *Cape Breton Ships and Men*, McGraw-Hill Ryreson Ltd., Toronto, 1967, Appendix.

Tucker, G., *The Naval Service of Canada*, Vol. II, King's Printer, Ottawa, 1952, App. XII.

Captain Max Reid entered the Royal Canadian Navy in 1943, serving as a naval gunner in merchant ships throughout World War II. He remained in the Navy after the war, serving in nine warships, including command of a frigate and a destroyer. After retiring from the Navy, he served as a convoy commodore in the Naval Reserve. He is now a foreign service officer with Canada's Ministry of External Affairs & International Trade.

The War Years: Canadian Combatants and Auxiliaries

W. J. Broughton

The Royal Canadian Navy (RCN), created in 1910, remained extremely small until World War II. Even at the height of World War I, its strength, including two submarines and various auxiliaries, was only 123 vessels and 5400 personnel. Between the wars the RCN averaged barely a thousand members and a handful of ships. Remarkably, by the end of the Second World War the RCN had grown to 775 vessels and more than 90,000 personnel. In 1945 it was the third largest western allied fleet, after the United States and Royal Navies.

Despite the inauspicious earlier decades of the RCN, the seeds were sown for future capabilities. In 1930 and 1931 the destroyers *Saguenay* and *Skeena* were built at the Thornycroft yard in the U.K. These were based on a British design, but with extensive modifications to suit Canadian conditions. These unique vessels were highlighted in Thornycroft's history:

> To meet the . . . exacting Canadian conditions, many interesting features were introduced, such as strengthened bows . . . as a protection against ice. They were also given exceptionally good metacentric height, as it was calculated that at times there might be 50 or 60 tons of ice on the upper decks and rigging. . . . Steam heat was provided . . . to maintain a comfortable internal temperature in sub-zero conditions. At the same time . . . good ventilation was . . . essential in view of the boats frequently having to pass through the Panama Canal and work from a base in the West Indies. Many improvements were introduced, as the Canadian Naval authorities were much more receptive to new ideas than the Admiralty. The bridges were exceptionally wide and partially streamlined. The bows were well flared, and this, in conjunction with special bridge screens, resulted in a very dry foredeck. . . . In fact so many improvements were introduced that . . . the ships were promptly dubbed the "Rolls-Royce destroyers" ([1], p. 101).

The hundreds of ships built in Canada during World War II tended to be anything but "Rolls-Royces." Workhorses would be more descriptive. Canadian corvettes, frigates, minesweepers, and motorboats became the backbone of convoy protection, antisubmarine warfare (ASW), and coastal defence. The common prerequisites were high quality, uniform standards, and mass production. In every case, the basic designs were British.

In 1940, to aid technology transfer, the British Admiralty Technical Mission (BATM) was established in Ottawa. Assistance to Canadian naval architects and shipyards was clearly in their mandate. But far more telling was the identification of Canadian sources for the myriad of equipments and systems to be fitted. Production was established for guns, mountings, sonar, radar, fire control, and compasses. By the end of the war, nearly all parts of Canadian ships were made in Canada ([2], p. 53).

The Canadian Government announced the initial shipbuilding program just 12 days after declaring war on Germany in September 1939. The 2-year program was to consist of 4 destroyers, 40 corvettes, 28 minesweepers, 32 motor torpedo boats (MTBs), and 2 sloops. In addition, three of Canadian National Steamships' *Prince*-class passenger ships were to be converted to armed cruisers. The sloops were subsequently dropped. When the RN could not provide the technical assistance required, the destroyer portion was changed, reducing the number to 2 and assigning them for construction in the U.K. in exchange for 10 corvettes to be built in Canada. In February 1940 funding of $54.25 million was approved, leading to the revised program for the construction of 64 *Flower*-class corvettes (10 to Admiralty account) and 28 *Bangor*-class minesweepers. All of these vessels but one corvette were delivered by the end of 1941.

Fig. 1 *Flower*-class corvette—HMCS *Sackville*: 950 tons, 205 ft, single-screw, 15.5 knots, 3450 n.mi, one 4-in. gun, one light AA gun, depth charges, minesweeping gear (Dept. National Defence photo CN-3557).

Fig. 2 *Bangor*-class minesweeper—HMCS *Cowichan*: 672 tons, 180 ft, 16 knots, 2950 n.mi, one 3-in. gun, one 12-pdr (or one 4-in. gun), minesweeping gear (Dept. National Defence photo L-5274).

During the course of the war the types of ships constructed changed to meet changing requirements. At the start, the main concern was the defence of Canadian coastal waters and the protection of convoys as they moved out toward mid-Atlantic. Small antisubmarine escorts, minesweepers, and coastal patrol boats were required in quantity, supported by a minimum number of destroyers.

For escorts, the initial thought was to build the RN's twin-screw *Bramble*-class sloop of 815 tons and 17-knot speed. Despite a slower speed of 15.5 knots, the smaller, British-designed, single-screw *Flower*-class corvette was substituted because it was cheaper and easier to build and yet had excellent characteristics for a ship of its size. The concept was based on whaling ships of the 1930s. These corvettes were designed to chase submarines, which, like whales, could maneuver in tight turning circles, surface rapidly, and make off at relatively high speed. They mounted a single 4-in. gun and carried minesweeping and ASDIC (sonar) gear. During the war, on both sides of the Atlantic, hundreds of corvettes were built, yet only one remains. After four decades of service as a warship and research vessel, the *Sackville* has been re-converted to her wartime configuration. As a naval war memorial, she is now the centerpiece of the Canadian Maritime Museum in Halifax, Nova Scotia.

While the exigencies of war led to reduced specifications for the corvettes, the opposite was true for the minesweepers. The coal-burning, slow (12.5-knot) *Basset*-class of the 1930s was dropped in favor of the newer *Bangor*-class. These vessels were oil-burning with higher speed and endurance. Although more expensive, they were better for the purpose and had wide producibility in Canadian industry.

Another need was for coastal patrol boats to monitor the thousands of miles of coastline, east and west, and to deter enemy landings, which were much feared after Pearl Harbor. That these fears were not out of place became evident some 30 years after the war, when evidence of a German weather station was discovered on the Labrador coast.

Both motor launches and motor torpedo boats (MTBs) were built. The Type B *Fairmile* was a 112-ft wooden motor launch of 79 tons powered by two 650-hp gasoline engines and equipped with ASDIC, radar, depth charges and light guns. It had an endurance of 1500 nautical miles at a speed of 16.5 knots, and an emergency speed of 20 knots. The MTBs were designed by the British Power Boat Company. They were considered excellent boats, with an economical endurance of 2000 nautical miles and a top speed of 38 knots from twin 1350 hp Packard gasoline engines. They carried two sets of twin half-inch guns in

Fig. 3 Algerine-class minesweeper—HMCS *Rockliffe*: 990 tons, 225 ft, 16 knots, 4500 n.mi, one 4-in. gun, other AA weapons, minesweeping gear (Dept. National Defence photo S-3403-R).

Fig. 4 River-class frigate—HMCS *Montreal*: 1445 tons, 301 ft, 19 knots, 7200 n.mi, one 4-in. gun, one 12-pdr (or two 4-in. guns), radar, ASDIC (Dept. National Defence photo CN-4026).

power-operated turrets, two 18-inch torpedo tubes, and four depth charges.

The middle years of the war brought submarine wolf-pack tactics in mid-ocean, while coastal areas had become too dangerous for submarines because of allied air patrols. This called for larger ships with more speed and endurance. Corvettes and minesweepers had to be augmented by frigates and destroyers. Although the destroyers were all built in the U.K., Canadian industry moved rapidly to the larger *Castle*-class corvettes, *Algerine*-class minesweepers and frigates. The newer corvettes were 45 feet longer than those of the *Flower*-class, with better accommodations and 3000 nautical miles added endurance. Thirty-six were ordered in March of 1943, of which twelve went to the Royal Navy as offsets for destroyers. The new minesweepers were 45 feet longer than the old, with 1500 nautical miles of added endurance at 12 knots.

A major improvement in capability came with the *River*-class frigates. Their endurance of 7200 nautical miles was double that of the *Flower*-class, and their top speed of 19 knots compared with 15.5 knots. One disadvantage was that they could not be built on the Great Lakes because their 301-ft length exceeded that of the then-existing locks in the St. Lawrence River. Thought was given to taking them out via the Chicago Drainage Canal and Mississippi River, but excessive complications with bridge clearances and shoal water led to the abandonment of that proposal.

It is appropriate here to note the origins of defence production sharing between Canada and the United States. This dates back to the Hyde Park Declaration of March 1941, which predated the U.S. entry into the war. A key provision stated that "each country should provide the other with the defence articles which it is best able to produce, and above all, produce quickly." Because Canada was buying more in the U.S. than vice versa, a remedy adopted in 1941 found the U.S. Navy ordering two frigates and fifteen corvettes from Canadian yards ([2], p. 68). Along with these awards came a large team of U.S. shipyard personnel to Canadian yards to help expedite the rapid expansion of the U.S. Navy's own convoy escort program ([1], p. 109).

The final phase of the war at sea for Canada shifted from mid-ocean escort service to offensive operations in European waters. This called for cruisers, light fleet carriers, destroyers, frigates, corvettes, minesweepers, and coastal defence vessels. Toward the end of the war and shortly afterwards, Canada acquired two cruisers and a fleet carrier from the Royal Navy. The steady buildup of expertise led to the construction in Canada of four British-designed *Tribal*-class destroyers. Although started during the war, they were not completed until late 1945 to 1948.

Table 1 World War II Canadian construction.

Type of Vessel	Total	Numbers Built: For Royal Navy	For U.S. Navy
Frigates	70	8	0
Corvettes	123	17	8
Algerine minesweepers	62	50	0
Bangor-class minesweepers			
Steam	50	12	0
Diesel	10	0	0
Western Isle trawlers	16	16	0
Motor torpedo boats	12	0	0
Fairmile B motor launches	88	0	8
126-ft minesweepers	30	21	0
105-ft minesweepers	42	32	0
Transport ferries	26	26	0
Total	529	182	16

Table 1 summarizes the extensive shipbuilding program completed by Canadian shipyards during World War II ([1], p. 111). In addition, many auxiliaries were built and major modifications carried out for troop carriers and armed cruisers and also to 19 aircraft carriers. Of equal importance was the massive development of support facilities on both coasts for the repair and improvement of ships and their equipment.

A 1977 study of World War II ship construction reported in the U.S. Naval Institute *Proceedings* notes, *inter alia*, that:

> . . . [T]he shipbuilding effort made in Canada was probably the most remarkable even of the three great naval powers, and delivery delays can take nothing away from the fantastic achievement. . . . The planning and production of [the RCN *River*-class of 70 frigates] is one of the more fascinating stories of the [allied] escort building programs. On the face of it, this was simply a carry on from the RN "River" class building program in the UK. But it was a magnificent achievement for the Canadian yards to switch from the much simpler "Flower" Class corvette program to these . . . sophisticated frigates, and if Canadian ships are examined closely a number of significant improvements can be found over the UK-built ships ([1], p. 110).

As another author has concluded, the one clear lesson was that "Canada need not and should not be dependent on others for future warships and their equipment" ([1], p. 115). The problems of overseas sources and non-Canadian standards had been overcome. The one ingredient missing, a warship design capability, leads off the Canadian naval postwar story to be found in Part VII, in the section Postwar Canadian Naval Vessels.

Acknowledgments

I want to thank Captain Keith Farrell for the invitation and encouragement to write the Canadian navy's technical story (as found above and also in Part VII). I greatly appreciate the word processing provided by the German & Milne Division of ECS-Power Systems Inc. of Ottawa. The critical comments of Rear Admiral Denny Boyle, Rear Admiral Mike Saker, Captain Jim Knox, and Messrs. Bob McNeilly and Don Nicholson have been instrumental in improving the accuracy and the telling. My thanks to all.

References

1 *RCN in Retrospect, 1910–1968*, Boutilier, J. A., Ed., University of British Columbia, 1982; the chapters by Captain J. H. W. Knox have been a major source for this Canadian section.

2 Tucker, G. N., *The Naval Service of Canada, Its Official History*, Vol. 2, King's Printer, Ottawa, 1952.

Commodore W. J. Broughton attended the Royal Military College of Canada, Queen's University, where he received a BSc degree in mechanical engineering, and then MIT, where he received both an MS degree in naval architecture and marine engineering and the Naval Engineer degree. During his naval career he worked, *inter alia*, on every Canadian destroyer program from the *Restigouche*-class conversion in the early 1960s to the *City*-class frigates currently under construction. He was Project Systems Engineer for the *Tribal*-class design and construction from 1970 to 1974. On his retirement in 1990, after 37 years of service, Commodore Broughton was the Canadian Navy's senior engineer at National Defence Headquarters, Ottawa.

Canadian Ship Construction

J. Y. Clarke

Origins of Shipbuilding in Canada

Few industries in Canada can trace their histories as far back as can the shipbuilding industry. From modest beginnings early in the 1600s, ships of increasing size took shape during the 17th century, first in shipyards along the Atlantic Coast, and later along the shores of the St Lawrence River. In the latter half of the century, shipyards sprang up in the Great Lakes area, delivering both cargo vessels and passenger ships of various configurations.

During the next two hundred years, a flourishing industry emerged, with merchantmen being completed for the French Government at the rate of about ten a year. Their quality was such that orders arrived for men-of-war, from 500-ton corvettes to a 70-gun ship of the line. An unlimited supply of high quality timber, coupled with competent, hard-working labour produced vessels that could compete with any in the world, either for trade or for naval purposes.

"Wooden Ships and Iron Men"

After 1759, when Britain acquired virtually all of New France, the industry made another leap forward. Ships were ordered not only to fill orders from Europe, but also to build a merchant fleet in Canada, which grew steadily until the late 1800s. By 1880, the Dominion's merchant marine numbered over 7200 vessels, and Canada ranked fourth among the shipping nations of the world. Canadian ships were sturdy, fast and well-manned and, together with those of the United States, all but overwhelmed European competition for world trade.

Also noteworthy is the construction in Great Lakes shipyards of both cargo and passenger vessels as far back as the late 1600s. During the War of 1812, several warships were built on the lakes, one of which was the *St Lawrence*, a triple-decked ship-of-the-line of 2300 tons displacement and mounting 102 guns. She was larger, with more fire-power than Nelson's *Victory*.

On Canada's West Coast, shipbuilding in any significant way developed at about the time the Canadian Pacific Railway reached Vancouver in 1885. Shallow draft stern wheelers were built to serve the Fraser River gold fields. Larger ships, however, were assembled from pre-fabricated parts from Scotland.

Another important phase of shipbuilding in Canada was the era of paddle steamers. The most famous of these was the *Royal William*, of 830 tons, built in Quebec in 1831. She completed a voyage, under steam power alone, from Nova Scotia to the Isle of Wight, probably ranking as the first merchant vessel to make the crossing without sail.

In the latter half of the 19th Century, stern wheelers operated in the lakes and rivers of the prairie provinces; and shipyards that sprang up in Winnipeg, Selkirk, and Grand Rapids produced eight to ten such vessels per year. They are believed to be the prototypes of the renowned steamers of the Mississippi–Missouri River systems, but their demise was signaled by the growth of the railways.

When iron and steel started to replace wood as a shipbuilding material, Canadian shipbuilders were slow to respond. A lack of government interest was a factor in this. From 1900 to to 1914 Canadian shipyards struggled to catch up, but their activities were largely confined to coastal waters, the Great Lakes, and other inland waters.

World War I

From the earliest days of World War I, it was clear that many new ships would be required to sustain transatlantic sea lines of communication. Orders arrived from Britain for oceangoing cargo ships, and the Government of Canada required submarine chasers, coastal patrol vessels, gate vessels (to open and close anti-submarine nets across harbour mouths), and tugs. These contracts provided the impetus for Canadian shipyards to modernize and to expand capacities. By mid-1917, yards were operating at peak capacities, and by war's end, over 50 cargo ships, displacing up to 8000 tons, had been completed.

Between the Wars

Shortly after the war ended, the Canadian Government Merchant Marine was formed, and during the next few years, orders for some 60 vessels were received for this service. Yards on each coast delivered oceangoing merchantmen of up to 8500 dwt, and the Great Lakes yards produced 2500 dwt cargo ships. These were placed in international trade, in competition with foreign fleets.

By the mid-1920s, the merchant marine requirements had been fully met, and shipbuilding in Canada entered another slump. Several yards closed and those that survived did so through ship repairing. The fortuitous completion of large graving docks on each coast during and immediately following the war, enabled Canadian yards to accept ships of all nationalities for high quality repair work.

In the early 1930s European merchant marines, now recovered from the war, provided competition that Canadian fleets were ill-equipped to handle. Perhaps the Canadian Government could have withstood this challenge to a large extent and kept the merchant fleet competitive, but the political will was lacking and Canadians were apathetic. Few seemed to understand either the advantages or costs of being a bona-fide maritime nation, hardly surprising with the country in the grips of the Depression.

In 1934 the Canadian Government Merchant Marine was disbanded, and its ships sold to any buyer at low prices, or simply for scrap. During the five years leading up to World War II, only 14 ships over 150 ft in length were built in Canadian yards, which struggled to survive with a few orders for small vessels, four 160-ft minesweepers, and repair work. At the outbreak of war in 1939, all sectors of the shipbuilding, repairing, and marine supply industries in Canada were run-down, and ill-prepared to respond to the massive demands that were soon to come.

The Second World War

The story of shipbuilding and ship repairing in Canada during the last war is a legend unto itself. In 1939, only 38 oceangoing cargo vessels were in the Canadian registry. With the devastating U-boat campaign increasing steadily throughout the early years of the war, there arose from Britain the most urgent demand for ships and more ships. In addition, the run down Canadian Navy badly needed patrol boats to protect transatlantic convoys. At the time, our shipyards were faced with the most challenging requirements to modernize, acquire skilled tradesmen and swing into full production in the shortest possible time.

One of the most critical issues was the availability of workers with the necessary skills, largely as a result of full mobilization of the armed forces. The selectivity base was increased by re-training older workers and by training large numbers of women in trades traditionally the purview of men.

Designs for cargo ships and tankers were obtained from Great Britain; and the Canadian Government set up a shipbuilding division within its recently established War Supply Board to accelerate the necessary major expansion of the shipbuilding infrastructure.

Canada's performance in delivering merchant ships and combatant vessels during the Second World War is covered comprehensively elsewhere in this book; however, Table 1, issued in 1949 by the Canadian Maritime Commission, is a good summary of total production. The delivery of almost 800 naval and merchant vessels to British, Canadian and U.S. requirements over a five-year period indicates the extent to which the industry exceeded all expectations, given the state of shipyards in 1939.

To reach these production levels, all phases of the shipbuilding industry, including the suppliers, had to be greatly enlarged and improved within urgent time limits.

Ship repair came into its own during the war, owing to the need for rapid repairs of enemy action and weather damage. Allied ships of all sizes, up to and including some of the largest capital ships, were repaired in Canadian shipyards on both coasts.

Refits, outfitting, and conversions were also a part of wartime shipyard life as, indeed, they are in peacetime. An unusual activity arose in West Coast yards, involving escort aircraft carriers being built for Britain in Washington State, then transferring to B.C. yards for installation of British weapon systems and final outfitting.

It is estimated that some 30,000 ships were refitted, converted or repaired in Canadian yards during the war; and peak shipyard employment exceeded 65,000.

Post War Achievements

From the cessation of hostilities in 1945 until well into the 1960s, Canada's shipyards remained actively

Table 1 Summary of steel naval and merchant vessels built in Canadian shipyards during World War II, 1939–1945 (from the 2nd Report of the Canadian Maritime Commission, June 1949).

Vessel Type	Naval	Merchant	Total
Atlantic Coast			
Corvettes	3	—	3
Destroyers[a]	4	—	4
Dry cargo ships, 4700 ton	—	32	32
Atlantic Coast total	7	32	39
St Lawrence River			
Ammunition lighters	2	—	2
Minesweepers, Bangor	16	—	16
Base supply ships	2	—	2
Corvettes	54	—	54
Frigates	53	—	53
Transport ferries	21	—	21
Water boats	2	—	2
Trawlers, Western Isles	4	—	4
Stores lighters, 79-ft	4	—	4
Diesel tugs, 1000 HP	5	—	5
Dry cargo ships, 10,000 ton	—	99	99
Tanker, 10,000 ton	—	1	1
Dry cargo ships, 4700 ton (Dominion, Grey, Grey-revised)	—	11	11
Tankers, 3600 ton	—	3	3
St. Lawrence River total	163	114	277
Great Lakes			
Minesweepers	84	—	84
Corvettes	51	—	51
Oceangoing tugs	18	—	18
Trawlers, Western Isles	12	—	12
Tankers, 3600 ton	—	3	3
Great Lakes total	165	3	168
Pacific Coast			
Minesweepers	22	—	22
Corvettes	14	—	14
Frigates	17	—	17
Transport ferries	5	—	5
Maintenance ships	—	16	16
Dry cargo ships, 10,000 ton	—	221	221
Tankers	—	12	12
Pacific Coast total	58	249	307
Grand total	393	398	791

[a] Under construction at cessation of hostilities.

employed in building several types of oceangoing ships of modern high-quality. Production and employment levels were lower than during the war, but orders continued to be placed by both domestic and foreign owners for general cargo ships, passenger liners, tankers, and colliers. Cargo vessels were generally of 4500 tons or 7500 tons and over 70 commercial ships were delivered during the five years following war's end.

In addition, large numbers of orders were placed for special purpose vessels during the same period. On the West Coast, work boats, fishing craft, powerful welded steel tugs, and boom boats to serve the rapidly growing forest industries, were built. These latter craft were unique. With 360° drives, they provided the high manoeuvrability required to control large log booms.

In Eastern Canada, the domestic demand was for coastal freighters, large "lakers" (i.e., bulk carriers and tankers for inland water service), ferries, coastal tankers, tugs, and trawlers.

Commercial contracts between 1945 and 1960 were well complemented by orders from the Federal Government. In 1948, as a result of a decision by Canada to contribute to NATO in anti-submarine warfare, the design and construction of a new class of destroyer escort was initiated. Known as *St Laurent*-class, the design was Canadian, and conformed to the following basic requirements for post-war ships:

- Canadian-designed with most, if not all, equipment built or supplied by Canadian companies.
- Quick construction in any yard in Canada in time of war. This was accomplished by unitized construction methods built across Canada, and assembled at a prime shipyard.
- Capable of handling any envisaged submarine, the ship had to be capable of operating in adverse North Atlantic conditions. This was done by rounding all hull contours, recessing the anchors behind heated doors, and locating the capstan within the forecastle.

Refer to the section Postwar Canadian Naval Vessels in Part VII for a detailed description of *St. Laurent*-class ships.

The initial order was for seven ships, but subclasses, involving installation of hangars, flight decks, and variable depth sonar systems were developed; and a total of 20 ships was subsequently ordered. The *St Laurent* was commissioned in 1955, and the twentieth ship in 1964.

Figure 1 shows the *Margaree*, one of the original seven, after her conversion for helicopter-carrying. Telescopic hangar doors, another Canadian innovation, were installed aft of the funnels. This photograph shows the ship under tow, entering refit, with the 3-in./50 gun mount removed. The *Margaree* was laid down in September 1951 and commissioned in October 1957.

Within this same 15-year time frame following World War II, ship repair activity was also substantial, involving refits, dockings, repairs, and conversions of

Fig. 1 HMCS *Margaree.*

Fig. 2 HMCS *Provider.*

ferries were built for coastal operators, and the demand for offshore semi-submersible drilling rigs and their supply and support vessels emerged and grew on both coasts.

Government orders were received from the Navy, Coast Guard, the fisheries protection fleet, and ferry corporations. Employment in Canadian yards was relatively stable at some 15,000 workers, and marine supply industries were able to continue their post-war development, striving to reach and sustain world-class standards for equipment and products.

Three naval contracts that were awarded during the decade are worthy of specific mention. (Complete descriptions of all these ships may be found in the section on Post-war Canadian Naval Vessels in Part VII.) The first of these was an order for an operational fleet support ship, designated AOR. This versatile vessel, commissioned in September 1963, was named the *Provider* (Fig. 2). Two follow-on ships of the same basic dimensions, *Protecteur* and *Preserver*, incorporated design changes stemming from *Provider*'s operational experience.

some 2000-plus commercial ships per year, on average. Added to this activity was the major conversion of 20 *River*-class frigates into enhanced anti-submarine warfare vessels, which became known as the *Prestonian*-class, after the first such frigate to be converted.

During the 1960s, shipyard activities involving new construction for both commercial and government accounts, as well as conversions, refits and repairs, remained at good levels. Several medium and large cargo ships and tankers were delivered to foreign owners, and domestic orders continued to be received. On the West Coast, 10,000-ton side-tipping log barges were developed and built; in Eastern Canada oceangoing cargo ships, and tankers up to 40,000 dwt were delivered to foreign owners. Passenger car

The second naval project was for a high-speed all-aluminum prototype hydrofoil craft, the *Bras D'Or*, capable of operating in the open ocean, and capable of providing a platform for anti-submarine warfare. This craft was of Canadian design. From the shipbuilder's perspective, many new aspects arose from the contract. Of these, the most significant were the construction tolerances, and stringent welding requirements, both of which stemmed from aircraft design standards. The builder (Marine Industries, Ltd., of Sorel, Quebec) elected to use an upside-down assembly concept, permitting a maximum amount of down-hand welding, which employed automatic metallic inert gas procedures. Needless to say, quality

Fig. 3 HMCS *Iroquois.*

control and assurance demands were comprehensive. In the 1960s, this project was considered a technical "first" in Canadian shipbuilding.

As the decade ended, the Navy placed contracts with two shipyards for two ships each of a new missile-equipped destroyer design, to be known as the *Tribal*-class. All four ships were delivered in 1972 and were, by far, the most sophisticated ships to be designed and built in Canada. Marine Industries, Ltd., of Sorel, Quebec was the lead yard, and built two of the class. The others were built at Davieship, Ltd., of Lauzon, Quebec. (Figure 3 shows the *Tribal*-class HMCS *Iroquois.*)

With considerable new building activity under way from government orders, perhaps this impact tended to disguise evolving trends. Orders from foreign owners diminished owing to the growth of lower cost shipbuilders, first in Europe and later in the Far East. By the early 1970s the commercial order book in Canada was all but entirely domestic. Notwithstanding this trend, shipyard activity remained at an average annual level well in excess of 150,000 G.T. delivered, with some 90 percent of the work being for commercial orders. This held until well into the early 1980s, with shipyard employment remaining relatively stable at 12,000 to 14,000 workers. Part of this work came from the design and construction of specialized icebreaking supply vessels for service in the Canadian Arctic, about which some specific information will be provided in later paragraphs.

Another example of Canadian innovation dating back to the 1970s is the Seabus pedestrian ferry service from downtown Vancouver across the harbour to North Vancouver. Boats and shore facilities are completely integrated so that exiting passengers leave by a set of doors on one side, while on-loading passengers are entering doors on the other side. Catamarans, powered by four diesel engines with propellers at each end of each hull, with well-found passenger accommodation, traverse the two miles across the harbour in a few minutes. At each terminal, a hydraulic fendering system permits the Master to dock his highly manoeuvrable ferry without shore-side assistance. He can then control the passenger doors and ramps from on board. Complemented by a fully automated ticketing system and connections with rapid transit systems at the terminals, the Seabus ferries provide efficient, economic public transport. Although all components of this system are of modern concept and design, it is their integration into a smooth, reliable total system, featuring frequent departures, low fares, rapid turnaround, and virtually no "down-time," that makes the Seabus popular. Figure 4 shows two of these craft on trials. These are another example of all-welded aluminum alloy fabrication.

The Canadian Coast Guard, realizing in the early 1970s that for their power and displacement, hovercraft were excellent icebreakers, initiated an R&D

Fig. 4 Vancouver Seabus ferries (Murray McLellan photograph).

program to identify the benefits of using air cushion technology to enhance the performance of standard icebreakers. As a result, the air-cushion icebreaking bow (ACIB) was developed, to be used in tandem with existing icebreakers in the Great Lakes. The concept was quite simple, namely that the power required for the ACIB to provide clear water through 75 cm of ice is far less than the power for the pushing icebreaker to do the same without the ACIB. Since the pushing vessel is, in essence, operating in open water, the only incremental power requirement to break ice is the provision of cushion pressure for the ACIB. The trials of the ACIB indicated increased icebreaking capability, increased speed of the ACIB/ship combination through consolidated ice, and greatly increased economy of operation of the combined unit over the ship itself. To date, however, this development has yet to be put to practical use in Canadian waters.

As mentioned earlier, a particularly Canadian expertise in the design and construction of icebreaking offshore support vessels was developed in the 1970s and early 1980s, largely in support of oil and gas exploration in the Beaufort Sea, at the western extremities of Canada's arctic waters. One of the most powerful of these is the *Miscaroo* (Fig. 5), built in Vancouver, to Canadian Coast Guard Arctic Ice Class 4 standards, and designed to operate in temperatures down to −50 °C. This was the first commercial icebreaker built in Canada to this standard.

The *Miscaroo*'s design contains several unique features to enhance her performance in multi-year ice. The hull form contains a forefoot wedge to deflect ice from the ship's path, and large bossings over the propeller nozzles and twin rudders to protect them from ice damage. The main propulsion system comprises four diesel engines delivering 14,900 bhp through two sets of reduction gears to twin shafts fitted with controllable pitch propellers housed in nozzles. Electrically driven bow and stern thrusters provide maximum manoeverability, and the most up-to-date communication and navigation systems have been provided. The spacious wheelhouse provides maximum visibility for icebreaking, cargo transfer, and anchor handling operations. It contains full ship control systems centre forward, at the bridge wings, and centre aft. Finally, through excellent thermal and acoustic internal insulation and HVAC systems, as well as first class outfitting, maximum comfort for six officers, sixteen crew and twelve passengers is provided. This vessel is an excellent example of innovative thought being focused on special operational requirements.

Fig. 5 Icebreaker *Miscaroo.*

The *Miscaroo* is 258 ft-8 in. in length, with a beam of 56 ft-6 in., a draft of 24 ft-7 in., and a full load displacement of 1898 tonnes.

During the early 1980s, our yards were kept fairly busy filling domestic commercial orders for coastal ferries, tugs, and offshore supply vessels as well as several large offshore trawlers. Government needs for Coast Guard icebreakers and small search-and-rescue craft were also forthcoming, while ship repair activities continued at a stable level.

It was about this time that the several Canadian yards strove to adapt their facilities to international state-of-the-art standards. Acquisition of CAD systems for ship design procedures; CAM systems for control of manufacturing and assembly processes; organization of planning into zone-by-stage modular construction techniques, in which units are fabricated and outfitted to a maximum by teams, drove the modernization process. In addition the yards effected major re-structuring of labour union agreements. These major changes are costly in both time and money, demanding good cash flows or significant investment of capital funds. During the 1980s, however, the worldwide shipbuilding slump hit the Canadian scene with a vengeance. Foreign commercial orders were no longer being received and domestic orders dwindled to a trickle. The overall effect meant that Canadian yards were almost entirely dependent on Government orders for new buildings and on refit and repair contracts. The devastating impact of this situation can be seen in Fig. 6. This resulted in our yards making very slow progress in adjusting to methods necessary to be competitive.

There were, however, a few exceptions, and these relate to those larger yards that won contracts for large passenger car ferries, Coast Guard icebreakers, and a series of twelve sophisticated frigates for the Canadian

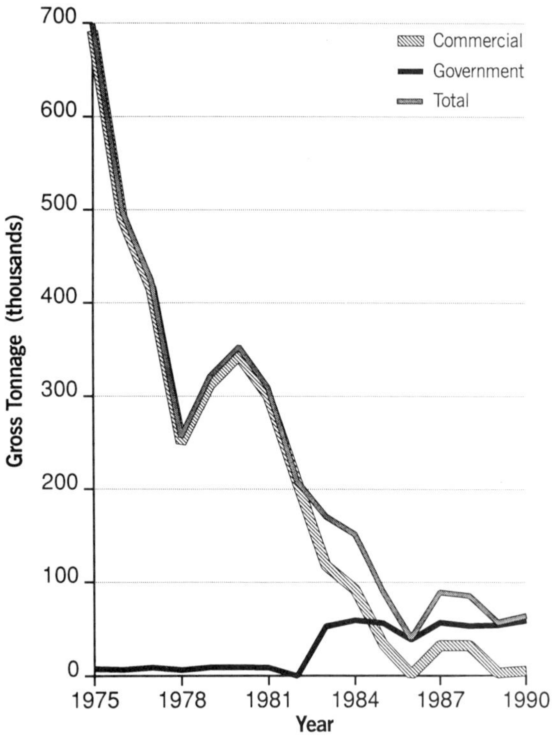

Fig. 6 Order book in CMIA shipyards, 1975–1990.

Navy. Each of these types of ship was built by final assembly of unit construction, and each has its own technical story to tell.

The *Caribou,* ordered by Marine Atlantic, was constructed in 180 unit blocks, as shown in Fig. 7. This procedure was engineered to reflect the yard's capacity and resources. A central planning department coordinated the efforts of six planning groups, which broke down all major planning objectives into specific work instructions. With maximum involvement of personnel, such instruction packages were developed for each unit. The end result of these efforts was the delivery of the ship on schedule. Moreover, the approach established an effective shipbuilding system that, with upgrading stemming from lessons learned, could be adapted to building virtually any type of ship. Indeed a second passenger car ferry, of the same class as the *Caribou,* has been delivered subsequently, and is shown in Fig. 8. Both ships were built by MIL Davie Shipbuilding Ltd. of Lauzon, Quebec.

The icebreaker story concerns the Canadian Coast Guard Ship *Henry Larsen,* an Arctic Ice Class 4 pure icebreaker. She was completed in 1987 at the now-defunct Versatile Pacific Shipbuilding Industries shipyard in North Vancouver B.C. The technical innovation implicit in this ship concerns the 16,000 shp propulsion system. The system comprises three 6000 kW diesels driving three 5100 kW alternators producing 4160-volt electrical power and feeding to a common bus. This power is then fed through cycloconvertors to two 6000 kW ac electric motors to drive two fixed pitch propellers at 1100 volts. A bubbler system is fed directly from the main bus and the ship's service supply transformed from the generated 4160 volts to 600 volts. An auxiliary generator of 840 kW provides power for use in harbour.

A machinery system of this sort had not previously been fitted in an icebreaker. The Finns were in the process of designing an icebreaker with this type of system; however, their vessel was using 20,400 shp from four generator sets to achieve an icebreaking performance of three feet. The Coast Guard designed the *Henry Larsen* with 16,000 shp using three generator sets to break four feet of ice.

The design of the machinery system incorporated both load and power management systems. The load management system required a fast, precise response capability. The authorities felt that a fully automatic system was necessary, one that would be operable during all working conditions. In addition the system would have to ensure that voltage and frequency levels would always be within a narrowly prescribed limit. An innovative power management system permits automatic control through a computer mimic program, which indicates the operating and readiness states of all components, and controls the on-line systems. For example, when operating in automatic mode on one main generator, with another on stand-by, if the load requirement exceeds 90 percent, the second generator will self-start, autosynchronize, connect

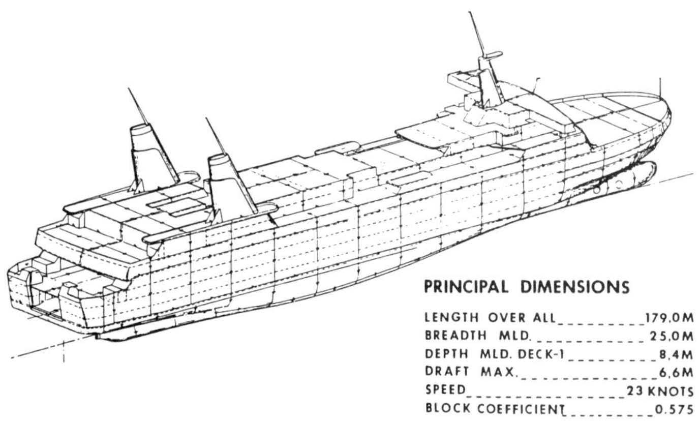

Fig. 7 MV *Caribou* block breakdown. (Source: MIA Technical Library.)

Fig. 8 Passenger ferry MV *Joseph and Clara Smallwood.*

on-line, and assume 50 percent of the load, all within 30 seconds.

Some other innovative features incorporated in this design include:

- monitoring of all machinery systems through microprocessors;
- vibration analysis equipment to detect any equipment deterioration;
- fibre optic inspection equipment between shaft liners to facilitate inspection without shaft withdrawal;
- centralized cooling system to minimize salt-water piping and lubricating oil detectors; and
- the highest standards of fire detection and fire fighting equipment, with full bridge monitoring.

The *Henry Larsen* has now been in service for several years. She met all design expectations, and exceeded many. Figure 9 shows this fine ship underway in open water.

The most significant naval project currently in the construction phase is the Canadian Patrol Frigate (CPF), of which twelve are under construction. The CPF is a modern monohull twin-screw anti-submarine frigate, with a primary role of maintaining sea lines of communication in wartime, and a peacetime role of surface and subsurface surveillance, including participation in NATO activities.

The CPF has an LOA of 134 m, a beam of 16.4 m and a design draft of 4.65 m. She displaces 3700 tonnes in light ship condition, is designed to have a maximum speed in excess of 27 knots from 45,500 hp provided by two gas turbines, and a cruising speed of at least 16 knots from 8700 hp provided by a single diesel engine. (Postwar Canadian Naval Vessels in Part VII describes these ships in more detail.)

Zone design and construction methods are being used in each of the two major shipyards building these warships. The prime contractor, Saint John Shipbuilding Ltd., developed a unit system whereby the CPF plan is divided into 57 assembly units, which are typically one deck high and contain at least one transverse bulkhead. These units are joined and pre-outfitted into two-, three-, and four-deck levels to form the 26 units shown in Fig. 10.

Powerful CAD/CAM systems were acquired and developed to integrate the contract design with unit plans, zone design, and stage design, and to control

Fig. 9 CCGS *Henry Larsen*, a Class 4 escort icebreaker (courtesy of the Canadian Coast Guard).

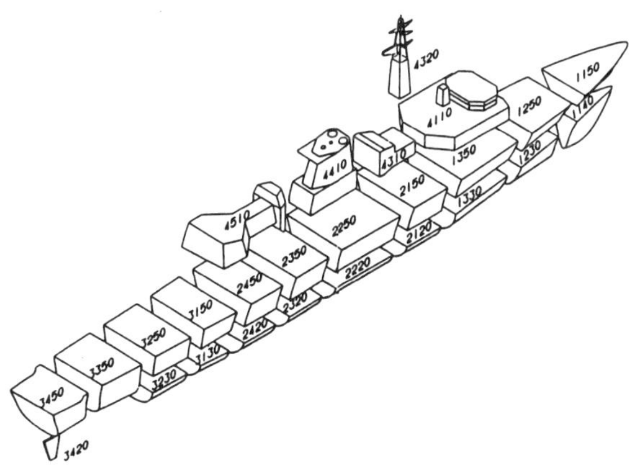

Fig. 10 CPF isometric erection unit breakdown (source: CMIA Technical Library).

construction stages, identify problem areas and monitor the assembly.

The first ship of this class, the *Halifax,* is shown undergoing sea trials in Fig. 11. The last of the twelve should be delivered in 1996.

To conclude this chapter, mention must be made of the application of modern technology within Canadian maritime industries within the past few years, in addition to those already noted. These include sophisticated model testing in water and ice tanks, particularly to identify optimum hull form compromises when conflicting requirements exist, as they usually do. To this we can add computer-controlled water-drenched plasma plate cutting, extremely accurate cold rolling equipment, self-unloading systems for bulk carriers, stealthy "super-quiet" propeller design and manufacturing for submarines and other warships, which virtually eliminate audio signatures; and unique polymer stern tube and rudder shaft bearings that are light, low in length-to-diameter ratio, and virtually impervious to wear. Internal to the ship, a computerized, fully integrated fire detection and suppression system, adaptable to all kinds of ship has been developed in Eastern Canada. Indeed, this system has been selected for the Channel Tunnel project connecting England and France.

Fig. 11 HMCS *Halifax* (courtesy Dept. of National Defence).

The future of Canada's shipyards and maritime industries in today's highly competitive world will depend upon many factors, not the least of which is the combination of quality and productivity. In this context, the technical sessions conducted by the Society of Naval Architects and Marine Engineers, the Canadian Institute of Marine Engineering, and the Canadian Maritime Industries Association play a most important role, as they always have, in advancing marine technology in North America.

Bibliography

Annual Reports of the Canadian Maritime Commission, 1949–1967.

Breton, D., "Gulfspan M.V. Caribou C.N. Ferry; Planning and Production Engineering," Davie Shipbuilding, Ltd., Feb. 1985.

Butt, M. N., "Newfoundland Dockyard Corporation—A Century of Service," March 1989.

The Canadian Coast Guard Fleet News, No. 16, 1987.

Canada's Navy, Commemorative Issue, 1985.

Clarke, J. Y., "Canada's Shipyards—Yesterday, Today and . . . Will There Be a Tomorrow?," *Maritime Affairs Bulletin,* No. 1/88, Navy League of Canada.

Clarke, J. Y., "Surge of High Technology Orders for Canadian Yards," *Marine Engineering Digest,* Vol. 8, No. 3, 1989.

CMIA Annual Statistical Reports, 1986, 1988, 1989.

Cosulich, N., "Historical Summary—West Coast Manly Shipyard," Feb. 1989.

Jaquith, P. E. and Snyders, G. J., "The Application of Advanced Design and Construction Techniques to the Canadian Patrol Frigate," Saint John Shipbuilding, Ltd., Feb. 1986.

McLaren, T. A., "Notes on Shipbuilding—Canada's West Coast," Allied Shipbuilders, Ltd., North Vancouver, Feb. 1989.

"Shipyard Approach to the Hull Construction of the FHE Hydrofoil Ship, *André Rochon,*" Marine Industries, Ltd., Sorel, Quebec, Feb. 1967.

J. Y. Clarke graduated from the Royal Canadian Naval College in 1947, and from the Royal Naval Engineering College (Plymouth, England) in 1951. He then served in the Canadian Navy as Engineering Officer aboard several ships. He was Executive Officer on the *Provider* and Commanding Officer on the *Athabaskan.* He joined the Canadian Coast Guard in 1970 and served as Director General Fleet Systems from 1979 to 1987, during which period 13 new ships were designed and built for the fleet. He retired from government in 1987 and assumed the presidency of the Canadian Maritime Industries Association, which represents shipyards and allied industries.

Part VI

Ships of Commerce

During the past half century we have seen a complete transformation in the ocean transport of dry cargo. We have seen the demise of the old-fashioned ship with a forest of masts and booms, whose cargo-handling inefficiency required it to spend at least half of its life in port. The advent of unitized cargo systems, on both ship and shore, has greatly increased marine transport productivity. Concurrent improvements in sea speed, weather routing, and shipboard automation have served further to reduce ocean transport costs. Unitized cargo systems have reduced insurance claims, packaging costs, and costs for the inland part of the complete door-to-door transport system.

In constant-value dollars, ocean transport costs are now but a small fraction of what they were fifty years ago. As a result, world trade has grown enormously and ocean shipping now represents only a tiny portion of the cost of imported or exported goods. (For example, the ocean transport cost of a Japanese auto selling for far more than $10,000 is only about $100.) Not a bad achievement for an industry that is sometimes looked upon as reactionary and slow to change.

Break-Bulk and Containerships

CHARLES R. CUSHING

Introduction and Overall Perspective

IN THE YEARS from World War I to World War II (approximately 1918 to 1939) cargo ship design and shipbuilding production in America saw two dramatic peaks. First, the Shipping Act of 1916 and World War I caused over two thousand ships to be built. Fifteen hundred mostly slow tramp-type cargo ships, were built for the government Emergency Fleet Corporation. Many other, smaller vessels, including 600 constructed of wood, were also built. On the Great Lakes, more than a hundred lake-bulkers were constructed. This surge of construction was completed by 1922. Unfortunately, however, these ships, many slower than 10.5 knots, proved unsuitable for competitive liner service. Later the Merchant Marine Acts of 1920 and 1928 encouraged owners to build new ships, but those measures proved ineffective.

The U.S. merchant marine, by the mid-1930s, comprised a 3000-ship, 12 million gross ton fleet. Most, however, were obsolete, over age and non-competitive.

The decade of the 1930s saw the Depression and ominous signs of an impending war, both of which were to have profound effects on U.S. cargo ship design and construction.

The Merchant Marine Act of 1936 formed the U.S. Maritime Commission and its ambitious subsidy programs for both ship construction and ship operation. The Commission's approach called for building many ships from a small number of standard designs. The Commission's naval architects favored welded hulls, more powerful steam plants, higher sea speeds, improved cargo gear, and higher safety standards. All these factors had a major favorable impact on the thousands of cargo ships built by the United States in the ensuing ten years.

While still newly formed, the Commission embarked on the design and development of a standard series of cargo ships. The first was the C2, followed by the slightly larger, 10,000-ton C3. A need was also perceived for a smaller ship. Accordingly, two versions of the C1 were developed. One was of full scantling design, the other, shelter deck.

In January 1938, twenty C2 and fourteen C3 ships were placed on order.

By August 1939, twenty-four standard ships were on order; and by the beginning of 1940 the number had increased to 145. Britain, already engaged in war and suffering severe shipping losses, ordered large numbers of its own designed standard ships from U.S. yards. Britain also put Canadian yards to work on standard ships, some similar to those in the United States. (See Max Reid's chapter on Canadian merchant ships.)

The attack on Pearl Harbor in December 1941, and subsequent entry of the United States into World War II, set the opening stage for SNAME's second half-century. The massive U.S. shipbuilding and ship design programs of the period 1941 to 1945 have been unequalled in history. Between 1938 and 1945, approximately 5171 cargo ships, with a cumulative deadweight capacity of nearly 55 million long tons, were built by or under the aegis of the U.S. Maritime Commission.

At war's end, the U.S. Government withdrew from merchant shipping, disposing of the cargo ships by selling them under the Merchant Ship Sales Act of 1946, or putting those that remained unsold by 1950 into the National Defense Reserve Fleets. The cargo ships in those reserve fleets have proven useful to the United States in the Korean War, Suez Crisis, Viet Nam, and Operation Desert Storm. During the 1960s, some of them were sold to owners, such as Sea-Land and Seatrain, for conversion to containerships and other types.

During the early post war years, efforts by the Maritime Commission to develop a new, competitive merchant ship centered on the C3-S-DX1, *Schuyler Otis Bland*, designed in 1948 and in service by 1951. She proved to be too small and too slow, however, and the C4-S-1A Mariner-class evolved, with 35 such ships ordered by 1951. These, together with other

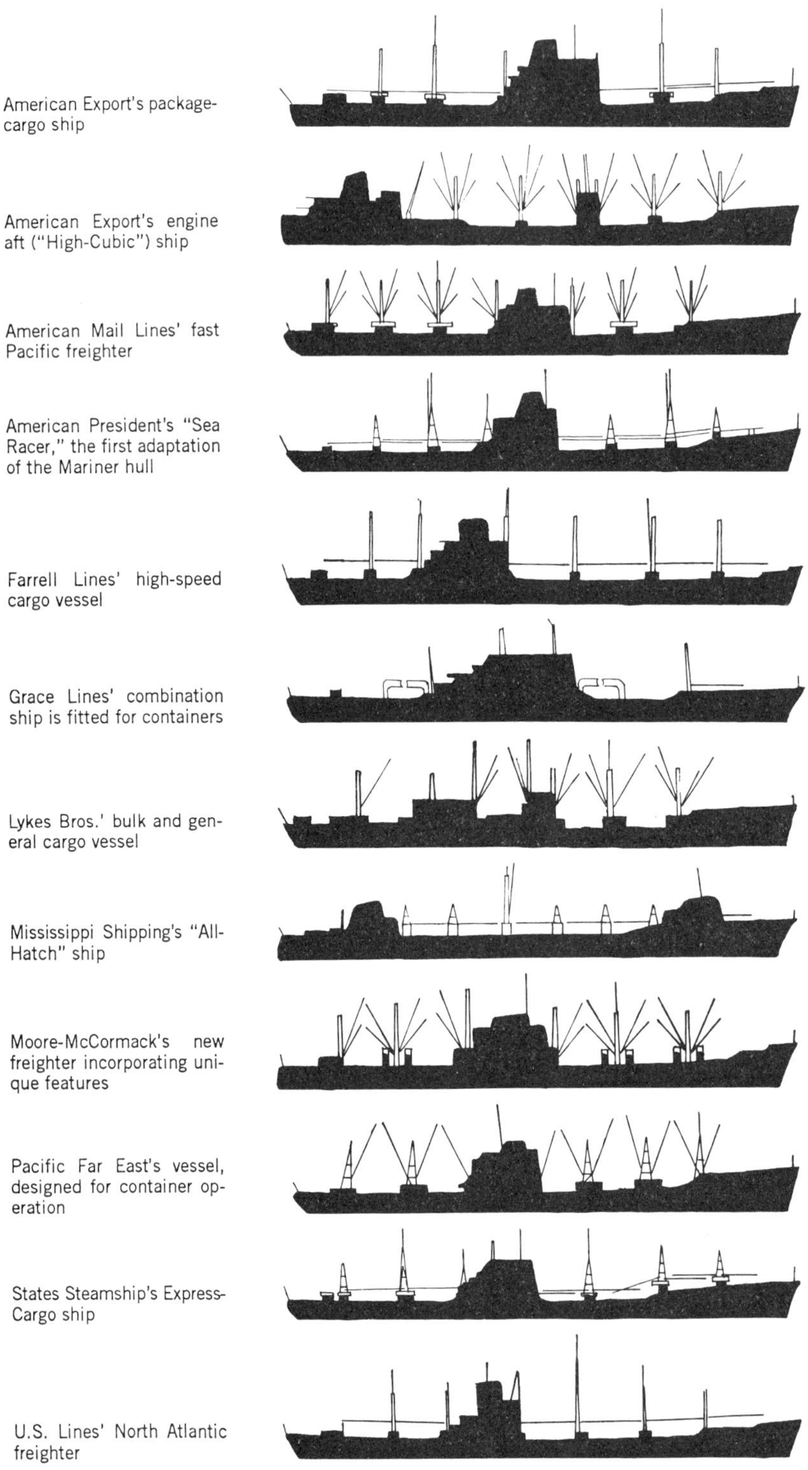

Fig. 1 World War II vessels were replaced by ships designed for their particular trade.

initiatives, such as Liberty Ship modernization program, and the nuclear ship *Savannah*, sustained U.S. progress in cargo ship design and construction until the later 1950s.

With the introduction of containerization in the late 1950s, the stage was set for a flurry of technical innovations in cargo ship design during the decades of the 1960s and 1970s. Containerships, roll-on/roll-off, and barge carrying ships, all vied for supremacy, each finding its special niche. Of these competing forms, containerization took a dominant lead. At first, owners were satisfied to convert conventional ships into fully cellular containerships, but by the 1970s containerships were being purpose-built.

The 1970s and 1980s proved to be an era of evolving technology, not only in cargo handling, but also in propulsion, automation, and shipboard work-rationalization. Notwithstanding the increased size and power of cargo ships during this period, crew sizes dropped dramatically from 45 to less than half that number. Pontoon covers, cellular holds, automated diesel engine rooms, superior coating systems, automatic mooring equipment, electronic navigation and communication equipment, and the shift of cargo handling equipment from the ship to the shore are but some of the reasons for these reductions in crew sizes.

The 1980s and beginning of the 1990s saw a steady erosion in the number of U.S.-flag companies and ships. Most of the U.S.-flag dry cargo ships built during this period were constructed abroad, principally in Japan, South Korea, and West Germany. The trend was toward large, more fuel-efficient containerships. The decade of the 1980s was definitely one of ship type-specialization.

Beginning in 1958, two dramatic activities were to dominate U.S. cargo ship design and construction for the next dozen years: the cargo ship replacement program, and the emergence of containerization.

Fig. 2 Cargo liner SS *Prudential Seajet*: length 544 ft, beam 75 ft, deadweight 13,450 tons, and service speed 20.8 knots. The *Prudential Seajet* makes use of automation in her propulsion, navigation, ship's services, and cargo-handling equipment and can accommodate general, refrigerated, and liquid—as well as containerized—cargoes.

Over the next 20 years, 1970–1990, the U.S. cargo fleet activity was characterized by the introduction of foreign built, U.S.-flag, diesel propelled, automated, large, fast containerships. While the number of U.S. dry cargo liner companies and U.S.-flag cargo ships and the share of U.S. cargoes carried in U.S. ships dwindled, the technical sophistication, efficiency, and total productivity of U.S. cargo ships increased.

Containerization has been, unquestionably, the most profound and far reaching advance in cargo transport during the 20th Century. It has revolutionized ship design and cargo handling. It has greatly shortened delivery times, eliminated unexpected delays, and permitted reductions in inventories. It has minimized packaging costs and significantly reduced damage and pilferage, and hence insurance costs. It has made it cheaper and easier to ship cargo and opened new markets for goods and products.

While the principles of unitization have long been understood (for example, CONEX boxes and Dravo Vans), and intermodal transport tried (for example, Seatrain's shipment of box cars), it wasn't until 1955 that the real benefits of containerization were demonstrated. Malcom McLean is universally recognized as the father of containerization. This trucker, on April 26, 1956, sent the converted T2 tanker, *Ideal X* to sea with fifty-eight 33-ft intermodal containers stowed on a spar deck.

What followed over the next 35 years was the introduction of some six million containers, millions of chassis, hundreds of thousands of tractors, thousands of straddle cranes, and large port gantry cranes. A group of mostly unsubsidized U.S. ship operators led the way. U.S. subsidized operators, skeptical at first, soon followed closely behind, and the U.S. merchant marine led the world into intermodal transportation.

Sea-Land (originally called Pan Atlantic Steamship Company) made the first moves. Matson, on the West Coast, quickly followed, especially with some innovative technology. Seatrain, on the East Coast, also undertook large scale containerization efforts. Several smaller operators, such as Bull Lines and Alaska Steamship, reacted to these improved container methods, but, unfortunately for them, not quickly enough.

Grace and American President Lines were earliest among the subsidized operators to see its benefits. Others took a wait and see position or experimented with barge-carrying ships or the RO/RO concept. Many suffered financially or went out of business by delaying. In some cases, new ships were put through major conversions shortly after their introduction into service.

The Era of World War II

Liberty Ship Class—EC2 Type. Max Reid's chapter on Canadian merchant ships outlines the story of how Canada and the United States worked together to build ships to help Britain during the early years of World War II. One of the series of ships built in those years was the Ocean-class, and that led to the start of our country's Liberty Ship program. The Liberty Ships were based on the British design; but extensive modifications and detailed plans were carried out by Gibbs and Cox.

The Liberty Ships were virtually identical to one another. Each had five cargo holds, three forward of the engine room and two aft. Each hold had a deep lower hold and a shallower 'tween deck. The lower portion of No. 1 hold was fitted with deep tanks.

Each Liberty Ship had a loaded draft limitation of 27 ft-9.25 in., which gave a displacement of 14,257 long tons and a deadweight of 10,920 long tons. With a full load of fuel, the vessel could carry about 9146 long tons of cargo. The gross and net tonnages varied slightly between ships, but were approximately 7500 gross and 4309 net registered tons. Each hatch was served by a pair of five-ton booms and double-geared steam winches. Hatch No. 2 had a 30-ton swinging boom and Hatch No. 4 had a 15-ton boom.

Most Liberties were powered by mass produced, 2500 ihp triple expansion steam engines.

The first Liberty Ship delivered was the *Patrick Henry*, in December 1941. Between September 1941 and October 1945, approximately 2750 Liberty Ships were delivered.

Victory Ship Class—VC2-S-AP1 to -AP5. While the Liberty Ships provided an excellent solution for emergency, it was clear that the United States needed a faster ship with the same carrying capacity. Consequently, in April 1943, the Maritime Commission carried out a preliminary design, and George G. Sharp, Inc., prepared the final design for a new class of vessel. The Victory Ship was originally designated EC2-S-AP1, but this was changed to VC2-S-AP1. As originally planned, several variations were envisioned. These included:

AP1 5500 hp Lentz engine version (not built)
AP2 6000 hp steam turbine propulsion
AP3 8500 hp steam turbine propulsion; this concept was developed into the U.S. Navy Attack Transport type A5
AP4 5850 hp diesel propulsion (not built)
AP5 8500 hp steam turbine configuration as an APA transport

In addition to 117 VC2-AP3 transports, 414 cargo type Victory Ships were built during the war.

These ships were designed from the keel up for welded construction.

Because they had been designed for a 15-knot speed, the Victory Ships were commercially viable in the years immediately following the war.

C1-Class. The smallest of the Maritime Commission's program of C-type vessels was the C1, which came in four models, C1-A (a shelter decker), C1-B (a full scantling ship), C1-M-AV1 (an engine-aft ship), and the C1-S-AY1 (converted to naval service). The C1's were both steam and diesel propelled. The numbers of each were as follows:

	Steam	Diesel
C1-A	19	46
C1-B	85	10

Between 1940 and 1945, 160 C1-A and C1-B vessels were delivered. In addition, 13 C1-S-AY1 were built.

The C1-A and -B dry cargo ships had five holds, three forward and two aft, with masts, booms, and electric winches between the holds. The cargo holds had three decks forward and two aft.

Crew accommodations were for 41 and 43 people on the diesel and steam vessels, respectively.

The C1-A ships were built as shelter deck, handy size, shallow draft break-bulk cargo ships. While some were built with steam turbine drive, by Pusey and Jones, Penn Shipbuilding built a larger number of diesel driven C1-A's.

The C1-B types were also intended as shallow draft ships and they were smaller than the C1-A's. As full scantling ships, they had low freeboards and, generally, wet decks. Federal built the initial C1-B's with steam propulsion, and Western Pipe and Steel built the later ones in this class with diesels. Consolidated Steel built some of the later C1-B's as troop ships (designated P1-S-AY1).

There was a definite need for small ships to assist in the South Pacific, i.e., vessels suitable for island hopping. Therefore, the Maritime Commission prepared the preliminary design for the C1-M-AV1. Between 1943 and 1945, 208 of these diesel ships were constructed. A number were built as refrigerated ships (R1-M-AV3), twin screw lumber carriers (C1-MT-BU1), and other types.

The C1-M-AV1 was an engine- and superstructure-aft dry cargo ship, with three holds forward and a small hold aft, two decks in each. The ship had raised forecastle and poop deck.

Many C1-M-AV1 vessels had names referring to knots, hence the nickname of the class "knot ships."

Two hundred eighteen of these slow (11-knot), general purpose cargo ships were built as 4000 dwt dry cargo ships. An additional 17 were built as refrigerated ships (designated R1-M- AVS). These C1 and R1 ships were arranged with holds forward and engine room aft. The six-cylinder, 180-rpm direct drive diesels were rated at 1700 shp.

The C1-MT-BU1 comprised four small ships, three for Sword Lines. These eventually (1964) became "box tops." (A "box top" was an industry term to describe an older ship to be traded in by a shipping company in exchange for a U.S. Government-owned surplus vessel in the reserve fleet.)

Owing to the shortage of steel, a Maritime Commission design group led by John Vasta designed a series of 24 small C1-S-D4 cargo ships to be built from reinforced concrete, adding 488,000 dwt of shipping to the urgently needed tonnage required during the war. These were propelled with 1300 ihp triple expansion steam reciprocating engines, which gave them a speed of ten knots.

C2-Class. The U.S. Maritime Commission's technical department carried out the basic ship design on the then-modern C2 class cargo ships before January 1938. The U.S. Navy participated in the preliminary design, especially with regard to the hull form. Federal Shipbuilding was awarded the first six contracts in January 1938.

The basic C2 was a single screw, five-hold, three-deck break-bulk cargo ship. Crew quarters were a decided improvement over those found on earlier ships. Higher speed was also an improved feature.

One unique concept was the contra-guide rudder, which was later used on all war-built cargo ships, but discontinued thereafter.

Most of the C2's were steam propelled, but eight were lengthened by 15 ft to accommodate Sun-Doxford diesels, and were designated C2-SU type ships.

The basic C2-class, first delivered in 1939, was a shelter decker intended to carry volumetric cargoes. Most of the variations on the basic C2's were also shelter deckers.

The exception was the series of C2-S-AJ1's built as full scantling ships, which loaded nearly two ft deeper in draft. Bulkheads were carried watertight up to the main deck. Passenger berths for eight and crew accommodations for over 50 were standard.

Fig. 3 The *American New York* in the Panama Canal. The ship carried mostly 40-ft containers. Four of the cargo holds were outfitted to carry 20-ft or 40-ft containers, and four stacks on-deck to hold 20-ft containers. (Courtesy American Bureau of Shipping.)

Late in World War II, a number of C2's were built specially for the trading needs of liner companies, e.g., C2-S1-A1 (Alcoa), C2-S (Robin), C2-S-E1 (Waterman), C2-G (Grace), etc.

The following list is a roster of the family of C2's:

C2	Basic 1939 vessel
C2-T	4 Tampa-built motor ships
C2-S	6 modified for Robin Lines 1940–1941
C2-SU	3 Sun-built motor ships
C2-SU (R)	5 C2-SU types with reefer spaces
C2-F	2 ammunition carriers, motor ships
C2-G	2 C2-S-B1's modified for Grace, with 500,00 cu ft of reefer space for bananas
C2-F	7 Federal steamships modified for Lykes
C2-S-A1	4 Bath-built export ships
C2-S1-A1	3 Moore-built Alcoa ships
C2-S-B1 (R)	6 C2-S-B1 types with reefer holds
C2-S-B1	Federal & Moore ships for Farrell and Grace
C2-S-AJ1	North Carolina-built shelter deckers
C2-S-AJ3	32 North Carolina-built shelter deckers
C2-S1-AJ4	6 Grace 52-passenger cargo liners
C2-S-AJ5	10 USL versions of the C2-S-AJ1
C2-S-E1	30 Waterman C2's
C2-S1-DG2	3 Grace 52-passenger C2-S-B1's
C2-N	3 naval ammunition carriers
C2-S-AJ2	5 AJ1-types with additional reefer space

C3-Class. The C3's were the largest, fastest, and beamiest class of cargo ships built up to the beginning of World War II. The first order for a C3 was let in 1938. Federal Shipbuilding was designated to build the steam versions, and Sun Shipbuilding the diesel versions. The vessels, because of their wide beam, exhibited good transverse stability characteristics. They proved useful to the Navy as auxiliaries and escort carriers. In fact, the *Tangiers*, a C3 that was anchored at the end of "Battleship Row" in Pearl Harbor on December 7, 1941, had the distinction of being the first naval vessel to return fire to the Japanese.

This useful class proved to be competitive after World War II as a general cargo carrier. In the 1950s, because of their superior stability, they were found useful as conversions to containerships.

The basic C3 was a shelter deck, five-hold, transversely framed cargo ship. The No. 2 and 4 lower holds were fitted as cargo deep tanks. While there were many variations in cargo gear, the basic C3 had eight sets of five-ton booms, two sets of ten-ton, and a single thirty-ton heavy lift boom.

Most C3's were 8500 shp, cross compound steam turbine-propelled ships. Some, however, were motor ships powered by four 2225 bhp, two-cycle geared diesels.

The C3-S-A1 was built in the early World War II years, initially for Isthmian Lines, but they proved to be of interest to the U.S. Navy for conversion to escort carriers. In the 1950s and 1960s many C3's were operated not only by Isthmian Lines, but by several other U.S.-flag operators as well.

The first ships built under the Merchant Marine Act of 1936 were a group for American Export Lines, called the Express-class, C3-E. The *Exemplar* was one of this group.

In 1941 and 1942, nineteen ships were built to a modified Express-class design, namely C3-S-A3. Four, including the *Extavia*, were built for American Export. These were smaller than the Express-class, being specially designed with shallower draft, capable of entering ports in the Mediterranean. They were shelter deck ships, with four hatches forward and two aft of the midship superstructure. They were readily identifiable because of their characteristics American Export knuckled, ducktail stern.

C4-Class. The C4-class was originally designed (in 1941) by George G. Sharp, Inc., for the American Hawaiian Steamship Company. When the need for troop ships and tank carriers became apparent, this class of vessel, with modifications, proved to be ideally fitted to those needs. Consequently, the following orders for 75 C4's were let:

- Thirty C4-S-A1 transports, Kaiser Shipbuilding (Richmond), a four-deck version, Navy manned,
- Fifteen C4-S-A3's, a four-deck version meeting U.S. Maritime Commission needs,
- Ten C4-S-A4's, Kaiser-built, two in Richmond and eight in Vancouver, four-deck versions that were general cargo ships,
- One C4-S-B1, Sun Shipbuilding, a three-deck version built as a tank carrier,
- Fourteen C4-S-B2's, Sun Shipbuilding, a three-deck version troop and hospital ships, and
- Five C4-S-B5's, Sun Shipbuilding, a three-deck version designed as a straight cargo ship.

The C4's were the largest of the standard cargo ships. The basic design had a split house; the forward house for deck officers and navigation bridge; the after house for engineers, unlicensed crew, engine casing, etc. A long, raised foredeck covered Holds 1 and 2. Aft of the after superstructure, a raised deck covered Hold 7 and the steering gear room. Four holds between the houses gave this class a total of eight. They were full scantling ships.

They had accommodations for 56 crew and 8 passengers. As troop ships, they carried about 3456 troops.

The C4-S-B1's had high 'tween decks and heavy structure and consequently were less than ideal as general cargo ships. Some were converted to hospital ships. On the other hand, up through the 1960s many commercial operators found the C4's useful for conversions to heavy lift ships, Great Lakes bulk carriers, and other miscellaneous types. Matson, Seatrain, Bull Lines and Sea-Land also converted a large number of them in the 1960s to containerships. Among other characteristics, because of their robust scantlings, they could be lengthened (jumbo-ized) without too much difficulty.

World War II Refrigerated Cargo Ships. In 1943, six C2-S-B1 ships were converted to transport frozen meat to troops overseas. These successful conversions were designated C2-S-B1(R). In 1945 an additional six were specially constructed as reefer ships and classed R2-S-RV1. They were powered and geared as C2's.

In 1943 six reefers were specially designed and ordered for the United Fruit Company. Designated R2-ST-AU1, the first in the series, the *Fra Berlanga*, was delivered in early 1945. Three more were ordered and delivered in 1947. These twin screw, 18-knot ships were powered by a pair of 6600 shp turbine sets. Each had three boilers. This class had four holds, served by four sets of booms.

To further augment United Fruit's "Great White Fleet," nine shorter (386-ft) *Yaque*-class types were ordered. These 5000 dwt ships operated at 16 knots. They were of full scantling design with flush deck. They had 195,000 cu ft of reefer space in four holds, and 209,000 cu ft of general cargo space. These 18 ships operated successfully for United Fruit for over 25 years.

Seventeen small reefer vessels were delivered in 1944–45. They were designated R1-M-AU3.

N3-S-A1 and A3. Over one hundred Baltic-type 250 ft coasters, designated N3-S-A1 and A3, were constructed during the war. These 1780 dwt ships had a somewhat outmoded appearance with three-island profile. Their shallow draft (a little over 17 ft) made them useful in some trades.

Immediate Post–World War II Period

The *Schuyler Otis Bland*—C3-S-DX1. After World War II, the Maritime Commission recognized that a larger, faster and more efficient type of cargo ship would be required. Their efforts centered on developing a design that would be an extension of the C3's. This led in 1948 to starting the design of a prototype ship, which was to become the *Schuyler Otis Bland*.

The ship was of all-welded design, except for four riveted crack arrestor straps. She had elaborate cargo handling gear of the king post-and-boom variety, and folding steel hatch covers.

The vessel's characteristics were:

Length, overall	478 ft-2 in.
Length, between perpendiculars	450 ft
Beam	66 ft
Depth	41 ft-6 in.
Draft, scantling	41 ft-2.5 in.
Draft, shelter deck	28 ft-6 in.
Deadweight	10,516 LT

Table 1 Some examples of U.S. and Canadian World War II built ships

Class	Designation	LOA	LBP	B	D	d	DWT	HP	V
Liberty	EC2-S-C1	441 ft-6 in.	416 ft-0 in.	56 ft-10¾ in.	37 ft-4 in.	27 ft-7 in.	10,865	2500	11.0
Ocean		441 ft-6 in.	416 ft-0 in.	56 ft-10¾ in.	37 ft-4 in.	27 ft-7 in.	10,500	2500	11.0
Fort/North Sands		441 ft-6 in.	416 ft-0 in.	56 ft-10¾ in.	37 ft-4 in.	—	9300	2500	11.0
Canadian Victory	VC2-S-AP3	455 ft-3 in.	436 ft-6 in.	62 ft-0 in.	38 ft-0 in.	28 ft-6 in.	10,861	8500	16.5
Park		328 ft-0 in.	315 ft-6 in.	46 ft-6 in.	—	—	4700	—	10.0
Victory	VC2-S-AP2	455 ft-3 in.	436 ft-6 in.	62 ft-0 in.	38 ft-0 in.	28 ft-6 in.	10,900	6000	15.5
	VC2-S-AP3	455 ft-3 in.	436 ft-6 in.	62 ft-0 in.	38 ft-0 in.	28 ft-6 in.	10,861	8500	16.5
	C1-A	412 ft-3 in.	390 ft-0 in.	60 ft-0 in.	37 ft-6 in.	23 ft-6 in.	7416	4000	14.0
	C1-B	417 ft-9 in.	395 ft-0 in.	60 ft-0 in.	37 ft-6 in.	27 ft-6 in.	9290	4000	14.0
	C1-M-AV1	338 ft-8⅜ in.	320 ft-0 in.	50 ft-0 in.	29 ft-0 in.	21 ft-¾ in.	5032	1750	11.0
Reefer "Yaque Class"	R1-S-DH1	385 ft-6 in.	365 ft-0 in.	56 ft-0 in.	35 ft-0 in.	26 ft-1 in.	5012	5500	16.0
C1-M-AV1 Reefer	R1-M-AV3	338 ft-8⅜ in.	320 ft-0 in.	50 ft-0 in.	29 ft-0 in.	21 ft-0 in.	4373	1700	11.0
Concrete	C1-S-D1	366 ft-4⅜ in.	350 ft-0 in.	54 ft-0 in.	35 ft-0 in.	26 ft-3½ in.	5004	1300	7.0
Shelter Decker	C2	459 ft-3 in.	435 ft-0 in.	63 ft-0 in.	40 ft-6 in.	25 ft-9 in.	9722	6000	15.5
	C2-S-B1	459 ft-2½ in.	435 ft-0 in.	63 ft-0 in.	40 ft-6 in.	25 ft-9 in.	9150	6000	15.5
Robin C2	C2-S	480 ft-0 in.	450 ft-0 in.	66 ft-0 in.	43 ft-0 in.	27 ft-0 in.	10,050	6300	15.5
American Mail C2	C2-SU	474 ft-1 in.	450 ft-0 in.	63 ft-0 in.	40 ft-6 in.	26 ft-6 in.	9620	7500	16.0
	C2-S-A1	420 ft-0 in.	400 ft-0 in.	60 ft-0 in.	39 ft-4 in.	27 ft-6 in.	8130	8800	16.5
Waterman C2	C2-S-E1	468 ft-8½ in.	445 ft-4 in.	63 ft-0 in.	40 ft-6 in.	25 ft-9 in.	10,565	6000	16.0
Full Scantling	C2-S-AJ1	459 ft-1 in.	435 ft-0 in.	63 ft-0 in.	40 ft-0 in.	27 ft-7 in.	10,775	6000	15.5
Fra Berlanga	R2-ST-AU1	455 ft-0 in.	431 ft-0 in.	61 ft-0 in.	35 ft-6 in.	27 ft-1⅞ in.	7067	12,000	18.5
Reefer	R2-S-BV1	459 ft-2½ in.	435 ft-0 in.	63 ft-0 in.	40 ft-6 in.	25 ft-9 in.	8133	6000	15.5
	C3	492 ft-0 in.	465 ft-0 in.	69 ft-6 in.	42 ft-6 in.	28 ft-6 in.	12,527	8500	16.5
	C3E	473 ft-1 in.	450 ft-1 in.	66 ft-0 in.	42 ft-3 in.	27 ft-0 in.	9660	8000	16.5
	C3-P&C	491 ft-10 in.	465 ft-0 in.	69 ft-6 in.	42 ft-6 in.	26 ft-6 in.	9937	8500	16.5
Del Mar	C3-S-BR1	494 ft-7⅜ in.	465 ft-0 in.	69 ft-6 in.	45 ft-6 in.	27 ft-9 in.	9290	8500	16.5
	C3-S-BH1	492 ft-0 in.	465 ft-0 in.	69 ft-6 in.	42 ft-6 in.	29 ft-1 in.	12,600	8500	16.5
	C3-S-A1	492 ft-0 in.	465 ft-0 in.	69 ft-6 in.	42 ft-6 in.	28 ft-6 in.	12,525	8500	17.0
	C3-S-A2	492 ft-0 in.	465 ft-0 in.	69 ft-6 in.	42 ft-6 in.	28 ft-6 in.	12,175	9350	16.5
Exporter Express (mod)	C2-S-A1	429 ft-0 in.	400 ft-0 in.	60 ft-0 in.	39 ft-4 in.	27 ft-7⅜ in.	7850	8800	16.5
Hospital Transport	C2-S1-A1	450 ft-2 in.	420 ft-0 in.	62 ft-0 in.	41 ft-6 in.	25 ft-0 in.	3647	5500	20.0
Lumber Carrier	C1-MT-BU1	317 ft-0 in.	309 ft-0 in.	48 ft-0 in.	26 ft-0 in.	17 ft-0 in.	3787	2400	12.0
Farrell	C3-S-BH2	492 ft-0 in.	465 ft-0 in.	69 ft-6 in.	33 ft-6 in.	29 ft-1 in.	12,031	9350	16.5
Exporter (C3-E)	C3-S-A3	473 ft-1 in.	450 ft-0 in.	66 ft-0 in.	42 ft-3 in.	27 ft-9½ in.	9902	8800	16.0
APL	C3-S-A4	492 ft-0 in.	465 ft-0 in.	69 ft-6 in.	42 ft-6 in.	28 ft-6 in.	11,665	—	16.5
Mooremac	C3-S-A5	492 ft-0 in.	465 ft-0 in.	69 ft-6 in.	42 ft-6 in.	28 ft-6 in.	11,167	—	16.5
A1=Coal/A3=Oil	N3-S-A1&3	—	250 ft-0 in.	42 ft-1 in.	20 ft-5 in.	17 ft-10 in.	2817/2760	1300	10/11
Troopship	C4-S-A1	522 ft-10½ in.	496 ft-0 in.	71 ft-6 in.	43 ft-6 in.	21 ft-6 in.	6094	9000	17.5
Cargo	C4-S-A4	522 ft-10½ in.	496 ft-0 in.	71 ft-6 in.	43 ft-6 in.	32 ft-9⅞ in.	14,863	9000	17.0
One of a Kind	C4-S-B1	522 ft-10½ in.	496 ft-0 in.	71 ft-6 in.	43 ft-6 in.	30 ft-0 in.	13,294	9000	17.0
	C4-S-B5	522 ft-10½ in.	496 ft-0 in.	71 ft-6 in.	43 ft-6 in.	32 ft-9⅞ in.	13,498	9000	17.0

Displacement	15,900 tons
Gross registered tonnage	8918
Net registered tonnage	5267
Bale cubic	561,290 cu. ft.
Shaft horsepower, normal	12,500
Shaft horsepower, max.	13,750
Speed	18.5 knots

She was powered by double reduction geared steam turbines (900F/860 psi) driving a single four-bladed propeller at 90 rpm.

The keel was laid May 1950 and she was delivered in July 1951, and chartered to American President Lines, Waterman and American Mail Lines. In 1959 she entered MSTS service where she worked until 1979, when she was scrapped.

While the ship featured substantial advances, the events of the Korean War and development of foreign merchant fleets indicated the need for something better. Consequently, the *Schuyler Otis Bland* represented an important but unique prototype.

Mariner-Class—C4-S-1A. In 1950 the Maritime Administration (successor to the U.S. Maritime Commission), surveyed the needs of the country's merchant marine in light of lessons learned during the Korean War. It then undertook the largest dry cargo shipbuilding program since the war. The liquidation of the war-built fleet was virtually complete. If we were to have a competitive merchant marine, there

Fig. 4 United States Lines' containership *American Legion*, one of six 700-ft, 22,225-dwt container liners built to American Bureau of Shipping class by the Sun Shipbuilding and Dry Dock Co. of Chester, Pennsylvania, for the North Atlantic run (capacity: more than 1200 twenty-foot containers). (Courtesy American Bureau of Shipping.)

was clearly a need for larger, faster, merchant ships. There was also a clear need for merchant ships convertible to modern naval auxiliaries.

The design, which was carried out by the Maritime Administration, recognized that a 20-knot service speed was necessary to compete with foreign fleets. Also, it was recognized that there was a trend toward higher stowage factors and hence a more volumetric ship was dictated. The essential trade routes dictated a 30-ft maximum draft. The ship was originally designed as a shelter decker, easily convertible to a full scantling ship by closing tonnage openings.

The Mariner was a seven-hatch (four forward and three aft) ship, which arrangement avoided long holds, and so minimized delays in loading and discharge. The arrangement of cargo gear and bale cubic in each hold was planned as a balanced system. The Mariner had a one-compartment standard of subdivision.

The steam turbine propulsion plant, turning a single four-bladed propeller at 102 rpm, was designed to provide a sustained sea speed of 20 knots at 17,500 shp (80 percent of maximum continuous shp of 19,250). A National Defense feature called for 22,000 shp.

The basic design was completed in April 1950, anticipating the Korean emergency, which occurred in the summer of that year. In August 1950 Vice Admiral E. L. Cochrane was appointed Maritime Administrator. He spearheaded what was originally intended as a one-ship prototype into a full shipbuilding program. In 1951 MarAd contracted for 35 Mariner-class ships in blocks of five from seven U.S. shipyards. This program was completed by 1954 and a number of U.S. liner companies enjoyed the benefits of competing internationally using the largest and fastest dry cargo ships built in the United States to that date.

The Mariners were usually given the nicknames of states, such as *Wolverine State.*

The cargo gear was based on the experience with the *Schuyler Otis Bland.* King posts with five- and ten-ton booms, together with electric winches and electric topping winches, easily met the three-ton burtoning capacity objective. Jumbo heavy lift booms with sixty-ton capacity were also provided. Steel pontoon folding hatch covers were also a notable feature. The advantages of folding covers were clear, but many early mechanical problems provided valuable lessons to the designers and operators. The biggest lesson learned was the need for sufficient lead time in developing new designs to meet national emergencies.

Mariner-class characteristics are

Length, overall	563 ft-7.75 in.
Length, between perpendiculars	528 ft-6 in.
Beam	76 ft
Depth	44 ft-6 in.
Draft	29 ft-10 in.
Shaft horsepower	17,500
Speed	20 knots
Deadweight	13,409 tons
Light ship weight	7684 tons
Cubic	736,723 cu ft

The chapter by Cherrix and Landsburg (The Maritime Administration and Antecedents) tells more about the Mariners, as well as about the *Savannah,* described next.

The Nuclear Ship *Savannah.* The construction of the world's first nuclear-powered merchant ship was proposed by President Eisenhower in 1955 to demonstrate peaceful uses of atomic energy. Congressional approval was granted in 1957, followed the same year by the Maritime Administration's award of the various component contracts. The ship design contract went to George G. Sharp, Inc., the nuclear reactor to Babcock & Wilcox, and the ship construction to New York Shipbuilding Co. of Camden, New Jersey.

Fig. 5 United States Lines stack two tiers of 40-ft-long containers five high and thirteen across the deck. The after deck mooring winches are located on the second deck so as not to sacrifice cargo space and to make it easier for crewmen to dock and undock the ship. (Courtesy American Bureau of Shipping.)

The *Savannah* was a single-screw, passenger/cargo ship, with a full load displacement of 22,000 long tons. Her normal service speed was 21 knots with her propulsion power plant operating at the normal rating of 20,000 shp and a capability of 22,000 shp at maximum output. The ship could carry 60 passengers, a crew of 110, and approximately 10,000 tons of dry cargo. She was fitted with a teardrop-shaped superstructure, set sufficiently aft to enhance the shape of the ship's fore section, which tapered gracefully to its well-raked bow. She had as attractive a profile as any ship entering service in the 1990s.

The ship was launched on July 31, 1959, and started her maiden voyage in August 1962, which included port visits on the East Coast, the traverse of the Panama Canal, and calls on the West Coast and Hawaii.

Savannah characteristics are

Length, overall	595 ft-6 in.
Length, between perpendiculars	545 ft
Beam	78 ft
Depth	58 ft-9 in.
Draft	29 ft-6 in.
Shaft horsepower	20,000
Speed	21.0 kts
Displacement	22,000 tons
Deadweight	9705 tons

Liberty Ship Modernization Program. The Maritime Administration, in an attempt to find use for the more than 1000 Liberty Ships in the reserve fleets, undertook a research/demonstration project. Four vessels were used to measure and demonstrate improvements in hull form, propulsion and cargo handling:

John Sergeant—A 6000 shp regenerative gas turbine and controllable pitch propeller were installed in this ship, giving speeds up to 17.2 knots. The ship proved the potential for such propulsion.

William Patterson—A 6000 shp (later downgraded to 4800 shp) free-piston engine was installed on this ship, but proved to be of little success.

Thomas Nelson—Two 3000 bhp geared diesels were installed in this ship, and run on Bunker C fuel. Unfortunately, the concept was ahead of its time in the United States, largely as a result of high maintenance and lube oil costs, and lack of interest by U.S. manufacturers.

Benjamin Chew—This vessel was equipped with a conventional, surplus 6000 shp steam turbine plant.

The hull forms of the *Sergeant, Patterson,* and *Nelson* were lengthened and bow lines made finer. The *Chew* remained unchanged.

The *Nelson* was used for experiments with crane-type cargo gear, which showed little advantage over burtoning gear when handling break-bulk cargo. When handling U.S. Army CONEX boxes weighing 2.3 tons, however, substantial gains were demonstrated, confirming the concept of containerization.

Post–World War II: 1958–1993

Cargo Ship Replacement Program. Except for the Mariner-class, there had been no major replacement ship programs between the end of World War II and 1958. Despite the utilitarian nature of the C-classes and Victory Ships, they were decidedly uncompetitive in the late 1950s, deficient in both size and speed. Advances in cargo gear, hatch cover design, refrigeration, the emergence of automation, the growing importance of fuel economy, the desire for improved cargo handling, and the need for improved habitability, all drove owners to undertake new construction programs.

Previously the Maritime Administration (Commission) had dictated the design of standard series, and owners were restricted to making minor design changes. Now, however, for the first time in many years, the owners had virtually a free hand in designing and specifying new cargo ships. A great variety of ships resulted. More than 25 different varieties of conventional break-bulk cargo ships were ordered in the short span of ten years. Most differed only slightly from the rest. The positive effect of this freedom was to stimulate innovation. Unfortunately, most of the innovative work was on a type of ship that was on the verge of obsolescence, soon to be replaced by specialized, intermodal ships: container, barge and RO/RO. Moreover, this plethora of designs denied

Fig. 6 *Sea-Land Defender* is one of twelve container carriers of 23,000 dwt built to ABS class for Sea-Land Service, Inc. This is the first of three container carriers built by Mitsui Engineering & Shipbuilding Co. (Courtesy American Bureau of Shipping.)

the potential economies of mass production because of the relatively small lot sizes ordered.

In retrospect, another remarkable feature during this period was the slow reaction of the dry cargo liner sector of the maritime industry to the emergence of containerization, except by a few operators, such as Grace, Seatrain, Matson, Bull and Alaska Steamship. McLean and his container were regarded as mavericks. The owners were not alone. In 1963 the N.Y. Port Authority, after an extensive study, predicted that not more than 50 percent of transatlantic cargo would ever be containerized.

In 1964, when intermodal containerization was first introduced in transatlantic service with the sailing to Europe of the *Fairland*, the shock and reaction was immediate. Many of the conventional cargo ships under construction were immediately modified to containerships. A number of the operators rushed to find alternatives that might improve on the container concept. Some selected pure barge carriers, others combination container and barge carriers. Some selected pure trailer ships, others combination container and barge carriers. One particularly poor compromise was the "multi-purpose" combination container/break-bulk ship, which emphasized the worst in both systems.

The period 1958 to 1968 could be characterized as one of emerging containerization among the unsubsidized owners and a search by the subsidized operators for the best and most competitive system.

Space limitations do not allow comprehensive accounts of how the various U.S. companies have gone about modernizing their fleets during the past few decades. What I have chosen to do is explain in some detail how five representative fleets (three subsidized, two unsubsidized) went about trying to survive (not always successfully) in the face of rapidly changing times. I have also included a short section on the specialized fleets of reefer ships.

The Subsidized Cargo Ship Companies. While the Mariners had served their purpose in providing the U.S. merchant marine with a modern and competitive break-bulk ship, the majority of cargo vessels in the subsidized fleet were World War II-built and were approaching 20 years of age, the statutory life of a subsidized ship. Therefore, around 1958 the many subsidized operators began to plan their replacement ships. Only Grace Lines and American President Lines (APL) seriously considered containerization. In 1960 Grace Lines converted two C2's, the *Santa Elena* and *Santa Leonor*, while in 1961 APL converted two Mariners to Sea Racer-class containerships.

The other subsidized operators resisted major changes, but rather opted for break-bulk ships. Many of these featured innovative details, but all were soon to become obsolete. In the period 1958 to 1968, there was a steady trend on subsequent ships in each company's reorders to include container capability. A few companies viewed barge carriers (Lykes and Delta), or RO/RO (States Lines and Moore McCormack), as the road to the future.

American President Lines. APL acquired 8 Mariners in the period 1952–1954 to augment their war-built ships. Their replacement program began with two Sea Racer-class (C4-S-1Q) built at Bethlehem San Francisco and three APL Mariner-class (C4-S-1S) ships.

Fig. 7 *Hawaiian Merchant,* a C-3 cargo vessel operated by Matson Navigation Company, outbound from San Francisco Bay for Honolulu to begin the first container service on the Pacific Ocean in 1958. (Courtesy American Bureau of Shipping.)

In 1964 an additional APL Mariner type (C4-S-1sa) was built by NASSCO, followed in 1966 by three Master Mariner-class ships also at NASSCO. All these replacement units were 20.5-knot break-bulk cargo ships. APL continued t o order general cargo ships with an order to Ingalls for five C4-S-69a class vessels delivered in 1967–68 and five C5-S-75a semi-containerships (332 TEU) at Newport News in 1968–69. The Sea Racers were traded to MarAd in 1979.

Recognizing the benefits of containerization, APL began converting most of its vessels to containerships in the 1970s, as follows:

1971–72	three C4-S-1 to C6-S-1x	1124 TEU
1972	one C4-S-1a to C6-S-1xa	1124 TEU
1972–73	three C4-S-1qa to C6-S-1qc	838 TEU
1972	five C4-S-69a to C6-S-69c	1094 TEU

Pacific Far East Lines (PFEL) built four LASH (C8-S-81b) ships at Avondale in 1971. These had a limited capacity of 335 TEU and 43 barges. When PFEL went into receivership in August 1978, APL bought and converted the LASH ships to C8-S-81e full containerships, with a capacity of 1856 TEU.

Neptune Orient Lines (Singapore) built two large 22.4-knot containerships in Japan in 1980, lengthened them the following year, bringing their capacity to 2838 TEU, and sold them to APL in 1984. APL reflagged them to U.S. registry.

The C6-S-69c's were chartered to Lykes in 1979. The Master Mariner conversions were sold to MARAD in 1979–82. The APL Mariner-class, which were operated by American Mail Lines from 1962 to 1978 (when they merged with APL), were chartered to Delta 1983–1985, and to U.S. Lines 1985–1987, and then turned over to MarAd. The American Mail Line C4-S-1xa, converted and acquired by APL in 1978, was sold in 1988 and later converted to a fish processing plant.

The decade of the 1980s saw a major, dedicated large containership building program beginning in 1980 in the U.S. at Avondale with the three C9-M-132b diesel-driven ships. It culminated in the 5-ship C10-M-F150a post-Panamax ships built in Germany in 1988. These latter ships have the largest container capacity of any ships afloat at the time I write this in 1992.

From 1980 onward, all new APL ships were diesel-driven and had speeds in the 22- to 24-knot range. The APL ships built and converted in the 1960s and 1970s (with the exception of the 23-knot Pacesetters) were in the 20- or 20.5-knot range.

The earlier containerships and conversions had provision for 20-ft containers, then later, 20- and 40-footers. The latest ships have capability also for 45- and 48-ft containers, in which APL has been a pioneer.

In 1973 and 1974 APL acquired its first four purpose-built containerships with the C6-S-85b Pacesetters, a design it shared with Farrell Lines, a class with a crew size of 47. Automation in the APL fleet started to reduce crew sizes during the 1980s.

It is worthy of note that in the 30 years 1961 to 1991, APL acquired a total of 41 ships. Of these, 18 were acquired from other companies. The remaining 23 were of seven different classes, plus four different conversion programs.

United States Lines. The vestigal beginnings of the United States Lines (USL) date back to 1921 when the U.S. Shipping Board took over the ailing U.S. Mail Steamship Co. During World War II USL acquired a large fleet of standard cargo ships, which were in need of replacement in the 1960s. They were, however, not quick to accept containerization when they planned their replacement program. In the period 1960 to 1964, eleven break-bulk Challenger-class (C4-S-57a) ships were ordered from Newport News and Bethlehem-Quincy.

Two years later five C4-S-64a American Rover-class break-bulk ships were ordered from Sun Shipbuilding & Dry Dock Co. These vessels were 22 ft shorter, but had 0.8-knot more speed.

In 1964, at Sun Shipbuilding, USL began construction of the new Lancer-class, as part of their faster (27.5-knot) replacement program. These were five

Table 2 American President Lines ships

Type	No.	Class	Built	Conv.	Year	LOA	LBP	B	Depth	Draft	Disp.	DWT	SHP	V	Prop.	TEU
Cargo	6	C3-S-AY														
Cargo	8	MSR[a]														
Cargo	2	C4-S-1Q[b]	Beth, SF	—	61	563-8	528-10	76-0	44-6	31-7	22,630	14,048	19,250	20.5	S	—
Cargo	3	C4-S-15 (M)[a]	Todd SP	—	61–64	563-8	528-10	76-0	44-6	31-7	22,630	14,048	19,250	20.5	S	—
Cont.	3	C6-S-1X[c]	—	Beth, SF	71–72	667-11	633-0	76-0	44-8	33-4	31,498	19,015	19,250	20.5	S	1124
Cargo	1	C4-S-1SA	Nassco	—	64	563-8	528-10	76-0	44-6	31-7	22,630	14,048	19,250	20.5	S	1124
Cont.	1	C6-S-1XA[d]	—	Todd SP	72	667-11	633-0	76-0	44-8	33-4	31,498	19,015	19,250	20.5	S	1124
Cargo	3	C4-S-1QA (MM)[a]	Nassco	—	66	563-8	528-10	76-0	44-6	31-7	22,630	14,048	19,250	20.5	S	1124
Cargo	3	C6-S-1QC[e]	—	Seattle Todd	72–73	667-11	633-0	76-0	44-8	33-4	31,498	19,015	19,250	20.5	S	838
Cargo	5	C4-S-69A	Ingalls	—	67–68	573-11	544-6	82-0	45-6	30-8	21,425	12,830	24,000	24.0	S	838
Cont.	5	C6-S-69C[f]	—	Todd SP	72	663-11	634-6	82-0	45-6	30-8	27,620	17,477	24,000	23.0	S	1094
Cargo	5	C5-S-75A[g]	NN	—	68–69	605-0	582-6	82-0	46-0	35-0	31,995	22,208	24,000	20.8	S	332
Lash	4	C8-S-81B	Avondale	—	71	820-0	724-0	100-0	60-0	40-10	54,460	39,602	32,000	22.5	S	335+43B
Cont.	4	C8-S-81E[h]	—	Beth	77–78	820-0	724-0	100-0	60-0	40-10	54,460	39,602	32,000	22.5	S	1856
Cont.	4	C6-S-85B (PS)[a]	Ingalls	—	73–74	669-3	625-0	90-0	53-0	33-1	30,490	18,540	28,500	23.0	S	1508
Cont.	2	C9-M-148A[i]	IHI	—	80	758-9	701-5	105-8	62-4	41-1	—	—	—	22.4	M	—
Cont.	2	C9-M-M8AC	—	IHI	81	861-8	804-6	105-8	62-4	41-1	63,989	45,864	40,200	22.4	M	2838
Cont.	4	C9-M-F151A[j]	Mitsubishi	—	87	849-9	803-9	105-8	61-8	39-1	61,156	44,256	28,800	21.0	M	2986
Cont.	5	C9-M-F150A	HDW BV	—	88	902-10	855-7	129-3	77-5	41-0	75,862	53,811	57,000	24.2	MXI	4300+
Cont.	3	C9-M-132B	Avondale	—	82	860-0	810-0	105-9	66-0	35-0	49,500	30,343	43,200	23.9	M	2500

[a] MSR = Mariner Sea Racer; M = Mariner; MM = Master Mariner; PS = Pacesetter.
[b] C4-S-1Q: MARAD 1979.
[c] C6-S-1X: AML 1962–78; APL 1978–83; Delta 1983–85; USL Truman/Builder/Banker 1985–87; Truman/Builder/Conv. to T-ACS.
[d] C6-S-1XA: AML 1964–78; APL 1978–88; sold and conv. to fish processing ship in 1988.
[e] C6-S-1QC: APL 1966–79/82; conv. by MARAD to crane ship (T-ACS) in 1984–86.
[f] C6-S-69C: APL 1968–87; chartered to Lykes, Grant Sunk 1976.
[g] C5-S-75A: AML 1968–78; APL 1978.
[h] C8-S-81E: AML 1971–79; PFEL bought/chart. 1979.
[i] C9-M-148A: NOL 1980–84; reflagged 1984 APL.
[j] C9-M-F151A: chartered from Lykes in 1987.

general cargo ships, with limited container capacity. Midway through this program, while two ships were well along in construction, it was recognized that containerships would be more desirable. The plans were changed to provide longer, wider (90-ft) midbodies, and other "container-friendly" features. The first was the *American Lancer*, and all five modified ships were delivered as full containerships (designated as C-S-68a) in 1968–69. Three more, slightly different versions, C7-S-68e, were built by Sun in 1969–71.

Table 3 shows the principal characteristics of the company's initial containership fleet.

The container competition that USL began to experience by the mid-1960s caused them in 1970–71 to convert the Challenger-class and some of their Mariners to full containerships. The latter became known as the Leader-class.

Through the 1970s, USL went through several changes of ownership and was acquired by McLean Securities in 1978. McLean undertook an expansion program, which included construction of 12 Jumbo Econships at Daewoo's shipyard in South Korea in 1984. These diesel-driven ships were the world's largest containerships. They were deployed in a round-the-world service. They had the following principal characteristics:

LOA	949 ft-10½ in.
LBP	915 ft-4½ in.
B	106 ft-0 in.
Draft, design	29 ft-7½ in.
Draft, max	38 ft-3⅝ in.
Brake horsepower	28,000
Speed	19.2 knots
Container capacity	4482 TEU

By 1985 the USL fleet included 44 containerships with a total capacity of just over 100,000 TEU, one LASH ship and a bulk carrier. In 1987, when the company was forced out of business, the 12 Jumbo Econships were acquired by Sea-Land and renamed the Atlantic-class.

Lykes Bros. Steamship Co. In 1900 Frederick and Howell Lykes opened an office in Havana to import cattle from Florida in schooners, including the three-masted *Dr. Lykes*.

Lykes Bros. Steamship Co. was formed in 1922 to separate the shipping from other interests of the brothers. In 1922 and 1923 Lykes bought ships from the U.S. Shipping Board, and also acquired other shipping companies and their ships, including Tampa Interocean Steamship Co., Lone Star Steamship Co., Reiley Steamship Co., and Dixie and Southern States Lines.

After passage of the Merchant Marine Act of 1936, Lykes committed itself to spend $30 million for

Table 3 Principal characteristics of United States Lines' initial containership fleet

	Leader 14-S-1A(C)	Challenger C4-S-57a	Rover C4-S-64a&b	Lancer C7-S-68c&d
LOA	661 ft-1¾ in.	560 ft-6 in.	544 ft-0 in.	700 ft-0 in.
LBP	625 ft-6 in.	529 ft-0 in.	507 ft-0 in.	670 ft-0 in.
B	76 ft-0 in.	75 ft-0 in.	75 ft-0 in.	90 ft-0 in.
D	44 ft-6 in.	42 ft-6 in.	42 ft-6 in.	50 ft-6 in.
d	29 ft-6 in.	32 ft-0 in.	32 ft-0 in.	32 ft-½ in.
DWT	15,627	10,804	13,057	21,665
SHP	19,250	16,500	18,750	27,300
Speed	20	21	20	22.25
Cont. cap.	1013 TEU	632 TEU	504 TEU	1270 TEU
Builder	various	Newport News	Sun/Beth	Sun
Date	1970–71	1962	1962, 1964	1964, 1968

replacing 28 ships, and by December of 1941 when the United States had entered World War II, Lykes had already received 16 new C type ships.

During the war Lykes lost 22 ships to enemy action at a cost of 272 lives.

After the war Lykes bought 21 additional new ships and by the 1950s had a fleet of 54 totaling 568,978 deadweight tons, the largest U.S.-flag cargo fleet under private ownership.

In 1958 Lykes again undertook a vessel replacement program which resulted in the construction of 26 C3's, 12 C4's and three unique barge carriers, each with a 2000-ton elevator on her stern. (See chapter by Stuart Thayer.)

In the mid-1980s Mitsui and Mitsubishi built four large 3000 TEU fully cellular containerships for Lykes.

In the half century 1942 to 1992, Lykes Bros. Steamship Co. has been one of the most important U.S. cargo ship operators. During the 1950s they operated one of the largest U.S.-flag fleets. Among subsidized companies, they have received more operating subsidy (1937 to 1989: $1.649 billion) than any other U.S. steamship company, in fact, one third more than the next most subsidized company, APL.

Lykes' base of operations has been the U.S. Gulf. The cargoes moving through that area did not appear to Lykes to be as readily containerizable as those moving through the East and West Coasts. Hence, in Lykes' replacement program during the late 1950s, they ordered large blocks of break-bulk ships, all of a fairly consistent size and speed. Between 1958 and 1961, Ingalls, Bethlehem, and Avondale built 21 C3-S-37a conventional cargo ships for Lykes.

Lykes remained committed to break-bulk, when in 1965 a series of 12 C4-S-66a freighters were delivered for their Far East service. In 1965, 60 percent of the fleet still comprised World War II break-bulk ships.

By the 1970s Lykes opted for the Seabee type barge carriers, with 38 lighters and 1000-TEU container capacity. Three such ships were delivered in 1972. At the same time, 18 of the C3's were lengthened as semi-containerships with a container capacity of 192 TEUs.

By the early 1980s, Lykes' mixed fleet also included two foreign (Hansa Lines) and one domestic-built RO/RO (988 TEUs), and four foreign-built containerships (1096 TEU), the C6-M-147a class. The decade of the 1990s has found Lykes planning a major fleet replacement program to substitute containerships for the 21 existing ships over 25 years old.

The Unsubsidized Companies. Although quick to see the potential benefits of containerization, the unsubsidized operators were hard pressed to finance such capital-intensive projects. Therefore, they turned to the expedient of converting older ships into containerships. For example, Sea-Land and Matson converted C2's, C4's, T2's and T3's; Seatrain converted T2's and C4's. Containerization was such a vast improvement that even inefficient, older ships, when converted, became competitive.

Sea-Land Service. In 1956 Sea-Land experimented with containerization by carrying 33-ft containers on the spar decks of three T2 tankers. Each carried 58 containers, which were loaded using shoreside cranes. The following year, the first of six converted C2's, the *Gateway City*, was introduced, carrying 226 thirty-five foot containers in vertical container cells. These ships established the technology-to-come, namely cellular structure, steel pontoon covers, automatic spreaders, gantry cranes, wider hulls, diagonal lashings, etc. Their hulls were sponsoned to provide greater stability.

In 1962 Sea-Land built four large midbodies in Germany, and joined them to T3 bows and sterns at Todd and Bethlehem shipyards, resulting in the *Elizabethport*-class, with 476 thirty-five foot container capacity.

Table 4 Lykes ships

Type	Class Name	Class	No.	LOA	LBP	B	D	d	DWT	SHP	V	Builder	Year	TEU
Cargo	Gulf Pride	C3-S-37C	8	495-8	470-0	69-0	41-7	30-0	11,368	11,000	18.0	Avondale/Beth	1963	—
Cargo	Gulf Andes	C3-S-37D	5	494-8	470-0	69-0	41-7	32-1	12,656	11,000	18.0	Avondale	1964	—
Semi.	Pacer	C5-S-37E	9	592-6	567-6	69-0	41-7	30-1	14,300	9,900	17.5	Todd[a]	1972[a]	192
Semi.	Pacer	C5-S-37F	4	592-6	567-6	69-0	41-7	30-1	14,300	11,000	18.0	Todd[b]	1972[b]	192
RO/RO	Maine Class	C7-S-95A	3	684-9	640-0	102-0	69-6	32-1	19,123	37,000	23.0	Bath Iron	1976	988
BC[c]	Seabee	C8-S-82A	3	875-11	740-0	105-11	74-9	39-1	38,410	36,000	20.0	General Dynamics	1972–73	730[d]
Cont.	Express	C6-M-F147A	4	659-8	607-7	80-9	48-1	25-10	15,770	15,750	18.0	Rheinstahl/Lishave[e]	1973[e]	1906
RO/RO	Pilot	foreign const.	2	633-10	584-0	88-7	57-8	29-10	14,319	18,972	19.0	HDW/SASEBO	1977	762
Cargo	Far East Clipper	C4-S-66A	5	540-0	514-0	76-0	42-6	33-4	15,295	15,500	20.0	Avondale	1965–68	62
Semi.	Constellation	C6-S-60C	3	665-9	635-0	75-0	42-6	31-7	15,288	19,000	20.0	Ingalls	1982	696
Cont.	Seamaster	C6-S-69C	3	663-11	634-6	82-3	45-6	34-11	22,307	24,000	20.4	Todd[f]	1972[f]	1112
Cont.	Pacific	C9-M-F151A	4	849-9	803-10	105-7	61-10	39-1	44,263	25,500	21.0	Mitsui/Mitsubishi	1987	3423
Semi.	C5	C5-S-75A	2	605-0	582-6	82-0	46-0	35-0	22,211	24,000	20.8	Newport News	1969	486
Cont.	Oceanus	C6-M-F146A	2	675-10	650-7	105-7	61-8	37-9	35,441	19,740	19.0	Samsung	1985	2409

[a] Jumboized from C3-S-37A built by Ingalls/Bethlehem during 1960–62.
[b] Jumboized from C3-S-37B built by Bethlehem during 1962–63.
[c] Barge carrier.
[d] 730 TEU + 38 barges.
[e] Jumboized from original built by Bremer-Vulcan and Blohn & Voss during 1968–69.
[f] Jumboized from C4-S-69A built by Ingalls in 1967.

Between 1963 and 1970 Sea-Land added 22 more converted ships to its fleet. These included the 2 Grace Line *Santa Elena*-class (which underwent further conversion from 17.5- to 35-ft cells), 2 Bull Lines C4-S-B2's, 4 Waterman C2's, and a number of C4 troopships, some of which received new forebodies. The earlier ships were equipped with traveling gantry cranes. The later ships were not self-sustaining, since, by 1970, most major ports had built shoreside gantry cranes.

In 1972 and 1973 eight European-built SL-7's entered service. They boasted 120,000 shp, speeds of 33 knots, and capacity for 1184 containers (35- and 40-ft). They quickly set speed records on the Atlantic and Pacific because they were the fastest cargo ships ever built. Eventually, because of a twenty-fold increase in the price of fuel, it became uneconomical to operate these high speed ships. In 1986 they were sold to the Navy and converted to fast naval auxiliaries, T-AKRs.

The decade of the 1970s saw Sea-Land build new diesel ships abroad rather than convert existing steamships in the United States, as it did during the 1960s.

In 1987, three D-7 class diesel powered containerships were built for Sea-Land's Alaskan service by Bay Shipbuilding, on the Great Lakes. The following year, Sea-Land acquired U.S Lines' 12 Jumbo Econships and renamed them the Atlantic-class.

Sea-Land has dominated the U.S. fleet and foreign fleets in terms of container import and exports to and from the United States. In its leading role over 35 years, it has changed hands, successively from McLean Industries to R.J. Reynolds to CSX.

Matson Navigation Co. A traditional and major Pacific shipping company, Matson in the past half century has traded mainly in the U.S. West Coast to Hawaii service. Matson was among the first to see the benefits of containerization, but had to overcome longshore labor union resistance before it could move ahead. The unions finally agreed to cooperate when Matson guaranteed job security. (They were aided in this by a study carried out by the National Research Council, which quantified the probable economic benefits of containerization.) Meanwhile, Matson's well organized research division had completed a careful study to choose the best container size. Based on the types of commodities and their densities moving in the U.S. and Hawaii trade, and existing high axle loads and trailer lengths, they concluded that the optimum container length, for them, was 24 ft.

Fig. 8 Matson's containership *Lurline*, built in 1973. (Courtesy American Bureau of Shipping.)

Table 5 Sea-Land ships

Class	Vessel Name	No.	Built Year	Built Yard	Converted Year	Converted Yard	FEU	LOA, ft	LBP, ft	B, ft	D, ft	T, ft	DWT, LT	Displ., LT	HP, BHP	Speed, kts	House[a]
T2	Ideal X	3	44	—	56	—	58	523.5	503.0	68.2	39.3	30.0	9400	16,200	7240	15	M,A
C2-C	Gateway City	6	42–43	Gulfship	57–58	Mobship	226	468.7	444.2	72.1	40.2	25.1	7785	13,125	6600	16	M
73-J	Elizabethport	4	42	Sun	62–63	Todd/Beth.	476	627.5	607.3	72.7	47.0	27.2	15,770	27,698	10,000	16	A
T2	Summit	1	43	Sun	63	ADDSCO	196	523.5	503.0	68.2	39.3	23.4	9394	16,240	7240	15.5	A
C4-M	Anchorage	2	45	Sun	65	Todd	272	520.0	496.7	71.6	45.5	32.8	10,000	22,034	9000	17	M,A
C2-C	Ponce	2	44	NCBB	60–65	MDDSBCO	274	504.5	480.0	74.1	39.7	25.3	7413	15,840	6000	15	M
C4-J	Long Beach	4	45	Kaiser	66	Ingalls	684	684.9	658.0	78.0	47.0	30.0	17,144	32,200	9900	15	A
C2-X	Wacosta	6	43–44	Chichasaw	67–68	Todd	225	468.5	445.3	63.3	40.5	27.6	6855	15,000	6600	16	M
T2-M	Houston	3	44	Marinship	67–68	Todd	360	523.5	503.0	68.0	45.3	30.0	10,400	21,700	10,000	16	A
C4-X2	Boston	3	44–45	Kaiser	68	Todd	366	522.9	496.0	71.5	43.5	30.5	10,200	19,800	9900	16	A
C4-JC	Pittsburgh	3	45	Sun	69–70	Todd/Wisco	699	695.3	658.0	78.0	47.0	30.1	17,904	32,900	9900	15	A
SL-7	Sea-Land McLean	8	72–73	AGW,RSNW	—	RDD	1184	946.1	880.5	105.5	64.0	34.7	29,300	51,815	120,000	33	F,M
T2/C4-M3	Seattle	1	44	Sun	70	Todd	354	494.1	471.1	71.5	43.5	30.5	9200	19,710	7000	15	A
MV	Aleutian Developer	1	60	MDDSBCO	75	Willamette	85	360.4	341.0	52.0	34.8	17.0	2296	5152	4200	15	A
C4-X	Charlestown	1	45	Kaiser	68	MDDSBCO	366	522.8	496.0	71.5	43.5	30.5	10,200	19,800	9900	16	A
C4-X3	Philadelphia	1	45	Kaiser	69	Beth.	366	522.8	496.0	71.5	43.5	30.5	19,800	19,800	9900	16	A
C4-C	Mobile	2	45	Kaiser	68	ADDSCO	366	522.8	496.0	71.5	43.5	30.5	8655	20,295	9900	16	A
T2-M1	Tampa	1	44	Marinship	69	Todd	360	523.5	503.0	68.0	45.3	30.0	13,544	22,890	10,000	16	A
T2-C4X	Baltimore	1	44	Sun	70	Willamette	360	497.4	471.1	71.5	43.5	30.5	9440	19,690	7240	15	A
Feeder	Desafio	1	79	Vulcano	—	—	193	339.2	307.3	60.7	27.2	21.0	6515	8649	6000	14.5	F
D6	Sea-Land Pacer	4	78	MHI-K	89	Jurong	589	661.9	623.9	78.1	47.0	27.3	15,241	28,301	17,400	18.0	A
D7	Sea-Land Anchorage	3	87	BSC	—	—	791	709.8	676.3	78.1	46.2	33.1	20,883	37,474	22,450	19.3	A
D9J	Sea-Land Patriot	12	80	MHI-K	85	MHI-Y	1343	845.6	798.7	100.4	54.1	32.9	30,776	47,182	30,150	20.7	A
Atlantic	Nedlloyd Holland	12	84–85	Daewoo	—	—	2177	949.6	915.1	105.6	70.5	38.3	57,949	81,284	28,000	18.0	A
C6	Sea-Land Expedition	2	73	Ingalls	—	—	723	669.1	624.8	89.9	53.0	36.5	21,694	34,694	28,500	21.6	A
C7 (Lancer)	Sea-Land Challenger	3	68–69	Sun	—	—	672	700.3	669.8	89.9	50.5	35.8	20,939	32,507	27,300[b]	20.9	A
C8	Sea-Land Navigator	4	72–80	Ingalls	83–84	Avondale	1025	813.1	768.8	89.9	53.0	36.5	30,810	47,792	28,500[b]	20.1	A
SL18P	Sea-Land Consumer	2	73–74	Beth.	89	MHI-K	832	396.2	363.8	95.1	54.0	34.1	25,247	38,860	32,000[b]	21.4	F,A
LASH	Sea-Land Reliance	2	80	Avondale	91	Lenac	1057	893.1	2615.1	100.2	60.0	40.8	40,238	62,314	32,000[b]	22.5	F
Panarea	Panarea I	3	80	MIE	—	—	288	396.6	363.4	68.2	34.4	25.9	9878	13,467	7040[c]	14.8	F
World	Sea Lotus	3	78–79	MIE	—	—	389	501.8	475.6	75.4	42.6	28.4	12,894	18,744	11,700[c]	16	A
Bay	Seward Bay	4	83–84	Imabari	—	—	284	450.3	419.8	68.9	35.1	26.0	11,875	16,551	7000[c]	15	A

[a] M= mid; A = aft; F = fwd.
[b] Units are SHP.
[c] Units are PS.

Matson's first container operations began in August 1958 by carrying containers on the deck of the *Hawaiian Merchant*, a C3.

Between 1958 and 1960, a series of six such vessels were put into service. In 1960 the *Hawaiian Citizen*, a fully containerized ship with a capacity of 436 twenty-four foot containers entered service. She was joined by two sisters, the *Californian* and *Hawaiian*, also capable of carrying bulk sugar. The *Hawaiian Motorist* and *Hawaiian Legislator*, also converted containerships, joined the fleet.

In 1963 a 300-ft cellular, self-sustaining, container barge, the *Islander*, with a capacity of 155 twenty-four foot containers, was put into the inter-island service in Hawaii. Two jumboized C4's, the *Hawaiian Queen* and *Hawaiian Monarch*, entered service in 1965 and 1966.

The 338-ft *Hawaiian Princess*, a specially built, cellular container carrier replaced the *Islander* in 1967. She carried 156 containers and 1600 tons of molasses.

In 1967 Matson started a Far Eastern service with the *Pacific Trader* and *Pacific Banker*, with 464 containers each; but Matson soon retired back into a solely domestic operation.

Four new ships, the *Hawaiian Enterprise* and sisters, were built, two at Bethlehem Sparrows Point and two at Bremer Vulkan in Germany. Sea-Land acquired these ships while they were under construction, and designated them their SL-18 class.

In 1973, Matson had built the RO/RO ships *Lurline* and *Matsonia*. In 1983, however, they were converted to container/RO/RO configuration with the insertion of a new 351-ft-long by 105-ft-wide midbody.

In 1990 the National Steel and Shipbuilding Co. began building a new combination container/RO/RO ship for Matson.

Refrigerated Cargo Ships. The long established United Fruit Company operated a special fleet of refrigerated break-bulk ships built just after World War II. Their principal trade was in bananas carried on the stem. This trade was shared with Grace Lines, Standard Fruit, and others. In the late 1950s, 44-lb cartons were introduced, and some operators began using conveyor systems for handling them. By the

Table 6 Matson ships

Type	Class	No.	Built	Conv.	Yard	LOA, ft	LBP, ft	B, ft	D, ft	d, ft	DWT	HP	V	Cont. (24-ft)	Names
Barges	—	2	—	—	—	106.7[a]	106.7[a]	19.7[a]	—	4.5[a]	—	—	—	216	Halea Kala, Maunaloa
Cont.	—	1	1992	—	NASSCO	217.0	217.0	32.0	—	10.51	27,000	—	22.5	1650	Hawaii II
Cont.	C3-S-A1C	2	—	1967	MHI	544-4	518-3	69-10	43-2	31-6	12,309	9350	16.5	464	Pacific Banker, Pacific Trader[b]
Cont./vehicle	C3-S-A1C	1	—	1963	—	491-7	465-3	69-10	42-6	29-5	—	9350	16.5	186	Haw. Fisherman[c]
Cont. barge	—	1	1963	—	Beth.	95[a]	95[a]	15.2	—	5.3	—	—	—	216	Islander
RO/RO	—	2	note *d*	—	—	—	—	—	—	—	13,600	3000	—	—	Kaimoku, Kain Alu[e]
Cont.	C7-S-88A	2	1978–80	—	Bath/Sun	219.2	—	28.9	—	10.4	26,600	32,000	23.0	1446	Mau, Kauai
Cont.	—	2	1970	—	—	219	—	28.9	—	10.4	27,100	32,000	23.0	1240	Manukai, Manulani
Bulk	T2-SE-A2C	1	—	—	—	—	—	—	—	—	24,611	10,000	17.0	—	Kopaa
Cargo	C4	1	—	1963	—	522-10½	496-0	71-6	43-6	32-10	14,800	9900	16.0	402	Haw. Legislator
RO/RO/cont.	—	2	—	1986	—	826	769.5	92	60.1	30.8	25,000	30,000	—	—	Lurline, Matsonia
RO/RO	—	2	1973	—	Sun	700	643	92	42.06	28	13,900	3000	—	—	Lurline, Matsonia
—	—	2	1970	—	Beth.	—	—	—	—	—	27,100	32,000	23.0	—	Haw. Enterprise, Haw. Progress[f]
Cont./bulk liq.	—	1	1967	—	Beth.	338-3	307-6	52-0	28-3	17-8	4443	1600	10.0	212	Haw. Princess[g]
Cont./auto	C4-S-B2C	2	—	1965	ADDSCO	630-4	606-0	71-6	52-0	32-10¾	17,642	9900	15.5	805	Haw. Queen, Haw. Monarch[h]
Cont.	C3-S-A2	1	—	1960	Willamette	492-0	465.3	69-6	42-11¼	29-4⅞	10,282	9350	16.5	408	Haw. Citizen
Cont./bulk	C4-S-A4C	2	—	1960	ADDSCO	632-11	606-0	71-6	43-6	32-11⅞	20,500	9900	15.5	436	Calif. Hawaiian

[a] In units of meters.
[b] Ex-Gulf SA.
[c] Haw. Motorist.
[d] 1976/1990, 1974/1976/1990.
[e] Ex-Cagans.
[f] Transferred to SL.
[g] Maunakea.
[h] Maunalli, Maunawili.

1960s some operators were carrying the bananas on pallets, for handling by forklift trucks.

The containership operators introduced refrigerated containers in the late 1950s. In 1968 a new concept was introduced. This consisted of a centralized shipboard refrigeration plant that blew cold air into insulated containers in the holds. Farrell Lines, in their Australian service, was the first U.S. company to use this method.

During the 1980s there was no widespread agreement among reefer operators as to the best method for carrying and protecting bananas or other kinds of fruit. Del Monte used pallet ships, Standard Fruit used highly mechanized conveyor systems, Reefer Express Lines introduced combination pallet reefer/car carriers, and United Fruit (renamed Chiquita Brands) experimented with reefer containers. By 1990, the trend had moved slowly toward containers with Del Monte building nine combination pallet/containerships in Spain, Dole building containerships in Italy, and Chiquita Brands building containerships in Japan.

Summary of U.S. Cargo Ship Progress: 1942–1992

I present here in outline form a summary of the most important changes that we have seen in cargo ship design in the past 50 years. I have deliberately omitted a goodly number of improvements that are well covered in other chapters of this book.

Speed. World War II–built cargo ships generally ranged in speed from 12 to 17 knots, with 15 knots being typical. The speed of U.S. cargo ships increased in one step to 20 knots with the advent of the Mariners in the early 1950s. Replacement cargo ships in the 1960s to 1980s varied from 18 to 33 knots, with 20 to 22 knots being typical.

Crew. The operation of ships built during the war was labor-intensive. Some of the manpower needs related to wartime duties, but many of them resulted from work practices prevalent from the previous century of operation of steam ships. Many of the crew members were required by government regulations and by union agreements. The most prevalent type of work was painting and chipping and, considering that coating materials were of limited capability, it is understandable that so much time was spent on this task. The peculiarities of the subsidy system also tended to encourage large crews.

It was commonplace for stewards departments on cargo ships to prepare all food from basic ingredients. Service in the mess rooms and officers' salon, as well as house cleaning functions, were extensive. The first major reduction in crew numbers, from the war-built sizes of 50 to 54 persons, occurred in the late 1960s with work rationalization and a modest amount of automation. Crew sizes in this period dropped to 30 to 36 men. This development was triggered by recommendations made by the Maritime Research Advisory Committee of the National Research Council.

In the 1970s and 1980s further crew reductions (to 18 to 22 men) followed the introduction of diesel propulsion, more bridge automation, and better communication equipment. By the early 1990s, international conventions and U.S. regulations, together with further automation and rationalization, made possible the reduction to twelve persons. Few U.S.-

flag operators have taken full advantage of this lower level, opting to keep personnel on board for maintenance, mooring, and other activities. Hence, crew numbers by 1992 were typically in the range of fifteen to twenty.

Cargo Gear. The cargo handling gear on break-bulk cargo ships built in 1942 through 1960 were almost exclusively of the boom and winch variety. During the early 1960s and through the 1980s, cargo gear on break-bulk and pallet ships gradually changed over to revolving deck cranes. But it was only with the emergence of containerization that cargo gear radically changed. Early containerships were equipped with portal-type gantry cranes. By the 1980s, most ports had built large shoreside gantries specially designed to handle containers, and it became more economical for the ships to operate without cargo gear.

Hatch Covers. 1942-built ships employed either hatch boards and tarpaulins, or small pontoons and tarpaulins, for hatch covering. In the late 1940s, the British-based MacGregor Co. introduced steel folding hatch covers in the U.K., and their use was quickly adapted to the Mariners in the early 1950s. After a period of developmental problems, they became more widely accepted and all break-bulk ships employed mechanized steel hatch covers after 1955. Containerships, on the other hand, found large, single piece pontoon covers to be more advantageous. These containership hatch covers were built as large as possible but at a weight not to exceed the capacity of container cranes (25 to 30 tons). By the 1990s the concept of eliminating hatch covers on containerships had appeared, and while no U.S. operator had built a "hatch coverless" ship by 1992, it was anticipated that many would do so in the future.

Cargo Tanks. Many of the war-built cargo ships incorporated cargo deep tanks in the lower portions of the cargo hold. These were suitable for carrying small lot sizes of liquid cargoes. Their use continued through the entire break-bulk era, to be replaced by the use of 20-ft and 40-ft intermodal tank containers. These greatly simplified cargo transfer, tank cleaning, and other problems.

Refrigerated Cargo. Many break-bulk cargo ships were constructed with large reefer rooms, which were loaded and unloaded by traditional break-bulk manual methods. In some cases entire decks were refrigerated, particularly for the carriage of fresh fruit. These break-bulk methods for carrying reefer cargo were gradually replaced by full reefer ships and by containerships carrying refrigerated containers. The reefer ships moved from carton-by-carton loading methods to conveyors and, finally, to efficient pallet operations.

Machinery Location. During the early 1940s, the engine rooms on cargo ships were usually placed in their traditional amidships position, with long shaft alleys running through the after cargo holds. With few exceptions, it was not until the introduction of diesel propulsion on a large scale that the engine rooms were moved aft. By now engine rooms are almost exclusively positioned all the way aft, except on fine lined ships, where they may be placed somewhat farther forward, but still well aft of amidships.

Superstructure Location. The traditional location of superstructures on 1940s-built cargo ships was amidships, over the engine room, except on some C4's that had split houses (amidships and aft). By the 1960s, a few ships had some superstructure forward and some aft. By the 1980s the most prevalent location was aft, but some containerships also had a superstructure above the collision bulkhead to protect the cargo boxes from green water coming aboard. Superstructures became smaller in size owing to the smaller crews. At the same time they grew in height in order to free deck area for cargo and to provide visibility from the bridge over the large deck loads of containers.

Beam Draft Ratios. There was little change in the first half of the 20th Century in beam draft ratios, with a bias against broad beamed ships. It was with the advent of containerships that high deck loads provided an economic incentive for designers to go to abnormally large beam draft ratios. The narrower converted containerships of the 1960s overcame stability problems by the inefficient expedient of using high density ballast. By the 1970s and 1980s, designers learned to live with broader beamed ships. This has extended to the introduction of the extra- or post-Panamax ship (beam in excess of 106 ft).

Draft. Because draft has always been viewed as a design limitation, and harbors seldom create dramatic increases in channel depths, the operating and scantling drafts of cargo ships have changed relatively little in 50 years. Any increase has been linear with length and displacement.

Block Coefficient. Higher sea speeds and dramatic increases in fuel costs in the 1970s and 1980s have led designers to finer hull forms and lower block coefficients, generally.

Bow Form. Traditional stem profiles and V-shaped raked bows gave way, in the 1950s, to bulbous bows. These were modest in size at first, but tended to increase as designers sought optimum forms. This growth did not stabilize until the 1970s. Higher speeds and the desire for drier decks led to more prevalent use of forecastle heads and extreme flare.

Stern Form and Rudders. Traditional cruiser sterns were commonplace from 1942 to the 1960s. As horsepower grew, so did problems with vibrations. The need for increased clearance between hull and propeller led to the popularity of the so-called clearwater stern, such as that used on the Mariners. This design featured an open-aperture with a horn-supported spade rudder.

Squaring off of the stern with a modest flat transom started to occur in the late 1960s. The pressure to maximize the deck area for containers, even at the stern, brought about wide, flat transom sterns.

Life Saving Gear. Life saving gear on cargo ships built in the early 1940s consisted of traditional open, oar-propelled, davit-launched, steel lifeboats. Few improvements occurred until the 1960s and 1970s, when lifeboat materials were changed to fiberglass, and self-unfolding rafts were installed on cargo ships. Immersion suits were introduced as a requirement in the 1980s, and free fall lifeboats in the 1990s.

Cargo Hold Outfitting. The World War II type cargo holds were equipped with wooden spar ceilings in the square of the hatch, sweat battens along the shell frames, and mechanically powered ventilators. More extensive use of mechanically powered ventilation occurred during the late 1940s and widespread use of cargo hold dehumidification was a feature of breakbulk ships of the 1950s and 1960s. Cluster lighting in the cargo holds was as typical on these earlier ships as was a fixed lighting system. The introduction of containerization eliminated the need for most of the cargo hold outfitting, particularly the wood battens and ceiling. (With the elimination of wooden hatch boards and other wood on board, the necessity for a carpenter and carpenter shop also disappeared.) Cargo hold outfitting on containerships is sparse, consisting usually of only cell guides and access ladders and platforms.

Closure

Cargo ship development in the United States during the second half of the 20th Century was distinctly defined by three events, namely World War II, the post-war government vessel replacement program, and the emergence of intermodal transportation, particularly containerization.

The evolution of more efficient and competitive cargo ships was largely a result of the application of work rationalization, mechanization, automation and the introduction of computers.

Surging energy costs, conservation, and a concern for the environment led to the development of more fuel-efficient propulsion plants, hulls, and appendages.

Improvements in material sciences, especially in metallurgy and metal joining techniques, resulted in lighter and stronger structures, with resulting savings in construction and fuel costs, and increased cargo capacity.

The computer revolution allows engineers to optimize cargo ship designs, leading to a dramatic increase in the efficiency of marine transport.

Charles R. Cushing is a graduate of the U.S. Merchant Marine Academy, MIT, and SUNY. He has sailed as an officer on U.S. passenger and cargo ships, and was a naval architect for Sea-Land for seven years in its container-pioneering days. For the past 25 years he has been president of the firm C. R. Cushing & Co., naval architects and marine engineers.

Development of Large Commercial Roll On/Roll Off Ships, Trailerships, and Car Carriers

Charles Zeien

Introduction

In a continuing effort to lower the cost of ocean shipping, ship owners have sought increasingly larger cargo ships. The larger ships can be propelled economically at higher sea speeds, but spend a greater part of their time in port for stevedoring operations. In an effort to reduce port time and the cost of stevedoring, the concept of unitization of cargo was born. (Unitized cargo is advantageous in that it allows mechanized handling.)

Early containerships used onboard cranes or other equipment to handle the containers in the same manner that break-bulk cargo ships used their own cargo gear. The addition of these massive cranes meant adding substantial topside weight and increasing the vessel's complexity. At about the same time, the concept of rolling cargo on and off the ship on wheels through structural openings began to look attractive. RO/RO's require minimal shoreside equipment and permit rapid loading and discharging of cargo, thereby shortening port time and reducing stevedoring costs.

Containerships and RO/RO ships developed side by side following World War II. These two types of ships frequently operated in direct competition on the same trade routes. Over the years, containerships evolved so that they used shore-based cargo handling equipment. Large containerships proved to be the most efficient carriers on longer high-density trade routes, while RO/RO's have found a niche in short-distance, specialized trades in high-value cargoes. RO/RO's frequently proved to be efficient in transporting cargo to less-developed ports, replacing break-bulk ships.

Definitions

There are three basic types of oceangoing RO/RO's:

- General purpose RO/RO's—ships or barges designed to carry palletized cargo and containers; fork trucks or other specialized equipment move cargo on and off the ship. Many of these RO/RO's have arrangements for carrying containerized cargo, sometimes with the use of special handling equipment.
- Trailerships—ships or barges that carry highway trailers, commercial vehicles, and automobiles; containers are carried on a wheeled chassis for the ocean voyage so that they can be handled like highway trailers.
- Pure car carriers (PCCs)—ships designed to carry newly manufactured automobiles by driving them to their position on the ship. Vessels are designed to be efficient in that service by generally excluding other types of cargo.

Early Commercial RO/RO Vessels

Large commercial RO/RO ships evolved from small harbor ferries that carried vehicles and railroad equipment. Shortly after World War II, larger ferries were built to transport vehicles between European ports on the North Sea. It wasn't long before truckers began to use these ferries to carry cargo in competition with small general cargo ships handling cargo in the conventional fashion with booms through hatches.

The earliest North Sea vessels loaded the cargo through stern doors fitted with hinged ramps. This required the vehicles to turn around in the ferry so that they could be discharged over the same ramps. Some later ferries were built with a turntable at the bow to aid in turning the vehicles. Later, in the 1950s, larger ferries were built with both bow and stern doors that permitted the vehicles to discharge without turning around in the ship. This speeded cargo handling considerably, especially on the larger ships.

Development of Large Military RO/RO's

During World War II, U.S. military operations, particularly in the Pacific, needed landing craft that enabled military forces to land onto the beach. Thousands

Fig. 1 The Greek-flag combination vehicle and container carrier *Hellenic Explorer* was built at Sasebo Heavy Industries Co., Ltd., Sasebo, Japan, for Hellenic Lines, Ltd., New York. The RO/RO is equipped with a stern ramp and a controllable-pitch propeller. (Courtesy American Bureau of Shipping.)

of craft, large and small, were built to permit discharge of troops, tanks and supplies onto the beach. The concept involved grounding the bow of the vessel and opening a large door that became the ramp needed to make a relatively smooth transition for the men and equipment moving ashore. Some of the earliest RO/RO commercial services involved war surplus landing ships or small craft adapted to particular requirements.

In the late 1940s the U.S. Navy recognized the military potential of the roll on/roll off concept; it enabled oceangoing ships to carry large, heavy military equipment and discharge the equipment quickly without requiring shore-based facilities or cumbersome cargo-handling gear on the ship. After testing and developing the concept by converting a few war surplus cargo ships, the U.S. Navy decided to take the next step.

In 1954 the Navy retained George G. Sharp, Inc., a naval architectural firm, to design the *Comet*, a large RO/RO ship, which was delivered by Sun Shipbuilding and Dry Dock Co. of Chester PA in 1959. This steam powered, twin-screw vessel loaded and discharged over the stern and through side ports and ramps; in many ways this was the real grandparent of large commercial RO/RO ships.

While the deck heights and basic design of the *Comet* made her unsuited to commercial operation, the specifications required hydraulic operation of the bulkhead doors, ramps and sideport doors. The development of those devices set the stage for subsequent development of two types of large RO/RO vessels defined above. The development of the car carrier was a later evolution.

The Navy prepared a performance specification for the *Admiral William H. Callaghan*, a larger RO/RO vessel patterned after the *Comet*. Sun Shipbuilding obtained the build and charter contract for the 26-knot, twin-screw, gas turbine-powered vessel. This RO/RO, delivered in 1967, had 167,537 sq ft of usable cargo space; cargo was loaded and unloaded through a large stern door that served as the ramp to shore. The ship had sideport doors and ramps, but these were seldom used in service. Over the next 20 years the ship operated on worldwide trade routes delivering military cargoes.

Recent Military RO/RO Programs

In the early 1980s, the U.S. Navy took a fresh look at RO/RO's when a need developed for prepositioning and rapid resupply of large amounts of military equipment. The prepositioned fleet was planned to be deployed in three groups at various positions in the world with equipment and ammunition to supply air-lifted Marines. The resupply RO/RO's were to be berthed near U.S. ports where military materiel is stockpiled.

For the prepositioned program, 13 large RO/RO's were chartered after modification to meet military requirements.

Three groups of ships made up the prepositioned fleet. One group of five commercial RO/RO's was acquired from Maersk Lines; three of these were

Fig. 2 The RO/RO vehicle and container carrier *Baltic Print* was built at Karlskronavarvet AB, Karlskrona, Sweden. She is fitted with a bow thruster, stern ramp, and a 40 metric ton cargo lift. (Courtesy American Bureau of Shipping.)

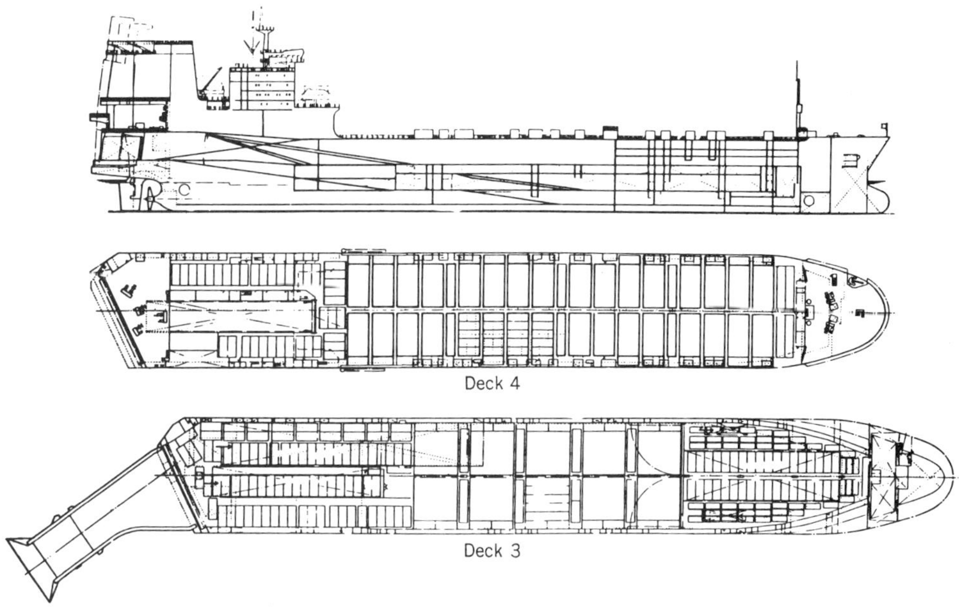

Fig. 3 The *Boogabilla*: 228.5 m LOA, 32.3 m beam, 10.8 m draft, 31,458 DWT, 20.5 knots, and 1707 TEU.

lengthened and modified for the military mission at the Sparrows Point Yard of Bethlehem Steel Co. and two were converted at the Beaumont Yard. Three Waterman Steamship Co. RO/RO's under construction at Sun Ship were taken to the National Steel & Shipbuilding Co., at San Diego CA, for lengthening and other modifications. Five new RO/RO's were built at the General Dynamics Yard in Quincy MA.

For the resupply mission, the Navy acquired Sea-Land's eight SL-7 ships. With speeds of 32 knots, these were the world's fastest containerships. They were converted to RO/RO configuration suitable for their military mission. Two angled side ramps were used to gain direct access to piers, barges, and causeways developed for the resupply program. Additional RO/RO's were added to the resupply fleet at a later date.

All of the ships for the prepositioned and resupply programs have a limited number of cranes for operation in the lift on/lift off mode. With maximum flexibility, these ships can be unloaded almost anywhere.

Again, in 1992, the Navy is embarked on a major Sea Lift Enhancement Program, which will involve ten or more large ships. More than half of them will be RO/RO's.

General Purpose RO/RO's

The first major commercial general purpose RO/RO ships were built in 1967 for the North Atlantic service. The *Atlantic Song* was built in France for Wallenius Lines while the *Atlantic Span* was built in Sweden for Transatlantic AB. Eight additional ships were built and a group of owners formed a consortium called Atlantic Container Lines (ACL). Each of these ships carried approximately three times as much cargo per voyage as the break-bulk ships with which they competed, and were able to make twice as many trips per year because of their slightly higher sea speed and much shorter port times.

Cellular containerships using shore-based cranes were beginning to make inroads in the North Atlantic service at that time, and the ACL RO/RO's carried similar containerized cargo in addition to their RO/RO cargoes. They achieved remarkable market penetration by combining the carriage of containers with the great flexibility inherent in the RO/RO concept.

The ACL ships used centerline stern ramps for cargo access. This necessitated special pier arrangements. Sometimes the stern ramp rested on a deck barge, which in turn had a ramp to shore. In 1975 the *Parella* was built with a stern ramp angled to starboard, and that permitted direct access to the shore. To increase the speed of loading and unloading, ramps were built wide enough for two lanes of traffic. The ramps were massive, and the winches that raised and lowered them were controlled by sophisticated equipment so that the ramp did not add excessive loads to the pier structure.

In a few years, slewing stern ramps were developed by the MacGregor Company of Sweden (currently the Swedish division of MacGregor-Navire of Piikkio,

Fig. 4 The RO/RO *Westward Venture*, the stretched sistership of *Ponce De Leon*, with a capability of carrying 390 forty ft trailers and 126 autos.

Finland); these provide access to the pier on either side or directly astern. In 1979 the Waterman Steamship Co. built three ships with slewing stern ramps as the first American flag vessels to include this feature. The slewing ramp usually requires the addition of a separate watertight door.

While the slewing ramp provided great flexibility, many ship owners continued to favor the fixed quartering ramp for simplicity, lower cost and the wider ramp that can be fitted as a result of that simplicity.

The RO/RO concept received a big boost following the 1973 oil price rise. This generated a heavy demand for industrial products in the OPEC countries, which benefitted from their greatly increased incomes. The prosperity created severe port congestion in 1975 and 1976 because the OPEC ports were as yet underdeveloped. Ship delays of 100 days or more became common. Since RO/RO ships could load and unload quickly at relatively undeveloped ports, these ships soon were given priority, creating an immediate demand for the redeployment of RO/RO ships to those trades. The resulting high freight rates influenced many owners to build RO/RO ships rather than cellular containerships.

By 1984, some 284 RO/RO's were operating in the world trade, with an aggregate capacity of 247,200 TEUs, an increase of 550 percent in vessel numbers and 640 percent in TEU container capacity as compared to 1975, when 43 ships were in service. These totals do not include some of the specialized ships such as pure car carriers.

In some trades, such as that between Europe and West Africa, RO/RO's were carrying 75 percent of the box traffic in 1984, with cellular containerships carrying the balance. On the other hand, RO/RO's never really penetrated the long transpacific trades. In the Pacific in 1984, only 3.5 percent of the boxes were carried on RO/RO ships. Since then one of the major carriers in that trade has withdrawn.

On high density trade routes, cellular containerships eventually come to dominate. They use efficient shore-based cranes, frequently provided at low cost to the ship operator by a publicly financed port authority. They prevail in competition with RO/RO's because a containership can carry roughly twice as much container-suited cargo as a straight RO/RO ship of similar size. Therefore, when new tonnage is planned for a well developed trade route, the lower cost per TEU of new cellular containerships becomes an overriding consideration. This chases the RO/RO to less developed trade routes.

In the midst of the intense competition between cellular containerships and full RO/RO's, some novel designs evolved combining the two concepts. In 1979 the *Booganbilla* was built as a combination ship in which the forward end was fitted with container cells. The aft portion was designed for RO/RO cargo; the center portion was convertible for carrying containers

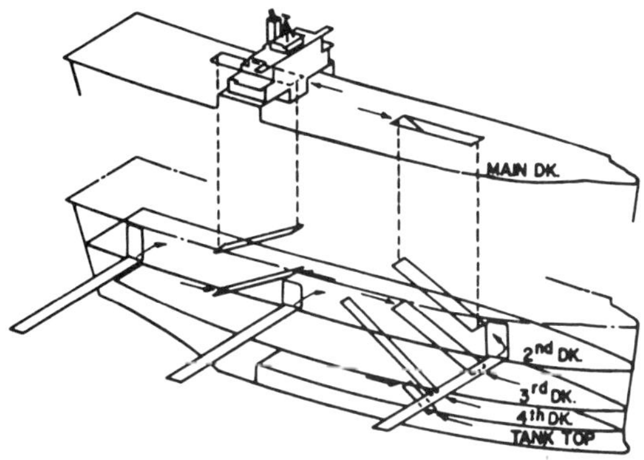

Fig. 5 Isometric arrangement of the *Ponce De Leon.*

Fig. 6 Roloc box deck fitting.

or RO/RO, depending on the cargo available for a particular voyage. Even some of the earliest RO/RO's carried containers on the top deck. These were placed on the ship with shoreside cranes. Later, some RO/RO's were built with a fixed guide system on deck to minimize the cost of lashing the containers.

Naturally, even in the most well developed trades, a significant amount of cargo remains for the RO/RO ships, especially after break-bulk ships are driven from the trade. RO/RO's are required to transport the cargo that cannot be readily containerized, such as agricultural machinery, heavy construction equipment, railway equipment, military equipment and other specialized cargo including industrial plant components. This portion of the trade has been variously estimated at three to six percent of the value of the container-suitable trade. RO/RO ships that are built for the special cargo usually fill out their cargo manifest by carrying containers by one means or another.

Trailership Development

The development of large trailerships began in the United States in 1965. The approach was to combine the military RO/RO experience with the extensive development of the detachable highway trailer that could operate on the highways in the continental United States and in the Commonwealth of Puerto Rico, which had the same highway regulations.

While the *Callaghan* was being designed and built, a commercial version was visualized by Sun Ship. The commercial version would use the same hull lines but would be single-screw and steam powered, with a service speed of 23.5 knots.

In cooperation with American Union Transport, Inc., the *Ponce De Leon* was built for weekly service between New York and San Juan, Puerto Rico. A new company named Transamerica Trailer Transport, Inc. was formed to own and operate the ship in weekly service. The vessel regularly sailed from New York on a Friday evening and arrived in San Juan on Monday morning, 60 hours later. The company was acquired by Puerto Rico Maritime Shipping Authority, Inc. in 1975, and the ship and her sister ships operated in this service until recently, when they were supplanted by cellular containerships.

Description of the *Ponce De Leon*

The *Ponce De Leon* loads through two large sideport openings on the main deck on the starboard side, fore and aft. The trailers and vehicles are moved over large shore-based ramps at each terminal by specialized tractors and some 40 longshoremen. The ship can carry 260 highway trailers and 292 automobiles. The automobiles use special ramps to reach lower levels of the ship unsuited for highway trailers. With careful planning, a relatively small stevedore crew regularly turns the ship around in eight hours.

The success of this approach to RO/RO service depends in large measure on the development of an efficient method of securing trailers on the ship for the 1400 nautical mile sea voyage. The concept, developed and patented with several unique features, improved upon a concept pioneered by Trailer Marine Transport, Inc. (TMT), a company offering barge service to Puerto Rico from Jacksonville FL using large barges.

When a trailer is delivered to the marine terminal it is supported temporarily on its own landing gear

Fig. 7 The *Matsonia.*

which, while suited to holding the trailer in a shoreside parking lot, is not strong enough when the ship is in a seaway. Therefore, a specialized structural steel support box was developed with a mechanical device on top similar to the "fifth wheel" on a highway tractor. The "fifth wheel" is a standardized mechanical device that is used in the United States almost universally; it allows any tractor to attach itself to any trailer.

The structural box with an integral "fifth wheel" is designed to lock onto a button welded to the deck of the ship at each location where a trailer would be parked. The box is designed so that it can be picked up on shore by the specially built tractor, which tows the trailer over ramps to where it is to be stowed on the ship. The box is then locked to the button by operating a hand lever. The back end of the trailer is held down by two simple wire rope devices installed by longshoremen, one at each corner. These lashings attach to D rings welded into the deck in a pattern that provides flexibility to accommodate trailers of various sizes. In summary, with the forward end of the trailer secured by the structural box to a button on the deck, and the back of the trailer tied down with lashings, the trailer is ready for sea.

Fig. 8 The 580 ft triple deck trailer barge *El Conquistador* was built for Trailer Marine Transport Corp. (Courtesy American Bureau of Shipping.)

Refrigerated trailers are plugged into the ship's electrical power for operating at sea. Unusual items, such as bulldozers and road building equipment, are lashed to the deck using gear similar to that used to tie down the back end of the trailer.

Recent Development of Trailerships

Later versions of the *Ponce De Leon*-class used elevators to provide trailer access to the tanktop when the trade route did not require extensive automobile space. Some of the later ships were stretched to 790 ft-9 in. LOA. These versions included partial superstructure decks for trailers with lightweight cargoes. In all, ten ships were built in various configurations.

Because of the previously mentioned Middle East port conditions, two Sun RO/RO's under construction were quickly finished and diverted to deliver containers, large construction vehicles, portable housing, and other equipment from the East Coast of the United States to ports in the Persian Gulf. When that trade returned to normal, the ships were redeployed to the U.S. domestic trades.

Matson ordered two ships from Sun Ship in 1973 for trade between the West Coast of the United States and Honolulu. In 1982 one of the ships was converted by Todd's San Pedro yard to carry containers in cells in the midship and forward portions of the vessel. A garage area was added aft of the deckhouse for automobiles. A portion of the container space was made convertible to trailers so the cargo space could be adapted to the needs of the trade on each voyage.

Fig. 9 The ABS-classed car carrier *Green Lake.*

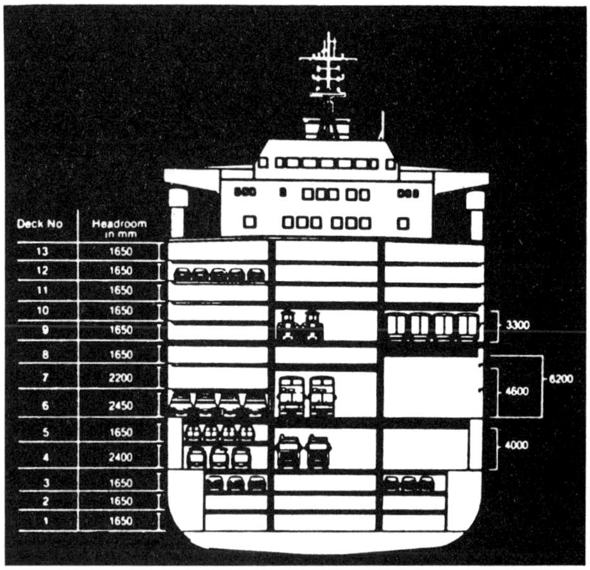

Fig. 10 Section through a typical 13 deck PC/TC showing level variations arranged for a mixed stow of cars, trucks, and buses. Very high loads can be accommodated by lifting two decks above the maindeck level.

Sun Shipbuilding & Dry Dock Co. delivered two of the ships to a wholly owned subsidiary, TOTE (Totem Ocean Trailer Express). This new company provides service between Anchorage AK and Tacoma WA in competition with containerships on the same route. TOTE has since been sold by Sun, but continues in successful operation.

Large Barge Trailerships

The use of large unmanned barges to carry trailers was pioneered by TMT, Trailer Marine Transport, of Jacksonville, FL. Initially, 400-ft barges were adapted to carry trailers on deck from Jacksonville to San Juan PR. Frequently, the tow boat would tow two or three barges. The trailers were loaded aboard by backing them into an aisle on the barge from a large shorebased ramp across the stern. The trailers were guided by high steel curbs on each side, which made the backing relatively easy for the driver. The front end of each trailer was supported by a steel box similar to those used on the *Ponce De Leon.* Because of the high curbing, it was not necessary to lash the back end of the trailer to the deck of the barge.

Later a second deck was added, requiring a double-level ramp at the stern in each port. By the late 1980s barges with three decks, 600 ft long by 100-ft beam, carrying 400 trailers were being used between various ports on the U.S. East Coast and San Juan. Initially, TMT would not carry refrigerated trailers because they could not be serviced. Later, however, they accepted refrigerated trailers and let them run unattended during the voyage, with acceptable results.

Pure Car Carriers (PCCs)

Automobiles have been carried on break-bulk ships since automobiles were first produced. The nature of the cargo-handling operations, however, resulted in high cost and damage to the vehicles. When automobiles started to move in large quantity between Europe and the United States in the late 1950s, bulk carriers were frequently modified to carry them more efficiently. Usually, a system of car decks was installed, and the cars were lifted into the vessel with cranes installed on the ship. Avoiding damage to the vehicles still depended on the skill of the crane operator, and so the appealing concept of a RO/RO ship, in which the car could be driven to the stowage position, was quickly adopted. Some of the early ships carried as many as 1000 cars.

In 1966 the Wallenius Lines, which had pioneered in the mass movement of cars, built the *Faust,* the first of a class of six ships, to carry 1700 cars by RO/RO. These ships could carry grain as well.

When the Japanese car export trade developed, the need for larger specialized ships became clear. These would carry vehicles exclusively and return in ballast, similar to crude oil carriers. In 1968, K Line built the *Toyota Maru No.10* as a pure car carrier to carry over 2500 cars.

In December 1974 the 6000-car *Jinyu Maru* was delivered. The size of this ship was constrained by the limits of the Panama Canal and various ports. The ship contained four holds and thirteen decks. Headroom on some of the decks was as little as 1.6 meters. Four ramps through the side of the ship provided access to the shore. Ships of this type require massive

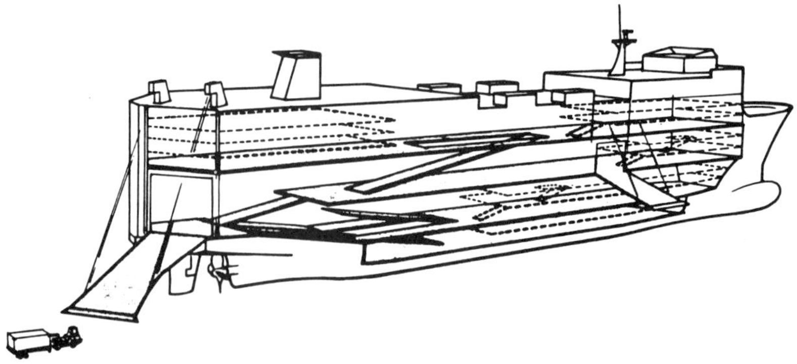

Fig. 11 Internal arrangement on one of the latest PC/TCs showing stowages suitable for containers and means of access to these strengthened decks. Dotted lines indicate cardecks in place.

ventilation systems since so many cars are operated simultaneously to achieve rapid loading and unloading. They will load or unload about 2000 cars per day.

The more recently constructed pure car carriers are no longer "pure." They may have several hoistable car decks, giving them the flexibility to handle trucks, buses, farm machinery, and even containers. Such cargo usually has a far higher freight rate than new automobiles in the Pacific trades, and is sometimes available as a back haul. They are usually designed for 18- to 20-knot sea speeds, carrying about 6000 cars.

In 1988 three American flag PCCs were built in Japan for the car trade between Japan and the United States. There are probably over 100 large pure car carriers in service today.

Bibliography

"Characteristics and Index of Maritime Administration Ship Design," U.S. Dept. of Transportation, Maritime Administration, Aug. 1987.

Dillon, E. S., Hoffmann, L. C., and Roseman, D. P., "Forty Years of Ship Design Under the Merchant Marine Act 1936–1976," *Trans.* SNAME, Vol. 84, 1976, p. 169.

"Innovation in the Maritime Industry," Appendix, National Research Council, Maritime Transportation Research Board, 1978, p. 73.

Koskivirta, V., Heikkila, V., Niini, M., and Harjuvaara, H., "New Finnish Barge Carriers for the U.S.S.R.," *Trans.* SNAME, Vol. 86, 1978, p. 304.

"The Market Role of Large Container RO/RO Vessels," (MarAd HE 731-D84-No. 130-c.3) Drewery Shipping Consultants.

The Motor Ship, Reed Business Publishing Group, 1970–1990.

Ro-Ro Ships and Shipping, Michael Grey, Ed., (MarAd-VM 393-R64R68) Maritime Administration, 1985.

"Ro-Ro Ships and Their Market Role—A Fairplay Report," Fairplay Publications Ltd., 1980.

Schorsch, J. J., Lawler, J. S., and Huntington, E., "Design and Operation of the *Ponce De Leon*," AIAA 2nd Advanced Marine Vehicle and Propulsion Meeting, May 1969.

Ship Design and Construction, SNAME, 1980.

Shipping World and Shipbuilder, Marine Publications International, 1978–1987.

Charles Zeien is a SNAME Fellow and former Vice President. From 1961 to 1972 he was Vice President of Sun Shipbuilding & Dry Dock Company, builder of many unusual ships, including RO/RO's and containerships and the Hughes' *Glomar Explorer*. The tanker *Manhattan* was converted to icebreaking service there, as well. From 1972 until 1985 he was President of the J.J. Henry Co., a naval architecture firm engaged in the design of innovative major commercial vessels, including the SL-7s. He is now the Principal of his own consulting firm, Charles Zeien Associates.

The LASH and "Seabee" Barge Carriers

Stuart W. Thayer

Introduction and Overall Perspective

There are many mystical stories in marine history—from the lost city of Atlantis to tall tales of the Bermuda Triangle. One of the most amazing events, however, was the conception and design of the barge carrying ships LASH (Lighter Aboard SHip) and "Seabee."

The two concepts originated in the same city, only blocks apart, for ships to load, carry and discharge barges, yet by methods that were entirely different. Most astounding! This occurred at about the same time without either party being aware of the other's ideas. LASH was the brainchild of Jerome Goldman, President of Friede and Goldman, an established firm of naval architects and marine engineers. "Seabee" on the other hand, was visualized by Frank Nemec, President of Lykes Bros. Steamship Co., Inc., one of America's major shipowners and operators. (Nemec invented the word "Seabee," stating that, "Like the Navy Seabees, this barge carrier can 'do anything, anywhere, any time'.")

The city of New Orleans, where barge carrying ships were conceived, is located not far from the delta of the Mississippi River, in the Gulf of Mexico. This is one end of the United States' central waterways system, which stretches thousands of miles from the Gulf to Chicago, Pittsburgh, Omaha, Minneapolis and cities in between. New Orleans is also at the hub of the Gulf Intracoastal Waterways System, which spans the coast from Brownsville, Texas to Apalachicola Bay, Florida. Moreover, Northern Europe also has many miles of navigable rivers and waterways suitable for barging. The availability of this tremendous source of barge-carried commodities inspired investigations into oceangoing barge-carrying ships from and to New Orleans and other Gulf of Mexico ports. The ability to move cargo in its original containment package (the barge), without rehandling, to inland facilities unreachable by ship was a great attraction to both shipper and transporter. Barge-carrying ships, because of their large unit loads, are able to load or discharge up to 3000 long tons per hour, and so carry the advantage of rapid port turnaround.

Fig. 1 The ABS-classed LASH vessel *Green Island*, one of the U.S.-flag ships in the ISC fleet, loaded with barges. The gantry crane, secured for travel, can be seen amidships.

The LASH ship is similar in concept to a self-loading/discharging containership with large floating containers (barges). Each loaded barge, weighing up to 376 tonnes (369 long tons), is lifted at the stern of the ship by a large gantry crane that then moves forward and lowers it into a vertical cell exactly like that in a containership. Barges can be stacked up to four-high in the hold and two-high on the hatch covers. LASH ships, in general, were constructed in long and short versions, with the former carrying a maximum of 89 and the latter 83 barges. There were also many combinations of barge and container combinations depending on the needs of the operator. LASH vessels have been used in international trade from the East, Gulf and West Coasts of the United States. The versatility of the LASH ships has been demonstrated by their use as bulk grain carriers and active operation as converted pure containers carriers.

There have been at least 29 LASH ships built, including one for the Soviet Union that was nuclear powered. (See Appendices A and C.) Avondale Shipyards, Inc. constructed eleven each of the LASH and jumbo LASH types. The building of foreign LASH vessels led to the design and construction of special feeder vessels, mostly of the float-on/float-off variety. (The feeder vessels were semi-submersibles and the LASH barges were floated either through the side or over the stern.)

The "Seabee" is not unlike a self-loading/discharging rail car carrier with long open horizontal decks. Its cargo handling equipment includes a stern elevator capable of raising or lowering two barges each weighing up to 2032 tonnes (2000 long tons), and two electrically powered transporters to move the barges along each deck. Two lower decks, each 183 meters (600 ft) long, and one weather deck 213 meters (700 ft) long, can hold a total of 38 "Seabee" barges. On the weather deck, container supporting adapter frames interface with the barge stowage positions to allow carrying 20- and 40-ft containers totaling 730 20-foot equivalent units (TEUs). Container support beams are also available for individual barges, allowing each to carry sixteen 40-ft containers above the hatch covers while they are in tow or stored on the ship's weather deck. The three American "Seabees" interchanged barges between the European and Mississippi River systems for more than 15 years and carried over 25 million tons of barges and cargo during that time.

Two "Seabee" barge carriers and two feeder vessels were constructed by Valmet Oy, Finland. (See Appendices A and C.)

LASH and "Seabee" barge characteristics are contained in Appendix B. LASH barge dimensions were dictated by certain European waterway widths, while "Seabee" barges are the same size as a half-length standard Mississippi River hopper barge.

Fig. 2 *Benjamin Harrison* is the first of two barge carriers built to ABS class at Avondale Shipyards, Inc.

Fig. 3 Handling containers with shipboard equipment or dropping lighters without tying up are options that LASH/Seabee operators value.

The three "Seabees" were bought by the Maritime Administration, Department of Transportation, in 1986 after completing 352 round trips between the Gulf of Mexico and Europe, averaging 42 days per voyage. In 1991 they took part in Operation Desert Storm, and at the time of this writing are in the Ready Reserve Fleet just south of New Orleans LA. The three diesel-powered LASH ships are still in commercial service, operated by Central Gulf Lines. Several of the Avondale-built, steam-powered LASH vessels are operating under charter to the Military Sealift Command.

(Bibliography appears on page 239)

Appendix A: LASH, Seabee, and Related Vessels

Type	Original Ship Name	Original Owner	Present Ship Name	Year	Present Owner	Builder	Hull No.	Del.
			American Seabee Ships					
1A	Doctor Lykes	Lykes Bros.	Cape Mendocino	1986	USA–DOT	Gen. Dyn. Quincy	17	1972
2A	Almeria Lykes	Lykes Bros.	Cape May	1986	USA–DOT	Gen. Dyn. Quincy	18	1972
3A	Tillie Lykes	Lykes Bros.	Cape Mohican	1986	USA–DOT	Gen. Dyn. Quincy	19	1973
			Foreign Seabee-Type Ships					
1B	Yulius Fuchik	Soviet Danube	Yulius Fuchik	—	Soviet Danube	Valmet Oy	290	1978
2B	Tibor Szamueli	Soviet Danube	Tibor Szamueli	—	Soviet Danube	Valmet Oy	291	1979
			American LASH Ships					
1C	Lash Italia	Prudential	Lash Italia	—	scrapped	Avondale	1184	1970
2C	Lash Turkiye	Prudential	Cape Florida	1987	USA–DOT	Avondale	1185	1970
3C	Lash Espana	Prudential	Austral Lightning	1976	USA–DOT	Avondale	1186	1971
4C	Lash Atlantico	Prudential	Lash Atlantico	—	Coastal Carriers	Avondale	1192	1974
5C	Lash Pacifico	Prudential	American Kestral	1988	Kestral Shipbldg.	Avondale	1193	1974
6C	Thomas E. Cuffe	PFE Line	President Hoover	1979	APL	Avondale	1187	1971
7C	Golden Bear	PFE Line	President Grant	1979	APL	Avondale	1188	1971
8C	Pacific Bear	PFE Line	President Harrison	1986	Calif. 1st IB	Avondale	1189	1971
9C	Japan Bear	PFE Line	President Tyler	1979	APL	Avondale	1190	1972
10C	China Bear	PFE Line	Austral Rainbow	1976	Farrell	Avondale	1191	1972
11C	Philippine Bear	PFE Line	American Veteran	1984	USA–DOT	Avondale	1194	1973
1D	Delta Mar	Delta	Cape Farewell	1987	USA–DOT	Avondale	1941	1973
2D	Delta Sud	Delta	Delta Sud	—	U.S. Lines	Avondale	1942	1973
3D	Delta Norte	Delta	Cape Flattery	1987	USA–DOT	Avondale	1943	1973
4D	Robert E. Lee	Waterman	Robert E. Lee	—	FNB Mobile	Avondale	1949	1974
5D	Stonewall Jackson	Waterman	Stonewall Jackson	—	FNB Mobile	Avondale	1950	1974
6D	Sam Houston	Waterman	Sam Houston	—	Waterman	Avondale	1951	1974
7D	Green Valley	Central Gulf	Green Valley	—	Central Gulf	Avondale	1952	1974
8D	Green Island	Central Gulf	Green Island	—	Central Gulf	Avondale	2258	1975
9D	Green Harbour	Central Gulf	Green Harbour	—	Central Gulf	Avondale	2257	1974
10D	Benjamin Harrison	Waterman	Benjamin Harrison	—	Sealand	Avondale	2307	1980
11D	Edward Rutledge	Waterman	Edward Rutledge	—	Sealand	Avondale	2308	1981
			Foreign LASH Ships					
1E	Acadia Forest	Central Gulf	Acadia Forest	—	Lash Carriers	Sumitomo	918	1969
2E	Atlantic Forest	Central Gulf	Atlantic Forest	—	Lash Carriers	Sumitomo	928	1970
1F	Bilderdyk	Holland America	Rhine Forest	1986	Lash Carriers	Cockerill	859	1971
2F	Munchen	Holland America	Lost At Sea	1978	Holland America	Cockerill	—	1971
1G	Aleksey Kosygin	USSR	Aleksey Kosygin	—	USSR FESCO	Kherson Sy	—	1984
2G	Indira Ghandi	USSR	Indira Ghandi	—	USSR FESCO	Kherson Sy	—	1985
1H	Sevmorput	USSR	Sevmorput	—	USSR Murmansk	Zaliv Kerch	—	1988
			Foreign Seabee Feeder Ships					
1I	Boris Polevoy	USSR	Boris Polevoy	—	USSR Danube	Valmet	314	1984
2I	Pavel Antokolsky	USSR	Pavel Antokolsky	—	USSR Danube	Valmet	315	1984
			Foreign LASH Feeder Ships					
1J	Stakhanovets Kotov	USSR	Stakhanovets Kotov	—	USSR Black Sea	Hollming Oy	217	1978
2J	Stakhanovets Yermolenko	USSR	Stakhanovets Yermolenko	—	USSR Black Sea	Hollming Oy	218	1978
3J	Stakhanovets Petrash	USSR	Stakhanovets Petrash	—	USSR Black Sea	Hollming Oy	219	1979
1K	Bacoliner 1	Bacoliner GMBH	Bacoliner 1	—	Bacoliner GMBH	Thyssen NSW	457	1979
2K	Bacoliner 2	Bacoliner GMBH	Bacoliner 2	—	Bacoliner GMBH	Thyssen NSW	458	1980
1L	Bacoliner 3	Bacoliner GMBH	Bacoliner 3	—	Bacoliner GMBH	Thyssen NSW	475	1984
1M	Mammouth Spruce	Central Gulf	Spruce	1978	Lash Carriers	Sumitomo	40100	1975
1N	Mammouth Oak	Central Gulf	Hong He Kou	1983	COSCO	Sumitomo	1063	1978
2N	Mammouth Willow	Central Gulf	Sha He Kou	1983	COSCO	Sumitomo	1064	1978

Appendix B: Barges

Barge Characteristics	LASH		Seabee	
		Standard Hopper Barge[a]		
Length overall (LOA)	18.74 m	61.50 ft	29.70 m	97.50 ft
Beam (B)	9.50 m	31.17 ft	10.67 m	35.00 ft
Depth (D)	3.06 m	13.00 ft	5.16 m	16.92 ft
Maximum draft SW (H)	2.66 m	8.73 ft	3.22 m	10.56 ft
Deadweight (DWT)	376 t[b]	369 LT[c]	847 t	834 LT
Lightweight	87 t	86 LT	169 t	166 LT
Displacement	463 t	455 LT	1016 t	1000 LT
Bale cubic capacity	563 m^3	19,900 ft^3	1091 m^3	38,546 ft^3
Number of barges constructed	6000		246	
		Heavy Lift Flatdeck Barge[d]		
Length overall (LOA)			34.29 m	112.50 ft
Beam (B)			10.67 m	35.00 ft
Depth (D)			3.35 m	11.00 ft
Maximum draft SW (H)			2.82 m	9.25 ft
Deadweight (DWT)			757 t	745 LT
Lightweight			259 t	255 LT
Displacement			1016 t	1000 LT
Number of barges constructed	N/A		3	

[a] Four Seabee hopper barges were modified for carriage of refrigerated chilled fruit. This reduced the cargo deadweight to 768 t (756 LT) and the bale cubic to 862 m^3 (30,446 ft^3).
[b] t is the abbreviation for tonne.
[c] LT is the abbreviation for long ton.
[d] Allowable deck loads of 337 kg/m^2 (8000 lb/ft^2).

Appendix C: LASH and Seabee Barge Carriers—Characteristics

SEABEE Barge Carrier

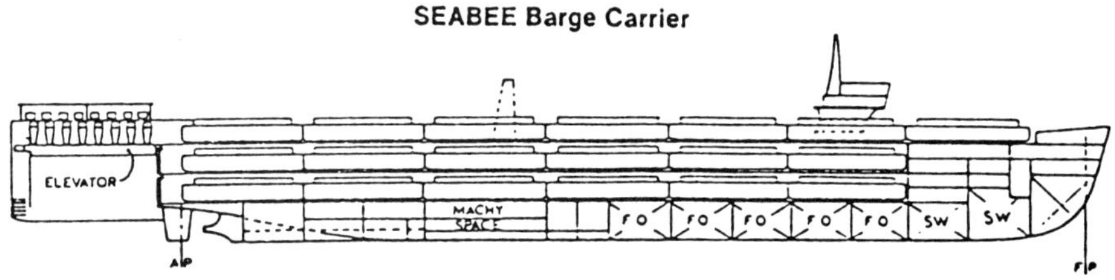

LASH Barge Carrier

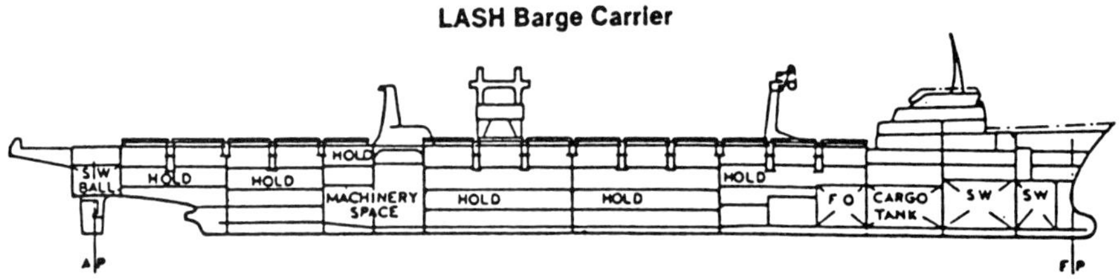

	Type	No. of Ships	LOA, m (ft)	LBP, m (ft)	Beam, m (ft)	Depth, m (ft)	Draft, m (ft)	DWT, t (LT)	Gross Tonnage	Net Tonnage	Speed, knots	Max. hp	Type Plant	No. of Props.
A	Seabee	3	266.99	219.93	32.40	22.80	11.92	39,010	21,667	12,024	20.0	36,000	Stm. turb./Gen. Elec.	1
			(876.00)	(721.60)	(106.30)	(74.80)	(39.10)	(38,400)						
B	Foreign Seabee	2	266.44	222.80	34.99	22.95	10.97	38,451	35,817	15,479	19.8	41,600	Diesel/4XPC2.5V4	2 CP
			(874.20)	(731.00)	(114.80)	(75.30)	(36.00)	(37,850)						
C	LASH	11	249.92	220.66	30.48	18.29	10.70	30,293	26,407	18,706	22.5	32,000	Stm. turb./DeLaval	1
			(820.00)	(724.00)	(100.00)	(60.00)	(35.10)	(29,820)						
D	Jumbo LASH	11	272.26	243.01	30.48	18.29	11.58	41,041	30,400	22,700	22.0	32,000	Stm. turb./DeLaval	1
			(893.30)	(797.30)	(100.00)	(60.00)	(38.00)	(40,400)						
E	Foreign LASH	2	261.41	235.20	32.61	18.29	12.13	43,683	33,221	22,846	19.1	26,000	Diesel/9RND90	1
			(857.70)	(771.70)	(107.00)	(60.00)	(39.80)	(43,000)						

	Type	No. of Ships	LOA, m (ft)	LBP, m (ft)	Beam, m (ft)	Depth, m (ft)	Draft, m (ft)	DWT, t (LT)	Gross Tonnage	Net Tonnage	Speed, knots	Max. hp	Type Plant	No. of Props.
F	Foreign LASH	2	261.41 (857.70)	230.97 (757.80)	32.28 (105.90)	18.29 (60.00)	11.28 (37.00)	45,510 (44,799)	31,386	20,940	18.0	26,100	Diesel/9RND90	1
G	Foreign LASH	2	262.91 (862.60)	232.00 (761.20)	32.19 (105.60)	18.32 (60.10)	11.64 (38.20)	41,530 (40,881)	37,464	11,239	22.0	33,580	Diesel/2X7DKRN80	2 CP
H	Foreign LASH	1	260.29 (854.00)	236.60 (776.30)	32.19 (105.60)	18.32 (60.10)	11.80 (38.70)	34,519 (33,980)	38,226	11,468	20.0	29,399	Stm. turb./nuclear	1 CP
I	Seabee Feeder	2	159.10 (522.00)	141.21 (463.30)	31.03 (101.80)	15.48 (50.80)	4.39 (14.40)	8909 (8770)	10,084	3206	13.5	3206	Diesel/2X ??	2 CP
J	LASH Feeder	3	139.41 (457.40)	117.19 (384.50)	20.24 (66.40)	12.59 (41.30)	6.28 (20.60)	5808 (5717)	4028	1200	14.0	6420	Diesel/2X6PC2.3L	2 CP
K	LASH Feeder	2	203.69 (668.30)	195.00 (639.80)	28.53 (93.60)	11.25 (36.90)	6.64 (21.80)	22,146 (21,800)	22,345	6703	15.0	9680	Diesel/8L55GF	1
L	LASH Feeder	1	204.85 (672.10)	195.06 (640.00)	28.53 (93.60)	14.20 (46.60)	4.60 (15.10)	22,117 (21,771)	22,528	6758	15.0	10,000	Diesel/7L55GFCA	1
M	LASH Feeder	1	112.47 (369.00)	103.75 (340.40)	34.23 (112.30)	8.11 (26.60)	6.49 (21.30)	8302 (8172)	1380	362	—	4300	Diesel/12-645	2
N	LASH Feeder	2	134.53 (441.40)	125.02 (410.20)	34.20 (112.20)	7.53 (24.70)	5.15 (16.90)	13,440 (13,230)	1856	807	10.3	5880	Diesel/6PC2.2L	2

Bibliography

"Characteristics and Index of Maritime Administration Ship Design," U.S. Department of Transportation, Maritime Administration, Aug. 1987.

Dillon, E. S., Hoffmann, L. C., and Roseman, D. P., "Forty Years of Ship Design Under the Merchant Marine Act 1936–1976, *Trans.* SNAME, Vol. 84, 1976, p. 169.

"Innovation in the Maritime Industry," National Research Council, Maritime Transportation Research Board, 1978, Appendix, p. 73.

Koskivirta, V., Heikkila, V., Niini, M., and Harjuvaara, H., "New Finnish Barge Carriers for the U.S.S.R.," *Trans.* SNAME, Vol. 86, 1978, p. 304.

"The Motor Ship," Reed Business Publishing Group, 1970–1990.

"Ship Design and Construction," SNAME, 1980.

"Shipping World and Shipbuilder," Maritime Publications International, Ltd., 1978–1987.

Stuart W. Thayer is a Life Fellow and Honorary Vice President of SNAME. He is Vice President-Engineering of the Lykes Bros. Steamship Co., Inc., of New Orleans. With Lykes he supervised the design and construction of more than 40 ships, including three "Seabees." He also served in the late 70s as a consultant to Valmet Oy (Finland) during construction of two "Seabees" for the Soviet Union.

Rebuilding the Great Lakes U.S. Cargo Fleet and the Metamorphosis of the Self-Unloader

George H. Palmer, Jr., Carlton E. Tripp, Richard Suehrstedt, and Joseph Fischer

Introduction

A design that becomes dominant through the years for vessels in a particular trade attains a certain dignity. The classic Great Lakes bulk freighter was easily recognized by its long, low profile and bluff bow, its prominent forecastle and forward deckhouse surmounted by its distinctive pilot house, and by its machinery spaces back aft topped by a tall funnel, which enhanced the draft through the coal burning boilers.

The professional career of SNAME Fellow Hugh C. Downer has encompassed responsible positions in all areas of the marine bulk transport industry: Great Lakes, inland rivers, and deep water service. In commenting on this chapter he said, "Over the years I have realized that practically every dry bulk handling innovation originated on the Great Lakes. The explosion in dry bulk transportation on salt water, beginning in the 1950s, was made possible by material handling methods developed on the Great Lakes. Even before that, when I was sailing salt water in the early 1930s, [I noted that] the Bethlehem ore loading facilities at Cruz Grande, Chile, were patterned after those on the Great Lakes."

These characteristics had been developed by the start of the 20th Century to suit the needs and constraints of the iron ore and coal trades. The basic design evolved from conventional wooden and iron sailing and steam propelled vessels used on the lakes in general cargo trade during the 19th Century. To adapt to bulk commodities, 'tween decks were omitted, double bottoms were added, and bilges were protected with ceiling or with side tanks. In addition, a large number of hatchways were installed in the weather deck to match the spacing of ore dock loading chutes and shore-based unloading rigs.

These vessels became known as "straight deckers," in contradistinction to the self-unloaders that evolved in more recent times, and the vestigial term "spar deck" still refers to the weather deck on Great Lakes bulk carriers. The "main deck" has shrunk to a shell stringer at mid-depth in the hold.

A pilot house with unimpeded 360 degree visibility at the bow gave the master a close view of locks and docks, and facilitated the lake practice of steering by looking astern to align the funnel with the wake.

These early lakers were also characterized by the considerable sheer and camber of weather decks, tumblehome (to reduce contact damage at locks and docks), and deadrise plus ample bilge radius (to reduce grounding damage in narrow, shallow channels). These characteristics, dear to the aesthetic eye of the naval architect, became less important as vessel mass and depth increased and all but disappeared as the design of the ultimate huge barge-like self-unloaders developed. Such features were costly to build and brought little practical advantage.

In the past half-century the most significant changes are not in individual special-purpose vessels, but in typical vessels of the Great Lakes bulk carrier and self-unloader fleets as a whole. The physical limitations of the channels and locks, and the economic and political constraints on transport of bulk materials within the Great Lakes watershed and St. Lawrence River defined the size and operating limits of these formerly landlocked ships. Major changes were gradually introduced in order for the fleet to meet economically a fairly constant demand, averaging 200 million tons per year in domestic trade.

These changes affected both ship designs and operating procedures and resulted in an increase in average trip capacity per vessel from less than 10,000 DWT in 1943 to over 60,000 tons in the latest self-unloaders. They occurred in five phases, each involving a major condition or building program. Each

Fig. 1 The tug/barge *Presque Isle* de-icing at the loading dock at Two Harbors, MN. (From the Great Lakes Historical Society Collection.)

phase was a consequence of at least one significant political or economic event and a multitude of minor developments. All phases overlapped. The one common inducement for change was any new opportunity to increase unit capacity. These events were varied, with political factors that affected trade patterns as in the opening of the St. Lawrence Seaway, economic pressures as in the reorganization of the railroads, labor unrest, and technological advances such as improvements in construction methods, powering and cargo handling. Discussions of cause and effect and typical characteristics of prototype vessels relating to each phase are presented below.

Owing to limits on space, we have made no attempt to discuss developments effected by our Canadian friends. Let us simply say that technology north of the border developed more or less in parallel with that in the United States.

Phase I (1941–1946): The War Years

At the beginning of 1943 there were about 350 active vessels in the Great Lakes bulk ore carrier and self-unloader fleets, with total trip capacity of about three million long tons. Of these, most were old vessels, some dating back to before the turn of the century. Limited in draft to 20 ft by the channels at the Soo locks, most were of 10,000 tons dwt or less, and were powered by steam reciprocating engines of between 1100 and 2600 ihp. There were a few exceptions such as the five 4400 SHP turbine driven 14,000 dwt ore carriers of the U.S. Steel Corporation's *Benjamin Fairless*-class, completed in 1942.

In 1943, to ensure wartime supply of ore to the steel mills in the Midwest, 16 government sponsored bulk carriers (U.S. Maritime Commission design L6) were delivered to the various major shipowners. Six were from American Ship Building Co. (AmShip) yards in Lorain OH and Cleveland. Ten were from the Great Lakes Engineering Works yards in Ecorse MI and Ashtabula OH. These ships followed traditional Great Lakes designs with a mixture of welded and riveted structural details. Two distinct hull and machinery designs were used. AmShip provided 2500 ihp four-cylinder compound steam engines with Lentz valves. Great Lakes Engineering Works used triple-expansion engines.

Typically, these war built ships had a combination of transverse framing on the side shell, with longitudinal framing on the deck and bottom, and on the sheer strake and upper side shell plating. Transverse side shell framing provided support for contact with dock and lock walls, a practice continued in postwar construction. In the earlier postwar hulls, the deck and shell butts were welded, while the seams and frames were riveted. Gradually, as owners and builders became more accustomed to welded construction (and after overcoming problems with fractures), riveted seams and frames were abandoned in favor of welded connections.

Since most of the welded repair work and much of the new construction in Great Lakes yards is done

Fig. 2 The *McKee Sons* is a C4 type ocean freighter converted to a Great Lakes self-unloader. (From the Great Lakes Historical Society Collection.)

under frigid winter conditions, considerable experimentation in welding procedure was needed to determine the lower practical limit on ambient temperatures. Consequently, substitution of welding in lieu of riveting was generally accepted somewhat later on the lakes than in more temperate parts of the nation. Moreover, owing to the great longevity of Great Lakes ships, riveters were needed to carry out repairs.

Phase II (1945–1961): Postwar Repairs and Interim Conversions of Ocean Vessels

After the end of hostilities (August 1945), the demand for steel remained near the production limit of the mills, and the Great Lakes ore fleet was operating at full capacity. After years of minimal repairs, the condition of the over-aged vessels (nearly 50 percent of the fleet) was critical, and repair facilities and drydocks were barely adequate. Ships were repaired in the winter when ice closed the lakes to navigation. Cargo docks became repair berths, and crew members became shipyard workers during that period, signing back on their individual ships in the spring. Consequently, shipyards were reluctant to expand their facilities and maintained only enough labor force to provide for emergency repairs during the sailing season.

Faced with the obsolescence of the older vessels, yet uncertain of the future effect of the proposed St. Lawrence Seaway and the adequacy of the domestic ore supply, owners undertook programs of necessary repairs to eke out a few more years of service from existing vessels. These programs included replacement of side tanks, tank top and hold bulkhead structures, replacement of inadequate machinery, and also reboilering. Upgrading of electrical equipment was prompted by the plentiful supply of war surplus and salvaged small turbo-generators, switchboards and pumps. Accommodations and galleys were improved to increase the comfort of the crew and encourage retention.

A replacement program was started by shipowners, but the capacity of the Great Lakes shipyards was inadequate to meet the immediate needs. As an interim measure, wartime built surplus ocean vessels were purchased and, of these, eight were converted in coastal shipyards to Great Lakes configuration and towed up the Mississippi River system to Chicago to be completed. After 1959 ocean ships reached the lake shipyards directly via the St Lawrence Seaway. These vessels, although over-powered and of sub-optimal capacity, were quickly available and augmented the fleet for many years and until they were superseded by more efficient units. These reserve fleet oceangoing ships included a VC-2 Victory Ship, six C4 cargo ships, three T-2 tankers, two T-3 tankers, and one C1-M-AV1 cargo ship. Hulls were altered to suit contemporary cargo hold and hatchway configurations of Great Lakes bulk carriers. Two were rebuilt as self-unloaders and the C1-M-AV1 as a self unloading bulk cement carrier. One of the C4's was completed as a passenger excursion ship.

Phase III (1950–1960): New Construction—Conventional Great Lakes Design

The decade starting in 1950 was not one of confidence for the shipyards on the Great Lakes. They were obsolete, and had only a small, skilled labor force, which was seasonally simultaneously employed in both new construction and repairs. This arrangement did not promote efficient cost control. The four major shipyards, Manitowoc Shipbuilding, Inc., American Ship Building Company, Defoe Shipbuilding Company and the Great Lakes Engineering Works had each developed its own standard details and designs over the years. Another major yard on the Lakes, Fraser Shipyards, Inc., located at Superior WI, chose to confine itself to upgrading and repowering contracts, and so did not participate in new construction. In the following decades Fraser became a major contractor for midbodies, as well as lengthening and converting to self-unloaders as described in Phase IV.

In 1953, there were 285 ships in the U.S. bulk fleet owned or operated by 23 domestic companies. The majority of these ships were over-aged and obsolete.

Owners, needing to keep in the shipbuilders' good graces because facilities were limited, tended to spread single-ship contracts among them in turn. Of these shipyards, Great Lakes Engineering Works was the most conservative, American Ship the most diversified, and Manitowoc in some ways the most innovative, specializing in self-unloaders. Defoe specialized in naval vessels. Unless there was good probability of continuing contracts for new construction, the shipyards were not disposed to make extensive plant improvements. Owners, however, were faced with major uncertainties, which made them reluctant to contract for multiple orders. New contracts were initially parcelled out to the major yards preserving the traditional status quo.

The Wilfred Sykes *boasted many unique features and "firsts" on the Great Lakes*

In size:

- *At 678 ft LOA, it was the longest ship on the lakes. In the entire U.S. merchant marine, only the 723-ft* America *was longer.*
- *First Great Lakes ship with a beam as great as 70 ft.*

In speed:

- *First lake ship designed for 16 mph speed.*

Machinery:

- *With 7000 shp it had the most powerful machinery on the lakes. The next most powerful plant was only 4400 shp.*
- *First lake steamer to be oil-fired, to be fitted with an AC electric plant, and to have switchboard and magnetic starters in a central location.*
- *First to have full mechanical ventilation in the machinery spaces.*

Hull form:

- *First with a full contra-guide stern.*

Arrangements and accommodations:

- *First on lakes with poop instead of after deckhouse.*
- *First with interior access to crew cabins.*
- *First with private toilet and shower for each crew member.*
- *First with mechanical ventilation throughout.*

On deck:

- *First ore carrier fitted with Kestner quick-acting hatch clamps.*
- *First unstayed masts.*
- *First unstayed stack.*

Primary uncertainties facing the bulk trades were the future of the Midwestern steel mills and the exhaustion of Minnesota's high-grade Mesabi Range ore. A secondary uncertainty, but no less foreboding to investors, was the inevitable competition from foreign steel and ore sources owing to the eventual opening of the St Lawrence Seaway.

The postwar new building on the lakes got off to a slow but significant start in January 1950 with delivery of the *Wilfred Sykes* from the American Ship Building Company (Lorain) to the Inland Steel Co. The dimensions of the ship were the maximum considered feasible, given the limitations at the loading and unloading ports, the locks and channels, the building facilities, and consideration of good proportions. The *Sykes* was 678 ft LOA, 70 ft in beam, and 37 ft deep. Its capacity and power were about half again greater than on any earlier lake carrier, and it set the pattern for the new construction in the 1950s.

The general arrangement and major structural details were similar to time tested prewar designs except for replacement of the after deckhouse with a poop superstructure, which reduced the freeboard in accordance with the current load line regulations. Anticipating improvements in the limiting dimensions of locks, channels, and terminal facilities, the *Sykes* was designed with deadweight capacity it could not immediately use at the prevailing drafts. Consequently the next vessels built were not true sister ships to the *Sykes*, but were 40 ft shorter and had slight changes of arrangement and structural details depending on owner's preferences and shipyard building practices.

During 1952 and 1953 eight ships were initially produced in three Lake Erie shipyards of the American Ship Building Company (the lead yard) and Great Lakes Engineering Works for five different owners. These followed a basic design used on the *Sykes*, but had principal dimensions of 620 ft LOA to suit prevailing port terminals and turning basins, 70 ft beam and 36 ft depth, giving about 19,500 dwt (long tons) at prevailing summer drafts. Concurrently, two smaller but similar vessels were built at the Defoe Shipbuilding Company in Bay City MI for two different owners active in the grain trade. These ships were of 17,500 dwt, with LOAs of 613 and 617 ft, beams of 67 ft, and depths of 35 ft.

A group of three 18,000 dwt ships were designed and built by Bethlehem Steel Corporation's Sparrows Point shipyard. These measured 610 ft LOA, 70 ft beam, and 37 ft depth. They were delivered from Maryland to the lakes by way of the Mississippi River system.

By 1953 owners had become accustomed to the 75-ft beam of the C-4 ocean vessel conversions and the unusual length of the *Sykes*. With concurrent enlargements of the building berths, the shipyards and owners altered the design of ships built in the following years to suit the desires of the owners for greater capacity and greater power. Overall lengths were increased progressively until 1960, when the last new conventional ship of this construction phase, the *Edward L. Ryerson*, was built. It had a deadweight of about 26,000 long tons. It had an overall length of 730 ft, a beam of 75 ft, and a depth of 39 ft. The power had by now increased to 9900 shp.

Because oil-fired boilers had become relatively trouble-free, three of the Cleveland Cliffs ore carriers had been fitted with but single boilers, thus gaining in thermal efficiency and ease of control. This was a radical departure from existing practice, but it worked well. John Horton, fleet engineer, was heard to remark to a boiler salesman, "One boiler is so much better than two that our next ship will have none at all."

At the time of this construction phase the U.S. merchant marine was still committed to steam propulsion. Moreover, following the massive wartime efforts, boilers, turbines and reduction gears were available in plentiful supply. These were domestically built and of proven design. Also Great Lakes engineers for generations had been accustomed to steam propulsion, albeit mostly with coal burning boilers and reciprocating engines. All of these factors came down in favor of steam turbine propulsion during this third phase.

Economic studies and model tests led the owners and designers of the *Wilfred Sykes* to select a speed of 16 mph and a deadweight of 21,000 tons. A single-screw power plant of 7700 shp was selected. It featured a geared cross compound steam turbine that powered the propeller at 100 rpm. Boilers were two-drum D type, oil burning, producing superheated steam at 450 psi and 750 °F. The auxiliary machinery was electric with 440-volt, three-phase power supplied by two 500 kW generators. Although coal was still a major cargo on the lakes it was decided that the new vessel would burn Bunker C residual oil, then the customary fuel for the U.S. merchant marine. Significantly, the *Wilfred Sykes*' base terminal was Inland Steel Company's mill at Indiana Harbor, which was adjacent to the major oil refineries in the Chicago area. Pipelines had been extended from the Gulf of Mexico and from western Canada to the midwest with major bunkering depots at Chicago, Toledo and Sarnia, Ontario. Considering that the coal mining industry had considerable financial interests in these shipping companies, some of the new ships of that time were fitted with boilers that could be converted to burn either coal or oil. Eight were initially fitted to burn coal; but, following labor difficulties at the coal mines in 1970, most lake vessels were refitted to burn oil. Ultimately coal bunkering facilities along the Great Lakes were phased out.

Construction of the *Wilfred Sykes* and its succeeding group of ships followed the customary Great Lakes practice of building on a level bank and side launching into a graving dock, slipway or river. Shell plating seams and frame connections were riveted and plate butts welded. Erection of shell plating was strake by strake, from the keel outward and upward. Internal structure was generally welded with size of subassemblies limited by the shipyard's generally modest crane capacity.

As the new construction program progressed, shipyards improved their facilities and equipment to handle larger assemblies. American Ship, for example, retired its three- and six-ton cranes in favor of 35-ton units. Ship dimensions were increased and arrangements were altered to suit individual owners' changing needs.

During this phase only four ships were built between 1952 and 1959 as self-unloaders. These ships, all built at Manitowoc, were similar in hull lines to the contemporary "straight deckers" of the period except they were two ft greater in beam and up to 4 ft greater in depth. Propulsive power was the same, but they were fitted with larger generators to provide for unloading demands. Arrangement of hoppered cargo holds, hold conveyors, bucket elevator and the conveyor boom and "A" frame forward were typical of pre-World War II design. No further new construction of self-unloaders was undertaken in this phase since the augmentation by converted ocean vessels, and small existing lake ore carriers provided adequate capacity in the stone and coal trades. This held true until the 1970s, when the development of taconite pellets made the general use of self-unloaders feasible in the ore trade, as described in Phases IV and V.

The *Cason J. Callaway* was delivered by the Great Lakes Engineering Works to the Pittsburgh Steamship Company in 1952, followed by the *John G. Munson*, the first post-war new self-unloader delivered by Manitowoc to the Michigan Limestone Division of U.S. Steel Corporation in 1952, and the *Richard M.*

Fig. 3 Prototype conventional ore carrier *Wilfred Sykes* at Lorain, OH. (From the Great Lakes Historical Society Collection.)

Marshall from Defoe to the Great Lakes Steamship Co. in 1953. The *Johnstown*, built for Bethlehem Steel's Great Lakes fleet by Bethlehem-Sparrows Point Shipyard in Maryland in 1952, was the first of three new hulls built at that yard, dismantled and towed up the Mississippi for completion at Chicago. These were the lead vessels from each of these yards in the Phase III group of 25 conventional ships built in U.S. yards for Great Lakes service from 1950 through 1960.

Phase IV (1950–1991): Alterations to Existing Vessels for Greater Capacity and Improvements to Self-Unloading Techniques

During the four decades since 1950 there occurred several unrelated events that completely upset the traditional operation of the bulk carrier fleet: the exhaustion of the rich deposits of hematite ore in the iron range in Minnesota; the opening of the St. Lawrence Seaway in 1959; and labor strife in the coal mines. The dilemma was that massive new investment was needed throughout the steel industry and the transportation chain. Mines, railroads, loading and unloading facilities all competed with ships and barges for limited funds. Industry leaders decided to invest their limited capital to rehabilitate the steel mills and mines rather than to continue the ship renewal program. This led the shipping lines to set about getting added capacity by lengthening existing vessels to take advantage of the 730-ft length to be allowed in the St Lawrence Seaway (anticipated to be open in 1958) and matching the 27-ft controlling depth of the locks and channels of the upper lakes.

The depletion of the rich open-pit hematite mines in the Mesabi Range prompted the development of the vast alternative deposits of the lower grade taconite and processing facilities for refining and concentrating the ore at or near the mines. That eased the iron reduction processes at the steel mill blast furnaces, requiring fewer tons of ore concentrate per ton of iron produced. This reduced the amount of ore that had to be moved down the lakes per ton of finished steel. Taconite pellets have another advantage in that they are easy to handle by conveyor, thus making them adaptable to transport in self-unloading ships. Moreover, they do not freeze at the loading docks like the wet, sticky hematite, thus permitting the shipping season to be extended into the winter, and further reducing the number of ships needed for the same volume of trade.

Great Lakes ore supply operations had been ideally balanced on a constant flow basis with iron ore and grain downbound and coal upbound. If there were enough hulls to meet the capacity of the unloading

Fig. 4 The *John G. Munson*, the first post-war new self-unloader delivered by Manitowoc to the Michigan Limestone Division of U.S. Steel Corporation. (From the Great Lakes Historical Society Collection.)

terminals, speed was of secondary consideration after fuel consumption. The power of the reciprocating steam engines fitted to ships in the ore trade was generally between 1100 and 2200 ihp, or just enough to maintain the normal fleet speed of 11.5 miles per hour. Simply put, owing to delays and congestion at terminals and locks, there was little point in having an open-lake speed any greater than that allowed in the rivers; and a ton of bunker coal saved was a ton more cargo carried.

Speed became more appropriate as the total number of ships decreased, the cargo capacity of each increased, and the turnaround times at the terminals shortened. A program of repowering the larger of certain prime ships built between 1910 and 1930 was undertaken. This involved upgrading the machinery to the technological level of the then-contemporary new construction. Uniflow reciprocating steam engines, steam turbines, and some diesels were installed, generally in the 2000 to 5500 horsepower range.

An increase in allowable draft for the existing vessels was obtained by adding to their longitudinal strength through the expedient of strapping deck, side and bottom shell plating. At the same time they overcame the freeboard penalties imposed by deficiencies in closures to hatchways and shell openings. These were penalties that had been applied when load lines had first been assigned to Great Lakes ships (about 1936). Several inches greater draft could be obtained by fitting gasketed steel hatch covers on higher coamings, and by other closure improvements. Capacity was also increased by lengthening existing ships to "Seaway size" of up to 730 ft, anticipating passage through the St Lawrence Seaway system when it would finally open in 1959.

Lengthening to over 600 ft had been planned into the conversion design of the ocean vessels for Great Lakes service. The *Cliffs Victory* (C2-S-VC2) conversion involved removing weather deck hatchways and hold structure and replacing them with lake-type hatchways, side tank, and tank top structure. The pilot house and deck officers' quarters were relocated on the forecastle deck. To give the maximum practical deadweight the ship was cut at the forward end of the parallel midbody and a 165 ft-6 in. midbody installed. For the initial C4 and tanker conversions similar alterations were made, except that the seagoing bow structures were discarded and new full sectioned lakes type bows were installed for greater buoyancy and transport capacity.

On the C1-M-AV1 and all subsequent lengthenings, new bows were not installed. These ships were towed up the Mississippi to Chicago between 1951 and 1953. Channel depth and bridge clearance limits in the Mississippi required that the ships be disassembled for the trip, with pontoons fitted aft to reduce draft, and reassembled and fitted out in the American Ship Building Company drydock in Chicago. Length and beam were similar to contemporary lake-built ships except for the *Joseph H. Thompson*, whose converted length of 696 ft was too great to negotiate the bends in the Chicago canal. The ship was delivered in two pieces to be joined in the drydock in Chicago.

The basic freeboard table of the Great Lakes load line regulations as established in 1935 applied to ships up to 620 ft in length. In 1966 the tables were extrapolated to the 750-ft length anticipated for the future. A more careful study of strength standards was deemed appropriate for the yet-longer ships that would be allowed by the 1200-ft Poe Lock at Sault Ste. Marie authorized by the Corps of Engineers in 1962.

Most naval architects believed that ships over 750 ft in length were not equitably covered by the existing Great Lakes regulations for freeboard and longitudinal strength. As a result, SNAME in 1964 undertook the coordination of hull stress data and the associated wave spectra. In 1966, the joint U.S. and Canadian Technical Committee on Great Lakes Load Lines was established by the Canadian Department of Transportation and the U.S. Coast Guard to study these matters, using the data being obtained and analyzed by the SNAME research program. Two working panels were established, one on freeboard, and the other on longitudinal strength. The proposed strength criteria that resulted were embodied in the report "Interim Strength Standards for Great Lakes Vessels," which was accepted by the Joint Technical Committee. The freeboard panel recommended extending the basic freeboard tables for ships over 750 ft up to 1000 ft in length by adopting the freeboard tables from the 1966 Load Line Convention for Type B ocean vessels. This allowed a lesser freeboard than would have resulted from straight extrapolation of the existing freeboard tables for lengths below 750 ft.

Revised load line regulations were published in 1971, containing the basic freeboard table extrapolated to 750 ft and also providing an alternative table for lengths from 440 to 1000 ft for ships of superior design, subject to certain design criteria and operational restrictions for controlling loading stresses.

By 1958, several of the "new" lake-built vessels and converted ocean ships were lengthened, adding new midbodies up to 72 ft in length. In 1961, after the opening of the Seaway, four more conversions from ocean tankers were made using European-built midbodies. After 1969, when new load line standards were accepted, several postwar ships were lengthened by as much as 120 ft. All in all, between 1950 and 1983 some 17 ships were lengthened. In cumulative transport capacity, this was the equivalent of three and a half new ships.

Self-unloading Great Lakes bulk carriers were well developed decades before World War II. Most were tramps and generally much smaller than ships in the iron ore trade, where greater capacity was an exploitable advantage in a dedicated trade between well equipped terminals on high-volume contracts. Self-unloaders carried most of the other dry bulk cargo trade, and this was largely between ports not on the main routes between Lake Superior and the lower lakes. One exception was the limestone trade, where the economies of scale led to self-unloaders comparable in size to the largest ore carriers. In total, there were about sixty self-unloaders in the 1950s. They were divided into two groups, one with traditional holds and two large revolving cranes fitted with magnets or clamshell buckets for use with pig iron or finished steel products as well as bulk commodities. The other type was fitted with a belt conveyor on a long boom on deck fed by a bucket elevator from the hold. Early vessels had a long single hold with a scraper to feed the elevator. Later improved arrangements had segregated holds with hopper-bottoms, which fed (by gravity) twin conveyor belts in tunnels beneath the hold.

Fig. 5 The post-war ore carrier *William Clay Ford* being lengthened 120 ft at Superior, WI. (Courtesy Fraser Shipyards.)

Conversion to gravity-type self-unloading, while decreasing the available cubic capacity of the hold, decreased the unloading time and so allowed more trips per year. The net result was a net increase in the ship's annual transport capacity. Significant evolution in self-unloading technology increased the unloading rate from 2000 tons per hour to 10,000 tons per hour.

The 1950s showed an extension of the prewar traditional approach with twin longitudinal conveyors carrying the cargo forward where it was transferred to a single centerline bucket elevator. The hold conveyors featured 25-degree idlers with fixed steel skirt boards and articulated gates with a drop lip intended to direct the cargo in the direction of belt travel and aid in centering the cargo on the belt. The return segment of the belt was completely covered by an idler deck to guard against spillage. Above deck, a long (200-ft) swiveling boom conveyor delivered the cargo ashore. The boom, consisting of a truss of rectangular cross section, was a typical civil engineering structure of steel angles and channels. Welding replaced the older riveted construction, but the design was similar. The topping lift, of multiple sheave wire rope tackle and winch, was supported by a massive "A" frame. Slewing was accomplished by winches with wire rope rigging and snatch blocks. Conveyor belts were powered by direct DC motors or stepped AC motors with wound rotors for speed control. The *Adam E. Cornelius* was a typical self-unloader of that period.

While the ore trade was dominated by "straight deckers" captive to the steel industry, the self-unloaders gained popularity in the coal and stone carriers that had no fixed routes. Independent carriers ran shorter routes to unimproved docks, up rivers and in remote harbors. They could put their cargo ashore while tied to a tree on the bank of a winding river. The electric generating capacity to power the unloading equipment was available to power side thrusters to maneuver the large "boats" without aid of tugs. These thrusters, which first appeared in the '50s had become standard in the '70s, with some ships featuring them at both ends.

An innovation appeared in 1958, when a 250-ft all-aluminum boom was made for the *J. R. Sensibar* to replace a shorter steel boom. The structural design was similar to past steel booms except the angle sections of 6061 T6 aluminum were fastened with cold-driven steel rivets.

In 1964 the *Richard Reiss* and *John T. Hutchinson* conversions each featured a multi-span boom with a steel base section supporting two hinged aluminum sections. Slewing was by a large chain drive with DC motors driving pinion sprockets on each side, located above deck.

In the conversions of the *Nicolet* and *Joseph H. Frantz* the high-maintenance bucket elevators were replaced by inclined belts that rose through the forward cargo holds in enclosures. To compensate for the loss of cargo cubic, a raised trunk was fitted above deck. A 250-ft boom truss was designed with a triangular cross section to reduce racking; and truss members were made of welded steel pipe. These vessels also pioneered the use of a hydraulic cylinder with fixed rigging for topping, and a hydraulic operated rack and pinion rotary actuator for slewing. These features have since become industry standards. The arrangement with the inclined belt passing through the hold had a single, rotating, cantilevered, kingpost design for the boom support, unique for these vessels, to reduce loading interference. Some conversions lacking electric capacity were fitted with steam engines to drive the belts.

During the '70s, many of the straight deckers built during the '50s and '60s were converted to self-unloading and lengthened to increase annual capacity. A single hold conveyor was fitted above the tank top and a Stephens-Adamson loop conveyor was located forward of the existing machinery space to avoid running new conveyor trunks through that space and the quarters. Diesel generators were added for unloading power. As these ships were primarily engaged in the ore trade, the reduction in cargo cubic was acceptable. The decrease in original deadweight capacity was offset by lengthening the hull.

The crane boat business was taken over by the railroads and mostly phased out by 1980. Since self-unloaders are expensive to maintain, most of the pre-war ships of the conveyor-boom type had been replaced gradually by conversion of a group of ore carriers built between 1920 and 1930. These conversions usually occurred when a ship's hold plating needed renewal. Eventually, after it had been demonstrated that self-unloaders were ideal for handling taconite pellets, most of the remaining ore carriers were converted. This development freed the ships from the frequent delays imposed by arbitrary constraints of shoreside labor disagreements. The next building program, described in Phase V, included self-unloaders designed and built in rather novel ways.

The technological advances seen in ocean shipping also occurred on the lakes during the construction or retrofit of ships described in Phases III and IV. These included innovations in navigation equipment such as radar and loran, engineering improvements such as oil lubricated stern bearings, controllable-pitch propellers on diesel driven conversions, and bow thrusters. Coal burning boilers in the ore trade were renewed or refitted for oil burning with automatic controls.

Although the U.S.-flag Great Lakes bulk fleet had been modernized during this fourth phase, the overall fleet had come down from 319 ships with three million tons total deadweight in 1943 to an apparently adequate fleet of 126 ships and two millions tons deadweight in 1973. Normal lake speeds, however, had increased from 11.5 to 15.5 statute miles per hour, and unloading rates had increased appreciably. Ship builders and owners were already contemplating further radical changes that would affect future construction.

Phase V (1969–1993): Refinements of the Supply System

Virtually no U.S.-flag Great Lakes vessels were built in the 1960s. In the first four phases of rebuilding for the iron ore trade, the basic objective was to maintain the ore supply to the midwestern steel industry. An ore carrier, impressive as it may be in physical aspect, is merely a unit of transport in the eyes of the steel industry. Except for the Maritime Commission involvement in the emergency building program during World War II, no direct federal funds were sought and government participation in Great Lakes shipping was resisted. Passage of the Merchant Marine Act of 1971, however, declared the Great Lakes as a fourth seacoast. As such, lake vessels became eligible for mortgage guarantees and tax incentives such as the Capital Construction Fund, all of which encouraged new construction. Nevertheless, shipowners hesitated to place orders for new ships until they could evaluate the cumulative effects of several important influences such as the opening of the Seaway, the growing obsolescence of the lower lake steel mills, and the impact of new oil and gas pipelines on the demand for coal.

By 1970 it was apparent that taconite pellets had solved the frozen cargo problem, and the mines, mills and shipowners saw that they could improve the system by extending the sailing season into the winter, thus reducing stockpiling inventory costs and maximizing the use of ships and terminal facilities. Of course that required government cooperation in maintaining aids to navigation, manning the locks, and providing icebreaker assistance. There were, in addition, shipboard personnel problems. Several meetings whimsically called "Ports and Shippers Conferences" were held with representatives of industry and government, principally the Corps. of Engineers and the U.S. Coast Guard. In time, Congress was influenced to authorize and fund a demonstration program.

This demonstration program extended operation of selected ships through January 1973. In succeeding years the operators proved that ships in Great Lakes ore service could operate year around. There were, however, negative political pressures advanced by environmentalists, owners of shore property, and island residents who found their winter bridge of ice destroyed. Moreover, shipowners heard complaints from shipboard personnel, most of whom were less than pleased about being so long away from home. There were also some high repair bills for bow shell damage. Overall, then, although year-round navigation was shown to be technically feasible, the consensus was that during normal winter weather, the maximum extension would be only into February.

From its conception, lake operators looked upon the St Lawrence Seaway as a Pandora's box. With its actual opening, deep water ports on the lower reaches of the river, as far east as Anticosti Island, became available to existing lake ships (with minor alterations for saltwater service). These ports provided loading terminals for iron ore coming out of Quebec and Labrador. American-flag ships had no advantage in such international trade and so Canadian-flag ships carried significant quantities of this ore into Lakes Ontario and Erie, in competition with ore from Lake Superior. The Seaway also made it possible for foreign operators to bring ore from various overseas sources. This latter competition, however, was minimal.

The first casualties of the Seaway were the small companies engaged mainly in the grain trade, from the head of the lakes to Buffalo, a generally seasonal trade suited for small vessels, which were otherwise available to supplement the main line ore trade. Traditionally, at the end of the shipping season after the last load of ore was delivered, those vessels not scheduled for repairs would load grain for holding and unloading during the winter at a lower lake port as elevator space became available. This brought in additional revenue for a hundred or more idle laid up ships each year, particularly at Buffalo, Toledo and Chicago. Other ships would hold grain at a loading port for delivery on the first trip in the spring.

The Seaway doomed Buffalo as a major terminal port for transfer of grain to the East Coast and overseas. Now it became possible for foreign deep draft ships to bring partial loads of overseas cargo to the lakes. Having discharged that cargo, they could load down to the controlling lake or harbor draft, and top off later to ocean draft with grain from elevators at Montreal or Quebec City. Domestic Great Lakes ships bypassed Buffalo to deliver overseas cargo to transfer ports on the lower St. Lawrence River, where they could load Canadian ore for Lake Erie and Lake Ontario steel mills and then load coal for the grain ports on upper lakes. This was a great advantage for Canadian owners who soon purchased most of the older, idled U.S.-flag hulls for operation until they could convert ocean tonnage or build new ships of greater size to meet the new demand.

Canadian owners on the lakes had faced the same phases of obsolescence as has been described for the U.S. fleet. They reacted promptly, however, and in a few years surpassed the U.S. owners in their number of large modern freighters and domestic tankers. Their new ships were sized to fit the Seaway and were modern in every respect. The incentive for this building program was the Canadian Government's "Angel" legislation, which allowed rapid depreciation of as little as three years. U.S. owners had no equivalent provisions. Canadians were thus able to dominate the grain trade, and also provide stiff competition in such trades as limestone, coal and salt.

The mainstream trade in iron ore was still protected in the United States and Canada by cabotage laws. There was thus little incentive for American companies to invest in new construction until the long term effects of the Seaway and other changes became evident. This was a turbulent period for the shipowners. In the next decade they would face many different concerns affecting their decisions about future requirements for ships, and where and how they would be built. In 1963 there were 21 U.S. companies with a combined fleet of more than two hundred ships. Most of these ships were old and inefficient, and it seemed likely that the smaller, less well managed companies might not survive. This also was a period of full employment, but with labor instability and inflation. The steel making facilities in the States were outdated in comparison to new European and Japanese plants, and were being evaluated for either major renewals or abandonment. Under these circumstances, shipowners were reluctant to place orders and shipyards were unwilling to spend money improving their facilities.

Rehabilitating the Midwest steel industry was gradual, extending over 30 years. Many mills, lacking room to expand, were not improved, but closed or perhaps reduced to specialized production. The Bethlehem Steel Co. built an entirely new, fully integrated mill on a large lakefront site at Burns Harbor IN, and in time reduced or phased out its mills at Lackawanna NY, and in Western Pennsylvania. The U.S. Steel Corp. also concentrated its improvements in its large lakefront Gary Works in Indiana, and eventually phased out mills in Ohio, Illinois, and around Pittsburgh, including its renowned Homestead Works. Ore traffic thus became concentrated at the large lakeside steel mills along Lake Michigan, also at Detroit and Cleveland, and the railhead transfer facilities along Lake Erie.

The new mills and transfer facilities were not equipped with unloading machinery, and so the need for self-unloaders became mandatory. Moreover, the economies of scale encouraged ships of the maximum possible size. Aside from the dimensional limits of the terminals, which would be improved by private investment, the major constraint was in the locks at Sault Ste. Marie, MI. Finally, after extensive feasibility studies, Federal approval and appropriations were obtained to enlarge the Poe Lock to accommodate vessels up to 1000 ft long and 105 ft in beam. Construction was started in 1963, and the work was completed and the lock dedicated on June 26, 1969.

U.S. shipowners and shipyard managers realized that the larger ships permitted by the new Poe Lock should be better able to compete with smaller Seaway-size ships. The question arose: Could the yards expand their building ways to handle the larger ships?

The Manitowoc yard had completed its last conventional lake freighter, the *Edward L. Ryerson*, in 1960; the Great Lakes Engineering Works had built the *Arthur B. Homer* in 1960; and the American Ship Building Co., at Toledo, the *Shenango II*, in 1959. All of these ships were at the physical limits of their respective shipyard facilities.

At Manitowoc, a bend in the river kept longer ships from having access to the shipyard. The parent company's crane business was expanding and needed the available property; so in 1968 they moved the shipyard business to Sturgeon Bay WI, where they bought the Sturgeon Bay Shipbuilding and Drydock Co. yard. They loaded their shipbuilding equipment onto their floating drydock and towed it to Sturgeon Bay. Two years later they bought the nearby Christy Corp. yard and formed the Bay Shipbuilding Corp. In 1976 they built a 1158-ft × 140-ft graving dock with a 200-ton bridge crane. They later bought the intervening land between the yards so they could build a large fabricating shop. This move placed the new Bay Shipbuilding Corp. in an excellent competitive position. They combined a skilled work force with adequate shorefront property, which would be built into an efficient production line shipyard.

Fig. 6 The self-unloader *Roger Blough*, demonstrating unloading into the war-built ore carrier *A. H. Ferbert* at Cleveland, OH. (From the Great Lakes Historical Society Collection.)

The Great Lakes Engineering Works, on the other hand, was not inclined to make substantial capital improvements at their Ecorse MI yard and so decided to go out of business.

The American Ship Building Co. by this time had begun to consolidate its Great Lakes hold-

Fig. 7 The *Stewart J. Cort*, the prototype 1000-ft self-unloader, on its maiden voyage on the Detroit River. (Courtesy Marine Consultants and Designers, Inc.)

ings. They eliminated their interests in shipyards in Superior WI, Buffalo, and Cleveland, and concentrated major repairs and new construction at Lorain and Toledo OH, with minor conversion and repair work in Chicago. To suit large vessels, a new building graving dock was built at the Lorain yard and some subassemblies were fabricated at their other yards.

At this time, the shipowners had evaluated the situation and concluded that the demands for iron ore were stable enough to justify replacing the interim postwar, and earlier, tonnage with new, highly efficient ships of maximum capacity. New concepts beyond established rules and experience were being developed, approval obtained, and financing arranged. About all that was not decided was how the pie would be divided.

In 1966 Litton Industries, realizing the potential of a sizeable new ship market on the Great Lakes, initiated a major construction program for the maximum size ships allowed. This program consisted of building a new ship fabrication assembly facility, with drydock, at Erie PA, and the prototype design of a new 1000-ft self-unloading iron pellet carrier. To complement their overall Great Lakes involvement, Litton purchased Wilson Marine Transit Co., an established lake carrier. They also bought Marine Consultants and Designers, Inc., a Cleveland-based design firm, for its domestic expertise. At the same time the Bethlehem Steel Co. was in immediate need of ships to supply its new, rapidly growing steel mill at Burns Harbor IN. This led in 1967 to a contract with Litton's Erie Marine for the construction of a 1000-ft lakes vessel, later named the *Stewart J. Cort.*

The *Cort*, completed in 1971, was the first ship that took full advantage of the maximum size allowed by the new Poe Lock, i.e. 1000 ft LOA and 105 ft beam. Its depth was 49 ft. The parallel middle body, constructed at Erie Marine, consisted of 17 identical 48-ft modules. The bow and stern, being unsuited to automated mass production, were built at Ingalls' conventional shipyard in Pascagoula MS. When completed, the bow and stern assemblies were temporarily joined as one hull, called "*Stubby*." This was sailed under its own power to Erie, where it was separated and joined with the midbody.

The *Cort* incorporated many innovative design features and set the pattern for the additional 1000-ft ships that were to follow. Longitudinal strength standards for ships of that length were not yet available, but the design team worked closely with the Great Lakes International Load Line Committee to develop interim criteria that revised the existing classification society strength requirements for vessels on the lakes. Higher strength steels were used in the upper and lower flanges of the hull girder to reduce the light ship weight. The *Cort* initiated the rudimentary cylindrical bow that maximized displacement and deadweight. It was the first to have both bow and stern thrusters, two forward and two aft.

A new concept in propulsion machinery used General Motors standard diesel locomotive components adapted to marine use. Four EMD Model

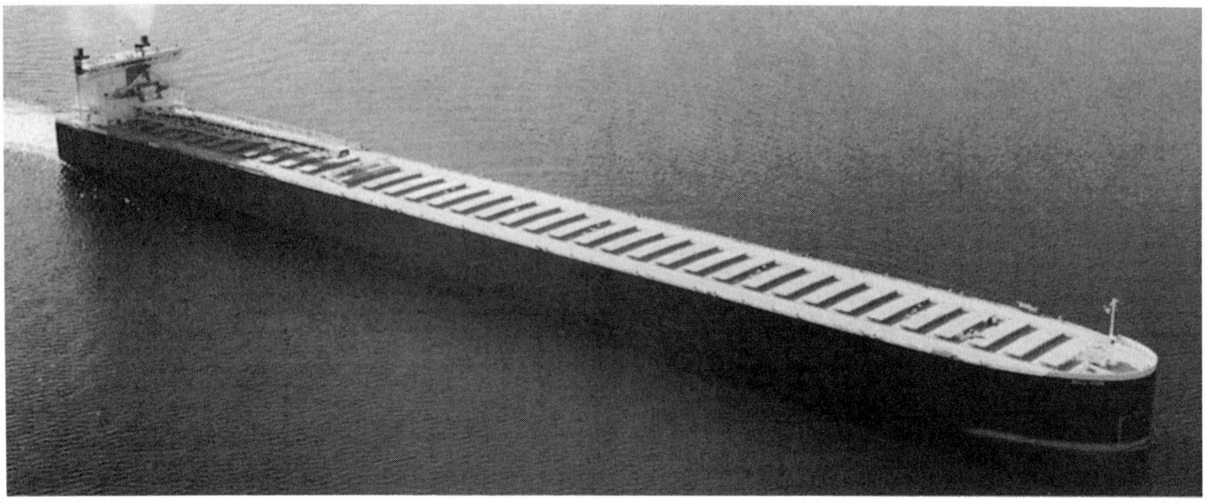

Fig. 8 The 1000-ft Great Lakes self-unloader *Belle River* is primarily a coal carrier, transporting up to 66,000 short tons.

20-E645-E7 engines each powered an EMD Model A-10 ac auxiliary generator. These engines, with a cumulative shaft horsepower of 14,000, were pneumatically clutched to reduction gears for the twin controllable-pitch propellers.

The ship had a unique, highly efficient ballast system. Each of the 18 ballast tanks was fitted with its own intake and discharge pumps. This eliminated the need for all valves except for one shut-off valve at each sea connection.

A fully automated unloading system with one-man control was installed. The main components consisted of 105 specially designed metering gates feeding a single 10-ft wide conveyor belt, which carried the cargo up over a 60-ft diameter wheel and dropped it onto a 98-ft long transverse cantilever boom conveyor. The system provided an unloading rate of 20,000 long tons per hour, which was then the highest in any dry bulk carrier worldwide.

The design and automation of the engine room, unloading gear, hydraulic actuated hatches, and other features allowed a crew reduction from the standard 35 to 26 despite the ship's increased size.

In 1966 U.S. Steel Corp. hired Marine Consultants and Designers, Inc. to design a new ship of maximum size for the new Poe Lock. This single-screw ship of 14,000 shp became the *Roger Blough*. Construction began in 1968 at the American Ship Building Co. yard in Lorain OH. The 858-ft × 105-ft × 41.5-ft dimensions required the construction of a new graving dock. The ship was to be delivered in 1971, but shortly before scheduled sea trials a fire destroyed the engine room and actual delivery was delayed until 1972.

Bay Shipbuilding Corp. at this time was not ready to build ships of this size since their building dock would not be ready until 1976, and they were developing their version of a new design for smaller vessels.

Both the *Cort* and the *Blough* were fitted with an unloading system that discharged at the stern over a transverse shuttle belt conveyor into a hopper on shore, which fed large capacity storage areas at a rate of 10,000 long tons per hour. The *Cort*'s nominal transport capacity was 58,000 tons of ore per trip, and the *Blough*'s was 44,500 tons. Both ships had their unloading transfer machinery aft in way of their machinery spaces. All crew accommodations in the *Cort* were located in the forward deckhouse. The pilothouse of each vessel was placed in the bow and equipped with remote controls for engine speed, propeller pitch, and bow thrusters.

While Great Lakes shipowners and builders sat back to evaluate the new high capacity ships, Litton brought out another innovation. This second vessel coming out of Erie Marine was a self-unloading barge pushed by a rigidly connected twin-screw tug, which was built down on the Gulf at Halter Marine, and delivered to Erie. The connecting system was developed by Breit Engineering, Inc. The combined tug/barge, of maximum size for the Poe Lock, was named the *Presque Isle*. The bow section was built at the Defoe Shipbuilding Co., while the rest of the barge, including self-unloading gear, was built at Litton's Erie yard. More versatile than the *Cort*, which was designed solely for iron ore pellets, the *Presque Isle* could handle limestone or coal, as well. The project was initiated in 1970.

The main propulsion plant in the tug consisted of two medium-speed Mirlees Blackstone geared diesel engines each rated at 7420 bhp, driving the twin controllable pitch propellers. Unloading, bow thruster and ballasting electrical power was provided by two 3000 kW gnerators driven off the propulsion engine, and auxiliary power from three 300 kW diesel generators located on the tug, and one 300 kW diesel driven generator on the barge. The tug was equipped with a large hydraulic cylinder designed to hold the tug in position in the notch with a constant force, or to move it in or out of the notch.

The cargo unloading system was designed to unload 10,000 long tons of iron ore pellets per hour over a conventional conveyor boom. A 67.5-ft diameter rotary wheel elevator was oriented transversely to accept cargo from the two 66-in. cargo hold belts and to be compatible with the notch arrangement. The length of the unloading boom was 250 ft. The assembled tug/barge was completed in late 1973 and operated by Litton Great Lakes Corporation.

The *Presque Isle* was designed with all crew accommodations on the tug. The intent was to operate the tug/barge with a crew of only 14, although various factors have so far prevented the crew numbers to be brought down to that level.

How do Great Lakes bulk carriers manage to be seaworthy despite their exceptionally high block coefficients? The answer lies in their proportions. A bow and stern of normal shape, when attached to an extremely long parallel middle body, lead to those surprisingly high coefficients. After all, a plank on edge is not a bad hydrodynamic shape, and yet its block coefficient is 100 percent.

Although the *Cort* and the *Presque Isle* had the same overall length and beam, the *Cort* had a deadweight advantage of about 2500 long tons. This was due primarily to the displacement lost in the notch in the barge. The tug, by regulation, must be ballasted to the load draft of the barge so that it exerts little or no force on the barge.

Repeat orders for 1000-ft hulls were not forthcoming. Not many ore loading docks were prepared to store and load 60,000 tons at one time, and not many unloading docks could handle that amount efficiently. Litton built no more vessels on the Great Lakes.

Over the next five years nine ships were built on the lakes. These were all self-unloaders with deadweights ranging from 20,000 to 40,000 long tons. These featured new designs in the *Charles E. Wilson*, at Bay Shipbuilding Corp., the *Roger M. Kyes* at American Ship Building Co., Toledo, and the *William R. Roesch* at American Ship Building Co.'s Lorain yard. These vessels followed the trend of the *Blough* and the *Cort*, but broke tradition by placing the pilothouse and all quarters in a single deckhouse at the stern. The cargo was unloaded onto a single belt that was carried up in a trunk within the engine room, transferred to an offset inclined belt running through the deckhouse, and dropped onto the boom conveyor. The swinging boom conveyor was secured to a tower at the forward side of the deckhouse—another innovation.

The cylindrical bow design for the *Charles E. Wilson* was adopted after extensive model testing at the Netherlands Ship Model Basin, sponsored by Bay Shipbuilding Corp. This form showed less resistance than a conventional pointed bow at the operating speed/length ratio of 0.504 and with a block coefficient of 0.89.

The depths of the ships were increased from past designs to compensate for the loss of cargo cubic with a single hold conveyor. Hopper slopes extended to ship's side at 35 degrees. The increased depth also increased the moment of inertia of the hull girder despite the smaller scantlings used with high strength steel (ABS Grade AH 36). The first and typical of such vessels was the *Charles E. Wilson*, built at Bay Shipbuilding Corp.

The *St. Clair*, from Bay Shipbuilding Corp., was designed exclusively for transporting western coal. Built as the largest ship possible for side launching on the available berth, 770 ft × 92 ft × 52 ft, the vessel had several unique features. Two cargo hold belts were used instead of a single belt to gain coal capacity. Both hold conveyors sloped upward through the engine room and used short transverse belts to feed an inclined center belt leading to the boom conveyor. A flat plate gate was used with an inflatable, retractable seal to make the cargo hold watertight from the conveyor tunnel. This allowed one-compartment subdivision without subdividing the conveyor tunnel. The vessel could unload 7000 short tons of coal per hour. Three locomotive type engines were used on a single shaft, two side by side forward of the reduction gear, and the third aft of the reduction gear, above the line shaft. Twin spade rudders were used with the single screw to shift the entire engine room aft, increasing the cargo volume and block coefficient without increasing the buttock angles of the run. Model tests were conducted at the University of Michigan.

Six 1000-ft LOA ships were built by Bay Shipbuilding Corp. between 1977 and 1981. Their hull design retained the cylindrical bow and featured a twin screw tunnel stern refined at the Netherlands Ship Model Basin. These vessels had a speed/length ratio of 0.43 and a block coefficient of 0.95. The CP propeller shafts were each enclosed by a pipe with V struts supporting

Fig. 9 The *James R. Barker*, a typical 1000-ft Great Lakes bulk carrier. (From the Great Lakes Historical Society Collection.)

the stern tube bearing. Each strut was raked and canted at an angle to minimize wake disturbance as determined by model tests. The ships showed little hydrodynamic vibration.

Five of the six ships were designed to carry coal, ore, or ballast water to their normal operating draft of 27.5 ft. A single hold conveyor with 35 degree hopper slopes extending to the ship's side was used to obtain the high ballast capacity. Cargo cubic volume was increased by a 56-ft depth, along with a unique arrangement of conveyor and machinery. The twin screw plan allows the centerline loop conveyor to pass between the engines and extend nearly to the transom stern. The Stephens-Adamson patented loop conveyor belt extends through the entire hold length without a transfer. The locomotive type engines, two per shaft, are tucked under the cargo hold slopes, making the hold overlap the engine room. This reduces the still water bending moment and maximizes the cargo volume. The high electrical requirements for unloading at 10,000 tons per hour are met by a 2500 kW, 4160 volt, AC generator driven by one of the propulsion engines. Symmetry provides a redundant generator on the other propulsion shaft. Bow and stern thrusters, each of 1500 hp are also powered by the unloading generators during maneuvering. (As you may note, those thrusters are as powerful as the typical main engine of 50 years ago.) The sixth vessel of this group has two main engines and is fitted with separately powered generators.

The smaller, exceptionally agile *American Republic*, from Bay Shipbuilding, was designed to combine the daily navigation of the Cuyahoga River in Cleveland with the normal Great Lakes service requirements. The vessel features an integrated hold conveyor and loop belt, with the latter in a casing extending aft between two diesel engines driving twin screws. The ship has Kort nozzles and twin flanking and steering rudders. It has, in addition, large bow and stern thrusters, with two sets of controls arranged in the pilothouse to facilitate operation while running either ahead or astern.

In 1973 and 1974, the American Ship Building Co. built four smaller vessels, one in Toledo and three in Lorain. In 1974, within a lengthened building dock at Lorain, the keel was laid for the *James R. Barker*, the first of a series of 1000-footers. A total of five such ships were built, with the last delivered in 1981. These were built at the Lorain yard, but some major subassemblies came by water from Toledo or Chicago. The ships were self-unloaders with up to three hold conveyor belts. They all had long cargo booms, quarters aft, and twin screws each driven by a medium speed diesel engine. The last of the series, the *William J. Delancey*, was also the longest on the Great Lakes at 1013.5 ft.

In 1984, subsequent to the delivery of that ship, the company sold the facilities and discontinued all operations on the Great Lakes.

Why have self-unloaders come to dominate the Great Lakes bulk trades? Economic logic originally had located unloading gear on shore where it would have a high use factor rather than on a ship where it would be used only once a voyage. Compared to a conventional bulk carrier, a self-unloader carries a large burden in capital-related costs and such annual costs as maintenance and repairs. On the other hand, the efficient recovery of capital and maintenance costs of a shore-side unloading rig depends on the intensity of use. Optimum utilization would come from a constant cargo flow, with rig capacity equal to blast furnace demand. Infrequent use becomes expensive, particularly in dockworkers' labor and in maintenance costs. From the dock superintendent's view real progress was to get unloading machines off the waterfront and off his short term budget and leave unloading costs to be borne by the shipowners where they would be buried in the freight rate. Thus the conventional American dry bulk freighter on the Great Lakes and the dockside unloading machines were interdependent, and beginning in 1971 were

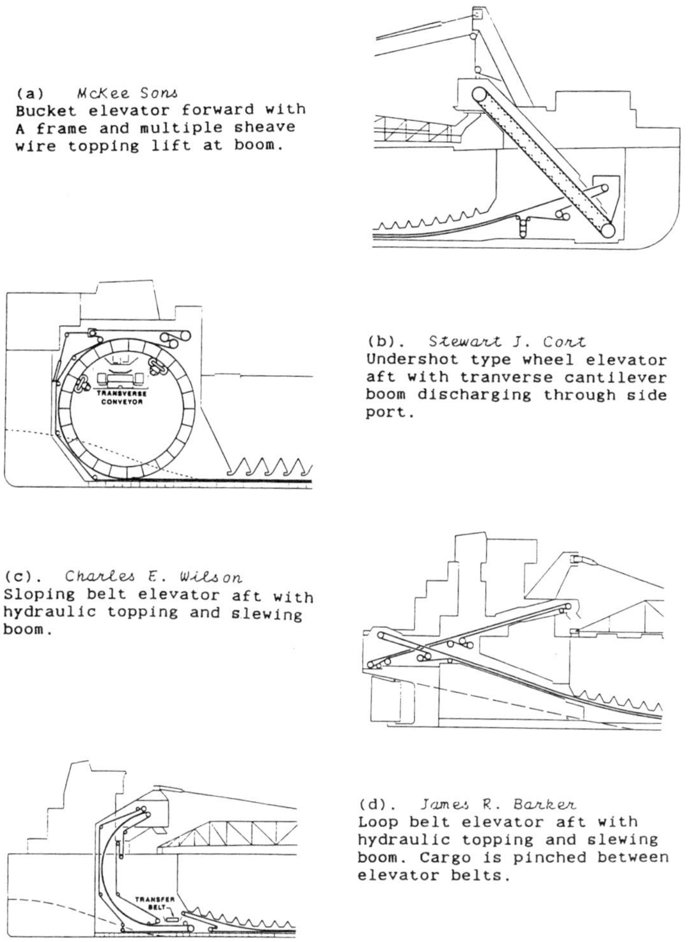

Fig. 10 Typical self-unloader conveying systems.

being phased out by competition with the more flexible self-unloaders.

Then a new problem arose. Delivery of large quantities of raw materials by the large new self-unloaders soon outgrew the limited capacity of storage pits at the blast furnaces and required extensive new storage and distribution facilities. Highly automated transfer systems using stacker-reclaimer machines and extensive belt conveyors both at loading docks and at unloading terminals gradually displaced labor intensive dockside operations. The operating economies of the 60,000-ton self-unloading, highly automated bulk carriers had made the old 10,000-ton conventional "straight deckers" uncompetitive. The same transition was occurring at the transfer terminals. As the "straight decker" fleet diminished, the cost of maintaining the shore gear and backup rail car systems became more expensive in inverse proportion, and so they were gradually phased out.

Throughout this narrative, we have stressed that major political events and long term economic forces affected the design and operation of Great Lakes ships. The opening of the Seaway and the enlargement of the Poe Lock caused significant changes in the domestic fleets; but it was the demise of the northbound coal trade that ultimately influenced the development of the 1000-ft self-unloaders suitable for carrying either ore or coal.

In earlier years coal had flowed largely from eastern or midwestern mines through Lake Erie ports northward throughout the Lakes. Coal thus provided a convenient backhaul cargo for the ore and grain carriers, allowing consequent reductions in freight rates. After World War II, however, a dramatic shift occurred. First, the demand for coal dropped off (owing to the introduction of alternative fuels), and then such demand as still continued called for low-sulphur coal from the Great Plains in place of the less clean-burning coal from the Midwest. Brought in by unit trains to Superior WI (at the western end of Lake Superior), this cleaner-burning coal introduced a demand for traffic down the lakes to power houses in the Detroit area. This traffic, of course, is alongside the ore trade, and so some of the newer ships are configured to be efficient carriers of either iron ore, coal, or grain.

Conclusion

Fifty years ago Great Lakes bulk carriers had evolved but slowly since the turn of the century. Riveted hulls, coal-fired boilers and steam reciprocating engines still dominated. Since then, however, rapid improvements have taken place. New economic and political pressures have been met through the introduction of improved technologies and managerial practices, which have affected the terminals as well as the ships. Labor intensity has been greatly reduced through increases in ship size, automated controls, and self-unloading arrangements. Diesel propulsion engines have markedly reduced annual fuel bills. Welded hulls have effected important savings in hull weight. Other improvements in design have increased the per-trip cargo capacity as well as the number of trips per year. The cumulative effect of these changes has been a significant increase in transport economies. Even the waterfronts are changing with the disappearance of magnificent, gigantic machines such as the coal dock car dumper and the Hulett unloaders. These, like the dinosaurs, have been driven out of existence by the changing environment. The belt conveyor and its host, the versatile self unloading bulk carrier, are the survivors.

Appendix

Here is a summary of data reported by shipowners, as compiled by the M. A. Hanna Company. The information pertains to new U.S.-flag Great Lakes dry bulk carriers built during each of the historic phases into which this chapter has been divided.

Phase I	Wartime construction	21 ships, 347,730 dwt total
Phase II	Ocean conversions	12 ships, 245,100 dwt total
Phase III	Conventional types	26 ships, 625,605 dwt total
Phase IV	(No new building)	
Phase V	Diesel self-unloaders	27 ships, 1,113,605 dwt total
Total 1941–1990		86 ships, 2,332,040 dwt total

Additional data from the same source show that in 1943 there were 319 ships with a total deadweight of 2,984,490 tons in the U.S. ore fleet. There were, in addition, about 30 self-unloaders. In 1983 there were 112 ships with a total deadweight of 2,752,225 tons in the U.S. fleet, including self-unloaders.

Bibliography

Bassett, N. L., Kline, R. G., and Miller, R. H., "Design and Operation of a Highly Maneuverable Great Lakes Self-Unloader," Great Lakes and Great Rivers Section, SNAME, Oct. 1981.

Benford, H., "Probable Size of Future Great Lakes Ships," Report for U.S. Army Engineers District, Detroit Corps. of Engineers, Jan. 1959.

Benford, Thornton, and Williams, "Current Trends in the Design of Iron-Ore Ships," SNAME Spring Meeting, 1962.

Bohn, W. E., "Great Lakes Load Line Rules," Eastern Canadian Section, SNAME, and Institute of Marine Engineers, Great Lakes Section, March 21, 1974.

Cashman and Ericson, "Great Lakes Ore Carriers S.S. *Johnstown* Class," Great Lakes Section, SNAME, Fall 1952.

"Conversion of C4 Class Ocean Vessels for Great Lakes Service," The Maryland Drydock Company, Great Lakes Section, SNAME, Oct. 2, 1953.

Cowles, W. C., "A Decade of Great Lakes Ships," *Inland Seas*, Fall 1989.

Cowles, W. C., "New Ore Carrier *Philip R. Clarke* Marks Advance in Pittsburgh Fleet," *Marine Engineering and Shipping Review*, July 1952.

Dennison, B. C., "Lake Freighter *E. J. Block* Converted to Diesel-Electric Propulsion," *Marine Engineering and Shipping Review*, Nov. 1946.

Duncan, Brissenden, and Bain, "Some Aspects of Current and Future Canadian Lake Ship Designs," Great Lakes and Great Rivers, and Eastern Canadian Sections, SNAME, Oct. 1962.

Earle, M. M., "The Conversion of T2 Tankers for Great Lakes and Seaway Service," Great Lakes Section, SNAME, Sept. 1959.

Greenwood, J. O., "Historical Perspective on Development of Great Lakes Bulk Freighters and a View Towards the 21st Century," Great Lakes and Great Rivers Section, SNAME, Jan. 24, 1991.

Kendall, H., "Bow Thrusters as Applied to Great Lakes Vessels," Great Lakes and Great Rivers Section, SNAME, May 1963.

Miller, R. H., "Design and Economics of Increased Capacity of the Steamer *William A. Reiss*," Great Lakes Section, SNAME, Oct. 1963.

Miller, R. H., "Designing and Building the Self-Unloader *Charles E. Wilson*," Spring Meeting, SNAME, 1974.

Miller, R. H., "Great Lakes Self-Unloader *Adam E. Cornelius*," Great Lakes and Great Rivers Section, SNAME, Sept. 1959.

Miller, R. H. "Great Lakes Thousand Footers," Great Lakes and Great Rivers and Eastern Canadian Section, SNAME, May 1979.

Pickersgill, T. C., Sharp, I. D., and Trunick, R. P., "1000 Ft. Tug/Barge Combination for Litton Great Lakes Corporation," Great Lakes and Great Rivers Section, SNAME, May 17, 1973.

Record, American Bureau of Shipping, various dates 1953 through 1991.

Reschke, G. W., "Design, Size, Construction and Operation of New Second Lock, Sault Ste. Marie, Michigan," Great Lakes and Great Rivers Section, SNAME, May 21, 1964.

"Self-Unloader *Detroit Edison*," *Marine Engineering and Shipping Review*, June 1955.

Snow, F. O., and Alman, P. R., "Loading of 1000 Ft. Ships at Two Harbors, Minnesota," Great Lakes and Great Rivers Section, SNAME, Oct. 11, 1979.

Sproul, G. "One Thousand Foot Great Lakes Self-Unloader *James R. Barker*," Great Lakes and Great Rivers Section, SNAME, Oct. 28, 1976.

Stearn, R. A., "Conversion of M/V *Joseph H. Frantz* and S/S *Nicolet* to Self-Unloader," Great Lakes and Great Rivers Section, SNAME, Oct. 1965.

Thornton, K., "Streamlining the Ore Carrier," *Marine Engineering and Shipping Review*, Sept. 1943.

Thornton, Meidlich, Douglas, and Butcher, "The Great Lakes Ore Carrier *George M. Humphrey*," Great Lakes Section, SNAME, Sept. 1955.

Tripp, C. E., and Plude, G. H., "Integrated Tug–Barge Versus Ship—An Economic Comparison for the

Great Lakes," SNAME National Spring Meeting, May 22–24, 1974.

Tripp, C. E., and Plude, G. H., "One Thousand Foot Great Lakes Self-Unloader, Erie Marine Hull 101," Great Lakes and Great Rivers Section, SNAME, Jan. 21, 1971.

Tripp, Rankin, Souris, and Miller, "Great Lakes Bulk Ore Carrier *Edward L. Ryerson*," Great Lakes Section, SNAME, Oct. 1960.

Wiening, P. G., "A Victory from the Wars," *Telescope* (Detroit Historical Society publication), March–April 1985.

Williams, Thornton, Miedlich, and Douglas, "Design and Construction of Great Lakes Bulk Carrier *Wilfred Sykes*," *Trans.* SNAME, Vol. 58, 1950.

Workman, J. C., "Shipping on the Great Lakes," SNAME *Historical Transactions*, 1943.

George H. Palmer, Jr., graduated from MIT in 1941 with a degree in Naval Architecture & Marine Engineering. After working with Bethlehem Steel Company's Shipbuilding Division and the U.S. Navy, he joined the American Bureau of Shipping in 1947, where he spent the rest of his professional career. From 1950 until he retired in 1983 he served the Great Lakes area as ABS Principal Surveyor for the Great Lakes. He has been a Member of SNAME since 1941.

Richard H. Suehrstedt earned a degree in Naval Architecture & Marine Engineering from the University of Michigan in 1951. His entire professional career has been spent with Marine Consultants & Designers, which organization he now heads as President. His company, based in Cleveland, was involved in the design of nearly all of the U.S.-flag ships now on the Great Lakes.

Carlton Tripp is a graduate of the Naval Academy (1942) and also of the University of Michigan (MSE Degree, 1950). After a year with Ingalls Shipbuilding, he joined Marine Consultants & Designers, advancing through the ranks of Chief Engineer, President, and now Chairman. In 1984 he received a SNAME certificate of appreciation for his work in setting strength standards for 1000-ft ships. In 1985 he was given the Elmer A. Sperry Award in recognition of engineering contributions in transportation.

Joseph Fischer spent four years as a loftsman and shipfitter before enrolling at the University of Michigan. He graduated from that institution in 1958 with a degree in Naval Architecture & Marine Engineering. He then went to work with R. A. Stearn, Inc., in Sturgeon Bay WI. From 1977 until 1986 he worked for the American Steamship Co. (in Buffalo), rising to the position of Vice President for Operations. In 1986 he returned to Sturgeon Bay to take his present position as Director of Engineering & Design at R. A. Stearn, Inc. He is a Member of SNAME, and Past Chairman of the Great Lakes & Great Rivers Section. He is on two ABS committees and has also done service with the National Research Council.

Bulk Carriers

Steven Spear

Historical Background

Ships have probably loaded bulk cargoes from earliest times. In the modern era, i.e., when sail was giving way to steam, the great Cape Horn square riggers carried South American nitrate to Europe as a bulk cargo. Salt was transported from England to Maine in large quantities for salting fish and preserving wooden hulls [1]. As steam came to dominate the oceans, the requirement for coaling stations around the world resulted in a demand for transporting coal to those stations. Much of this low value cargo was carried in the remnants of the sailing fleet as well as in traditional three island tramp freighters. Along the U.S. East Coast, coastal schooners served to carry coal from Chesapeake Bay to New England. These were the precursors of a fleet of steam powered coastal colliers.

During 1929–1930 Bath Iron Works obtained contracts for two 13,000-ton colliers through the Emergency Fleet Corporation. These were destined for the Crowell and Thurlow firm of Boston, which once operated a fleet of schooner colliers. Photographs show those vessels to have been of typical three island construction.

The navies of the world's sea powers, of course, also had to provide fuel for their coal-fired ships. Colliers such as the U.S. Navy's ill fated *Cyclops* and *Jupiter* class were perhaps among the first ocean bulk carriers to adapt the characteristic machinery aft arrangement, which ultimately became the norm for bulk carriers.

The modern era of ocean bulkers might be considered to have begun in 1917 with the construction by and for the Bethlehem Steel Corp. of five ocean going ships incorporating many features of Great Lakes ore carriers. Built to deliver iron ore from Cruz Grande, Chile, to the company's mills at Baltimore MD and Claremont NJ, the 20,000 dwt ships featured large steel hatch covers [2]. All these ships were lost by submarine action during World War II. Consequently, plans were instituted even during hostilities for their replacement, initially by the *Venore*-class.

Almost parallel to the development of the *Venore*-class in 1945 was the collier version of the World War II Liberty ships. Designated EC2-S-AW1, 23 of these were built by the Delta Shipbuilding Company in New Orleans. Although based on the basic Liberty hull and fitted with the Liberty triple expansion engine, the machinery was located aft. The hatch covers were of steel, and the holds and internal structures were arranged for efficient handling of coal, together with provision of suitable ballast capacity for the return voyage without cargo. Built as an urgent replacement for the aging coastwise collier fleet, these ships were a considerable upgrade in arrangement and fittings from the standard Liberty. Most served efficiently in their designated trade for up to 20 years [3,4].

During World War II, the United States expended vast quantities of material resources to supply the Allied armed forces. High grade domestic reserves of bauxite (aluminum) and iron ore were reduced substantially. Importing these minerals became economically feasible. War-built Liberty ships of about 10,000 dwt and 11 knots speed were initially used for the trade. About 2700 were built during the war, and the major portion were declared surplus at war's end.

The Late 1940s

The Bethlehem Steel Corp. designed and built the first post-war ore carrier, the *Venore*, delivered in 1945.

Fig. 1 The collier USS *Jupiter* (1908). (Courtesy Alex Shaw.)

Five other ships of this class were delivered for trade between Chile and Sparrows Point, MD [5]. Two additional ships of the *Baltore*-class, with dimensions and power plant similar to that of the *Venore*, were delivered in 1948 [6].

The 1950s

By the early 1950s many purpose built bulk carriers for iron ore, bauxite and gypsum were under construction worldwide. American shipowners such as Hendy International and Universe Tankships had ships built for the Venezuela to U.S. iron ore trade. M. A. Hanna Co. and the C-4 conversions of Ore Transport, Inc. were engaged in the Seven Islands (Canada) to U.S. East Coast trade. Iron ore carriers of this period ranged from about 18,500 to 31,500 long tons dwt. A notable exception was the *Ore Chief*-class built for D. K. Ludwig interests. At 60,400 dwt, these were the largest commercial ships in existence at that time. They had a beam of 116 ft, a length/beam ratio of 6.52, and a block coefficient of 0.828. These proportions set the trend of future tanker and bulk carrier hull forms [6].

Reynolds Metals acquired a fleet of self-unloading bulk carriers to transport bauxite from Jamaica to near Corpus Christi. Alcoa and Alcan moved bauxite in shallow draft (20 ft) ships from Surinam and British Guiana to a transshipment port in Trinidad. There alumina was loaded into deeper draft ships for delivery to Mobile, AL (Alcoa) or Port Alfred, Quebec (Alcan). Ships built for these trades are described in [6]. A significant publication of the period [7] clearly showed the economies of scale in the bulk trades.

The 1960s

During the decade of the '60s world shipments of coal and grain reached major proportions. Hampton Roads was the major U.S. port for coal with trades to Rotterdam and to Japan via the Panama Canal. Ports in the U.S. Gulf, Seattle/Tacoma, and British Columbia handled the grain trade. The "Panamax" vessel was developed to serve these trades. In these ships the beam is limited to 106 ft by the Panama Canal. Cargo capacity is 60,000–80,000 tons. The holds have a relatively large cubic capacity for cargoes of coal (42–48 cu. ft per ton) or grain (45–65 cu. ft per ton) [8].

During this period the "handy size" bulker evolved. They are in the 25,000 to 40,000 dwt range. Many are fitted with hydraulic jib cranes of 5 to 25 tons capacity and are known as "geared bulkers."

A new type of bulk carrier was developed to carry forest products such as wood chips and logs. These are light density, high volume cargoes; hence, these ships are unusually deep. Some have deck-mounted cranes with special grabs to load and discharge logs, or to discharge chips [9].

A unique type of ship, the newsprint carrier, was introduced on the West Coast during this period. These have deck-mounted gantry cranes equipped with a patented device to lift four 1100-lb newsprint rolls at one time. The devices employ clamps and vacuum to do their work.

Also introduced during this period was another special purpose bulk carrier designed to deliver salt from Mexico to the Pacific Northwest. During the voyage the salt's crystalline structure nearly solidifies. Large bulldozers are used in the holds to assist grab buckets lowered from deck-mounted cranes. The buckets dump onto a conveyor system arranged to discharge overboard by shuttle conveyors fore and aft.

The 1970s

The international shipment of bulk commodities had grown to major proportions during the 1950s and

Fig. 2 The SS *Venore*, built at the Bethlehem Sparrows Point Yard and placed into service in 1945 by Ore Steamship Co. (Courtesy Alex Shaw.)

1960s. Most of the mineral resources were in remote areas of the World, far from the major manufacturing centers in Europe, Japan and the United States. As the population of the world increased, shipments of grain from Canada, the U.S., Argentina, and Australia provided employment for many bulk carriers. Since this was international commerce, most of the cargoes were carried under foreign flag. A percentage of the U.S. aid shipments to foreign nations, however, was reserved for U.S.-flag ships.

The 1970 amendments to the Merchant Marine Act of 1936 stimulated an interest in U.S.-built bulk carriers. The National Steel and Shipbuilding Company (NASSCO) designed and built two OBOs: the *Ultramar* and the *Ultrasea.* The Overseas Shipholding Group ordered two 24,500-ton ships. A ship of interesting design for the Jones Act trade was the 30,000 dwt *Sugar Islander,* built to carry bulk sugar from Hawaii to the U.S. West Coast [10,11].

The 1980s

During the 1980s eight bulk carriers were built for the U.S.-flag fleet. Levingston Shipyard in Texas built three 36,500 dwt ships. Liberty Marine contracted for two 63,000 dwt diesel powered ships, and Ogden Marine had two of 48,000 dwt built at the Avondale Shipyard.

An unusual ship for a unique trade was the *Energy Independence,* a 38,300 dwt self-unloading collier to carry coal from Chesapeake Bay to electric power plants in Massachusetts. She was built by General Dynamics, Quincy MA. The power plant has the only marine, coal fired boilers built in the United States in recent years [12].

Particulars of the U.S.-flag bulk oceangoing bulk fleet are listed in Table 1. The last three on the list are steam powered, the rest are diesel powered.

The 1990s

The economic recession of the early 1990s has reduced the international bulk trades. However, projections indicate that world-wide orders of replacement tonnage for dry bulk carriers will rise substantially through the 1990s. Nevertheless, it is doubtful that U.S.-flag carriers will be built, except for the protected Jones Act trades [13].

The Unique Problems of Bulk Carriers

Steel Hatch Covers. Post-war bulk carriers have all been fitted with quick acting steel hatch covers to minimize the time to open and close them in port. A good description of the types, operation and details of these covers is contained in [6].

Tank Top Reinforcement. For iron ore the buckets may weigh ten tons and have a capacity for fifteen tons of ore. If the tank top plating is of minimum classification scantling, the buckets will cause indents and possible penetration. To minimize such damage, plating 1.25 in. (30 mm) thick should be fitted, along with longitudinals spaced about 15 in. (380 mm). Since the plating and longitudinals are continuous throughout the hold space, the material and its vertical location have a definite effect in bolstering the ship's section modulus.

Clean-up of the hold spaces is generally done by bulldozers. The blades of the bulldozers may shear-off

Table 1 Bulk carriers in the U.S.-flag fleet in 1990.

Ship Manager	Ship's Name	Year Built	dwt ($\times 10^3$)	HP ($\times 10^3$)
Falcon Shipping Group	Pride of Texas	1981	36.5	15.6
	Spirit of Texas	1981	36.5	15.6
	Star of Texas	1982	36.5	15.6
Liberty Maritime Corp.	Liberty Sea	1984	63.7	12.3
	Liberty Wave	1984	63.4	12.3
Marine Transport Lines	Marine Princess	1967/1982	51.7	13.8
OMI Corp.	OMI Sacramento	1983	48.9	11.0
	OMI Missouri	1983	48.9	11.0
OSG Bulk Ships	Overseas Harriet	1978	25.5	11.2
	Overseas Marilyn	1978	25.5	11.2
Pacific Gulf Marine	Sugar Islander	1973	30.0	12.0
Sealift, Inc.	Inger	1945/1962	24.0	7.2
American Maritime Transport	Ultramar (OBO)	1973	82.2	24.5
	Ultrasea (OBO)	1974	83.4	24.5
Keystone Shipping Co.	Energy Independence[a]	1983	38.3	12.0

[a] Self-unloading collier.

metal from the plating and any projections such as access hatch bolts.

Alternate Hold Loading. Bulk carriers may be designed with a hold cubic capacity sized for coal (42–48 cu. ft per ton) but also operate in the iron ore trade (13–19 cu. ft per ton), which requires considerably less cargo hold cubic for a full deadweight of cargo. To expedite cargo discharge the ore should be loaded as deeply as possible to give maximum "free-digging" time and minimize bulldozer clean-up. Hence, some holds will be empty, while others are loaded to full depth. In way of the empty holds, large buoyancy forces acting upward are not balanced by the weight of any cargo acting downward. Therefore, large shear stresses are induced at bulkheads between the empty and loaded holds. To resist these forces, shear material in the form of intermediate girders extending one transverse space fore and aft of the bulkhead should be fitted at the deck and shell.

Stability. Full block bulk carriers tend to be "stiff" (have a large metacentric height), particularly in ballast. High wing ballast tanks help alleviate this condition, provided the tanks have appreciable capacity; otherwise the crew may not bother to use them.

Ore carriers may also be stiff in the loaded condition, depending on the cargo's center of gravity. Ore carriers generally have a high tank top to raise the cargo and its center of gravity. Alternate hold loading on bulk carriers also raises the cargo center of gravity.

Main Deck Plating. Classification society rules have in the past permitted minimum deck plating thickness between hatches. Cracks have occurred inboard of the line of hatch side girders. Plating of 0.5 in.(13 mm) thickness is therefore recommended, with well rounded corners at the hatch openings.

References

1 O'Leary, W. M., *The American Neptune*, Vol. 47, No. 2, Spring 1987, p. 83.

2 Snow, R. L., *Bath Iron Works, The First Hundred Years*, Maine Maritime Museum.

3 Sawyer, L. A., and Mitchell, W. H., *The Liberty Ships*, 2nd ed., Lloyd's of London Press.

4 Williams E. B., "Adaptation of Basic Liberty Designs," *Trans.* SNAME, Vol. 53, 1945.

5 Robinson, H. F., and Worthen, E. P., "The Ore Carrier S.S. *Venore*," *Trans.* SNAME, Vol. 53, 1945.

6 Henry, J. J., "Modern Ore Carriers," *Trans.* SNAME, Vol. 63, 1955.

7 Benford, H. "Ocean Ore Carrier Economics and Preliminary Design," *Trans.* SNAME, Vol. 66, 1958.

8 Dorman, W. J., "Combination Bulk Carriers," *Marine Technology*, Oct. 1966.

9 Jacobs, C. L., "Development of the Specialized Dry Bulk Carrier," *Marine Technology*, Jan. 1983.

10 "U.S. Flag Oceangoing Merchant Fleet," *Maritime Reporter and Engineering News*, June 1991.

11 "U.S. Flag Fleet," *Marine Log*, July/Aug. 1991.

12 Baseler, R., "Design, Construction and Operation of a 36,250 ST Self-Unloading Collier with a Coal-Fired Steam Plant," *Marine Technology*, Oct. 1984.

13 "World Shipbuilding & Repair," *Maritime Reporter and Engineering News*, June 1991.

STEVEN SPEAR* graduated from the University of Michigan in 1951 with a degree in Naval Architecture & Marine Engineering. During his professional career he was engaged almost exclusively with the design and construction of tankers (including LNG) and bulk carriers. His work took him to Japan and Taiwan, as well as to the United States. While employed by National Bulk Carriers, he worked on the design of what were then the world's largest tankers and bulk carriers. Other employers have included Esso International and Burmah Oil Company.

*The editors note with deep regret the death of Mr. Spear on February 25, 1993.

Tankers

STEVEN SPEAR

Historical Background

THE MERCHANT MARINE ACT of 1936 helped to revitalize tanker construction in the United States. The country was in the depths of the Great Depression. There was massive unemployment and little capital to invest in new ships. The U.S. Maritime Commission administered the Act. Details of many of the ships built under this program are listed in [1] and [2].

By today's standards, tankers of 10,000–12,000 tons deadweight and 10–13 knots speed seem small and slow. They were, however, economically suited to deliver cargoes from the oil fields of Texas, Louisiana, California, and Venezuela to the major markets on the U.S. East and West Coasts. In those days coal was the major source of energy in the United States, so the domestic demand for petroleum products was much less than it is today. Those were the days when the United States actually exported oil to Europe and the Far East.

Tankers of 22,600 dwt were built in 1921 at Newport News for the Standard Oil Company [3]. Twelve *Cimarron*-class tankers of 18,300 dwt, 17 knots service speed were built for Standard Oil of New Jersey and the U.S. Navy in 1939–1940.

By the 1940s the conventional arrangement of tankers had evolved with machinery aft and longitudinally framed structure. Cargo spaces were divided by two longitudinal bulkheads to provide center and wing tanks. The vessels had "three island" superstructures, i.e., poop, bridge, and forecastle.

In January 1941 the "Emergency (shipbuilding) Program" was instituted by the U.S. Maritime Commission. Standard tanker designs, such as the T-1, T-2, T-3, and Liberty Ship, Z-ET1-S-C3, were built in large numbers. Between 1939 and 1945 an amazing 669 tankers were built, of which 481 were of the T2-SE-A1 series. The number of tankers built during each year of this period is tabulated below [4].

Year	Number
1939	7
1940	5
1941	2
1942	33
1943	175
1944	266
1945	196
1946	5

Four shipyards built the T2-SE-A1 tankers during the war years. They were:

Sun Shipbuilding & Drydock Co.	198
Kaiser Swan Island Yard	147
Alabama Drydock and Shipbuilding Company	102
Marinship Corporation	34

Marinship also built 44 vessels of the T2-SE-A2 and T2-SE-A3 designs.

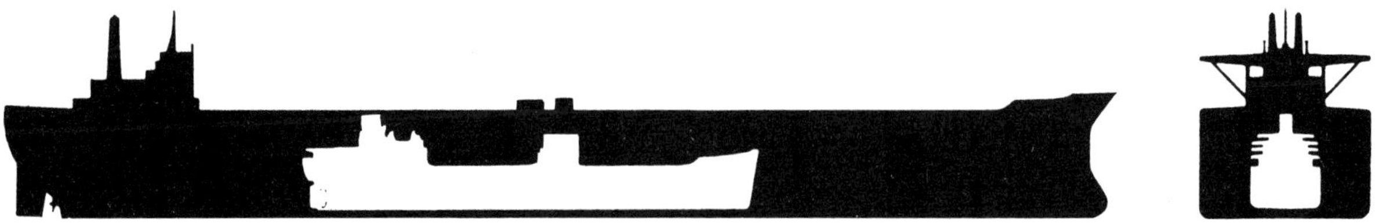

1942 1975

Fig. 1 The transformation of oil carriers to their present proportions appears at first more unbelievable than it does after careful consideration. It is more than coincidence that the plans of large tankers, drawn to a reduced scale of 1 to 100, are easily confused with those of T-2s drawn at one-half inch scale (1 to 24). The casual observer has difficulty grasping the difference in size and assessing the implications. When a VLCC with a length of 1000 ft is compared with a T-2 tanker of 503 ft, it seems at first that the one is twice as large as the other. Yet just doubling all dimensions increases the volume *eight* times and the cargo capacity more than eight times. In fact, the comparative beam of these vessels is 164 ft versus 68 ft, or a 240 percent increase, and the depths about 82 versus 39 ft-3 in., an increase of 210 percent. Small increases in linear dimensions beyond these figures yield much larger increases in deadweight capacity.

The war-built tankers provided vital petroleum supplies to our military forces, as well as to those of our allies, in North Africa, Europe, and the Pacific. At the end of the war the United States had the largest tanker fleet in the world.

Fig. 2 The *Gettysburg*, hull number 241, the first T-2 tanker. (Courtesy Sun Shipbuilding Collection, Philadelphia Maritime Museum.)

Tanker trading patterns changed in the post war years. The vast oil fields of the Middle East were discovered during the 1920s and 1930s. World War II, however, delayed their exploitation. The post war demand for oil increased dramatically in Europe and the United States. The Middle East oil fields were about 8300 miles from the U.S. East Coast via the Suez Canal. Before World War II, tanker routes from the Gulf of Mexico or Venezuela to our East Coast were in the 2000-mile range. By 1948 the United States had become an importer of crude oil. Hence, larger and somewhat faster tankers were required for economical transport of oil from the Middle East to Europe and the United States. The era of the "Supertanker" began.

Supertankers of the 1940s

The first supertankers were larger (about 600 feet) and faster (16–17 knots) than the pre-war tankers. Many built for the major oil companies were designed to serve a dual role:

(1) Larger deadweight, speed and bunker capacity for the long haul of crude oil (a relatively heavy cargo) from the Middle East. Draft was limited by the Suez Canal to 33 ft.

(2) Large cubic capacity to carry gasoline and heating oil (relatively light cargoes) on the shorter coastwise and Caribbean runs.

In the late 1940s U.S. shipyards were internationally competitive and available to construct these new ships. Major U.S. shipyards had multiple orders for tankers built to standard designs. The shipyards and some of their designs are tabulated in the Appendix.

The 1950s

During the 1950s the economic advantages of large tankers for the long crude oil trades had become apparent. The U.S. coastal trade in refined products also required larger ships to satisfy the increasing demand for gasoline and heating oil.

Two canals; the Suez and the Panama, imposed limits on the draft of new tankers. The permissible Suez Canal drafts and corresponding approximate deadweights in the various years are listed below [4]:

Year	Draft	Dwt
1954	35 ft (10.7 m)	32,000
1961	37 ft (11.3 m)	45,000
1964	38 ft (11.6 m)	65,000

The Panama Canal limited beams to 106 ft and drafts to 39 ft. These dimensions correspond to about 60,000 dwt for the Panamax tankers and bulk carriers.

In October 1956 Egypt nationalized the Suez Canal. The canal was blocked by Egypt when Britain and France tried to take it over by military force. It remained closed until March of 1957 [4]. Tankers had to sail around the Cape of Good Hope to deliver cargoes to Europe and North America, an 11,000-mile, 30-day voyage each way. Before that, voyages to Europe ranged from 4500 to 6500 miles while the distance to the U.S. East Coast was about 8300 miles [4]. To satisfy the demand for oil and to provide economic transport, tankers of 100,000 dwt were required. These large vessels had dimensions that far exceeded the limitations of the Suez and Panama Canals.

During the 1950s each year saw a new claimant to the title "largest tanker in the World" (based on deadweight). Ships owned by the American Daniel K. Ludwig held the title for several years starting in 1952 with a class of 38,000 dwt ships. The 44,600 dwt *Phoenix* followed in 1954, and then came the combination ore/oil self unloader *Sinclair Petrolore* of 56,000 dwt in 1955, and the 85,500 dwt tanker *Universe Leader* in 1956. These were built at Ludwig's National Bulk Carriers shipyard in Kure, Japan. That yard also delivered the 103,000 dwt tanker *Universe Appollo* in 1959.

Major U.S. shipyards enjoyed a boom in tanker construction with multiple orders and series production. The major U.S. oil companies contracted for ships to meet the needs of the U.S. coastwise trade as well as the long hauls from the Middle East. Foreign owners also placed orders for tankers to enter the latter trade. Shipyards involved were Bethlehem Steel Corporation's yards at Quincy MA and Sparrows Point MD, with tankers ranging in size from 25,000 to 106,600 dwt [12]. New York Shipbuilding entered the picture with a 30,000-ton class, and Sun Shipbuilding with a series of about 31,000 tons. Newport News Shipbuilding delivered 43 tankers in the period from 1949 to 1962 [13]. These ranged in deadweight from 19,000 to 66,500 tons [14].

The American built tankers were all steam turbine driven with higher horsepowers and greater speeds than those built in the rest of the world. The high speeds were justified economically by the high wages of the U.S. crews, which meant that the ship had to deliver more cargo in a given time. Suitability for national defense was also a consideration. The service speed of U.S.-built tankers ranged from 16 to 18.5 knots, while those built worldwide during the same period were generally designed for the 14- to 16-knot range.

The SNAME *Transactions* of 1957 included a landmark paper by Benford: "Engineering Economy in Tanker Design." His paper was unique in that it gave naval architects tools with which to make design decisions based on economic analysis, rather than on strictly technical characteristics. Among other things, the paper made clear the advantages of scale in tankers (or other bulk carriers).

The 1960s

The decade of the 1960s was a period of dramatic change for tankers in size, arrangement and details. Tankers tripled in size from 100,000 tons to over 300,000 tons. But the size of crews was reduced from about 43 to about 25–30 men. Crew accommodations improved with individual, air conditioned staterooms, and expert thought given to interior decoration.

Tankers in the short haul trades were fitted with bow thrusters to assist in docking and maneuvering. The mooring of large ships required considerable changes from earlier practice. Studies of the mooring forces indicated the need for larger capacity winches, different arrangements of mooring lines and deck fittings and the use of wire rope to withstand the forces. "First line ashore" rope reels were provided.

Electronics helped the navigator with such equipment as radar, Loran, Decca, speed logs, weather facsimile and quartz crystal chronometers and, later, with collision avoidance systems and satellite navigation.

Crew reduction in the engine room was justified by automation and remote control equipment. Pressures and temperatures were sensed remotely and recorded on a data logger. Standby pumps and generators could be started and brought on line automatically. Boiler automation was improved with wide range burners and steam atomization. Bridge control of the main engine became standard. Large machinery control boards with multiple pushbuttons, lights, and alarms became the norm and were eventually housed in air conditioned control rooms. Unattended engine rooms were approved by the classification societies.

The International Maritime Consultative Organization (IMCO) developed safety recommendations that were implemented by the laws of the various flag states. These included requirements for emergency firepumps located outside the machinery spaces. Steam smothering of cargo tanks was replaced by

Fig. 3 The *Esso Zurich* (hull number 566), 26,000 dwt, launched in 1948—one of the first post war supertankers. (Courtesy Sun Shipbuilding Collection, Philadelphia Maritime Museum.)

Fig. 4 The tanker SS *Manhattan*: length 940 ft, beam 132 ft, deadweight 108,590 tons. Delivered in 1962, the *Manhattan* has twin propellers and rudders, a cargo capacity of more than 38 million gallons, and a top speed of 19 knots.

foam systems. Emergency generators were required to provide power for emergency lighting, fire fighting and steering capability.

Concern for the environment became a factor in the 1960s. For years standard tanker practice was to wash the cargo tanks with hot saltwater and discharge the residue and muck (sand, sediment, rust, oil) into the sea. IMCO established recommendations that the tank residues not be discharged into the Mediterranean Sea or similar confined bodies of water or within 50 miles of shore. The slops should be retained on board. Tankers were fitted with slop tanks, i.e., small, deep tanks that permitted the oil and water to separate by gravity. After separation, the water was pumped overboard and the oil retained to mix with the next cargo. Oil/water separators were tried, but proved inadequate to handle the large quantity of slops involved.

In March 1967 the 118,000 dwt tanker *Torrey Canyon* ran aground off England's southwest coast, creating the largest oil spill to that date. As a result, more stringent regulations were developed for tanker construction and operation [4].

In one week in December 1969 four large tankers suffered explosions while cleaning tanks in ballast. The major oil companies and independent tanker owners immediately responded by studying possible causes and solutions to the problem. The solution was to install inert gas plants to supply washed boiler flue gas under pressure to the cargo tanks, and so exclude oxygen from those spaces.

During the 1960s world demand for oil increased approximately seven to nine percent per year. In June 1967 the Israeli–Egyptian war broke out. The Suez Canal was closed until June 1975 [4]. Tanker charter rates were relatively high during 1966 and 1967. These factors led to a rash of orders for large tankers from shipyards worldwide. In 1968–69 six 326,000 dwt tankers were built for D. K. Ludwig and Gulf Oil interests. By 1969 the tanker tonnage on order was more than double the tonnage of the existing world tanker fleet.

The situation in America, however, was different. The demand for larger tankers in the Jones Act (coastwise) trades had been pretty well satisfied. The T-2 tankers were still relatively young and many had been "jumboized" to carry 20–25,000 tons deadweight. The long haul cargoes from the Persian Gulf were generally carried under foreign flags. In 1959, however, the U.S. Government introduced oil import quotas, freezing the portion of oil imports to the then current level. The intent was to keep the portion of the country's oil consumption covered by "unreliable sources" from rising too high in this Cold War era. One reaction to the quotas was the formation of OPEC, the Organization of Petroleum Exporting Countries [4].

In the late 1960s the discovery of oil on the Alaskan North Slope provided a demand for large new tankers to carry this product to the lower forty-eight. This, of course, was domestic trade and so was covered by the Jones Act. Also, replacements were required for the pre-World War II tonnage still trading.

Deliveries of tankers from U.S. shipyards in the decade of the 1960s are tabulated in the Appendix [1,14–16].

The Maritime Administration (an outgrowth of the former U.S. Maritime Commission) provided Title XI assistance in the form of government guarantees of construction loans and mortgages for several of the tankers listed in the Appendix.

The 1970s

The decade of the 1970s saw drastic changes in the oil industry and the World economy and, as a result, the tanker industry as well. The decade started with world

Fig. 5 The tanker SS *Sinclair Texas*: length 736 ft, beam 102 ft, deadweight 48,563 tons. Delivered in 1963, the *Sinclair Texas* was the first large bridge-aft tanker to be built in the United States.

oil demand increasing each year. U.S. oil imports increased from 176 million tons in 1970 to 316 million tons in 1973 [4]. Much of this oil came from the Persian Gulf. Tanker spot market rates in 1970 were Worldscale 280. By the summer of 1973 the going rate was Worldscale 350. Since the shipowners' voyage costs were equivalent to about Worldscale 30, there were enormous profits to be made in the spot market. Many shipowners invested their profits in more and larger tankers. Special shipyards had been constructed in Europe and Japan to deliver these larger tankers. By 1973 most major shipyards offered standard designs in the 300–450,000 dwt range. A series of three 550,000-ton tankers were built in France and still remain the largest in the world. A conference was held in New York to discuss the possibility of a million-ton dwt tanker. These were exciting times in the tanker industry.

In October 1973, however, all of this changed owing to OPEC and the "Yom Kippur War" between Egypt and Israel. When diplomatic measures failed, Saudi Arabia imposed an oil embargo on the United States and Holland because of their relationships with Israel. OPEC increased the price of oil from $3.00 per barrel to $5.12. By January 1974 OPEC had raised the price to $11.65 per barrel and reduced total production of oil by 5 to 10 percent. OPEC discovered that it could control the price of oil [4,17].

The World economy based on cheap oil changed almost overnight, with a drastic effect on the tanker industry. Charter rates dropped from Worldscale 440 to Worldscale 50 between October and December of 1973. By April 1975, rates hit bottom at Worldscale 18. In 1976–77 rates averaged about Worldscale 40 [4]. The drastic change was affected by several factors:

Fig. 6 Powered by a slow-speed, 17,000-horsepower diesel and moving at an average speed of 16 knots, the *Exxon Charleston* (1973) carries petrochemicals in as many as 43 tanks from Exxon refineries in Louisiana and Texas to the U.S. East Coast.

- The high price of oil caused stagnation in the World economy and reduced the demand for oil.
- The high price of oil made exploitation of the North Sea and Alaskan oil fields economically viable. The distance to transport this oil was, however, considerably reduced when compared with a Persian Gulf voyage. Hence, the demand for tankers was drastically cut.
- By 1978 there was a 30 to 40 percent oversupply of tankers.
- Bunker prices increased from $10 per ton pre-1970 to $75 per ton by 1975, and $180 per ton in 1980.

Many tankers went into lay-up; some were used for floating storage tanks. Tankers on long term charter were "slow steamed" at 10–12 knots instead of the 14–16 knot design speed. This conserved bunkers, which had become a major expense to the charterers.

The technical efforts of shipowners, shipyards, and machinery manufacturers were concentrated on improving the propulsive efficiency and the fuel efficiency of the power plants. For example:

- Exxon set up a program to "tune" and adjust their steam plants to near-trial trip conditions.
- Mobil scrapped a steam plant in favor of diesel propulsion.
- Manufacturers modified existing steam turbines to operate more efficiently at the slow steaming speeds.
- Diesel engine manufacturers developed long stroke engines turning 70 to 80 rpm (versus normal 110–120 rpm) to improve propeller efficiency.
- Diesel main engines and generators were designed to use cheaper heavy fuels (6000 Redwood).
- Ducted propellers were installed to improve ship's speed or fuel economy.

Worldscale: What Does It Mean?

A Worldscale freight rate of 100 is obtained on any given tanker trade route by applying standard assumptions affecting transport economics on that route. These imaginary figures are found in a book called Worldscale, published by Worldscale Association (NYC), 17 Battery Place, New York, 10004. For various reasons, people engaged in tanker chartering find it convenient to base charter rates on a standard voyage adjusted for distance, ports, and vessel size rather than in actual dollars per ton.

During the 1970s a good deal of research and technical studies were devoted to tankers. The Society was fortunate to be a forum for technical papers in the following areas:

- The maneuvering characteristics of large, full block vessels studied by means of model tests, computer programs, and actual sea trials [21–24].
- Studies of structural members in the cargo tanks to determine loads and verify computer calculation methods [25–28].
- Model testing and studies to understand and minimize propeller induced vibrations [29,30].
- Investigations of new tanker operating procedures for preventing pollution [31,32].

By the decade of the 1970s it was apparent that the United States was not going to have deep water ports. This led the major oil companies to develop lightering techniques for unloading VLCC's in the Gulf of Mexico. An offshore terminal was eventually built with pipelines to shore and underground storage. At the same time, American naval architects recognized that restricted draft tankers could operate economically into some U.S. ports [33,34]. Chevron Corporation, for example, "de-jumboized" one of their 200,000 dwt tankers to about 150,000 tons to suit their limited draft trade.

Oil pollution by tankers was a worldwide concern highlighted by the stranding of the *Torrey Canyon* in

Fig. 7 The 188,500 dwt tanker *Arco Alaska* (1979), built by National Steel and Shipbuilding Co. for Atlantic Richfield.

1967 and the *Amoco Cadiz* in 1978. The *Argo Merchant* stranding instigated the 1978 Port and Tanker Safety Act. In the same year the IMCO Conference on Tanker Safety and Pollution recommended that vessels be equipped with inert gas plants and oil/water monitors. Segregated ballast tanks or crude oil washing systems, or both, were required to minimize the amount of oil residue discharged overboard. Crude oil washing proved economically feasible on existing tankers because (1) the process reduced the amount of oil clinging to the surfaces within the tanks, and oil was now a valuable commodity and (2) the modifications could be carried out during any extended shipyard overhaul period.

IMCO recommendations required sufficient segregated ballast to achieve minimum draft, limits on trim, and required propeller immersion. The Conference also recommended that a second radar be fitted and improvements be made to increase the reliability of the steering systems [35].

There was considerable demand for new construction of tankers in the United States during the 1970s. This was owing to three factors:

- The need to replace World War II tankers with larger ships of more economic and efficient design.
- The Alaskan oil trade.
- The 1970 Amendments to the Merchant Marine Act of 1936. The changes provided Title V (Construction Subsidy) and Title XI financing (government guaranteed mortgage bonds) for U.S.-flag tankers in both foreign and domestic trades.

The result was that the major shipyards and some "not-so-experienced" shipyards built considerable new tonnage. Over 50 tankers were delivered during the 1970s [1,36]. Some of the vessels are listed in the table in the Appendix.

Several innovative designs were developed for the U.S.-flag fleet. Avondale produced a wide-beam (L/B = 4.99) shallow draft design of 173,380 dwt. Chevron Shipping had six 39,200 dwt product carriers with gas turbine/electric main propulsion, built by the FMC Corporation. Ingalls Shipbuilding built the first post-war diesel powered tankers, with Colt-Pielstick engines. National Steel designed and built the first 82,000 dwt OBOs for the U.S. fleet. Sun Shipyard developed the 124,000 dwt "ecology class" tankers.

Several of the tankers listed above have double hulls or double bottoms to carry segregated ballast and minimize oil pollution. The designs include the Chevron gas turbine vessels and the "*Tonsina* ecology class" tankers. The ships were built about 15 years before U.S. laws required such construction [37].

Crude Oil Washing

"Crude oil washing" is a procedure for cleaning cargo tanks while discharging cargo. One cargo pump takes oil from a given tank and sends it to tank cleaning machines located on the deck and in the cargo tank being emptied. Other pump(s) take oil from that same tank and discharges it ashore. This procedure removes muck (sand and rust) from the tanks and reduces the amount of oil clinging to the internal structure. Crude oil washing has proven to be more effective than the hot (175 °F) salt water washing system of former days.

The 1980s

The decade of the 1980s started with the World economy in recession. Part of the cause occurred in late 1978, when the Shah of Iran was deposed. Oil production in Iran was shut down. OPEC could not increase oil supply to meet the demand. Spot trading of oil in Rotterdam and Geneva bid up the price. By May of 1979 crude oil was selling at $30 to $34 a barrel. The effects on the international tanker market were drastic [17]. By 1982 sixty to seventy percent of the world tanker fleet was surplus to demand. World tanker requirements dropped from 280 million tons deadweight in 1979 to 200 million tons in 1982 [4]. Major independent shipowners went bankrupt. Banks and governments attempted to refinance ships, while the major oil companies reduced their tanker fleets. The rate of scrapping vessels increased [39].

By 1985 there was a surplus of oil in the world since Iran and Iraq were producing to capacity to finance their war requirements. Astute financial interests such as Loews Corporation (theaters and hotels) began purchasing large surplus tankers at bargain prices. By May of 1986 the price of oil had bottomed out at $10 per barrel, which stimulated the world economies to produce the boom of the 1980s.

Tanker demand immediately increased. Ships came out of lay-up. Slow steaming became a thing of the past. Orders were placed for new tankers, first in Yugoslavia, then in Korea and Japan. The SUMAX tanker of 140–145,000 dwt was offered. Its 53-ft draft and dimensions were developed to suit the enlarged Suez Canal.

In the late 1980s Lloyd's Register, the American Bureau of Shipping and other major classification societies developed rules and requirements for "Vessel

Life Extension" programs. The large tankers built in the early to mid-1970s were approaching 15 years of age or Third Special Survey. The intent was to perform a rigorous survey of the vessel, to carry out maintenance and repair to permit it to continue active trading for another 10 to 15 years [38].

The IMCO Tanker Safety and Pollution Prevention Specification of 1978 required the installation of inert gas systems. The installation on a U.S. built ULCC and the test results are described in [40]. The IMCO specifications also required Crude Oil Washing (COW) [41] and segregated ballast tanks [42] or clean ballast tanks [43]. SNAME concurrently provided a continuing open forum for technical discussions related to these emerging international requirements.

Size Definitions

Within the marine industry there are a number of loosely used terms categorizing tanker sizes. During World War II the standard war-built T-2 tanker (of about 17,000 dwt) was considered large. Then, however, in the late 1940s the term "supertanker" was initially applied to ships of about 28,000–30,000 dwt. During the 1950s, as tanker deadweights went up in five- to ten-thousand-ton steps such as 45, 60, 80, and eventually 100,000, the term "supertanker" still applied. (By then many owners thought of these monsters as being the equivalent of so many T-2s.)

In the 1960s, tankers designed to go around the Cape of Good Hope grew to 200,000 to 280,000 dwt; these were commonly called VLCCs (very large crude carriers). "Supertankers" had in effect now become "Inferiortankers." The 26,000–60,000 dwt tankers of the '40s and '50s are now considered handy size, or "handies."

During the 1970s we saw the introduction of tankers ranging in size from 300,000 to 550,000 dwt. These were called ULCCs (ultra large crude carriers). What next? Quick, Henry, Roget's Thesaurus!

Now, to add to the confusion, we find that modern chartering arrangements call for "parcels" of one million barrels of oil (135,000–140,000 dwt) or two million barrels (260,000–280,000 dwt).

With the price of a barrel of oil approximately $20, the accuracy of cargo tank calibration [44] and tank gauging systems [45] became important considerations. The presence of inert gas in the cargo tanks made remote reading tank level indicators necessary.

Research continued on the sloshing phenomena in slack tanks to determine whether the loads imposed and the structure of the long cargo tanks required additional reinforcing. The results of these studies are reported in [46] and [47]. The problem of surplus tankers was taken up at symposiums and in the technical journals [48].

A paper on shallow or restricted draft tankers offered a solution to the lack of deep water ports in the United States and other parts of the world [51].

The concern for the environment fostered in the 1960s continued through the 70s and 80s. Large oil spills gave the press and television a dramatic opportunity to focus on the problem, but did not aid in the clean up. Each large spill brought new regulations for the operation and equipment required for oil tankers. Some solutions to the oil spill problem were offered in the Society's technical forums [52–54].

Laws were passed requiring vapor recovery systems to control hydrocarbon vapors escaping into the atmosphere. Reference [55] provides guidance and experience on the installation of such a system.

In the United States, the tanker building programs started in the 1970s to renew the domestic fleet continued into the 1980s. From 1980 to 1986, 27 major tankers and two small coasters were built. During this period there was a major change from steam turbine propulsion to diesel engines. Of the 29 tankers built, 4 had steam propulsion and 25 had diesel main engines. Many of these vessels are listed in the Appendix.

Sun Refining and Marketing Company had two small coastal tankers built for delivering products. The 2,650 dwt *Northern Sun* was built by Mangone Shipyard in 1981 [57]. Based on the experience of the *Northern Sun*, the *Eastern Sun*, of 3,400 dwt, was built at Jeffboat, Inc., in 1985.

The 1990s

In March 1989 the *Exxon Valdez* became stranded in Prince William Sound, Alaska. This major oil spill

Fig. 8 The 1143-ft tanker U.S.T. *Pacific*, built at Newport News Shipbuilding and Dry Dock Co. in 1979.

Fig. 9 The 855-ft, 90,393-dwt tanker *Overseas New York*, built at National Steel and Shipbuilding Co. (1977).

and the resulting publicity had a significant effect on tanker construction and operation worldwide. Responding to the hue and cry of the press, television and the public, the U.S. Congress passed the Oil Pollution Act of 1990. The Act mandates that any ship or barge, of whatever flag, carrying oil as a cargo and operating in U.S. waters (and the U.S. Exclusive Economic Zone) be built to double hull standards if contracted after June 30, 1990. The Act provides a phase-out schedule for older single hull tankers to begin in 1995 [59].

The 1990 Oil Pollution Act goes well beyond previous legislation, such as the 1976 requirement that tankers of 70,000 dwt or more have "protectively located segregated ballast tanks" (PL/SBT). This requirement was incorporated into the Ports and Waterways Safety Act (1972). Annex I of MARPOL 73/78 extended the PL/SBT requirement to vessels of 20,000 dwt. The 1990 Act requires protection along the entire bottom and sides of the cargo block. In response to OPA90, the international community, led by IMO (formerly IMCO), has adopted double hulls or equivalents, such as the mid-deck tanker, as the standard for new construction.

Legislation by the U.S. Federal Government and some states raised the dollar limits per gross ton of legal liability in the event of an oil spill. The new legislation will have definite effects on the world tanker industry in future years:

(1) Some tanker operators may choose to keep their vessels out of U.S. waters because of the liability requirements. This will affect charter rates to U.S. ports. Nonetheless, the United States is the largest oil consumer in the world, and so many shipowners will comply.

(2) The conventional design of oil tankers and barges will change to meet the new double hull standards.

During 1991 the U.S. Coast Guard requested comments and studies from the industry to assist in determining the specific requirements of the double hull standard [59].

Reaction to the new legislation and liability requirements was forthcoming in various Society technical forums. A panel discussion was held at the 1990 Annual Meeting in San Francisco. One speaker proposed a "mid-deck" tanker, a concept developed by Mitsubishi Heavy Industries, to minimize oil spillage

Fig. 10 The *Frank Rockefeller*, built in 1896 for the carriage of iron ore.

in event of a casualty. Another explained his experience with 39,200 and 130,000 dwt double hull vessels concerning access, inspection, tank coatings, damage stability, flooding and gas freeing. An insight was given into the background and framing of the legislation for the 1990 Act [60].

At a SNAME Philadelphia Section meeting a scheme for short loading single hulled tankers was proposed. The method would balance the level of oil to the hydrostatic pressure of the water outside the hull to minimize oil spill leakage. Another suggested method involved curved shell plating that would take hydrostatic loads as balanced membrane stresses [61].

Not All Tankers Look Alike

Built in 1896 by the American Steel Barge Company in Superior, Wisconsin, for the carriage of iron ore, the Frank Rockefeller *was a typical example of a remarkable new technology in hull form. The concept of Captain Alexander McDougal, these "whalebacks" were designed to go through the waves rather than over them, in effect to "adapt to the sea state" rather than fight it. During her almost century-long career, the* Frank Rockefeller *first served her namesake firm in the trade for which she was designed. In 1926 she became the* South Park. *In 1936 her configuration was adapted for the carriage of automobiles. In 1941 she was reconverted to a bulk freighter to serve wartime requirements for the transport of iron ore, and in 1943 she was again converted, this time to a tanker configuration and became the* Meteor. *She continued in the tanker trade until her final and continuing service as a museum ship at Superior, the city of her birth. In a remarkable parallel to industrial society and to her namesake family she served first as a carrier of ore, then in perhaps the heyday of automobile developments, she became an adjunct to that industry. Finally, in her last days as a tanker, she was a tranporter of aircraft jet fuel! As a unique representative of our own Society, very nearly paralleling our own organization's 100-year history, she is on permanent display to bring to the public some sense of what our industry is all about.*

A noteworthy paper presented at the Northern California Section analyzed 20 years of crude oil supply and demand to compare with new tanker orders and construction costs, second hand prices, tanker tonnage built, spot market rates and tonnage laid up. The age profile of the tanker fleet, shortages of seafaring personnel, reduced shipyard capacity and the "Life Extension Programs" were all considered in an attempt to determine when tanker replacement programs should start [62].

The salvage and repair of the *Exxon Valdez* provided three valuable papers [63–65].

During the late 1980s and early 1990s new construction of tankers was greatly reduced. There was still much surplus tonnage available. One special ship, however, was built for the specific task of lightering the VLCC's [66].

The Persian Gulf War of 1990–91 caused oil spot market prices to reach the $30–$40 per barrel range, but the impact was brief because most of the major industrial countries had plentiful oil in storage. Oil from Iraq and Kuwait was embargoed, but the other oil producing nations made up for that. The price of oil increased, but not drastically, and tankers continued to enjoy a reasonably good market.

The long term effect of the global economic downturn of 1991 cannot be determined at this time (January 1992).

General Configuration

The oceangoing tanker of the 1940–1950s was generally a "three-island" vessel with forecastle, midship, and poop superstructures. Catwalks provided access between the superstructures, and carried hull piping and electric cables. The navigation bridge and deck officers' quarters were located in a midship deckhouse. Engineer's and unlicensed crew's quarters, the galley and mess rooms were in the poop deckhouse. Machinery spaces were fitted aft with relatively large machinery casings to provide a 32 percent propelling machinery space deduction from gross tonnage. (The tonnage laws have since been changed.)

In the late 1950s the first large tankers were built with the navigation bridge and accommodations aft. This change reduced the cost of construction,

Fig. 11 The *Meteor*, the result of the *Frank Rockefeller* undergoing a conversion to an automobile carrier, a reconversion to a bulk freighter, and a final conversion to a tanker configuration, as well as two name changes.

increased deadweight and reduced maintenance costs of the catwalk, piping and electrical services to the midship house. It became standard when it was adopted as part of the International Convention for Safety of Life at Sea (SOLAS). Designers were quick to follow this practice with a combined deckhouse, machinery casing and tall stack. Then they found that the machinery casing and stack transmitted fore and aft vibrations to the deckhouse, which was detrimental to the bridge electronics and operations. Separation of the deckhouse and machinery casing/stack evolved in the 1960s with stores handling cranes for the engine room and accommodations placed between the deckhouses.

From the 1940s to 1960s tankers were designed to be aesthetically pleasing. Deckhouses had rounded corners, curved fashion plates, some degree of streamlining and a good looking stack. Over the past fifty years, however, functional design has prevailed in order to reduce construction costs. Many tankers are now built with no sheer, square stacks, square deckhouses and transom sterns. Streamlined masts have been replaced with pipe tripods.

The design of cargo manifold arrangements and cargo hose handling was generally left up to each shipowner until OCIMF (Oil Companies International Marine Forum) developed standard arrangements in the mid to late 1960s.

Post World War II tankers were generally fitted with a dry cargo hold, serviced by kingposts and booms. This was possibly the last remnant of the time when oil was carried in barrels or cases. As tankers were designed specifically for long haul crude oil service, this feature was deleted.

During the period 1940 to 1960 cargo tanks were about 40 ft long, and arranged three abreast (i.e. port, centerline and starboard) with two longitudinal bulkheads. In the 1960s designers reduced steel weight by increasing the tank lengths, reducing the number of tanks and thus eliminating some bulkheads. This practice was restrained by two international conventions: the 1966 Loadline Convention and the 1973 International Convention for the Prevention of Pollution from Ships. The authors of the 1966 Loadline Convention realized that large vessels with tanker freeboard no longer had tanker sub-division, i.e., numerous tank compartments and bulkheads. The 1966 Convention permitted deeper drafts, but required damaged stability calculations, engine room flooding calculations, loading manuals, stability booklets and protection against down flooding in the arrangement of vents, closures and watertight doors.

The 1973 Pollution Prevention Convention imposed limits on tank size to restrict the hypothetical oil outflow. Also slop tanks were required to contain cargo tank washings.

Early tankers were generally designed with forward and aft pumprooms. Modern practice has one pumproom aft with vertical pumps driven by steam turbines located above in the engine room.

Hull Form

The T2-SE-A1 design had a notably efficient hull form. A small bulbous bow was fitted. The vessel's characteristics are listed below:

LOA	523 ft-0 in.
LBP	503 ft-0 in.
Beam	68 ft-0 in.
Depth	39 ft-3 in.
Draft	30 ft-3 in.
L/B	7.4
L/D	12.8
B/H	2.25
SHP (normal)	6000 @ 90 rpm
Propulsive coeff.	~0.78
Service speed	14.5 knots
Speed/length	0.65
Block coeff.	0.740

During the next 20 to 30 years tanker hull forms changed considerably. Service speeds generally were in the 14- to 16-knot range regardless of ship size. As a result the block coefficient progressively increased from about 0.75 to 0.85. The bow lines were filled out to provide a fuller hull form and to move the longitudinal center of buoyancy forward of amidships. The stern lines, however, had to be kept relatively fine in order to provide good flow to the propeller. Tanker hull forms had considerable parallel middle body. This hull form proved satisfactory for low speed/length ratios of 0.4 to 0.5. Experience showed that the full bow could not be driven into heavy seas without severe slamming damage. In moderate sea conditions, however, the form proved to be entirely satisfactory. The primary advantage of a full block coefficient was increased deadweight on a given set of vessel dimensions.

Between the 1950s and 1970s the length to beam ratio (L/B) gradually decreased from about 7.5 to 5.5. Vessel length is a major determining factor in structural design. A shorter, wider vessel requires less steel, hence is less costly to build.

In the mid-1960s large bulbous bows were fitted to the full bodied tanker hull forms. Bulbous bows had been used on high speed vessels for many years. Model tests showed that large bulbous bows could be designed to give an increase in speed at constant power

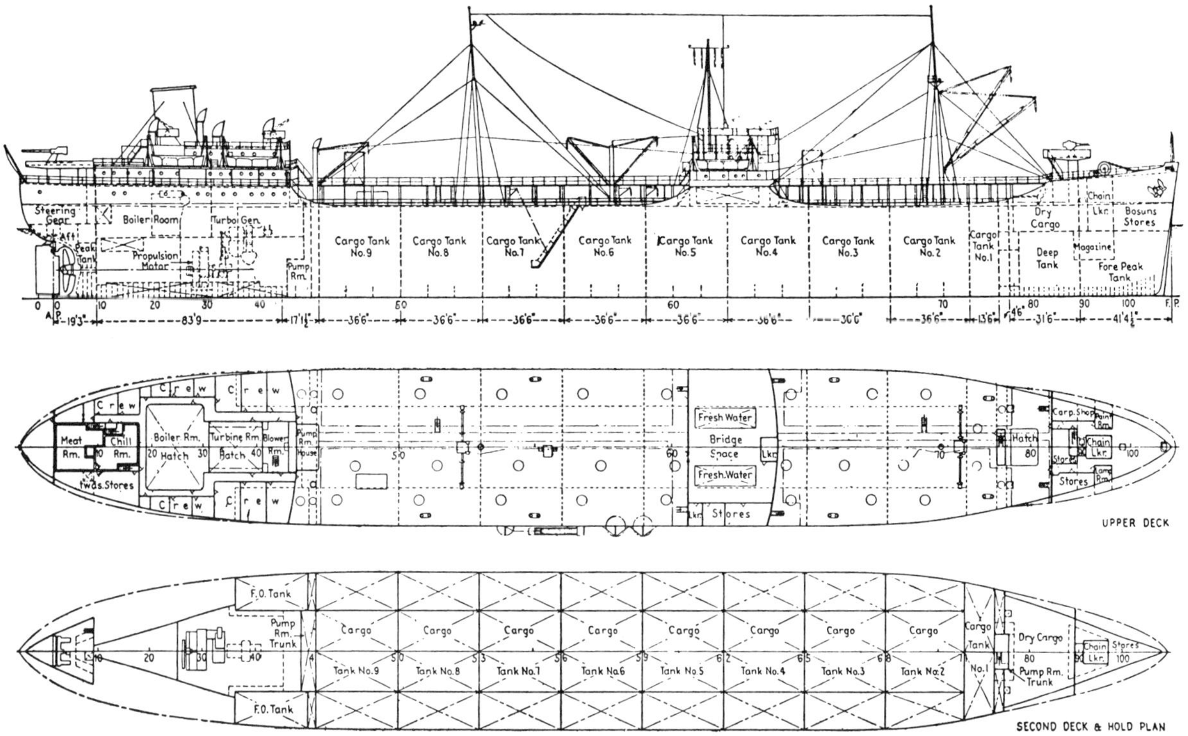

Fig. 12 General arrangement of the *James Island*, a typical T2-SE-A1 turbo-electric tanker designed by Sun Shipbuilding and Dry Dock Co. in collaboration with the U.S. Maritime Commission.

for either the full load condition or the ballast condition. Speed increases of about a knot were not unusual. Bulbous bows, however, required considerable shaping of plates by furnacing or plate rolls. As a compromise, more easily shaped cylindrical bows were developed to provide the same increase in speed.

Steam turbine plants were designed to run at about 100 rpm. Slow speed direct drive diesels operated in the range of 110 to 120 rpm. The oil crisis of 1973 caused marine engineers to develop diesel engines and geared turbines operating in the 70–90 rpm range to improve propeller efficiency.

Hull Structure

Over the past 50 years there has been a major change in the steel structure of tankers brought about by the switch from riveted to welded construction. The first tanker to employ extensive welding in the cargo spaces was the *J. W. Van Dyke*, built at Sun Shipbuilding and Drydock Company in 1937 [9]. During World War II construction methods changed dramatically. The war created the need for great numbers of ships within a short time. Many welders were available from other industries, hence their training was easier and quicker than training riveters, chippers and caulkers. Fabrication of large subassemblies was generally simplified by the welding process. Building ways were at a premium and building time could be reduced by the use of large subassemblies. Most of the war-built ships were all-welded, and welding was used extensively in a wartime construction.

In post-war years U.S. shipyards had large multiple ship orders for tankers. The numerous orders provided the incentive for considerable research on ship structures by the major shipyards, the American Bureau of Shipping, the National Research Council, and the David Taylor Model Basin. Some of the results were presented in the Society *Transactions*, Vol. 57, 1949. Research findings included the following:

- Studies of the reactions of riveted versus welded ships proved that rivet slip was not a major factor in the deflection of tanker hulls [5].
- Studies of the distribution of nonhomogeneous cargoes showed that cargo distribution has an important influence on bending stresses in a tanker hull [7].
- Welded construction requires specific details to eliminate "hard spots" and notches in the struc- ture [6].
- The chemical composition of ships' steel has a definite effect on the material's ability to resist brittle fracture at cold temperatures [8].

The need for loading instructions for the proper distribution of cargo in tankers was recognized by the

Maritime Administration. An influence diagram showing the tension and compression in the deck per change of 100 tons added load per tank was developed for the T2-SE-A1 tankers [7].

The decision in a major law case obligated U.S. ship operators to provide their ships' masters with means to properly load their vessels. This hastened the development of loading manuals, which ultimately became a requirement of the 1966 Load Line Convention.

Several war-built T-2 tankers broke in half in service. The American Bureau of Shipping in concert with the tanker operators required that riveted straps be installed on the deck and shell to serve as "crack arrestors." Alternatively, some owners chose to increase the hull's section modulus by about 15 percent by means of deck and shell girders. Both methods seemed to be effective in reducing the incidents of failure in service.

The American Bureau of Shipping revised its steel specifications in 1948 to require Class B steel for plates over 0.50-in. to 1.0-in. thickness inclusive and Class C steel for plates 1.0-in. to 1.375-in. thickness [9].

The structural details of tankers changed considerably between 1945 and 1950. The war-built T-2s had flanged web frames and transverses. Connecting brackets were also flanged. Lapped joints were the rule. By the 1950s transverses and brackets were of single plane construction with sections butt welded together, although some erection joints were still lap welded. Large radiused corners were provided at the junction between horizontal and vertical members. Welded face

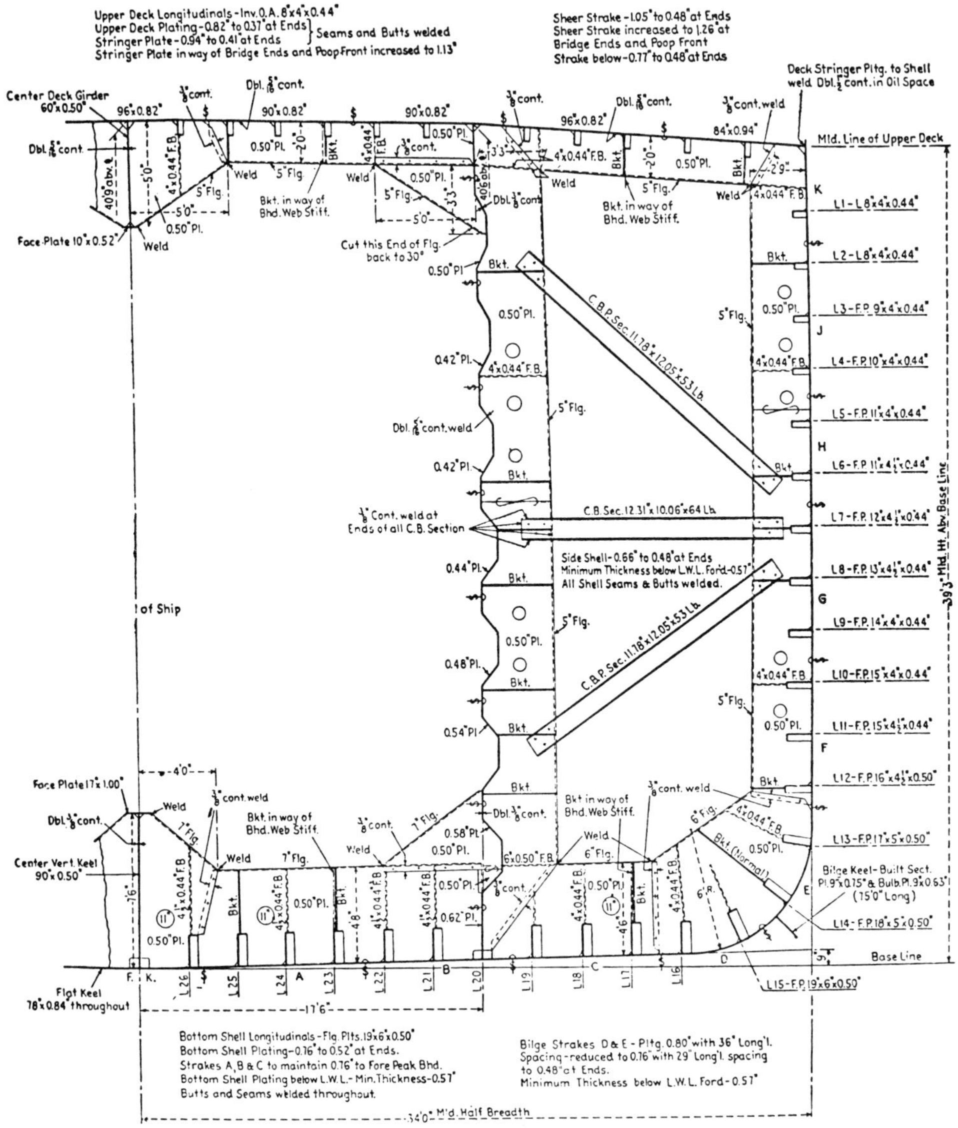

Fig. 13 Midship section of the tanker *James Island.*

plates were fitted in lieu of flanges. Brackets were designed with a "soft toe." H-beams fitted in the wing tanks were replaced by fabricated struts in the plane of the transverse web frames [10].

The section modulus requirements for all vessels up to 600 ft long were included in the 1930 Load Line Convention. Before the 1960s, the "standard" bending moment for structural design was based on the vessel loaded with a homogeneous cargo and balanced on a trochoidal wave of the same length as itself. The assumed wave height was 1/20th of the vessel's length. For vessels with machinery aft, the greatest bending moment was generally in the sagging condition. As a precaution against deck buckling, the effective longitudinal deck area was increased 20 percent over that set by normal classification rules. This resulted in increased section modulus requirements for tankers.

As tankers increased in lengths over 600 ft, studies showed that the L/20 wave criterion was too severe. Research efforts produced the following results:

- Data obtained from weather ships and ocean stations provided information on actual wave proportions.
- Instruments installed on a number of ships measured the actual stresses at sea.
- Model testing programs established the correlation pattern between model response to simulated wave conditions and the measured response of ships in actual waves.

Based on this research, computer programs were developed to predict ship motions and wave loadings. The result was a reduction in the required section modulus for tankers over 600 ft in length. The first reduction occurred in the early 1960s, based on less severe effective wave height criteria. About 1980 a second reduction in tanker section modulus was permitted, based on the research findings listed above. This provided a significant increase in the deadweight capacity owing to the reduction of major scantlings.

During the 1960s tanker steel weight was also reduced by the following design changes:

- Tank lengths were increased from about 40 ft to over 100 ft. Light weight swash bulkheads were fitted at structural points of support.

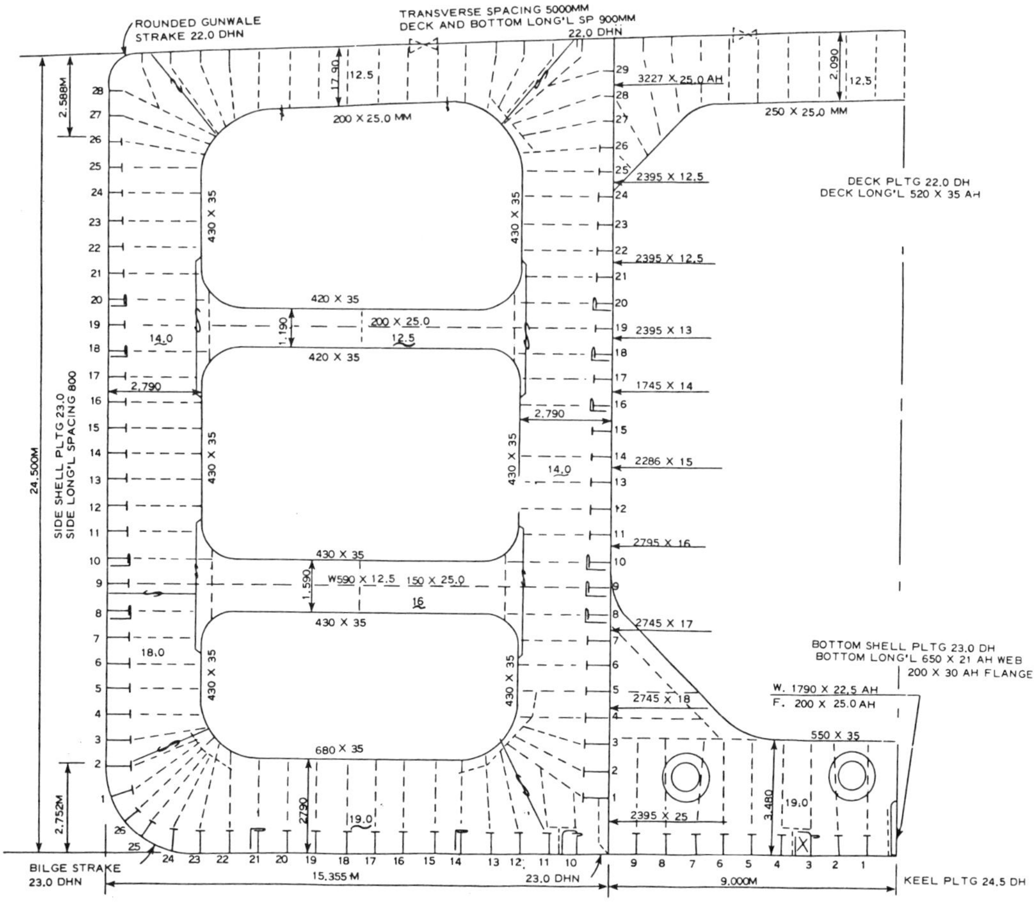

Fig. 14 Midship section design based on ABS Rules 1980.

- The spacing of transverse web frames was increased from about 10 ft to 15 or 16 ft, thus reducing the number of frames.
- Longitudinal frame spacing was increased from a standard 2.5 ft to 3 ft or more.

The classification society rules provide a corrosion allowance on steel structure. The allowance is based on normal corrosion rates in the marine environment. But, the internal structure of tankers is subject to much more severe corrosion when tank cleaning is performed after carrying refined products such as gasoline, heating oil, or jet fuel. In those trades it was normal to "Butterworth" cargo tanks with hot (175 °F) saltwater before loading a different cargo. Tankers in the crude oil trade experienced severe pitting brought on by sulphur in "sour crudes." During the early to mid-1960s tank coatings such as inorganic zinc, epoxy and vinyl showed promise of providing good protection against corrosion. Cathodic protection in the form of anodes placed in the tanks also reduced the rate of corrosion.

Prior to these developments it had been necessary to make major structural renewals in the cargo tanks between the 8th and 12th years, and again in about the 15th year [11]. The sophisticated coatings required blasting the steel to bare metal to ensure good surface preparation. Nevertheless, the costs of blasting and coating proved to be less expensive than the periodic renewal of steel. As tank coatings and cathodic protection proved to be effective, the classification societies allowed a reduction in scantlings.

Computers gave ship designers a more accurate means of predicting stresses and strains on ship structure. Scantlings required by classification society rules were based on experience, i.e., what had worked successfully over the years. But, the rapid increase in tanker size outpaced experience. Fortunately, computer programs permitted the rapid calculation of forces, moments, and deflections of complex structures. Grillage analysis programs allowed the designer to determine the interaction of plates, stiffeners and web frames when subjected to hydrodynamic forces and cargo loadings.

Finite element analysis permitted the structural designer to model the entire cargo tank space using a "coarse model." Hull bending moments and shear forces could be imposed on this model to determine their effects on the major structural members. A "fine model or mesh" was used in areas where structural behavior was complex [10].

Now the classification societies require scantlings derived from rule equations or tables. In addition, computer analysis is required to verify the imposed loads and stress distributions.

High strength steels have been used in the deck and bottom shell plating of some tankers as a means of further reducing the weight, although they allow greater hull deflections. However, the fatigue characteristic of this steel is similar to that of mild steel. The higher stresses, combined with relatively greater corrosion effects, has contributed to fatigue cracking problems for some design details.

Historical Overview

In the past 50 years the design, construction, and demand for tankers has been affected by world economics, politics, wars, and environmental concerns.

Tankers are no longer pleasing to the eye. They are strictly functional. Perhaps this is logical because they spend so much time at sea, and their ports of call tend to be rather remote.

The arrangement and number of tank spaces has changed over the past 50 years to increase vessel deadweight capacity. This will change again to suit the double hull configuration.

New equipment, such as inert gas plants, foam fire extinguishing systems and crude oil washing is now installed to meet the safety and environmental requirements that were not considered 50 years ago.

The hull form of tankers has been pushed to the limits by increasing block coefficient and beam, while at the same time reducing bow wave resistance and refining the lines through model testing. These changes have increased deadweight and reduced both fuel consumption and construction costs.

Computers have played a major role in the hull and structural design of tankers. They have made possible the mammoth tankers of today.

Research in model tanks and at sea has provided knowledge for structural design, wave induced vibration, propeller vibration, and the maneuvering characteristics of large tankers. This knowledge has largely replaced the empirical methods of design used 50 years ago.

The past 50 years have seen major technological change, but the basic function of a tanker remains the same, i.e., transporting petroleum or products economically from the producer to the marketplace.

References

1 Dillon, Hoffman, and Roseman, "Forty Years of Ship Designs Under the Merchant Marine Act, 1936–1976," *Trans.* SNAME, Vol. 84, 1976.

2 Robinson, Roeske, and Thaeler, "Modern Tankers," *Trans.* SNAME, Vol. 56, 1948.
3 Dunn, L., *The World Tankers*, Adlard Coles, 1956, p. 65.
4 Ratcliffe, M., *Liquid Gold Ships—A History of the Tanker 1859–1984*, Lloyd's of London Press, 1985, p. 140.
5 MacCutcheon, E. M., "Rivet Slip, Stress Distribution and Deflection of Ship's Hulls," *Trans.* SNAME, Vol. 57, 1949.
6 Brown, D. P., "Structural Design and Details of Longitudinally Framed Tankers," *Trans.* SNAME, Vol. 57, 1949.
7 McDonald, J. H., and McNaught, D. F., "Investigation of Cargo Distribution in Tank Vessels," *Trans.* SNAME, Vol. 57, 1949.
8 Acker, H. G., "Highlights of Welded Ship Research," *Trans.* SNAME, Vol. 57, 1949.
9 *Design and Construction of Steel Merchant Ships*, SNAME, 1955, pp. 128–129, 151–153.
10 *Ship Design and Construction*, SNAME, 1980.
11 "Symposium on Control of Internal Corrosion of Tankers," *Trans.* SNAME, Vol. 60, 1952.
12 Nichols, W. O., Rubin, M. L., and Danielson, R. V., "Some Aspects of Large Tanker Design," *Trans.* SNAME, Vol. 68, 1960.
13 Long, C. L., Stevens, J. L., and Tompkins, Jr., J. T., "Modern High Speed Tankers," *Trans.* SNAME, Vol. 68, 1960.
14 Fox, W. A., *Always Good Ships*, Histories of Newport News Ships, 1986.
15 *The Tanker Register*, H. Clarkson & Co., Ltd., London, 1972.
16 *Seatrade North American Yearbook*, Seatrade Publications, Calchester, Essex, U.K., 1985.
17 Sampson, A., *The Seven Sisters*, rev. 4th ed., Bantam Books, June 1991.
18 Narita, H., Kunitake, Y., and Yagi, H., "Application and Development of a Large Ducted Propeller for the 280,000 DWT Tanker MS *Thorsaga*," *Trans.* SNAME, Vol. 82, 1974.
19 Meyerhoff, L., Hill, J. G., and Meyerhoff, S., "Ducted Propeller Applications for Modern Ships," *Trans.* SNAME, Vol. 80, 1972.
20 Laredo, A., Beghin, D., and Garguet, M., "Design of the First Generation of 550,000 DWT Tankers," *Trans.* SNAME, Vol. 85, 1977.
21 van Berlekom, W., and Goddard, T., "Maneuvering of Large Tankers," *Trans.* SNAME, Vol. 80, 1972.
22 Crane, Jr., C. L., "Maneuvering Safety of Large Tankers: Stopping, Turning and Speed Selection," *Trans.* SNAME, Vol. 81, 1973.
23 "Effect of Rudder Rate on Maneuvering Performance of a Large Tanker," SNAME T&R No. R-22, 1976.
24 Eda, H., Falls, and Walden, "Ship Maneuvering Safety Studies," *Trans.* SNAME, Vol. 87, 1978.
25 Little, R., Lewis, E., and Bailey, "A Statistical Study of Wave-Induced Bending Moments on Large Oceangoing Tankers," *Trans.* SNAME, Vol. 79, 1971.
26 Mansour and Jones, "Elastic and Plastic Analysis of a Tanker Web Frame," *Journal of Ship Research*, Sept. 1973.
27 Finifter and Mansour, "Finite-Element Analysis and Optimization of a Web Frame of a Tanker with Isolated Ballast System," *Journal of Ship Research*, June 1974.
28 "Dynamic Slosh-Induced Loads on Liquid Cargo Tank Bulkheads," SNAME T&R No. R-19, 1975.
29 Dyne, G., "A Study of the Scale Effect on Wake, Propeller Cavitation and Vibratory Pressure at Hull of Two Tanker Models," *Trans.* SNAME, Vol. 82, 1974.
30 Dyne, G., and Hokstra, "Propulsion, Cavitation and Propeller Induced Fluctuations of a Tanker," SNAME Spring Meeting Papers, 1976.
31 Overaas, S., and Solum, E., "The Load on Top System for Crude Oil Tankers—Experience and Possible Design Improvements," *Trans.* SNAME, Vol. 82, 1974.
32 Bass and Morrow, "Improvements in Tank Atmosphere Control for Liquid Bulk Cargo Carriers," *Trans.* SNAME, Vol. 87, 1979.
33 "Economics of Wide Shallow Draft Bulk Carriers by Nickels," SNAME Los Angles Section Paper, Sept. 1974.
34 Gertler, M., Kohl, R., and Roseman, D. P., "Characteristics of Bulk Products Carriers for Restricted-Draft Service," *Trans.* SNAME, Vol. 82, 1974.
35 Snider, W. D., "IMCO Conference on Tanker Safety and Pollution Prevention," *Marine Technology*, July 1978.
36 "U.S.-Flag Oceangoing Fleet," *Maritime Report and Engineering News*, June 1991.
37 Information from the private files of D. P. Roseman, Jan. 1991.
38 "Guide for Vessel Life Extension," American Bureau of Shipping, Sept. 1989.
39 Hattori, K., and Perakis, A. N., "The Overtonnage of Large Tankers: Status and Prospects," *Journal of Ship Production*, Aug. 1986.
40 Bass, Morrow, and Magott, "Design and Operational Considerations for Liquid Bulk Carrier Inert Gas Systems to Provide Safe Tank Gas Concentrations," *Trans.* SNAME, Vol. 89, 1981.
41 Cowles, W., "A Practical Guide to the Design and Installation of Crude Oil Washing Systems," *Trans.* SNAME, Vol. 88, 1980.
42 Czimmek, D., and Jordan, C., "Optimization of Segregated Ballast Distribution and Its Impact on Tanker Economics," *Marine Technology*, April 1981.
43 Daidola, J., Graham, D., Jovino, V., and Hogg, W., "Subdivision of Crude Oil and Product Tankships," *Marine Technology*, April 1982.
44 Morrell, R. C., "Calibration of Tanks on Tankers and Oil Barges," *Marine Technology*, April 1984.
45 Johnson, J., Rogers, A., and Bass, R., "An Assessment of Shipboard Tank Level Indicating Systems," *Marine Technology*, July 1984.

46 Hamlin, N. A., Lou, Y. K., Maclean, W. M., Seibold, F., and Chandras, L. M., "Liquid Sloshing in Slack Ship Tanks—Theory, Observations and Experiments," *Trans.* SNAME, Vol. 94, 1986.

47 Dillingham and Falzarane, "A Numerical Method for Simulating Three Dimensional Sloshing," STAR Symposium, 1986.

48 Noble and Swift, "Innovative Uses for Surplus Tanker Tonnage," STAR Symposium, Sept. 1984.

49 Podenzana-Bonvino, C., Tedeschi, R., Boote, D., and Ferrando, M., "Shortening Afoot of a Very Large Crude Carrier," *Marine Technology*, July 1988.

50 Buetzow, "De-Jumboizing of Four 212,000 DWT VLCC's to 150,000 DWT," SNAME Northern California Section, March 1981.

51 Ono, Takekuma, and Kawaguchi, "The Design of Tankers for Restricted-Draft Service," STAR Symposium, 1985.

52 Bianchi, R. A., "The Economics of Oil Spill Cleanup," *Marine Technology*, Oct. 1981.

53 "Spill Containment and Cleanup—Research and Development, *Marine Technology*, Oct. 1981.

54 Levine, R. A., "Contingency Planning for Oil Spills: An Update Based on Experience Gained from the Port Angles Spill," *Marine Technology*, April 1988.

55 Dielsi, G. J., "Principles of Marine Vapor Recovery," *Marine Technology*, Jan. 1989.

56 Koops, Caldwell, R. X., and Gordon, "Two State-of-the-Art Specialty Product Ships: Design, Construction and Operation," *Trans.* SNAME, Vol. 92, 1984.

57 Mazzel and Terry, R. G., "Revival of the Coastal Tanker: Development and Construction of a U.S.-Flag 2500 DWT Product Carrier," *Trans.* SNAME, Vol. 89, 1981.

58 Terry, R. G., and Lore, D. N., "Design and Construction of the Coastal Tanker MT *Eastern Sun*," *Marine Technology*, Oct. 1986.

59 "Double Hull Standards for Tank Vessels Carrying Oil," Proposed Rules 33 CFR Part 157, *Federal Register*, Vol. 55, No. 234, Dec. 5, 1990.

60 Hagner, T. R., Corbett, C., Kimble, J., and Thompson, T., "The Reconstruction of the *Exxon Valdez*—Panel Discussion," *Trans.* SNAME, Vol. 98, 1990.

61 Tornay, E. G., "Tanker Design for the '90s," SNAME Philadelphia Section Meeting, Oct. 18, 1990.

62 Bishop, B. R., "Tankers in the 1990's: Legacies of the 80's and Opportunities of the 90's," SNAME Northern California Section Meeting, March 8, 1990.

63 Leitz, M., "Salvage of the *Exxon Valdez*," SNAME Pacific Northwest Meeting, April 7, 1990.

64 Hudson, D., "Technical Aspects of the *Exxon Valdez* Salvage," SNAME Pacific Northwest Meeting, May 12, 1990.

65 McNeill, D. C., Malmquist, D., Walter, D., Grothen, D., and Londot, K., "The Reconstruction of the *Exxon Valdez*," *Trans.* SNAME, Vol. 98, 1990.

66 Ternus, R., Buetzow, M., and Selle, E., "Chevron's 78,000 DWT Lightering Tankers," *Marine Technology*, Jan. 1990.

67 Wildenberg, T., "The Origins and Development of the T-2 Tanker," *The American Neptune*, Vol. 52, No. 3, Summer 1992.

Appendix

Shipyard	No. Built	DWT	Horsepower	Ship Manager
		The Late 1940s and 1950s		
Bethlehem Steel Co.	24	28,170	13,750 T	various major oil companies
Quincy, MA, and Sparrows Point, MD	—	45,000	13,600 T	
Newport News Shipbuilding & Dry Dock Co.	16	26,800	12,500 T	Esso Tankers
Newport News, VA	4	42,000	26,500 T	Esso Tankers
New York Shipbuilding Camden, NJ	3	30,000	16,000 T	Atlantic Richfield Co.
Sun Shipbuilding & Dry Dock Co.	5	18,500	13,500	USMC
Chester, PA	7	16,500	7500	USMC
	2	12,700	5000	Standard Oil of California
	1	17,000	7000	Standard Oil of California
	11	26,800	12,500	various
	9	30,200	13,500	Sun Oil Co. and various others
	3	25,000	18,600	USMC
	2	30,100	16,000	Socony Mobil Oil Co.
	3	60,000	18,500	Sun Oil Co.
	1	80,000	24,000	Millcreek Leasing Co.
Welding Shipyard Norfolk, VA	5	30,000	12,500 T	Universe Tankships

Shipyard	No. Built	DWT	Horsepower	Ship Manager
		The 1960s		
Avondale Shipyards Avondale, LA	3	75,600	19,000 T	Humble Transport Co.
Bethlehem Steel Co. Sparrows Point, MD	—	49,800	15,000 T	Maritime Overseas Corp.
	5	25,500	13,600 T	Texaco Marine Services
	11	37,800	15,000 T	Keystone Shipping Co., Maritime Overseas Corp., Ogden Marine
	2	62,000	20,000 T	Maritime Overseas Corp.
	1	62,000	20,000 T	Keystone Shipping Co.
Newport News Shipbuilding & Dry Dock Co. Newport News, VA	2	48,000	19,000 T	Humble Transport Co.
	—	32,600	12,500 T	Atlantic Refining Co.
	2	71,000	19,000 T	Humble Transport Co.
New York Shipbuilding Camden, NJ	—	50,560	18,750 T	Maritime Overseas Corp.
	—	38,200	18,750 T	Maritime Overseas Corp.
Sun Shipbuilding & Dry Dock Co. Chester, PA	—	53,400	18,500 T	Sun Transport, Inc.
	—	80,700	24,000 T	Trinidad Corp.
		The 1970s		
Avondale Shipyards Avondale, LA	4	173,380	27,000 T	Standard Oil Co., Ohio
	2	173,380	27,000 T	Exxon Co., U.S.A.
Bath Iron Works Bath, ME	—	27,300	14,000 D	International Marine Carriers, Inc.
Bethlehem Steel Co. Sparrows Point, MD	1	37,500	15,000 T	Keystone Shipping Co.
	1	37,500	15,000 T	Union Carbide
	2	70,200	20,000 T	Arco Marine, Inc.
	3	70,200	20,000 T	Chevron Shipping Co.
	3	120,000	26,000 T	Arco Marine, Inc.
	1	120,000	26,000 T	Maritime Overseas Corp.
	2	262,000	35,000 T	Gulf Transportation & Trading Co.
	3	264,000	35,000 T	MFC-Boston Carriers
FMC Corp. Portland, OR	6	39,200	12,500 GT/E	Chevron Shipping Co.
Ingalls Shipbuilding Pascagoula, MS	3	37,250	15,000 D	Falcon Shipping Group
National Steel and Shipbuilding Co. San Diego, CA	2	82,000	24,500 T	American Maritime Transport, Inc.
	3	91,000	24,500 T	Apex Marine
	2	91,000	24,500 T	Keystone Shipping
	4	91,000	24,500 T	Maririme Overseas Corp.
	3	39,200	15,000 T	Keystone Shipping
	3	39,200	15,000 T	Mormac Marine Transport
Newport News Shipbuilding & Dry Dock Co. Newport News, VA	2	398,143	45,000 T	Interocean Management Corp.
Seatrain Shipbuilding Brooklyn, NY	4	225,000	50,000 T	Marine Transport Lines
Sun Shipbuilding & Dry Dock Co. Chester, PA	3	80,700	24,000 T	Trinidad Corp.
	1	130,000	30,000 T	Mobil Oil Corp.
	3	124,000	30,000 T	Keystone Shipping
Todd Shipyards Corp. Los Angeles, CA	9	27,300	14,000 D	International Marine Carriers, Inc.
		The 1980s		
Avondale Shipyards Avondale, LA	2	50,850	15,000 D	OMI Corp.
	2	48,000	17,000 D	Exxon Shipping Co.
	1	57,100	17,000 D	Exxon Shipping Co.
Bath Iron Works Bath, ME	2	33,500	14,700 D	Falcon Shipping Group

Shipyard	No. Built	DWT	Horsepower	Ship Manager
Jeffboat, Inc. Jeffersonville, TN	1	3396	2960 D	Sun Transport, Inc.
National Steel and Shipbuilding Co. San Diego, CA	2	188,400	28,000 T	Arco Marine, Inc.
	2	188,400	28,000 T	Marine Transport Lines
	2	37,500	11,244 D	Ingram Tankships
	2	39,990	13,000 D	West Coast Shipping
	2	42,600	15,000 D	Maritime Overseas Corp.
	3	50,000	11,400 D	American Transport and Trading
	2	211,400	31,650 D	Exxon Shipping Co.
Sun Shipbuilding & Dry Dock Co. Chester, PA	2	34,000	14,200 D	Sun Transport, Inc.
Tampa Shipyards Tampa, FL	5	30,100	15,300 D	Ocean Shipholdings, Inc.

Steven Spear* graduated from the University of Michigan in 1951 with a degree in Naval Architecture & Marine Engineering. During his professional career he was engaged almost exclusively with the design and construction of tankers (including LNG) and bulk carriers. His work took him to Japan and Taiwan, as well as to the United States. While employed by National Bulk Carriers, he worked on the design of what were then the world's largest tankers and bulk carriers. Other employers have included Esso International and Burmah Oil Company.

*The editors note with deep regret the death of Mr. Spear on February 25, 1993.

Refined Product and Specialty Tankers

Maurice Gordon

Petroleum Product Tankers

During World War II there was a considerable movement of refined products from the Gulf Coast refineries to the East Coast and to Europe. This trend continued until the late 1950s. Most of the tankers built in that period were similar in arrangement to crude oil tankers and, in many cases, they were used interchangeably. Cargo in those ships was carried in 24 to 30 tanks arranged in four isolated systems with double valved crossover piping. This permitted the ship to handle up to four grades of cargo with a minimum possibility of contamination. Large quantities of gasoline, diesel oil, aviation fuel, kerosene, heating oil, and similar products were moved on those ships. Some built during that period had dry cargo holds forward. These were used for carrying specialty cargoes such as lube oils in drums.

From 1943 to 1970, product tankers increased in size from the 16,000 dwt MarAd T-2 design to the 75,000 dwt *Exxon San Francisco*-class. As requirements for larger tankers grew, many ships built during the 1940s and 1950s were enlarged by adding new fore- or mid-bodies, or both. Another trend during that period was a reduction in the number of cargo tanks—from 30 to about 15. That simplified cargo handling significantly.

Also during that period, the oil companies and independents undertook some major pipeline construction. Many of the pipelines extended from the Gulf Coast to the northeast. Advances in pipeline technology permitted different products to be sent through the same pipeline with minimal mixing. Also, the discovery of oil on Alaska's North Slope shifted the focus of the tanker owners to crude oil. As a result, most of the U.S.-flag tankers built or converted after 1968 were intended solely for crude oil.

Today, in 1993, the main uses of product tankers are to:

- Transport clean products to terminals that are not yet served by a major pipeline.
- Transport products (such as heavily leaded aviation gasoline) that cannot be sent through a pipeline.
- Carry products to alleviate spot shortages. This is most notable in the winter when tankers carry heating oil to the northeast terminals.

Fig. 1 The chemical tanker SS *Marine Dow-Chem*: length 551 ft, beam 68 ft, deadweight 16,200 tons. Delivered in 1954, the *Marine Dow-Chem* was the first oceangoing vessel specifically constructed for the transportation of liquid chemicals in bulk.

Fig. 2 The liquid sulfur carrier SS *Marine Floridian*: length 612 ft, beam 42 ft, deadweight 24,412 tons. The *Marine Floridian*, a jumboized vessel, has a cargo capacity 50 percent greater than that of the T-2 tanker from which it was built (in 1967). Construction of the new tanker required extensive modification work on the ex-*Paoli* stern and ship's systems.

Specialty Product Tankers

Solvents, lube oil blending stocks, waxes and other specialty products are also transported by tanker. These specialty tankers or "drugstore ships" carry a wide variety of products, perhaps as many as 30 or 40 different ones on a trip. The tank arrangements on these ships vary widely but there are generally no special containment or isolation requirements as there may be on chemical ships. Also, cargo compatibility is such that some of the cargoes can be discharged sequentially through the same cargo pumping system.

Only a few ships of this type have been built new, but several have been converted from product tankers. The most recent new buildings were the 42,000 dwt *Exxon Charleston* and *Exxon Wilmington* built by Avondale Shipyards in 1983–84. These ships have 42 cargo tanks, each capable of carrying a separate grade of cargo [3].

The main reason for the existence of these ships is that the products they carry require more protection from contamination than a pipeline can provide, or they require other special treatment such as heating. These ships are classified as product tankers, and transport their cargoes in compliance with 46 CFR Subchapter D ("Tank Vessels") of the Coast Guard Rules.

The Noxious Liquid Substance (NLS) provisions of MARPOL Annex II have shaded the division between specialty tankers and chemical tankers. Annex II classifies the pollution and toxic potential of many products and chemicals. Some solvents that had been carried as petroleum products under Subchapter D were labeled NLS and are now transported under the provisions of 46 CFR Subchapter O ("Hazardous Cargoes").

Chemical Tankers

In the past half century, the transport of chemicals by water has increased dramatically. At this writing, however, the largest chemical tanker fleets are not under U.S.-flag. Chemical tankers have some of the most sophisticated tank and pumping arrangements. Special coatings and linings are required for some products. In addition, many of the chemicals react with others and with water. The violence of one of these reactions resulted in the loss of the *Puerto Rican* in 1983. In that case, caustic soda corroded a damaged area in its tank and leaked into an adjacent cofferdam coated with inorganic zinc. The caustic soda and the zinc reacted to form hydrogen gas, which subsequently exploded.

Chemical tankers have three basic types of containment, which are mainly related to damaged stability. The most stringent system, Type I, requires protection from grounding and collision damage. Type II requires protection only from grounding damage. Type III requires only protective positioning of cargo tanks. Type I and II systems also require that the tank bulkhead be at least 76 cm (30 in.) away from the ship's shell. Most chemical tankers have a combination of containment systems.

Cargo handling arrangements vary, but the current standard arrangement is to have a cargo pump, either deepwell or submerged type, in each tank with a dedicated line to the loading manifold. A highly effective stripping arrangement is also required, so as to discharge the maximum amount of high value cargo, and to meet the requirements of MARPOL Annex II.

Some chemicals have properties requiring unusual discharge arrangements. On the *Marine Dow-Chem*, for example, six rubber lined pressure tanks were installed for hydrochloric acid [1]. The tank surface was pressurized with compressed air to force the liquid through the ship's piping to the suction of a pump ashore.

In 1954, Bethlehem Steel's Quincy yard delivered the 16,625 dwt *Marine Dow-Chem* to Marine Transport Lines and Dow Chemical. This was the first tanker built from the keel up as a chemical tanker. Until that time, tankers and barges had been converted from other uses to carry chemicals. The *Marine Dow-Chem* had 24 chemical tanks served by 11 totally isolated cargo systems. In addition, there were 18 wing tanks that could be used for petroleum products, and five double bottom tanks. The ship carried perchlorethylene, methylene chloride, styrene, glycol, carbon tetrachloride, and chloroform. Six of the cargo tanks were 100 psig pressure vessels for hydrochloric acid. Four other tanks, clad with 50 tons of nickel, were installed for caustic soda.

Another notable U.S.-built chemical tanker was the 37,100 dwt *Edgar M. Queeny*, built in 1970 by Bethlehem Steel, Sparrows Point for Keystone and Monsanto. The ship was capable of carrying up to 26 separate cargoes, including liquid naphthalene and caustic soda. In 1971, the same yard delivered the 34,600 dwt *Puerto Rican* to Hendy (now West Coast Shipping) and Union Carbide. This ship included cryogenic tanks and handling arrangements for uninhibited vinyl chloride and butadiene. It also carried such products as caustic soda, ethylene glycol, trichloroethylene, pylene and methyl-chloroform.

In 1983, Newport News delivered the 35,000 dwt *Chemical Pioneer* to Marine Transport Lines and Union Carbide for service between Texas and New Jersey. This ship was, in fact, a conversion using the afterbody from the *Sea Witch*, which had been all but destroyed in a fiery collision with the *Esso Brussels* in 1973. Newport News built a new forebody, which incorporated the latest chemical tanker technology. The ship has a double hull and 48 cargo tanks. Each tank has a submerged hydraulic pump with a dedicated discharge to the loading manifold. All piping is self draining to the cargo pump. Product is stripped from each tank using compressed nitrogen through an eductor system. The ship also has a liquid nitrogen system to supply nitrogen for tank atmosphere padding, stripping and instrumentation.

Fig. 3 The *Chemical Pioneer*. Cones that resemble large eyes protruding port and starboard from the bow are inexpensive alternatives to flaring the upper hull enough to protect the bulbous bow against damage when the anchors are lowered.

Edible Liquid Tankers

While not a mode for transporting the finest wines, one tanker was built in the United States to carry wine as its main cargo. In 1957, the Bethlehem Steel San Francisco yard converted the 15,343 dwt T-2 *Angelo Petri* by building a new midbody. The ship was arranged to carry 26 different wines in stainless steel vats. Another product transported by this tanker is orange juice. This requires stainless steel containment and special inert gas arrangements.

Elevated Temperature Products

Two elevated temperature products are notable for being transported by tanker. These are asphalt and liquid sulfur. Asphalt is a residual of the petroleum refining process. It must be maintained at about 300 °F. But, even at this temperature it is quite viscous and generally requires positive displacement pumps, special seals and heat tracing. It has generally been carried in single hull vessels and the temperature differential with the seawater has made thermal hull stress a serious consideration.

Liquid sulfur also requires special consideration. The main reason for carrying sulfur as a liquid is that the end user needs it in that form. Moreover, storing it as a liquid eliminates a number of hazards that arise when storing it in solid form. Liquid sulfur is carried at 260–280 °F [4]. It solidifies at 238 °F.

The U.S. Coast Guard requires that liquid sulfur tanks be independent of the hull. They are also insulated. The vapor space above the cargo is ventilated to maintain hydrogen sulfide concentration at half its 3.7 percent lower explosive limit. In addition, sulfur reacts with copper, so no copper bearing materials are used in the cargo system and special provisions are made in the machinery space to protect equipment from damage.

References

1 Danielson, R., and McDonald, A., "The Design of the S.S. *Dow-Chem* Tankship for Special Commodities," New England Section, SNAME, Sept. 1954.

2 Price, R. I., Goldbach, R. D., Dyer, A. F., and Wheeler, F. J., "Sending Chemicals to Sea," Metropolitan Section, SNAME, May 1967.

3 Koops, Coldwell, and Gordon, "Two State-of-the-Art Product Specialty Ships," *Trans.* SNAME, Vol. 92, 1984, pp. 51–83.

4 Johnson, L., "Sulphur Vessels," Gulfcoast Section, SNAME, Feb. 1972.

Maurice Gordon graduated from Case Institute in 1970 with a degree in mechanical engineering. Following eight years of tanker design experience with Bethlehem Shipbuilding, he spent two years managing maintenance and repairs for Atlantic Richfield and El Paso Marine. Since 1980 he has been with the Exxon Shipping Company, largely as a project engineer for new construction. In 1984 he was given the SNAME Linnard prize as co-author of the best annual meeting paper (on specialty product ships). He is a life member of SNAME, a fellow of the Institute of Marine Engineers, a member of ASME, and a registered Professional Engineer.

Liquefied Natural Gas Carriers

EDMUND G. TORNAY

THE MARINE TRANSPORTATION of LNG is an industry and technology that was conceived in the United States in the 1950s, developed in Europe in the 1960s, reentered by American shipbuilders in the 1970s, and finally dominated by the Japanese in the 1980s. The principal reason for this worldwide flow of venue has been the corresponding shift of market for natural gas. Japan, without its own energy resources, is by far the greatest importer of LNG and therefore has extensive control of shipbuilding contracts.

The first commercial shipment of LNG was made by the *Methane Pioneer* in February 1959. The cargo was loaded at Lake Charles LA and discharged at the Canvey Island terminal in England after a voyage of about 26 days [1]. In spite of severe weather and heavy rolling, the cargo was delivered safely, proving the feasibility of ocean transport of bulk cargoes at cryogenic temperature.

The *Methane Pioneer* was a converted diesel powered C1-M-AV1, a dry cargo ship with machinery aft. The conversion was by the Alabama Shipbuilding and Drydock Co. to a design prepared by the J. J. Henry Company using specialized technical input from many sources, assembled by Constock. Constock was a joint venture of Continental Oil and the Chicago Stockyards. After the initial voyages of the *Methane Pioneer*, Shell Oil joined with Constock to form Conch Methane Services, and further development shifted to London. Development and construction of commercial LNG ventures and ship construction continued in Europe throughout the 1960s, the only American involvement being Exxon's construction of four LNG ships to its own design in Italy and Spain.

Fig. 1 The *Methane Pioneer* at dock in Lake Charles.

As a result of the development of several LNG import projects in the United States in the early 1970s, the construction of a total of sixteen LNG tankers was undertaken by three American shipyards during the decade. Of these, only thirteen were successfully completed and placed in service. Three suffered from a flawed insulation and secondary barrier system and were eventually scrapped or converted to carry more prosaic cargoes. Of the remaining thirteen, five entered service from Algeria to receiving terminals newly built at Cove Point MD, Savannah GA, and Lake Charles LA. Together they carried some 59 cargoes before being taken out of service and laid up owing to a suspension of the commercial agreement between the Algerian gas monopoly (Sonatrach) and the American gas utilities. The remaining eight ships have operated almost continuously under American registry, carrying about 120 cargoes a year from Indonesia to Japan. These eight vessels comprise, to this day, the largest single fleet of LNG carriers in the world.

The three shipyards that entered the LNG vessel market in the 1970s each chose a completely different system of cargo containment. The Quincy Shipbuilding Division of the General Dynamics Corp. built ten ships using the Moss–Rosenberg design of spherical aluminum cargo tanks. See Fig. 2. Each ship has five tanks, which are nearly perfect spheres of a diameter roughly equal to the height of a 12-story building (120 ft). Each sphere is continuously welded at its equator to a vertical cylinder, the lower end of which is directly welded to the hull structure. This provides an ingenious combination of rigid connection and mechanical and thermal isolation between the extremely cold tank and the relatively warm ship. Because of the inherent strength and

Fig. 2 The *LNG Gemini*.

Fig. 3 The 71,495 dwt gas carrier *LNG Taurus* can transport approximately 125,000 m^3 of liquefied natural gas.

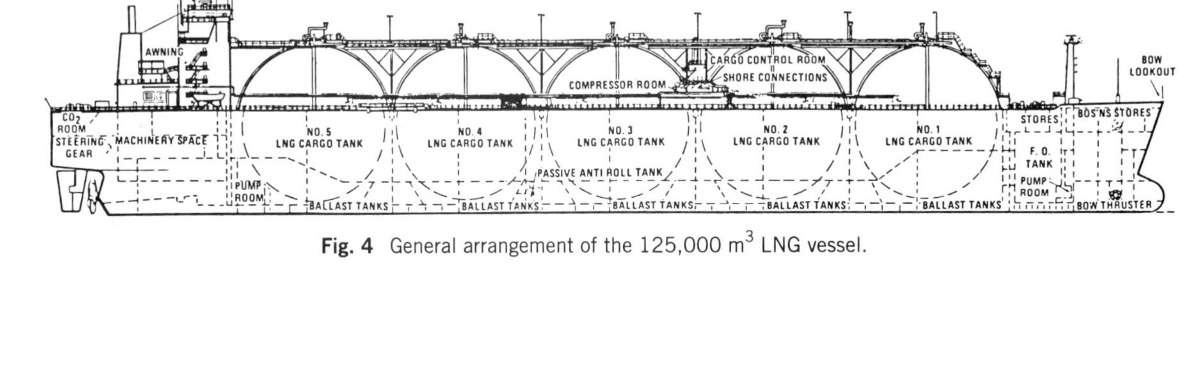

Fig. 4 General arrangement of the 125,000 m^3 LNG vessel.

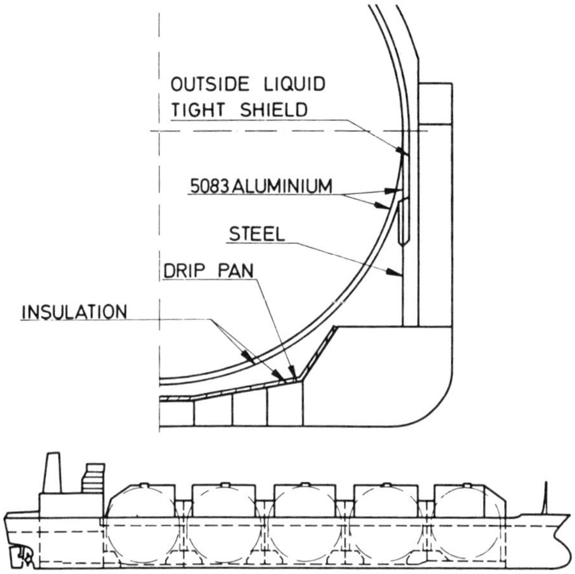

(*a*) Moss-Rosenberg design with spherical independent tanks of type B.

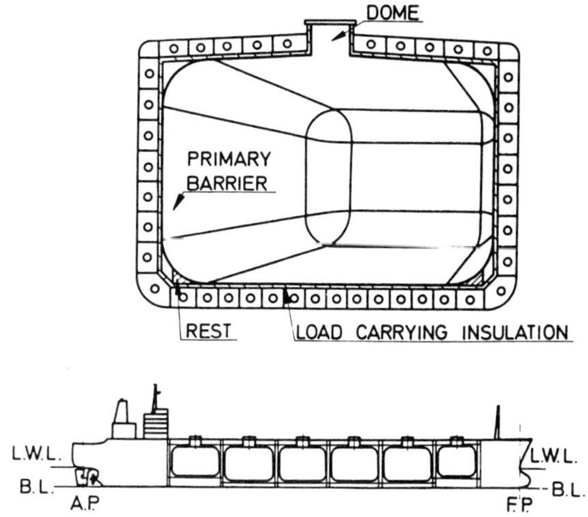

(*b*) IHI design with semi-membrane tanks of aluminum.

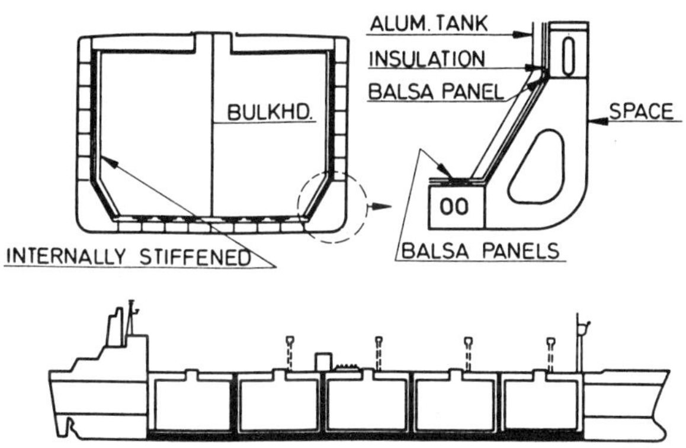

(*c*) Conch design with rectangular independent tank of type A.

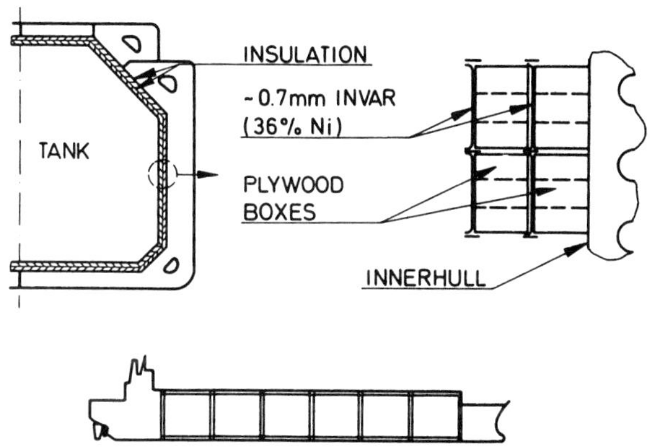

(*d*) Gaz-Transport design with membrane tanks of Invar.

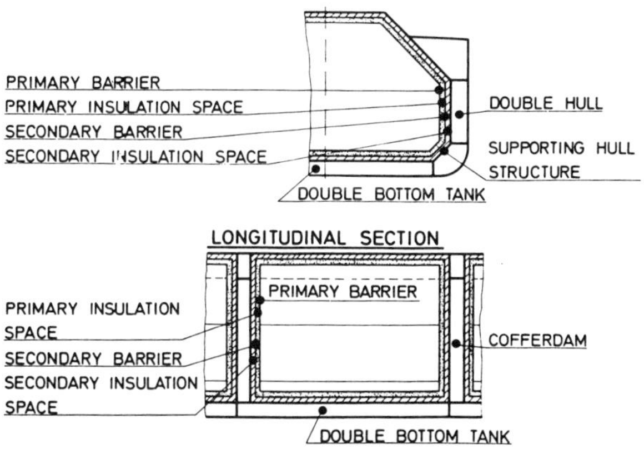

(*e*) Principal configuration of membrane tank.

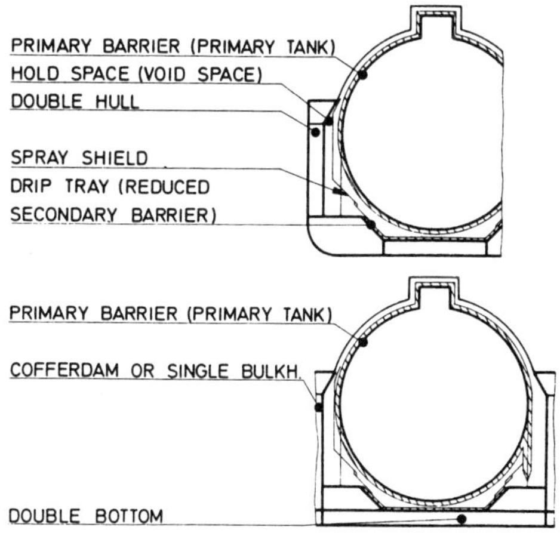

(*f*) Principal configuration of independent tank type B (body of revolution).

Fig. 5 Tank design in LNG carriers.

Fig. 6 *Lake Charles*, the first of two independent tank carriers completed in 1980.

simplicity of the spherical shape, only a limited secondary barrier is required.

In over 100 ship-years of service, not one ounce of cargo has leaked from these tanks. They are heavily insulated with panels of foam, and extend well above the main deck to create a distinctive profile. The main operational advantage of the spherical tank design is the large space between the inner hull and the cargo tanks, which permits easy inspection and repair of any corrosion or cracks in the hull.

Unfortunately, the history of the construction of these ships was flawed by allegations of fraudulent subcontracting practices by the management of the shipyard and a major subcontractor, leading to indictments and convictions of some of the participants.

Newport News Shipbuilding and Dry Dock Co. built three ships of the membrane tank type to a French system designed by Technigaz. Their double-hull ships were constructed with smooth-sided cargo holds. Precisely fitted balsa and plywood insulation panels were attached to the inner hull and to each other by a series of bonded plywood splices. The entire inner surface of these panels was lined with special thin stainless steel sheets with corrugations to allow for thermal shrinkage. These sheets were welded together to form the primary cargo containment surface. The plywood surface of the insulation panels formed a secondary barrier capable of containing the cargo and protecting the ship's hull against low temperature in case of a primary barrier leak. The complexity of such construction combined with a tank area of about six acres per ship is a measure of the accomplishment of building three such ships, each of which proved successful in carrying forty-seven cargoes.

The last three ships of the program were built to the Conch design, which was a development of the original Constock system used on the *Methane Pioneer.* The Conch system had been used in the early 1960s for two of the first LNG ships, *Methane Princess* and *Methane Progress,* which were built by two separate yards in the United Kingdom. By the time this system had been selected for the Avondale ships in 1973, it had been modified and developed to allow it to remain commercially competitive for the large increase in size from the 27,000 m^3 ships of the 1960s to the 125,000 m^3 ships of the 1970s.

The principal development was the extensive use of an insulation system comprising sprayed-in-place foam connected at the edges to a "picture frame" of balsa and plywood panels, similar to the panels used

Fig. 7 The *El Paso Southern*, now the *Southern*, was built to Technigaz Mark I design.

on the Technigaz ships. The Conch cargo tanks were free-standing aluminum rectangular prisms resting on load-bearing wooden bottom panels. Unlike the spherical tanks, a full secondary barrier was required by the regulatory bodies. This meant that the sprayed foam and balsa panels would have to be capable of containing the cargo for an extended period in case of total loss of containment of the primary tank. The insulation system had been carefully tested at low temperatures in a laboratory to demonstrate its containment capability. Unfortunately, when the first completed ship underwent gas trials, extensive cracking of the inner surface of the foam insulation occurred. Certification for LNG service could not be obtained and construction of the remaining ships was suspended. Eventually, two of the three ships were converted to combination oil–ore carriers; and the other, after sustaining grounding damage, was scrapped.

Since the completion of these three large construction programs, no LNG tanker has been built in the United States. With the 126,000 m^3 size as a standard, many spherical tank ships have been built in Japan. Recently European yards have re-entered the market with membrane type gas tankers. The possibility of future design and construction in the United States is favorable owing to the sophisticated nature of the ships and the improving competitiveness of American shipyards and supporting manufacturing industries.

Reference

1 Ffooks, R., *Natural Gas by Sea (The Development of a New Technology)*, Gentry Books, Ltd., London, 1979.

Edmund G. Tornay is a graduate naval architect and marine engineer who has been professionally engaged in all aspects of liquefied gas overseas transport for over 30 years. His LNG-related work has taken him for prolonged periods to Japan, Italy, and England. He returned to the United States to serve as owner's representative for the eight LNG ships built at Quincy MA. Since their completion in 1979 he has been responsible for the technical aspects of their operation. Mr. Tornay is a member of ABS and chairman of a committee for the Society of International Gas Tanker and Terminal Operators. He is the author of numerous SNAME and gas industry papers.

LPG Tankers and Barges

James E. Steele

The Product

Liquid petroleum gas (LPG) consists primarily of propane and butane. These are petroleum compounds that are produced from the well along with the crude oil and natural gas. The properties of propane and butane require that they be handled differently from other petroleum products.

Propane is chemically C_3H_8. It boils at a temperature of −44 °F if it is held at atmospheric pressure. If the confining pressure is increased, the boiling temperature also increases. At a temperature of 100 °F the pressure rises to 172 psig.

Butane is chemically C_4H_{10}. It boils at a temperature of +31 °F if it is held at atmospheric pressure. At a temperature of 100 °F the pressure rises to 38 psig.

Commercial LPG is frequently a combination of propane and butane.

Both of these products may be handled at atmospheric temperature as either a gas or a liquid, depending on the confining pressure. This is in contrast to natural gas (Methane, CH_4), which boils at a temperature of −259 °F and which, at atmospheric temperature, remains a gas regardless of the confining pressure.

Another product that is similar in physical characteristics to propane and butane is anhydrous ammonia (NH_3). This boils at a temperature of −28 °F if it is held at atmospheric pressure. At a temperature of 100 °F the pressure rises to 198 psig.

Ships and barges designed to handle propane are generally suitable for the carriage of butane and anhydrous ammonia.

Propane and butane are produced in three different types of plants:

- In natural gasoline plants in which the natural gasoline and the propane and butane are separated out from the crude oil.
- In recycling plants in the field in which the LPG is separated out of the natural gas before the natural gas is recycled into the formation.
- In oil refineries.

In the early days of oil production, LPG was looked upon as a waste product and was burned. It was soon found that it could be transported by truck as a liquid in small pressure vessels and that it was useful for heating and cooling in rural areas where manufactured gas or natural gas was not available. Consumption in the USA went from about 5000 barrels in 1922 to 44 million barrels in 1947, and to 480 million barrels in 1990. World consumption of LPG was 1560 million barrels in 1990. For comparison, in 1990 world consumption of crude oil was about 22 times that of LPG, while world consumption of natural gas was about 12 times that of LPG; both comparisons are based on BTU content.

Moving the Product

As consumption increased transportation was primarily by pressure vessels on trucks and then on railroad cars. Rail cars generally have pressure vessels that contain about 10,000 gallons. Storage tanks at terminals tend to be somewhat larger; one standard size holds 30,000 gallons in a cylindrical pressure vessel that is 103 in. inside diameter by 70 ft-10 in. long.

Shortly after World War II the Warren Petroleum Corp. of Tulsa, a leading producer and marketer of LPG, decided that their business between Texas and New York had grown to the extent that water transport was warranted. Up until that time the only waterborne LPG had been in a few small individual pressure vessels mounted on deck or in the holds of freighters or tankers. Warren wanted a ship devoted entirely to the carriage of LPG. This was an engineering project of considerable novelty. To accomplish it they entered into an agreement with the Pennsylvania Shipyards of Beaumont TX. Two approaches were possible. One was to install pressure vessels that would store the LPG at atmospheric temperature and at the corresponding pressure. (Nominal design pressure, which includes some margin, is usually about 250 psig for propane or 100 psig for butane). Alternatively, it would have been possible to use insulated rectangular

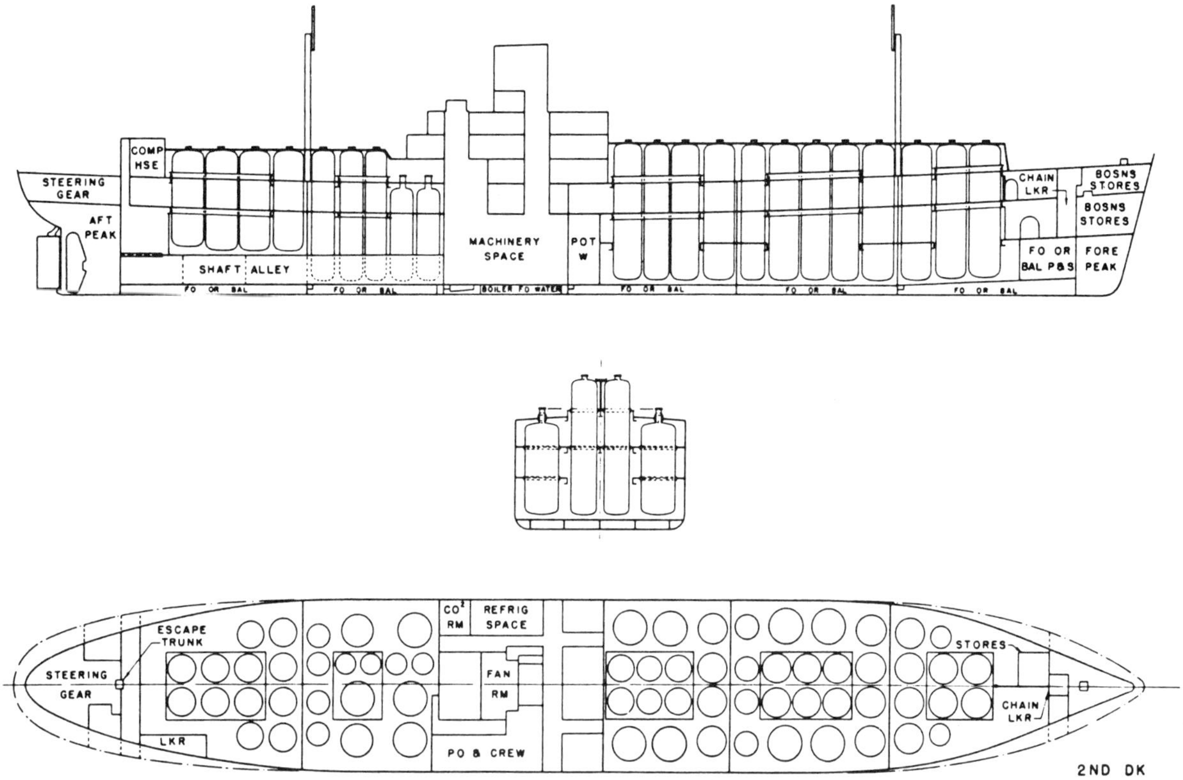

Fig. 1 General arrangement of the *Natalie O. Warren.*

tanks, which would carry cargo at atmospheric pressure and refrigerated temperature. At that time, all of the industry's experience in trucks, in railcars, and in storage terminals was confined to pressure vessels; so that was the course followed.

The *Natalie O. Warren*

The availability of war surplus ships made it desirable to convert an existing ship rather than to build a new one. The ship selected was the *Cape Diamond*, a standard C-1A cargo ship that had been built at Pusey and Jones in Wilmington, DE. One of the considerations that led to its selection was that it was steam propelled and therefore had steam available for running the large LPG compressor installed for unloading the cargo.

After conversion the ship was renamed the *Natalie O. Warren* in honor of the owner's wife. During conversion a total of 68 vertical pressure vessels were installed in four rows. See Figs. 1 and 2. The maximum

Fig. 2 The *Natalie O. Warren.*

Fig. 3 The *Ultragaz*.

diameter of 13 ft was controlled by clearances on the railroad from the tank construction site in Alabama to the shipyard in Texas.

Some specific considerations related to the nature of the cargo were:

- Tank size and cargo capacity tended to be controlled by volume and stability rather than displacement. Additional capacity was achieved by increasing the height of the tanks while maintaining stability with fixed ballast on the double bottom.
- Pressure vessels sat on structural foundations on the tank top and had watertight sliding connections where they penetrated the upper deck of the ship.
- All pressure vessels, or pressure vessel necks, extended above deck so that no LPG piping was below deck.
- All holds and machinery spaces had mechanical exhaust ventilation taking suction from just above the tank top to guard against accumulation of propane or butane vapor, which is heavier than air.
- Cargo was unloaded by a ship mounted compressor which drew vapor from the shore tanks and discharged it into the ship's tanks, thereby raising the pressure in the ship's tanks and decreasing pressure in the shore tanks. Steam driven pumps augmented the cargo flow. No vapor was discharged to the air. The system was reversed for cargo loading.
- Piping attachments to the tanks were fitted with "excess flow" valves, which shut down in the event of a broken pipe.
- Relief valves on all tanks were set at 250 psig, which was the design rating of the pressure vessels. Discharge was led to the tops of the masts.

Fig. 4 A seagoing barge.

Fig. 5 An inland barge.

The *Ultragaz*

In 1948 the Beaumont yard converted a second ship to carry LPG. The yard, in the meantime, had been bought by the Bethlehem Steel Co.

The *Rio Novo,* a smaller British built cargo ship with Scotch boilers and a reciprocating steam engine, was owned by Oivind Lorentzen of Oslo, who had developed a trade by shipping LPG from Houston to Brazil in small portable pressure vessels. By converting a ship to the exclusive carriage of LPG in fixed pressure vessels the business would be made more efficient.

The conversion was similar to that of the *Natalie O. Warren.* The name of the ship was changed to *Ultragaz.* See Fig. 3.

Other Early LPG Ships

Table 1 describes a number of LPG ships that were converted or built new in the next few years. Except for

Table 1 Early LPG ships

Name	New or Convert	Date	Builder	Owner	Product	Tanks	Press., psig	LOA, ft	LBP, ft	Beam, ft	Depth, ft	Draft, ft	LPG, m^3	Mach.	HP	Speed, knots	Service
Natalie O. Warren	conv.	1947	Penn. Shipyard, Beaumont, TX	Warren Petroleum, Tulsa	propane	press.	250	412	390	60	37	24	6050	steam	4000	14	Houston to New York
Esso El Salvador	conv.	1947	—	—	—	press.	—	—	—	—	—	—	650	—	—	—	Aruba to Caribbean
Ultragaz	conv.	1948	Bethlehem, Beaumont, TX	Oivind Lorentzen, Oslo	propane	press.	250	—	310	47	29	—	2720	steam	—	—	Houston to Brazil
Heroya	new	1949	Marinens, Norway	Norsk Hydro, Norway	ammonia	press.	—	252	228	38	26	15	2800	diesel	1500	13	N. Norway to S. Norway
Rasmus Tholstrup	new	1954	Marstrands, Sweden	Kosangas, Denmark	propane	press.	255	165	150	33	14	12	1350	diesel	520	11	Holland to Denmark
Gasbras Norte	conv.	1954	Kieler, Howaldtswerke	Oivind Lorentzen, Oslo	propane	press.	250	412	390	60	37	26	7650	diesel	4150	—	Houston to Brazil
Neviges	new	1955?	Peute, Hamburg	Petrobras	propane	press.	—	310	285	41	21	15	1900	diesel	2000	14	Brazil
Butagaz	new	1955	Chantiers, Navals	Soc Maritime Shell	propane	press.	227	279	265	39	21	15	1350	steam	1500	11	N. Africa to France
Marian P. Billups	new	1957	Bijker's, Holland	Marine Caribbean Lines	propane	press.	250	285	262	39	22	15	2000	diesel	1580	13	—
four ships	new	1959	Leghorn, Italy	Marine Transport Lines	propane	refrig.	atmo.	—	—	—	—	—	46,000	steam	16,000	17	Persian Gulf to Germany
Esso Puerto Rico	new	1959	Monfalcone, Italy	Esso Transport	propane	press.	—	690	660	90	47	37	35,000	steam	16,000	16.5	—

the *Natalie O. Warren* and the *Ultragaz*, these were all done in European yards, a reflection of the difference in construction costs. Two of the designs were for anhydrous ammonia rather than propane and one of these used refrigerated tanks rather than pressure vessels.

LPG Barges

After the conversion of the *Natalie O. Warren* it became apparent that water transport of LPG and similar products was practical and economical. This led to barges that were equipped with pressure vessels and piping systems. During the period from 1950 to 1971 some 55 such barges were built at the Bethlehem yard at Beaumont. These had a number of variations:

49 were for inland service
6 were for ocean or coastwise service
38 were for propane/butane/anhydrous ammonia
17 were for chlorine
50 had pressure vessels
5 were refrigerated

Figure 4 shows a typical oceangoing barge with a capacity of 3400 m^3. Figure 5 shows a typical inland barge with a capacity of 1800 m^3.

Bibliography

Blachy, D. B., and Shaw, H. F., "Design of Tanks and Cargo Handling System for "5500 m^3 LPG/NH_3 Carrier," New England Section, SNAME, March 1983.

BP Review of World Gas, Aug. 1991.

"From All Points—Leghorn Yards Propane Carriers," *Tanker Times*, 1959.

Marlow, H., and Phelps, R. V., "The Carriage of Liquefied Petroleum Gas on Ships," unpublished, 1948.

"Motor Tanker for the Carriage of Butane," *Shipbuilding and Shipping Record*, March 14, 1957.

"M/S *Gasbras Norte* Delivered," *Norwegian Shipping News*, No. 1, 1954.

"Petroleum Gas in Tankers," *The Shipping World*, Jan. 25, 1956.

"The Waterborne Transportation Of Special Liquid Cargo," Bethlehem Steel Co., Shipbuilding Div., May 1950.

"World's First Ammonia Tanker Powered by Nordberg Diesel," *Nordberg Progress*, 4th quarter, 1949.

"World's First Propane Tank Ship," *Maritime Reporter*, Nov. 15, 1947.

Appendix

Pressure Vessel Barges—Bethlehem, Beaumont

Year	Length	Beam	Depth	Cargo	Service	Owner
1950	195	44	10.5	propane	inland	Warren Petroleum
1951	195	44	10.5	propane	inland	Warren Petroleum
1951	210	44	10.5	butane	inland	Warren Petroleum
1953	242	48	16.5	propane	ocean	Warren Petroleum
1954	180	30.3	16.2	propane	coastwise	International Petroleum Co.
1954	252	48	18	ammonia	ocean	Tidewater Shaver
1954	245	48	17.5	propane	ocean	Warren Petroleum
1955	175	26	9.5	chlorine *Total of 3*	inland	Marine Transport Co.
1955	245	48	17.5	propane	ocean	Warren Petroleum
1955	210	44	15.6	propane	inland	Warren Petroleum
1957	252	48	18	propane	ocean	Warren Petroleum
1957	210	44	15.5	propane	inland	Warren Petroleum
1957	195	35	11.5	chlorine	inland	Columbia Southern
1957	175	26	9.5	chlorine *Total of 2*	inland	Diamond Alkali
1958	210	44	15.6	ammonia	inland	Commercial Transport
1958	210	44	15.8	propane *Total of 2*	inland	Tennessee Gas Transport
1960	210	44	13	ammonia	inland	Mid-South Chemical
1960	195	44	11	ammonia	inland	Mid-South Chemical
1960	210	44	16.9	propane	inland	Warren Petroleum
1960	195	50	12	propane	inland	Alamo Chemical Transport
1961	195	45	11	ammonia *Total of 2*	inland	Dixie Carriers
1963	380	53	12.8	propane *Total of 3*	inland	Alamo Chemical Transport

Year	Length	Beam	Depth	Cargo	Service	Owner
1964	195	35	12	chlorine	inland	Pittsburgh Pl. Glass
				Total of 3		
1964	255	50	12	chlorine	inland	E. I. duPont
				Total of 2		
1965	225	52	12	propane	inland	E. I. duPont
1965	380	53	12.8	propane	inland	Alamo Chemical Transport
1966	195	35	12	chlorine	inland	Pittsburgh Pl. Glass
				Total of 4		
1966	195	35	12	chlorine	inland	Diamond Alkali
				Total of 2		
1968	355	53	12.8	propane	inland	Alamo Chemical Transport
1970	180	52.5	12.5	propane	inland	Union Carbide
				Total of 2		
1971	175	42	12	propane	inland	Warren Petroleum
				Total of 3		
1971	175	42	12	butane	inland	Warren Petroleum
				Total of 2		

Refrigerated barges

Year	Length	Beam	Depth	Cargo	Service	Owner
1967	250	52	11	ammonia	inland	Port Arthur Towing
				Total of 2		
1967	270	53	14	ammonia	inland	E. I. duPont
1968	270	50	14	ammonia	inland	Tuloma Gas Product
				Total of 2		

James E. Steele graduated from Webb Institute in 1941 and set to work building destroyers at Consolidated Steel in Orange TX. In 1946 he shifted to the Pennsylvania Shipyard (later acquired by Bethlehem Steel Corp.) at Beaumont TX. There he was engaged in designing LPG barges, in developing LPG ship conversions, and in designing offshore drilling rigs. In 1958 he transferred to the Quincy shipyard, where he helped design the power plants for nuclear cruisers. He moved to Bethlehem Steel's home office in 1964; he is now retired.

Passenger Ships

JOHN W. BOYLSTON

Introduction

PASSENGER SHIPS were covered in the SNAME *Historical Transactions of 1893–1943* under the following Sections of Part 2 entitled "Special Types of Vessels":

- The Pacific Coast Steam Schooner
- Early American Steamship Lines
- History of United States Coastwise Steamers
- Development of Steam Navigation on Long Island Sound

In reviewing these transactions and other material for this chapter, it appears to me that during the past fifty years passenger ships (as opposed to ferries) have been confined to transoceanic service or, more recently, to cruises in which the destination, if any, is not necessarily important.

While the previous *Historical Transactions* took history to 1943, the discussion of offshore passenger ships ended in about 1939. Therefore, this chapter will deal with the period of 1943–1993, but will include ships built as early as 1938.

Capt. Leback's chapter, which follows immediately after this one, is dedicated to a historical chronicle of Maritime Commission standard cargo ship designs modified to cargo/passenger service from 1940 through 1947. As these are not considered strictly passenger ships by design, no discussion of hull form, arrangements, stability, and subdivision is presented in his chapter.

Historical Background

To understand developments over the past half century, we must first consider the period's immediate antecedents. Perhaps going back for a brief survey since 1920 will suffice.

During the 1920 period, new ships were ordered by almost all of the American coastwise steamship lines. The Savannah Line acquired the *City of Chattanooga* and *City of Birmingham*, two new 380-ft ships in 1923. In the same year, the Merchants and Miners Line followed suit with similar vessels, the *Berkshire* and *Allegheny*. Practically a whole new fleet was ordered by Eastern Steamship Lines. Their *Acadia* and *St John*, built at Newport News from designs by Theodore Ferris, were particularly handsome ships of over 6000 gross tons and 11,400 shaft horsepower. East Coast, Gulf, and Caribbean ports saw a reasonable number of new combination passenger and freight ships between 1924 and 1932. Most of them were geared-turbine ships of moderate powers, moderate speeds, and moderate fuel economies, with steam pressures within the 200 to 250 psi bracket, and steam-driven auxiliaries.

The *Dixie*, of 1927, built by Federal Shipbuilding and Dry Dock Co. for Southern Pacific Lines, pioneered in higher pressures with 350 psi and 100 °F superheat.

The first large transoceanic liner since the war-time building program was the *Malolo*, built by Cramps in 1927 for the Matson Line's Pacific service. Shortly thereafter came the *President Hoover* and *President Coolidge*, ordered by the Dollar Line from Newport News; then came two more Matson ships, *Mariposa* and *Monterey*, from Bethlehem's Fore River yard; and two Atlantic liners, *Washington* and *Manhattan*, from New York Ship. All these gave our merchant marine a series of modern ships badly needed to replace older craft with many years of service to their credit. These were intermediate-class passenger and freight liners with sustained sea speeds just over twenty knots. The *Mariposa* was a particularly good looking ship at 631 ft-6 in. length overall. The *President Hoover* had a length of 659 ft-3 in., and the *Manhattan* 705 ft. All were twin-screw ships of generally similar appearance with conventional straight stems, extensive superstructure and two stacks.

It was not until 1939, however, that definitely fireproof construction of oceangoing passenger steamers was inaugurated in the *Panama*, the first of three similar ships designed by George G. Sharp for the Panama Railroad Steamship Co. and built by Bethlehem's Fore

River yard. The *Panama*, *Ancon*, and *Cristobal* were 471 ft-6 in. × 64 ft × 46 ft-9 in., carrying 202 passengers in a single class. These three ships may well be regarded as examples of the comfortable, economical, efficient, safe type of ship that American designers and builders had developed just before World War II (Fig. 2).

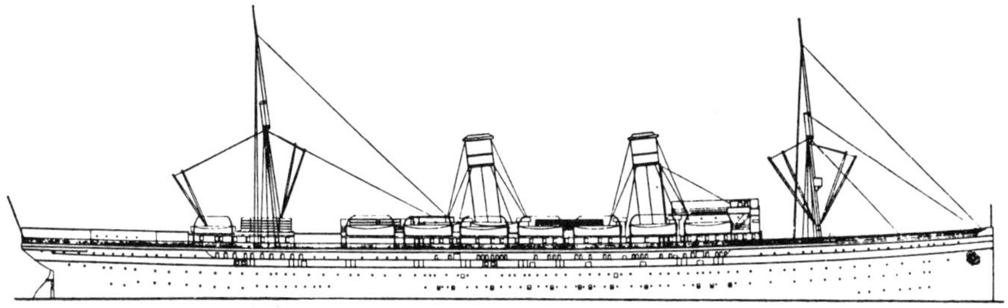

Fig. 1 *Saint Louis* (1895).

Developments in Transoceanic Passenger Ship Design (1943–1993)

Introduction. The development of transoceanic passenger ships from 1893 to 1943 was steadily toward larger and faster vessels culminating with the foreign built *Normandie* in 1935. She had a length of 1029 ft-4 in. and a speed of 29.5 knots. The trend from 1943 and on, however, was toward smaller size and slower speed. With the exception of the great liner *United States* (the last North Atlantic liner built in the United States) the trend that started in 1940 with the *America* at 723 ft and 22 knots ended in 1963 with a passenger/cargo ship, the *Santa Magdelena* of 546 ft-8 in. and a speed of 20 knots.

The battle between the United States and various European countries for Atlantic liner supremacy produced probably the best designed, best built U.S. liners in the late 1930s up to the early 1950s. By 1965, however, the airplane and economics had forced most of the great liners off the seas.

The Mediterranean passenger ships, *Constitution* and *Independence* (1951), the Caribbean passenger liners *Santa Rosa* and *Santa Paula* (1952), and the South American liners *Argentina* and *Brazil* (1962) were modern, efficient ships; nonetheless, the criteria to get from point A to point B as fast as possible, and in as luxurious a setting as possible, was not used for these vessels. These were semi-cruise ships that carried modest amounts of cargo and transit passengers who had the time and perhaps an aversion to flight. R. L. Sanford in discussing George Sharp's 1947 SNAME paper [3] said, "There has been widespread speculation among those interested as to the ultimate effect of overseas air travel as a competitor of passenger ships, even to the extent, in some cases, of predictions that air travel eventually will supersede sea travel. I, for one, refuse to accept such a denouncement."

Within the next ten years, the trend would become so clear that Mr. Sanford and the marine industry would have to accept the passing of an era.

The rest of this section is broken down into a number of technical subsections as follows:

Hull Form and Powering
Machinery
Arrangements
Crew Accommodations
Structures
Subdivision
Stability
Joiner Work
Fire Detection and Extinction
Electrical
Lifesaving Equipment
Ecological Requirements

These sections will deal with large offshore passenger ships built in the United States through the 1960s.

A final section will deal with smaller passenger cruise vessels built generally after the 1960s.

Hull Form and Powering. In comparing hull form ratios and coefficients, speed, power, and displacement, I have selected the passenger ships *Manhattan* (1932), the *Normandie* (1935), and the *America* (1940) as base comparison vessels to U.S. passenger ships built after 1943. Those vessels in Table 1 built after 1943 comprise almost all of the major passenger ships built during that period. Table 1 contains two passenger/cargo ships, the nuclear powered ship *Savannah* (1962) and the *Santa Magdelena* (1963). As this

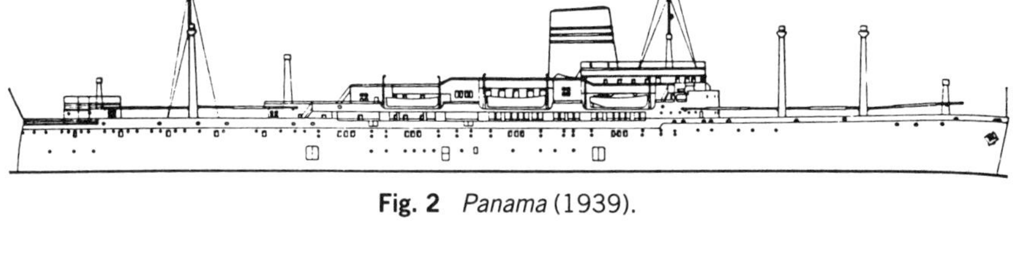

Fig. 2 *Panama* (1939).

Table 1 A comparison of vessel characteristics

Vessel	Year Built	LOA, ft-in.	LBP, ft-in.	B, ft-in.	Depth, ft-in.	Draft, ft-in.	Δ Tons	Speed	SHP	No. Pass.	LWL, ft-in.	C_P	L/B	B/H
Manhattan	1932	705-3	666-0	86-0	75-0	30-6	33,250	20	30,000	1224	683	0.659	8.00	2.37
Normandie	1935	1029-4	962-0	117-9	91-10	36-7	68,350	29.5	160,000	1972	962	0.578	8.18	3.22
America	1940	723-0	660-7	93-3	73-5	32-6	35,440	22	34,000	1202	690-3	0.600	7.50	2.83
LaGuardia	1944	622	—	75-0	—	26-1	—	19	—	609	—	—	8.3	—
President Wilson	1948	609	—	75-0	—	30-2	—	20	—	832	—	—	—	—
Independence	1951	682-6	632-0	89-0	70-10	30-0	30,090	22.5	37,000	1003	650	0.637	7.30	2.96
United States	1952	990-0	905-3	101-6	75-0	31-3	45,400	38.32	240,000	1725	940	0.559	9.26	3.25
Santa Rosa	1952	583-7	534-9	84-0	51-3	26-0	19,364	20	20,000	300	550	0.59	6.54	3.23
Argentina	1958	617-0	510-0	84-0	63-0	27-0	22,700	23.5	28,000	553	590-0	0.62	7.02	3.11
Savannah	1962	587-6	545-0	78-0	50-0	29-6	21,840	20.25	22,000	60	550	0.62	7.05	2.64
Santa Magdelena	1963	546-8	508-6	79-0	48-1	29-1	19,799	20	19,800	119	508-6	0.6014	6.44	2.93

chapter is organized, these vessels might have been excluded; nevertheless, their historical significance and passenger capacity were sufficient to include them in this section.

There is a distinction between U.S. passenger ships designed before the mid-1950s and those that came after. Before the mid-1950s, U.S. passenger ships were competing with airplanes on fixed, two-port routes where dependable, rapid transits were required year round. In my opinion, after the late 1950s, most U.S. operators knew the battle was lost and while transatlantic liners continued to be built, they were all built overseas. The challenges to U.S. naval architects in the 1950s to produce attractive, fast, dependable passenger ships at minimal construction and operating cost produced some remarkable hull forms.

While there are not enough ships to indicate trends in Table 1, it is worthwhile to compare three passenger ships of almost identical size, the *Manhattan* (1932), the *America* (1940), and the *Independence* (1951). With similar lengths, the length/beam (L/B) ratio becomes slightly smaller as time goes on. The prismatic coefficient drops for the higher speed *America* as compared to the *Manhattan.* It increases somewhat for the *Independence,* however, with a speed equivalent to that of the *America.* Power over a 20-year period modestly increases from 30,000 to 37,000 shp. Those who may conclude that power was limited in early years by the types of machinery available should recall that the *Europa* (1930) had 105,000 shp and the *Rex* (1932) had 120,000 shp. Both of them were 27-knot ships; thus clearly the lower speeds of the three American ships were driven by economics.

The *Normandie* built in 1935 with a LWL of 962 ft and the *United States* built in 1952 with a LWL of 940 ft again are similar; the *Normandie,* however, was built to be the largest, fastest transatlantic liner at the time, while the *United States* was finally conceived for reasons other than strictly passenger transport. A short history of the ship is in order to understand her hull form.

The independent naval architect William Francis Gibbs had been obsessed with building the ultimate U.S. liner since the 1920s. His plans for an American superliner were developed initially to compete with foreign liners for passenger business. After World War II, however, he modified the concept to make the ship quickly convertible to a troop transport. In the late 1940s the U.S. Navy and Maritime Administration, collaborating with United States Lines, accepted Gibbs's concept and a graving dock was prepared at the Newport News Shipbuilding to build a ship whose lines, power, and potential speed would be kept secret for over 25 years. John Kane's 1978 SNAME paper [4] would finally reveal the awesome potential of the ship. Her record breaking 35.59 knot transatlantic passage, which stood to the 1990s, was made at about 158,000 shp; but, she attained 38.32 knots on trials using 241,785 shp! This speed, achieved at full design draft, was nearly ten knots faster than the *Normandie* at about 50 percent more power. For those who would say, "well, no wonder," let me remind them that speed/power curves for displacement hulls are fairly unforgiving at speed length ratios above 1.0. (For the general reader: speed length ratio equals the speed in knots divided by the square root of the length in ft.)

Returning to Table 1, it is not surprising that the *United States'* prismatic coefficient is significantly below the *Normandie's.* It is worthy of note that the *United States* had an L/B ratio in excess of the *Normandie's* by some 13 percent, which is counter to the declining trend of the chronologically arranged ships in Table 1. The 33 percent reduction in displacement of the *United States* as compared to the *Normandie,* the slender proportions, and the fine ends clearly indicate a criterion for top end speed.

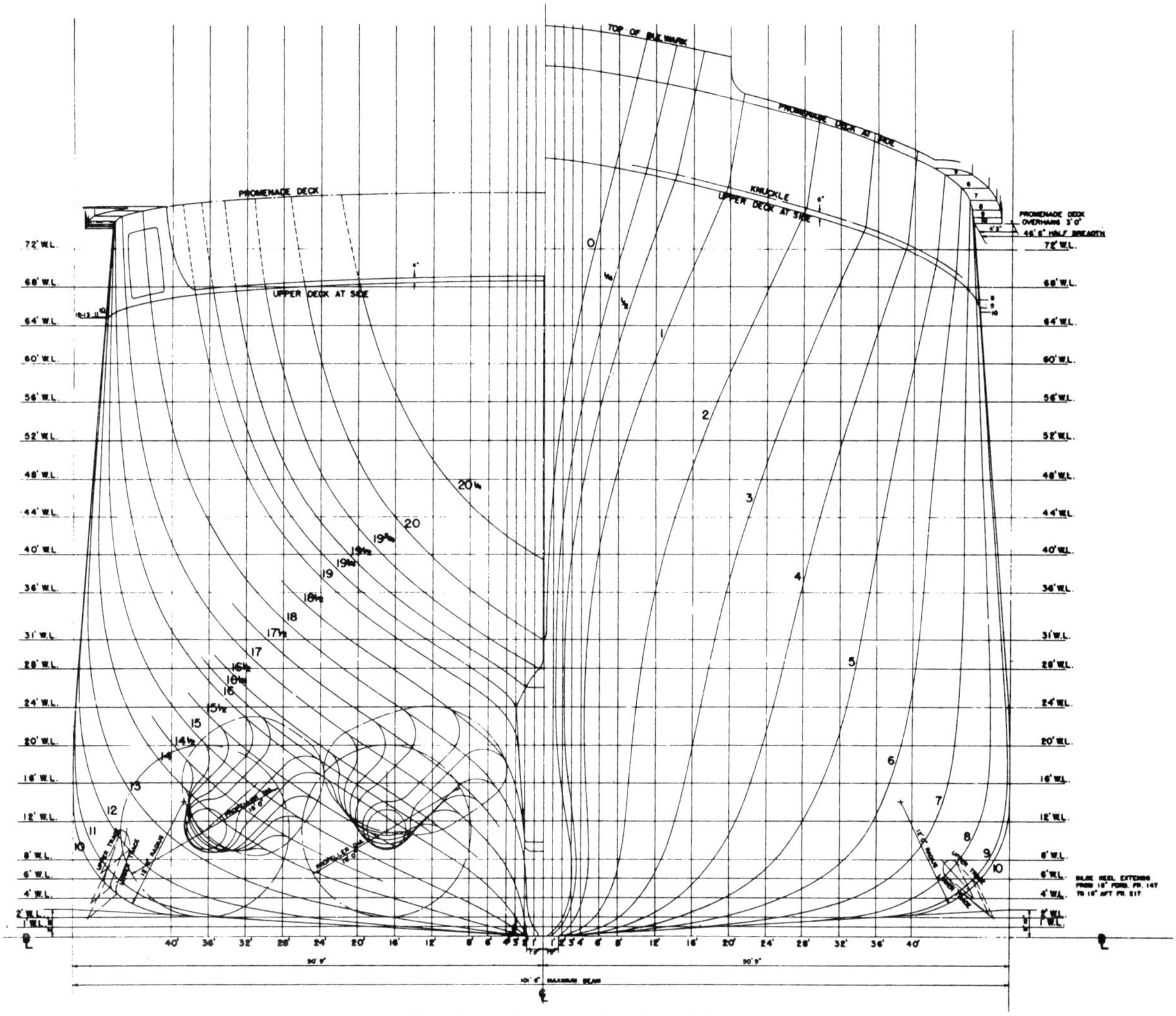

Fig. 3 Body plan of SS *United States.*

The body plan of the *United States* is presented as Fig. 3. It is worthy of note that such a high speed ship had only a slight bow bulb with no forward protrusion past the stem [4]. Gibbs was evidently concerned that the bulb would cause spray, leading to wet decks, or that in heavy weather the bulb would cause slamming. As the *United States* was known as a fairly dry ship in heavy weather, and was able to maintain fairly high speeds without slamming, Gibbs was obviously right.

The arguments over U form and V form forebodies started in the late 1940s and early 1950s, and still persists today. Most of the passenger ships built until the late 1940s employed U forms. A P3-S2-DA1 proposed by MarAd in 1949 was the first V form to be proposed by that agency. The ship was never built and surprisingly none of the remaining vessels in Table 1 employed a bulbous bow.

In reviewing all of the lines of ships shown in Table 1, the lines of the *Savannah* are the most striking. Designed by George Sharp, this modest speed ship did not have to carry too much (a cargo deadweight of 9250 tons on a length of 587 ft LOA) and, with her reactor located deep within the vessel, probably had no stability problems. The designer, thus cast free of normal design constraints, produced a handsome underbody for a single screw ship.

Another continuing debate during the period was the consideration of twin screw shafts running in bossings or in shaft brackets. The *Savannah* and the *Santa Magdelena* are the only single screw ships shown in Table 1. All of the rest are twin screw ships with shafts encased in bossings. Everyone agreed that shaft struts produced less drag than bossings, but no one was willing to step up to shaft struts. Gibbs and his team in designing the *United States* considered them but dropped the idea as they were concerned with longitudinal vibration at such high shaft power, squatting of the ship at speed, protection of the shafts, and pitching motion in a seaway. Gibbs felt that bossings would dampen pitching; and that, with careful fairing and attention to flow lines, a relatively uniform flow to the propellers could be obtained.

The combined disciplines of propeller design and underwater form started using model testing to a greater degree after 1943. Greater confidence in the results and better cavitation tunnels allowed the designers to optimize hull forms, appendages, and propellers. While none of the vessels in Table 1 had any significant hull or propeller problems, the higher powered *United States* had some trailing edge propeller curling in service. To allow proper hull clearance, the ship had relatively small propellers of only 18-ft diameter and, consequently, a developed area ratio of 0.904. The thin trailing edges of the original five-bladed wheel curled owing to cavitation. They were replaced in 1958 with propellers having a heavier trailing edge. The practice of truncating the trailing edge was discovered shortly after by the Navy.

Machinery. All of the ships in Table 1 employed steam turbine propulsion with steam coming from boilers, except for the *Savannah*, where steam came from a pressurized water nuclear reactor. Steam pressures hovered around 600 psi, with superheat temperature as high as 850 °F.

The *Normandie* used auxiliary boilers to produce power for hotel services and low pressure steam. From the *America* on, U.S. practice was to use high pressure evaporators based on crewing and machinery cost economics. Another U.S. departure, starting with the *Manhattan* and continuing with the *America*, was the use of scoops for the supply of condenser sea cooling water as opposed to the European reliance on pumps.

The *Laguardia* and *President Wilson* were both originally P2 class troop ships, and were powered by two T2 tanker steam turbo-electric plants totalling 19,800 shp. While the turbo-electric plant was less fuel efficient than the straight steam turbine plant, the degree of maneuverability and unlimited electric power made these plants popular.

Today diesel electric propulsion using cycle converters are becoming popular in passenger ships owing to the ability to share hotel and propulsive loads powerhouse style.

Arrangements. George Sharp's SNAME 1947 paper [3] noted how passenger ships had become progressively beamier over time. The *City of Rome* in 1881 had a length/beam ratio of 10.72, while the *America* in 1940 had a ratio of 7.10. (As you may note in our table, the *Santa Magdelena* had a ratio of 6.44.) Stability, both intact and damaged, lower speeds, and shallow draft all forced wider beams as time went on. Mr. Sharp wrote in that paper:

> In my experience with the problems of ship design, I have come to observe that a great deal of unnecessary trouble arises out of a tendency to continue the use of certain methods long after substantial changes have occurred in the nature of the problem for the solution of which those methods were first developed.

The fruitful field he thought needed further study in this regard was passenger vessel arrangements. Three trends were identified, in addition to wider beam, as bearing on modern passenger ship arrangements:

- The desirability and salability of outside rooms.
- The diminished size of machinery casings.
- The trend toward private toilet/shower facilities and thus the diminished use of the center of the ship for public facilities.

The desirability of outside cabins continues to this day and needs no explanation. Casing and uptake sizes diminished as steam reciprocating engines and coal-fired boilers were replaced by steam turbines and oil-fired boilers before 1943 and by progressive improvements in steam plant design after 1943. Higher pressure, higher temperature steam plants reduced casing size. Moreover, the machinery itself, coupled with improved engine room ventilation systems, dramatically reduced the size of the engine room.

Fig. 4 *Santa Rosa*, 1958. (Courtesy Newport News Shipbuilding.)

Despite all of this increased space available, two dramatic accidents would restrict forever the naval architect's freedom in passenger ship arrangements. In 1934 the *Morro Castle* suffered a fire that gutted the entire ship and killed many passengers. In 1942 the liner *Normandie* caught fire while docked in New York. The weight and free surface of the firefighting water caused the ship to roll over on her side. It is worthy of note that two of the three passenger vessels shown in the table that were built before 1943 had disastrous fires in 1942 (*Normandie* and *Manhattan*) whereas none of the vessels built after 1943 had any fires of significance. The *Morro Castle* forced standards for fire rated bulkheads, fire zones, fire doors, access routes, automatic sprinkler system, and automatic fire detection systems. The *Normandie* fire pointed out that U.S. regulations for vertical zoning for fire and subdivision not only contained the damage, but minimized the effects of firefighting water.

The *America* (1940) had nine watertight compartments carried nearly to the weather deck whereas the *Manhattan* (1932) had but four. While remote operating watertight doors helped keep some passageways open on upper decks, tourist class and crew cabins on lower decks would be confined to short watertight compartments requiring "up and over" diversions to go forward or aft. Many alternative arrangements were proposed from 1943 on to reduce the claustrophobic disadvantages of inner cabins. There were, for example, cross passages with courts or verandas onto which the cabins faced. None of these ideas caught on, however. To this day the twin fore and aft passageway with one room extending to the side of the house or hull outboard of the passageway, and two rooms, public spaces or utility spaces extending between the passageways has been the established norm. Attempts at all outside cabins with long windowless public spaces between the passageways met with adverse passenger reaction. Surveys showed that passengers, particularly on cruise ships of the 1960s and beyond, did not mind the interior cabin (if it was inexpensive) as they spent the majority of their day in public spaces or on deck.

While interior arrangements were influenced by regulatory constraints and passenger desires, exterior arrangements have changed dramatically primarily based on economics. The aesthetic appeal of the proportioned passenger ship with a fore deck and some aft deck and modest deck house between has given way to cruise ships with house from stem to stern of almost continuous height. A dramatic comparison is made between Figs. 4 and 5. The former shows the *Santa Rosa* as delivered in 1952 and as configured for 300 passengers. The latter shows the ship converted in 1992 on the same original hull to carry 1000 passengers.

Crew Accommodations. Figure 6 [11] shows the trend in crew size versus passenger capacity up through the 1950s. An incredible linear relationship exists for all U.S. ships, as the only ones off the straight line plot (Points 8, 12, 13, 16) are foreign vessels. It is also significant that the foreign ships with their lower crew wages still had smaller crews—a testimony to the strength of U.S. maritime labor unions. This arithmetic progression of about 1.7 passengers per crew could not be sustained with U.S. crew costs and was, in part, the reason for the demise of the U.S.-flag cruise industry. While from the 1960s on, all passenger ships were foreign, it is worth noting that the passenger to crew ratio in the 1970s and 1980s varied from 2.15 to 2.90. With automation and enhanced steward facilities today, the ratio has exceeded 3.0 on ocean cruisers and well exceeds 4.0 on the smaller domestic cruise vessels described in the closing section of this chapter.

Through the 1960s, arranging passenger ship crew accommodations was no small matter. While foreign passenger ships through the 1960s averaged 20 to 30 sq ft of crew space per person, U.S. practice (again forced by U.S. regulations and unions) averaged nearly 40 sq ft. A foreign commanding officer rated about 300 sq ft of cabin while his U.S. counterpart commanded 400. Up into the 1950s foreign passenger

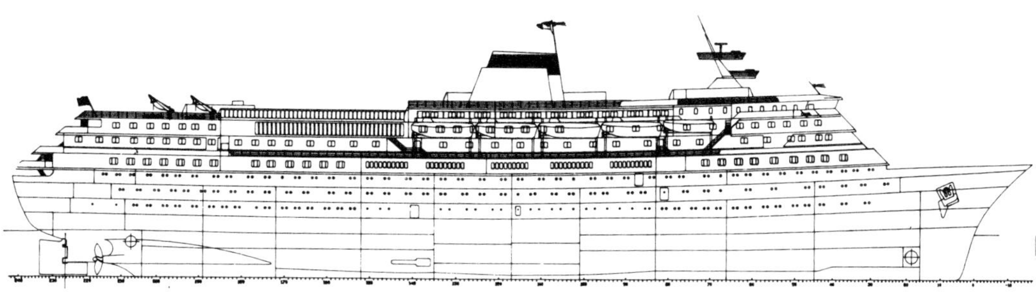

Fig. 5 Profile of *Santa Rosa* (1992).

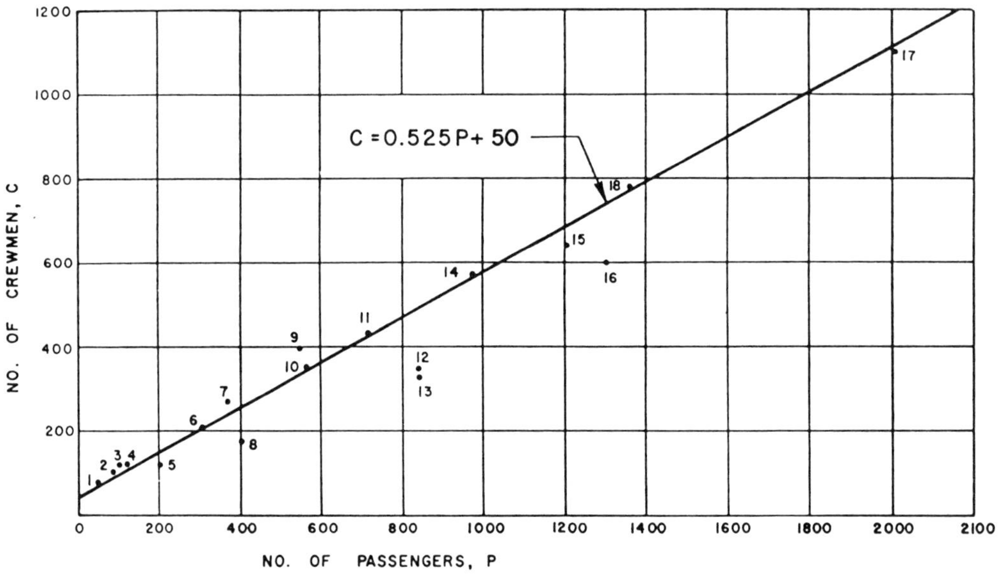

1	SS *Santa Barbara*	7	SS *Mariposa*	13	MS *Bergensfjord*
2	SS *Alcoa Courier*	8	MS *Stockholm*	14	SS *Independence*
3	SS *African Comet*	9	SS *Brasil*	15	SS *America*
4	SS *Del Norte*	10	SS *President Cleveland*	16	SS *Leonardo Da Vinci*
5	SS *Panama*	11	SS *Lurline*	17	SS *United States*
6	SS *Santa Rosa*	12	MS *Gripsholm*	18	SS *Rotterdam*

Fig. 6 Approximate total of crew for passenger ship.

ships would accommodate up to twenty crew in a single room, whereas the U.S. maximum was ten and averaged about six. Foreign ships in that era averaged one toilet for each 16 crew, and one shower for 27, as opposed to the U.S. standard of one T/S space to each eight crew.

The pre-war and post-war trends in crew quarters were dramatically illustrated when the *America* (1940) was rebuilt after the war (1946) to then current U.S. crew standards, losing some 151 passenger spaces in the process.

The *Santa Rosa* and *Argentina* started the trend toward semi-private crew cabins with two crew members to a room; but the U.S. merchant ship standard of individual crew rooms with shared T/S spaces was never attained.

Structures. When passenger ships were still being designed in the United States, naval architects could use a standard longitudinal bending moment calculation assuming a trochoidal wave, a standard bending moment formula, or detailed bending moment calculations. Table 2, reprinted from [12], shows the comparison of two of these methods of bending moment calculations and the resultant predicted stresses.

In small and medium size passenger/cargo ships, the designer had to be careful of increased passenger capacities, the resultant longer accommodation block, and the tendency to push cargo to the ends of the ship. This practice would increase the severity of the traditional hogging moment. The *Santa Rosa* and *Argentina*, built in the 1950s, used side ports and casings through the house to locate cargo spaces closer to amidships. For large passenger ships carrying small amounts of cargo, the designer had greater freedom to distribute weights more favorably. As a result, designers usually carried out detailed bending moment calculations considering all conditions of loading, and thereby maximized the efficiency of the structures.

It was generally considered (and still is) that stress levels on the order of nine tons psi are acceptable. Table 2 shows general agreement with this criterion. The *Rex* and *Conte di Savoia* used high-tensile steel and thus their higher permitted stress levels.

Overall, passenger ships have significant depths to the strength deck, with length/depth ratios on the order of eleven. They are, thus, rather stiff structures in longitudinal bending.

Subdivision. The subdivision of passenger ships is regulated by the International Convention of Safety of Life at Sea (SOLAS). At the beginning of the period this chapter covers, passenger ship subdivision was based on the International Convention for Promoting Safety of Life at Sea of 1929. Of the large passenger ships in Table 1 built before 1940, most were two-compartment vessels based on the floodable length criterion. Only the U.S. designed and built *America*

Table 2 Longitudinal strength characteristics of passenger lines

Vessel	Dimensions, ft-in. LBP	B	d	Section modulus, in.^{2}·ft	Calc. bending moment on standard trochoidal wave, ft·tons	Standard bending moment, ft·tons $\frac{L^2Bd \times 0.75}{35 \times 35}$	Stress on calc. bending moment, tons/in.^{2}	Stress on standard bending moment, tons/in.^{2}
Queen Mary	965-0	118-0	38-10	279,100	2,602,500	2,615,000	9.32	9.37
Leviathan	907-0	100-0	39-1	209,300	2,514,000	1,968,000	12.00	9.40
Bremen	888-1½	101-8½	33-7½	180,800	1,733,000	1,650,000	9.60	9.13
Aquitania	865-0	97-0	36-0	181,000	1,512,700	1,600,000	8.36	8.84
Rex	817-3¼	96-9⁷⁄₁₆	32-9	128,300	1,216,000	1,295,000	9.48	10.10
Conte di Savoia	800-10½	95-9½	31-0	121,200	986,000	1,170,000	8.13	9.65
Manhattan	666-0	86-0	32-0	89,200	734,000	748,000	8.23	8.39
America	660-6	93-3	32-0	93,800	682,000	797,000	7.27	8.50
Independence	632-0	89-0	30-0	78,100	724,000	653,000	9.27	8.36
Mariposa	605-0	79-0	28-0	63,400	486,000	496,000	7.67	7.82

exceeded the rule requirement, as she was easily a three-compartment (nearly four-compartment) ship.

In addition, European designers tended to step bulkheads to fit the desired arrangement. In some cases some good sized passenger ships might have three compartments damaged within the length of only 45 ft. Subsequent to SOLAS 29 came Senate Report 184 upon which the *America* was designed, then came SOLAS 1960 and SOLAS 1974. SOLAS 1974 still used an updated formula involving the number of passengers, length of ship, and crew size; but IMCO Resolution A.265 (VIII) was the first convention to recognize additional factors such as probable statistical occurrence of damage (probable length of damage), data from previous casualties, and validation of formulas by model tests.

It had already become apparent by the 1980s that increased numbers of bulkheads (and thus smaller spacing between bulkheads) was not improving damaged stability. IMCO A.265 (VIII) was the beginning, in my opinion, of an enlightened approach to subdivision in which substantially safer passenger ships were designed.

Stability. To see how passenger ship stability has changed over time, compare the *Panama* profile (Fig. 2) with the modified *Santa Rosa* (Fig. 5). While damage stability is certainly part of the stability question, clearly the sail effect of the superstructure, and the U.S. Coast Guard wind heel criterion for intact stability, have played an important role in passenger ship design.

Bulkhead spacing, the setting of the bulkhead deck and margin line, and detailed damage stability calculations allow some latitude in solving the damage stability problem in design; owners, however, want all outside rooms, a high deckhouse running the length of the ship, and a hull with minimal draft and displacement. Consider the cumulative effect of these trends on the required metacentric height (GM): in the numerator of the equation we find an increasing trend arising from longer, higher deckhouses. At the same time, the denominator is decreasing because of diminishing displacements. Thus the required GM continues ever higher.

Reference [12] recommended a large passenger ship GM of not over six percent of the beam. This is not easily attained, however, as the trend from 1943 to 1993 has been toward greater sail area without much increase in beam. Thus we find that relative GM values are running about ten percent. The designer, who could at the beginning of this era assume the depleted fuel oil tanks would be filled with ballast water to maintain stability and draft, must today double his tankage to provide separate clean ballast and fuel tankage. A careful balance always had to be maintained between GM and good seakeeping, as too much GM would cause a stiff rolling, cranky vessel clearly unsuitable for passengers; whereas too little GM caused lolling, also discomforting to passengers. The development of stabilizer fins and anti-roll tanks in this era have improved passenger ship seakeeping to the point where the larger required GMs can now be tolerated.

Joiner Work. From 1930 to 1945, the passenger ships *Georges Philippar*, *Atlantique*, *Morro Castle*, *Lafayette*, *Paris*, *Normandie*, *Manhattan*, *Alcantara*, and *Empress of Russia*, all had significant fires, the worst of which was the *Morro Castle*. The joiner work in these ships was largely of wood and plywood, and where the joiner bulkheads were stopped below the deck beams to permit easy access for piping, wiring, and ducting.

The wooden encased rooms would burn easily; the real horror, however, was the rapidity with which fires would spread from room-to-room, generating a wind-driven firestorm that was nearly impossible to stop other than by flooding.

The *Morro Castle* disaster prompted the U.S. Government to make a complete investigation of the incident, focusing on joiner work and fire. Two SNAME papers in 1937 and 1938 presented the results of these studies. As a result of the investigations, the United States mandated changes to the International Convention of 1929 to the extent that passenger and crew rooms and all public spaces were to be constructed of fire resistant (fireproof) materials. One of the principal partition panels was developed by Johns Manville. It was an asbestos product called Marinite. Alternative products were double faced steel bulkhead panels with the center filled with rockwool. It was found that this latter panel reduced sound transmission between rooms to a greater extent than did Marinite. Nevertheless, Marinite became the leading product used in U.S. ship construction.

The weight of joiner material and its supporting steel framework dramatically increased the joiner weight component of passenger vessels. This trend, in conjunction with higher deck houses, would lead to lower length/beam ratios in passenger ships.

While the product Marinite and other incombustible panels surrounding rooms almost negated major fires in the 1943–1993 era, the Marinite panel would be found to have created another hazard. As already noted, these panels contained asbestos; and in the late 1970s it would be discovered that merchant seamen working on ships (and particularly in engine rooms where great amounts of asbestos insulation were used), were afflicted with various diseases emanating from asbestos exposure. While shipyard workers had been previously thought to be at risk as they worked with the product during construction, no one thought seamen would be. With a twenty-year gestation period, seamen in the 1970s working in ships built in the 1940s to 1960s, started experiencing a startling incidence of afflictions. This culminated in the prohibition of asbestos products in U.S. ship construction in the late 1970s, early 1980s.

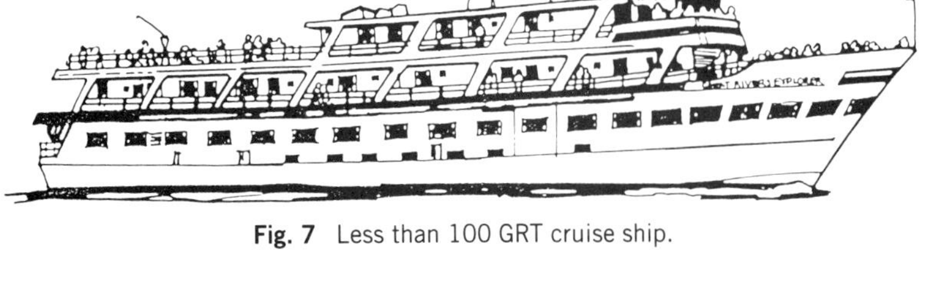

Fig. 7 Less than 100 GRT cruise ship.

Thus, the double steel bulkhead panel with rockwool or fiberglass core is now the preferred method of joiner construction.

Fire Detection and Chemical Extinguishing Systems. The United States in this era was the leader in fire detection, with all manner of cleverly engineered detection systems, which in some cases were coupled to automatic extinguishing systems. In 1935 U.S. passenger ships had to be fitted with either an automatic sprinkler system or automatic fire detection system with manual extinguishing. Even in the early 1940s fire detection systems would immediately pinpoint the space where the fire was located.

While the United States was proceeding full ahead on fire extinguishing and detection, much of Europe was on a slow bell. Reciprocal agreements allowed foreign passenger ships into U.S. ports without the fire protection required on U.S. vessels. U.S. passenger ships, either new or refitted, therefore cost more than their foreign counterparts. Thus while U.S. ships were safer, passenger traffic had already started moving toward foreign ships with their better service and lower fares.

Engine room fire protection was another matter. As we turned to oil fired plants, old steam and foam smothering systems were considered inadequate. In the late 1920s, carbon dioxide was used to extinguish fires and this became the standard for engine room protection to this day. Carbon dioxide has not been without hazard, as accidental releases have killed engine room and shipyard personnel. In the 1970s and 1980s, Halon was thought to be the ultimate answer, where one could walk into a compartment with a fire, have the compartments flooded with Halon and calmly walk out while still breathing. This miracle is now about to come to an end as Halon is considered to be partly responsible for depletion of the ozone layer. So CO_2 still seems to be the answer for machinery spaces.

Electrical. Electrical systems in this era have possibly made the greatest strides of any of the shipboard systems. From dc, open-front breakers to modern ac cycloconverter propulsion systems, the technological advance of electrical systems has been phenomenal. More importantly, although higher voltages and amperages are carried in shipboard systems, personnel safety and the reliability of electrical systems have progressed to near perfection. Electronic systems guide and operate ships better and more efficiently than ever before. Crews have become hi-tech engineers using computer systems to catch faults before failure, to optimize operations, and to transmit data throughout the ship and to the shore. These changes now provide a level of efficiency beyond belief at the beginning of the era.

Lifesaving Equipment. As incredible as it may seem, while dramatic advances have been made in electrical, air conditioning, and sewage disposal systems aboard passenger ships, lifesaving equipment during this period, generically, has remained almost unchanged.

Even though cork and kapok filled lifejackets have been replaced with synthetic foam lifejackets, a 1943 passenger would have no trouble identifying a 1993 lifejacket. Similarly, he might not notice that his lifeboat was made of GRP instead of metal, but he would certainly recognize it as a lifeboat. Inflatable life rafts are an improvement over the cork "donuts" used at the beginning of the era; but a life ring is still a life ring.

The new "free-fall" lifeboat used on tankers and cargo ships today certainly causes the seaman to ponder his two fates: that of being consumed by fire (or sea) or hurtling a hundred feet or so through space for a 1-g impact with the sea. These units have not been used, to date, on passenger ships undoubtedly owing to arrangement problems, and possibly the reluctance of passengers over the age of 12 to enter any such device.

While passenger ships have become many times safer during this era, the few disasters we have witnessed have pointed up how ineffective lifeboat systems are in safely deploying all passengers off the ship. As ships grow in size and complexity, this archaic system must be replaced with some new technology.

Ecological Requirements. As the airplane laid to rest the transatlantic liner during this era, the requirements of preserving the world's ecology may injure the modern day cruise ship. Today the tons of sewage on a cruise ship must be totally contained while within U.S. waters (the limit of which continually expands). Moreover, before discharging sewage at sea it must be reconverted in a process seemingly more complicated than the process that produced the sewage in the first place.

Incinerators now burn everything that cannot be otherwise "processed." In some cases, certain waste products must be bagged to be taken ashore to an ever-decreasing number of dump sites.

In short, the passenger ship has not escaped the world's problems of pollution; and ecological considerations will play an increasing role. The volumetric, deadweight, and power requirements will have a marked effect on the economics and designs of the future ships.

U.S. Coastal Cruise Trade

To a modern group of naval architects involved in the U.S. coastal cruise trade, it does not matter that transoceanic liners are no more. These creative people have discovered the aging U.S. tonnage laws, some written at the time SNAME was founded and still in existence today. Truly incredible feats of ingenuity have produced fairly sizeable passenger vessels whose gross tonnage is miraculously below one hundred. Figure 7 shows one such ship attaining a one hundred passenger capacity, and sailing from the States to Alaska, Tahiti and other exotic locations. The difference is that the vessel goes to the remote locations without passengers and waits there for passengers who fly in to cruise the local area.

Why is the one hundred gross ton limit sought? The answer lies in a multitude of costs affected by regulation. The requirements for lifesaving, firefighting, communication/navigation equipment, electrical systems, piping systems, and the regulatory inspection criteria for vessels under one hundred GRT are all dramatically different from those applying to larger oceangoing ships.

A crew of 10 or 15 can legally operate a one hundred GRT passenger ship, with only a few licensed individuals required. Their licensing requirements are considerably less than those for oceangoing officers. A crew of 20 to 25 can operate a coastal cruise ship with a hundred passengers, where offshore regulations would force a crew of over 50.

In a complete reversal, we now see modern sailing cruise ships where the passengers can assist in the operation. For less eager passengers, these modern sailing marvels (Fig. 8) can be remotely worked from

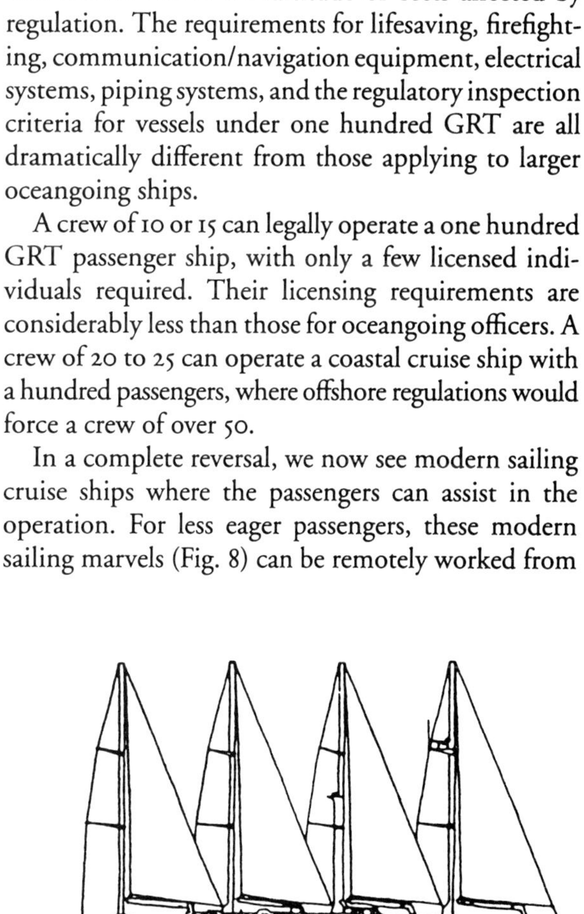

Fig. 8 110-passenger Windcruiser proposal

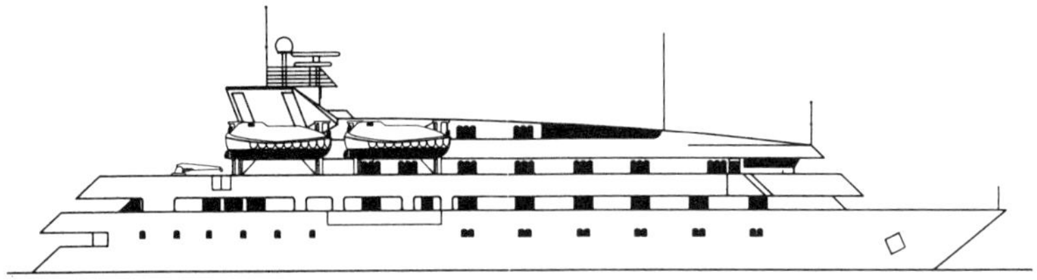

Fig. 9 Yacht-like small cruise/passenger ship

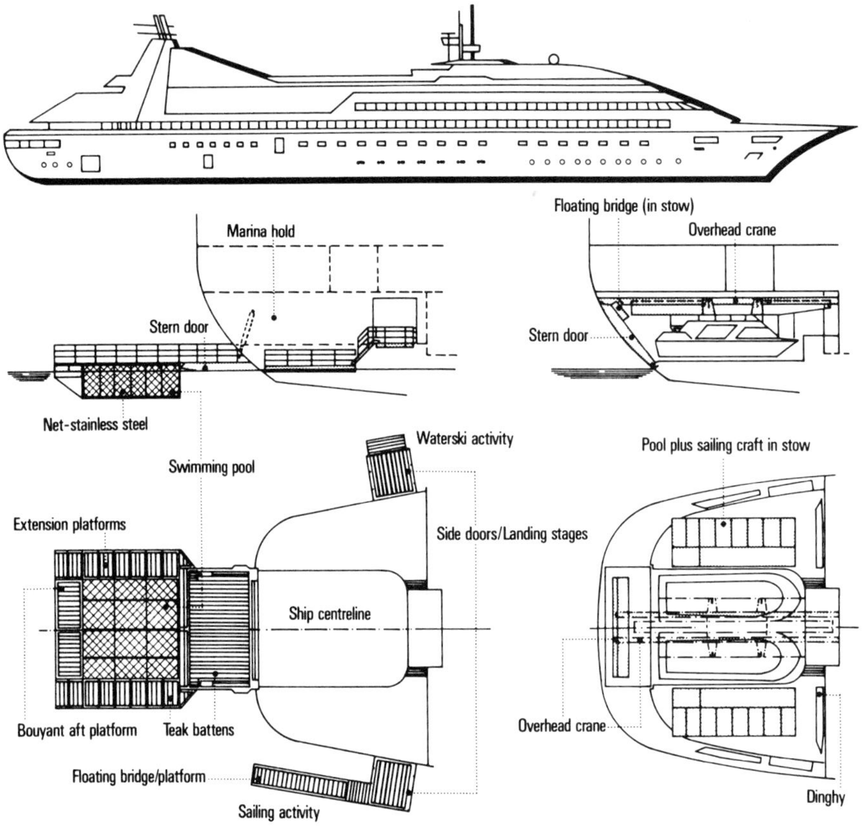

Fig. 10 Intermediate cruise ship with swimming platforms, windsurfers, and diving.

a one-man central console. The automatic sailing ship! One would question if this is truly progress.

Yacht-like small passenger ships catering to a free spending class of clientele cruise to Monte Carlo and similar locations, providing their passengers with an opulent, waterborne hotel, and a grand entrance wherever they go (Fig. 9).

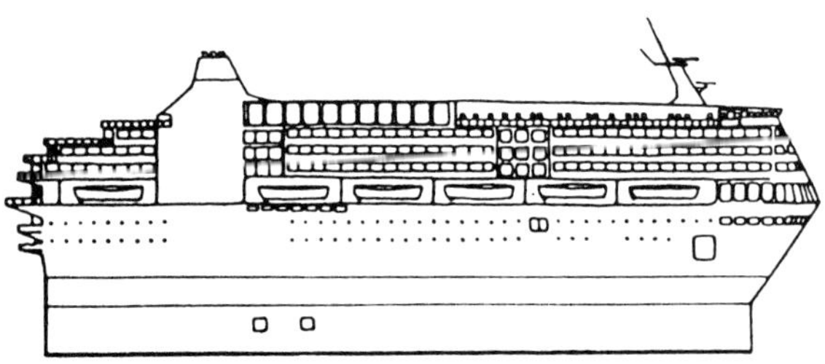

Fig. 11 2000-passenger SWATH cruiser.

A large and growing segment of the cruise ship industry operates out of the United States, but generally under foreign flags. The luxurious amenities and comforts of these ships are such as to make the destination immaterial. On them you may find in addition to the traditional enticements, swimming platforms, recreational boats, windsurfers, and scuba diving equipment (Fig. 10). And finally, to be delivered in late 1992 or early 1993, at the end of this era, is the first SWATH cruise ship (Fig. 11). A floating hotel like no one has ever seen, with complete facilities such that the cruise passenger does not even care where he is going. This gives only a hint of the future. I suppose this is progress.

References

1 "Large Passenger-Carrying Ships for Certain Essential Trade Routes," *Trans.* SNAME, 1948.
2 "Aspects of Large Passenger Liner Design," *Trans.* SNAME, 1946.
3 "Some Considerations in the Design of Modern Ships," *Trans.* SNAME, 1947.
4 "The Speed of the SS *United States*," *Marine Technology*, Vol. 115, No. 2, April 1978, p. 119.
5 "Principles of Naval Architecture," SNAME, 1967.
6 "Design and Construction of Steel Merchant Ships," SNAME, 1955.
7 "N.S. *Savannah*," *Marine Engineering Log*, May 1958.
8 "Main Particulars of SS *Brazil*," *Marine Engineering Log*, Dec. 1958.
9 "Main Particulars of *Santa Rosa*," *Marine Engineering Log*, Sept. 1958.
10 *Marine Engineering Log*, March 1951, pp. 60, 61.
11 *Ship Design and Construction*, SNAME, 1969.
12 *Design and Construction of Steel Merchant Ships*, SNAME, 1955.

John W. Boylston, after graduating from Kings Point, served as a deck officer on the *America*, *Argentina*, and *Santa Rosa*. He then entered the University of Michigan, where he earned a BSE degree in naval architecture and marine engineering. He then became Naval Architect for Sea-Land during a major construction program. He is currently President of Argent Marine, operating a fleet of LNG tankers. He is an active member of SNAME and the author of several highly regarded papers. He is also known for his energetic efforts in restoring the Liberty Ship *John W. Brown* to an active floating museum.

U.S. Maritime Commission Cargo/Passenger Ship Conversions

WARREN LEBACK

THE UNITED STATES MARITIME COMMISSION was established with the passage by Congress of the Merchant Marine Act of 1936, The Commission was charged, among other things, to rebuild the American-flag merchant fleet. In 1937 conceptual designs for two somewhat different cargo ships were distributed to the maritime industry for constructive criticism. These were intended for the U.S. overseas liner trade and were what we now call break bulk general cargo carriers. From comments and suggestions by shipowners, shipbuilders, and naval architects there emerged the final designs for the standard C-2 and C-3 cargo ships. The general characteristics for these vessels are shown in Table 1.

The first C-2 to be delivered was the *Donald McKay* in June 1939, while the first C-3 was the *Exporter* in September 1939. Parallel with the building and delivery of the standard design C-2 and C-3 types, such companies as American President Lines, Mississippi Shipping Co., American South African Line, Inc., Moore-McCormack Line, Inc., and United States Lines were actively interested in replacing their cargo/passenger ships. American President Lines was interested in replacing their 502- and 535-class vessels built at the end of World War I. Mississippi Shipping and United States Lines were both interested in replacing their Hog Island cargo/passenger ships, while America South African Lines wanted to augment their South African service based on the success of their 1930-built *City Of New York.* Moore-McCormack Lines wanted to establish a secondary cargo/passenger service to South America. The Maritime Commission, in conjunction with operators, developed four cargo/passenger variants using the standard C-3 type hull and machinery. A total of 24 ships were ordered and laid down.

Following World War II, seven additional C-3's were converted to cargo/passenger service. Four were for American Export Lines, Inc. as explained below.

Grace Line, Inc. bought nine C-2's during construction and converted them to cargo/passenger vessels in 1946 to replace war losses for their services to the Caribbean and the West Coast of South America. Alcoa Steamship Company contracted to convert three Victory type hulls and machinery into cargo/passenger ships for their Gulf/Caribbean Service.

Table 1 General characteristics for C-2 and C-3 cargo ships

	C-2		C-3	
	Diesel	Steam	Diesel	Steam
Tonnage, gross	6160	6085	7886	7773
Tonnage, net	3653	3573	4702	4635
Deadweight, tons	8656	9758	11,975	12,595
Displacement, tons	13,876	13,893	18,215	18,215
Length OA	459 ft	459 ft	492 ft	492 ft
Length BP	435 ft	435 ft	465 ft	465 ft
Breadth, molded	63 ft	63 ft	69 ft	69 ft
Depth, molded	40 ft-6 in.	49 ft-6 in.	42 ft-6 in.	42 ft-6 in.
Draft, loaded	25 ft-9 in.	25 ft-9 in.	28 ft-6 in.	28 ft-6 in.
Bale capacity, cu ft	523,131	562,849	701,873	730,549
Horsepower	6000	6000	8500	8500
Service speed, knots	15.5	15.5	16.5	16.5
Cruising radius, nmi	12,500	13,000	14,500	12,000
Passengers	12	12	12	12

A total of 40 standard cargo ships were converted to cargo/passenger service. Additionally, in 1941 Alcoa Steamship Co. contracted for three modified C-2 hulls using C-3 type machinery. These, when modified, did not resemble either C-2 or C-3 hulls. But we shall discuss only the 36 standard hulls.

C-3 Conversions

American Export Lines. In 1943 American Export Lines contracted with Bethlehem Steel Company, Sparrows Point MD, to build fifteen export type C-3's. Ten were completed as either attack or troop transports. At the end of World War II all fifteen were returned to American Export Lines. Four (*Dauphin, Duchess, Queens,* and *Shelby*) were converted into cargo/passenger ships, replacing the original "Four Aces," three of which were lost during the war and the fourth found to be unfit for service. The vessel characteristics were:

Tonnage, gross	9644
Tonnage, net	5313
Deadweight, tons	7231
Displacement, tons	14,893
Length OA	470 ft
Length BP	450 ft
Breadth, molded	66 ft
Depth, molded	42 ft-6 in.
Draft, loaded	27 ft-9 in.
Horsepower	8800
Service speed	16 knots
Passengers	125

The four conversions were completed in 1948. The ships were renamed *Excalibur, Excambion, Exeter,* and *Exchorda.* Two were eventually sold overseas. In 1967 the *Exchorda* was sold to Stevens Institute as a floating dormitory, but was dismantled in 1979. The *Excambion* has, since 1965, been used as a training ship by the Texas Maritime Academy.

American President Lines. In 1939 the Newport News Shipbuilding and Dry Dock Co. was given an order to construct seven cargo/passenger ships using standard C-3 hulls and machinery. These were for American President Lines' Round-the-World service replacing the World War I 535- and 502-type cargo/passenger vessels. All seven were delivered between October 1940 and November 1941. Five were acquired by the U.S. Navy for conversion to attack transports while two were time-chartered to the U.S. Navy for service as transports. After the war the *President Monroe* and *President Polk* were returned to their original owners for their Round-the World service, which they continued until 1965, after which they were sold overseas and eventually scrapped. The major characteristics were:

Tonnage, gross	9255
Tonnage, net	5157
Deadweight, tons	9947
Displacement, tons	6175
Length OA	492 ft
Length BP	465 ft
Breadth, molded	69 ft-6 in.
Depth, molded	42 ft-6 in.
Draft, loaded	28 ft-6 in.
Horsepower	8800
Service speed	16.5 knots
Passengers	96

American South African Line. In October 1939 the United States Lines awarded to Ingalls Shipbuilding Corp. a contract for four C-3's with accommodations for twelve passengers. In September 1940, three additional ships for American South African Lines, Inc. were awarded. United States Lines ceased its North Atlantic services in 1940 owing to the Neutrality Act. The first three ships ordered were transferred to American South African Line, and the contracts called for modifications to bring the passenger capacity up to 116. The three vessels were taken over by the U.S. Navy in 1942 before completion. They had all been scrapped by 1973. The general characteristics were:

Tonnage, gross	12,069
Tonnage, net	6577
Deadweight, tons	5278
Displacement, tons	15,450
Length OA	492 ft
Length BP	465 ft
Breadth	69 ft-6 in.
Depth, molded	45 ft-6 in.
Draft, loaded	25 ft-6 in.
Horsepower	8800
Service speed	16.5 knots
Passengers	116

At the end of World War II, Farrell Lines (American South Africa Line) purchased from the government two cargo/passenger ships originally laid down for the Mississippi Shipping Co. These were two from a group of six ordered. Their general characteristics were:

Tonnage, gross	7922
Tonnage, net	4327
Deadweight, tons	9021
Displacement, tons	14,247
Length OA	468 ft
Length BP	465 ft
Breadth	65 ft-6 in.
Depth, molded	30 ft-9 in.
Draft, loaded	25 ft-8 in.
Horsepower	8800
Service speed	16.5 knots
Passengers	67

The *African Endeavor* was laid down as the *Delbrazil* and delivered in May 1940. The U.S. Navy requisitioned her in August 1943 and converted her to a transport renamed *George F. Elliott.* In June 1946 she was returned to her owners. Farrell acquired her in 1949, and in 1960 she was transferred to MarAd. She was scrapped in 1972.

The *African Enterprise* was laid down as the *Delargentino* and was delivered in November 1940. In June 1941 she was acquired by the U.S. Navy and converted to a transport (troop capacity 1891). The Navy renamed her *J. W. McAndrew.* She was acquired by Farrell in 1949. She was placed in their South African service until 1960, and was then transferred to MarAd, and scrapped in 1972.

Mississippi Shipping. The Mississippi Shipping Co. gave the Bethlehem Steel Co., Sparrows Point, a contract for building three cargo/passenger ships to be delivered in the second half of 1940. An additional three were contracted for delivery in the second half of 1942. These were for service between U.S. Gulf ports and the east coast of South America. The general characteristics were:

Tonnage, gross	7922
Tonnage, net	4327
Deadweight, tons	9021
Displacement, tons	14,247
Length OA	468 ft
Length BP	465 ft
Breadth	65 ft-6 in.
Depth, molded	30 ft-9 in.
Draft, loaded	25 ft-8 in.
Horsepower	8800
Service speed	16.5 knots
Passengers	67

The ships were all taken over by the Navy before, or shortly after, delivery and saw but little civilian passenger service. For the historic record, their original names were *Delbrazil, Delorleans, Deluraguay,* and *Delorleans II.* The *Delorleans* is now in service as a training ship for the California Maritime Academy. The others have been scrapped.

At the end of the war, three basic C-3's were ordered by the Mississippi Shipping Company for their East Coast to South America service. They were modified to carry 121 passengers. These were the last standard C-3 hulls to be converted as cargo/passenger hulls for American operators. Their general characteristics were:

Tonnage, gross	12,069
Tonnage, net	6577
Deadweight, tons	5278
Displacement, tons	15,450
Length OA	496 ft
Length BP	465 ft
Breadth	69 ft-6 in.
Depth, molded	45 ft-6 in.
Draft, loaded	25 ft-6 in.
Horsepower	8800
Service speed	16.5 knots
Passengers	121

These ships were named *Delnorte, Del Sud,* and *Del Mar.* They entered service in late 1946, early 1947, and were scrapped in 1972.

Moore-McCormack. Moore-McCormack Lines, Inc. ordered four cargo/passenger diesel driven ships from Sun Shipbuilding and Dry Dock Co. at Chester PA. These were to be delivered during the second half of 1941. They were basic C-3 hulls, but they were the first large American ships to be diesel driven. Their passenger capacity was 216 with a number of single-berth cabins measuring 12 ft by 6 ft, with private baths. These were indeed on the small side. All passenger spaces were to be air conditioned. The ships were contracted for in February 1939 and all were launched during the winter of 1940–41. Owing to the war, they were never completed as contemplated, but were converted to aircraft carriers. Their original general characteristics were:

Tonnage, gross	9255
Tonnage, net	5157
Deadweight, tons	9947
Displacement, tons	16,176
Length OA	492 ft
Length BP	468 ft
Breadth	69 ft-6 in.
Depth, molded	42 ft-6 in.
Draft, loaded	28 ft-6 in.
Horsepower	9000
Service speed	17.5 knots
Passengers	216

These four ships were destined to lead adventuresome lives. The *Rio Hudson* was launched in November 1940, delivered in July 1941 to Bethlehem Steel Corporation, Staten Island, and completed as an escort aircraft carrier and renamed *Avenger.* She was loaned to the British Navy. On November 11, 1942, she was sunk by a torpedo west of Gibraltar.

The *Rio Parana* was launched in December 1940, and delivered to Atlantic Basin Iron Works, New York, for conversion and completion as an escort aircraft carrier. She was leased to the British Navy and renamed *Biter.* In April 1945 she was transferred to France and renamed *Dixmude.* She was returned to the U. S. Government and subsequently sunk as a test target.

The *Rio De La Plata* was launched in March 1941, and delivered to Newport News to be completed as an escort aircraft carrier. In September 1941 she was delivered on a lend/lease basis to the British Navy and renamed *Charger.* She was used to train British naval air crews. In July 1943 she was returned to the U.S. Navy and operated throughout the remainder of the war. In March 1946 she was stricken from the U.S. Navy and transferred to the Maritime Commission. She was sold in 1949 to Alvion SS Corp., Panama, for conversion by Bethlehem Steel Corp., Hoboken, to a passenger ship having a capacity of 1800 passengers. She was renamed *Fairsea.* She was broken up in August 1969.

The *Rio De Janiero* was launched in April 1941, and delivered to Tietjen and Long Drydock Co., New York, for completion as an escort aircraft carrier. She was completed and delivered in July 1942, to the British Navy and renamed *Dasher.* On March 27, 1943, she suffered a tremendous explosion on her hangar deck, tearing a hole in her hull. She sank within five minutes in sight of Ardrossan, just south of the Cumbrae Islands. An investigation concluded the explosion resulted from a leaky fuel valve that permitted aviation fuel to drain below the mess deck and become ignited from a local source. Subsequent carriers were fitted with return lines.

United States Lines. In October 1939 the United States Lines ordered four C-3's carrying twelve passengers from Ingalls Shipbuilding Co., Pascagoula. In September 1940, three additional ships were ordered for American South African Line. Because the Neutrality Act prevented U.S. Lines from serving on the North Atlantic, the first three vessels were transferred to American South Africa Line. Their passenger capacity was increased to 116. The remaining were completed as troop transports for the U.S. military. In the end, all seven did yeoman service during the war, and came through unscathed. Their general characteristics were almost exactly the same as the ships described earlier, under the section on the American President Lines.

All of these ships experienced numerous changes in name, ownership, and configuration. Perhaps the extreme example was the *American Banker.* She was launched as *Biloxi* in February 1943 and completed as a U.S. Army transport, being renamed *Henry Gibbons.* She had a troop capacity of 1948. In January 1946, she was rebuilt as a servicemen's dependent ship to carry 273 adults and 179 children. In March 1950, she was transferred to the U.S. Navy, and in December 1959 she was transferred to MarAd. She was immediately loaned to the N. Y. Maritime Academy as a training ship and renamed *Empire State IV.* In 1974, she was reassigned to the Massachusetts Maritime Academy and renamed *Bay State.* She was scrapped in 1984.

Summary. In summary of our history of the C-3 conversions, it is worth noting that all of them were built between September 1939 and May 1947. The series started with the *Exporter* and ended with the *Del Mar.* Thirty-one were converted for American owners and six for foreign owners. They were designed to serve on a wide range of trade routes and kinds of service. The C-3's served in the U.S. Navy as troop transports, repair vessels, seaplane tenders, and in both the U.S. and British Navies as escort aircraft carriers. In the post war period several were converted to containerships.

C-2 Conversions

Grace Line. Nine C-2 ships were completed as cargo passenger vessels for the Grace Line between June 1946 and January 1947. Of these, six were completed by the North Carolina Shipbuilding Co. at Wilmington NC. Three were built by Federal Shipbuilding and Drydock Company. The six built at North Carolina Shipbuilding were the last ships built at that yard. The three at Federal also closed out that yard's activities. The vessels' general characteristics were:

	North Carolina	Federal
Tonnage, gross	8327	8610
Tonnage, net	4919	5891
Deadweight, tons	9652	8735
Displacement, tons	14,945	13,907
Length OA	459 ft	459 ft
Length BP	435 ft-6 in.	435 ft-6 in.
Breadth	63 ft	63 ft
Depth, molded	40 ft-6 in.	40 ft-6 in.
Draft, loaded	27 ft-7 in.	25 ft-9 in.
Horsepower	6600	6600
Service speed	15.5 knots	15.5 knots
Passengers	52	52

The six North Carolina-built vessels operated on a weekly service from New York to the west coast of South America. The remaining three were assigned to a weekly service from New York to the Dutch West Indies, Venezuela, and Columbia. Most of them underwent occasional name changes and some changes of ownership throughout their long careers. The last was scrapped in 1970.

Alcoa Steamship Company. In 1941 Alcoa laid down three cargo/passenger ships at Moore Drydock Company in Oakland CA. Although designated as C-2 types, they little resembled the standard C-2. They were somewhat yacht-like in appearance, with raked bow and stern, and with the main deck house overhanging the hull. Prior to launching, they were taken over by the U.S. Navy and converted to hospital transports. Their characteristics were:

Tonnage, gross	7486
Deadweight, tons	3452
Displacement, tons	10,900
Length OA	429 ft
Length BP	425 ft
Breadth	62 ft
Horsepower	9350
Service speed	20 knots
Passengers	700

Their original names were *Alcoa Courier, Alcoa Corsair*, and *Alcoa Cruiser*. Their war and peacetime services saw them undergoing changes in ownership, name, and service. The last was scrapped in 1971.

Victory Ship Conversions

In 1945 Alcoa decided to order three ships to replace the three C-2's that had been taken over by the Navy. They selected one AP5 and two AP3 Victory Ships as their basic designs, and had them redesigned as cargo/passenger vessels. They were redesignated A57-class Victories. The Oregon Shipbuilding Corp., at Portland OR, did the conversions. The *Alcoa Cavalier* was launched in September 1946 and delivered in March 1947. The *Alcoa Clipper* was launched in September 1946 and delivered in April 1947, with the *Alcoa Corsair* following in October 1946 and delivering in April 1947. These were the last ships built at Oregon Shipbuilding. Their general characteristics were:

Tonnage, gross	8481
Tonnage, net	4588
Deadweight, tons	9733
Displacement, tons	15,199
Length OA	455 ft-3 in.
Length BP	436 ft-6 in.
Breadth	62 ft
Depth, molded	38 ft
Draft, loaded	28 ft-8 in.
Horsepower	9350
Service speed	17.5 knots
Passengers	98

Two of these ships had rather nugatory careers. The *Alcoa Corsair*, however, finally lived up to her name. On October 22, 1960 she was outbound from New Orleans on her regularly scheduled run (which was to have been her last voyage) when she collided with the Italian ship *Lorenzo Marcello*. Five of the 50 passengers aboard and five of her crew of 98 died in the collision. About a dozen more were injured. She was beached to prevent sinking. She was subsequently refloated, discharged, drydocked, and sold at public auction in March 1961. She was repaired and entered cargo service in March 1962. In 1963, she became the *Rye*. In June of that year, while on a voyage from Houston to Bombay, she developed boiler trouble. She sailed from Bombay in September 1963 for scrapping in Japan.

It is historically significant that sixteen Victory Ships were sold to foreign interests for conversion to cargo passenger/vessels; they are listed in Table 2.

Closure

The cargo/passenger ships that were converted from standard cargo hulls filled an interim demand during the period from 1945 through 1960. By the end of that period trade patterns had changed, and air travel reached all points in the world. Cargo ships were becoming specialized in such configurations as containerships, bulk carriers, car carriers, etc., thus closing the era of the cargo/passenger ship.

Table 2 Victory ships sold to foreign interests for conversion to cargo/passenger vessels

Original Name	Name after Conversion	Flag	Passengers
Costa Rico Victory	*Groote Beer*	Dutch	900
La Grande Victory	*Waterman*	Dutch	900
Cranston Victory	*Zuiderkruis*	Dutch	900
Smith Victory	*Buenos Aires*	Argentine	800
NYU Victory	*Cordoba*	Argentine	800
Rock Hill Victory	*Entre Rios*	Argentine	800
William & Mary Victory	*Mendoza*	Argentine	800
Gustavus Victory	*Santa Fe*	Argentine	800
La Crosse Victory	*Tucuman*	Argentine	800
Vassar Victory	*Begona*	Spanish	940
Wooster Victory	*Montserrat*	Spanish	825
Medina Victory	*Laurentia*	British	55
Taos Victory	*Lismoria*	British	55
United Victory	*Khedive Ismail*	Egyptian	100
Atchison Victory	*Mohamed Ali el Kabir*	Egyptian	100
St. Lawrence Victory	*Hrvatska*	Yugoslavian	60

Outboard Profiles: Four Aces

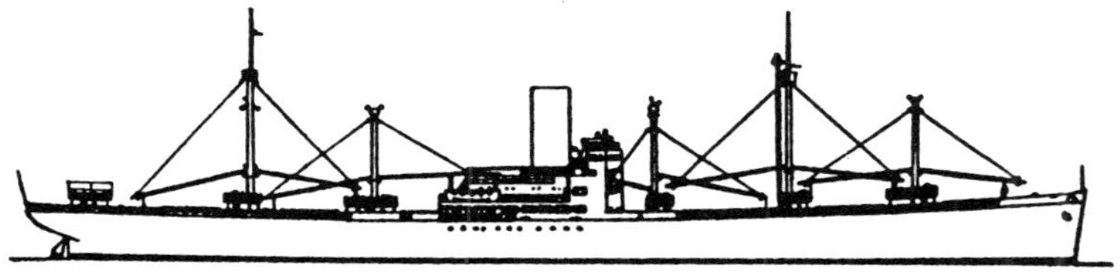

Exporter

Exeter, Excambion, Excalibur, and *Exochorda*

Outboard Profiles: President Jackson

Mormacsea

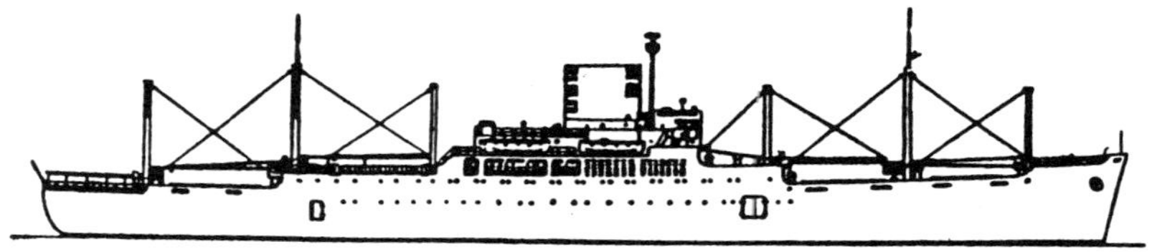

President Monroe and *President Polk*

Outboard Profiles: African Comet

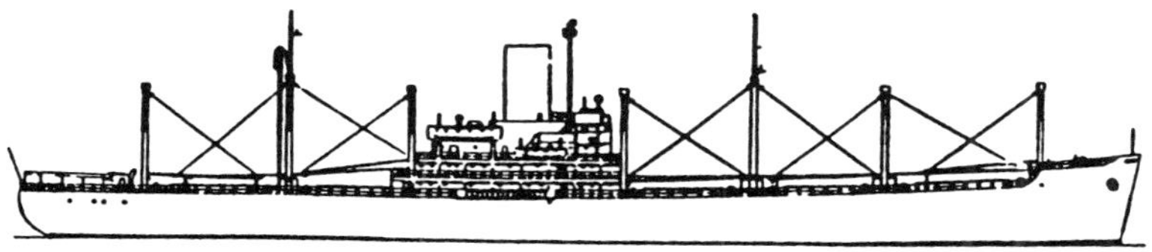

President Grant

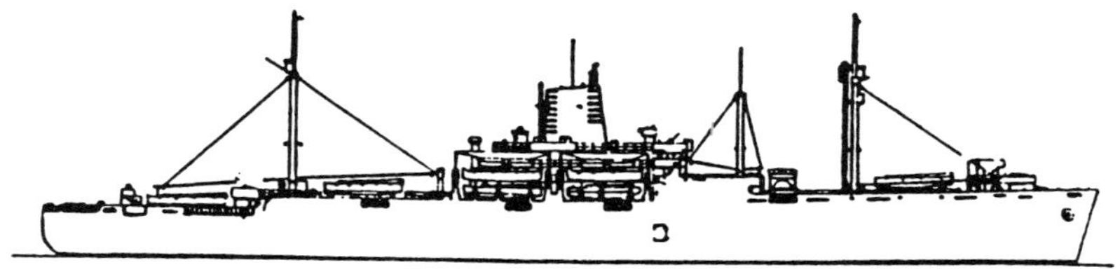

African Meteor

Outboard Profiles: Delbrazil

Mormacsea

African Enterprise and *African Endeavor*

Outboard Profiles: Rio Hudson

Mormacsea

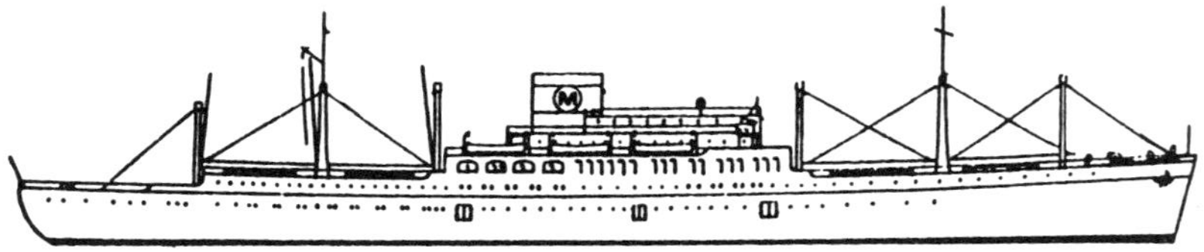

Proposed design of the C3-P & C type

Outboard Profiles: American Merchant

Mormacsea

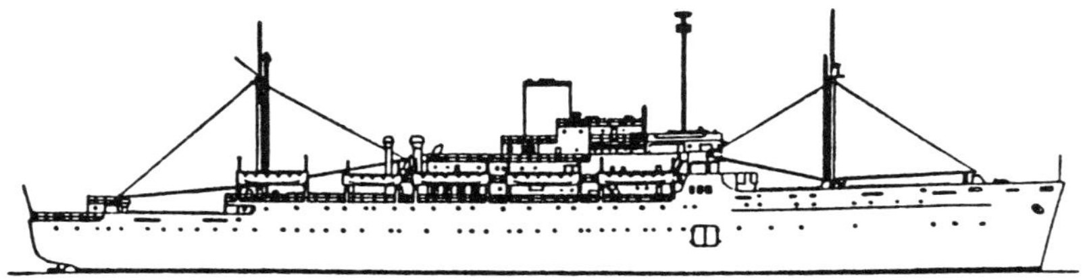

Henry Gibbins

Outboard Profiles: C3 Passenger/Cargo Vessel Conversions (Foreign Flag)

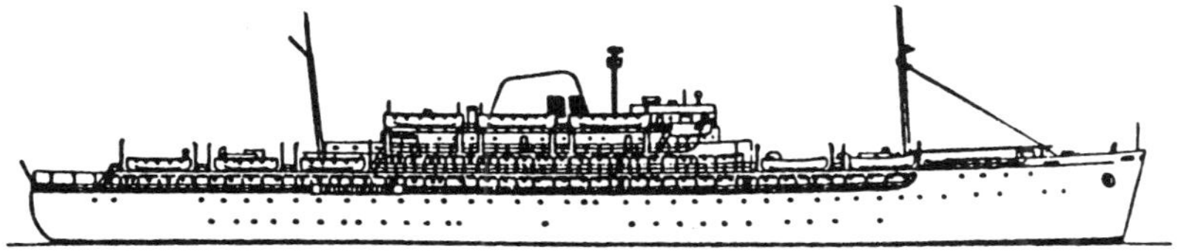

Corrientes and *Salta*

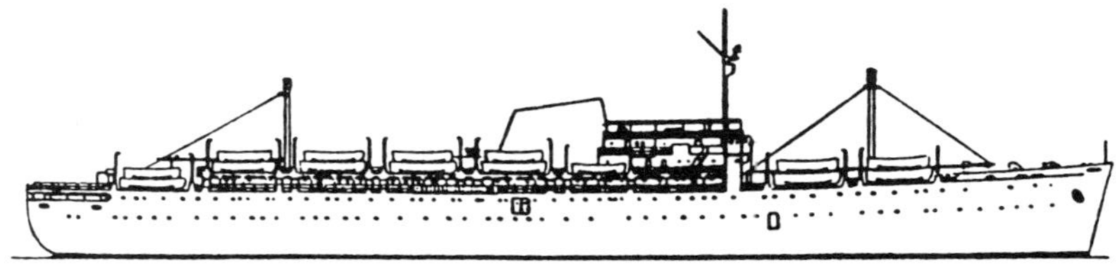

Fairsea

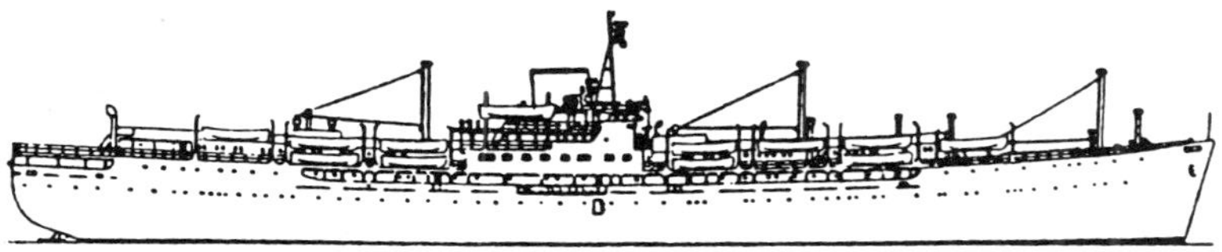

Seven Seas

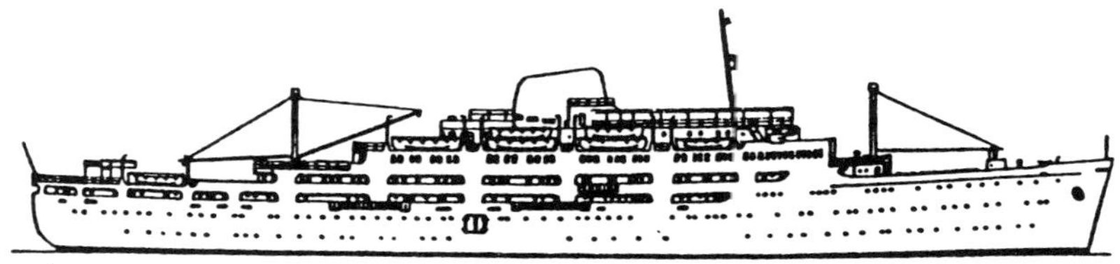

Sydney (C3-S-A2) and *Roma* (C3-S-A1)

Outboard Profiles: C2-P & C, Santa Barbara

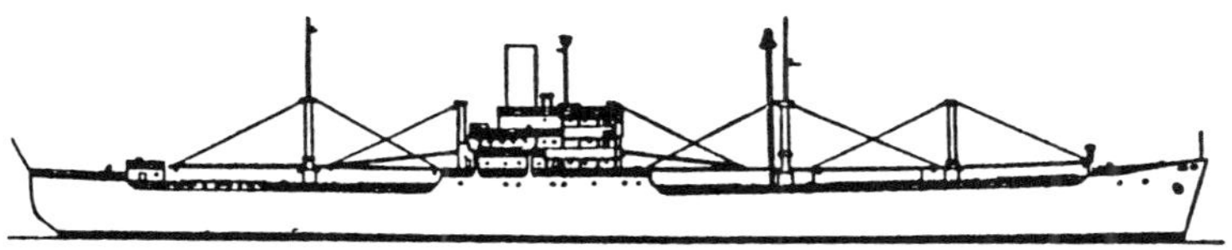

American Clipper

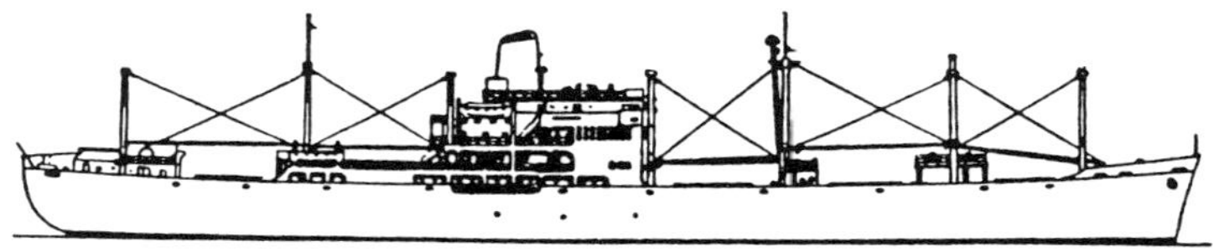

Santa Cecilia

Outboard Profiles: Victory Ships Passenger/Cargo Conversions (Foreign Flag)

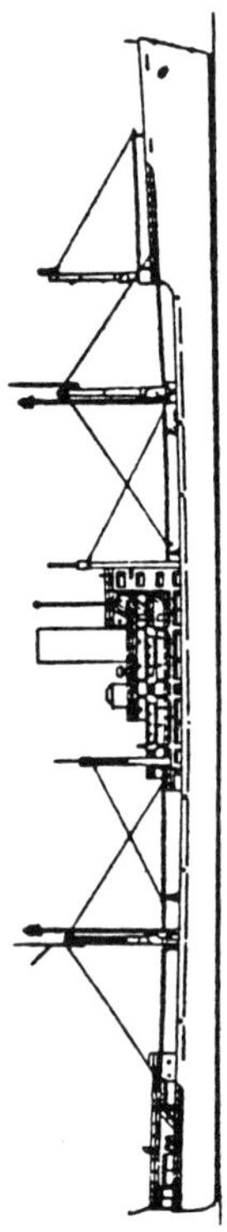

VC2-S-AP3

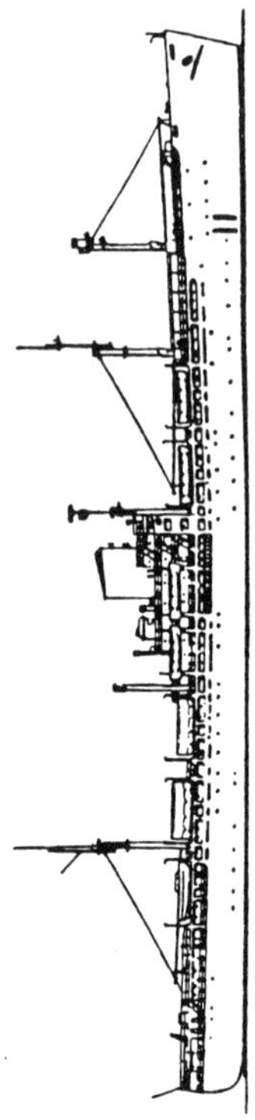

Castel Verde

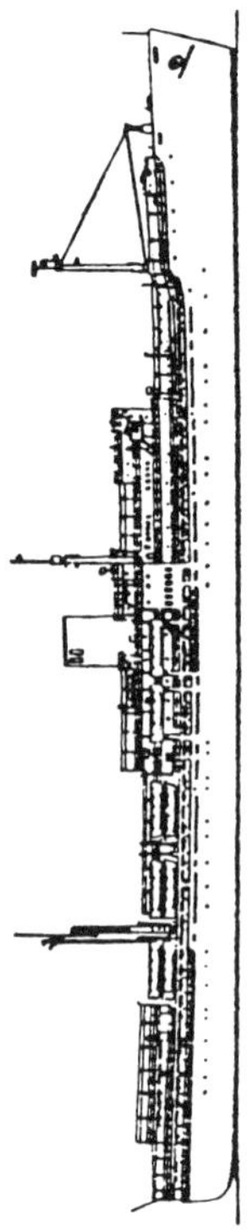

Zuidereruis and *Waterman*

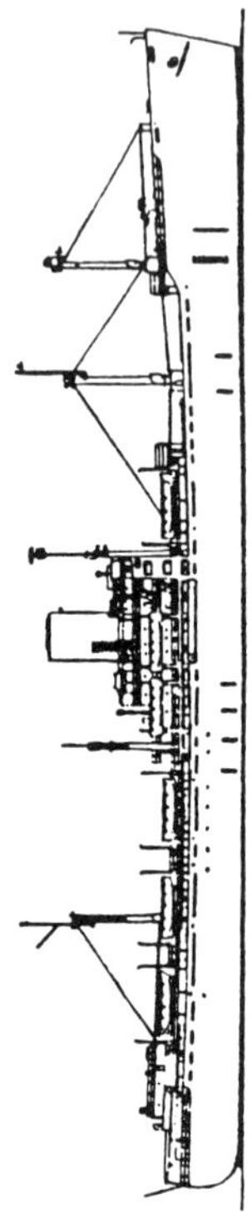

Cordoba Mendoza and *Entre Rios*

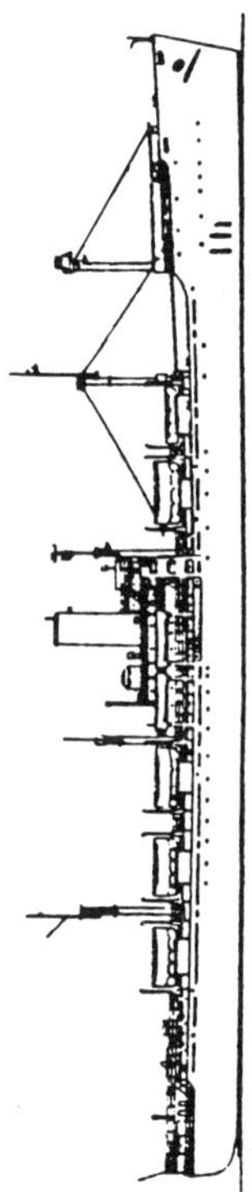

Castel Bianco

Outboard Profiles: Alcoa Cavalier

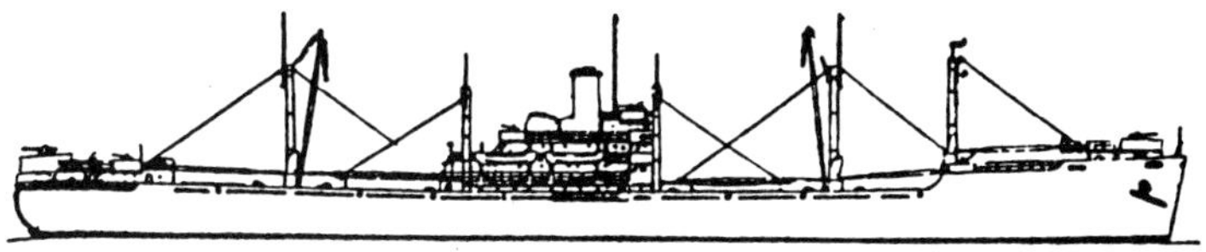

VC2-S-AP2

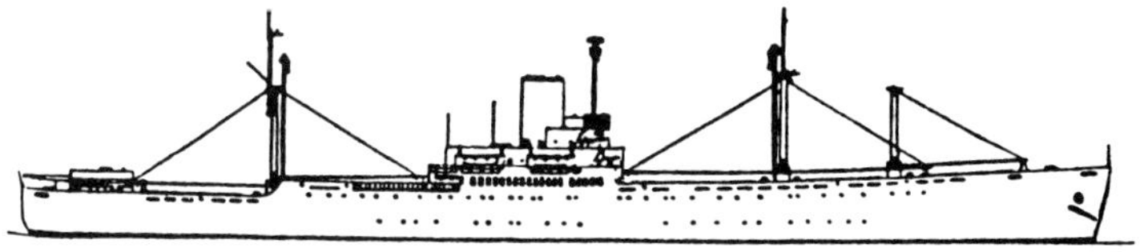

VC2-S1-AP7: *Alcoa Cavalier, Alcoa Clipper,* and *Alcoa Corsair*

Pvt. William N. Thomas

CAPTAIN WARREN LEBACK has had a distinguished career in the marine industry, both in operations and design. An active member of SNAME, he is well known for his several stimulating papers dealing with the managerial aspects of engineering. He is currently serving as Administrator of the Maritime Administration, U.S. Department of Transportation.

Part VII

Ships of War and Deterrence

Aircraft Carriers

B. F. TIBBITTS and ROBERT J. RIGGINS

Introduction

THE EVOLUTION of U.S. aircraft carriers from the end of World War II to the present is the subject of this chapter. Six major classes of ships form the principal milestones of the development: *Essex* (CV 9), *Midway* (CVB 41), *United States* (CVA 58), *Forrestal* (CVA 59), *Enterprise* (CVAN 65), and *Nimitz* (CVAN 68). The significant features of these six transition classes will be discussed.

The never-built *United States* is included here because its history is key to understanding modern carriers. Note that the "B" (for heavy) and the "A" (for attack) in aircraft carrier designators were later dropped, and these became all-purpose carriers (CV or, if nuclear powered, CVN).

Various public and private shipyards, some no longer in existence, built these ships. The Newport News Shipbuilding Company, however, played a unique role. In each case it built the lead ship; and is today our nation's only builder of nuclear carriers.

There are many factors that drive aircraft carrier design. The characteristics of naval aircraft, and the missions assigned to them, obviously are of overriding concern. The ship designer will sacrifice almost anything to increase the effectiveness of embarked aircraft. One writer has likened the aircraft carrier to the first stage of a three-stage rocket, with the aircraft being the second stage, and its payload being the final stage. Note that the aircraft carrier, in turn, has an impact on the design features of carrier aircraft. These considerations are not minor, which is why there are few examples of modern aircraft designed to operate from land bases that can perform satisfactorily as carrier aircraft. By considering both the ship and its aircraft, the more successful carrier designers adopted a true "systems engineering" approach long before that expression came into vogue. On the other hand, since the life of the ship (45 years) is much longer than the life of its aircraft (previously 5–10 years, more recently 10–20 years), a successful design is one that is flexible and adaptable.

Lessons learned from war damage and peacetime accidents, and the nature of potential enemy threats have also been decisive factors in aircraft carrier design. Since carriers are few in number, much greater attention is paid to their survivability than to other, more numerous, types of ship. Like other warships, aircraft carrier design has been driven by advances in technology. Before describing the specific ship designs, this chapter will outline the changing influences, and the resulting aircraft carrier design trends.

Principal Features of Evolution

Perhaps the most hotly debated subject of aircraft carrier design has been the question of ship size versus cost and effectiveness. Ship size is directly related to the size, speed, number and variety of aircraft carried. Since the 1930s there has been a continuing debate over whether it is more effective to invest in a large number of small carriers or a small number of large carriers. The proponents for small carriers have generally argued that:

- Dispersion of a given number of aircraft to more ships would make the total force less vulnerable to enemy attack (the "don't put all your eggs in one basket" argument).
- A smaller ship would present not only a more difficult target to hit, but would force the enemy to spread his offensive resources more thinly.
- A larger number of carriers could be geographically dispersed to a larger number of trouble spots.

Proponents for larger carriers have argued that:

- The ship cost per aircraft carried is less for large carriers if other variables remain constant. This is true because the number of support systems such as catapults, arresting gear, and maintenance shops per aircraft is proportionally less.
- Modern air wings include a large number of aircraft that can be viewed as "overhead." These include ASW (antisubmarine warfare), AEW (airborne

early warning), ECM (electronic counter measures), rescue and tanker aircraft. The number of such aircraft is relatively fixed, rather than being a function of total air wing size. Thus a smaller carrier's air wing contains a smaller percentage of offensive aircraft than a larger carrier's air wing.

- Larger carriers are inherently more survivable (owing to sheer size), have better seakeeping qualities (can safely operate aircraft in higher sea states), have greater range, etc.
- The marginal increase in horsepower to obtain the required speed range is small compared to the size differential.

Essentially the same arguments have been offered for the two viewpoints over the entire history of aircraft carrier development, although with markedly different constraints. Until World War II, the overriding concern was treaty limitations. Small carrier proponents won because this enabled more carriers to be built for a fixed limit on total displacement. Since World War II, the issue has primarily been cost. History shows that the large carrier proponents are now in control, but this has by no means occurred automatically. There have been many side studies, and even a few full fledged designs of smaller alternatives. More details of this will come out as we discuss specific designs and studies.

The design of flight decks has gone through major changes with respect both to ship structure and to arrangements. The flight decks of earlier carriers were merely platforms erected above the hulls and did not add to the ship's longitudinal strength. Many of the earliest carriers were in fact conversions of existing hulls, but the same approach was also used on carriers designed from the keel up. This was changed after World War II because of many factors. There was a recognized need to armor the flight deck, and to increase longitudinal strength.

Flight deck arrangements went through a significant evolution intricately linked to:

- Canted decks, i.e., angling of the landing area outboard to port to improve safety, and permit true simultaneous launch and recovery.
- Aircraft elevators located at the deck edges rather than inboard, primarily to improve aircraft handling patterns, but with secondary survivability benefits.
- Landing and launching lengths increased for heavier and faster aircraft. The canted deck reduced the total impact to a degree because the forward portion of the landing area (to port) could overlap with the after portion of the catapult launch area (to starboard).
- Provision of blast deflectors for jet exhausts.
- Changes in aviation weapon elevators and handling procedures, to accommodate bigger, heavier, and more sophisticated weapons.
- Automated launching and landing aids to handle faster, heavier aircraft.
- Improved catapults and arresting gear.
- Armored flight decks based on World War II experience with dive bombers, and later, Kamikazes.

Like other warships, aircraft carriers have achieved significant improvements during the post-war period from the addition of a wide range of electronic sensors, directors, controllers, improved communications and processors. There is a continuing element of improvement both from one class to another, and in modernization of commissioned ships. Not only have ship command, control and communication systems become more capable, but aircraft electronic systems have proliferated and so require increased ship support.

Nuclear power became an alternative to oil-fired boilers for aircraft carrier propulsion in the 1950s. Nuclear propulsion offers several advantages for large aircraft carriers. Possessing essentially unlimited endurance it allows sustained operations at high speed, a major military advantage, and one that enhances the carrier's own survivability. Eliminating bunker fuel increases the allowable weight of combat load (aviation fuel and ordnance). While nuclear ships cost more to construct than oil-fired ships, many have argued that the proper comparison is of the relative life cycle costs. Studies have over the years produced mixed results. It is now generally considered that the life cycle costs are essentially equivalent, but that the nuclear carrier has significantly improved fighting capability.

Aviation weapons stowage and handling have undergone significant changes because of changes in the weapons themselves. The "iron" bombs and anti-ship torpedoes of World War II have been replaced by less dense smart weapons and missiles. Longer weapon elevators traverse the full ship depth and thereby influence every deck. "Universal" magazines must accommodate multiple weapon sizes and configurations rather than simply provide for bombs stowed like cord wood. Aircraft carrier magazines must accommodate a wide variety of ordnance because of the many different missions that may be assigned to embarked aircraft. Those who plan the aviation ordnance load face a bewildering array of competing assignments because

Fig. 1 *Essex* (CV 9) with canted flight deck and "hurricane" bow.

of the many potential targets: armored ships; unarmored ships; "hard" land targets (industrial facilities, bridges); "soft" land targets (troops and vehicles); submarines; aircraft (both in the air and on the ground); enemy radar and flak batteries.

A clearly visible evolution in aircraft carrier design has been the nature of the self-defense weapons. A World War II carrier bristled with 5-in./38, 40mm, and 20mm anti aircraft (AA) guns to protect against subsonic aircraft. Since attacking aircraft had to close to short ranges before launching weapons, the effectiveness of the carrier's own self defense weapons was of critical importance. Those guns have now been replaced with fewer, faster, more accurately controlled missile launchers and 20mm close-in weapons systems (CIWS) to protect against much faster incoming missiles. The range of engagement (point blank to 10,000 yards) is little changed from World War II, although reaction times may now be much shorter. Near the end of the war the carrier's own aircraft became its first line of defense, and this has continued. Fighter aircraft, under close radar control, patrol 100–200 miles from the ship and attack enemy aircraft before they are within missile-launching range. The second line of defense is provided by the surface Navy: currently Aegis cruisers and destroyers, which can engage targets at 50–100 miles.

During World War II both the carrier and its supporting escorts (battleships, cruisers and destroyers) were equipped with the same anti-aircraft weapons (as described above). This is no longer true, and from time to time the issue is raised of installing long range missiles (and possibly the Aegis radar) on carriers.

Another design aspect that has been significantly modified is that of aviation maintenance. Like modern warships, modern warplanes are full of electronic systems. In order to keep their aircraft fully operational, greatly expanded and sophisticated avionics (aviation electronics) maintenance facilities have been added. In addition, a modern carrier operates more kinds of aircraft. As a result, as many as fifty different aviation maintenance shops may be required.

Essex (CV 9)—World War II State of the Art

No discussion of the post-war history of U.S. aircraft carriers would be complete without some mention of the *Essex*-class. If success in battle is the ultimate test for a warship, the *Essex*-class ships were arguably the most successful aircraft carriers ever built. Two dozen were built during, or right after the war, and they formed the largest single class of fleet carriers ever built by the United States. They formed the backbone for the fast carrier task forces that proved victorious in the great Pacific air battles of 1944–1945. At war's end the Navy possessed many of these ships, all of which had many years of service life remaining. *Essex*, the lead ship, was only three years old. In the budget-cutting climate immediately following the war, no new aircraft carriers would be authorized, and the *Essex*-class ships would dominate the postwar carrier Navy for years. As a result, the Navy's early transition to jet aircraft was primarily played out on the axial decks of the *Essex*-class carriers.

The design of this class had been strongly influenced by the limitations of the Washington Naval Armament Treaty of 1922 even though those limitations were no longer binding when construction started. The treaty was signed by the five leading naval powers of the day: the United States, Britain, France, Italy, and Japan. It limited the warships each nation would maintain by establishing total standard displacement limits. Standard displacement equated to full load displacement (the characteristic in common use today), minus fuel and reserve feed water. The total weight of aircraft carriers for the United States was limited to 135,000 tons. The maximum displacement of individual aircraft carriers was limited to 27,000 tons. The treaty also limited carrier gun armament to 8-in. guns. Some exceptions were stated for each country to accommodate special situations. For example, we were allowed to convert two partially built battle-cruisers (*Saratoga* and *Lexington*) to 33,000-ton aircraft carriers. Thus we had 69,000 tons to "expend" on additional aircraft carriers.

By 1939 when the *Essex* design was begun, the treaty had expired, but the Vinson–Trammell Act effectively extended it for the United States by authorizing only enough new tonnage to match the treaty limit. During the years of the 1920s and 1930s there was continuing debate over the most effective aircraft carrier size to maximize capability within treaty limits. Several classes of either 15,000 or 20,000 tons displacement were built. The argument for the 15,000-ton ships was that four of them could be built for the same total tonnage as three of the larger ones. This was not a trivial matter since it meant the difference between having five or six carriers within treaty limits. By 1939, however, the success of the 20,000-ton *Yorktown* provided evidence that the heavier types of aircraft, such as torpedo-bombers, required at least a 20,000-ton standard displacement carrier. The scheme initially adopted for the *Essex* was for a ship of 20,400 tons. Design development, however, increased this to 27,100 tons by the time contract design was completed in 1941. By the time construction got under way the start of the war had abrogated all treaty limitations, but the design had nevertheless been strongly influenced by those limitations. The *Essex* was in many regards an improved *Yorktown*, with a longer flight deck, better sub-division and side protection system, a 25 percent increase in aviation gasoline, more guns, and increased propulsion power.

Despite its wartime success, by war's end the *Essex*-class ships were viewed as deficient in protection against both bombs and torpedoes, with insufficient aviation fuel, and cramped for space. As air wings started comprising more and more jet aircraft, with their weight and speed seemingly increasing with each new model, the old ships became less and less suitable. Although eleven ships of the class served tours during the Korean war, the days of air wings comprising piston engine and subsonic jet aircraft were numbered. It was no longer a question of minor changes being satisfactory; a major upgrade, or conversion, was needed.

The major problem was the flight deck, which was of the traditional axial variety. That is, a single rectangular shape, parallel to the longitudinal centerline of the ship, which served as both the landing and launching strip. At a given time, aircraft could either be launched or landed, but not both at the same time. Aircraft were parked at the aft end for launching over the bow, and at the forward end for landing over the stern. This arrangement was acceptable only for relatively light, slow aircraft, which did not require long accelerating and decelerating distances.

Old Ships—New Technologies

Coal-fired, paddle wheels and high-tech—well at least some major innovations. Yes, two of the most significant contributions to the war effort were made by the two side-wheel training carriers operating on Lake Michigan. Converted from the great passenger ships Seeandbee *and* Greater Buffalo, *the* Wolverine *and* Sable *not only served as training platforms for over 17,000 carrier pilots but also served two other innovative functions.*

The Sable *was the first carrier to be fitted with a steel flight deck. Indeed, as the prototype for such an installation, the deck consisted of alternating sections of more-or-less typical plate and stiffener construction and a patented design of formed steel sections normally intended for use as warehouse flooring. For whatever reason, the Navy apparently decided that the warehouse design was not feasible.*

In another development relative to advanced technologies, the Wolverine *spent some time in the rather sparsely populated Grand Traverse Bay area of northwestern Michigan, where she participated in the testing of radio-controlled drone aircraft (from the Traverse City* Record-Eagle, *Aug. 24, 1990).*

Several published articles on these conversions include a comprehensive story by Commander John D. Alden appearing in a 1961 issue of Naval Institute Proceedings *and an article in the Winter 1988 issue of* Naval History. *I must, however, express a rather special interest in the conversions, having been involved in some of the design aspects, including the preparation of the flight deck plating plan for the Sable.*—W. C. Cowles

Major refits included replacing the axial deck with a stronger, canted deck; incorporating bigger and stronger aircraft elevators, steam catapults, more robust arresting gear, the Fresnel lens landing system, an enclosed "hurricane" bow, additional aviation fuel, 3-in./50 AA guns in place of 40mm guns, and improved command, control and communications systems. By any definition, these were major modifications, and the final product bore little resemblance to the original ships. Many of the lessons learned in these ships can be found in the *Nimitz* today.

Those *Essex*-class ships that received all of the above modifications were designated as CVAs, and they formed the basis of the U.S. carrier striking forces for a number of years. Nine ships of the class served during the Vietnam war although, as the duration of that war extended and more modern aircraft came into action, their shortcomings became apparent. Because of the availability of the *Midway*- and *Forrestal*-classes (and eventually the *Enterprise*), the *Essex*-class carriers were no longer "first line." The last to serve as a true CVA was the *Intrepid.* Her carrier air group commander (CAG) during the Vietnam war was Commander (later Admiral and Chief of Naval Operations) Tom Hayward.

Owing to budgetary restrictions, some ships of the class received less extensive modifications and were therefore unsuitable for use as attack aircraft carriers. They were redesignated as antisubmarine carriers (CVS). They carried about 45 aircraft, a mixture of fixed wing and ASW helicopters. A bow sonar dome was added. In fleet ASW operations they proved effective. By the late 1960s, under the budgetary pressures of the Vietnam war, a proposal to further modernize these ships or to construct new CVSs to keep pace with advances in Soviet submarines was not approved. When the CVSs were finally phased out, their function was absorbed by adding a small number of ASW aircraft to the air wings of existing CVAs (primarily *Forrestals*), which were then redesignated as CVs.

Midway (CVB 41)—Transition Class

The *Midway* design represents the transition between World War II and post war aircraft carriers. She was designed in 1941–42, and was partially influenced by the Washington Treaty. The design profited from lessons learned in early World War II combat. (Note, however, that design lessons may not be perfectly learned until the end of hostilities, when enemy records are available for examination.) The *Midway*-class was originally designed for relatively light, slow, propeller driven aircraft. The primary design threat was viewed as air dropped bombs, air dropped or submarine launched torpedoes, and surface gunfire (at that time, assumed to be Japanese heavy cruisers armed with 8-in. guns). She was strongly built along the lines of the earlier *Essex*-class, but with more protective armor, more planes, more aviation gasoline and more aviation weapons. While the *Essex* was 27,000 tons standard displacement, the *Midway* was 45,000 tons—a 65 percent increase. As mentioned, the *Essex* was originally designed to meet the Washington Treaty limitations. The *Midway* continued the operational theory and design practices of the *Essex*, but abandoned the Treaty limitations.

Studies of a proposed 45,000-ton aircraft carrier had begun before the start of World War II. This was a bold step in light of the size of existing carriers, and also because much of the Navy was still battleship-oriented. These studies centered on significantly greater protection, more planes, more gasoline, and more ammunition compared to the *Essex*-class. These early studies eventually resulted in the *Midway*-class. Thus, even though this was a much larger ship, the same planes, weapons, operational concepts and naval theory went into the early design as those that formed the Treaty ships. Some wartime operational experience eventually influenced the design—in particular, confirmation of the need for armoring the flight deck—but it was truly a transition between the pre-war Treaty designs and the post war designs intended for a new era.

The *Midway* was designed with protection as a primary consideration. The flight deck was armored to protect against 8-in. shell fire and against dive bombers. There was extensive debate over the question of armor thickness on both the flight and hangar decks. Early wartime experience of both British and U.S. carriers contributed heavily to the final determination. Detractors of heavily armored flight decks argued that, because of the increased weight, the number of aircraft would have to be drastically reduced. This had indeed been the experience with the British, whose heavily armored ships carried substantially fewer aircraft than U.S. carriers of comparable size. For example, the 23,000-ton *Illustrious* carried only about half the aircraft of our 20,000-ton *Yorktown.* Some argued that the reduced freeboard would lead to wet decks and less reserve buoyancy. In addition, aircraft elevators would be slowed by the added weight (inboard elevators would require armor equivalent to that on the flight deck), and there was uncertainty

Fig. 2 *Midway* (CV 41), as built.

over the armor effectiveness. Regardless, the fact remained that British carriers had survived bombing attacks that would have put ours out of action. Eventually, the proponents of flight deck armor won, and a flight deck consisting of 3.5-in. ballistic armor over the vitals and a hangar deck of 2-in. armor were adopted.

In addition to overhead protection, the *Midway* had extensive below-decks protection. To guard against heavy cruiser gun attack, a 7.6-in. armor belt was provided at the waterline. (Note that, except for the unarmored escort carriers at the Battle of Leyte Gulf, no U.S. carrier was ever subjected to major caliber gunfire during World War II. Ironically, because these escort carriers were unarmored, many armor piercing shells passed harmlessly through them without detonating.) Much of the *Midway*'s side armor was later removed to compensate for weight growth during the post war conversions.

A five-bulkhead torpedo protection system guarded the machinery rooms and magazines. Extensive subdivision of the machinery spaces provided significant redundancy in propulsion components. There were four propeller shafts, each served by a separate engine room. Three separate boiler rooms served each engine room. Thus, there were in total twelve separate spaces to isolate the major propulsion components.

This emphasis on protection, and the desires for increased planes, weapons and gasoline resulted in a balanced design at 45,000 tons, significantly bigger than the *Essex*. In March 1942 the air wing designated for the *Midway* consisted of 120 aircraft.

The *Midway*'s propulsion plant was of standard design for that era. Geared steam turbines drove the four propellers. Standard design steam power plants available in the early 1940s were of 150,000 shp, 172,000 shp, and 212,000 shp. High speed was considered essential in order to generate sufficient wind-over-the-deck to launch and recover aircraft safely and then rejoin the formation. Propeller-driven aircraft were routinely deck launched, and rarely catapulted until near the end of the war. Thus, a design speed of about 33 knots was desired. To achieve this on the 45,000 ton design, the 212,000 shp plant was needed.

As noted, the *Midway*'s aviation features were designed for the relatively light weight, piston engine, propeller aircraft of the early World War II era. Flight deck arrangement, hangar, aircraft elevators, arresting gear, catapults and aircraft support equipment would later prove to be less than optimal for the post war faster and heavier jet aircraft. The flight deck was of the traditional axial variety.

The aircraft's piston engines had to be warmed up before launching. To minimize delays, engines were started while the aircraft were still in the hangar. This required large vent openings in the side shell. There were also openings for engine uptakes and aircraft elevators. Thus the structure supporting the flight deck could not be made to take the shear loads. As a result, the flight deck (like those on all earlier U.S. carriers) was not designed to add to the ship's longitudinal strength. In fact, two transverse expansion joints in the deck allowed it to flex without becoming over-stressed.

With respect to armament, like other contemporary carriers, the *Midway* was originally built with an extensive suite of self-defense weapons. Eighteen (of the then new) single 5-in./54, and 21 quad-40mm gun mounts and numerous 20mm mounts were placed on sponsons around the flight deck periphery at and below the gallery deck level. In addition to acting as self defense weapons, the larger mounts supplemented the protection in way of the hangar, particularly against Kamikaze aircraft.

Fig. 3 *Midway* (CV 41) with canted flight deck and "hurricane" bow.

Owing to the ending of World War II, only two more of the *Midway*-class were built: the *Franklin D. Roosevelt* (CVB 42) and the *Coral Sea* (CVB 43); three sister ships were cancelled. Subsequently, the *Midway*-class carriers were all extensively modified to incorporate angled flight decks, hurricane (enclosed) bows, steam catapults, deck-edge elevators, new radars and electronics, and modern self-defense systems. These improvements, however, brought added weight, over-crowding and reduced stability. This was compensated for by adding hull blisters to the class in the 1960s. The *Midway* received an additional blister in the late 1980s. It is worthy of note that both the *Midway* and the *Coral Sea* operated for over 45 years. Fifty years after the *Midway* was first designed, she served with distinction in the recent war in the Persian Gulf. This is testimony to the soundness of the basic design of this class, and the fact that aircraft carriers are amenable to incorporating major design modifications during their service lives.

United States (CVA 58)—The First Super Carrier Design

Aircraft carriers had proven decisive during the war in the Pacific. Ironically, they were soon threatened with technological obsolescence because of the advent of the atomic bomb and long range bombers. The Japanese navy had been sunk. The Soviet surface navy was primarily a coastal defense force and highly subservient to the Red Army. Not even the most ardent carrier advocate could envision a future Battle of Midway or the Coral Sea. It was commonly accepted that the next war would be nuclear. If the Navy was to play an important role, it needed to develop a nuclear delivery capability.

While carrier aircraft could carry 2000-lb bombs, the early atomic bomb weighed 9000 lbs. A new aircraft was needed, and it would be too big and too heavy to operate off existing decks.

The Bureau of Ships and the Chief of Naval Operation's (CNO's) Ships Characteristics Board (SCB) were already investigating a number of aircraft carrier concepts. Their aim was to define the next generation carrier to support new aircraft, weapons and aviation tactics. Alternatives included variations in mission, aircraft size, ship size, weapons size, passive protection, fire protection, aircraft elevators, and catapults. By mid 1946 the need for larger, longer range aircraft, heavy aviation weapons and heavy passive protection focused attention on a carrier significantly larger than the *Midway*. A new Navy heavy bomber was under development with an estimated fully loaded weight of up to 100,000 lbs. This was an enormous increase, since the heaviest U.S. carrier aircraft during World War II was the 15,000-lb *Avenger* torpedo bomber (the kind George Bush piloted). The carrier design to operate a wing of these bombers was established in 1947 as CVBX. This was later designated the CVA 58 and given the name *United States*. This concept was carried through the contract design stage and construction was begun at Newport News.

This activity was going on against a backdrop of controversy within the armed forces. Fleet Admiral Nimitz spoke of a "fleet of super carriers capable of bombing the interior of the Soviet Union." Navy spokesmen argued that the Navy could do this more

Fig. 4 The never-built USS *United States* (CVA 58), artist's impression.

efficiently than the Air Force. The Navy pointed out that land attack was far from a new mission since during the war in the Pacific 77 percent of bombs dropped by carrier aircraft had been against land targets. It is unclear if giving aircraft carriers a strategic nuclear capability was of concern to the Soviets. There is no doubt, however, that it did not find favor with the U.S. Air Force, which argued that aircraft carriers were vulnerable because of the need to launch at relatively short distances from the Soviet Union. The Navy countered that the Air Force B-36 was so large as to be easily detected and shot down. The debate was both fierce and public.

The Air Force argument was accepted. Secretary of Defense Louis Johnson viewed the proposed carrier as an unnecessary duplication of the Air Force mission, and in April 1949 he cancelled construction nine days after the keel had been laid. Though the ship was cancelled, the investigations and deliberations that constituted its development became the foundations for the next successful U.S. aircraft carrier design, that of the *Forrestal* (CVA 59).

The *United States* was meant to be a truly single mission ship. Its objective was to launch 18–24 long range attack aircraft equipped with atomic bombs. The design featured a flush flight deck with no protrusions to interfere with flight operations. This had been a goal of naval aviators for decades, and was deemed particularly important because of the wing span of the projected new bomber. When flight operations were not being conducted, a retractable island could be raised for conning the ship. The hangar and elevators were designed to support the bomber, which was estimated to weigh 100,000 pounds when fully loaded. The ship would have an overall length of 1088 feet and a full load displacement of 79,000 tons. Although the bombers were the controlling factor, the resultant ship had room for 50 fighter planes to provide ship self-defense and bomber escort. There were four catapults; two in the bow and two amidships angled outboard to port and starboard. They were to be of a new type powered by an explosive charge with enough thrust to launch the heavy bombers.

The 280,000 shp propulsion plant included 1200 psi boilers. Designed sustained speed was 33 knots. Exhaust gas from the boilers was trunked outboard, port and starboard, below the flight deck. This feature was reminiscent of some 1920s and 1930s designs, although they had not proved terribly successful. The many air search radar antennas, which typify carriers, were simply eliminated because there was no suitable place to put them. Accompanying escort ships would provide the necessary radar coverage.

Forrestal—First Major Post War Stride

Meanwhile, the Navy started deploying carriers to the Mediterranean, and technological improvements started to shrink the size of atomic bombs. At that time with NATO in its infancy, there were few land bases for U.S. aircraft along NATO's southern flank. The Navy won the right to use the atomic bomb in this area, but only against land targets that could threaten sea control, such as naval installations or coastal airfields. These limits were never precisely defined, and in the final analysis the Navy's nuclear strike role was limited by available technology.

A flush deck carrier continued to be a goal of Navy aviators. And a new generation of long range heavy bombers for carrier operations was still an expectation. However, by mid 1949 the projected weight of the bomber had been reduced to about 70,000 lb. There were other changes as well. By 1950 the primary mission of carrier based aircraft was centered on tactical air strikes by a large number of aircraft, rather than long range strategic strikes by a few super-carrier bombers. This also was of concern to the U.S. Air Force, although the organization affected was the Tactical Air Command as opposed to the Strategic Air Command. By this time, the size and weight of atomic bombs promised to become smaller and suitable for smaller aircraft not expressly designed for atomic weapons. These trends, plus the desire to bring down the cost of large carriers, led to the creation of an informal target displacement for the next carrier. Carl Vinson, a strong Congressional supporter of the Navy, suggested the limit of 60,000 tons standard displacement (13 percent lighter than the *United States*, but 33 percent heavier than the *Midway*). This became a real objective for the *Forrestal*'s design.

I remember when National Steel and Gibbs & Cox won the competition to design and construct the Sea Control Ship, but Congress did not support the acquisition.

After a bit, as the technically responsible person for all Gibbs & Cox work, I ended up in Madrid as a member of a team intent on "selling" the Spanish Navy the Sea Control Ship. They "bought" the idea—or ironically, with a $150 million encouragement from "somebody" in the U.S. of A., they agreed to build an SCS. The first modification: this was now a Spanish aircraft carrier!

The major difference between the SCS and the Spanish Aircraft Carrier (SAC) was a matter of 3000 tons displacement, from 14,000 to 17,000 tons. This was the combined result of an optimistic contract design weight estimate and capability improvements by the Spanish Navy. A ship floating 3000 tons deeper in the water can exercise the naval architect doing the damaged stability, particularly in way of the aft elevator, where freeboard is reduced. However, he—we—prevailed and the last calculations showed everything O.K. Anyway, I doubt if theory will be tested. I don't think SAC will go far enough from port to be hurt.

*The SCS/SAC is also a lesson for the marine engineers. Gibbs & Cox had recently designed the FFG-class (*Perry*) frigates: 4000 tons do 29 knots on one shaft and two LM 2500 gas turbines. SAC does about 26 knots at 17,000 tons with a pair of LM 2500's. Length is the key factor. The slower rpm resulted in a heavier propulsion system with added weight for the larger propeller shaft and gears.*

The flagship of the Spanish Navy (SAC) is about 643-ft long with an 80-ft beam. The flight deck is wider and has a 12 deg "ski-jump" forward. The hangar deck can provide for a mix of twenty V/STOL aircraft and/or Helos.

No other country in our world builds and operates nuclear carriers. The countries that provide for aircraft at sea use some form of mini-carrier. It has been variously estimated that one could build fifteen or twenty Sea Control Ships for the price of one nuclear Nimitz. *A very interesting trade-off!*—Thomas M. Buerman

There was no formal program for developing a new carrier design in early 1950. However, the Bureau of Ships continued a low level effort to define a design in the Vinson target range with the capabilities generally understood from earlier studies. The focus was still on the *United States*, and an intensive effort was undertaken to reduce the size of that design. The war in Korea, which began in June 1950, revalidated the utility of aircraft carriers. This spurred efforts to design a new carrier. During the initial stages of the conflict, when it appeared that all of South Korea might be overrun, all U.S. Air Force units had to be relocated to bases in Japan. Several years later, when the Chinese entered the war, the Air Force had once again to relocate. Carrier-based aircraft ensured that the U.N. forces retained local air superiority during those critical periods when ground forces were outnumbered and local air bases were unavailable. The increasing awareness of the Soviet threat to Europe also played a role. In addition, the threats to Greece, Turkey and Iran, and the overall fragility of the Middle East all helped demonstrate that aircraft carriers could play a cold war role. The budget submitted in October 1950 contained a new carrier to be awarded in 1952. This was an extremely tight schedule.

The initial objectives for the *Forrestal* included operation of the 70,000-pound bomber, flush flight deck for safety of air operations, four catapults, passive protection, armament and aviation weapons equivalent to those of the *United States*, increased aviation fuel capacity, and an enclosed bow for all-weather operations. All this was to be provided on a 60,000-ton ship, or about 8000 tons lighter than the *United States*. Because of the recent focus on the *United States* design, the *Forrestal* was approached as a *scaled down* version of that, rather than a *scaled up* version of the *Midway*.

At the start of construction the design for the *Forrestal* retained the flush deck, four-catapult arrangement and retractable island of the *United States*. The waist catapults were angled outboard port and starboard on flight deck overhangs supported by sponsons. A driving consideration of this arrangement was a clear deck with no island protruding as a hazard to landing the new heavy jet bombers.

Development of the canted deck occurred late in the *Forrestal*'s design. This approach solved several difficult flight deck arrangement problems, allowed return to a fixed island without undue hazard, and became standard in all future U.S. aircraft carriers. Essentially, the concept was to place the landing strip at an angle to the ship's centerline so that the forward end projected outboard to port on a large sponson. The concept was generated by the British and first applied by the United States in an *Essex*-class conversion. Aircraft landing on an axial deck carrier cut their engines when they came over the stern ramp, as directed by the Landing Signal Officer, and thus they were committed to landing. Planes failing to engage an arresting wire (there might be as many as a dozen) would crash into a barrier to protect the aircraft parked on the forward end of the flight deck. If planes landing on a canted deck missed a wire, they would merely "bolt" off the port side. This was possible because aircraft landing on a canted deck retained flying speed throughout. Indeed, as soon as the aircraft touches down, the throttle is fully advanced just in case the hook does not catch. Only when a sudden deceleration alerts the pilot to a successful "trap," does he throttle back.

The decision to change to the canted deck and fixed island occurred in May 1953 after start of CVA 59 construction. Those readers familiar with detail design and construction will appreciate the major difficulties associated with incorporating such fundamental changes into a design after contract award. History records, however, that the schedule impact was minor.

Another British innovation adopted for the *Forrestal* was steam powered catapults. Earlier U.S. carriers had hydraulic catapults. First intentions for CVA 59 were to develop explosive powered catapults based on the *United States* design intended to achieve the significantly increased thrust required to launch 70,000-pound aircraft. Development of the explosive catapults did not go well and an alternative was needed. The British steam catapult was quickly tested in early 1952 and accepted in lieu of the explosive concept. There were two sizes installed. Two bow catapults were of the higher thrust C7 model needed to launch heavy bombers. Two waist catapults of the lower thrust C11 model were located at the forward end of the landing strip on the port flight deck overhang.

An unusual aspect of the *Forrestal*-class design was the placement of pilot ready rooms below the hangar deck, five levels below the flight deck. That put the rooms in a more protected location. Escalators to the flight deck provided rapid movement of pilots (wearing bulky flight suits).

Aviation payloads received differing degrees of attention. There was no rigorous analysis of aviation ordnance capacity. Stowage space for ordnance was retained at about the same level as on the *Midway* and *United States*: about 2000 tons. There was, however, extensive debate over aviation fuel capacity because of the newly developing mission for aircraft carriers. During the design period, the principal mission changed from long range strategic bombing to shorter range tactical attack. Sortie range, number of sorties per day, and frequency of carrier replenishment were heavily debated. Also, the mix of aircraft between gasoline driven piston engine and jet powered aircraft was a new variable. Both gasoline and jet fuel would be required for some time. Jets burned fuel at a substantially greater rate (particularly at low altitudes), but there were still more gasoline powered aircraft in operation. No one foresaw the increased prevalence of jet aircraft in the near future. The balance between fuel types was almost equal, with provision for 750,000 gallons of gasoline and 789,000 of jet fuel. In later ships of the class, the amount of gasoline stowage was reduced by half in favor of increased jet fuel capacity.

The size of the air wing was not a determinant of the size of the *Forrestal*, although the design of the aircraft did significantly affect the design of the ship. The original Vinson target of 60,000 tons maximum standard displacement continued to be a constraint on the size of the ship. As a historical aside, note that standard displacement was still being used as a criterion years after the treaties had expired. Perhaps this reflected Carl Vinson's great longevity. The size of the air wing would be whatever could be accommodated on a ship of that size. Designing a ship to a displacement limit is both historic and pragmatic. Purists might argue that the first order of business is to set the size of the air wing based on mission requirements. In fact, balancing cost vs military effectiveness is a highly iterative process. Flight and hangar deck strength, catapult thrust (size), flight deck arrangement, aircraft elevator capacity, hangar height, and arresting gear were driven by aircraft characteristics. Hangar height was increased to 25 ft by the clear height requirement of the new A-3D attack bomber. This 7.5-ft increase over the *Midway*'s hangar height was enormous, and raised the flight deck higher above the waterline. This created an adverse impact on ship stability. The solution was thinner flight deck armor and increased

Fig. 5 *Forrestal* (CVA 59).

waterline beam compared to the *Midway*. The 70,000-pound bomber weight drove elevator, catapult, arresting gear and deck parking designs. Both the flight and hangar decks were designed as longitudinal strength decks. This was a first for U.S. aircraft carriers, which has been followed ever since.

Although the *United States*'s design propulsion plant was based on 1200 psi steam, the urgency of the *Forrestal* program led to reliance on the proven 600 psi plants. Eight boilers at 600 psi and 850 °F delivered 260,000 shp to four propellers. This provided a design sustained speed of 30 knots. Fuel provided an endurance of 12,000 miles at 20 knots.

The *Forrestal*'s self defense armament was composed of guns as in the *Midway*. There were eight 5-in./54 caliber guns on sponsons below the flight deck. There were no bow mounts as in earlier carriers because the ship was built from the keel up with a totally enclosed bow to prevent heavy weather damage.

The *Forrestal* design was basically retained for all following carriers built up to the mid-1960s, except for the *Enterprise* (CVAN 65). There were four ships of the *Forrestal*-class, CVA 59 through CVA 62. CVA 63 and 64 constituted a new design class, but were essentially an evolutionary development of the CVA 59, emphasizing enhanced operability on the flight deck. These ships had two starboard elevators forward of the island and one aft, versus one forward and two aft on the *Forrestal*, and the single port elevator was moved to the stern quarter.

A significantly different nuclear powered aircraft carrier, the *Enterprise* (CVAN 65) followed the CVA 63/64. The *America* (CVA 66) design was much the same as that of the *Kitty Hawk* (CVA 63). The *Kennedy* (CVA 67) was a new design, but eventually came to be similar to the *America*. The *Kennedy* design started with nuclear power and was to be ordered in FY 1963. Changes in administration, advances in technology, and increasing operational experience with the *Enterprise*, however, caused the design to be changed from nuclear to oil-fired, back to nuclear, and finally to oil-fired again. The contract for the *Kennedy* was awarded in FY 1964. All of these fossil fueled carriers were therefore evolutionary advances of the basic *Forrestal* design.

There were changes within classes, as well as between classes. The *Forrestal*'s full load displacement was about 78,000 tons (59,900 tons standard). This was increased somewhat in later oil fired carriers, but dimensions remained much the same through the *Kennedy* design.

Some of the more significant evolutionary changes in carriers during this period included the propulsion plant, armament, aviation support systems, and command, control and communication systems. Steam conditions changed from 600 psi to 1200 psi in carriers subsequent to the *Forrestal*. Increasing aircraft speeds on takeoff and landing required longer runs, increased catapult length and longer arresting gear run-outs. These requirements demanded new design, higher powered catapults and more robust arresting gear which, together with larger aircraft, caused changes to flight deck arrangement, and aircraft elevator size and locations. Five-inch guns were reduced in number, and ultimately eliminated altogether. Terrier missile launchers were installed, and later replaced with Sea Sparrow missiles and 20mm Close-In-Weapon Systems (CIWS). Aviation gasoline eventually disappeared and was totally replaced by JP-5 jet fuel. Aircraft electronic systems became more extensive and sophisticated, which resulted in greatly expanded aviation maintenance shops on the carriers. Sensor, intelligence, and other electronic equipment was updated continuously during this period. The torpedo defense system became more compact, but remained equally effective. Pilot ready rooms moved to the gallery deck and escalators were deleted. The multitude of evolutionary changes that took place resulted in many changes to the internal arrangement of aircraft carriers. But, the hull and overall design of the oil fired aircraft carriers from CVA 59 in 1952 to CVA 67 in 1964 were fundamentally the same. There were many studies for alternatives during this period, but the nuclear powered CVAN 65 was the only other design actually constructed.

Seven of the eight *Forrestals* or "improved" *Forrestals* served during the Vietnam war. In July 1967, while preparing to launch air strikes off Vietnam, serious misfortune overtook the *Forrestal*. An accidental firing of a missile from one aircraft into another started a fire. Eventually a number of aircraft were engulfed in flames, and bombs began to "cook off" on the flight deck despite heroic fire fighting efforts. The armored flight deck was holed repeatedly, and personnel casualties were high. Meanwhile, down on the hangar deck a large number of bombs that had been removed from their protective magazines were in danger of being set off by fragments coming down from above. Detonation of such a large body of ordnance would have been truly catastrophic, and many credit the armored flight deck with preventing high velocity fragments from reaching this "bomb farm."

Enterprise (CVAN 65)—Applying New Technology

The potential advantages of equipping aircraft carriers with reactor-powered steam boilers had been discussed within the Navy frequently in the post World War II years. Three major advantages that nuclear propulsion promised were:

- Almost unlimited ship's fuel endurance, even though other supplies would still have to be replenished.
- Elimination of propulsion exhaust smoke from the flight deck, which would greatly diminish corrosion damage of topside sensors and aircraft. This would also relieve visibility problems caused by smoke; and it would reduce turbulence, thereby enhancing flight safety. The size of the island could be reduced, and this would create more deck area for parking aircraft handling equipment.
- Elimination of space required for ship's propulsion fuel, which would allow carrying significantly greater amounts of aviation fuel, enabling aviation operations to continue for longer periods before having to replenish aviation fuel.

The most important disadvantages of a nuclear propelled aircraft carrier were the higher ship acquisition and crew training costs.

Nuclear propulsion had been considered in the *United States*'s design, but development of the technology was insufficiently advanced to make it a serious consideration. The same held true during the design of the *Forrestal*. Nevertheless, expectations of eventually developing a suitable nuclear plant did influence the decision to provide the *Forrestal* with a conventional, largely "off the shelf" oil-fired plant rather than making the investment required to develop a new conventional plant.

Fig. 6 *Enterprise* (CVAN 65).

The following are a few things that come to mind regarding aircraft carriers:

Midway: *The 3.5-inch flight deck had much to do with the requirement for a hull beam of 113 feet. This made the class the first U.S. warships unable to transit the Panama Canal.*

Forrestal: *The night before the ship was to go to sea on her builder's trials, radio commentator Walter Winchel predicted* Forrestal *would capsize because she was so topheavy. Several of us who had worked on her design and were aboard for the trials wished we could have had a bet with Walter. We knew we couldn't lose.*

Enterprise: *In those hectic days when BUSHIPS didn't know whether the design would end up with nuclear or conventional power, one of each was underway concurrently. In-house staff couldn't handle two jobs of this size, so a team of designers was farmed in for several months from the Newport News Shipbuilding and Drydock Co.*

To demonstrate the practicability of nuclear power in a large surface ship, a half-carrier plant was built and sent to sea in the cruiser Long Beach *(CGN 9).*—George A. Dankers, Jr.

Studies of a carrier nuclear propulsion plant were conducted in the early 1950s, and the Navy proposed building a land based prototype in 1952. This was not approved because defense funds were dwindling and the available nuclear elements were needed for weapons. Indeed, there were then few Navy proponents of nuclear power for aircraft carriers. Many felt that the advantages were minor compared, for example, to the enormous impact it had in submarines. Nevertheless, the proposal was made again in 1954, and approval was given to build a prototype in 1955.

Design studies of an aircraft carrier to mate with the new nuclear plant were started in 1954 with the issue of tentative characteristics for the ship. Many ship sizes and flight deck configurations were examined. The absence of smoke stacks led to a sense of greater freedom in deck arrangement, and a wide variety was considered. The heavy weight of reactor shielding and the large volume of side protection tanks quickly led to the conclusion that a nuclear carrier, to be cost effective, must be relatively large. Other aspects of a nuclear ship tended to drive it to the large end of the spectrum. The amount of aviation fuel carried was large because of the elimination of bunker oil. To provide balance, the aviation magazines were made about 50 percent larger than in earlier carriers. Balance meant providing aviation fuel and aviation ordnance in the amounts needed to support the same number of days of aircraft operations. A heavy bomber was still a Navy expectation, and this pushed up flight deck size, hangar size, and ship displacement. In fact, the decks and aircraft handling systems were eventually designed for a maximum aircraft weight of 80,000 pounds, based on projections of the actual weight of the heavy bomber. Flight deck arrangement followed the general layout of the *Forrestal*, with a few exceptions. The island was moved aft, with two starboard aircraft elevators forward and one aft. The port side elevator was moved aft from the forward end of the landing strip. These changes improved aircraft handling and, as mentioned earlier, had been included in the preceding *Kitty Hawk*-class.

The power level of the carrier nuclear plants under development was such that eight reactors would be required in order to make the required ship speed.

Nuclear propulsion was not the only innovative feature of the *Enterprise*. Technical developments were at a stage that made CVAN 65 a ship of opportunity for several new systems. The most obvious was the phased array electronically scanned radar, which created the distinctive dome shaped island. The long flight deck allowed for installation of four of the longer, higher thrust C-13 steam catapults for launching heavy aircraft. The *Enterprise* had the first carrier-installed Terrier self defense missile system. A new Fresnel pilot landing aid was introduced, allowing a reduction of arresting wires from six to five.

Construction of the *Enterprise* was started in 1958 and completed in 1961. It was the only ship of this design to be built. The Navy then reverted to construction of modified *Forrestal*-type carriers. This continued until more experience could be gained from at-sea operations of the nuclear carrier. In addition, time was needed to muster political support for the additional cost.

The *Enterprise* served several tours off Vietnam. In January 1969, while in waters off Hawaii, it experienced a flight deck mishap similar to that on the *Forrestal* described earlier. Once again, the value of armored flight decks was demonstrated, albeit in situations that the original designers had likely not anticipated.

Although the *Enterprise* was one of a kind, there were continuing studies and attempts to acquire a follow-on nuclear carrier. Initially, the Navy had a repeat *Enterprise* in the FY 1959 program, but it did not survive budget cuts. Studies examined the feasibility of smaller nuclear carriers. A 60,000-ton version using the cruiser size nuclear plant was a promising concept, but the size of its air wing was reduced as compared to that on the *Enterprise*. The price tag did not drop as rapidly as the capability. Since none of the smaller nuclear carriers appeared to be cost effective, the Navy requested a repeat fossil fueled carrier for FY 1960, but it was not approved. An improved CVA 64

Fig. 7 *Nimitz* (CVAN 68).

conventional powered carrier was requested for FY 1961 and approved for construction as the *America* (CVA 66). But, the Navy continued design studies to define the next generation nuclear powered aircraft carrier. The objective was to provide aviation facilities equal to those of the *Constellation* (CVA 64), but at considerably lower cost than the *Enterprise*.

As mentioned earlier, the performance of carriers during the Korean war garnered political support for the *Forrestal*. Likewise, carrier performance in the Vietnam war garnered support for the *Nimitz*.

Nimitz—Standard of Nuclear Carriers

Reducing the size of the power plant (largely the number of reactors) was seen as the key to achieving a smaller, fully capable nuclear carrier. Strides in this direction were being made in the early 1960s, and a four-reactor propulsion plant appeared promising (as opposed to eight in the *Enterprise*). But, this would still be insufficient to reach the necessary cost reduction.

The next intended nuclear carrier design was already on the Bureau of Ships' drawing boards in 1961 when the CVA 66 contract was awarded. The new design was designated CVA(N) 67. In fact, the CVA 67 was partially designed three times before it was finally completed as a fossil fueled ship. Some participants might even object to the word "partially" as understatement. The design started as a nuclear ship in 1959. It was shelved and restarted as a conventional design; then stopped with resumption of the nuclear design. It was finally completed and authorized for construction in FY 1964 as a conventionally powered ship. It became the *Kennedy* (CVA 67). It was largely an improved CVA 64, but there were some innovative features that had been developed for the nuclear design. Also, many of the improvements that first appeared in the *Kennedy* were carried over to the *Nimitz* design. The CVA 67 full load displacement was over 87,000 tons, which was 7000–9000 tons greater than earlier ships of the CVA 59 and CVA 64 classes. The hull dimensions, however, were essentially the same.

A significant innovation in the CVA 67 was the compact torpedo side defense system. This grew out of efforts to increase the ship speed achievable with four reactors. To get increased speed, the naval architects decreased the bow waterline entrance angle. As ship designers know, most changes in one system influence other systems. The finer hull made it impossible to install the then state-of-the-art torpedo side protection system. This spurred development of a new system to achieve equal protection in less space. Though the need for this no longer applied to the conventionally powered *Kennedy*, the configuration was retained and became the standard for the *Nimitz* and later designs.

Design of a nuclear carrier to succeed the *Kennedy* continued to evolve in the early 1960s. Objectives were to increase the aviation fuel and ordnance, and decrease the nuclear plant size. There was political resistance both to the increased size of carriers and to nuclear power. The Navy was accused of "gigantism," and in truth the growth in the size of aircraft carriers had been both consistent and large for many years. See Table 1. Growth, however, has leveled off with the

Table 1 Growth of aircraft carrier classes.

Class		Commissioning Date	Full Load Displ., tons	Percent Increase Over Previous Class
CV 8	*Hornet*	1941	26,000	44
CV 9	*Essex*	1942	35,000	35
CV 41	*Midway*	1945	60,000	71
CV 59	*Forrestal*	1953	78,000	30
CVN 65	*Enterprise*	1962	91,000	17
CVN 68	*Nimitz*	1975	95,000	4

Nimitz as a result of reaching the limits imposed by graving docks and harbor depths.

The virtually unlimited propulsion endurance of nuclear power provided an advantage that would be lost (or diminished) if the ship had frequently to replenish aviation fuel and ordnance. Increased aviation endurance could help justify the high investment cost of nuclear propulsion; but higher aviation endurance required a larger ship. Thus, the equation showed an economy of scale in which the advantage of nuclear propulsion could be realized only in a relatively large carrier.

Continuing studies and debate over the size, cost and propulsion plant delayed the *Nimitz* authorization until FY 1966 and until a smaller (yet equally efficient) reactor was available. The perfected power plant used only two reactors, required significantly less space than the *Enterprise* plant, and pushed the ship at well over the 30-knot minimum required speed.

The *Nimitz*'s hull had the same waterline length as the *Enterprise*'s, but had greater internal volume because of a 1-ft greater beam and a fuller hull form. Even so, the *Nimitz* was extremely volume-limited. Her design called for a complement of 439 officers and 5182 enlisted crew, compared to 402 officer and 4980 enlisted crew on the *Enterprise*. Aviation fuel capacity was only slightly greater on the *Nimitz*, at 2.7 million gallons of JP-5; but aviation ordnance, at about 3000 tons, was 65 percent greater. Moreover, aviation shops required significantly greater room on the *Nimitz*. The net result was much tighter space. The Bureau of Ships made a final effort to lengthen the ship by 40 ft to relieve the problem, but the request was denied. Note that a 40-ft increase is the maximum feasible based on existing graving docks. The resulting space shortage primarily affected habitability standards.

Principal innovations in the *Nimitz*, other than the nuclear power plant, included the flight deck arrangements and magazine arrangements. The long flight deck accommodated four of the new, long C13-1 steam catapults with sufficient thrust to launch the developmental F-111B aircraft. Catapults and aircraft elevators were rearranged to reduce the angle of the canted deck and thereby reduce air turbulence along the landing strip.

To reduce vulnerability, aviation weapons magazines were arranged in a manner different from those in earlier carriers. Weapons elevators had to be enlarged over those of previous carriers because new, longer missiles were being carried by aircraft.

Local air defense weapons for the ship were consistent with what was being supplied to other carriers at the time. Three Basic Point Defense Missile Systems (Sea Sparrows) were initially installed.

Another innovative development for the *Nimitz* was the retractable catapult control stations. One was located in the flight deck between the two bow catapults. A second was located at the port edge of the flight deck outboard of the waist catapults. Water cooled jet blast deflectors were provided for all four of the C13-1 catapults. A new design wet accumulator was installed in the catapult system to boost the reactor-generated steam pressure to that needed for maximum catapult thrust.

The extensive studies, development, and debate that led to the *Nimitz* design were apparently worthwhile, because the result has been a lasting design. From 1966 to the time of this writing, the *Nimitz* and follow ships have been the only U.S. aircraft carriers acquired. A total of eight ships of the class have been authorized to date, and a ninth is in the FY 1995 budget. Because of the extremely long time span in constructing this class (30+ years) there have been many changes and improvements in the design. Nevertheless, it remains the same basic configuration that was started in construction in 1966. The most significant of these updates came in the *Theodore Roosevelt* with the addition of anti-ship cruise missile protection for the magazines and machinery box, plus state-of-the-art combat systems improvements, including updated computers and standardized display consoles (primarily in the combat direction center). This latter improvement is being retrofitted in all

Fig. 8 *Nimitz*—battle ready.

in-service carriers as well. Later, in the *Abraham Lincoln*, new, larger diameter, low pressure catapults were installed, providing the same launch capability but at less demand on the steam plant. Numerous studies and designs for competitive carriers have been developed during those years, but none of them would have been as cost effective. Some of them are described in the next section.

Post-*Nimitz* Designs

Though the *Nimitz* design was highly successful in filling the needs of carrier based aviation, the high acquisition cost has caused continued interest in developing a lower cost solution. There was divided opinion within the Navy as well as within Congress. In general, the naval aviation community favored the large carriers because they provide maximum capability and the large deck provides safer launching and landing under a far greater range of sea conditions.

There was, however, a large group among the debaters who believed that a larger number of smaller carriers would allow carrier aviation to be dispersed to a wider range of trouble spots around the world. They argued that a single ship loss would not be as great a catastrophe. There was essentially a continuum of aircraft carrier studies and designs between about 1970 and 1985. There were periods of renewed or flagging interest as budget cycles evolved, and as personnel changes occurred. Although some of the designs progressed to the contact design level of detail, none was authorized. Most of the concepts were nuclear powered, although there were a few with oil fired boilers or gas turbine plants. Carriers designed to operate aircraft that could take off vertically or with short deck runs were examined in detail. Many felt that the studies validated the economy of scale advantages that large carriers enjoy. For example, a carrier that cost two-thirds as much as the *Nimitz* could support an airwing only half the size of the *Nimitz*'s.

Because of the many uncertainties involved, the Congress has been slow to authorize new construction. This led the Navy to establish a Service Life Extension Program (SLEP) to provide 15 years of additional service life to the *Forrestal*- and improved *Forrestal*-class ships. This included extensive overhaul of existing systems, upgrading or replacing equipment no longer logistically supportable, and installation of new electronics and weapons systems.

Since the end of World War II aircraft carriers have been involved in two lengthy conflicts (Korea and Vietnam), and numerous smaller ones. In addition, there have been countless "show the flag" operations where the presence of a carrier battle group was used to advance U.S. foreign policy objectives merely by the threat of force. In all cases, including the recent Gulf war, their primary mission has been to achieve local air superiority and to attack land bases. Owing to time and distance considerations, tactical aircraft from an offshore carrier were frequently the first U.S. aircraft available on the scene. As events in the Gulf of Sidra demonstrated, land based air is frequently too distant, and gaining approval for overflights may not be politically achievable. While the former Soviet worldwide threat has virtually disappeared, the proliferation of third-world nations with sophisticated weapons (and the willingness to use them) continues. Because of a general reduction in defense spending, big ticket items such as aircraft carriers will come under increased scrutiny. As this chapter has shown, this is not a new experience!

Recommended Reading

Chesneau, R., *Aircraft Carriers of the World*, Naval Institute Press, 1984.

Friedman, N., *U.S. Aircraft Carriers*, Naval Institute Press, 1983.

St. Denis, M., "The Strike Aircraft Carrier: Considerations in the Selection of Her Size and Principal Design Characteristics," *Trans.* SNAME, Vol. 74, 1966.

Robert J. Riggins held degrees from the University of Florida (mechanical engineering), University of Michigan (naval architecture), and George Washington University (engineering administration). He also completed special studies at the Naval War College and the Defense Systems Management College. He worked as a civilian with the U.S. Navy from 1959 to 1987, rising to the rank of division director. During most of that time he concentrated his attention on aircraft carrier design. The final years of his life were spent with JJH Inc.

Capt. B. F. Tibbitts, USN (ret.), graduated from the U.S. Naval Academy in 1956. After six years of sea duty (including gunnery division officer on an attack aircraft carrier), he attended MIT where he earned an MS degree in mechanical engineering and a Naval Engineer degree. He was then assigned to a series of increasingly responsible positions in a wide variety of naval establishments. He was, for example, Commander of the David Taylor Naval Ship R&D Center and, later, director of The NAVSEA Ship Design Group. He retired from the Navy in 1987, but was recalled to teach ship design courses at MIT for three years. He retains an association with MIT as a senior lecturer, but is also employed by John J. McMullen Associates as chief systems engineer.

American Submarine Development, 1944–1990

GARY E. WEIR

This chapter is dedicated to the late Rear Admiral Andrew I. McKee, USN, gentleman, acknowledged dean of American submarine engineers, and author of the essay on American submarine development in the fiftieth anniversary volume of the SNAME Historical Transactions.

IN DECEMBER 1944, four naval veterans of the ALSOS mission to explore German progress on the atomic bomb suggested that the Navy Department embark upon a comprehensive investigation of German wartime developments in science and technology.[1] Heeding their advice, in January 1945, the Navy Department commissioned the Naval Technical Mission in Europe to follow the advancing Allied Forces for the purpose of gathering scientific and technical intelligence for the postwar American Navy. The four "founders" were Commodore Henry A. Schade, leader of the mission and later Director of the Naval Research Laboratory, Captain Albert G. Mumma, future Chief of the Bureau of Ships, Captain J. P. Den Hartog, a reserve officer and Professor of Mechanical Engineering at Harvard, and Captain Wendell Roop, who later did ground-breaking work on structural materials at the David Taylor Model Basin. They spearheaded a team composed of nearly 900 officers, civilians, and enlisted men on a year-long pursuit of the best in German research and development. What they found changed the history of the American submarine.

The Naval Technical Mission discovered a series of revolutionary U-boat designs developed by the German Navy in direct response to the effectiveness of Allied antisubmarine warfare (ASW) in the Atlantic. The intelligence provided by the breaking of the German naval codes made the efforts of RAF Coastal Command and the American Escort Carrier Groups particularly effective. By May of 1943, Admiral Karl Dönitz, commander in chief of the German Navy, virtually abandoned his North Atlantic submarine campaign because of the incredible losses suffered by his forces. Only a submarine capable of much greater submerged speed and endurance offered the Germans some hope of success against Allied ASW. These dire circumstances gave birth to the Type XXI.

As with all submarine designs, the Type XXI was an engineering compromise with a long history. For the Germans the possibility of a submarine capable of high submerged speed originated with a 1933 proposal by Professor Helmuth Walter for a U-boat capable of 25 to 30 knots. He planned to achieve this goal by employing a closed-cycle diesel propulsion system that employed scrubbed exhaust gases enriched with oxygen for combustion. Rather than storing the oxygen on board, the Walter system derived the gas from the chemical breakdown of concentrated hydrogen peroxide.

The professor soon discovered that the chemical reaction, which separated the hydrogen peroxide into its component parts, generated considerable heat. Taking advantage of this knowledge, Walter eventually set aside the diesel and developed a closed-cycle submarine steam turbine system. He first demonstrated this new mode of submerged propulsion in 1939 with the experimental, 80-ton, V-80, secretly built behind wooden screens at the Krupp Germania shipyard in Kiel.

The V-80 provided a foundation for a series of variations on the high-speed U-boat. Although it was small, the 28.1-knot submerged speed achieved in its 1940 trials gave the vessel's designers a reason to feel that the Walter propulsion system and its sleek, unadorned hull form held promise for the future.

[1] ALSOS was the Greek word for "grove," chosen to emphasize the link between this mission and the head of Manhattan Engineering District, Lt. General Leslie Groves.

In response to the emergency conditions of 1943, the German Navy concentrated its design and engineering talent at Blankenburg in the Harz Mountains to give them the time, resources, and opportunity to conceive an operational, high-speed U-boat. The German Navy needed a fast submarine capable also of prolonged submergence. This combination might conceivably checkmate the highly successful Allied ASW effort. The outcome of their combined effort was the series of advanced submarines designs that included the Type XXI.

The Type XXI represented a compromise between the diesel, direct drive submarines that formed the backbone of Germany's U-boat force during World War II and the Walter V-80's promise of greater speed and endurance. The Blankenburg design team did not incorporate the teardrop hull form in the Type XXI. Instead, they selected a nearly conventional hull that revealed an inverted figure-eight shape in cross section. This form had less in common with the V-80 than it did with the workhorse of the German U-boat fleet, the Type IXC. To reduce submerged resistance they dramatically streamlined the hull. In part, this entailed removing the deck gun and any external fixtures that would heighten resistance, as well as reducing the size and profile of the conning tower.

The still experimental nature of the Walter propulsion system and the scarcity of concentrated hydrogen peroxide forced the Blankenburg designers to adopt enhanced diesel-electric power to propel the Type XXI. The ship's 372 battery cells allowed 18 knots submerged speed, which would enable it to outpace any Allied ASW vessel that might offer a challenge. In addition, the Germans equipped the Type XXI with a snorkel to provide for both battery charging and greater endurance while submerged.

Although no Type XXI U-boat ever saw action, the technologies which gave birth to this and other experimental U-boats profoundly affected worldwide postwar submarine development.

By late February and early March of 1945, the U.S. Naval Technical Mission initiated its field operations. The Ships Division of the Mission's Technical Branch took responsibility for exploring German advances in submarine technology. Staffed by experts in materials, hull design, propulsion and other specialties, this division performed the first analysis of German advances in hydrogen peroxide propulsion, periscopes, batteries, snorkels, and closed cycle diesels.

The scores of technical reports and the products of German wartime research and development sent back to the United States between December 1944 and November 1945 had an immediate effect on the Navy's submarine force. This was especially true of the Mission's analysis of the Type XXI, the snorkel, and Walter propulsion technology done by officers like Captains Albert G. Mumma and William R. Ignatius, Commander Henry A. Arnold, and Lieutenant Commander Ralph A. Smith.

Fig. 1 Floating crane Schwimmkran 2 lifts the stern section of Type XXI submarine U-3001 during construction at Deschimag Shipyard, Bremen, Germany, ca. April 1944. (Courtesy Capt. Henry A. Arnold, USN (Ret.) and U.S. Naval Historical Center.)

With the end of the war in August 1945, the U.S. Navy needed to redefine its mission and strategy. Although its most likely adversary, the Soviet Union, posed no immediate naval threat, the Red Army had as much success in capturing German submarine technology in the closing months of the war as did the Naval Technical Mission. Since the Type XXI represented a quantum leap forward in submarine development, the Navy Department had to assume that the Soviets would capitalize on the same technology that interested Commodore Schade and his associates.

Under Allied postwar agreements the United States Navy acquired two captured Type XXI U-boats, the U-2513 and U-3008, for examination and operational testing. The performance of these vessels while in the custody of the Portsmouth Naval Shipyard confirmed the impressive reports filed by the Naval Technical

Fig. 2 Hull sections of two Type XXIs in position for joining, on the ways at the Schichau Shipyard, Danzig, ca. 1944. Note large lugs welded to each section for preassembly handling.

Mission on capabilities. They submerged in approximately 20 seconds and could provide a sustained underwater speed of 17 knots for over an hour.

The advances in design and propulsion uncovered by Portsmouth and the Naval Technical Mission prompted the Bureau of Ships to pursue two avenues in the development of postwar attack submarines. The first took some of the wartime fleet submarines and converted them to Type XXI facsimiles, called "Guppies" because of their "greater underwater propulsive power." The other, which found expression in the *Tang* class (SS-563), sought to imitate the Type XXI from the keel up.

The postwar Guppy program drew its inspiration from German Type XXI design and technology. The Bureau of Ships adapted the extremely successful American "fleet submarine" of World War II for greater underwater speed by streamlining the hull and conning tower, removing all appendages—including the guns—and dramatically increasing the battery capacity.

Less than a decade earlier, the original fleet submarine was derived from the *Salmon* (SS-182), *Sargo* (SS-188), and *Tambor* (SS-198) classes. Designed by the Bureau of Ships, Portsmouth Naval Shipyard and Electric Boat Company, the Navy Department introduced these vessels between 1936–1940. General Motors or Fairbanks–Morse engines in a diesel-electric reduction gear arrangement propelled them at 20 to 21 knots on the surface. Submerged, fleet submarines operated at a maximum depth of 250 feet at 8.75 knots, powered by General Electric or Elliott motors. While the speed of these vessels changed little before the war ended, thicker hull plating increased the maximum diving depth to 400 feet. These years also witnessed the introduction of high tensile steel (HTS) for submarine hulls. But wartime shortages made HTS more significant for the postwar submarine fleet.

Less than one year after the Naval Technical Mission was decommissioned in November of 1945, Portsmouth Naval Shipyard converted a *Tench*-class fleet submarine, the *Odax* (SS-484), into the first "Guppy." Since the focus was on increased submerged speed, the *Odax* lost its deck guns, the hull was streamlined, and she received a "sail" to replace the conning tower. These changes increased the submarine's speed by reducing its submerged drag. After doubling the usual *Tench*-class complement of two 126-cell batteries, this Guppy I conversion attained a submerged speed of 18.2 knots, exceeding its surface speed by 0.4 knots.

The Guppy II program added a snorkel to the converted characteristics of the *Odax*. The Bureau of Ships chose to limit the Guppy I conversions to *Odax* and *Pomodon* (SS-486), the latter accomplished at Mare Island, while the Navy's Engineering Experiment Station at Annapolis and the Portsmouth Naval Shipyard busied themselves perfecting an American snorkel. After this device made its debut on the *Irex* (SS-482) in 1947, the Bureau of Ships transformed 24 fleet submarines of the *Balao* and *Tench* classes into Guppy IIs. This number included the *Odax* and *Pomodon*, each upgraded with the addition of a snorkel.

The expense of these conversions limited the number the Navy could accomplish. The advanced battery developed for the Guppy IIs illustrates this point well—it offered only one extra knot of submerged speed for a dramatically increased cost. Thus, the Navy Department chose to perform less expensive and drastic alterations after the completion of the Guppy II series. The Guppy IA and the Guppy IIA employed a cheaper, standard-size, battery of higher capacity called the Sargo II. Other variations in these conversion plans included a variety of machinery and sonar room arrangements as well as better living conditions.

The least expensive fleet submarine conversion plan barely changed the *Tench*- and *Balao*-class submarines. These "fleet snorkel conversions" provided only

Fig. 3 USS *Nautilus* (SSN-571), the Navy's first atomic-powered submarine, on her initial sea trials, January 1955. (Courtesy U.S. Naval Institute.)

rudimentary exterior streamlining, additional sonar equipment, and a snorkel. Between 1947 and 1952, 19 fleet submarines were converted in this manner.

Unlike the Guppies, the *Tang*-class was designed from the keel up as America's version of the German Type XXI. Congress authorized the *Tang* (SS-563) in May of 1947 and the workers at Portsmouth Naval Shipyard laid her keel in April of 1949. Like their German progenitor, the *Tang*s employed a streamlined hull and snorkel, but not the Walter propulsion system. Instead, a greatly increased battery capacity allowed the six vessels of the class to reach 17 knots submerged, a speed they could match on the surface using the new General Motors model 184 radial "pancake" diesels. The latter possessed less than half the weight of the regular 1000 bhp opposed piston diesels. Unfortunately, the new engines proved far too difficult to service. The *Tang*-class vessels eventually received new propulsion plants in order to avoid extended overhaul periods and the occasional need to be towed back for repairs.

In the effort to anticipate the type of challenge posed by the Soviets at the opening of the Cold War, the Navy also adapted American fleet submarines to a variety of other uses. The idea of employing fleet submarines for radar picket duty extended back into the war years. The original project, code named "Migraine," called for 24 SSRs, or radar picket submarines. Between 1946 and 1953, the Bureau of Ships took delivery on ten of these conversions. Other converted fleet submarines functioned as oil transports (SSO) and cargo (SSA) and troop (SSP) carriers.

At the same time, the Bureau of Ships investigated employing submarines to launch guided missiles patterned after the German V-1. The *Cusk* (SSG-348) and *Carbonero* (SS-337) were modified to deliver the American version of the German V-1, called the Loon.

When its Trounce Program produced the larger and more powerful Regulus I to replace the Loon, the Navy Department authorized fleet submarine conversions to carry the system. Between 1953 and 1955 the Bureau of Ships transformed the fleet submarines *Tunny* (SS-282) and *Barbero* (SS-317) into guided missile platforms (SSGs). The modifications included a missile hanger, retractable launch platform, and missile fire control equipment. The conventional submarines *Greyback* (SSG-574) and *Growler* (SSG-577) and the nuclear ship *Halibut* (SSGN-587) carried a second generation of the Regulus system, first launched in 1957. Regulus remained operational until replaced by Polaris during the next decade.

Like some of these specialized Guppy conversions, the Navy also experimented during the early 1950s with a smaller, mission-controlled design for the early postwar hunter-killer submarines or SSKs. A very different type of submarine, the hunter-killer lingered at slow speed in the shallow depths near harbors and at the geographic choke points through which enemy forces would have to pass. The first of these vessels, the K-1 through 3, each displaced only 765 tons on the surface with a length of 196 feet. The real advantage of these small, cramped, uncomfortable vessels was their large AN/BQR-4 sonar. Like the Guppies and *Tangs*, this technology also traced its roots to the discoveries made in Germany by the Naval Technical Mission. The Navy derived the AN/BQR-4 from the German Gruppenhorchgerät, or GHG, hydrophone system.

The necessity to do combat against other submarines as well as surface vessels led the Navy to develop multipurpose attack submarines, which performed the hunter-killer role as part of their regular mission. Rather than continuing with the smaller K-series, the Navy spent the balance of the decade arming seven

Fig. 4 USS *Seawolf* (SSN-575), August 1955.

old *Gato*-class fleet submarines with the more advanced BQR-4 sonar to take over the hunter-killer mission. These ships provided increased speed, better habitability, and the capability of firing pattern-running and acoustic-homing torpedoes like the Mark 37.

Two other developments of far-reaching significance took place during this first postwar decade. In 1946 the Bureau of Ships, under the direction of Vice Admiral Earle W. Mills, embarked upon a program to develop nuclear propulsion for naval ships. In June, Mills' choice to direct that project, Captain Hyman G. Rickover, began his nuclear training at Oak Ridge TN. In the second of these developments, the Bureau of Ships received congressional authorization to build the experimental submarine, *Albacore* (AGSS-569) on November 24, 1950. This vessel gave the Bureau the opportunity to explore the promise of a teardrop hull configuration. Together these two influences determined the future of American submarines.

The roots of the *Albacore* hull design extend back to 1948, when the Committee on Undersea Warfare of the National Research Council initiated an effort to accumulate the collective wisdom of the naval and scientific communities on the hydrodynamics of submerged bodies. The various research institutions consulted by the committee included among others, the David Taylor Model Basin, Brooklyn Polytechnic Institute, Portsmouth Naval Shipyard, and the Woods Hole Oceanographic Institution. The scientists at these and other research centers were investigating the performance and control characteristics of high speed submarines either independently or under government contracts. In one case, a group at Brooklyn Polytech, working under a contract from the Office of Naval Research, developed a method for determining the precise shape of cylindrical pressure hulls for high speed submarines and the framing required to build them.

The selection of the best hull form was the critical issue. Reflecting upon his contribution to *Albacore*'s design during an interview with historian David Allison, Dr. Louis Landweber, physicist and head of the Hydrodynamics Division of the David Taylor Model Basin, recalled his research into Series 58, a collection of 24 different variations on the body-of-revolution submarine hull form. Landweber recalled that the goal of the research was to find the body-of-revolution that provided maximum performance, "from the point of view, first, of resistance and pressure distribution; and then of stability and control." The best of the series would become that used for the *Albacore*.

From the beginning the expectations for *Albacore* were high. Since the vessel's designers and engineers expected a submerged speed of at least 25 knots, this vessel would provide an opportunity to experiment with single screw propulsion, control surfaces and ship stability at high speed, and hull coatings to reduce frictional resistance. *Albacore*'s double hull and electric drive also promised to make it one of the quietest submarines in the Navy, a fact which interested Dr. Marvin Lasky and other physicists at the David Taylor Model Basin interested in underwater acoustics.

Albacore was launched on August 1, 1953, commissioned by December, and completed a year of shakedown and acceptance trials by February of 1955. According to Dr Morton Gertler, who participated in designing *Albacore* and tested the vessel at the David Taylor Model Basin's Hydrodynamics Laboratory, this submarine "served the twofold purpose of providing fundamental hydromechanic research data and providing a means for evaluating new and unusual concepts of control, propulsion and quieting." The tests that produced this data were divided into four phases determined by alterations made to the stern control surfaces. These variations included both cruciform and "X" stern planes combined with either the *Albacore*'s single screw or a set of contra-rotating

Fig. 5 USS *Albacore* (AGSS-569), off Isle of Shoals, May 1954. (Courtesy U.S. Naval Institute.)

screws. In her unmodified form *Albacore* proved capable of 26 knots submerged and eventually reached 33 knots with the help of silver–zinc batteries and highly efficient contra-rotating propellers.

This research program provided the Navy with the opportunity to improve significantly the performance of later classes of American submarines. In conjunction with the scientific community, the officers and crew of the *Albacore* demonstrated to the fleet that a streamlined, single-screw submarine could produce high submerged speeds and hydrodynamic efficiencies under the control of a single operator. Work was also done on depth-keeping without the benefit of sail planes or bow planes, arresting snaproll, and placing sonar in the bow of the submarine.

The Navy applied the revolutionary results of its research with *Albacore* to an operational vessel for the first time with the three submarines of the *Barbel* class. The *Barbel* (SS-580), *Blueback* (SS-581), and *Bonefish* (SS-582) represent the transition to the modern American submarine. Save for nuclear power, these diesel-electric vessels had everything that the latest knowledge of hydrodynamics and modern technology could provide. *Barbel* was the first operational submarine to have the *Albacore* hull design, a single propeller on the axis, an advanced ballast control panel, and HY-80 hull steel. She displaced 2145 tons and could make 21 knots submerged.

Nuclear power provided the perfect complement to the high speed of *Albacore* and *Barbel*. In 1939, seven years before Captain Rickover began his nuclear training at Oak Ridge, Dr. Ross Gunn of the Naval Research Laboratory in Anacostia made the first recorded suggestion that nuclear power "would enormously increase the range and military effectiveness of a submarine." A full decade after Gunn's insightful comment, Rickover, then head of the Bureau of Ships' Nuclear Power Division (Code 390), initiated his effort to create just such a vessel.

In March of 1949, the Submarine Officers Conference (SOC), under the direction of Rear Admiral Charles B. Momsen, finished a comparative study of the potential advantages of both closed cycle and nuclear propulsion. The SOC gave its complete support to nuclear power in a report issued on May 18, 1949. Less than three months later the hard work of the Nuclear Power Division of the Bureau of Ships turned this vote of confidence into a commission from the Chief of Naval Operations, Admiral Louis E. Denfeld, for a nuclear propulsion system ready for installation into an operational submarine by January 1955.

The Navy's interest in closed-cycle propulsion did not end with the SOC's recommendation and Denfeld's decision. The staff at the Navy's Engineering Experiment Station in Annapolis continued exploring this alternative for nearly three years. In Project Wolverine, they designed a semi-closed cycle gas turbine employing liquid oxygen as the oxidant to burn diesel fuel. The initial investigation of the original Walter closed cycle propulsion plant, called Project Hill, was 70 percent complete by October 1946. Its successor, Project Alton, was scheduled to "fire up" a final product by mid-October 1952 when the promise of nuclear propulsion ended this line of investigation.

Initially, Rickover sought to create two partnerships for the production of nuclear submarines. General Electric Company, accustomed to supplying Electric Boat Company (EB) with motors, generators, and other components for conventional submarines, would work with this shipbuilder once again. In this case, EB was enthusiastic and Rickover convinced General Electric to assume responsibility for the liquid metal-cooled SIR, or submarine intermediate

Fig. 6 USS *Shark* (SSN-591) under construction at Newport News Shipbuilding and Dry Dock Co.

reactor. In West Milton, New York, EB built a mock-up of the submarine compartment for the SIR reactor prototype, called the Mark A. If the SIR system proved successful, the company also agreed to build the vessel in which this propulsion plant would eventually go to sea.

The admiral's effort to form a second partnership, this time between Westinghouse and the Portsmouth Naval Shipyard, met with less success. In a meeting with Rickover and his associates from Naval Reactors, Portsmouth officials declined to serve as a partner with Westinghouse to produce a submarine propelled by a pressurized water reactor. As a naval shipyard, Portsmouth felt the economic pressures of the postwar slowdown less than the private sector. Commander of the shipyard, Rear Admiral Ralph E. McShane, did not want to draw design talent and building skills away from the SSRs, the Guppy conversions, and the *Tang* class, then under construction at Portsmouth. He offered to give Westinghouse advice on the project, but would not commit himself to the same role EB had assumed for General Electric.

Rickover immediately went to EB and received assurances of the company's willingness to build both vessels. In addition, the company agreed to work closely with Bettis Laboratory at Westinghouse on reactor development. In the end, Electric Boat built not only the hull sections to house the prototype reactors at Arco ID and West Milton NY, but also laid the keel of the *Nautilus* (SSN-571) at Groton on June 14, 1952. Portsmouth would not begin working on a nuclear submarine until about 16 months after the commissioning of the *Nautilus.* The *Swordfish* (SSN-579) of the *Skate* class was the first Portsmouth-built nuclear submarine.

Shortly after the Mark I pressurized water reactor was completed at Arco in 1953, Rickover authorized an ambitious trial run that would far exceed those tests originally planned. The Idaho prototype had first gone critical at 11:17 pm on March 30. On June 25, in an effort to demonstrate his conviction that the Mark I would perform reliably, Rickover proposed a 100-hour test run simulating a "voyage" across the Atlantic. Although the operating staff experienced considerable trouble with the steam plant and mechanical equipment after the sixtieth hour, the reactor performed extremely well and the test proved to be a milestone in American nuclear development. As historians Richard Hewlett and Francis Duncan pointed out many years later:

> The test run had been an extraordinary achievement, with a significance extending far beyond the Navy and the United States. The Mark I was the world's first fully-engineered nuclear reactor capable of producing practical amounts of energy on a sustained and reliable basis.

As the *Nautilus* neared completion at Electric Boat an accident occurred that demonstrated the need for absolute care in the procurement, storage, and marking of essential construction materials. In the course of propulsion plant hot water pressure tests, a pipe connecting the two main steam generators burst. Engineers from the Navy and Electric Boat determined that the fracture occurred along a seam in the pipe. Since the designs and orders specified only seamless, cold drawn steel tubing all of the pipes became suspect. The Navy canvassed its own laboratories, private industry, and experts in the academic community for a non-destructive means of testing the pipes already installed in the *Nautilus.* For the safety of the ship and its crew the engineers had to determine the extent to which "seamed" tubing was mistaken for the proper seamless material in work already accomplished. No means of testing then known could perform the task without damaging the installed pipe. Thus, the Navy ordered Electric Boat to replace all of the suspect piping with seamless material procured by placing emergency orders with the contractor. According to Captain William Roseborough, then inspection officer for the Supervisor of Shipbuilding at Groton, "It was almost four months and $4 million later before the job was completed and testing resumed, even though we worked 21 shifts per week with essentially unlimited overtime."

Since Electric Boat Division had to pay for the lost time and materials, this incident precipitated major improvements in the process of pipe procurement, marking, storage, and handling at this yard as well as others. EB went so far as to forbid the use of seamed pipe anywhere in the yard for any purpose. Nuclear technology proved an unrelenting taskmaster, demanding complete fidelity to specifications for the safety of the ship and crew as well as for the survival of the infant nuclear program.

On the morning of January 17, 1955, the *Nautilus* left the dock at Electric Boat Company and actually got underway on nuclear power. Westinghouse had successfully prepared the improved version of the pressurized water reactor, now called the Mark II, for installation in the vessel.

Six months after the *Nautilus*'s revolutionary maiden voyage, the *Seawolf* came down the ways at Electric Boat. This vessel's reactor presented General Electric with significant and complicated problems. The Mark A, renamed the S1G, took longer to develop than the *Nautilus*'s S2W pressurized water reactor because of the strongly corrosive effect of the sodium–potassium coolant. Both the S1G, and its more advanced successor the S2G, experienced leaks in the cooling system, which advanced the corrosion of the superheater as well as the pipes in the steam plant. Although the *Seawolf* passed its initial sea trials in February of 1957, Rickover, now a Rear Admiral, had already informed the Atomic Energy Commission of his decision to convert the *Seawolf* to a pressurized water system at the first available opportunity.

The four-ship *Skate* class (SSN-578/579; SSN-583/584) which followed *Nautilus* and *Seawolf* represented more of a shift in mission than a change in technology. The Chief of Naval Operations, Admiral Robert B. Carney, decreed that in the design of these vessels speed should take second place to the capability to wage antisubmarine warfare (ASW). Thus, the Navy built the *Skate* class as a smaller version of the *Nautilus*, designed for serial production and ASW. The lead ship of the class became the first submarine to surface at the North Pole, a particularly important region for ASW operations. Since the lessons learned from the *Albacore* were not yet ready for application to an operational vessel when this class was designed, the marriage of nuclear propulsion and the teardrop hull design had to wait for the *Skipjack* in 1958.

In the *Skipjack* (SSN-585) the Navy combined nuclear power and the *Albacore-Barbel* hull form to create the fastest and most maneuverable submarine in history. The noted submarine engineer and design expert Captain Harry Jackson has described this ship

Fig. 7 USS Sam Houston (SSN-609). (Courtesy Newport News Shipbuilding.)

Fig. 8 USS *Skipjack* (SSN-585). (Courtesy Nautilus Memorial Submarine Force Library & Museum.)

as the fifth and latest submarine design revolution of this century.

According to historian Francis Duncan

> For the first time nuclear propulsion and a streamlined hull had been combined in a submarine. Beneath the surface the *Skipjack* had behaved like an airplane, banking and rolling as she maneuvered at high speed.

Equipped with the new S5W reactor, this ship displaced 3070 tons over its 252-foot length. Like the *Albacore* and *Barbel, Skipjack* was small. But the success of this vessel set the stage for all of America's future nuclear submarines, whether guided missile or attack types.

The success of nuclear power also prompted the Navy to apply this technology to unique, mission-controlled designs which would address specific strategic and tactical problems. The role of the *Halibut* (SSGN-587) in the Regulus program is one of the best examples. Others include the construction of the *Triton* (SSN-586) for duty as a radar picket submarine. Commissioned in 1959, this vessel had two S4G nuclear reactors to give her sufficient surface speed to conduct surveillance and direct air strikes, submerging only to avoid detection. With the launching of the *Tullibee* (SSN-597) in 1960 the Navy enhanced the effectiveness of the hunter-killer submarine or SSK with nuclear power. The *Tullibee* operated in the gap between Greenland, Iceland, and the United Kingdom, with a quiet nuclear turboelectric power plant and exceptional sonar capability fitted tightly into its small 2460-ton submerged displacement hull. The Navy later employed this type of propulsion once again in the *Glenard P. Lipscomb* (SSN-685) in a bid to reduce drastically the noise generated by submarine nuclear propulsion plants. The same motive gave birth to the *Narwhal* (SSN-671), laid down by Electric Boat Division in January 1966. With this vessel the Navy tried to combat noise generated by the coolant pumps and the reduction gears with a natural circulation reactor (S5G). In this case convection would insure the proper flow of the reactor coolant rather than a noisy mechanical pumping system.

In each case these technically successful designs outlived their purpose. Regulus was displaced by Polaris and the quiet and versatile *Thresher-Permit*-class fast attack submarines took over both the surveillance and hunter-killer missions. In spite of their image as ideas that did not work, these ship designs and technologies were not technical blind alleys. In many cases they provided new submarine systems and valuable engineering experience that contributed significantly to the success of the *Skipjack, Thresher*, and *Sturgeon* attack submarines.

Thus the fundamental design revolution which determined the configuration of current American submarines took place in the late 1950s. Over the next 30 years the Navy and private industry displayed the adaptability and versatility of the *Skipjack* formula with missile and torpedo-equipped submarines that went faster and deeper than any ships in history.

Only seven months after Electric Boat Division laid the keel of the *Skipjack*, the Navy took steps to accelerate its development of a new generation of submarine launched missiles. On December 8, 1956, the Navy's Special Projects Office (SP) embarked on the development of Polaris. Under the direction of Rear Admiral William F. Raborn, and with the strong support of Chief of Naval Operations Admiral Arleigh Burke, SP took a considerable risk and embraced the new solid-propellant ballistic missile concept as the type of missile best suited for launching at sea. This system would not use the highly volatile fuels so hazardous to handle on board ship but demanded by the air-breathing guided missiles.

Motivated by the success of the soviet satellite Sputnik in October 1957, Admiral Raborn accelerated the Polaris project by two years, moving the target date for the first successful submerged launch up to 1963. The planned range of the missile was reduced from 1500 nautical miles to 1200 and SP, with support from the Bureau of Ships, received authorization to convert the partially completed hull of the last *Skipjack* class submarine under construction at EB into the first of the Polaris ballistic missile submarines, or SSBNs.

Contrary to popular myth, this ship was not simply cut in half to admit the 180 foot section which would accommodate the 16 polaris missiles. Already under construction as the *Scorpion* (SSN-589), the Bureau of Ships reordered the vessel as the *George Washington* (SSBN-598) and took advantage of the long lead time in procurement and design afforded by using a vessel already on the ways to complete the construction of the ship in 27 months. After successful test firings of the polaris missile from Cape Canaveral and from the missile test ship *Observation Island*, the *George Washington* made the first two successful submerged Polaris launches on July 19, 1960, nearly two years ahead of schedule.

When President Kennedy took office in January of the following year, he asked for an acceleration of the program, which would eventually provide a fleet of over 40 SSBNs of the *George Washington*, *Ethan Allen*, and *Lafayette* classes. The capability of the missiles increased along with the number of submarines able to fire them. By the time the development of the three-type Polaris series was completed in the autumn of 1964, the missile's range had increased from 1200 to 2500 nautical miles.

Fig. 9 The nuclear-propelled attack submarine *San Francisco* (SSN-711) underway in the James River, Virginia, during sea trials. The submarine was built by Newport News Shipbuilding and Dry Dock Co. (Official U.S. Navy photograph.)

In the years since the first launching of Polaris, the Navy's submarine-launched missiles have undergone considerable changes. Poseidon (C-3), introduced in 1971, had a range of approximately 2000 nautical miles. The Trident I (C-4) model, which first went to sea in 1979, increased that limit to 4000 and its successor, the Trident II (D-5), can reach targets 6000 nautical miles away.

Early in 1958, as SP raced to develop Polaris and *Skipjack* astounded the naval world with its performance, the Bureau of Ships under Rear Admiral Albert G. Mumma, assigned the construction of the *Thresher* to the Portsmouth Naval Shipyard. This vessel would lead a new class of submarines that would test a variety of innovations in materials, sound detection, and quieting. With the *Thresher* the Navy explored the effectiveness of flexible pipe joints as well as sound isolation mountings for machinery to reduce the major sources of noise aboard the vessel. Slightly slower than *Skipjack*, the *Thresher* would dive deeper thanks to the advances made in working with the HY-80 in its pressure hull and frames. In addition, the new ship would carry the bow-mounted AN/BQQ-2 sonar

Fig. 10 The nuclear-powered strategic missile submarine USS *Ohio* (SSBN-726) as it transits the Hood Canal. (Official U.S. Navy photograph.)

system complemented by the capability to deploy and retrieve towed array sonar. Once he detected a foe, the commander of this vessel could dispatch his adversary with torpedoes or the newly deployed SUBROC missile. After a torpedo tube launch, this ballistic missile emerged from the sea to kill an enemy submarine operating 30 miles away. The Navy adapted later submarines to employ the Tomahawk and Harpoon cruise missiles.

The loss of the *Thresher* on April 10, 1963, during depth trials and the subsequent investigation raised many questions about the materials and procedures used to build the vessel. After the inquiry into the disaster, Chief of the Bureau of Ships, Rear Admiral William A. Brockett, initiated the *Subsafe* program to recertify the Navy's submarines as safe to operate at their test depth. Subsafe remedial measures included an automatic emergency seawater valve closing system, a revised ballast-blowing system, and a procedure for identifying and tracing critical materials from fabrication to installation in the submarine. Critical pipe joints were universally welded and radiographed, and the yards ultrasonically tested all silver-braised joints. Skilled technicians responsible for working with these materials had to obtain proper training and certification that guaranteed their familiarity with current procedure in their respective fields. Rear Admiral Willis C. Barnes, former Commander of Mare Island Naval Shipyard, noted in a recent publication that Admiral Brockett himself set the tone for the Subsafe Program by insisting that "Specifications are not to be considered as desirable goals, but rather as minimum standards which must be met or exceeded." Careful scrutiny became more regular and a stricter adherence to safety standards demonstrated that the *Thresher* disaster would have a lasting effect on the submarine community and its industrial support.

The Navy followed the 13 *Thresher-Permit* submarines with 37 vessels of the larger *Sturgeon* class. The *Sturgeon* (SSN-637) and its sister ships combined the best characteristics and attempted to repeat the operational performance of their predecessors. While carrying the same weapons payload, the heavier *Sturgeons* placed a greater burden on the 15,000 hp S5W nuclear power plant they shared with the *Skipjack* and *Thresher-Permit* classes, and thus surrendered the superior speed characteristics of the latter vessels.

In 1968 an unexpected event suddenly intensified the debate within the Navy and private industry over the type of ship that would succeed the *Sturgeons* as the submarine force's front line attack vessel. While pacing the *Enterprise* (CVAN-65) on January 5, a Soviet *November*-class submarine demonstrated sustained speeds far in excess of anything the American intelligence

Fig. 11 USS *Kentucky* (SSBN-737). (Courtesy General Dynamics, Electric Boat Division.)

community had predicted. As a result it quickly became clear that the next generation of American submarines had to regain and surpass the loss in speed accepted by the Navy in building the *Sturgeon* class.

This imperative as well as budget considerations precipitated a conflict within the Navy over the type of submarine to build. Rickover and his Naval Reactors Division of the Naval Ships Systems Command (NAVSHIPS) wanted to build a single submarine prototype to test the new 30,000 shp S6G reactor. Naval Reactors had pursued a similar course in the past. The *Nautilus, Seawolf, Triton, Narwhal, Tullibee,* and *Skipjack* had all provided valuable test platforms for new reactor plants and Rickover initially wanted the SSN-688 to do the same.

However, the submarine design group at Naval Ships Systems Command had their own ideas. Employing methods developed through decades of submarine design and engineering experience at the Bureau of Construction and Repair, the Bureau of Engineering, and the Bureau of Ships, NAVSHIPS developed a new-generation submarine design from the best characteristics of the 13 different types then at sea. As head of the NAVSHIPS submarine design division, Captain Donald Kern worked closely with the Electric Boat Division of General Dynamics Corporation to develop a matrix of these favored attributes. *Albacore's* best features formed the core of this design which acquired the name CONFORM as an acronym for Concept Formulation. The *Albacore's* single screw and the body-of-revolution hull developed at the David Taylor Naval Ship Research and Development Center provided the best possibility for maximum submerged speed. Other *Albacore* features that found their way into CONFORM were X-stern control surfaces and single-stick control. Kern and his designers also preferred contra-rotating propellers which had demonstrated a high propulsive coefficient on the *Jack* (SSN-605) as well as the 20,000 shp, quieter natural-circulation reactor (S5G), which first went to sea in the *Narwhal* (SSN-671).

Unfortunately, as Kern later recalled, "the CONFORM design became very competitive with his [Rickover's] very conservative 30,000 shaft horsepower [SSN-] 688." Captain Kern felt that with contra-rotating screws the NAVSHIPS design could make the same speed as a 688 submarine but with 10,000 less shaft horsepower and in a smaller, deeper-diving, and potentially cheaper ship.

Just as these facts and the competitive potential of CONFORM began to emerge, Dr John S. Foster Jr., Deputy Secretary of Defense for Research and Engineering argued forcefully that the Navy could no longer afford to build expensive prototypes. The next submarine proposed by the Department of Defense under Secretary McNamara would have be the first of a new class. Thus the submarine community could have the 688 or CONFORM but not both.

Before the debate had begun Rickover and Kern had agreed to support one another so the Navy could benefit from both ideas. Kern would support the admiral's new nuclear power plant and in return Naval Reactors would support the new submarine proposed by the NAVSHIPS design team. When he discovered that CONFORM had become a formidable competitor to the 688 and the S6G, Admiral Rickover

Fig. 12 USS *Maryland* (SSBN-738). (Courtesy General Dynamics, Electric Boat Division.)

Fig. 13 USS *Annapolis* (SSN-760). (Courtesy General Dynamics, Electric Boat Division.)

used his considerable political and naval clout to kill Kern's project and commit the Navy to build an entire run of SSN-688 submarines, beginning with the *Los Angeles*.

From NAVSHIPS' viewpoint a 688 prototype was acceptable. It would allow thorough testing of the new reactor and the results, combined with NAVSHIPS data on CONFORM, would allow the Navy and Electric Boat Division to design the best possible submarine for a long production run as mandated by the Navy Department and Department of Defense. Faced with the possibility that CONFORM would replace the 688, Rickover discarded his memorandum of understanding with Kern as well as the plan to build a single test platform for the new reactor. In the end there was little the NAVSHIPS submarine designers could do.

With a 360-ft length, the *Los Angeles*-class submarines displace roughly 6900 tons. Owing to the increased size of the S6G reactor, these ships are larger than the *Sturgeons*. The 30,000 shp generated by the propulsion plant allows the 688 class to exceed 30 knots submerged speed. Since NAVSHIPS decided to reduce the thickness of the pressure hull to conserve weight and enhance speed, the 688 class cannot operate as deeply as some of the Navy's other submarines and has to exercise greater care in deep diving and maneuvering at high speed. These exceptionally quiet submarines carry enhanced sonar and an improved fire control system for SUBROC and Mark 48 torpedoes. They also have the capability to launch Tomahawk and Harpoon missiles. To date the Navy has 48 of these vessels in commission, built by Electric Boat Division and Newport News Shipbuilding Company.

While the fine officers and crews of the 688 class have been able to keep their ships marginally better that the best the Soviets have to offer, American submarine designers and engineers do not hesitate to point out the significant flaws in this class. As Captain Joseph F. Yurso, former Supervisor of Shipbuilding at Groton, has recently pointed out,

> the *Los Angeles*-class submarine violated the submarine "holy trinity" of speed, depth, and stealth. Although her design was adequate in the stealth area, speed and depth were inferior to the best Soviet designs.

Anticipating the long production runs, complexity, and cost of the Navy's future submarine construction plans, Electric Boat Division, under the leadership of Joseph D. Pierce, broke ground in 1973 on a $140 million renovation of its facilities in Groton. Completed in 1976 in time for the first Trident-class SSBNs, the Land Level Submarine Construction Facility, or LLSCF, made better use of the eight acre site on the Thames River than the old building ways. The company now assembles the Navy's submarines in the construction shed and the vessels emerge from the enclosure as they lengthen and reach completion. When finished each submarine is moved outdoors on rails recessed into the shipyard floor and then transferred to a pontoon launch facility that operates in a fashion similar to a floating drydock. Instead of sliding down the ways, the submarine is floated and released into the Thames.

To reduce costs and construction time further, in 1978 Electric Boat Division also purchased the site of the old Naval Air Station at Quonset Point RI and built a $120 million automated submarine frame and cylinder manufacturing facility. Here the company prefabricates hull sections of up to 42 ft in diameter.

Fig. 14 USS *Springfield* (SSN-761). (Courtesy General Dynamics, Electric Boat Division.)

The cylinders are manufactured on four sets of machines. The first completes the frames, the second fabricates the cylinder shells, a third installs the frames, and the last pairs the cylinders. The company has automated the entire assembly as well as any required welding. Once the sections are complete and filled with as many pipes and components as possible they are floated down river on a barge and delivered to the LLSCF at Groton.

While certainly the most advanced and ambitious, this effort at sectional construction was not the first. During World War II the Manitowoc Shipbuilding Company of Wisconsin built fleet submarines in this manner as did the Germans with their Types XXI and XXVI. The Navy also investigated this construction technique in the early postwar years. Electric Boat received a contract from the Navy to explore the feasibility of this technique and submitted its preliminary report to BUSHIPS in October of 1953. The study recommended the creation of assembly sites on the Gulf coast which would take advantage of the excellent river network in the south and midwest for the transport of prefabricated hull sections. Although BUSHIPS investigated the industrial potential of the region, the Navy never implemented the plan. As the Guppy Program ended, the need to invest in new facilities and expand existing ones for the mass production of submarines seemed extravagant.

The improvement of construction facilities at Groton came primarily in response to the advent of a truly exceptional class of SSBNs. Designed primarily by Electric Boat Division, the *Ohio*-class Trident fleet ballistic missile submarines surpass their *Lafayette*-class (SSBN-616) predecessors with enhanced survivability, noise reduction, and hull strength characteristics. For its time the SSBN-616 class represented a considerable increase in size, exceeding the dimensions of its forbearers by 40 ft in length and 1500 tons in displacement. Displacing 18,750 tons, the 560-ft *Ohio*-class submarines dwarf the 8250 ton *Lafayette* and compare more closely in size to a World War II light cruiser. The Trident design employs a variation on the body-of-revolution hull employed in American submarines since *Albacore* and *Skipjack*. Propelled by the S8G reactor turbine system driving a single 26-ft propeller, these quiet giants boast the most advanced digital computers available and an extremely sensitive sonar system. The Trident design also anticipated future changes in weapons and vital systems by including factors that would make replacing components, sonars, and new weapons far easier. This eliminated the need for cutting the hull open at some future date or performing drastic alterations to the internal spaces. With the launching of the *West Virginia* (SSBN-736), the Navy's fleet of Trident submarines currently stands at twelve.

The Navy has also sponsored the construction of many submersibles for underwater salvage and rescue, deep diving experimentation, and oceanographic exploration. A few of these vessels have contributed significant service to the operational Navy and the submarine force in particular. The *Triest II*, with a collapse depth of nearly 40,000 feet, built by Mare Island Naval Shipyard to surpass the best features of its predecessor, the *Triest I*, helped in the search for the *Thresher* in 1963. The Navy also built two Deep Submergence Rescue Vehicles (DSRVs) in response to the *Thresher* disaster. These vessels were built for Submarine Development Group One by Lockheed Missile and Space Corporation in 1970 and 1971. The DSV *Alvin*, operated by Woods Hole Oceanographic Institution, is characteristic of Navy-owned submersibles, which have a wider variety of missions. *Alvin* has helped civilian science explore the geology, marine life, and physical oceanography of deep ocean trenches. In one of many projects for the Navy, *Alvin* played a key role in the successful search for a 20 megaton hydrogen bomb lost in 1966 near Palomares, Spain in the Mediterranean Sea.

The future of the American submarine is both exciting and uncertain. The relationship between the U.S. Navy and the submarine industry has improved dramatically since the conflicts of the 1970s between General Dynamics, Admiral Rickover, and the leadership of Electric Boat Division. The Trident- and *Los Angeles*-class submarines have served the Navy and the nation well. But in a world of limited budgets how will the submarine fare? Furthermore, have the Navy and private industry exhausted the potential of the *Albacore–Skipjack* formula for improving the performance of American submarines? Is it still wise and practical in terms of size and available technology to build a single all-purpose attack submarine capable of a wide variety of missions? Are more mission-controlled designs in order? Hopefully, unrestricted approaches to design, like concept formulation, will combine with a fresh naval strategy in the nineties to reveal new missions and design configurations for the future based on a diverse and stormy, but extraordinarily successful past.

Suggested Reading

Alden, J. D., *The Fleet Submarine in the U.S. Navy: A Design and Construction History*, Naval Institute Press, Annapolis, 1979.

Arentzen, E. S., and Mandel, P., "Naval Architectural Aspects of Submarine Design," *Trans.* SNAME, 1960.

Dallinger, F., and McGill, J. E., "Submarine High-Pressure Air System," *Trans.* SNAME, Vol. 70, 1962.

Duncan, F., *Rickover and the Nuclear Navy: The Discipline of Technology*, Naval Institute Press, Annapolis, 1989.

Friedman, N., *Submarine Design and Development*, Naval Institute Press, Annapolis, 1984.

Hewlett, R. G., and Duncan, F., *Nuclear Navy, 1946–1962*, University of Chicago Press, Chicago, 1974.

McKee, A. I., "Development of Submarines in the United States," *Historical Transactions of the Society of Naval Architects and Marine Engineers (1893–1943)*, pp. 344–355.

McKee, A. I., "Recent Submarine Design Practices and Problems," *Trans.* SNAME, Vol. 67, 1959.

Naval Engineering and American Seapower, R. King, Ed., Nautical and Aviation Publishing, Baltimore, 1989.

Polmar, N., *The American Submarine*, Nautical and Aviation Publishing, Baltimore, 1981.

Roddis, L. H., and Simpson, J. W., "The Nuclear Propulsion Plant of the USS *Nautilus* SSN-571," *Trans.* SNAME, Vol. 62, 1954.

Rössler, E., *Geschichte des deutschen Ubootbaus*, J. F. Lehmanns Verlag, Munich, 1975.

Schade, H. A., "German Wartime Technical Developments," *Trans.* SNAME, Vol. 54, 1946, pp. 83–111.

Weir, G. E., *Building American Submarines, 1914–1940*, Naval Historical Center, Washington, D.C., 1991.

Weir, G. E., *Forged in War: The Naval-Industrial Complex and American Submarine Construction 1940–1961*, Naval Historical Center, Washington, D.C., 1993.

Born in New York City in 1951, **Gary E. Weir** graduated from Manhattan College with a BA in History in 1973. After studying German history and Naval–Industrial Relations under Prof. John H. Morrow at the University of Tennessee, Knoxville, he received his doctorate in History in 1982. After a year on the faculty of the U.S. Naval Academy, Dr. Weir joined the newly established Contemporary History Branch of the Naval Historical Center in 1987. In this capacity he is preparing a history of naval–industrial relations in the construction of the American submarine fleet for the period 1914 through 1960. The first portion of this study, entitled *Building American Submarines, 1914–1940*, appeared in June 1991 as part of NHC's Contributions to Naval History series. In the Spring of 1992, Naval Institute Press published Dr. Weir's history of the construction of the German High Seas Fleet under the title *Building the Kaiser's Navy: The Imperial Naval Office and German Industry in the Tirpitz Era, 1890–1919*. His work has also appeared in *Military Affairs*, *Naval War College Review*, *International History Review*, *U.S. Naval Institute Proceedings*, *Naval History*, and the *Naval Engineers Journal*.

Surface Combatant Ships

William H. Garzke, Jr.

During World War II the aircraft carrier emerged as the capital ship, and the U.S. Navy surrounded it with a number of ships with differing functions to form a task force. Even the battleship, which had been the nucleus of the prewar fleet, now performed a vital defense role for the aircraft carrier. The destroyers, cruisers, and battleships formed a screen against air attacks by the piston-driven propeller planes of the 1940 era, surface ships,and submarines. They could also conduct shore bombardments and fight enemy surface forces in bad weather or darkness.

The Japanese suicide air attacks during World War II demonstrated the need for better fleet defense against such attacks. The Kamikaze's sudden emergence at high altitude, and high-speed dives made it imperative to develop more sophisticated radar for early detection and tracking of targets. This was underscored with the emergence of the atomic bomb, the jet plane, the guided missile, and the high-speed nuclear submarine. Jet aircraft, with fast rates of closure, posed a serious threat to the fleet since gun antiaircraft batteries of World War II could not destroy them before they released their weapons. Although the 5-in./54 and 6-in./47 guns were perfected by the U.S. Navy as good anti-aircraft weapons, it was obvious to the naval high command that missiles were the best able to destroy such threats at long range.

The *Iowa*-class battleships, as well as the light and heavy cruisers, were to form the aircraft carriers' inner defense, but these ships were also given independent missions such as shore bombardment when amphibious operations were mandated. Some World War II cruisers were armed with anti-aircraft missile systems in the postwar era because they had the platform and displacement to support them. Cruisers like the *Boston*, *Canberra*, and others, were converted to single-ended missile ships with guns and missiles. The cruisers, *Albany*, *Chicago*, and *Columbus* were double-ended missile ships. But, these conversions were only interim solutions to fleet defense.

One of the most severe post-war threats was the Soviet submarine force. These submarines were faster, dived deeper, had improved weapons, and ultimately became quieter than those deployed by the Axis forces in World War II. The need to equip surface combatants with weapons adequate to meet this menace as well as with anti-aircraft (AAW) systems, was a major design development. World War II destroyers were too small to accommodate the new weapons systems and the requisite electronics, so a whole new type of destroyer force had to be developed. An operational doctrine evolved calling for a greater interdependence between major ships (cruisers, aircraft carriers, etc.) and smaller ships (destroyers and frigates) in the task force. Some of these ships were specifically designed to be convoy escorts and to defend against submarines, while others provided anti-aircraft defense and the ability to conduct anti-ship missions. A limited gun capability was retained on the latter to counter small and fast missile-equipped patrol craft that could pose a threat to the task force.

A most significant development was the rapid advance in electronics, both radar and sonar, that enabled detection of threats at much longer ranges. All of the electronics and weapons had to be controlled by the ship's command and control system. This greatly increased the required number of spaces for related electronic equipment, particularly in the Combat Information Center (CIC). This was important because the newer threats all had increased speed, and so had to be identified at greater distance. This also required correct placement of this equipment and antennas to minimize mutual interference. The weight and topside location of all this gear led to problems with stability, leading to aluminum superstructures in cruisers, destroyers, and frigates. The disadvantages of aluminum were that it was expensive to fabricate and more vulnerable to fragmentation damage, gun blast, or fire than equivalent steel structures. Fire damage to the aluminum superstructure of the *Belknap* (after colliding with the carrier *Kennedy*) emphasized this vulnerability. The ultimate result was that the *Arleigh Burke*-class would feature the first steel superstructure in a modern destroyer, with many of the important electronic spaces located within the hull.

During the immediate postwar period, and during the Korean conflict, Navy planners could see that habitability standards needed to be upgraded. This was to attract and retain the necessary skilled manpower. In 1955, for example, crew berthing spaces were freed from main traffic routes. Pullman-style berths were provided with lockers beneath, instead of pipe and rack berths. In 1970, a new CNO policy proposed habitability standards that provided, among other things, greater privacy in dressing, recreation, and berthing areas. Mess rooms and sanitary spaces were improved and made more spacious and attractive. This was important in the post-Vietnam era with the all-volunteer military force. Some ideas were borrowed from the Royal and Canadian Navies where enlistments were also down and who were employing similar techniques to entice men to join. These new standards increased ship volume, increased damage-control complexity, increased weight, raised deck heights, and forced auxiliary machinery to be combined into spaces called auxiliary machinery rooms (AMRs). When the *Perry*-class frigates and *Virginia*-class nuclear cruisers were built, these standards were fully implemented and were major factors in these ships' cost and size.

Another important consideration in surface combatants was the choice of the propulsion plant: conventional steam, nuclear power, diesels, or gas turbines. High pressure, high temperature steam offered the advantages of reduced machinery volume and an improved fuel rate. These plants, however, required special crew training and were intolerant of below-par maintenance.

Aircraft-derivative gas turbines offered reductions in weight, volume, and manpower. They were also easily automated. The problems with gas turbines were their high fuel rates when operated at off-rated power, the large intakes and uptakes, and the need to use controllable pitch propellers for reversing. Unlike steam plants, which could be made in a continuous range of powers, gas turbines were available only in discrete powers.

During this period marine engineers faced constantly increasing demands for electrical energy arising largely from new air conditioning and electronic requirements. Thus, any volume saved in the main machinery spaces was taken up by auxiliary machinery rooms scattered throughout the ship. The reduced weight of the gas turbine plants, which were low in the ship, raised the overall center of gravity, necessitating larger beams for a given length to achieve the requisite stability.

The gas turbine was easy to silence since there was no direct fluid (noise) path to the sea. This was extremely important in ASW operations, allowing sonar detection of quiet, high-speed nuclear submarines.

Another factor that made gas turbines competitive with steam plants was the Navy's decision to employ Navy distillate fuel in all its ships to simplify the logistics of delivering fuel at sea. The Navy wanted gas turbines in its new warships and these burned only distillate fuel. This decision immediately removed the steam plant's economic advantage arising from its ability to burn less expensive fuels.

Ocean (Destroyer) Escorts or Frigates

***Dealey* (DE-1006 Class) Ocean Escorts (Frigates).** A 1947 Anti-Submarine Warfare Conference concluded that there was a need for small patrol craft to escort convoys. Producibility was important so that they could be built quickly by a number of different yards. Discussions before the Ships Characteristics Board led to a proposal for a ship of 1000 tons with a speed of 25 knots and an endurance of 3000 miles at 20 knots. The armament was to be basically an ASW type with fixed torpedo tubes, depth charge throwers, trainable hedgehogs, and twin 3-in./50 guns. The ships had to be highly maneuverable with a turning circle of 300–350 yards. Design work began in 1949. This would be the forerunner of the ocean or destroyer escorts.

In 1950 the Ships Characteristics Board authorized further studies with various combinations of 3- and 5-in. guns and a depth charge projector of the Mk 17 type. This armament forced a small and concentrated superstructure, which was in any event a means to achieve a low silhouette. Unfortunately, it led to poor visibility from the pilot house. A twin Squid projector, similar to a British model, was chosen for the lead ship and Weapon A in all follow ships of the FF-1006 (*Dealey*-class) ocean escorts. Two 3-in./50 twin mounts were positioned forward and aft with no 5-in. guns being mounted. The major characteristics of these ships are summarized in Table 1.

The real debate in the design centered on single versus twin shafts. A major issue was that the propulsion plant, rather than the hull, was the bottleneck in building escorts. Single screw plants would speed procurement, yet be more vulnerable to damage or breakdown. In the end the ships were built with a single screw, but a debate had been started on single versus twin shafts in ocean escorts.

Table 1 Characteristics of frigates

	Dealey FF 1006	Claud Jones FF 1033	Bronstein FF 1037
Displacement: (tons)			
Full Load	1,877	1,678	2,723
Design	-	-	2,695
Dimensions: (feet)			
LWL	308.0	301.0	350.0
Beam (DWL)	36.4	38.56	39.5
Depth	20.5	21.25	30.25
Draft (design)	11.75	12.0	13.5
Armament:			
AAW/ASU	3"/50 twins (2)	3"/50 single (2)	3"/50 twin & single (3)
ASW	Weapon A (1) Squid (Dealey)	Torpedo tubes, triples (2)	ASROC (8 missiles) Mk 32 TT (2) DASH Helos (2)
Propulsion:			
Type	Steam	Diesel	Steam
Shafts	1	1	1
Shaft horsepower (SHP)	19,800	9,600 (BHP)	20,000
Speed (knots)	26.8	22.0	26.0
Endurance @ 12 knots	6,000 mi @ 12 knots	7,000 mi @ 15 knots	4,000 mi
Radar & Sonar:			
Radar - Surface search	AN/SPS-10	AN/SPS-10	AN/SPS-10
- Air Search	-	-	AN/SPS-40
- Gunfire	-	-	Mk 56
- Missile	-	-	-
Sonar	AN/SQS-4	-	AN/SQS-26
ECM	-	-	-

	Garcia FF 1040	Brooke FFG 1	Knox FF 1052
Displacement: (tons)			
Full Load	3,403	3,426	3,939
Design	-	-	3,763
Dimensions: (feet)			
LWL	390.0	390.0	415.0
Beam (DWL)	43.75	43.75	45.8
Depth	30.0	30.0	28.3
Draft (design)	14.5	14.5	14.15
Armament:			
AAW/ASU (2) (1)	5"/38 single (1)	5"/38 single	5"/54 single
		Mk 22 Tartar (16 missiles)	Sea Sparrow (Space & Wgt)
ASW	Mk 32 TT (2) DASH (later, SH-2B Helo) ASROC (8 missiles)	Mk 32 TT (2) DASH (later, SH-2B Helo) ASROC (8 missiles)	ASROC (16 missiles) Mk 32 TT (2) DASH Helos (2) (later SH-2B Helo)
Propulsion:			
Type	Steam	Steam	Steam
Shafts	1	1	1
Shaft horsepower (SHP)	35,000	35,000	35,000
Speed (knots)	26.0	26.0	27.0
Endurance @ 20 knots	4,000 mi @ 20 knots	4,000 mi @ 20 knots	4,500 mi
Radar & Sonar:			
Radar - Surface search	AN/SPS-10	AN/SPS-10	AN/SPS-10
- Air Search	AN/SPS-40	AN/SPS-40	AN/SPS-40
- Gunfire	Mk 56	Mk 56	Mk 68
- Missile	-	SPG 51C	Mk 115 (Sea Sparrow)
Sonar	AN/SQS-26	AN/SQS-26	AN/SQS-26
ECM	AN/ULQ-6	AN/ULQ-6	AN/ULQ-6

	Perry FFG 7
Displacement: (tons)	
Full Load	4,002
Design	3,400
Dimensions: (feet)	
LWL	340.8
Beam (DWL)	45.17
Depth	30.0
Draft (design)	14.33
Armament:	
AAW/ASU	3" single (1) 20-mm Phalanx CIWS (1)
AAW	Mk 13, Mod 4 TARTAR
ASW	Mk 32 TT (2) LAMPS III SH-60B Helos (2)
Propulsion:	
Type	Gas Turbines
Shafts	1
Shaft horsepower (SHP)	40,000
Speed (knots)	28+
Endurance @ 20 knots	4,500 mi
Radar & Sonar:	
Radar - Surface search	AN/SPS-55
- Air Search	AN/SPS-49
- Gunfire	-
- Missile	SPG-60 (STIR)
-	
Sonar	AN/SQS-56
- ECM	AN/SLQ-32

Fig. 1 The *Garcia* (FF-1040) was to be the prototype for a class of 62 ocean escorts. Equipped with the AN/SQS-26 sonar, two single 5-in./38 enclosed mounts, an ASROC launcher with reloading capability, and a large DASH helicopter platform aft, these ships were to be the classic ASW escort equipped with high-pressure/high-temperature pressured fired boilers. They were not successful, as their construction costs exceeded budgets and the pressured fired boilers were a maintenance problem until the late 1970s, when the problem was finally resolved. A variant of these ships was the *Brooke* class, in which a TARTAR Mk 13 launcher replaced the after 5-in./38 gun mount. The cut out in the stern is for long torpedo tubes. (Courtesy U.S. Naval Institute.)

The first ships in the class had steel superstructures, but this was changed to aluminum for the later ships.

The propulsion plant featured 600 psi/850 °F boilers, which were then standard for U.S. destroyers. It was decided to use such steam conditions for simplicity in operation and mass production of units.

These ships were later modernized and equipped to operate the DASH unmanned helicopter and FANFARE—a torpedo countermeasure device. The aft 3-in./50 gun was removed to make room for the DASH.

***Claud Jones* (DE-1033 Class) Ocean Escorts (Frigates).** Despite the design intention, the DE-1006 class ocean escorts were not inexpensive, with the lead ship costing $12 million and follow ships $10.6 million. The Navy wanted a less expensive escort. Diesel propulsion was attractive if the speed requirement could be reduced to 21 knots. Special attention was to be paid to seakeeping, as this vessel would be sailing with convoys.

Four new ocean escorts arose from this program. A slower firing 3-in./50 single mount and two hedgehogs were selected in an effort to reduce the size of the ship. The SQS-4 sonar was retained along with depth charge racks. Design studies indicated that it was feasible to achieve a smaller displacement, but the twin 3-in./50 gun was no longer in production, so a single rapid firing 3-in. gun was used instead.

The class was powered by four 2400 bhp diesels geared to a single shaft. The problem with this plant was the selection of a proper cruise speed so that the diesels could operate efficiently. A speed of 12 knots was chosen, even though there were some who felt a higher cruise speed was necessary.

These ships had generous freeboard with good righting arms and were for their size very seaworthy vessels. Their characteristics are summarized in Table 1.

These ships ended their service in the U.S. Navy as electronic intelligence platforms, and in 1973–1974 all four were transferred to the Indonesian Navy.

***Bronstein* (DE-1037 Class) Ocean Escorts (Frigates).** Long-range detection of submarines at safe distances from a convoy or fleet became important in the post-Korean war era. The Navy believed that submarine speeds would reach 30 knots. Merchant ship speeds were also increasing. The obvious conclusion was that ocean escorts would need increased speed and enhanced ASW capability. This was, however, (in the late 1950s) a time of budget cutbacks. The Navy's approach was to use more sophisticated technology, but in a smaller number of ships.

The plan was to use an AN/SQS-26, a powerful new sonar capable of overcoming the thermal gradient problem in detecting deeper diving submarines. To match the sonar detection ability, an ASROC missile could be teamed with the AN/SQS-26, making ship speed less important.

Studies were begun using the *Dealey*-class frigate as a parent ship with a helicopter added to give extended range against submarines. The resultant design was for a ship with a speed of 24 knots and a more limited endurance of 4000 miles at 15 knots.

More speed and better endurance were needed for ocean escorts that would be equipped with AN/SQS-26 dome. Further studies in 1957 indicated that a 350-ft ship with a power plant similar to that in the DDG-2 would yield 24 knots on a single shaft. This design proved feasible and led to the two *Bronstein* (DEG 1037) class ocean escorts (later frigates) with the characteristics summarized in Table 1.

Although pressure-fired boilers were available for these ships, they were given the same type of plant that was being used in the DDG-2 class, but modified for a single shaft. The propulsion plant occupied a good portion of the hull, which was the smallest that could accommodate the SQS-26 dome. These ships were extremely maneuverable. The modest endurance was a fall-out from the hull specified, which led to limited fuel volume. Although these ships were austere, the Navy believed they could be a prototype for a mass producible convoy escort.

The air- and surface-search radars were mounted atop a special "mack" stack structure, which was a combination stack and mast that was so configured that electronics mounted on it would not be affected by flue gases. In the mid 1970s a Towed Array Surveillance System (TASS) was installed at the stern for additional submarine passive detection capability.

***Garcia* (DE-1040) and *Brooke* (FFG-1) Classes.** In 1959 the Ship Characteristics Board asked BuShips to investigate an ocean escort that would provide close escort to a relatively high speed convoy. Based on trends in merchant ship speed, it seemed reasonable to expect that 20 knots would be the average speed of merchant ships by 1969. Costs of a 25-knot escort were astounding. Alternatives, such as using a specialized cargo ship or submarine, were equally costly. Finally, the Navy decided to build a ship of 4000–5000 tons, but with the minimum armament of a destroyer—a "low mix" ship. Some of these ocean escorts would protect primarily against submarines, the rest against aircraft.

Admiral James, head of BuShips, pushed the concept of an ocean escort equipped with SQS-26 sonar, pressurized boilers, TARTAR or TERRIER missiles, and a single shaft. This was based on his belief that future destroyers would be armed primarily with missiles. In March 1960 he proposed an improved *Bronstein*-class escort with a TARTAR launcher forward, a DASH (Drone Anti-Submarine Helicopter) facility, twin 3-in./50 guns, and an ASROC launcher.

The pressure-fired boiler had been under development by Foster Wheeler and Babcock and Wilcox since World War II. First introduced in the French *Richelieu*-class battleships in 1937, the reduced size of the Indret–Sural boiler allowed a much more compact fireroom. Although the pressurized boiler design had progressed significantly by the time of the DE-1040 design, it had to burn a more refined fuel—diesel oil or JP-5. Engineers estimated that the fireroom volume of the *Bronstein* could be cut in half using pressurized boilers of the type under development.

Until this time, the Navy had imposed an arbitrary 3000-ton upper limit on the displacement of ocean escorts. By the time Admiral James's specifications were met, however, this new class of escorts weighed almost 3500 tons. Their characteristics are summarized in Table 1.

These ships were built of high tensile steel with aluminum superstructures and a mack stack.

The ships were equipped with the AN/SQS-26 sonar in a bow dome and, to provide better air search capabilities, a 3-D radar was used in some of the ships.

These ships had severe problems with their pressure-fired boilers until the latter part of the 1970s when the problems were corrected.

An unusual variant was the *Glover* (AGDE-1, later FF-1099). She was equipped with a water-jet propulsion system. With reduced propeller noise and a special plastic SQS-26 dome, she was intended to detect and track submarines at great distances. She had the same weapons and electronics as the *Garcia*.

***Knox* (DE-1052 Class) Frigates.** The U.S. Navy's quest for the ultimate ocean escort continued with design studies in the fiscal year program of 1964. This was despite misgivings about high costs on the part of Secretary of Defense McNamara. These proposed ships would need to have improved habitability, reduced topside clutter, reduced manning, and space for a variable depth sonar (VDS) capability, yet retain features like the SQS-26 sonar, DASH, and a 5-in. gun. These requirements were in conflict with McNamara's cost-cutting objectives. Nevertheless, the Navy convinced him that a class of ships with such features was needed. The new escort would have ASROC with reloading capability. The endurance was to be 4500 miles at 20 knots, dictating a larger ship than the DE-1040, not withstanding reductions in other features.

The new design included pressure-fired boilers along with a large cylindrical mack stack that would become a distinctive feature of these ships. This stack supported the increased complement of radars and

Fig. 2 The *Knox* (FF-1052) class ocean escorts were the largest and the last of their type to be equipped with steam propulsion, and the biggest postwar construction program undertaken by the U.S. Navy. Designed to be anti-submarine vessels, they carried a 5-in./54 gun, an ASROC launcher, and a helicopter. Fitted with the AN/SQS 26 sonar dome, they were capable of detecting Soviet submarines at long distances. This view of the *Knox* was taken after spray rails had been installed and the bow bulwark was extended upwards to improve seakeeping performance. (W. H. Garzke Collection.)

other sophisticated electronic countermeasure equipment. The electrical plant, too, contained several advances that enabled a consolidation of the machinery plant to reduce watch standing.

In 1963 the detailed design of the DE-1052 class was awarded to Gibbs & Cox, Inc. Mr. Gibbs was aware of the difficulties of the pressure-fired boilers in the DE-1040 and felt it unwise to use them in a class where large numbers were to be built. This led him to recommend that these ships be redesigned to use the standard 1200 psi/950 °F steam plant. The change called for a 20-ft increase in ship length, but the hull form was made finer by decreasing the midship coefficient from 0.837 to 0.810.

These became the *Knox*-class frigates. With 46 units completed, these became at the time the largest postwar ocean escort class. Although they were later surpassed by the FFG-7 class in numbers, they were considered to be mass producible, and a number of yards built them until it was felt more economical to confine them to only one shipyard. The last ships, starting from DE-1078, were built by Avondale Shipyards. The yard perfected a technique for mass production of prefabricated sections so that the hulls could be constructed upside down and then rotated for launching. The characteristics of the *Knox*-class are summarized in Table 1.

These ships have been equipped with a number of different radars and sensors, which are summarized in Table 1.

***Oliver Hazard Perry* (FFG-7 Class) Destroyer Escorts.** The FFG-7 concept was born in 1970 when the CNO, Admiral Zumwalt, asked the Naval Ships System Command to investigate the feasibility of a single-purpose AAW or ASW escort ship costing between $30 and $40 million. Within a few months NAVSHIPS proposed a destroyer escort weighing 3500–4000 tons, but costing between $40 and $50 million. These results were presented to OPNAV early in 1971, and it was decided to proceed with the design.

The initial concept of single-purpose ASW or AAW ships with a common hull and machinery was discarded in favor of a single dual-purpose design. Because of a misunderstanding, OPNAV imposed an arbitrary displacement limit of 3400 tons, an unreasonably low figure. A follow-ship cost limit of $45 million in fiscal year 1973 dollars was also imposed. Characteristics were approved in May 1971 for a lead ship to be purchased in 1973. Follow ships were not to be procured until 1975 in order to maximize learning from the lead ship. These became the FFG-7 class destroyer escorts.

Much of the design effort was devoted to considering an array of new and different general arrangements, most of them aimed at minimizing first cost. The propulsion plant was located well aft permitting the central location of accommodations, minimizing piping and shafting runs, and putting maximum distance between the propulsion plant and the SQS-56 sonar. The resulting high-shaft angle was a concern of the hydrodynamicists, but it turned out not to be a problem. Similar careful thought was given to locating the various weapon systems. A long slender hull was selected to maximize the speed potential since

Fig. 3 This bow view of the *Perry* (FFG-7) was taken during her acceptance trials. One of the largest peacetime construction programs in the U.S. Navy, 55 ships were built to this design, which was the first ocean escort equipped with gas turbine propulsion. This program included four ships for the Australian Navy; but later, Spain, Australia, and the Republic of China (Taiwan) built ships to the same basic design with modifications to carry specific weapons and sensors required by their navies. Armed with the Mk 13 missile launcher, a 3-in. gun, and two helicopters, these ships were intended for convoy escorts but have served a dual role as fleet escorts as well. (W. H. Garzke Collection.)

these ships were being equipped with a single shaft powered by two LM-2500 gas turbines, which had fixed power outputs.

The use of gas turbines in a relatively small helicopter-carrying ship made it necessary to consider the flow of exhaust gases under varying conditions to ensure minimal interference. The number and size of generators was studied thoroughly. It was decided to use diesel rather than gas turbine generators for reasons of economy, although the gas turbine generator would have been a better choice for silencing.

Another new problem was that of pollution abatement. It was no longer acceptable to pump oily ballast water into the oceans except in emergencies. Heretofore, the stability requirements at some conditions of loading had required ballasting some fuel tanks. The provision of clean ballast tanks equivalent to full fuel capacity would require significant increases in displacement and cost. After careful analysis of operating practices, the engineers concluded that clean ballast capacity need come to only about one third that of full fuel capacity. The coincidence of severe weather and low fuel was deemed sufficiently rare that it could qualify as an emergency.

Roll stabilization was given attention as a means of increasing the safety of helicopter landings and of improving crew effectiveness. The extent of the problem became apparent only after the first ships in the series were in service. The design was then modified, enlarging the stern, fitting a special hauldown device, and adding fin stabilizers.

There was difficulty, too, with the question of crew complement. Estimates of required numbers varied from 185 to 230 (officers plus enlisted men). OPNAV dictated the lower figure, but this eventually proved to be woefully inadequate.

During the Functional Baseline phase of the design, it was decided that the hull material would be medium steel with HY-80 crack arresters, primarily for cost reasons. An aluminum superstructure was selected.

Many other cost-saving measures, such as the elimination of one anchor, were adopted. At the end of this phase, a formal review was conducted including reviews by two contractors, both fleets, INSURV, and the usual in-house review by NAVSEC. This review took several months and the next design phase did not start until April 1972. During one of the reviews, Gibbs and Cox recommended a two-foot increase in beam, but this was rejected as being too costly. As a result every FFG required fixed ballast to meet the Navy's stability requirements.

Gibbs and Cox was named the principal design agent for the FFG-7 design and assisted the Navy in the construction of 56 ships in three shipyards. As the program progressed, changes were made in later ships to accommodate new and modified weapon systems.

The principal characteristics for the latest FFG-7s are shown in Table 1.

The 3400-ton displacement was a design objective never achieved when the ships were delivered. As the class evolved, weights inevitably increased. With the addition of LAMPS III helicopters at one point, for example, it was necessary to increase the displacement to 4100 tons, which was the limit of the last ship, FFG-61.

These ships were planned for mass production and modular construction. This proved beneficial when the *Roberts* (FFG-58) was damaged by a mine in the

Persian Gulf in April 1988. The blast destroyed the propulsion plant, but through innovative repair and the modular construction techniques, the ship was repaired rather quickly.

The *Perry*-class frigates are one of the most successful modern combatant designs. They have been adopted by three foreign navies, so there are now 65 of them serving in four navies.

Destroyers

The destroyers of the postwar fleet evolved into two principal types: one to screen the task forces against air threats, the other to provide protection against submarines. The latter, which evolved into the destroyer escort or frigate, served with the fleet or convoys. A destroyer development conference was held at BuShips in October 1946 to chart the paths that these two types would take. Destroyers had to be capable of speed and yet have endurance that would enable them to keep up with the fleet. The slower destroyer escorts or frigates, although primarily for antisubmarine warfare, would also be equipped to handle a limited antiaircraft role.

During the war in the Pacific, the U.S. Navy recognized the need for a faster, more seaworthy, and longer-range destroyer. With aircraft carriers and battleships operating at speeds near 30 knots, the pre-war types were hard pressed to continue their screening duties. Studies were begun, but it was not until after the war that a new class of destroyers was delivered.

***Mitscher* (DL-2 Class).** The first of the new postwar destroyers was the *Mitscher* (DL-2 class), which emphasized speed and endurance. These factors dictated increased displacement (to 4700 tons), which made them almost double the size of some of the pre-war destroyers. Four were built and were classed as destroyer leaders during their construction; they could make 36.4 knots and had the endurance of 4500 nautical miles at 20 knots. These characteristics, however were far less than the design objectives of 39 knots and 6500 miles at 20 knots. They were considered overly large at the time, and there was much opposition within the Navy to their construction. They were the first destroyers to carry the 5-in./54 rapid fire single-mount gun, which took up as much space as the twin-mount 5-in./38, but fired a heavier shell, had a faster rate of fire, and a longer range. To provide antisubmarine support to the fleet, two Weapon A's, (Able) were mounted. These were single barrel rocket projectors, which could fire a lead-weighted depth charge ahead of the ship. These were later replaced by homing torpedoes and the antisubmarine rocket torpedo (ASROC), which had greater range. Although 3-in./70 guns were specified, they were still being developed, so 3-in./50 twin-mounted guns were substituted. These ships, though lightly armed for a U.S. destroyer, had nearly 360-degree coverage for all weapons systems.

The main characteristics of the DL-2 Class are summarized in Table 2.

With an eye to saving weight, the steam machinery chosen for these ships was of advanced design: 1200 psi/950 °F. These plants were apparently ahead of their time, and caused many difficulties in trials and in operations.

The postwar electronic revolution significantly affected destroyer design. The importance of early detection of enemy aircraft had been demonstrated off Okinawa in 1945. As a result, air-search and height-finding radars were to become important features in postwar destroyers. These ships were provided with the AN/SPS-8 height-finding radar, which added as much weight as a 3-in./50 twin gun mount and severely influenced stability. This is one reason why the armament was so limited in comparison to World War II destroyers.

One important objective in the DL-2 design was good seakeeping. In heavy seas, prewar destroyers could not keep up with the carriers and fast battleships. In the postwar period, speed and seakeeping were important attributes in countering submarines, which had reached 19 knots submerged in the German Type XXI. Undoubtedly, this technology had fallen into Soviet hands. To improve seakeeping, our Navy considered flush-deck destroyers with increased freeboard or forecastle-deck destroyers. The increased freeboard, flush-deck design was chosen for the DL-2 hull.

All four ships were scrapped by 1970.

***Forrest Sherman* (DD-931 Class).** The DL-2 class was not altogether satisfactory, so the Navy sought a smaller destroyer that could be produced in greater numbers. To accomplish this objective, a lower speed had to be accepted; the endurance of the DL-2, however, was maintained, as the staying power of the destroyer with the task force was critical. In 1951, the Ship Characteristics Board (SCB) issued preliminary characteristics for a destroyer with a standard displacement of 3500 tons (maximum), which had to support an armament of three 5-in./54 single-mounted guns, four twin 3-in./50 twin mounts, and four fixed torpedo tubes. Fixed hedgehogs and depth charge racks were to be provided. The elimination of Weapon A in favor of a heavier gun armament was a tacit admission

Table 2 Characteristics of destroyers

	Mitscher DL 2	Forrest Sherman DD 931	Charles Adams DDG 2
Displacement: (tons)			
Full Load	4,758	4,139	4,588
Design	4,485	3,685	4,440
Dimensions: (feet)			
LWL	476.0	407.0	420.0
Beam (DWL)	49.63	44.25	39.5
Depth	27.88	25.25	30.25
Draft (design)	14.0	14.5	13.5
Armament:			
AAW/ASU	5"/54 single Mk 42 (2) 3"/50 twins (2)	5"/54 single Mk 42 (3)	3"/50 twin & single
ASW	Weapon A (2)	Fixed TT (4) Hedgehogs (2) Depth Charge Racks (2)	ASROC (8 missiles) Mk 32 TT (2)
Propulsion:			
Type	Steam	Steam	Steam
Shafts	2	2	2
Shaft horsepower (SHP)	80,000	70,000	70,000
Speed (knots)	34.8	33.9	33.0
Endurance @ 20 knots	4,500 mi @ 20 knots	4,500 mi @ 20 knots	4,500 mi
Radar & Sonar:			
Radar - Surface Search	AN/SPS-10	AN/SPS-10	AN/SPS-10
- Air Search (2D)	-	-	AN/SPS-40
- (3D)	-	-	
- Height Finding	AN/SPS-8		
- Gunfire	Mk 67	Mk 67	Mk 68
- Missile	-	-	AN/SPG-51 AN/SPG-53
Sonar	-	AN/SQS-23	AN/SQS-23
ECM	-	-	AN/SLQ-6

	Spruance DD 963	Arleigh Burke DDG 51
Displacement: (tons)		
Full Load		8,232
Design		
Dimensions: (feet)		
LWL	529.0	466.0
Beam (DWL)	55.0	59.0
Depth	42.0	41.83
Draft (design)	20.5	19.68
Armament:		
AAW/ASU	5"/54 single Mk 45 (2) Sea Sparrow Point Defense System	5"/54 single Mk 45 (1) Mk 41 VLS 32 Cell (1) 64 Cell (1) HARPOON SSM (8) 20-mm CIWS, Phalanx (2)
ASW	ASROC (24 missiles) Mk 32 TT (2) Helicopters SH-2F LAMPS I (2)	Mk 32 triple TT (2) AN/SQR-19 Towed Array
Propulsion:		
Type	Gas Turbines	Gas Turbines
Shafts	2	2
Shaft horsepower (SHP)	80,000	100,000
Speed (knots)	32.5	30+
Endurance @ 20 knots	6,000 mi	-
Radar & Sonar:		
Radar - Surface Search	AN/SPS-10	AN/SPS-67
- Air Search (2D)	AN/SPS-40 AN/SPS-49 (DD 997)	AN/SPY-1D
- (3D)		
- Height Finding	-	
- Gunfire	Mk 86	Mk 86
- Missile	Mk 91	Mk 99 (3)
Sonar	AN/SQS-26 (modified to AN/SQS-53)	AN/SQS-53C
ECM	-	AN/SLQ-32

that these ships were actually anti-aircraft destroyers. Within a month or two the SCB had decided in favor of what was to become the *Forrest Sherman* (DD-931 class) destroyers with the characteristics listed in Table 2.

These were the last of the all-gun destroyers and eighteen were built. Modifications were made as the ships were constructed, but the armament was basically the same. As the result of NATO exercises in the North Atlantic, the forward freeboard was increased on ships of the class not yet complete. Therefore, in DD 936 to 944 freeboard was increased by simply raising the sheer line three feet at the bow. The last seven destroyers in the DD-931 class (DD-945 to 951) were built to a slightly modified design, which involved relocating gun directors to improve fire-control. All the ships had aluminum superstructures and made extensive use of high tensile steel in their hulls.

Fig. 4 *Forrest Sherman*-class destroyers were designed to be close escorts to the fast carriers and battleships. Armed with three dual-purpose 5-in./54 gun mounts, they also carried an assortment of anti-submarine weapons, including fixed torpedo tubes for homing torpedoes, depth charges, and Weapon A (a trainable Hedgehog). During the course of their careers, some of these were modified to carry TARTAR missiles and ASROC. DD-945 and later members of the class were completed with more freeboard at the bow. (Courtesy U.S. Naval Institute.)

These destroyers were to provide radar picket duty as well as search for submarines so they were equipped with the sensors tabulated in Table 2.

The propulsion plant was an improved 1200 psi/950F type, which enabled them to avoid the problems of the DL-2, class and they were reliable ships. Because the speed requirement was reduced to 33 knots, the shaft horsepower was reduced from 80,000 to 70,000. A speed of 33.93 knots was achieved on trials with 71,500 horsepower at a displacement of 3650 tons.

***Charles Adams* (DDG-2 Class) Guided Missile Destroyers.** The design of the U.S. Navy's first true guided-missile destroyer began as the DD-931 ships were nearing completion. The *Gyatt* of the *Gearing* class had been converted to carry guided-missiles on a trial basis. She was equipped with a TERRIER twin-missile launcher. Only fourteen missiles could be fitted in her magazine, and missile guidance was provided by a modified Mk 25 fire-control radar system. It became obvious that the TERRIER missile was not suitable for ships the size of the *Forrest Sherman*-class and smaller, but that the TARTAR missile could become so.

The Navy decided to develop a guided-missile destroyer on a *Forrest Sherman*-type hull with the latest long-range antisubmarine weapons. To achieve the design objectives, the superstructure remained aluminum and the ships grew in length and beam, offsetting the increased weight and higher center of gravity brought on by the requisite electronic gear.

The first flight of the DDG-2 class carried a twin TARTAR launcher (Mk 11), but this was unsuccessful

Fig. 5 The *Charles Adams* class were the first American destroyers to be designed from the keel up to carry missiles to sea. Although having the lines of the *Forrest Sherman* class, these ships were longer, but had less beam and more depth, the latter for better seakeeping capabilities. Armed with two 5-in./54 gun mounts, an ASROC launcher, and a TARTAR missile launcher, these ships were the escorts for carrier task forces until the 1980s. Several ships saw service in the German and Australian Navies. (Courtesy U.S. Naval Institute.)

and was later replaced by the single-arm Mk 13. The magazine, however, remained at a capacity of 42 missiles.

The principal characteristics of the *Charles Adams* class are shown in Table 2.

One of the essential requirements of the new guided-missile destroyer was to detect targets at long range so they could be tracked and locked on the missile guidance system for the TARTAR launcher. Both the TARTAR and TERRIER missiles are beam riders to their air target, with radar tracking an important function for a successful intercept.

Fig. 6 The *Spruance* class was the first destroyer type conceived outside the U.S. Navy establishment. These ships were designed for modular-type construction and oversized to allow for updates in the combatant system and sensors. The propulsion plant featured gas turbines—the first American combatant to be so equipped. Although first equipped with two 5-in./54 lightweight gun mounts, an ASROC launcher, Sea Sparrow, and a dual helicopter facility three-quarters of the way aft, updates to these ships have included replacement of the ASROC with the vertical launch missile system. (Courtesy U.S. Naval Institute.)

Volumetric needs presented a severe problem from the beginning. In particular, improved habitability was proving difficult to achieve. Increasing space demands for electronics caused further difficulties, and because these ships had austere supporting systems and limited access for maintenance, modernization was difficult.

The hull envelope was constructed of HY 80, high strength, and mild steels. The HY 80 plating was used in the sheer strakes and outboard strake on the main deck as ballistic protection and crack arresters. This type of construction was to be used in the DLG-9 class and follow-on destroyer leaders, frigates, and cruisers. Riveted seams in way of gunwale and sheer strakes, also provided insurance against the brittle fractures that had plagued the all-welded World War II merchant ships. This combination permitted several of these ships to survive severe structural damage.

Six DDG-2 class ships were constructed in U.S. shipyards for foreign export. Three were sold to Australia and three to Germany. These ships all had American weapons and electronics at the time of their transfer.

***Spruance* (DD-963 Class).** By the late 1960s most of the destroyers built during, or immediately after, the war were nearing obsolescence. A fleet repair and maintenance program extended their service lives, but new concepts in habitability, electronics, and weapons demanded more suitable platforms. One of the severest constraints was that of high costs, particularly for the new electronic and weapons systems. Until 1967 the Navy had set its own procurement policies, but the then Secretary of Defense, in the name of economy, stepped in and specified mass procurement of the replacement fleet from a single shipyard. Moreover, whichever yard was chosen would be responsible for design, construction, and eventual maintenance of the ships. There were to be two variations in the design, some destroyers being fitted for antisubmarine warfare, the others for anti-air. Both types, however, were to share a common hull and modular features. Basic specifications called for a 30-knot speed in sea state four, and an endurance of 6000 miles at 20 knots.

A request for proposal for concept formulation was sent out, and six shipyards responded. The contract definition was innovative. The contractors were asked to bid on 30, 40, or 50 DXs (ASW version) and 25 or 35 DXGs (AAW version). A contract for 30 ships was awarded to Litton Industries in June 1970. These were to become the *Spruance*-class destroyers. Particulars for the winning design are shown in Table 2.

Of particular note is that these were the first gas turbine powered destroyers in the U.S. Navy. Four turbines are geared to two shafts with controllable/reversible pitch propellers. The electrical generators are also gas turbine powered.

The relatively narrow beam dictated an aluminum superstructure and a compensating fuel system that allows adding water ballast as fuel is consumed.

These ships carry a number of electronic arrays suited to the weapon systems and missions for which they have been designed. As first completed, these ships were outfitted with the sensors summarized in Table 2.

All thirty of the *Spruance*-class ordered were built in commonality with the lead ship (DD-963). The ships were designed for mass production in Ingalls' new shipyard. They were built in prefabricated units delivered to the building site for final assembly.

During the late 1980s a number of the *Spruance*-class ships were modernized. Some of them have had the new vertical launch Mk 41 guided missile launcher added in place of the ASROC launcher. This change gave them a Tomahawk strike capability and also permitted their future upgrade to a vertical-launched antisubmarine weapon. The sonars were upgraded from the SQS-26 to the SQS-53 system to increase effectiveness in antisubmarine warfare. Included in the modernization was a helicopter hauldown and traverse system to handle two LAMPS III helicopters.

***Kidd* (DD-993 Class).** In 1974, the Royal Iranian Navy contracted to buy six (later changed to four) DD-963 destroyers modified to a DDG configuration with Mk 26 guided missile launch systems and 5-in./54 guns at both ends. When the Shah was overthrown, the new government canceled the contracts. The U.S. Navy purchased them in July 1979; and those four destroyers were the only true DDGs in the 35 ships of the *Spruance* type constructed by Ingalls. Since the hull had been designed with space and weight for future weapons systems, the only major changes were in complement, sensors, and weapons. These differences in features in the *Kidd*-class resulted in the following differences in particulars from those of the *Spruance* shown in Table 2:

Displacement	9590 tons (full load)
Armament	Missiles 2 GMLS Mk 26 (Mod 1 + Mod 0) (ASROC and surface to air missiles); 2 quad Harpoon launchers
Radars	2 MK74 missile fire control system; 1 AN/SPS - 48C air search radar (3-D)
Sonar	AN/SQS 53A

***Ticonderoga* (CG-47 Class) Cruisers.** Originally designated DDG-47, the first of the *Ticonderoga*-class cruisers was changed to CG-47 during construction. These were to be the first warships designed to carry the new AEGIS radar system. This multi-function or phased array system traces its beginning to the early 1960s when it was called the Advanced Surface Missile System (ASMS). The AEGIS system scans the skies by directing a radar beam electronically rather than by physically rotating the antenna. This allows higher scan rates and permits detecting and tracking hundreds of targets simultaneously. Following land-based testing, AEGIS went to sea for further tests on the *Norton Sound* (AVM-1) in 1975.

In mid 1974, the Ingalls Shipbuilding Division of Litton Industries proposed the installation of the AEGIS radar system on a *Spruance* (DD-963) class hull as an alternative to a proposed nuclear-powered AEGIS strike cruiser (CSGN) that was in the process of design. This proposal was favorably received by the Navy. An attempt was made to develop a system suitable for either kind of ship, but some compromise was needed. As a result, the AEGIS combat system design for the CSGN became the "super set," while that for the DDG-47 was a downsizing of the "super set."

A thorough feasibility study and concept design for the AEGIS destroyer was accomplished during 1975. The contract design began in 1976. It included certain objectives dictated by the Secretary of the Navy. Among these was a requirement that the new destroyer had to have maximum commonality with *Spruance*-class (DD-963) in the hull, mechanical, and electrical areas. This allowed only those changes to the DD-963 as-built configuration that were directly required to:

- Install and support the AEGIS weapon system.
- Accommodate the AEGIS combat system.
- Improve the accessibility, supportability, and maintainability of the DD-963 class subsystems and equipment.

In the end, the naval architects were forced to undertake massive changes in the original *Spruance* design as dictated by the exigencies of the AEGIS hardware. For example, they had to raise the pilot house by two decks, modify the helicopter hangar and landing platform, and then add an AEGIS deckhouse atop the hangar. The stacks had to be raised to clear the higher deckhouse, and new masts had to be designed to carry additional radars and other sensors. They settled on a forward/starboard and aft/port ar-

Fig. 7 The *Ticonderoga* (CG-47) was the largest possible variant of the DD-963 class. With a symmetrical missile and gun battery plus a hangar for two LAMPs III helicopters and the AN/SQS-53 sonar dome, this ship was some 2000 tons heavier than a standard DD-963 hull. A later variant of these ships, the CG-52 class is equipped with vertical launch missile systems instead of the Mk 26 Guided Missile Launching System, which proved to have some mechanical problems. The important feature of CG-47 and later ships is that they are equipped with the AEGIS phased radar system and are the first combatants to be so equipped. In addition, they carry the AN/SPS-49 air search radar and the Mk 86 gunfire control system. (Courtesy U.S. Naval Institute.)

rangement of the AEGIS arrays. This was necessary to keep exhaust gases and their heat away from the sensitive radar antennas. It also resulted in an athwartships orientation of the forward illuminators to minimize topside interference and increase coverage. Internally, the CIC area was increased in size to accommodate the displays for an embarked unit commander and additional weapons and electronic systems. Communication spaces had to be increased and additional living spaces had to be found for officer, CPOs and crew. In addition, significant survivability features were added to the design, including the addition of fragmentation protection and fire insulation to the superstructure.

Many other changes were found necessary. These included larger generating capacity, additional air conditioning, and several changes in the propulsion plant and uptakes.

All approved changes to the basic *Spruance* design were reviewed for possible inclusion in the then DDG-47 design. In addition, an ongoing analysis of the *Spruance* operating deficiencies was performed, and design solutions were incorporated where applicable.

At the end of contract design, the *Ticonderoga*, DDG-47 (later, CG-47) had the characteristics summarized in Table 3.

In November 1976 an industry briefing was held to explain the DDG-47 acquisition program and to solicit competitive bids. In the end Ingalls won the contract.

As they developed the design, the naval architects discovered rather serious weight problems. This led to a strict program of weight control and ultimately the follow ships were modified. These changes included redesigning the masts, limited re-straking of the upper hull with HY-80 steel, redesign of the uptakes to use an all welded CRES honeycomb, and the first use of light-weight, low-smoke jacket electrical cable on a Navy ship.

In the early 1980s the Navy decided to use a vertical launching guide system, VLS Mk 41, for the missiles. This was to replace the mechanical Mk 26 system. This change began with the CG-52. The new missile launchers would increase the ship missile capacity from 88 to 122 and also permit the use of the Tomahawk cruise missile. The change in launchers increased the weight and raised the center of gravity. Although those changes were small, their impact was large, requiring major changes in internal arrangements and kind of steel in the hull. After this redesign, the CG-52 showed a full load displacement of 9410 tons and a KG of 23.21 ft.

Planned improvements in the sonar and AEGIS systems in the class necessitated further weight control measures. As a result, the CG-60–73 are now delivered with 51 tons of solid ballast to account for in-service weight growth.

CG-60–73 have the following differences from the CG-47 lead ship:

Displacement	full load 9583 tons vice 8910 tons
Draft	22.2 vice 21.5 ft
Armament	2 VLS, Mk 41 launchers versus the Mk 26, Mod 5 GMLS
Sensors	1 AEGIS Radar (AN/SPY-1B)
Sonar	1 Sonar set AN/SQS-53C

In September 1981, the Navy directed that a second source be developed for FY 82 AEGIS cruisers. In May 1982, Bath Iron Works was selected as this second source and awarded a contract to build CG-51. As a result of this contract award, Ingalls has built 19 of the 27 ships authorized for the class. Eighteen ships were delivered as of February 1991.

The principal sensors of the CG-47 class are as follows:

Surface search	AN/SPS-64
Air search	AN/SPS-49 (3-dimensional); AN/SPY-1A (B in CG-55–73)
ECM	AN/SLQ 32(V)3
Missile Guidance	Mk 99
Gun director	Mk 86
Sonar	SQS-53A (B in CG-55–64; C in CG-65–73); AN/SQR-19 towed sonar in CG-47–53

***Arleigh Burke* (DDG-51 Class).** The DDG-51 *Arleigh Burke*-class guided missile destroyers are intended to provide general defensive service in the multi-threat environment of the 1990s. They will replace the aging *Adams* (DDG-2) and *Coontz* (DDG-40, ex-DLG-9) classes. The DDG-51 class are the first destroyers to incorporate the AEGIS phased-array radar system, which, in conjunction with the SM-2 standard missile, can meet the threat of high performance aircraft, air-to-surface, and surface-to-surface missiles.

Interest in a general-purpose destroyer designed from the keel up for AEGIS was rekindled in the early 1980s with the beginning of design for a concept referred to as DDGX. The requirements, issued in November 1980, were intended to improve on the *Ticonderoga* (CG-47), particularly in survivability.

The DDGX preliminary design baseline issued in March 1982 differed from CG-47 in a number of ways in addition to the propulsion plant and improved survivability characteristics. These included a single deckhouse incorporating all four AEGIS SPY-1 array faces, rather than split fore and aft. This arrangement

Fig. 8 The *Arleigh Burke* was one of the largest destroyer building programs undertaken by the U.S. Navy since World War II. These ships feature reduced radar reflectivity, vertical launch systems forward and aft, the AN/SQS-53 sonar system, and a 5-in./54 lightweight gun forward. The forward deckhouse contains the AEGIS phased radar system. Equipped with gas turbine propulsion, these ships are the fastest destroyers in service. Later units of the class have been modified to service and land helicopters, but are not equipped with hangars. Starting with DDG-78 a twin hangar facility was added aft with the elimination of the HARPOON and TACTAS systems. (W. H. Garzke Collection.)

permitted the use of a single transmitter. It also included a vertical launch system (VLS) for firing long-range standard anti-aircraft missiles, TOMAHAWK cruise missiles, or vertically launched ASROC ASW missiles.

The DDGX design evolved into the current configuration of DDG-51 with the beginning of a new design effort in March 1982. The principal change from the DDGX configuration was a mechanical propulsion system similar to CG-47 (four LM 2500 gas turbines, operating up to 100,000 shp, and driving two controllable pitch propellers via double reduction gears). This change was made to reduce technical risk, development time and cost.

The characteristics of the *Arleigh Burke* (DDG-51) are summarized in Table 2.

The ship incorporates a number of the survivability improvements originally introduced in the DDGX. These include a steel superstructure, extensive fragmentation protection, nuclear blast resistance, vital combat spaces located in the hull rather than in the deckhouse, collective protection system (against chemical, biological, and radiological contamination), reduced radar cross section, infrared signature suppression, and separation or redundancy of vital systems.

The Navy planned to procure the DDG-51 class in flights, with each new flight introducing major combat system upgrades. Flight I includes DDG-51 through -71. DDG-72, the last FY 92 ship, will be the first Flight II ship. While identical in size to Flight I, it includes a larger limiting displacement for potential upgrades. Improvements in exterior communications, information processing, and electronic countermeasures were also introduced.

With the collapse of the Soviet Union in 1991, the U.S. Navy initiated a number of DDG variant studies. The selected configuration eliminated the AN/SQR-19 towed array and the HARPOON launchers, which would be installed later. The addition of a twin hangar structure around the aft vertical launch missile launcher allowed the use of antisubmarine or surface warfare helicopters. This configuration started with DDG-79.

***Norfolk* (DL-1) Destroyer Leader.** After World War II the U.S. Navy perceived a need to counter a threat from Soviet submarines capable of 25-knot speeds and diving depths of 1000 ft. This called for a "killer" ship that could operate in severe seas, possess sufficient endurance, have a speed in excess of 25 knots, and be equipped with deep-diving homing torpedoes. To meet these requirements, such a ship should have an elaborate sonar that would operate on low- and medium-frequencies. It should also have a sonar equivalent of the Identification Friend or Foe (IFF) equipment used with air- and surface-search radars.

In December 1946 the design of this super submarine-killer had already begun in earnest at BuShips. The design concentrated on a ship with an endurance of 6000 miles at 20 knots and a standard displacement of some 5500 tons. The proposed speed of 33 knots was that of the *Iowa*-class battleships and *Midway*-class carriers, which were the nucleus of the postwar American fleet. The hull form of these killer ships was to be based on the successful *Atlanta* (CLAA-51 class) anti-aircraft cruisers. It could easily achieve a 30-knot speed in rough weather, providing it with a five-knot advantage over the high-speed submarine. Such a large hull would provide the necessary topside area for the weapon systems desired, particularly the many deck-mounted torpedo tubes.

In February 1947, the designation CLK for cruiser, light, killer was given to this concept and certain changes were made in the proposed weapon array.

The need to operate in rough weather at high speed made it necessary to increase freeboard so CLK-1 had more bow freeboard than the *Atlanta*-class hull from which it was derived. Because of the need to operate in the Arctic in conditions of severe icing, the beam was increased to provide a metacentric height of 4.97 ft. The characteristics of the CLK-1 (later changed to DL-1, Destroyer Leader), which was to be named the *Norfolk*, are tabulated in Table 3.

Table 3 Characteristics of destroyer leaders or cruisers

	Norfolk DL 1	Coontz DLG 9	Leahy CG 16	Belknap CG 26	Ticonderoga CG 47
Displacement: (tons)					
Full Load	7,280	5,648	7,590	7,960	9,959
Design	6,456	4,856	6,158	7,914	-
Dimensions: (feet)					
LWL	520.0	490.0	510.0	524.0	529.0
Beam (DWL)	53.5	49.63	53.5	54.0	55.0
Depth	34.5	30.25	38.5	38.44	42.0
Draft (design)	16.0	15.75	16.5	18.83	21.5
Armament:					
AAW/ASU	3"/70 twins (4)	5"/54 single Mk 42 (1)	3"/50 twins (2)	3"/50 twins (2)	5"/54 single Mk 45 (2)
AAW		Mk 10 TERRIER (1)	Mk 10 TERRIER (2)	Mk 10 TERRIER (1)	Mk 26, Mod 5 (2)
ASU					HARPOON (8)
ASW	Weapon A (4) Mk 37 triple TT (2) Mk 24 fixed TT (8)	Mk 37 triple TT (2) ASROC (8 missiles)	ASROC (8 missiles) Mk 32 TT (2)	Mk 32 triple TT (2) Drone Anti-submarine Helos	Mk 35 triple TT (2) LAMPS I Helos (2)
Propulsion:					
Type	Steam	Steam	Steam	Steam	Gas Turbines
Shafts	2	2	2	2	2
Shaft horsepower (SHP)	80,000	83,350	85,500	85,850	80,000
Speed (knots)	32.8	33.7	32.1	32.2	30.0
Endurance @ 20 knots	6,000 mi @ 20 knots	5,000 mi @ 20 knots	8,000 mi	11,000 mi @ 20 knots	6,000 mi
Radar & Sonar:					
Radar - Surface Search	AN/SPS-39	AN/SPS-10	AN/SPS-10	AN/SPS-10	AN/SPS-64
- Air Search (2D)	AN/SPS-6B	AN/SPS-29	AN/SPS-27	AN/SPS-37*	AN/SPS-49
- (3D)	AN/SPS-26	AN/SPS-39	AN/SPS-39	AN/SPS-48	AN/SPY-1A
Height Finding					
- Gunfire	Mk 56 (2)	Mk 68	-	Mk 68	Mk 86
- Missile	-	AN/SPG-55 (2)	AN/SPG-55 (4)	SPG-53A	Mk 99 (4)
SPG-55B					
Sonar	QHB/SQG-1	AN/SQS-23	AN/SQS-23	AN/SQS-26	AN/SQS-53A
ECM	-	AN/SLQ-6	AN/SLQ-6	AN/SLQ-6	AN/SLQ-32

*AN/SPS-40 in CG 29-34

The design of the *Norfolk* reflected findings of the 1947 Bikini Atoll nuclear tests. Deck camber was increased to facilitate wash-down, and the superstructure was reinforced.

The *Norfolk* was built to modified cruiser, rather than destroyer, standards. She had a doublebottom that extended to the bilge. To provide splinter protection, the shell from the main deck to the waterline was of STS plate, as was the main deck itself. This also provided a notch-tough steel to guard against crack propagation.

Two ships were authorized under the FY 48 program, but only the *Norfolk* was built because of excessive cost. The designation of the *Norfolk* was later changed to a destroyer leader (DL). Being a prototype, she was used primarily as a test platform for new antisubmarine equipment. A large SQS-26 sonar was fitted in 1958, replacing the QHB/SQG-1 initially installed. An ASROC launcher was also added to evaluate this new system to detect and destroy submarines at longer range. During this same conversion, she was also equipped with a prototype AN/SPS-26 three-dimensional radar.

The *Norfolk* had a 1200 psi/950 °F steam plant of the type fitted to the *Mitscher*- and *Forrest Sherman*-class destroyers. The power plant would serve as the prototype for the *Coontz*-, *Leahy*-, and *Belknap*-class destroyer leaders that would follow.

***Coontz* (DLG-9 Class) Large Carrier Escorts.** The emergence of Admiral Arleigh Burke as the CNO in the 1950s brought about a renaissance of the naval task force. With the retirement of all the battleships by 1958, the central ships were aircraft carriers of the *Forrestal*-type. These required escorts with anti-air and antisubmarine capabilities, since they had few weapons to defend themselves other than their own fighter squadrons. The TERRIER missile launcher was an ideal long-range weapon of that period but required too much space for a destroyer.

A 1954 study had shown that an improved DL-2 type ship with emphasis on speed and four single 5-in./54 dual-purpose guns would need a length of 530 ft and a displacement of 5500 tons to attain a speed of 38 knots. Such a ship required an 80,000–100,000 shp power plant. A more modest 3600-ton destroyer was desired. Unfortunately, to carry the intended weapons, the projected displacement grew from 3600 to 5000 tons. To reduce this displacement, an aluminum superstructure was proposed, and various alternative armament schemes were explored.

During this design process, the *Gyatt* completed her preliminary trials with her new TERRIER armament. Some schemes involving these missiles were studied. With six ships of this class now authorized in the FY 56 program, the Navy thought it prudent that three of them be gun ships and three be gun and missile ships. Two Mk 32 triple torpedo tubes were proposed. The big change was a TERRIER launching facility fitted to replace the two aft 5-in./54 guns. To reinforce the ASW role, an ASROC launcher with eight missiles was fitted. The missile ship design showed so much promise that the gun ships (DLG 6–8) were reordered as sister ships to the DLG-9s, which accounts for the odd numbering system in this class. These were the first large carrier escorts and were the size of World War II light cruisers. Their characteristics were as tabulated in Table 3.

The DLG-9 class was unique in that the entire hull envelope over the midship half length was of HY-80 steel. This also provided a degree of splinter protection similar to the STS steel that had been used in World War II destroyers.

The endurance of these ships was slightly more than destroyers since they were to be close support ships for the carriers. Like their predecessors, DD-931 and DDG-2 class destroyers, these ships were equipped with 1200 psi/950 °F steam plants. The big difference was in electrical load with the inclusion of the TERRIER missile launchers and their associated guidance radars, which required larger generators. With the substitution of the ASROC launcher for a forward gun, there was a need for sonar, so an AN/SQS-23 bow dome was included. The ASROC required a Mk 111 fire-control system. The bow dome and larger electrical load reduced the endurance from 6000 to 5000 miles.

The real difference between these and earlier ships was in the electronics. The numerous radars created a large increase in superstructure and more windage area than in previous destroyers. Stability became marginal.

***Leahy* (DLG-16 Class) Guided Missile Cruisers.** The futuristic-looking *Leahy*-class cruisers were the first double-ended guided-missile launching surface ships in the U.S. Navy. One of the principal missions of these ships, like that of the DLG-9 class, was to form part of the anti-air and antisubmarine screen. They were also expected to control aircraft from the carrier, vectoring them to their assigned targets.

Their antisubmarine weapons system featured the Mk 32 triple torpedo tubes and an ASROC launcher with no reloads for either. The ASROC required the AN/SQS-23 sonar system mounted in a bow dome.

The major characteristics of the DLG-16 class are shown in Table 3.

The increased endurance was one of the major reasons for their growth in size over the DLG-9 class. Although the propulsion plant of the DLG-9 was retained, the electrical plant showed major growth because of the increases in missile launchers and their requisite electronics.

The inclusion of the second TERRIER missile launcher and the provision of protection against nuclear blast brought about a change in the scantlings of the superstructure. The area requirements for the Combat Information Center (CIC) were doubled, and the new air search radars required more electrical power and better air sweeps. The increased space requirements led to the long raised deck configuration. These ships had an aluminum superstructure. They introduced the split or mack stack on which the new radars could be mounted without smoke interference.

In later upgrades to this class, the AN/SPS-48 air search radar replaced the AN/SPS-52, a helicopter landing platform was added, and the Naval Tactical Data System was added during an AAW modernization.

***Belknap* (DLG-26 Class) Guided Missile Cruisers.** In the next acquisition of guided missile cruisers, several new weapon systems were available that altered the characteristics of the follow-on designs. The growth of electronics and guided-missile systems (too numerous to enumerate) overtook the DLG-16 design, which had already exceeded its design displacement by some 400–500 tons. One rather odd development in these ships came about because this class was basically a lengthened DLG-16. To meet the need for more displacement, the naval architects enlarged the hull through the expedient of inserting a 14-foot extension between the forward engine room and aft fire room. Since the hull had significant curvature at the point of insertion, the enlarged hull was somewhat less than fair. This makeshift approach did, however, reduce design and building costs.

The engineering plant was the same as in the previous cruiser class, with 1200 psi/950 °F steam conditions. The maximum power was the same but the displacement was greater. Therefore, maximum speed was 0.1 knots slower in the DLG-26. The rpm of the shafts were decreased to reduce propeller noise and achieve greater efficiency. The most significant difference was in the electrical generators where a 50 per cent increase in power was provided for the larger combat systems.

The major characteristics and sensors of the DLG-26 class are shown in Table 3.

Nuclear Cruiser *Long Beach* (CGN-9). Discussions about using nuclear power in surface combatants started in 1956. Although the installation in the submarine *Nautilus* was successful, no nuclear plant was then available to power destroyers and larger combatants. In January 1956, the chief of BuShips stated that nothing would be gained by placing nuclear power plants in cruisers or battleships; but by May Rear Admiral Rickover proposed a nuclear power plant for almost any size naval combatant. Informal work had been underway at BuShips on a nuclear cruiser, but it was not until Admiral Burke became CNO that the program had a focus. He wanted to investigate the feasibility of a nuclear cruiser to support a nuclear carrier. Conceptual studies considered various complements of military equipment, but problems arising from the great weight of the then-available nuclear plants plagued the studies.

The debates continued on the ship's characteristics. The Ship Characteristics Board wanted armor protection but was unwilling to pay for the resulting loss in offensive capability and the increased cost.

Fig. 9 The third series of the fast task force escorts were units of the *Belknap* (CG-26) class, which were 14 feet longer than predecessors of the *Leahy* (CG-16) class. Designed to be close escorts to attack-carrier battlegroups, these ships were armed with the TERRIER ship-to-air missile system forward, a 5-in./54 mount aft, and a large helicopter platform at three-quarters aft for two Drone Anti-Submarine Helicopters (DASH). The helicopter facility was large enough for one manned helicopter, as DASH did not prove to be feasible in operation. *Belknap*-class ships were also equipped with the large sonar, AN/SQS-26, so that this warship was a complete ASW and AAW ship and had a very large cruising radius of 8000 miles at 20 knots. (Courtesy U.S. Naval Institute.)

Fig. 10 The nuclear cruiser *Long Beach* (CGN-9) was the only one of a type built. She is shown here with six of her phased array missile guidance radars, a pair of 5-in./38 enclosed mounts, a TALOS missile launcher aft, two TERRIER missile launchers forward, an ASROC launcher just aft of the forward deckhouse, and the huge AN/SPS-33 phased array radar atop the forward deckhouse. The *Long Beach* was a test platform for missile and radar systems until the mid 1970s, when the ship was equipped with HARPOON and Tomahawk missile systems. A proposed AEGIS installation was canceled. The large space amidships was to be for Polaris missile tubes, which were never fitted. (Courtesy U.S. Naval Institute.)

Although shipbuilding appropriations were at a low level in the late 1950s, the authorities felt that a nuclear cruiser should have both ASW and AAW capabilities. A proposed armament of Regulus or Polaris would increase construction costs. The Navy decided, however, to provide space and weight for eight Polaris midship launching tubes, which were never fitted. All the proposed concepts featured the AN/SQS-23 sonar dome and an ASROC launcher. Two TERRIER missile launchers were provided forward and a TALOS missile launcher aft to give this nuclear cruiser more anti-aircraft capability. With the missiles and electronics specified, they could engage four air targets simultaneously. Admiral Burke set a priority on speed. The new ship had to make 30 knots; thus a longer hull and relatively small beam were necessary. Some variations to achieve additional power were examined using gas turbine booster power but were not pursued. The final design of the nuclear cruiser, now named the *Long Beach* (CGN-9), was an all-missile ship with no 5-in. guns. Because this ship had no protection against surface-ship attack and low-flying aircraft, a ship alteration in 1962–1963 added two 5-in. gun mounts. The cruiser *Long Beach* had the characteristics summarized in Table 4 when first completed in 1961.

She was:

- First U.S. Navy nuclear surface ship
- First phased array radar ship (SPS-32/33)
- First all-missile ship (completed before CG-16s)

In short, this cruiser design was a major departure from tradition, being a concept with an all-missile armament that could deal with both air and sub-surface threats.

In 1977 the *Long Beach* underwent a major overhaul during which her 32/33 radars were replaced by 48/49 radars, and the ability to land helicopters was upgraded. The conversion to a more modern missile system and the inclusion of AEGIS was prohibitively expensive, so the Mark 12 TALOS system was removed along with the Mk 77 guided missile fire-control system and replaced with Harpoon canister launchers and Tomahawk armored box launchers.

The Navy recognized that the *Long Beach* was less than a complete success, being simply too big, too slow, and too expensive. The nuclear frigates that would follow were somewhat more affordable and better suited to task force defense. She was decommissioned in 1992 and later sold for scrap.

***Bainbridge* (CGN-25) Nuclear Guided Missile Cruiser.** Despite nuclear power's obvious drawbacks (mostly related to cost and weight), many in the Navy were convinced that it was the propulsion system of the future. Attempts back around 1955 to squeeze a nuclear plant into a DLG-9 (*Coontz*) hull

Table 4 Characteristics of nuclear cruisers

	Bainbridge CGN-25	Truxtun CGN 35	California CGN 36
Displacement: (tons)			
Full Load	7,250	8,927	10,096
Design	-	-	-
Dimensions: (feet)			
LWL	540.0	540.0	570.0
Beam (DWL)	56.0	57.83	60.0
Depth	-	-	-
Draft (design)	19.42	19.9	20.5
Armament:			
AAW/ASU	3"/70 twins (2)	5"/54 single Mk 42 (1)	5"/54 single Mk 42 (2)
AAW	Mk 10 TERRIER (1)	Mk 10 TERRIER (1)	Mk 13 TARTAR
ASW	Mk 32 triple TT (2)	Mk 32 Triple TT (2)	Mk 32 triple TT (2) ASROC (16 missiles)
Propulsion:			
Type	Nuclear	Nuclear	Nuclear
Shafts	2	2	2
Shaft horsepower (SHP)	-	-	70,000
Speed (knots)	-	-	30+
Endurance	Unlimited	Unlimited	Unlimited
Radar & Sonar:			
Radar - Surface Search	AN/SPS-10	AN/SPS-10	AN/SPS-10
- Air Search (2D)	AN/SPS-29	AN/SPS-40D	AN/SPS-40
- (3D)	AN/SPS-39	AN/SPS-48C	AN/SPS-48
Height Finding			
- Gunfire	Mk 68	Mk 68 (Mod B)	Mk 86
- Missile	SPG-55 (2)	AN/SPG-55B (2)	AN/SPG-74D Mk 73 Dir
Sonar	AN/SQS-23	AN/SQS-26	AN/SQS-26
ECM	AN/SLQ-6	AN/SLQ-6	-

	Virginia CGN-38	Long Beach CGN 9	CSGN
Displacement: (tons)			
Full Load	10,271	17,525	17,210
Design	-	-	-
Dimensions: (feet)			
LWL	560.0	666.0	666.0
Beam (DWL)	63.0	73.21	77.0
Depth	-	45.0	-
Draft (design)	19.8	29.7	22.0
Armament:			
AAW/ASU	5"/54 single Mk 42 (2)		8"/55 single Mk 71 (1)
AAW	Mk 26 Mod 0 Mk 26 Mod 1	Mk 10 TERRIER (2) TALOS (2)	Mk 26, Mod 2 (2)
ASW	Mk 32 triple TT (2) Helo (2)	ASROC	Mk 32 triple TT (2) ASROC (16 missiles)
Propulsion:			
Type	Nuclear	Nuclear	Nuclear
Shafts	2	2	2
Shaft horsepower (SHP)	70,000	70,000	85,500
Speed (knots)	30+	30	32.1
Endurance	Unlimited	Unlimited	Unlimited
Radar & Sonar:			
Radar - Surface Search	AN/SPS-55	AN/SPS-10	AN/SPS-55
- Air Search (2D)	AN/SPS-40B	-	AN/SPS-49
- (3D)	AN/SPS-48A	AN/SPS-32 AN/SPS-33	AN/SPY-1D
Height Finding			
- Gunfire	Mk 86	-	Mk 86
- Missile	AN/SPG-74 (2) Mk 73 Dir	AN/SPG-55	MK 99 (4)
73 Dir			Mk
Sonar	AN/SQS-26	-	AN/SQS-53 -2
ECM	-	AN/SLQ-6	

Fig. 11 The *Bainbridge* was the nuclear version of the *Leahy* class. This 1968 view shows her improved planar array SPS-39 air search radar. Equipped with two TERRIER missile launchers, the *Bainbridge* also carried an ASROC launcher (with no reloads), two twin 3-in./50 gun mounts (later removed), and the AN/SQS-23 sonar system. The four conical antennas are the AN/SPS-55 missile guidance radars which emits a beam to guide a TERRIER missile to its target. (Courtesy U.S. Naval Institute.)

found the 490-ft ship expanded to 540 ft. Nuclear power seemed to demand displacements of not less than 6900 tons. Worse yet, the nuclear ships promised to cost $20 to $30 million more than those with conventional power.

Despite these discouragements, the Navy pressed ahead with the design of the nuclear guided missile cruiser that was to become the *Bainbridge.* Her armament was patterned after the DLG-16 class with a few changes.

Since the *Bainbridge* had no stacks to interfere with her radars (smoke or heat), the exhaust of the auxiliary boiler was made integral with her foremast, upon which was mounted the surface search and 3-D air search radars.

The major characteristics of the ship are shown in Table 4.

During 1974–1976 the *Bainbridge* underwent a major mid-life modernization in which her anti-aircraft combat system was upgraded. In a later overhaul (in 1983–1985), other major improvements were effected.

***Truxtun* (CGN-35) Guided Missile Cruiser.** The cost of the *Bainbridge* was $60 million more than the $108 million initially estimated to build the Navy's first nuclear cruiser. This development eroded support for any follow ships of her type. During discussions on the Navy's FY 62 budget, however, Congress arbitrarily substituted a nuclear frigate for one of the ten ships the Navy had projected for the *Belknap*-class. A repeat of the *Bainbridge* design was sought at first, but since the large AN/SQS-26 sonar was being introduced in the CG-26 class, the Navy decided that the nuclear version of these ships should be also equipped with that system and the hull design would not be a repeat of the *Bainbridge.*

The *Truxtun,* like the *Bainbridge,* was one of a type, in effect a nuclear version of the *Belknap.* There was

Fig. 12 The *California* (CGN-36), as completed. She has a symmetrical gun and missile armament of a 5-in./54 lightweight gun and a Mk 13 TARTAR missile system with 40 missiles in each silo. Equipped with the AN/SQS-26 sonar system, these ships were also fitted with an ASROC launcher with reload capability (the small house forward of the launcher). No helicopters were provided in these ships, but LAMPS data link antennas were provided. (W. H. Garzke Collection.)

the absence of stacks and no smoke or temperature interference with the radar. With only a single missile launcher, only two missile directors were required. These and latticed masts gave the *Truxtun* a significantly different topside arrangement from the *Bainbridge*.

The major characteristics and electronics for the *Truxtun* are summarized in Table 4.

During her mid-life modernization in the early 1980s various new offensive weapons were added. She was redesignated CGN-35 in June 1975.

***California* (CGN-36 Class) Guided Missile Cruiser.** The design of the CGN-class guided missile cruiser was begun by the Ship Characteristics Board in mid-FY 64. The design went along the path Navy began in the Coontz (DLG-9 class) to provide the fleet with high-performance surface combatants. These units carried a guided missile main battery and accompanying advanced electronic suits of sensors for search, fire control, electronic warfare, and communications on a platform powered by nuclear reactors. (Congress, in its wisdom, had dictated that any naval ship over 8000 tons displacement must be nuclear powered, saving naval personnel the trouble of debating the merits of alternative choices.) A major design decision involved the choice of guided missile systems: TERRIER versus TARTAR. The latter was fairly quickly chosen on the basis of its relative simplicity and smaller size.

The *Truxtun* (DLGN-35) was chosen as the baseline design. There were many combinations of weapon systems from which to choose. As those decisions were made, the resulting ship outgrew the parent design by 30 ft in length, 2.5 ft in beam, and 900 tons in displacement. The propulsion system retained the same D2G pressurized water reactor, but was improved to increase endurance. Habitability standards were also raised.

These ships have an anchor stowage integral with the sonar dome to reduce excessive flare and prevent anchor fluke damage to the sonar dome. This installation followed the concept first introduced in the *Knox*-class frigates (FF-1052).

The preliminary design ended in September 1965 and resulted in a ship with the principal characteristics shown in Table 4.

In 1975 the Navy changed their designation from frigates to cruisers. Originally, three ships were authorized, but the third was overtaken by technical developments and became the CGN-38.

These ships had their electronics and weapons systems upgraded during their service life and by 1990 the following added:

Fig. 13 The *Virginia* (CGN-38) class was conceived as a class of eight, but their cost and the emergence of the *Ticonderoga* (CG-47) resulted in only four ships being completed. These ships were provided with the new Mk 26 Guided Missile Launching Systems, with the Mod 0 (24 missiles) forward and the Mod 1 (44 missiles) aft. The ship was also equipped with the AN/SQS-26 sonar and a below-deck helicopter hangar in its stern. The helicopter facility was later removed because of mechanical problems and leakage of the sliding deck cover; it was replaced with a Tomahawk missile battery. These vessels were designed to take the AEGIS radar system to sea, but this proved to be too expensive to accomplish. The *Virginia* class are versatile combatants with an unlimited cruising radius and a symmetrical missile and gun battery forward and aft. (Courtesy U.S. Naval Institute.)

Harpoon	2 quadruple barrels added amidships
CIWS	2 20mm 6-barrel Mk 15 mounts amidships
Torpedoes	4 Mark 32 torpedo tube (2 fixed)
Countermeasures	4 SRBOC 6 barreled (Mk 36) AN/SLQ 32V (3) radar warning and jammer
Radars	AN/SPS 48C 3D Air search; AN/SPS 40B 2D Air search; AN/SPG 51 D (Four sets); AN/SPQ 9A; AN/SPG 60

Virginia (CGN-38 Class) Guided Missile Cruisers. The *Virginia* (CGN-38) guided missile cruiser was designed during the late 60s and early 70s. The decision makers were well aware of the rapid technological progress being made in all aspects of naval weaponry and were anxious to design a ship best suited to support those systems. This meant that the hull design was in a constant state of flux. The *California* (CGN-36) design was used as starting point, but the final product was about 20 feet shorter, 3 feet wider, and 200 tons heavier. (More than 10,000 tons displacement tended to distort the design.) The draft remained nearly the same, but there was an increase in trim aft. Principal improvements over the CGN-36 class included the Mk 26 guided missile launcher, a new replenishment-at-sea system, lightweight guns, extensive use of HY-80 steel in the hull, and elimination of the ASROC launcher.

The preliminary design was concluded in the fall of 1969 and represented a ship with the characteristics summarized in Table 4.

During contract design the decision was made to accommodate the LAMPS helicopter. There was some unassigned space in the stern and it was proposed that this be used as a below-deck hangar for the helicopters plus the twelve associated magazines, stowages, and ammunition. The concept was based on the seaplane hangar arrangements used in the *Baltimore* (CA-68) class cruisers. The helicopter would land on the raised elevator platform; with the platform lowered the hangar would be enclosed by sliding doors or hatches. In seemed an excellent concept, but in practice it proved unsuccessful. The seals to the hatches enclosing the helicopter hangar leaked seawater onto the helicopters, necessitating extensive freshwater wash-downs. The helicopter facility was later removed and Tomahawk box launchers were installed on the main deck.

The CGN-38 class had a distinctive bow anchor arrangement unlike their predecessors. Because of bow trim problems in early contract design, the sonar dome-stowed anchor was eliminated and replaced with a bow-stowed anchor and an on-deck stowed lightweight anchor. The bow flare was increased slightly to allow these anchors to clear the AN/SQS 53A bow sonar dome and improve dryness.

These guided-missile cruisers are the latest nuclear-powered surface combatants, other than the nuclear carriers of the *Nimitz* (CVAN-68) class. Four ships were completed, although the last two came along much later than their sisters owing to substantial cost increases of $140 million. The fifth ship, which would have featured the AEGIS weapons system, was canceled as the result of costs and the emergence of a new strike cruiser (CSGN) concept.

The electronics and weapons of these ships have been improved since their completion, but owing to costs not all of the changes are uniform. As of early 1991 these ships had the following additions:

Armament	2 Quadruple Tomahawk missile launchers; 2 Quadruple Harpoon canister launchers; 2 Mk 15 200 mm 6 barreled Phalanx machine guns

Countermeasures	4 Mk 36 SRBOC fixed launcher SLQ 32(V)3
Electronics	1 AN/SPS-48D Air search; 2 SPG 51 Guided missile fire control

CSGN Proposed Strike Cruiser. The cancellation of the fifth *Virginia*-class cruiser (CGN-42) was brought about by the emergence of the AEGIS radar system. Because the ship would have missions somewhat independent of the task force, she was to be equipped with nuclear power and flag facilities. The first design cycle ran into difficulty because the ship was to be appreciably heavier than the previous guided missile cruisers but retain essentially the same power. She would be hard pressed to make 30 knots. A fresh start produced in July 1975 a design featuring a slightly more powerful D2G reactor, which would allow a speed of 30 knots. The real strength of this ship was to be the AEGIS system,and a tactical command center. The ship was to feature protection of vital components, and all vital communication and control spaces were to be located in the hull to reduce the size of the superstructure. These ships included hangar stowage for the LAMPS or VSTOL aircraft and full rearming facilities. The design characteristics for the FY 78 program was as tabulated in Table 4.

A total of eight ships was contemplated and they would have cost approximately $1.2 billion each, with the nuclear propulsion and AEGIS being the primary cost items. These high costs escalated as the AEGIS system developed, and the Ford administration decided to return to an AEGIS cruiser developed from the *Virginia* CGN-38 class. This was abandoned in favor of the *Ticonderoga* CG-47 class AEGIS cruisers, which were much less expensive.

Battleships

***Iowa* (BB-61 Class).** The *Iowa*-class battleships were the last of the breed and have been the subject of conversion and modernization studies ever since the end of World War II. The availability of the ships in the reserve fleet, their size, and their excellent attributes of gun power, speed, and protection have contributed to the popularity of the *Iowa*-class as candidates for resurrection.

The immediate postwar plans were to keep the *Iowa*-class in commission with the *South Dakota*- and *North Carolina*-class in standby reserve as they did not have the speed to keep up with the *Essex*- and *Midway*-class carriers. Manpower and cost considerations, however, forced three of the *Iowa*-class to be decommissioned, leaving only the *Missouri* in service by 1949.

During World War II and the Korean War, the *Iowa*-class ships were used with fast carrier task forces for anti-aircraft support and independent missions of shore bombardment. They carried a large number of 20mm and 40mm guns, which by 1945 were considered to be of limited value against Japanese suicide planes. To destroy such planes, it was important to do so at ranges that these batteries could not reach. During refits from 1946–1948, almost all of the 20mm guns were removed because they were prone to weather damage, and entailed high maintenance. Plans were made for significant improvements, but the only change authorized was to add six Mk 56 directors to the *Iowa* and *New Jersey*. These new directors were to control the 5-in./38 as well as the 3-in./50 mounts. When the four battleships were mothballed from 1954–1958, all of their 40mm armament still remained aboard.

Three of the class were later fitted to fire nuclear shells, but none were ever fired.

The *Kentucky* was scrapped in 1959.

Proposals were made to convert the four *Iowa*-class ships to guided missile capital ships that would feature anti-aircraft missiles and the Polaris ship-to-shore missile. A critical problem was the removal of the barbette armor, a very costly operation. The key argument used against this proposal was the high cost, which seemed ill advised because hull and machinery were over 20 years old. For a slightly higher cost it seemed prudent to build an entirely new missile ship. There was lingering interest in such conversions until 1960.

In August 1967 the *New Jersey* was recommissioned for Vietnam service. The austere reactivation was accomplished at the Philadelphia Naval Shipyard. All the 40mm guns were removed, and the electronics were considerably changed. The surface-search radar was improved by the addition of the AN/SPS-10B and the air search with the AN/SPS-6C. A large radio mast was placed on the bow and her foremast was considerably changed with the addition of an early version of SHORTSTOP, an anti-missile jamming system. This was required because the Soviets had given the North Vietnamese 22-N-2 (STYX) missiles. To complete her protection against such attack, four specially converted Zuni Chaff rocket launchers were added. New communication systems were added and the helicopter landing platform was made to conform with the latest NAVAIR standards. Habitability changes were made and seven air conditioning units were added for crew and electronic needs.

The *New Jersey* remained in commission until September 1969. Her deactivation was forced by the

Fig. 14 The *Iowa* is shown in her 1980 modification, which saw the installation of 8 quadruple Tomahawk missile launchers, 16 HARPOON canister launchers, the AN/SPS-49 air-search radar, and four CIWS mounts, which replaced the four 5-in./38 twin mounts. The *Iowa* and her three sisters carried one of the most successful bagged guns afloat, the 16-in./50 gun, capable of hitting targets as far away as 50 miles. These battleships were the fastest of their type ever constructed, though not the most heavily armored or armed. The Japanese *Yamato* class had an armor side belt of 410 mm and nine 460 mm guns in three triple turrets on a displacement of 69,100 tons full load. However, the American 16-in./50 gun was almost as powerful as the Japanese 460-mm gun because it was designed to fire the heavy weight 2700-lb armor-piercing shell. Defense budget cuts in 1990–1992 and the collapse of the Soviet Union led to the layup of all four *Iowa* class, the most successful battleship ever built. (Courtesy U.S. Naval Institute.)

North Vietnamese representatives in Paris who would not continue negotiations until she was deactivated. Her inactivation was ironic because her toughest critic during congressional debate on her funding had been Representative Melvin Laird, who now had become the Secretary of Defense.

In 1977 the Naval Ship System Command examined all four ships and found that their hulls had a minimum service life of 15 years. During this period, a retired Air Force pilot and defense consultant, Mr. Charles Myers, started a publicity campaign on the usefulness of the *Iowa*-class battleships in coastal zone warfare. He cited the *New Jersey*'s service in Vietnam and the fact that a number of aircraft were lost in areas where 16-in. gunfire could have produced equal or superior performance in all weather. Moreover, the same defensive fire that brought down aircraft could not deal with the destruction of a 16-in. shell.

Although there was opposition in Congress, reactivation of the *New Jersey* was approved in 1981. In 1940 the cost of these ships was approximately $71 million each, but the cost of reactivation of the *New Jersey* in 1982 was $325 million, with $170 million of that cost in the modernization of the hull systems. The *Iowa, Missouri*, and *Wisconsin* were also reactivated, but at higher costs (of about $470 million) largely because they had been out of service longer.

The modernized version of the *Iowa* class featured the following:

- nine 16-in./50 caliber guns in three turrets (some firing arc restrictions were imposed),
- twelve 5-in./38 caliber guns in six twin mounts,
- thirty-two BGM-109 Tomahawk cruise missiles in eight armored box launchers,
- sixteen RGM-84 Harpoon anti-ship missiles in four quadruple canister launchers,
- four 20mm/76 caliber Mark 15 CIWS six-barrel Phalanx Gatling guns,
- improved air-search radar AN/SPS-49, and
- the AN/SLQ(V)3 defensive ECM system with its automated signal processing and analysis capability.

The reactivation also included improved accommodations, a sewage and holding system, a modern fuel system, and improved communications.

The replacement of the air-search radar was one of the most important changes. The AN/SPS-49 radar is a large system, with the antenna requiring its own supporting structure. The existing pole mast from the 1968–69 modernization was removed and replaced with a tripod mast just aft of the conning tower. This location provided a better view of the horizon and permitted adequate structural support for the radar antenna as well as shielding the radar from stack gas interference. The new air-search radar has a range of approximately 100 miles and is capable of detecting oncoming missile warheads.

The modernized *Iowa*-class ships have the following characteristics:

Displacement	45,280 tons (light ship); 57,256 tons (full load)
Dimensions	
LBP	860 ft
Beam	108.25 ft
Depth	52.99 ft
Draft	34.77 ft (design)
kW capacity	10,500
Shaft horsepower	212,000
Speed	33 knots
Endurance	15,000 miles at 15 knots

All four battleships were modified to the above characteristics, although the displacement of each of the four is slightly different. It should be noted that these ships have a limiting draft of 37.17 feet, which corresponds to a displacement of 59,250 tons. There is still considerable growth potential for armament and electronic changes.

All four battleships were equipped to land and refuel a variety of helicopters, but no hangars were provided to protect them from the ravages of sea spray. They were also equipped with five remote pilotless vehicles for gunfire spotting and reconnaissance. During Operation Desert Storm these tiny craft with their video cameras were invaluable aids to the gunners on the *Wisconsin* and *Missouri.*

During 1991–1992 all four *Iowa*-class ships were placed in the reserve fleet once again as a matter of budget reductions. With the ending of the cold war, it was felt that these ships with their large crews and steam power plants were not needed, so they are inactive once again awaiting a future call if the need should ever arise.

Acknowledgments

Several highly qualified friends were kind enough to render me considerable advice and assistance as I was creating this chapter. Prominent among them were Robert J. Scott, Robert Staiman, Thomas Wickert, William Rogalski, and George Kerr.

Suggested Reading

Abbott, J. W., "Modular Payload Ships in the U.S. Navy," *Trans.* SNAME, Vol. 85, 1977.

Brand, R. A., Broderick, T. E., and Mayer, J. J., "Design Criteria of U.S. Naval Ship Electrical Systems," *Trans.* SNAME, Vol. 66, 1958.

Ferrerio, L., and Stonehouse, M. H., "A Comparative Study of U.S. and U.K. Frigate Design," *Trans.* SNAME, Vol. 99, 1991.

Garzke, Jr., W. H., and Kerr, G., "Major Factors in Frigate Design," *Trans.* SNAME, Vol. 89, 1981.

Garzke, Jr., W. H., and Kerr, G., "A New Warship Design Strategy—A Perspective," *Trans.* SNAME, Vol. 93, 1985.

Kennell, C. G., and White, B. L., "Innovative Naval Designs for North Atlantic Operations," *Trans.* SNAME, Vol. 93, 1985.

Leopold, R., and Reuter, W., "Three Winning Designs—FDL, LHA, DD-963: Method and Selected Features," *Trans.* SNAME, Vol. 79, 1971.

Meigs, C. H., "Recent Naval Steam-Plant Design," *Trans.* SNAME, Vol. 62, 1954.

Sarchin, T. H., and Goldberg, L. L., {Stability and Buoyancy Criteria for U.S. Naval Surface Ships," *Trans.* SNAME, Vol. 70, 1962.

Sherwin, Jr., S. A., and Miller, R. T., "The Impact of Electronics on Warship Design," *Trans.* SNAME, Vol. 70, 1962.

Thornbury, J. W., "Current Electrical Systems and Equipment for Warships," *Trans.* SNAME, Vol. 69, 1961.

Warren, G. B., "Development of Steam Turbines for Main Propulsion of High-Powered Combatant Ships," *Trans.* SNAME, Vol. 54, 1946.

William H. Garzke, Jr. is a graduate of the University of Michigan with a degree in naval architecture and marine engineering. He also attended MIT for graduate studies in naval architecture, and received a masters degree in applied mathematics from Adelphi University. He is currently employed by Gibbs & Cox, Inc. as their staff naval architect. He has done research work at Webb Institute, participated with Dr. Robert Ballard of Woods Hole Oceanographic Institution in his discovery and analysis of the *Bismarck* wreck, and was an Adjunct Professor of Naval Architecture at the U.S. Merchant Marine Academy. A member of SNAME he is the recipient of both the Cochrane Award and the Linnard Prize. He is noted for his world-renowned trilogy of books, *Battleships,* published by the Naval Institute Press on the design, construction, and operational histories of World War II battleships and battle cruisers.

Amphibious Ship Evolution

CHARLES H. PIERSALL

Introduction

THIS CHAPTER reviews briefly the evolution in the composition of U.S. amphibious forces since World War II to see where the force is, how it got there, and where it is heading in the year 2000 and beyond. Fig. 1, which depicts the projection of naval power ashore, contains the four major elements essential for the interdiction of amphibious assault forces and the sustainability of those forces.

The amphibious ships and advanced air cushion landing craft (LCAC), as well as the maritime prepositioning ship force, which carries prepositioned supplies for the U.S. Marine Corps, and the sealift enhancement ships are discussed in this chapter. The ships for the assault follow-on echelon (AFOE) are naval auxiliary and commercial ships used to provide the sustainment and re-supply requirements of the amphibious assault. The ships of the AFOE are discussed in other chapters.

Background

The general mission of amphibious ships is to transport elements of landing forces, troops and their equipment and supplies, and to land these elements in amphibious assault. The characteristics of these elements and certain naval elements needed to support the mission, as well as the characteristics of the amphibious functions, determine the specialized characteristics needed in amphibious ships.

A brief look at the first-generation amphibious force, that was built before and during World War II, indicates about six major amphibious functions that led to a large number of substantially different types of amphibious ships. These functions are troop transport, vehicle/cargo transport, landing craft transport/support, tank/heavy vehicle transport, reconnaissance, and embarked staff support. There are many other amphibious functions, some of which call for specialized characteristics in ships, but these six functions appear to be the ones that demand different types of ships.

First Generation

Troops/Staff. The six major amphibious functions and the first generation of amphibious ships used to support those functions are shown in Fig. 2. The troop transport function led to a specialized troop transport ship called LPA (earlier APA). These ships were designed to transport and land battalion-size landing forces, and could lift about 1200–1500 troops plus various quantities of vehicles, cargo and bulk POL (petroleum/fuel). Deck-loaded landing craft (primarily LCPR and LCVP with some LCM-3/6) were used for ship-to-shore movement. Some LPAs were configured to support embarked Navy staffs and were designated flag ships. Some LPAs were converted merchant ships and others were variations on U.S. Maritime Commission hull designs. These ships were the mainstays around which amphibious task forces were configured.

Vehicle/Cargo. The vehicle/cargo transport function led to a specialized transport called LKA (earlier AKA). These ships were the major vehicle transports and could handle vehicles weighing up to 30 tons.

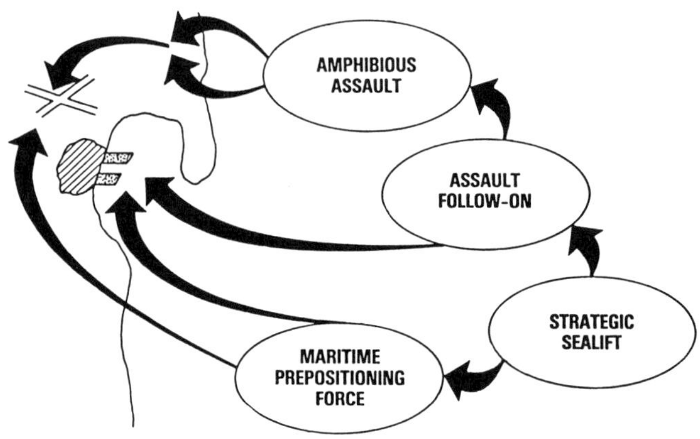

Fig. 1 Projection of naval power.

Vehicles and cargo were the primary LKA load, but LKAs also carried 200–300 troops (vehicle/equipment operators) and various quantities of bulk POL. As in the LPA, deck-loaded landing craft (LCPR, LCVP, LCM-3/6) were used for ship-to-shore movement, but the LCMs were the primary craft carried in LKAs. LKAs, like LPAs, were acquired through converting merchant ships and modifying Maritime Commission hull designs. Speeds were about 14–16 knots.

Landing Craft. The landing craft transport/support function led to the landing ship dock called LSD. These ships could transport large landing craft (LCUs) to the amphibious objective area (AOA) and provide drydock support there for landing craft as well as for small beaching ships such as Landing Ship Infantry (LSI) and Landing Ship Mechanized (LSM). Landing craft were the primary LSD load, but LSDs also carried vehicles, troops (landing craft and vehicle operators), and various quantities of cargo and bulk POL. Unlike LPAs and LKAs, the LSDs were specially designed from the keel up. Speeds were 12–14 knots.

Tank/Heavy Vehicle. The tank/heavy vehicle transport function led to the Landing Ship Tank (LST) and Landing Ship Mechanized (LSM). A need existed to load, transport and put ashore main battle tanks weighing approximately 50 tons, and LKA boom capacity was generally limited to 30 tons. The numbers of tank capable landing craft (LCUs) and ships for transporting such craft (LSDs) were also limited. Further, a need existed for a capability to launch and recover amphibious vehicles in the near offshore area. These heavy vehicle requirements led to the design of the highly specialized LST and LSM to provide ocean transports with direct beaching capability. Vehicles were the primary LST/LSM cargo load, but these ships also carried troops (vehicle operators) and various quantities of cargo and bulk POL. LSTs could lift approximately 500 tons of vehicles and cargo at speeds of 8–10 knots. The LSM was smaller and more powerful but only about two knots faster.

Reconnaissance. The reconnaissance function calls for a ship with a substantial speed advantage over the task force to facilitate beach surveys and other reconnaissance activities before the assault. This led to the LPR (earlier APD), which was a small troop transport for Underwater Demolition Teams (UDT) and other reconnaissance teams. The design was based on a destroyer escort (DE) hull. Its speed was about 20 knots.

Embarked Staff. The embarked staff function led to a specialized staff support ship called LCC (earlier AGC). These ships were designed to embark and support major Navy and landing force staffs with heavy emphasis on command, control and communications (C^3). Like the LPAs and LKAs, the first-generation LCCs were converted merchant ships or modified Maritime Commission hull designs. Speeds were about 14–16 knots.

First Generation Summarized

The planning done in the 1930s was primarily by the Marines, with some Navy participation. It was done on a shoestring budget and concentrated largely on prototypes such as the LCPR, LCVP, and LCM. The demands of World War II exploded the growth of amphibious ships. This placed the naval engineers in a catch-up role, but they rose to meet the challenge.

In summation, the first-generation amphibious force was characterized by

- many types of highly specialized ships,
- ship-to-shore operations limited to surface means and short ranges, and
- landing craft largely deck-loaded.

Second Generation—First Approach

Troops/Staff/Helicopters. The second generation of amphibious ships (see Fig. 3) began in the early 1960s following the development of the vertical envelopment concept. Note the addition of helicopters to what was the troop/staff function. Introduction of

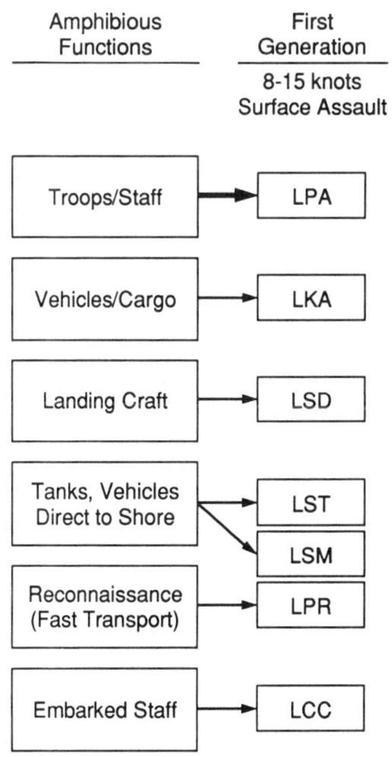

Fig. 2 Amphibious ship evolution—first generation.

helicopters, a major technological breakthrough, into the ship-to-shore movement called for ships that could transport, maintain and operate such aircraft. Two new types of troop transports were introduced into the fleet, one with aviation facilities (LPH) for vertical assault and one with docking facilities (LPD) for surface assault.

The LPH was designed using a single-screw merchant ship hull (similar to the then newest LKAs) as the basic platform. Troop capacity was increased by several hundred over the LPAs, providing space for a battalion plus an aviation detachment. These ships also carry vehicles, cargo and bulk POL, but they carry no landing craft and for practical purposes have no surface ship-to-shore capability. Speed was increased to 20 knots.

The LPD is a derivative of the then newest LSD, the LSD-28 class. The wet well deck of the LPD is less than half the size of that in the LSD, a large part of the wet well and ballast system volume having been configured for vehicles, cargo and POL. The superstructure and wing walls were modified to increase troop capacity to 800–900, and a helicopter deck was constructed over the well deck to provide a modest helicopter capability. The LPD is far more general purpose in character than any of its predecessors, carrying some of all lift components (i.e., troops, vehicles, cargo, helicopters, boats, and POL), and has both surface and vertical assault capabilities. The use of wet wells in transport ships provided a major improvement in debark capabilities, providing for underway launch of preloaded landing craft instead of alongside loading from ships at anchor. Speed is 20 knots.

Vehicle/Cargo. The post-World War II LKAs, both the LKA-112 (mid-1950s) and the LKA-113 class (late 1960s) are larger and faster than their predecessors. The LKA-113 class also carries LCM-8 (tank capable) as well as the LCM-6 landing craft, has more surface debark stations, has faster and safer cargo handling gear, and has a helicopter platform, giving it a substantial capability to debark vehicles and cargo via helicopter. These ships, like their predecessors, carry troops (vehicle operators) and bulk POL. Speed is 20 knots.

Landing Craft. The post–World War II LSDs, both the LSD-28 class (mid-1950s) and LSD-36 class (late 1960s and early 1970s) are larger and faster than their predecessors. The wet well of the LSD-28 class is sized for three LCU-1466 series, and the LSD-36 class is sized for three longer LCU-1610 series. Both classes have portable mezzanine decks for use in the forward part of the well deck to permit conversion of landing craft transport capacity to vehicle transport capacity, a feature that facilitates adjustment to wide variations in lift requirements for different operations. The ships have helicopter platforms, although helicopters are not permanently embarked. Both ships have speeds of 20 knots.

Tank/Heavy Vehicle. Of the two tank/heavy vehicle transports developed during World War II (LST and LSM), only the LST survived to the second generation. Several new classes of LSTs were constructed after World War II, each larger and slightly faster than its predecessor, culminating in the LST-1179 class in the late 1960s and early 1970s. To achieve the 20-knot speed capability, this class of LST used a completely different hull form. It has a bow ramp for discharge that extends from the main deck, rather than from the tank deck via bow doors, as in earlier classes. A stern ramp is used for LVT launch and recovery. LVTs can be launched with the ship under way. These ships carry substantial quantities of vehicles, several hundred troops (vehicle operators), small quantities of cargo, and substantial quantities of bulk POL. These are the only ships that have the tankage, fuel piping, deck connections and bow and stern anchors needed to pump bulk POL directly to landing force POL facilities ashore. As stated earlier, their speed is 20 knots.

Reconnaissance. The second generation has no ship specifically designed for the reconnaissance mission, all LPRs having been retired without replacement.

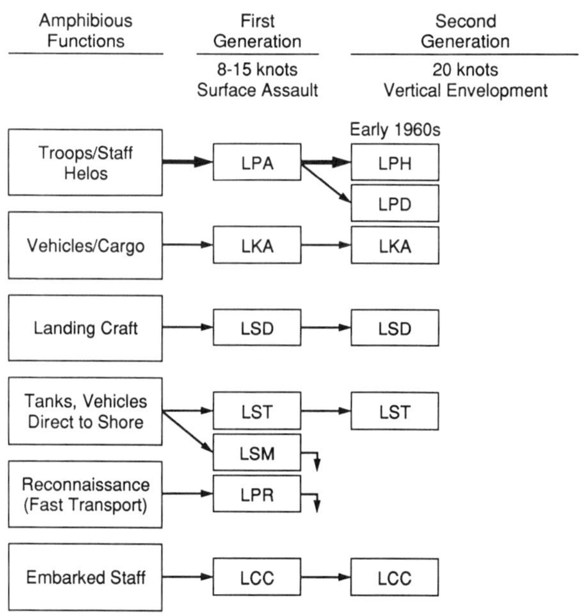

Fig. 3 Amphibious ship evolution—second generation, first approach.

Embarked Staff. One post-World War II-class LCC has been introduced into the fleet. This is the LCC-19 class, with two ships, commissioned in 1970 and 1971. These ships use a hull form similar to the LPH, and therefore are larger and faster than their predecessors. The role of these ships in future operations is not clear, since they have been used as fleet flagships from the beginning. Speed is 20 knots.

Second Generation—Second Approach

In the early 1960s, the early years of vertical envelopment and the second generation of amphibious ships (see Fig. 4), a study (NAVWAG-44) was conducted that indicated that the cost of the amphibious capability could be reduced substantially through the use of fewer, larger ships and fewer kinds of ships. This led to the development of the LHA, a troop transport with aviation facilities, as in the LPH, but also including a large wet well and substantial quantities of vehicles, cargo and POL. The LHA was procured under the auspices of Total System Responsibility, where the mission requirement was stated by the government, but the total design and construction was accomplished by industry. This ship, with size constrained only by the Panama Canal, takes advantage of the economies of scale and provides both vertical and surface assault capabilities in the same hull. It was envisioned that one LHA, assisted by one or two well deck or tank deck ships, could conduct BLT (Battalion Landing Team) -size operations, and that multiples of these ships could conduct larger operations. Subsequent studies support the conclusion that the use of fewer but larger ships and fewer kinds of ships reduce overall force costs. The LHA speed is 20 knots.

The remaining ships shown in the second generation, LPD, LKA, LSD, LST, and LCC all have 15 or more years of useful service. They will therefore be used for a number of years in amphibious task forces built around LHAs as well as LPHs. The LHA has substantial command ship capabilities, which brings into question the future use of the LCC in amphibious operations, as mentioned earlier.

Second Generation Summarized

In summation, the second-generation amphibious force is characterized by the following:

- seven ship types, including general and special purpose types,
- ship-to-shore via vertical as well as surface means,
- surface ship-to-shore limited to short ranges,
- landing craft largely in wet wells, preloaded with vehicles or cargo,
- substantial capabilities for underway launch of amphibian vehicles and landing craft, and
- 20-knot speed.

Third Generation

The third generation of amphibious ships (see Fig. 5) has arrived; it is being ushered in by the development of the Landing Craft Air Cushion (LCAC) and Vertical/Short Take Off and Land (VSTOL) aircraft. The LCAC provides a standoff capability for surface assault comparable to the current vertical assault capability, and VSTOL provides amphibious forces with some air support capabilities. The LHA will be the mainstay of this generation for a number of years. It is a major troop transport and helicopter aviation ship. It has capabilities to debark troops, vehicles and cargo via vertical or surface means. It can debark via LCAC but can transport only one LCAC because of the "pier" in the center of the LHA well deck. Follow-on to the LHA is the LHD, which is a modified LHA. It replaces LPHs as they retire and provides additional capacity that may be needed. The LHD has improved compatibility with LCAC and with VSTOL. The LHA and LHD are the centerpiece of amphibious warfare shipping. The outstanding command and control capabilities suggest that the LCC is not needed

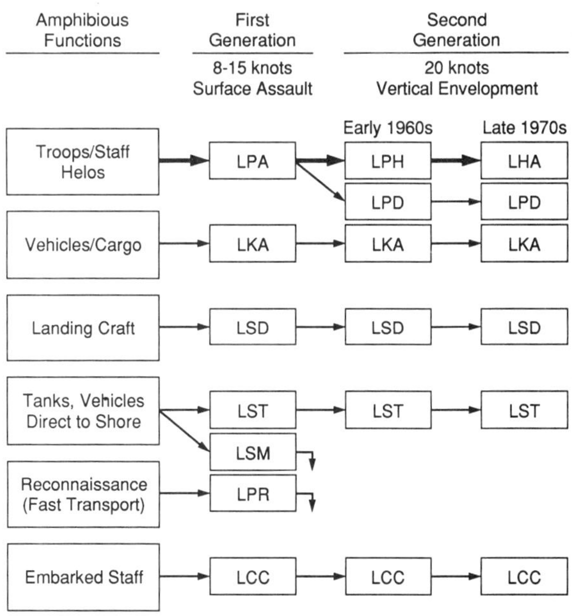

Fig. 4 Amphibious ship evolution—second generation, second approach.

The 1171 class was the last of the WWII style LSTs, that is, vessels fitted with bow doors and a main deck bow ramp. These followed the 1156 class. The 1156s, following a more or less traditional hull form but with higher power, suffered such severe propeller-induced hull vibration that it was necessary to entirely rebuild the stern portions.

The re-designed form was a semi-tunnel stern and resulted in ships that were almost free of propeller-induced vibration. Dick Couch was at the Taylor Model Basin during this period and was much involved in developing this highly successful design.

There were six ships in the 1171 class. The Navy had determined that it would be desirable to develop a facility on the Great Lakes with the capability and experience to build modern Navy vessels. The American Ship Building Company was the successful bidder for the two ships assigned to the Lakes. Newport News and Avondale were each assigned two. It was also decided that economies could be achieved with a joint procurement program. Accordingly, material procurement was divided between the three yards, each responsible for about one-third of the material purchasing.

These ships, although of the same general configuration as the LSTs of WWII, were a good deal more sophisticated. The lead vessel, 1171, was built at the Boston Navy Yard with production drawings produced by Gibbs & Cox (1172 was cancelled). Avondale had 1173 and 1174, Newport News had 1175 and 1176 and American Ship 1177 and 1178. Although they started out to be sisters, they of course incorporated a variety of design changes along the way. The basic design provided for powering with six engines divided in two engine rooms with twin screws. The Newport News ships ended up with four engines and were equipped with controllable-pitch propellers. Midway in construction, the American Ship vessels were modified by the addition of a 'tween deck to serve as troop carriers. They were also fitted with cargo tanks and a cargo pump room in the lower hull.

Although scheduled for delivery in 1957 via the Illinois–Mississippi waterway, only the first made this trip. During tank testing, it developed that the very sophisticated tank coating used in the ballast tanks suffered seriously when exposed to fresh water. Indeed, the material that had not come off on its own, had to be physically removed and the tanks recoated. To transit the Chicago Sanitary and Shipping Canal, it was necessary to remove the upper deckhouses and stow them on the main deck and also to precisely

for an amphibious mission. Furthermore, the LHD has a secondary mission capability as a sea control ship by maximizing the AV-8 Harrier load carried.

The payload flexibility of the LSD type of ship makes it a key player in the third generation. The number of landing craft needed in amphibious operations varies widely with the scenarios. This is true whether using conventional craft, LCAC, or a mix. Therefore, the force needs capabilities to handle large numbers of craft, and it also needs capabilities to convert any unneeded landing craft space for other users. The LSDs, particularly those with portable mezzanine decks, provide these capabilities. Unlike its predecessors, the LSD-41 class does not require mezzanine decks to meet its cargo, vehicle and landing craft capacity requirements. The LSD-41 class is the principal LCAC carrier, and is capable of embarking four of them in the well deck.

The LPD is a modern, general-purpose amphibious ship with both vertical and surface debark capabilities and is expected to be able to transport two LCACs. It is a relatively new class of ship from the second generation, which will most likely be a major factor in the third generation for many years. A new "class of LPD," the LSD(CV), is a cargo variant of the LSD-41 class, which evolved from the LPDX studies. The LSD(CV) has the same diesel propulsion system as the LSD-41. The well deck has been shortened to permit greater cargo capacity than that in the LSD-41. It carries two LCACs. The approach of modifying the LSD-41 to increase cargo capacity at the expense of vehicle capacity provides an LPD type of ship while retaining the same basic LSD hull.

As a follow-on to the LSD-41 and the LSD(CV), the Navy is studying the LX, which is envisioned at

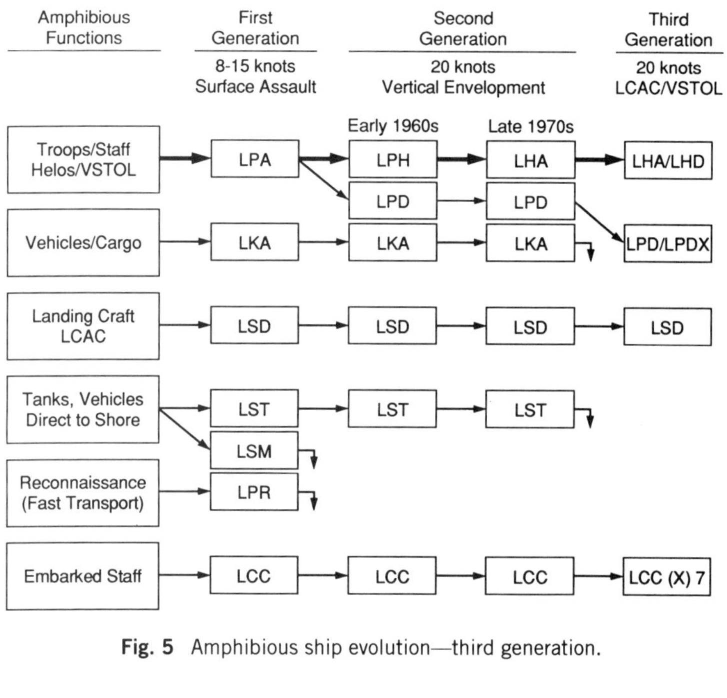

Fig. 5 Amphibious ship evolution—third generation.

ballast the vessel. As fresh water could not be used for ballast due to the tank coating problem, the vessel called at Port Huron to load salt brine available from the salt wells in the area. The transit of the canal allowed only about two feet of clearance between the bottom of the canal and the lowest of the fixed bridges. Even in the ballasted condition, it was necessary to drop the water level in the canal about five feet to permit passage. This was the last major vessel to transit this route. After the canal passage, the vessel was integrated into a river tow and delivered to New Orleans for final outfitting and sea trials.

The second vessel held builders and acceptance trials on Lake Erie—an operation that resulted in an embarrassing situation. Lake Erie being very shallow, it was difficult to find a shore where the bottom gradient was steeper than the keel drag of the ship. An area was finally located in a cove of Kelly's Island, one of the group in the western part of the lake. The ship left the yard at Lorain and headed immediately for the beaching area. The operation was carried out with the ship under command of an experienced Navy beaching officer. An underwater cable near the mouth of the cove prevented the dropping of the stern anchor as far out as would be normal practice. The affair was carried out as though the vessel were making a combat landing, charging up on the beach at nearly full speed. Beached, the bow doors were opened, the ramp dropped and somewhat of a general exodus onto the beach for picture taking, etc., took place. In due course, the ramp was raised, the doors were closed, the engines reversed and the stern anchor winch activated. The anchor pulled in nicely, but the vessel didn't budge! There had of course been no unloading of cargo and there was no ballast to discharge. All other minor efforts such as dropping the bow anchor were of no avail, even the most unique engine order—"Give her all she's got, Gus!"—had no effect. The morning came and went, lunch was served and most of the afternoon was gone before a passing tug with a barge in tow was hailed. After parting two or three hawsers the ship was finally freed and headed home. Aside from this incident, the trials were highly successful and the delivery voyage to Norfolk completed via the not-yet formally opened Saint Lawrence Seaway. Incidentally, SNAME members participating in the 1959 Spring Meeting cruise on the SS Tadousac *were thrilled to be called on deck during an early morning passage through the Thousand Island area to view the* Wood County *as she charged past.*—W. C. Cowles

this time to be an LPD kind of ship. The notional LX is expected to carry about 700 troops and will have a capacity for 25,000 cubic feet of cargo. Recognizing the current emphasis on regional contingencies and rapid deployment, the LX will provide the versatility of the LPD, LHA, and LHD with its well deck and flight deck. It will carry the LCAC, but the quantity has yet to be determined. As many as eighteen LX ships will be required to support the specified three Marine Expeditionary Brigades. The initial deliveries should occur in the years 2000 or 2001.

The LKA appears to have been overtaken by events. The vehicle and cargo functions are being filled to a large extent by LHA and LPD types of ship. Further, even though the current LKAs are substantially improved over their predecessors, they lack underway launch capabilities and LCAC compatibility. Although it is a low-cost ship, in terms of cost per unit of lift, it is likely that any cost savings that might result from its use would be more than offset by the cost of acquiring and maintaining an additional kind of ship.

Like the LKA, the LST appears to have been overtaken by new technology. The tank/heavy vehicle function can better be filled by tank-capable craft and well deck ships. Tanks can be put ashore by craft early in the assault under circumstances in which the force commander would be unwilling to risk ships such as the LST-1179 class. Underway launch of LVTs can be accomplished with less difficulty and at higher speeds from small wet well ships (LPD/LSD) than from LSTs. There is substantial risk in beaching the LST. The classic World War II photos of beached LSTs sitting side-by-side show what easy targets they are. Further, the LST is the most costly ship of the amphibious force in terms of cost per unit of lift. We can not only launch underway now, but with the helicopter and the LCAC we can launch "from over the horizon."

As stated earlier, the future role of the LCC is not clear. It may or may not be a major player in the third generation of amphibious ships. While Fig. 5 shows an LCC(X), the attraction of using such a ship for amphibious operations is small, based on the tremendous capabilities of the LHD.

Third Generation Summarized

In summation, the third-generation amphibious force is characterized by the following:

- three types of ships: one large troop transport/aviation and two smaller wet well types,
- over-the-horizon ship-to-shore, vertical and surface capability,
- LCAC or LCAC plus large conventional craft preloaded in wet wells,
- underway launch capability, and
- greater than 20-knot speed.

The Maritime Prepositioning Force Concept

The Maritime Prepositioning Force, conceived in the Marine Corps Mobility Enhancement (MCMOBE)

Fig. 6 USS *Lenawee* (APA-195).

Study of late 1979, was based on the procurement or charter of merchant ships to be manned by civilian crews of the Military Sealift Command. These forward deployed logistics ships were to enhance the mobility of the Marine Corps to respond to any contingency in the Indian Ocean, principally for the Middle East and Persian Gulf. Two potential scenarios for comparison with other deployment capabilities were used. One called for rapid deployment of a Marine Brigade into a permissive, non-hostile environment. The other called for the deployment of a Marine Brigade to achieve rapid force buildup after seizure of a port by other amphibious forces. Thus was born the concept of prepositioning supplies at sea.

The ship program consists of thirteen combination RO/RO/containerships. Five were built new, eight were conversions. Each carries a balanced spread load of unit equipment, ammunition, supplies, water, and fuel. The mission requirement was that each ship be capable of self offloading in a benign environment. Each squadron of Maritime Prepositioning Ships (MPS) is capable of providing needed supplies to a Brigade for thirty days. The number of ships in each squadron varies because of differences in size. The five new ships were built at General Dynamics, Quincy Shipbuilding Division. Five of the conversions (from Maersk Line) were done at Bethlehem Steel's Sparrows Point MD and Beaumont TX yards. The other three conversions (from Waterman Steamship Co.) were done at National Steel and Shipbuilding Co., San Diego CA. All contracts were let in August 1982 and the entire program was complete by early 1986. The first Maersk conversion was completed in August 1984, the first Waterman conversion came along the following month, and the first of the new ships (*Braintree*-class) was completed in March 1985. Thus, the first of each conversion was completed within two years of contract award, and the first of the new ships within three. By the summer of 1986 all three MPS squadrons were operational. One squadron comprises one *Braintree*-class and three Waterman-class ships. The second comprises five Maersk conversions. The third comprises the remaining four *Braintree*-class ships.

A second Mobility Enhancement Program provides two Aviation Logistics Support Ships (TAVBs). These care for the Marine Corps' landing force aviation support, either to the Brigade associated with the Maritime Prepositioning Ship squadron, or the amphibious assault ships. The TAVB ships are Seabridge-class RO/RO/containerships converted at Todd Shipyards, Galveston TX during 1984–1986. In peacetime, these ships are in port in a reduced operating status under aegis of the Military Sealift Command. During operational scenarios these ships embark and transport aviation stores and maintenance support to the Marine Corps' air element associated

Fig. 7 USS *Dawson*.

Fig. 8 LST-125.

Fig. 9 LSM-253.

Fig. 10 LSM(R)-512.

Fig. 11 USS *Lorain County*, LST-1177 (1958).

with the Marine Brigade. The TAVB provides for the aviation support offshore until a base is established on land.

The Mobility Enhancement Programs, both the MPS and the TAVB, of the 1980s add a new dimension and flexibility to the amphibious capabilities of the United States. While these are commercial types of ship, not intended for hostile environments, they complement the amphibious assault forces. In the era of low- to mid-intensity conflicts, the marrying up of afloat prepositioned supplies to Marines flown to the area offers yet another flexible response capability to meet our national objectives.

Summary

From the development of the first-generation (World War II) amphibious force, we have discussed the evolution to the current force as well as the beginning of a future, third generation amphibious force. Several observations worthy of emphasis are the following:

- The development of general-purpose amphibious ships, such as LPD, LHA, and LHD has reduced the number of types of ship needed.
- The use of fewer, larger hulls substantially reduces the cost of acquiring and maintaining amphibious forces.
- The use of wet wells in troop/vehicle transports substantially improves debarkation (underway launch of preloaded craft).
- Wet wells plus helicopter pickup spots in all amphibious ships improve ship-to-shore flexibility.
- Two events having major impact on amphibious forces are the introduction of helicopters in the 1960s and the introduction of LCAC in the 1980s.
- Future amphibious ships may use dry wells if the force uses the LCAC landing craft (or other ACV) exclusively. The ship does not need to fully ballast down for the LCACs to enter or exit the well. LCACs fly into and exit the well of the amphibious ships with the ships underway.

As we look at Operations Desert Shield/Desert Storm in the summer of 1990, we see the obvious role of the Maritime Prepositioning Ship (MPS) program in support of the projection of amphibious forces. First on the scene, the MPS force from Diego Garcia in the Indian Ocean married up with the fly-in echelon of Marine Corps troops, using roughly 250 airlift sorties, to provide the early presence of U.S. sustained capability in the Persian Gulf in response to Iraq's invasion of Kuwait. All three MPS squadrons supported Desert Storm. Each provided thirty days of supplies, ammunition, unit equipment, water, and fuel for a 16,000-man Marine Expeditionary Brigade. The performance of the MPS squadrons validates the effort and resources devoted to the concept. Future use of the MPS is being considered for other deployment options of less than brigade size. The U.S. Army, based on the utility of MPS for the Marine Corps during Desert Storm, is considering afloat prepositioning to support its own requirements.

Additional Reading

Homer, A., Humphreys, G. C., and Swiderski, F. E., "Propulsion Diesel Engines for Landing Craft and Small Boats," *Trans.* SNAME, Vol. 56, 1948.

Glossary

ACV—Air Cushioned Vehicle—a type of advanced hovercraft. LCAC is an ACV.
AE—Assault Echelon—that element of a force that is scheduled for initial amphibious assault.
AFOE—Assault Follow-On Echelon—troops, vehicles, and supplies not required to initiate the assault. Embarked in merchant ships providing reinforcement/sustainability.
AOA—Amphibious Objective Area.
BLT—Battalion Landing Team—the ground combat element of a marine expeditionary unit (MEU).
CNA—Center for Naval Analyses—the Navy's "think tank."
LCAC—Air Cushioned Landing Craft.
LCC—command ship (earlier known as AGC).
LCPR, LCVP, LCM—small landing craft.
LCU—Landing Craft, Utility—a large landing craft.
LHA—a large, air capable, amphibious assault ship.
LHD—a ship similar to the LHA, with significant improvements; next generation LHA.
LKA—attack cargo ship.
LPA—troop transport with supplies and equipment.
LPD—troop carrying ship with docking facilities.
LPH—amphibious assault helicopter carrier.
LPR—a small troop transport for underwater demolition teams (UDT).
LSD—Landing Ship, Dock.
LSM—Landing Ship, Mechanized.
LST—Landing Ship, Tank.
LX—the notional next generation amphibious ship planned for the year 2000 and beyond.
MCMOBE—Marine Corps mobility enhancement study —joint Navy/Marine Corps study conducted together with CNA which developed the MPS concept.
MEB—Marine Expeditionary Brigade—4000 to 18,000 troops for amphibious ships; 16,000 troops for MPS. Regimental size with 30 days sustainability.

MEF—Marine Expeditionary Force—30,000 to 60,000 troops. Division size with 60 days sustainability.
MEU—Marine Expeditionary Unit—1000 to 4000 troops. Battalion size with 15 days sustainability.
Mobility enhancement—MPS and TAVB ships.
MPS—Maritime Prepositioning Ships—Predeployed commercial type ships with unit equipment/supplies.
MRS—Mobility Requirements Study—a 1991 study conducted by the Joint Chiefs of Staff establishing the amphibious force and sealift sustainability requirements for the 21st Century.
NAVWAG 44—a study performed by the Center for Naval Analyses. This led to development of LHA.
POL—petroleum/fuel.
RO/RO—Roll On/Roll Off—a commercial ship that can load and offload by driving cargo aboard/ashore.
Sealift enhancement—FSS (Fast Sealift Ships-TAKR, ex-SL-7s) and TACS (auxiliary crane ships).
TAVB—civilian manned aviation support ship. Part of the mobility enhancement program.
VSTOL—Vertical/Short Take-Off and Land aircraft—current Marine VSTOL is the AV-8B Harrier.

Capt. Charles H. Piersall (USN ret.) was the U.S. Navy's program manager for Amphibious Warfare and Strategic Sealift from 1978 through 1985. During that period he presided over the rebirth of amphibious warfare, turning out 28 ships valued at a total of $24.5 billion. These included ships of seven different designs. He holds the ASNE Gold Medal and the Navy's highest noncombat award: the Distinguished Service Medal. He serves on several national committees, including three with SNAME.

Fig. 12 *Tuscaloosa*, LST-1187.

Fig. 13 *Tuscaloosa*, LST-1187.

Fig. 14 LSD-48.

Fig. 15 LSD-1.

Fig. 16 *Mount Whitney*, LCC-20.

Fig. 17 USS *Tawara*, LHA-1.

Fig. 18 LSD-41 with 3 LCACs.

Fig. 19 *Wasp*, LHD-1.

Fig. 20 TAVB-15.

Postwar Canadian Naval Vessels[1]

W. J. Broughton

Development of a Canadian Warship Design Capability

At the close of World War II, Canada had the third largest western allied fleet after the United States and Royal Navies. Despite Canada's remarkable ship construction achievements during the war, Canada lacked the capability to design its own warships. Immediately after the war, Canada rapidly reverted to its traditionally small military posture. Some 500 commissioned warships and 90,000 naval personnel were reduced to 44 ships and 7000 sailors by 1948. Nevertheless, key lessons learned were not altogether forgotten. Steps were taken early on to retain the more highly capable vessels and to develop a design team. While four wartime Tribal-class destroyers were being completed, plans were made to develop a new design under the technical leadership of the Chief of Naval Technical Services (CNTS), RAdm(E) J. G. Knowlton. He clearly was the father of the new RCN, in a managerial sense, as the CNTS from 1945 to his retirement in 1956. From the design viewpoint, the RCN owes an enormous debt to the U.K. Royal Corps of Constructors, and in particular to Constructor Commodore Sir Rowland Baker, who was on loan from 1948 to 1956, first as the Director of Naval Construction, and from 1953, as Constructor-in-Chief.

Key decisions made by these two officers gave the Canadian Navy excellent warship design capability that continues to this day. Initially, they recruited large numbers of well-trained civilian naval architects, engineers, and technologists, largely from the U.K., as well as Royal Navy technical officers (on loan). They also established educational programs, which involved sending Canadian officers to both Britain and the United States for graduate-level training in naval architecture, marine engineering, and combat technology.

Real urgency to this endeavour occurred in 1948 with the announcement of a new destroyer-building program. The ships were to be both capable and suitable for rapid production in a variety of Canadian shipyards. When the first of the class, the *St Laurent*, was commissioned in 1955, the staff of CNTS had grown to about 400 from the 84 in 1948 [1]. (This figure had shrunk to about 230 in 1975 and is back to 400 with the current City-class construction, of which more later.)

But there was much more technical expertise involved than just the design offices of the CNTS. Traditionally, the Navy prepared the specifications for the ship and its equipment, plus a preliminary design. Then industry prepared the detailed design, bills of material, and purchase orders before proceeding to acquire the material and construct the ships. Although familiar with the process, Canadian industry had a limited capability to handle the technical requirements of warships. In order to overcome this deficiency, arrangements were made to establish the Naval Central Drawing Office (NCDO) and the Naval Ship Construction Purchasing Agency (NSCPA). These were industry manned and operated in the lead yard, Canadian Vickers, Ltd. of Montreal, but were under the cognizance of an on-site Principal Naval Overseer. These arrangements remained in place for four destroyer programs plus several conversions over 25 years.

In 1952 a second technical agency was created, the Naval Engineering Test Establishment (NETE). It is sited in Montreal. Its central mission was to test (and perhaps modify) Canadian-made naval auxiliary equipment for shock, vibration, humidity, and the like. Now its work also covers control systems, computer technology, and health monitoring. As with NCDO and NSCPA, the agency is government owned but industry operated, in this case by Peacock Brothers, Inc.

NEDIT, the Naval Engineering Design Investigation Team, had a shorter history as a separate technical unit. It was established in 1953 to provide a design

[1] The views expressed are solely those of this author and do not necessarily represent those of the Canadian Department of National Defence nor those of the authors of the referenced articles.

assessment capability for marine propulsion systems. It was decided to have a team of civilian engineers, with a civilian employer (again Peacock Brothers) reporting through a naval officer to the Engineer-in-Chief. The employees of NEDIT were transferred to the Federal Public Service in 1957, but the team was disbanded in 1966, being absorbed by the naval engineering division at National Defence Headquarters.

Thus it was through the rapid expansion of the technical staff at Headquarters, the establishment of NCDO, NSCPA, NETE, and NEDIT, and the initiation of on-going post-graduate technical education that Canada moved successfully into the business of warship design in the early 1950s. Similar expansion of commercial design capability occurred in the private sector. Clearly there was a need for a more permanent local forum for exchange of professional information. Largely under the leadership of Mr Richard Lowery, later President of Canada Steamship Lines, the Eastern Canadian Section of SNAME was established in 1952. Later additions have been the Canadian Atlantic Section (1962), and the Arctic Section (1981). In addition, Canadian members are regularly active in SNAME's Great Lakes and Great Rivers Section, and in the Pacific Northwest Section.

The requirements for expertise and organization in warship design are well known and understood. But the proof is in the results. It is now appropriate to turn to Canadian programs for warship design and construction since World War II.

The Canadian *St Laurent*-Class and Follow-On Destroyers

The newly developed technical expertise was first applied to the design of destroyer escorts (DDE). These were built in four groups numbering seven, seven, four, and two ships each. Named after Canadian rivers, each group was identified by the name of its first ship. The lead-off ship was the *St Laurent*, so the first group of seven was known as the *St Laurent*-class.

From the staff requirements to the design concepts, to the execution of details, the *St Laurent*-class project was ambitious and innovative. Many new features were incorporated, some based on the recent experience of the Battle of the Atlantic. The twenty ships were commissioned from 1955 to 1965, followed later by three different, major modifications.

The basic mission was to be anti-submarine warfare for the protection of shipping, but in an environment that had vastly changed with the development of long-range aircraft and nuclear bombs. Accordingly, the staff sought ship characteristics that included a sustained speed of 17 knots in a seaway, excellent seakeeping, low self-noise, nuclear, biological and chemical defence features (NBCD), and improved damaged stability and damage control. Improved habitability and amenities were also seen as a needed response to an all-volunteer force drawn from a nation with a high and rising standard of living. Lastly, Canada's decision to sustain its move to "made-in-Canada" called for relatively small size and mass producibility.

As noted earlier, Commodore Baker, who was on loan from the Royal Corps of Naval Constructors, led the project team. More significantly, he brought a personal imprint to the design itself. His new ideas and approaches found much fertile ground. Starting with the hull, he changed the traditional warship approach from design for top speed in calm water to design for sustained speed in a seaway. The result was fine lines forward below the waterline but large flare above, with high freeboard. A standard was adopted that there was to be no negative righting arm (GZ) due to damage caused by holing the hull. This last feature complemented Baker's approach of liberal design margins to cater for future growth and weapons to be added in time of conflict [2]. So farsighted were the hull characteristics, that the last follow-on class and the *St Laurent*-class conversions accepted a hangar, hangar deck, and the large Sea King helicopter without altering the hull, adding significant ballast, or lowering the damage control standard.

In terms of operational service, these ships have been employed in a variety of duties spanning over forty years, including the recent Persian Gulf War.

The *St Laurent* design was the first NATO ship with the citadel concept of closing down with some internal over-pressure to keep out airborne contaminants for NBCD purposes. Gas mask filters in the air intakes solved the air replenishment problem. Other firsts included a saltwater washdown system for the upper works, assisted by a reverse forecastle sheer on the flush deck and rounded deck edges, the removal of numerous upper deck fittings, enclosed anchor windlasses, anchors stowed behind flush anchor pocket doors, and the capability to close down rapidly with ingress decontamination centres.

That the design incorporated features that added weight, such as air conditioning, extra insulation and full NBCD provisions, special attention was given to saving hull weight. This was achieved by adopting a weight-efficient egg-box construction of longitudinals and frames. This also helped to retain the shape of the sub-units of the hull. (Construction of the hull in units

Fig. 1 *St Laurent* class—HMCS *St Laurent*: 2700 tons, 366 ft, 28 knots, 30,000 shp, 4750 n.mi, two twin 3 in./50 guns, two Mk 10 triple mortars, two triple torpedo mounts, AN/SQS-502 sonar. (Dept. National Defence photograph.)

facilitated sub-contracting steelwork as a procedure that could be useful in an emergency building program.)

Again, as a weight saving measure, the ship's superstructure, masts, and much of its equipment were made of aluminum alloy. In all of this, the Canadian industry provided great assistance.

To help in developing the *Laurent*-class destroyers, a private Quebec company, Sorel Industries, was encouraged to establish a capability to produce such sophisticated military equipment as rapid fire gunmounts. Similarly, in 1955 Sperry of Montreal developed a synchro gyro repeater and plotting table incorporating the first transistor amplifier.

At the start of the *St Laurent*-class project, there was virtually no Canadian capability to produce warship propulsion systems. A version of the British Y100 steam plant (550 psi, 850 °F) was developed to meet specific Canadian requirements. Production capability was developed for boilers, turbines, gearing, controls and auxiliaries. These were all changed from British to Canadian standards for such basics as dimensions and screw threads. A fundamental change was to hardened and ground reduction gearing.

It was not always easy. Practical production difficulties arose from the lack of centralized facilities. Take steam turbine rotors for example. The ingots were poured from furnaces in Montreal, then shipped 800 miles east at 1600 °F (in insulated boxes) to a 7000-ton forging press in Nova Scotia, followed by machining and assembly 1200 miles west in Toronto [3].

Perhaps the most striking fundamental change was in habitability. In Baker's own words,

> The welfare of our men is our paramount concern. The Naval Service is a fighting Service, but in peace the hardship of seagoing must be reduced as much as possible so that in war the added hardship will not of itself make the struggle for survival in the ships so great that men have no margin to struggle for victory . . . The ships must be regarded as home [4].

and

> It cannot be denied that at first the general idea was that recognized military characteristics . . . had a priority in the design field. By accident, partly, it was discovered that habitability, if it was to be satisfactory, had to be treated just as seriously as these other items. In fact, it seems to [me] that the whole secret of design lies in the layout—to be blunt, anyone can (with the help of model basins) get the correct speed, anyone can get the stability right, and nine times out of ten no one will ever know or care whether the constructor succeeded or not—but if the ship is hard to live in, everyone will know and say so.

In order to meet the foregoing aims, the *St Laurent*'s accommodations were accorded many improvements over those in earlier destroyers. To begin with, Baker would not allow a crew complement that could not be comfortably housed in a hull of a given size. Other features included:

- Main passageways were not allowed through accommodation spaces.
- Eating and sleeping spaces were kept separate.
- Sanitary spaces were distributed in numerous, convenient locations.

Fig. 2 Improved *Restigouche* class—HMCS *Restigouche*: 2900 tons, 371 ft, 28 knots, 30,000 shp, 4750 n.mi, twin 3 in./70 gun, ASROC, Mk 10 triple mortar, two triple torpedo mounts, AN/SQS-505 hull-mounted sonar and variable depth sonar (VDS). (Dept. National Defence photograph.)

- Hammocks were replaced by bunks.
- Items of equipment were isolated in their own enclosed spaces. This greatly reduced the problem of noise in the accommodations.
- Full air conditioning was used in all living and operational spaces.
- Careful attention was given to making the ship easy to maintain and keep clean.

To those who argued about the cost of these innovations, Baker responded that they came to only five percent of the total, when "the men themselves are the most valuable equipment of all."

Seven ships of the *St Laurent*-class were built. The ample margins and sound basic design of these ships eased the task of introducing variations in follow-on ships. A second group of seven destroyer escorts, the *Restigouche*-class, featured some improvements in gunmounts, fire control, and air conditioning. A planned six-ship repeat, the *Mackenzie*-class, was reduced to four in number, the final two being changed in design from destroyer escort (DDE) to helicopter-carrying destroyer (DDH) of the *Annapolis*-class.

The *St Laurent*-class and its follow-ons are almost history now as they are being replaced by the new City-class, although some are still serving. The design

Fig. 3 *Annapolis* class—HMCS *Annapolis*: 2600 tons, 371 ft, 28 knots, 30,000 shp, 4750 n.mi, CH-124 Sea King helo, twin 3 in./50 gun, Mk 10 mortar, two triple torpedo mounts, AN/SQS-503 hull-mounted sonar, AN/SQS-504 VDS. (Dept. National Defence photograph.)

was innovative and has proved capable and enduring. It deserves to be included among Canada's great achievements in engineering.

Canadian System Developments in the 1950s and 1960s

Even as the first Canadian warships were being designed and built, a number of R&D projects were initiated that were to put the RCN in the forefront of anti-submarine warfare (ASW). The more significant of these were the variable depth sonar (VDS), propellers with high cavitation inception speeds, integrated command and control, and the helicopter hauldown and rapid securing device (HHRSD).

Scientists at the Naval Research Establishment (NRE) in Dartmouth, N.S., proposed a way to overcome the two chief causes of sonar performance degradation: ship self-noise and ocean temperature gradients. The concept was to separate the transducer from the ship by towing it in a pod below the temperature inversion layer. The idea was simple but the execution was fraught with difficulties, largely with the towline. Round lines led to high drag and vortex instability. Faired cables were developed, but only after overcoming severe problems with streaming, retrieving, and stowing. Ship motions jerked and broke cables, requiring the development of a special boom-bobbing mechanism. By 1959 the problems had been overcome and system fitments had begun.

In 1961 the development of a new generation VDS was begun, leading to an order of magnitude increase in body size to the seven-ton *Diana One*. It was the RCN's first completely solid state system, and it was retrofitted on existing ships in the late 1960s and specified for the upcoming Tribal-class.

Concurrent developments led to significant reductions in ship self-noise. Chief among these were improvements in propeller cavitation inception speeds arising from work at the Hydrodynamics Laboratory at the National Research Council in Ottawa. Coupled with this were developments leading to better quality control for shape accuracy in the manufacture of propellers. In due course, the new propellers were retrofitted on all twenty River-class ships.

The potential benefits of integrated command and control were recognized early. By the early 1960s, development in Canada was advanced sufficiently to incorporate central control of combat systems into a new ship design, the General Purpose Frigate [5]. Although this project was cancelled by the government at the contract design stage in 1963, Canada did not lose its momentum. Attention was shifted to the experimental 200-ton seagoing hydrofoil boat *Bras D'Or*, commissioned in 1968. The experience learned with this small system (of which more later) was instrumental in establishing the data flow and interface standards that were set for the Tribal-class destroyers, then in the preliminary design stage.

But of all the successful system developments in the 1950s and 1960s by the RCN, the most ambitious by far was to marry the large ASW helicopter to the small destroyer. With the purchase of Sikorsky HO4S helicopters in 1955 the possibility of the frigate-with-helicopter concept was revived. A temporary flight deck was constructed on the *Buckingham*, a World War II *Prestonian*-class frigate. Three months of rough weather trials were conducted in 1956 using the smaller Sikorsky S58 helicopter. More trials followed on the destroyer *Ottawa*. It became clear that the limiting factor to operations was not the helicopter or the pilot, but the ability of the ship's crew to secure and traverse the helicopter safely.

A solution to the landing problem was the "beartrap" developed by Fairey Aviation of Dartmouth, N.S. The system consisted of an open box on deck fitted with jaws between which the ship's crew could feed a wire dropped by the helicopter. This was led to a constant tension winch, which then proceeded to pull the helicopter down to the deck. The jaws were then snapped shut on the helicopter probe; tailguide bars on deck aligned the helicopter fore and aft, and it was moved into its hangar on a special track.

The first such full fitment was aboard the *Asssiniboine* in 1964–65. In order to make the upperdeck space available for the hangar and flight deck, the funnel was split port and starboard, and one gun and one mortar were removed. In 1962 Canada ordered the Sikorsky Sea King helicopter with dunking sonar and ASW torpedoes. Trials began in 1964 and proved the system met all specifications including all-weather operations, day or night in Sea State 5 giving up to 31 degrees roll, 9 degrees pitch, heaving at 20 ft per second, and in winds up to 50 knots. Canada's Navy was the first to operate ASW helicopters from destroyers. Subsequently the Italian, Japanese, and U.S. navies bought helicopter hauldown systems based on the Canadian beartrap design. Now, however, modern electronics are changing the concept to signal-controlled fly-ons.

The Canadian Hydrofoil Program, the *Bras D'Or*

Canada has long been associated with hydrofoil craft. It started with that famous Canadian-American-Scot,

Fig. 4 Canadian ocean hydrofoil HMCS *Bras D'Or*: first time foil borne, 200 tons, 150 ft, 2000 HP diesel hull borne, 22,000 hp gas turbine foil borne, over 60 knots. (Dept. National Defence photograph.)

Sir Alexander Graham Bell, whose ladder-foiled boat set a world water speed record of 70.86 mph back in 1919. It remained a record for hydrofoil craft until the late 1950s [6]. As subsequent weight-saving advances in hull and machinery made the concept more practical, Canadian naval authorities kept the potential in mind. The demands of antisubmarine warfare seemed to exploit the unique characteristics of hydrofoil craft [7]. This was finally realized in the 200-ton *Bras D'Or* (FHE 400), designed by the de Havilland Aircraft Company of Toronto and built by Marine Industries, Ltd. in Sorel, Quebec.

Bras D'Or had many novel features. The unique Canadian tail-first (canard) configuration, with surface-piercing foils, was inherently stable when foil borne, and gave excellent seakeeping performance both foil borne and hull borne. The 150-ft hull was built of aluminum alloy and the foils were of very high yield (250 ksi) maraging steel. The supercavitating bow foil rotated to provide steering either hull borne or foil borne. The main foil unit carried the driveshafts and propeller pods. Propulsion consisted of two separate systems. Hull borne, a Paxman 16YJCM diesel of 2000 shp drove two 7-ft diameter, fully reversible, controllable-pitch propellers. Foil borne, a Pratt & Whitney FT4A-2 gas turbine of 22,000 shp drove two fixed-pitch, 3-ft diameter supercavitating propellers. All ship and combat systems were bridge-controlled.

Bras D'Or went on trials in 1968. In many respects she was a great success, attaining speeds over 60 knots and showing comfortable riding characteristics in 15- to 20-ft swells. In time, however, stress-corrosion cracks appeared in the foils, a problem never satisfactorily resolved. Meanwhile, changes in the Canadian Navy's basic role called for more versatile craft, and so the project was placed on hold in 1971 and eventually abandoned.

Despite the problems, the *Bras D'Or* proved a valuable test bed for some advanced technologies that were applied to other, more conventional ship programs. Notable among these advances were gas turbine propulsion, centrally controlled command, bridge control of main machinery, high speed faired cables for sensor pods, and improved filters for gas turbine intakes in saltwater atmosphere. In addition, valuable experience was gained in welding special steels and large aluminum alloy structures.

Mention should also be made of some of the *Bras D'Or*'s other pioneering features. Among the western navies we find the first supercavitating foil, the first CODOG plant (combined diesel or gas), and central bridge control of ship and combat systems.

Canadian Replenishment Ships

In 1957 the Navy started to design its first replenishment ship, the *Provider*. The requirement was for a ship to maintain a squadron of destroyers on station in support of a task force. It would be the fleet's one-stop supermarket at sea.

After examining various existing designs, the (American) Maritime Administration's *Mariner*-class cargo liners seemed a most suitable starting point for the replenishment ship. The lines drawing used, however, turned out to be a variant on the original (apparently modified in the afterbody to make room for nuclear machinery) [8]. This led to severe vibration problems, but these were overcome by switching to a six-bladed propeller. Upon investigation, it was found that the lines drawing was unsigned and without a known author.

The 20-knot *Provider* was built by Davieship in Lauzon, Quebec and delivered in 1963. With state-of-the-art equipment, she was capable of replenishing ships port and starboard with both fuel and stores. By and large, she was a successful ship and continues to serve the Navy well.

Continuing staff studies demonstrated that additional replenishment ships would do much to increase the Navy's ability to stay on station. Accordingly, two additional ships were approved in the mid-1960s. These were called Operational Support Ships (OSS). Experience gained on the *Provider* led to many improvements in the new ships. A sealift capability was added, as was a capacity to do helicopter maintenance and act as a command ship.

Given the *Provider's* initial vibration problems, the OSS design involved careful development of the hull form using a modified Hogner stern [9].

The steam machinery plant of the *Provider* was similar to that on the Mariners, and, having been found generally good, was used as a starting point for propulsion of the new ships. Some of the new ships' special requirements (such as unlimited bridge control) required some alteration in the main components.

Even more radical changes were effected in the overall arrangements. The additional operational requirements dictated such additions as a helicopter flight deck and hangar, more guns, improved communication and navigation systems, four landing craft, and a hospital complex. Lessons learned from the *Provider* suggested major improvements in the arrangement (and fittings) of the new ships' replenishment at sea gear [10]. Four dual stations were spaced 105 ft apart longitudinally to suit the majority of receiving ships, and 30 ft apart athwartships to satisfy equipment placement for leading the lines and for working areas.

In selecting and arranging the replenishment gear, particular attention was paid to safety. Special fittings included quick release mechanisms on hoses and rigging, emergency shut-off valves for fuel lines, and fail-safe arrangements on winches in case of loss of power. Proper control of the operation dictated careful attention to good visual and verbal communications. For example, the central control officer had an internal communication system with the liquid (fuel) control officer, who was equipped with a fuel distribution calculator for hull stress and trim management.

These new supply ships featured portable dunnage with modular shelving for the dry cargo, and various features to make them self-sufficient for loading in harbour. Lastly, the helicopter facility was arranged so as to allow vertical replenishment.

The new supply ships, named *Protecteur* and *Preserver*, were commissioned at the Saint John Shipbuilding and Drydock Company in Saint John, New Brunswick in 1970 and 1971. With the exception of the auxiliary diesels, there have been few problems and the OSSs continue to serve the fleet admirably well. Most recently, *Protecteur* performed the tactical coordination role for all allied seaborne logistic support groups in the Persian Gulf War [11].

Tribal-Class Destroyers

The 1964 Canadian government's White Paper on Defence stated the following as policy:

> To preserve peace by supporting collective defence measures, to support foreign policy, and to provide for the protection and surveillance of our territory, our air space, and our coastal waters [12].

As noted earlier, the impact on the Navy was to enhance general purpose capabilities while retaining the pre-eminence of ASW. After considering a series of costed operational analyses, the government confirmed the building of the two OSSs, the acquisition of three submarines, the improvement of the seven *Restigouche*-class destroyers and the construction of four destroyers to carry helicopters. The last item was to become the Tribal-class, but only after a tortuous period of project development.

In 1963, the previous government had cancelled the general purpose frigate. At about the same time as the new government's issuance of the White Paper, the shipbuilding industry was lobbying the politicians for orders to assist during its badly depressed state. The urge for a plan that could be rapidly implemented resulted in a 1965 proposal for four, "stretched-*Nipigons*" at an estimated cost of $146 million. It would have an updated Y100 steam plant, and the

Fig. 5 Operational support ship HMCS *Protecteur*: 25,000 tons, 546 ft, 20 knots, 21,000 shp, 7500 n.mi, twin 3 in./50 gun (later removed), three CH124 Sea Kings for maintenance. (Dept. National Defence photograph.)

same Sea King helicopter, but would be stretched to cater for missile systems, the larger 505 VDS and a command capability. This proposal, although approved, did not proceed on the proposed quick schedule. It would appear that several factors were in play. Even the best possible schedule would not give immediate relief to the yards. The resulting design was not popular with either the naval staff or the engineers. Concurrently with the development of the stretched-*Nipigon* preliminary design, two seminal studies were completed. One was operational and demonstrated a large increase in ASW cost-effectiveness for a two-helicopter destroyer. The other was technical and it expounded the savings in weight and cost for gas turbine propulsion. Finally, the stretched-*Nipigon* would simply not be big enough. The net result of all of these considerations was the abandonment of the stretched-*Nipigon*.

The preliminary design of the GP frigate was re-worked and became the DDH 280 Tribal-class. The basic hull form of the GP frigate was found to be feasible with an additional two feet of beam. Otherwise stability would be unsatisfactory because of the reduced machinery weight with gas turbines, and the increased topside weight and windage area with two helicopters, a missile system house and increased combat system electronic equipment.

It is a great credit to the engineers of the day that this massive change from the equipments that were to have been fitted in the stretched-*Nipigon* was so rapidly re-worked in the DDH 280 design, leading to a revised approval in 1967 at a cost of $220 million plus a $5 million contingency. Given the nature of defence procurements, it seems inconceivable that a contingency margin for a major development project such as a new warship should be only 2.25 percent. This project has had a poor reputation in some quarters based on the inappropriate comparison of the final ships with the initial stretched-*Nipigon* at $146 million. The ships, which were commissioned in 1972 and 1973, actually cost $242 million, or ten percent over the $220 million approval. Clearly this more proper comparison, showing an overrun of only ten percent, was an outstanding achievement given all of the developmental "firsts" that were achieved with this design.

The DDH 280 design had many innovative and successful features [13] that made it a truly remarkable success story [14]. These included this being the first destroyer in the free world to have all-gas turbine propulsion, to operate two ASW helicopters, to have bridge control of the propulsion plant, to have a centrally controlled, federated command and control system, an integrated interior communication system and to have such extensive noise-reduction measures. In addition, further improvements were made to the high Canadian warships standards for NBCD and ship habitability.

The hull form was typical of destroyers towards the lower end of length-to-beam ratios. The forefoot shape was re-worked from the GP frigate's to use elliptical shapes for both the sections and lower stem profile in order to reduce vortex shedding affecting performance of the hull-mounted sonar. A long roll period of twelve seconds was achieved in favour of helicopter operations. Active fin stabilizers were considered but rejected in favour of the Muirhead Brown controlled passive flumetank system. Fin stabilizers were deemed to be a potential noise source at higher cruising speeds and less effective below twelve knots. Later in service, the passive system was de-activated in favour of tankage.

The all-gas turbine propulsion system was unique in destroyers for its day. It was a combined gas or gas (COGOG) arrangement with two Pratt & Whitney FT4 main engines, each rated at 25,000 horsepower, and two Pratt & Whitney FT12 cruise engines of 3700 horsepower each. Top speeds were 30 and 18 knots, respectively. The physical installation was particularly unusual and achieved a significant reduction in vibration and noise transmitted through the hull into the water. The requirement to resiliently mount the main machinery required the port and starboard gearboxes to be married together in order to counteract significant movements under torque reactions that had been witnessed and measured in the *Columbia* noise trials. Therefore, the main gearboxes (MAAG hardened and ground, double reduction helical gearing) were bolted together. Bolted to the joined gearboxes and cantilevered forward was a large steel raft structure on which all four gas turbine engines, complete with acoustic enclosures, were hard-mounted. This total rigid structure was then resiliently mounted on the hull seatings. Of course, all pipes connected to the rafted equipment were fitted with flexible joints. The two main shafts used Vulcan flexible couplings that could accept raft movements as they transmitted the required torque. A unique alignment procedure was developed by Mr. Don Nicholson of the Directorate of Marine & Electrical Engineering. Because the gearboxes were bolted together, the normal alignment procedure had to be reversed and aligned from gearbox to propeller. Furthermore, the requirement was to achieve proper alignment in the average waterborne condition. This involved docking, floating and re-docking the ships to measure hull deflections and account for the waterborne settling of the raft and flexible couplings.

The large volume of high-velocity hot exhaust gases presented a difficult problem to abovedeck electronic equipment and helicopter operations. Wind tunnel tests were conducted at the National Research Council to examine different funnel heights and configurations. The original low funnel was found to completely envelope the flight deck with smoke at certain wind angles. As a result of these tests, a compromise increased height was adopted using a Y-configuration and giving the ships their characteristic "bunny-eared" funnels. It gave satisfactory flight deck operations but still resulted in hot gases impinging on the lower mast structure under certain situations. Accordingly, the mast was designed in T1 steel instead of aluminum (steel for hot gas impingement, T1 for high strength and reduced topside weight). The flight deck was also of T1 steel to reduce weight.

The machinery control system had to meet requirements for bridge control of propulsion, unmanned main and auxiliary machinery spaces, and minimized complement [15]. For main propulsion, these requirements meant pitch control of the inward-turning (for better steering when going astern), controllable pitch and reversing propellers, rpm control above 12 knots, and automatic control of fuel, cooling, lubrication and other auxiliary services. The system adopted was electro-pneumatic with monitoring functions processed electrically by a Bailey Meter 760 digital system and with control functions transmitted pneumatically, the only evidence of a degree of conservatism in the design. Main propulsion, auxiliary machinery, and the electrical power generation and distribution system were all monitored and controlled from the machinery control room.

Two difficulties with the main propulsion plant and its controls arose on sea trials. A problem with the cruise gas turbine's combustion chamber, and required modifications to the machinery control system pneumatics (to reduce the standard feedback error) were accomplished with virtually no impact on operations. In large measure this was owing to the installation and testing of one complete shaft line at the U.S. Navy's Naval Ship Systems Engineering Station in Philadelphia [16]. With the use of the large water-brake, full power could be generated to test most components except for the propellers. More importantly, the logic, response and alarm aspects of the machinery control system could be tested and debugged. The necessity of incurring this added expense can always be debated, but it unquestionably prevented unknown difficulties and delays later.

Electric power was generated by three 750 kW Solar gas turbine driven generators and one 500 kW diesel generator. The thermal efficiency of the Solar generating systems was enhanced by exhausting the two units in the auxiliary machinery room through waste heat boilers. In due course, the waste heat boiler arrangement was found in service to be too complex and a maintenance headache. Accordingly, it was removed at the mid-life update.

Special firefighting arrangements were required. A centrally controlled "light water" (an aqueous film forming foam) system served the engine room, auxiliary machinery room, helicopter hangar, JP5 pump room and forward gas turbine generator room. A separate "light water" system served the flight deck. The air conditioning system represented further advances on the all-air conditioned citadels of the River-classes and Operational Support Ships noted earlier. Heating and cooling were distributed by a network of hot or chilled water feeding small local fan coil units with individual temperature controls in each space.

The Canadian naval emphasis on habitability has been a continuing story and the DDH 280 design was no exception. An improved bunk standard (6 ft-6 in.) and a new, larger kit locker were designed. A minor bulkhead system of extruded aluminum sandwich construction was developed and was clad with decorative plastic sheathing giving easy cleaning and reducing painting. Cleanliness around furniture in accommodation spaces was enhanced by the use of explosion-bonded bimetallic pads that could be welded to the steel deck and to aluminum furniture legs and supports. These also used less production hours and have not been a source of corrosion problems.

The core of the Tribal-class combat system was the development of a centralized command and control system, the CCS 280. It operated on a federated concept by which each sub-system could be operated on its own or from the CCS 280. Therefore, before explaining the CCS 280 further, it is appropriate to outline each major sub-system. The helicopter system with its beartrap, flight assists, firefighting and operational envelope was described earlier under system developments of the 1950s and 1960s. It is the primary ASW weapon system.

But, before the helicopters are given target detection information and launched, the ship's sonar sensor systems must be capable of making that detection. The DDH 280 design was fitted with both hull-mounted and VDS sonars of the latest AN/SQS 505 design, as well as the AN/SQS 501 sonar for classifying "bottomed" targets. An entirely new ASW data system performed integrated signal processing from the three sensors and provided target information to both the CCS 280 and the ASW mortar as well as for torpedo settings. Similarly, the electronic warfare (EW) systems were designed to operate on an integrated basis from a sub-system console with the operator as the integrating element. All of the received radiated emissions could be analyzed and presented independently for active jamming or use of chaff launchers, or could be passed directly to the CCS 280 for centralized control.

The surface and air weapon system was also an integrated sub-system comprising radars, Sea Sparrow point defence missile launcher and a 5 in./54 Oto Melara gun. This system's computer was programmed to allow simultaneous surface and air operations. The missile system was a unique Raytheon Canada design, which provided retractable launch arms port and starboard capable of single or salvo launchings to give 360 degree coverage and automatic re-load capability. The rapid fire gun could be operated locally or from the surface and air weapons system in the operations room against both surface and air targets.

Returning now to the CCS 280, how was integration achieved? Each sub-system contractor was given the data transfer specification and required to include an interface unit as part of his sub-system. The contractor could set things up within his sub-system however he chose to meet the sub-system performance requirements. But he had to convert at the interface as necessary to match the CCS 280. Radar, sonar, fire control, electronic warfare, log and gyro were all interfaced. A program generation centre was built at Litton Canada, the CCS 280 supplier, and co-manned by naval personnel. In this way both technical and tactical expertise could be coordinated. Fail-safe features were a prime consideration. The failure of a subsystem did not prevent other sub-systems, or the CCS 280, from continuing to function without further degradation.

Marine Industries Ltd of Sorel, Quebec, was the lead yard, commissioning the *Iroquois* on July 29, 1972. They also built the *Huron*. The follow yard was Davieship Ltd of Lauzon, Quebec who built the *Athabaskan* and *Algonquin*. As with most of the River-classes before them, the Tribals were to see a major modernization and update. Beginning in 1988 they were back in the hands of Davieship (now MIL Davie) as a subcontractor to Litton Canada. Major changes included replacement of the OTO Melara 5 in./54 gun with the OTO Melara 76/62 Super Rapid Firing Gun, replacement of the Sea Sparrow point defence missile system with vertical launched SM2 Block III area air defence missiles (same as AEGIS), replacement of the CCS 280 with the new Ships Integrated Processing and Display System (SHINPADS, of which more later), replacement of the machinery control system with an integrated all-digital system, replacement of the complete EW suite, new cruise engines and a host of other improvements.

The DDH 280 Tribal-class turned out to be the last and most sophisticated warship to be designed largely in-house by the Canadian Department of National Defence. As will be recounted later, the next major warship procurement was not to occur for a decade. An industrial, total responsibility approach would be adopted.

Fig. 6 Tribal-class destroyer HMCS *Iroquois*: 4700 tons, 426 ft, 28 knots, 50,000 shp, 4500 n.mi, two CH124 Sea King helos, Sea Sparrow S/A missiles, 5 in./54 gun, Mk 10 triple mortar, two triple torpedo mounts, AN/SQS-505 hull-mounted sonar and VDS. (Dept. National Defence photograph.)

Fig. 7 Modernized Tribal-class HMCS *Algonquin*: 5000 tons, 426 ft, 28 knots, 50,000 shp, 4500 n.mi, two CH124 Sea King helos, SM2 Block III vertical launch missiles, 76 mm/62 gun, Mk 10 triple mortar, two triple torpedo mounts, AN/SQS-505 hull-mounted sonar and VDS. (Dept. National Defence photograph.)

Canadian Auxiliaries

A number of major, special purpose ships have been acquired since 1950 by the Canadian Navy. Although they will not be treated as extensively as the destroyer programs, brief accounts are in order.

The first special purpose ship was the icebreaker *Labrador*, built by Marine Industries, Ltd., at Sorel, Quebec, and commissioned on July 8, 1954 [17]. Her design was patterned closely after the Wind-class of the U.S. Coast Guard. A major addition was the installation of retractable, Denny Brown stabilizers [18]. As is well known, the extremely rounded hull form without bilge keels, beneficial for operating in ice, causes severe rolling in the open ocean. Forty and fifty degree heel angles had been reported. These stabilizers were of new design with a tail flap to give the rapid response necessary to be effective. Fin control was based upon two gyroscopes, one for ship rolling velocity and the other for measuring the angular departure from the horizontal. The gyro outputs could be combined in any desired ratio to produce the best result for the particular direction and severity of the sea. The stabilizers effectively reduced rolling and had little effect on ship's speed or manoeuvrability. The retractable feature protected them when the ship was in ice.

The *Labrador* transitted the Northwest Passage in the summer of 1954 carrying scientists in cosmic ray research, oceanography, meteorology and terrestrial magnetism. They also conducted trials on a new RN gyro-magnetic compass (AGM5). She returned via the Panama Canal to Halifax in November and became only the second ship to circumnavigate North America. (The first was the *St Roch* of the Royal Canadian Mounted Police in 1940/42, taking 27 months for the first arctic passage west to east [19].) The following year the *Labrador* assisted with transport and surveys for the construction of the DEW Line in coordination with the U.S. military. She was transferred on April 1, 1958 to the Ministry of Transport.

Two specialized research vessels were designed and built by the Canadian Navy, one each for the Defence Research Establishments Pacific and Atlantic. The *Endeavour* and *Quest* were built by Yarrows, Ltd. of Victoria and Burrard Dry Dock Ltd of North Vancouver, and entered service in 1965 and 1969, respectively. The ships are of moderate size at 2200 tons and 2900 shp, with a speed of 15 knots.

The primary areas of research for these ships were to be ocean acoustics and sonar systems, with secondary capabilities in the related fields of ocean wave spectra, seakeeping, general oceanography and atmospheric studies. Accordingly, much design emphasis was placed on achieving quiet operations, maintaining station with good seakeeping characteristics, suitable laboratory facilities, and high standards of habitability in both the living and working spaces.

Fig. 8 Icebreaker HMCS *Labrador*: 6490 tons, 369 ft, 10,000 shp, 16 knots. (Dept. National Defence photograph.)

For example, the master of the *Quest* was given the choice of two modes of propulsion and three modes of electrical power depending upon the speed and quietness required for various experiments [20]. Normal propulsion and electric power was by a diesel-electric system with two main diesel generators. A gas turbine driven generator provided propulsion during quiet operations and extended periods of slow running. When extreme quietness was needed, electronic equipment and lighting loads of up to 30 kVA could be run for an hour from battery-supplied inverters.

The best possible scientific working conditions were incorporated. The hull shape itself was model tested extensively for seakeeping and a roll stabilization system was installed. Twin screws and a retractable bow thruster gave maximum manoeuvrability and position keeping. A unique feature was a complete ship control console amidships on 01 Deck to assist in keeping the ship close to a suspending wire for deep experiments without drifting and streaming the wire horizontally.

Experience had indicated that frequent and significant temporary wiring changes are associated with different experiments conducted over the life of the ship. To facilitate such changes, the *Quest* was equipped with an extensive system of raceways plentifully fitted with access panels in which temporary wiring could be laid. These raceways ran between electronic rack positions throughout the ship and included special packing blocks to seal watertight bulkhead penetrations. Finally, a wide range of winches and cranes of varying capacities and locations were fitted. The largest was a 370 hp winch below the quarterdeck (fantail) that could hold 15,000 ft of 1.5-in. diameter, electrically-cored cable with a haul speed of 350 ft per minute.

These ships have supported successfully a wide variety of research and development projects. One of the more noteworthy has been the Canadian towed array described in a later sub-section.

Fig. 9 Oceanographic research vessel CFAV *Quest*: 2200 tons, 253 ft, 3550 shp, 16 knots, 10,000 n.mi. (Dept. National Defence photograph.)

Submarines

Although Canada has never designed its own submarines, it has a long history of building and operating them that should be noted [21]. The first two were acquired at the outbreak of the First World War. They were built by the Seattle Construction and Drydock Company for the Chilean Navy. A disagreement between the parties made them available. In an unprecedented move, the provincial government of British Columbia acquired them on August 4, 1914. The Federal Government was thereby "forced" to underwrite the purchase three days later. The two submarines (313 tons surfaced) served in the Pacific until 1917 when they were transferred to Halifax for service until scrapped in 1920. Also in World War I, Canadian Vickers of Montreal built twenty American H-class submarines (364 tons surfaced) for the RN, putting the first four into service in seven months. They subsequently built fourteen more, eight for Italy and six for Russia. Two of the British boats were given back to Canada and served in the RCN until paid off in 1922. The RCN did not operate submarines again until after World War II, nor have submarines, other than hull sections for the USN, been built in Canada since World War I.

During World War II, Canadian personnel did serve in RN submarines, including some stationed in Canadian ports. In the 1950s an agreement with the RN provided a squadron of three of their "A"-boats, comanned and stationed in Halifax to provide ASW training to the RCN surface fleet. In 1961, the *Burrfish* (a World War II snorkel submarine) was acquired from the U.S. Navy. Renamed *Grilse*, she provided similar ASW training on the West Coast. She was replaced in 1969 by the U.S. Navy's Fleet Snorkel class *Argonaut*, renamed *Rainbow*, and served until paid off on December 31, 1974. The West Coast has been without a submarine ever since, but not so for the East Coast.

The acquisition of new submarines to replace the three RN "A"-class boats serving on agreement with the RCN was studied extensively in the 1950s. Both nuclear-propelled and conventional submarines were considered. Nuclear propulsion was not accepted because of costs and the technical risks associated with what was then a new technology. Three RN "O"-class submarines were contracted for construction in Her Majesty's Dockyard Chatham. HMC Submarines *Ojibwa*, *Onondaga*, and *Okanagan* were commissioned in 1965, 1967 and 1968, respectively. For their day they were deep-diving, quiet-running, high underwater speed, conventional boats with diesel-electric propulsion and of 2030 tons surface displacement. When ordered, *Ojibwa* was under construction and was built entirely to first-of-class RN specifications. Some changes to the general arrangements were made for the other two. *Onondaga* included RN changes to the galley location to place it nearer to the messes and away from the work places. *Okanagan* introduced an entirely Canadian change, a cafeteria, so that food was consumed in one location rather than being distributed from the (relocated) galley to the messes. With the acquisition of these submarines, new facilities were added to HMC Dockyard Halifax, including a battery shop, a clean shop for hydraulics, a specialized welding school for the hull and piping, a quality assurance section, and a docking facility. The latter was an elevator type Synchrolift complete with rails to transfer the docked submarine into a new covered shed for refits.

By the late 1970s, the combat system of the "O"-boats was becoming dated and support problems were becoming difficult and costly to resolve. A submarine operational update project was approved in 1979 that replaced the fire control system, the passive ranging sonar and the interior and exterior communications systems, and added a low light modification and television system to the attack periscope.

In the 1980s the Canadian Navy began staff studies for a submarine replacement program. Initial project approval was for an undetermined number of conventional boats. The Defence minister subsequently ordered an investigation into nuclear propulsion. Specific studies related to submarine requirements clearly demonstrated that the now-mature technology of nuclear propulsion was more cost-effective than conventional submarines (essentially three times as effective for twice the cost) and had the added advantage of under-ice capability in the Arctic. After some two years of project development, including the submission of industrial proposals based on the RN's *Trafalgar*-class and the French *Amethyst*-class, the project was cancelled. The government, in making this decision as part of its budget presentation to Parliament in February 1989 cited the growing Federal deficit and the need to reduce and stretch out government programs, including defence. A strong antinuclear lobby by peace and environmental groups undoubtedly was also a major factor. At the time of writing, a replacement conventional submarine program remains in the Canadian Navy's plans for the 1990s, but its future is uncertain.

Canadian System Developments in the 1970s and 1980s

The three Canadian services were unified into the Canadian Forces effective February 1, 1968 by an Act of Parliament. An important fallout was a major study of how the National Defence Headquarters should be organized. One of its concerns was the management of defence acquisitions. The earlier-noted management difficulties with the Tribal-class were influential in the organizational outcome for Research and Development. The Defence Research Board, as a separate agency reporting to the Minister, was disbanded and incorporated into a new single departmental headquarters that integrated the former civilian and military organizations. Specifically, the engineering, construction and properties, research and development, and supply functions were all placed under a single assistant deputy minister.

New policies were developed for Research and Development (R&D) and for major weapons systems acquisition. In effect, R&D with the potential to meet stated Canadian operational and technical requirements would be supported on a competitive basis as funds permitted. However, the major weapons platform contracts normally would not specify what system developments were to be fitted. The prime contractor would make the choice as part of his total systems responsibility. Hence, a supported R&D project would have to be competitively successful to be chosen by a prime contractor and thereby go into production. What follows is an overview of some dozen such R&D projects, all of which were significant Canadian naval advancements of international quality. This continued record of success has meant that the smallest service of the Canadian Forces has continued to have the largest share of the Defence R&D budget.

Although Canadian destroyers had an enviable record for good seakeeping, developing technology in the 1970s lent itself to the generation of a destroyer/frigate hull form for superior seakeeping [22]. The approach taken was to conduct, in parallel for comparison and design synergism, theoretical parametric studies and model series testing. The model series scope was a potential total combination of 96 hull forms by combining separate forebody and afterbody shapes. This arrangement allowed trends to be determined as the bow form was changed from extreme V to extreme U sections, while the sterns varied between wide shallow and deep narrow sections. "Success" was determined by comparing the vertical acceleration, deck wetness and slamming results of the optimized form with those of existing Canadian and foreign designs. It is these three seakeeping parameters that affect ship and crew performance. The conclusion was that significant improvement could be made to seakeeping qualities without compromising power requirements.

The specific hull form characteristics are of particular interest. It was found that the bow sections should have a fine V-form near the keel, but with a relatively full waterplane. The waterplane coefficient forward (CWF) should be high, whereas CWA (aft) should be determined by powering and general arrangements. A low block coefficient would be beneficial for seakeeping and also for lower power requirements at intermediate speeds. The studies confirmed the well known rule that increasing ship's length improves seakeeping, but added the important refinement that equivalent percentage changes in waterplane and block coefficients would have far greater effect. Lastly, it was found that the breadth to draught ratio (B/T) would have the least effect on seakeeping for normal ranges required for adequate stability, although increased B/T would increase slamming somewhat.

Of course, the real proof is in the final product. Initial sea trials and operations of the Canadian patrol frigate indicated that significantly improved seakeeping has indeed been achieved.

Turning to the structural side of warship hull design, the Canadian Navy became particularly interested in long-term performance with delays in ship replacement during the late 1970s and 1980s. A thorough history of ship structural failings and repairs was documented [23]. The River-class destroyers built from 1955 to 1964 experienced a number of failures over the years traceable to the use of poor materials, poor fabrication techniques, rough plate edges, misalignments, incorrect selection and use of welding rods, and poor detailed design in high-stress areas. All of these were isolated situations either in structural location or particular ship. None was catastrophic or greatly affected ship operations, and most had been recognized and avoided in the subsequent Tribal-class vessels. However, as the ships were by then expected to be in service for ten or more years beyond their nominal design life of 25 to 30 years, concern mounted over the long-term fatigue life of the overall hull structure in bending, as well as the identified local problems.

Two shortcomings were recognized. There was room for improvement in both structural analysis capability and knowledge of actual seaway loadings. The Navy's ability to analyze complex structures was significantly augmented by a suite of finite element programs developed by the Defence Research Establishment Atlantic under the name VAST (Vibration Analysis of STructures). It provided a high-powered capability to analyze complex details and nicely complemented the overall hull computer code called MAESTRO that was developed in Australia about the same time. In order to get a better understanding of the history of ship loadings, the Canadian Navy began in the late 1980s to permanently instrument fourteen ships of various classes on both coasts. Both strain gauges and pressure transducers have been installed. In time it is hoped that the data collected will increase the naval engineers' ability to assess fatigue and other structural responses.

The advent of high-volume, hot exhaust gases from gas turbine powered frigates and the introduction of heat-seeking missiles accentuated the need to develop funnel designs that would lower exhaust gas temperatures and obscure them from infrared sensors. A concept that has achieved excellent results was pioneered by the Defence Research Establishment Suffield (DRES) in the province of Alberta. The so-called DRES Ball combined the effects of film-cooling, cooling air entrainment, and optical blockage to reduce infrared emissions from the exhaust stack metal surfaces and from the exhaust plume.

The design, construction and testing for shipboard use was conducted by W. R. Davis Engineering, Ltd. of Ottawa. The arrangement consisted of a large, teardrop-shaped centrebody inside the uptake. Its primary function was to block the view down the exhaust duct, but it also was used to add cooling air to reduce the plume temperature. It provided "full" protection in that all visible metal surfaces from any viewing angle were cooled. The development program involved both computer simulation and scale model testing that demonstrated the DRES Ball to be extremely effective at reducing both metal surface and plume infrared emissions. In the 3–5 micrometer and 8–14 micrometer atmospheric windows these were reduced by 90 to 95 percent when compared to an unprotected stack.

Given these promising results, design of full-scale infrared suppression devices was applied to the new Canadian patrol frigates and the modernized Tribal-class vessels. The former was a full DRES Ball installation and the latter an eductor-diffuser owing to weight and space considerations in the existing ships. (A DRES Ball arrangement is slightly heavier than a standard uptake, whereas an eductor-diffuser gives substantial weight savings.) At the time of writing the full scale trials have not yet occurred, but little departure from analytical and large scale modelling results are expected.

An earlier section highlighted the work in the 1950s and 1960s on low-cavitation propellers at the Canadian National Research Council. Theoretical work has continued at the Defence Research Establishment Atlantic (DREA), leading to three different designs for special applications. In all cases verification by model tests was conducted at the Maritime Research Institute Netherlands. Advances in sonar technology required additional silencing of the ship propellers on the acoustic research vessel *Quest*. The original propellers had five narrow blades. The new propellers were similar except the skew was increased dramatically from a few degrees to 60 degrees. These propellers have shown reductions in both cavitation and noise. The Tribal-class needed to retrofit new controllable-pitch reversing propellers as part of its modernization. Changes to the propellers in this case were more marginal. Far more dramatic was the propeller redesign for the two River-class destroyers *Annapolis* and *Nipigon* being modified for towed array passive sonar operations. A new seven-bladed design replaced the earlier five-bladed propellers. At the time of writing, sound ranging has not been completed to determine the degree of improvement actually achieved.

Of course, sonar operations and ASW have been the primary open-ocean role of the Canadian Navy for over fifty years. It is not surprising, then, to note several major developments in the sonar field. A study by DREA in 1976 concluded that the greatest potential gains with the 505 sonar could be realized by replacing the analogue receiver with a digital one incorporating advanced techniques for signal processing and computerized detection aids. System development was contracted to Computing Devices of Canada in Nepean, Ont., leading to fitting an advanced development model in the *Nipigon* in 1985. As expected, in poor water conditions no significant difference could be measured between the new 510 sonar and the 505. However, nominal water conditions gave range improvements of 37 to 50 percent. With full scale production development still underway in 1989, including all software in the ADA language, regretfully the timing was too late for application in the Canadian patrol frigate. Nevertheless, future fleet installation is planned. In the meantime, initial naval experience will be obtained on the Portuguese *Meko*-class frigates for which the 510 sonar was selected as part of a NATO assistance program.

Just as several NATO navies did during the 1970s, the Canadian Navy also moved into the development of large, passive, towed sonar arrays [24]. Experimental work was initiated by the Defence Research Establishment Pacific at Esquimalt, BC. The success of the experimental systems led to a full scale development project based on purchasing the USN's AN/SQR-19 "wet end" and using the excellent Canadian capability in signal processing and display equipment to develop the "dry end" to meet specific Canadian requirements. Computing Devices of Canada built the advanced development model, which was extensively tested in 1988 aboard the *Annapolis*. The experience gained was used to incorporate higher throughput computing resources, a more streamlined computer system architecture and the ADA programming language. The CANTASS system subsequently went to full scale production for the twelve new Canadian patrol frigates and for retrofitting on HMC ships *Nipigon* and *Annapolis*.

In response to an operational requirement for an automated system, Canada embarked in the late 1960s on the development of electronic warfare systems (EW) for the first time. As in so many instances of Canadian development, specific aspects were selected for concentrated effort. In this case, it was the so-called Electronic Support Measures, or ESM. More precisely, Canada addressed automatic signal processing, classification, display, and overall system integration with antennas, active EW jammers and the ship's command and control system.

Once again, the engineering development model of the Canadian EW System (CANEWS), as tested aboard the *Terra Nova* in 1981, was a multi-sourced system. Westinghouse Canada had the prime contract with the receiver from MEL in the UK and the display system from the Loral Corporation in the United States. The combined technical and operational evaluation produced good results. The system was further evaluated in a multinational exercise in the Pacific Ocean in 1982. Subsequently, contracts were let to MEL Defence Systems, Ltd. of Stittsville, Ont. for twelve systems, four for the DDH 280 Tribal-class, four for the improved *Restigouche*-class, two for the *Annapolis*-class, one for the Fleet School and one as a third line repair test bed. CANEWS also became the system of choice by the prime contractor for the Canadian patrol frigate.

Even as the CANEWS system was going into production, a requirement was established to counter the expanding frequencies of electronic emissions and to identify and classify them automatically even more rapidly. In 1990 a definition contract was let to Lockheed Canada, MEL Division for the development of CANEWS II to meet this requirement. It is expected that it will lead to fleet application in the late 1990s.

The Canadian Navy for a long time has sponsored another passive detection technology, that of infrared sensors. Although the development became a joint project with the USN, the roots of the system concept were in experimental research at the Defence Research Establishment Valcartier in Valcartier, Quebec, followed by early system development at SPAR Aerospace of Toronto in the early 1960s. After a series of parallel developments in Canada and the United States, a memorandum of understanding and joint project agreement was signed in July 1976. It resulted in an advanced development model called the IR Search and Track (IRST) System that was successfully tested on the *Algonquin* (1978) and the U.S. Navy's *Kincaid* (1979). A subsequent joint agreement was signed leading to a prime contract in August 1984 with SPAR for engineering development of the IR Search and Target Designation (IRSTD) System.

The SPAR contract called for designing, building, and testing two full scale engineering development models primarily to warn ships against attacking missiles and aircraft, and to provide infrared imaging for navigation and station keeping. The first was delivered in September 1989 to the U.S. Naval Ship Weapons System Engineering Station in Port Hueneme CA for land based test and trials. The second system was integrated into a 30-ft container and installed aboard the *Kincaid* for sea trials in 1991. At the time of writing results are encouraging and within performance specifications. There is reason to hope that they will lead to installations in warships of both countries.

It will be recalled that the CCS 280 command and control system in the Tribal-class was a federated system providing overall combat system coordination, yet allowing independent sub-system operation such as ASW and AAW. One of the disadvantages of this concept was that a single point of failure still existed in terms of the overall ship system; namely, the CCS 280 computer. Other factors, such as the proliferation of sub-system computers within the ship, and the concentration of vital cabling which could be severed, but which could not be duplicated in a cost effective manner, also posed major handicaps. These considerations, and others, led to an examination in 1975 of the overall integration problem. Officers representing all aspects of the entire ship system (not just combat engineers and operators, but also naval architects, marine engineers and logisticians) were involved. After months of gradually breaking down traditional parochialism, a paper was produced proposing a new electronic architecture for an entire ship based upon a distributed processing approach.

The SHipboard INtegrated Processing And Display System (SHINPADS) was not really a system, but rather a concept of using a distributed architecture for ship integration [25]. In centralized systems, large processors run all the sub-programs for the sub-systems. In federated systems, a central processor coordinates the sub-systems, which run on their own processors. In a distributed system, small processors throughout the ship can run any sub-system or do the coordination. The SHINPADS concept proposed three forms of distribution: geographic distribution, functional distribution, and distribution of control. It became rapidly evident that the key elements to maximizing the concept's potential benefits were standardization of equipment (notably the processors, the displays and the software), and a data bus for high speed data exchange.

The potential benefits were many. These included reduced life-cycle costs (from standardization), replacement of tons of point-to-point wiring by the data bus, and system redundancy and survivability from geographic and functional distribution. Finally, SHINPADS represented the single, largest step towards synergism of the entire ship design, i.e. "co-operative action of discrete agencies such that the total effect is greater than the sum of the [two] effects taken independently" [26]. The conceptual purview of SHINPADS included all combat, marine, hull and administrative support systems and equipment.

For the standard minicomputer, the U.S.N.'s militarized AN/UYK-20 was chosen. For interfaces, a NATO standard STANAG 4153 was developed and adopted. The standard language became CMS2, a high-level language. A standard display was developed by Computing Devices of Canada to meet the SHINPADS requirement. Sperry Univac developed an AN/UYK-20 compatible microprocessor (the UYK-502) and the serial data bus [27]. Although data busses existed, increased capacity and passive couplers that would not disturb the information flow were required. The estimated peak data rate for a frigate was two million bits per second. In order to provide low access time and room for growth, the overall throughput for the bus was set at eight million bits per second. A successful concept demonstration in 1979 led to the adoption of SHINPADS for both the Canadian patrol frigate and the modernization of the Tribal-class.

The acronym SHIN grew into rapid usage in the Canadian Navy. It was applied by the marine engineers to their own development of a digital electronic MAchinery Control System (SHINMACS) and a fully integrated COMmunications system (SHINCOM), although the latter did not use a data bus. (The naval architects felt that they had been into SHip INtegration for decades and shunned adopting the term SHINSHIP!)

SHINMACS was a microprocessor-based control system that provided automatic control and monitoring of the ship's machinery and systems [28]. Based on a requirement identified by the Navy's marine engineers, the Canadian Defence and Civil Institute for Environmental Medicine (DCIEM) in Toronto undertook in 1977 a human engineering investigation of a ship propulsion control console design for a microprocessor-based, distributed control system. Through their work, it became evident that the elimination of the dials and gauges of the current systems and their replacement with graphic video displays of the plant information database, could present more useful information to the operator. Development progressed from a wooden mock-up to a demonstrator model in 1983 built by CAE Electronics Ltd. of Montreal. The specific system characteristics that were emphasized were survivability, reliability, redundancy, modularity, standardization, and graceful degradation. A remarkable measure of the human engineering success of the demonstrator model was that fleet operators ably handled simulated emergency drills after a four-hour training period.

This success led to a contract for CAE to build and test an advanced development model. It consisted of operator and supervisor consoles, a triplicated data bus, dual redundant digital propulsion controllers, two remote terminal units, a health monitoring and diagnostic system, and two printers. As a result of its demonstrated success to control the monitor, the main propulsion system, ancillary and auxiliary machinery, electrical power, and damage control, the SHINMACS concept was selected by the prime contractors for both the Canadian patrol frigate and the modernization of the Tribal-class. The contractor's sea trials of the Canadian patrol frigate in the summer of 1990 had the distinction of being the first fully integrated, digitally controlled machinery control system taken to sea in a warship.

The second "SHIN concept" developed by the Canadian Navy's marine engineers was for communications, or SHINCOM. It, too, represented state-of-the-art digital technology to meet both operational and secure voice communications requirements. This fully digitized system integrated some dozen and a half independent systems into a single system of all the internal and external telephones, broadcasts, intercoms, and radios, i.e., everything except flags, lights, and the underwater telephone.

The development of SHINCOM generally paralleled the SHINMACS time-frame, in this case carried out by Leigh Instruments of Kanata, Ont. As developed, the system contained two central switches interconnected with two high-speed data links to provide full redundancy. No single point of failure or cable cut would interrupt the communications. The flexible software and database provided special features such as override, voice call, auto answer, meet-me-conference, and pre-arranged conference nets for operational and routine communications.

The SHINCOM system was also selected by the prime contractors for the Tribal-class modernization and the Canadian patrol frigate. The latter installation in the *Halifax* was fully operational and passed sea trials in 1990. The commonality of the software, hardware, database, and fleet training will result in significant savings in life-cycle support costs and operational effectiveness. By the time the first system put to sea, development contracts were already in place for such improvements as a modular expansion switch (with reduced power, weight and size), an embedded processor using ADA language, and intelligent terminals with multi-channel communications capable of concentrating a number of terminals by means of a high-speed fibreoptic cable.

Some Canadian developments have been less esoteric but of equal importance. An example would be the engineering development of a reverse osmosis desalination system that could satisfy the stringent naval environmental requirements for performance, shock, noise, vibration, and corrosion [30]. Evaporators had long been the Navy's mainstay for the production of both potable water and boiler feedwater. This was the case for both the River-class "steamers" and the gas turbine Tribals via waste heat auxiliary boilers. The principle of reverse osmosis had been known for some time. However, the inability of semipermeable membranes to withstand high pressure, and difficulties with mechanically holding them in place, had deterred any serious move to adopt this potentially superior technology.

By the 1980s, the development of more suitable synthetic membranes and improved mechanical attachments prompted the letting of a contract to Seagold of Vancouver to build a unit to the Canadian Navy's design. Polymer ester membranes, which provided for higher reject rates and were less susceptible to biological fouling, had replaced the earlier cellulose acetate. The requirement was for a reverse osmosis desalination system, incorporating a high-pressure (800 psig) energy recovery pump, that would generate at least 8800 U.S. gallons of fresh, potable water daily, 20 percent of which was to be of feedwater quality from a second stage. It was anticipated that the energy recovery pump would save about fifty percent of the energy costs. The supplied unit was exhaustively tested at the Naval Engineering Test Establishment. The development process proved lengthier than expected, requiring five progressive prototypes. But all deficiencies were overcome in 1987, and manufacturing contracts were placed with Zenon of Burlington, Ont., who had bought Seagold.

The accepted equipment has resulted in an overall reduction in energy and maintenance costs, less demand on space, a safer ambient temperature, and nonpolluting operation. Use in the Tribal-class has meant operation on one auxiliary boiler instead of two in all but the most extreme conditions of steam demand for heating and the galley. Although the development time-frame missed the first six ships, the second batch of six Canadian patrol frigates will be fitted with the world's first reverse osmosis desalination system fully qualified for naval use.

The last Canadian naval development of the 1970s and 1980s to be mentioned was a continuation of earlier work on helicopter hauldown and handling systems. As the pioneering navy in the operation of large ASW helicopters from small warships, Canada maintained a continuing interest in this field in an attempt to apply new technologies to give similar or better performance with reduced weight, space and maintenance requirements. Although the "beartrap" system installed in the fleet continued to meet the need fully, the hauldown cable, tail guide cables and associated hydraulic equipment represented several tons of topside weight and a heavy preventive maintenance workload for the crew.

In the early 1980s, an entirely new concept was postulated in which photogrammetry would be used to locate the helicopter and manoeuvre a new, more compact trap to meet the helicopter's probe. The RAST MK III project (a Recovery, Assist, Secure and Traverse system) originated in September 1984 with an unsolicited proposal from Indal Technologies Inc. of Mississauga, Ont. [31]. Their proposal resulted from their experience delivering over 200 RAST MK I systems to the U.S. Navy and another dozen contracted for the Canadian patrol frigate. (The RAST MK II version was to be for the Australian Navy but was not completed.) Within the company, a careful analysis of RAST operations led to the conclusion that the recovery assist function was a drawback and that its elimination could result in significant gains without prejudicing the existing operating envelope.

The most significant feature, other than the compact trap itself, was the helicopter position sensing equipment. This used two cameras to sense the helicopter's position based on the reception of a laser light source from four beacons on each side of the helicopter. This new approach allowed the elimination of the hauldown gear and the yaw restraint system of tail-guide bars and winches. As a result, there was no below-deck equipment requiring valuable space. A weight saving of five tonnes was realized. The helicopter associated equipment was also reduced. In addition, RAST MK III offered several operational advantages. The required helicopter over-deck hovering time was reduced substantially for improved flight safety and less time required for the ship to be confined to its flying course. There was no longer a requirement for flight deck personnel during recovery and traversing, or for voice communication between the pilot and the landing signal officer. Finally, the life-cycle cost was estimated to be reduced by sixty percent.

After completion of a feasibility study, a full scale engineering model of the system was contracted with Indal Industries. It was successfully tested in-plant and led to a contract for the development and shipboard evaluation of an advanced development model. At time of writing, these trials were scheduled for late 1991 aboard the *Ottawa*. The designated replacement for the Sea King, the EH 101, will be fifty percent heavier with a nose-wheel rather than a tail-wheel, thereby creating a completely new hauldown requirement, which will be fully met by RAST MK III.

It has been evident throughout this section on Canadian naval system developments in the 1970s and 1980s that the majority have been (or likely will find themselves) incorporated into the new Canadian patrol frigate. It is now time to describe Canada's largest ever, single defence project.

The City-Class (Canadian Patrol Frigate)

This City-class project had its beginnings in 1973 just as the last of the four Tribal-class destroyers, the *Algonquin*, was being completed and delivered. Once again the Canadian Navy was taking a broad long-term look at its future fleet requirements. The operational studies [32] identified two major missions: anti-submarine warfare (ASW), and fleet air defence. Engineering concept designs confirmed that it would be more cost effective to build separate classes rather than one larger class. A key factor in this determination was the proposed new ASW operational concept using passive towed array sonars deployed for maximum effectiveness well away from the task force or convoy. In program terms, this conclusion translated into the Canadian Patrol Frigate (CPF), initially six ships, designed primarily for ASW, to be followed by two further batches for a total of twenty frigates, and the Tribal-class Update and Modernization Program, or TRUMP project, giving these four ships an area air defence capability.

Commensurate with its size and expense, the CPF project definition was far more elaborate than any previous Canadian defence project. And it must be made clear that this was a project involving far more than just the ships. They were to be delivered fully supported including training facilities and documentation, training of the crews, new dockyard maintenance facilities as required by new systems, a lifetime buy of repairable spares (ten years for combat systems, thirty for platform), plus two years of consumable spares and tons of support documentation in Canada's two official languages. The ratio of the total estimated cost to the six ship sailaway cost was 1.8 to one. Preliminary government approval was received in December 1977 at a total project cost of 1.785 billion (1977) dollars.

For the first time in warship acquisition, the Canadian government intended that a prime contractor have total systems responsibility. Although initially the prime contractor did not have to be Canadian, another project objective was the establishment of a Canadian industrial capability for combat system integration. Needless to say, before the detailed contract package was issued, many reviews and preliminary designs (with estimated costs) were completed in-house to ensure as much as possible the overall compatibility and cost efficacy of the Navy's specifications. A measure of that success is that the project was still within the original budget (including contingency margin and escalation for inflation) when the first of class was delivered nearly fourteen years later.

Five industrial organizations submitted, without remuneration, their proposed approaches. After consortium realignments to meet a change that the prime contractor must be Canadian, two were contracted to prepare and submit preliminary designs with supporting calculations and documentation for the detailed design, construction, and delivery of the six, fully-supported ships. After another evaluation (and about an eighteen-month delay caused by a change in government), an implementation contract was awarded to Saint John Shipbuilding in July 1983.

As was noted in the previous section, the completed ship at 4750 tons deep displacement incorporated many recent Canadian developments. The maximum speed exceeded 27 knots with two LM2500 General Electric gas turbines (23,000 shp each). The range was over 4500 nautical miles at 15 knots on the Pielstick PA6 diesel engine (8,800 shp). A unique propulsion component was the combined port and starboard cross-connected gearbox built by Royal Schelde in Holland. Space, accommodation, and support facilities were provided for a complement of 225, with adaptability (after a 1986 design change) to accommodate a 10, 25 or 50 percent mixed gender crew. Storage held fresh provisions, frozen provisions, and general stores for 30, 45 and 60 days at sea, respectively.

The fully integrated, SHINPADS-controlled combat system provided the latest in ASW capability and self-defence in depth. The main sub-contractor for the combat system was Paramax Electronics, Ltd. of Montreal. The primary weapons system was the Sea King helicopter with its sonobuoys, dipping sonar and MK 46 torpedoes. The ship was equipped with the CANTASS towed array, the 505V hull-mounted sonar, and its own MK 46 torpedo launchers. Air defence was provided by search and tracking radars, vertical launch Sea Sparrow missiles and the U.S.N.'s PHALANX close-in weapon system. Surface defence was provided by radars, the U.S.N.'s Harpoon missile and a 57 mm Bofors, rapid-fire gun. Passive defence included the CANEWS ESM equipment for electromagnetic emission detection, the DRES Ball uptake infrared (IR) suppression system, radar and IR chaff launchers, RAMSES jammers and reduction of the ship's radar cross-section by careful selection of the ship's above-water geometry.

New features of these ships were not limited to the contract approach and the ship itself. The method of construction also deserves mention. As one author noted, the basic principles of unit construction go back to the *Warrior*, the world's first ironclad, when she was built in 1860 [33]. But the extent and refinement of the concept have improved dramatically. In the case of the CPFs, the steel, machinery, computers and weapons have been "built" into operating ships in just over three years. At her naming ceremony, the *Halifax* "was 55 percent complete, the highest percentage completion at launch for a lead ship of any naval combatant construction program in the West" [34]. This was achieved by a high degree of pre-outfitting of erection units. The combat system hardware as delivered from PARAMAX for installation had, short of weapons firing, already been fully tested [35]. The combat system software will be fully tested ashore prior to each sea trial.

At the time of writing, the first-of-class, the *Halifax*, had completed contractor's platform sea trials, been provisionally accepted on June 29, 1991, and had begun operating with a navy crew. Eighteen months of extensive combat system trials and operational evaluations will follow before full acceptance into the fleet in 1993.

Canadian Naval Procurement Strategies

It seems appropriate to close this section on warship design and construction in the Canadian Navy with a few remarks on how ship procurement strategies have changed over the years. Before World War II, the small Canadian Navy's ships were designed and built in the UK. During World War II, most Canadian warships were built in Canada, but to Royal Navy designs. With the postwar decision to maintain a modest fleet designed and built in Canada, procurement strategies became an issue with each ship program. Perhaps because of the significant time-lapses between the River-classes, the Tribal-class and the City-class, in each case the strategies were different. But more likely, many of the changes were brought about by managers in central government agencies who had insufficient faith in the Navy's engineers, and a desire to reduce the government's risk through a clearer, more legalized, more business-like allocation of responsibilities.

The River-class contracts were cost-plus on an allocation basis. No fewer than seven yards built the twenty destroyers. By the late 1960s, competitive bidding had become *de rigueur*. As a result, an attempt was made to have the lead shipyard (two of four ships) for the Tribal-class undertake the responsibility for the development and performance of the detailed design, which it was to develop from the Navy's contract design using a government designated sub-contractor. However, virtually every major equipment was separately contracted and delivered to the shipyard as government supplied material. Needless to say, even when equipment delivery dates were met, all too frequently the technical data package for the equipment was received far too late to meet the detailed design drawing schedule. In some cases it appeared as

Fig. 10 City class Canadian patrol frigate HMCS *Halifax*: 4750 tons, 440 ft, 46,000 shp, 27 knots, CH124 Sea King helo, vertical launch Sea Sparrow missiles, two quad harpoon anti-ship missiles, PHALANX close-in weapon system, 57 mm gun, two triple torpedo mounts, AN/SQS-505V hull-mounted sonar and CANTASS towed sonar array. (Dept. National Defence photograph.)

if contractors held back the data to assist themselves in other objectives. In the end, the Navy exercised most of the responsibilities and accepted most of the (limited) risks that were supposed to have been transferred to the lead shipyard.

By the time the CPF project was out to industry in the early 1980s, a new, major objective had been added to all Canadian government procurements over $2 million: the promotion of Canadian industrial benefits. Contrary to some prevalent popular opinion, it was this change more than any other factor that moved design responsibility to industry, and certainly not a lack of capability on the part of naval engineers. Because of regional economic considerations, the industrial benefit plan for each procurement had to be approved by the Cabinet. If the Tribal-class procurement strategy had been repeated, well over 200 separate contracts would have had to be urged through the full government approval process with all the attendant potential for lobbying, interference, and delay, to say nothing of the time demands that would have been placed on senior officials and ministers. The prospect of ever getting a ship could probably be described as moot. Under these circumstances, it was far better to let a single, prime contractor deal with the Pandora's box of individual equipment selection and industrial benefits within the envelope of protection provided by their mandated acceptance of total systems responsibility.

Total systems responsibility was examined as practised in the U.S. Navy's *Spruance*-class program. A fairly similar approach was adopted with two important differences. First, the competitive phases were carried through to a fuller completion of preliminary design. More importantly, the competing industrial groups were given far more than the operational requirement. They were also given detailed functional specifications (not written around specific equipments) and all the technical standards for every system in the ship. These took the form of mandatory and desirable requirements, leaving trade-offs and equipment selections to the competing groups.

At time of writing it is too early to draw a final conclusion about the CPF procurement strategy. In the opinion of at least one, well-positioned naval engineer, the prospect on the whole looks good [36]. He concluded:

> In a nutshell, our bold experiment with TSR (Total Systems Responsibility) so far has worked. The process, as imperfect and frustrating as it is, is not endangering our likelihood of acquiring a fully compliant ship with CPF. The sceptics can maintain their guarded optimism for now, but they may be pleasantly surprised at the results.

I would support such an opinion.

Acknowledgments

I want to thank Captain Keith Farrell for the invitation and encouragement to write the Canadian Navy's technical story (as found above and also in Part V). I greatly appreciate the word processing provided by the German & Milne Division of ECS-Power Systems Inc. of Ottawa. The critical comments of Rear Admiral Denny Boyle, Commodore Mike Saker, Captain Jim Knox, and Messrs Bob McNeilly and Don Nicholson have been instrumental in improving the accuracy and the telling. My thanks to all.

References

1 *RCN in Retrospect, 1910–1968*, J. A. Boutilier, Ed., Univ. of British Columbia, 1982, p. 318.
2 *RCN in Retrospect, 1910–1968*, J. A. Boutilier, Ed., Univ. of British Columbia, 1982, pp. 319–320.
3 Galbraith, D. E., "Turbine Rotor Forgings for DEs Manufactured in Canada," SNAME, Eastern Canadian Section, 12 Nov. 1957; *RCN in Retrospect, 1910–1968*, J. A. Boutilier, Ed., Univ. of British Columbia, 1982, pp. 320.
4 Baker, R., "Habitability in the Ships of the RCN," *Trans.* SNAME, Vol. 64, 1956, p. 69.
5 Personal notes of the author taken at a presentation in 1964 of the General Purpose Frigate design.
6 Lynch, T. G., *The Flying 400—Canada's Hydrofoil Project*, Nimbus, Halifax, N.S., 1983.
7 Eames, M. C., "A Review of Hydrofoil Development in Canada," First International Hydrofoil Society Conference, July 1982.
8 *RCN in Retrospect, 1910–1968*, J. A. Boutilier, Ed., Univ. of British Columbia, 1982, p. 325.
9 Shenker, H. A., et al., "Operational Support Ship for the Royal Canadian Navy," SNAME, Eastern Canadian Section, March 1966.
10 Farrell, K. P., Judge, R., and Engh, I. B., "RCN Application of Method Study in Design," SNAME, Eastern Canadian Section, 12 Jan. 1965.
11 "Starshell," The Naval Officers Association of Canada, March 1991, p. 3.
12 *White Paper on Defence*, Queen's Printer, Canada, 1964.
13 Farrell, K. P., et al., "DDH 280 Class Design," SNAME, Eastern Canadian Section, 23 Feb. 1971.
14 Edwards, G. L., "The DDH 280 Revisited—An Operational Review," SNAME, Eastern Canadian Section, 15 Feb. 1977.
15 Sunley, A. N., and Patterson, G. A., "DDH 280 Control System," SNAME, Eastern Canadian Section, 3 Feb. 1970.
16 Le Gallais, G. A., and Healey, E. J., "Description of the Shore Test of Propulsion Plant for DDH 280 Class Gas Turbine Destroyers," SNAME, Canadian Maritime Section, 25 March 1971.
17 *RCN in Retrospect, 1910–1968*, J. A. Boutilier, Ed., Univ. of British Columbia, 1982, pp. 286–294.
18 Matthews, F. W., "Stability and Control of HMCS *Labrador*, *Trans.* SNAME, Vol. 67, 1959, p. 70.
19 Tranter, G. J., *Plowing the Arctic*, Wrigley Printing, Vancouver, 1945.
20 "QUEST Commissioning Booklet," Defence Research Establishment Atlantic, Dartmouth, Nova Scotia, August 1969.
21 Bowness, G., McFarlane, J. R., and Sargent, W. C., "Submarines in Canada," SNAME, Eastern Canadian Section, 21 Jan. 1969.
22 Schmitke, R. T., Glen, I. F., and Murdey, D. C., "Development of a Frigate Hull Form for Superior Seakeeping," SNAME, Eastern Canadian Section, 17 April 1979.
23 Hussey, D. J., "DDH 280 Structural and Corrosion Problems," SNAME, Eastern Canadian Section, 16 Feb. 1982.
24 Marchand, R., "CANTASS—Bringing ASW into the 21st Century," *Maritime Engineering Journal* (published by Director General Maritime Engineering and Maintenance at National Defence Headquarters), Jan. 1989, pp. 9–14.
25 Carruthers, J. F., "SHINPADS—A New Ship Integration Concept," *Naval Engineers Journal* (American Society of Naval Engineers), April 1979, pp. 155–163.
26 *Webster's Dictionary.*
27 Kuhns, R. S., "The SHINPADS Serial Data Bus," *Naval Engineers Journal*, April 1979, pp. 164–172.
28 Baxter, B. H., Rhodenizer, R. J., and Penny, P. V., "Shipboard Integrated Machinery Control System (SHINMACS)—A Canadian Forces Concept," *Proceedings*, Sixth Ship Control Systems Symposium, Vol. 5, 1981, pp. M2-1 to M2-9.
29 Khan, R., "Canadian Patrol Frigate Machinery Control System (An Update)," *Proceedings*, Eighth Ship Control Systems Symp., Vol. 1, 1987, pp. 1.51 to 1.66.
30 Shak, M., and Thibault, R., "Development of a Reverse Osmosis Desalination System for the Naval Environment," *Maritime Engineering Journal*, Jan. 1989, pp. 15–21.
31 Johnson, R., "RAST MK III A New-Generation Helicopter Handling System," *Maritime Engineering Journal*, Sept. 1989, pp. 6–10.
32 Archer, R. F., "CPF Operational Capability," *Maritime Engineering Journal*, Feb. 1984, pp. 17–21.
33 Payne, R., "The Product Work Breakdown System—Blueprint for CPF Construction," *Maritime Engineering Journal*, Sept. 1988, pp. 17–20.
34 McCullough, B., "New Frigates a Shipbuilding Success Story," *Maritime Engineering Journal*, Sept. 1988, p. 21.

35 Hemming, D., "The CPF Combat System Development Process," *Maritime Engineering Journal*, Sept. 1987, pp. 7–11.
36 Chiasson, R. E., "Total Systems Responsibility—Will It Work?" *Maritime Engineering Journal*, Sept. 1989, pp. 24–26.

Additional Reading

Baker, R., "Habitability in Ships of the Royal Canadian Navy," *Trans.* SNAME, Vol. 64, 1956.

Commodore W. J. Broughton attended the Royal Military College of Canada, Queen's University, where he received a BSc degree in mechanical engineering, and then MIT, where he received both an MS degree in naval architecture and marine engineering and the Naval Engineer degree. During his naval career he worked, *inter alia*, on every Canadian destroyer program from the *Restigouche*-class conversion in the early 1960s to the City-class frigates currently under construction. He was Project Systems Engineer for the Tribal-class design and construction from 1970 to 1974. On his retirement in 1990, after 37 years of service, Commodore Broughton was the Canadian Navy's senior engineer at National Defence Headquarters, Ottawa.

The Military Transport Fleet

ALAN R. REID and EDWARD BARRETT

Introduction: Historical Development

DURING WORLD WAR II, our government's sea transportation was controlled by four separate federal agencies. That was changed in 1949 when the Military Sea Transportation Services (MSTS) became the single managing agency for Department of Defense ocean transport.

The immediate post war era saw the constriction of the *Barrett*-class (P2-S1-DN1) ships and a number of troop ship adaptations of Maritime Administration (MarAd) designs (T-AP). MSTS acquired many of its shipping assets from the Maritime Commission reserve fleet. Many of these ships were employed in transporting troops during the Korean War. During that war, the MSTS inventory consisted of 226 MSTS ships and 212 time chartered ships. The MSTS fleet included 64 troop transports, 2 hospital ships, 42 cargo ships, 81 tankers and 37 miscellaneous ships. Time chartered ships included 190 cargo ships and 22 miscellaneous ships.

The types of ships in the MSTS special projects fleet ranged widely and included C1-M-AV1s, C2s, Victory ships, C4s, T2 tankers and specially designed and constructed survey ships. In 1957, six C1-M-AV1s were modified for Air Force missile range support missions by receiving electronic equipment and antennas for telemetry and communications. Five ships were converted to support the Apollo Man-in-Space/Man-to-the-Moon programs.

The Service also converted and operated ships to support programs of the Office of Naval Research, U.S. Naval Oceanographic Office, National Science Foundation, and the Naval Research Laboratory. The Service was heavily involved in the development of the initial forward floating depot concept by converting three Victory ships to carry prepositioned Army armored infantry division equipment. During the Viet Nam war, the MSTS fleet was expanded to provide support to allied troops. Additionally, the Service converted and operated an ex-Navy seaplane tender for the Army as a floating helicopter repair facility.

In 1970, MSTS was renamed the Military Sealift Command (MSC). In 1972, its mission was expanded to include Naval Fleet Auxiliary Force (NFAF) responsibilities after shipboard tests showed that civilian crews could operate the Navy's fleet auxiliary ships efficiently and effectively, thereby freeing military crews for other responsibilities.

Between 1981 and 1984, MSC's responsibilities were expanded to include strategic sealift ship management. This included the acquisition of eight high-speed (33 knots) containerships that were converted to roll on/roll off, or sea shed-flat rack fast sealift ships. Additionally, the capability to preposition three Marine expeditionary brigades around the world was accomplished with build or convert-and-charter contracts for aquiring thirteen maritime prepositioning ships (MPS).

By 1984, MSC's role received additional prominence when the Secretary of the Navy designated strategic sealift as one of the Navy's four primary functions. It achieved equal status with the Navy's other three functions, namely, sea control, power projection, and strategic deterrence.

Yet another milestone was reached in 1987, when MSC became one of three component commands reporting to the newly established joint service U.S. Transportation Command (USTRANSCOM).

Current Composition

MSC executes its mission within three broad operational categories: Naval Fleet Auxiliary Force (NFAF), Special Mission Support Force, and Strategic Sealift Force. The following sections will address these forces' compositions and responsibilities.

Naval Fleet Auxiliary Force. The Naval Fleet Auxiliary Force, NFAF, (Table 1) provides direct support for Navy combatant ships at sea, allowing them to remain on station for extended periods. Some of these ships replenish supplies for our forces afloat; others conduct underwater surveillance; and still others provide lifting and towing services to our ships at sea.

Table 1 Naval fleet auxiliary force (49 total ships)

AO (Oilers) — 17 total	TAGOS (Surveillance) — 18 total	AE (Ammo) — 1 total
USNS *Hassayampa*	USNS *Stalwart*	USNS *Kilauea*
USNS *Kawishiwi*	USNS *Contender*	ATF (Fleet Ocean Tug) — 7 total
USNS *Mississinewa*	USNS *Vindicator*	USNS *Apache*
USNS *Navasota*	USNS *Triumph*	USNS *Catawba*
USNS *Neosho*	USNS *Assurance*	USNS *Mohawk*
USNS *Passumpsic*	USNS *Persistent*	USNS *Navajo*
USNS *Ponchatoula*	USNS *Indomitable*	USNS *Powhatan*
USNS *Truckee*	USNS *Prevail*	USNS *Sioux*
USNS *Kaiser*	USNS *Assertive*	USNS *Narragansett*
USNS *Humphreys*	USNS *Invincible*	AK-FBM (Fleet Ballistic Missile) — 2 total
USNS *Lenthall*	USNS *Audacious*	USNS *Marshfield*
USNS *Higgins*	USNS *Bold*	USNS *Vega*
USNS *Diehl*	USNS *Adventurous*	AFS (Combat Stores) — 3 total
USNS *Grumman*	USNS *Worthy*	USNS *Sirius*
USNS *Pecos*	USNS *Titan*	USNS *Saturn*
USNS *Ericsson*	USNS *Capable*	USNS *Spica*
AF (Fleet Stores) — 1 total	USNS *Tenacious*	
USNS *Rigel*	USNS *Relentless*	

A review of Table 1 indicates the largest share of NFAF ships are oilers or ocean surveillance ships. Of the 49 NFAF ships, as of January 1992, sixteen ships were transferred to MSC. More information on these ships follows.

Fleet Oilers (T-AO 143 and T-AO 105 Classes) (9 Ships). Both of these classes of ships were transferred to MSC. They are steam propelled, 180,000-barrel capacity, fully capable oilers. In view of their age and budget considerations, they were retired as the new construction Fleet Oil T-AO 187 class ships were delivered. Principal characteristics of T-AO 143 class ships are

Length Overall	655 ft-0 in. (199.6 m)
Maximum Beam	86 ft-0 in. (26.2 m)
Displacement, Full Load	11,600 LT (11,786 mt)
Draft, Full Load	35 ft-0 in. (10.7 m)
Rated Horsepower	14,000 (each shaft)
Speed	20 knots
Endurance	20,000 n. mi
Accommodations	125
Propulsion Type	Steam turbine

Fleet Oiler (T-AO 187 Class) (18 Ships). Oilers such as the *Kaiser* are the newest additions to the fleet. These ships incorporate the Navy's latest cargo transfer equipment to provide underway replenishment of fuel oil and dry cargo to Navy ships at sea. In addition, they provide jet fuel for aircraft assigned to our aircraft carriers. Each ship has a crew of about 100 civil service mariners and 20 Navy personnel, and they cruise at up to 20 knots. They incorporate technological state-of-the-art features such as large bore, medium speed, twin diesel engines with an integrated electrical plant,

Fig. 1 Fleet oiler USNS *Mississinewa.*

Fig. 2 Fleet oiler USNS *Henry J. Kaiser* (T-AO-187) in the Gulf of Mexico, Sept. 1986. (Official U.S. Navy photograph.)

hardened and ground reductions gears, and controllable reversible pitch propellers. The machinery plants are designed for unattended operation. Principal characteristics of T-AO 187 class ships are

Length Overall	677 ft-6 in. (206.5 m)
Maximum Beam	97 ft-6 in. (29.7 m)
Displacement, Full Load	40,700 LT (41,351 mt)
Draft, Full Load	36 ft-0 in. (11 m)
Rated Horsepower	16,000 (each shaft)
Speed	20 knots
Endurance	6000 n. mi
Accommodations	137
Propulsion Type	Diesel/CRP

Fleet Stores (AFS 1 Class) (6 Ships). Fleet stores ships of the Mars class are being turned over to the MSC Fleet for operation, with the first, *Concord*, turned over in October, 1992. The remaining five ships will be turned over in a phased manner through 1995. These ships provide underway replenishment of freeze, chill, and dry cargo to the fleet. The *Concord* has a crew of about 135 civilian mariners and 80 Navy personnel. The ships are single-screw, three-boiler, steam turbine propelled, with a displacement of 16,500 tons.

Undersea Surveillance Ships (T-AGOS Monohull) (18 Ships). As potentially hostile submarines deploy around the world, MSC's undersea surveillance ships, designated T-AGOS, operate a towed array sensor system to gather submarine acoustical data, specially designed to locate new and quieter submarines. All T-AGOS ships propelled by diesel electric motor (dc) so as to reduce noise and provide effective operations over a wide range of propulsion and shipboard electrical system requirements. The T-AGOS ships are the most rapidly increasing segment of NFAF. Despite their small size, they can accomplish their mission in high sea states with deployments up to 93 days. Principal characteristics of T-AGOS 13 class ships are

Length Overall	224 ft-0 in. (68.3 m)
Maximum Beam	43 ft-0 in. (13.1 m)
Displacement, Full Load	2262 LT (2298 mt)
Draft, Full Load	14 ft-11 in. (4.6 m)
Rated Horsepower	1600
Speed	11 knots
Endurance	3000 n. mi (+90 days at 3 knots)
Accommodations	33
Propulsion Type	Diesel electric, ac/dc SCR

Undersea Surveillance Ships (T-AGOS 19-class) (SWATH) (4 Ships). A newly designed twin-hulled T-AGOS ship provides excellent stability, even in the rough seas encountered at high latitudes. This latest advance in Navy surveillance ships is called SWATH, which stands for Small Water Plane Area Twin Hull. Initial at-sea performance of the first ship of this class, the *Victorious*, demonstrates the excellent stability and maneuverability characteristics of the SWATH hull. Principal characteristics of T-AGOS 19 class ships are

Length Overall	234 ft-6 in. (71.5 m)
Maximum Beam	93 ft-6 in. (28.5 m)
Displacement, Full Load	3380 LT (3434 mt)
Draft, Full Load	24 ft-9 in. (7.6 m)
Rated Horsepower	1600
Speed	9.6 knots
Endurance	3000 n. mi (+90 days at 3 knots)
Accommodations	38
Propulsion Type	Diesel electric, ac/dc SCR

Undersea Surveillance Ships (T-AGOS 23-class) (SWATH) (1 Ship). A construction program is now

Fig. 3 T-AGOS 13 (M230) on acceptance trial, July 1988.

Fig. 4 SWATH T-AGOS 19, May 1990.

underway to build this larger SWATH, the T-AGOS 23, which can accomplish additional mission requirements in high latitudes. Principal characteristics of T-AGOS 23 class ships are

Length Overall	281 ft-6 in. (85.8 m)
Maximum Beam	95 ft-9 in. (29.2 m)
Displacement, Full Load	5370 LT (5456 mt)
Draft, Full Load	26 ft-0 in. (7.9 m)
Rated Horsepower	5000
Speed	12 knots
Endurance	3000 n. mi (+90 days at 3 knots)
Accommodations	45
Propulsion Type	Diesel electric, ac/dc SCR

Fleet Ocean Tugs (T-ATF) (7 Ships). Another workhorse of NFAF is the fleet tug. MSC operates seven of these powerful units, which can tow even the largest Navy ship, fight fires, pump out other ships, and provide other limited salvage capability. Principal characteristics of T-ATF 166 class ships are

Length Overall	226 ft-0 in. (68.9 m)
Maximum Beam	42 ft-0 in. (12.8 m)
Displacement, Full Load	2260 LT (2296 mt)
Draft, Full Load	16 ft-0 in. (4.9 m)
Rated Horsepower	3600 (each shaft)
Speed	14 knots
Endurance	10,000 n. mi
Accommodations	41
Propulsion Type	Diesel, geared

Kilauea (Ammunition Ship) (T-AE). In addition to ammunition, the T-AE ships can provide at-sea replenishment of fuel oil and dry cargoes. Principal characteristics of T-AE 26 class ships are

Length Overall	564 ft-0 in. (171.9 m)
Maximum Beam	81 ft-0 in. (24.7 m)
Displacement, Full Load	20,500 LT (20,828 mt)
Draft, Full Load	25 ft-9 in. (7.9 m)
Rated Horsepower	22,000 (total)
Speed	20 knots
Endurance	9384 n. mi
Accommodations	190
Propulsion Type	Steam turbine

Sirius (Stores Ship) (T-AFS). Food, spare parts, and limited ammunition can also be transferred to ships

Fig. 5 Artist's concept of T-23.

while underway. The *Sirius* is one of four MSC-operated stores ships supplying the fleet. Previously, this diesel-powered ship was a British Royal Fleet Auxiliary replenishment ship. It was acquired and modernized by the U.S. Navy in the early 1980s. Principal characteristics of T-AFS 8 class ships are

Length Overall	524 ft-0 in. (159.7 m)
Maximum Beam	72 ft-0 in. (22 m)
Displacement, Full Load	16,680 LT (16,947 mt)
Draft, Full Load	26 ft-6¾ in. (8.1 m)
Rated Horsepower	11,520 (total)
Speed	17 knots
Endurance	13,870 n. mi
Accommodations	195
Propulsion Type	Diesel, slow speed

Special Mission Support Force. MSC's second major force of ships is the Special Mission Support Force. This force carries out a variety of highly specialized missions, including oceanographic research, range tracking, oceanographic and hydrographic surveys, as well as cable laying and repairing. In addition, the Pacific Launch Area Support Ship (LASS), *Point Loma*, and a previous frigate, the *Glover* (with pressure fired boilers), supporting sonar research were

Fig. 6 Fleet ocean tug USNS *Powhatan* (T-ATF 166).

Fig. 7 Ammunition ship USNS *Kilauea* (T-AE 26).

transferred to MSC in the 1980s. Table 2 shows a summary of MSC Special Mission Support Ships.

Outlined below are some of the ships that make up the Special Mission Support Force.

Observation Island (Missile Range Instrumentation Ship). Missile range instrumentation ships, such as the *Observation Island*, monitor missile launches to collect data that can be used to improve missile efficiency and accuracy. They also track foreign missile launches to ensure compliance with arms limitations agreements. Principal characteristics of the T-AGM 23 class ships are

Length Overall	563 ft-0 in. (171.6 m)
Maximum Beam	76 ft-2 in. (23.2 m)
Displacement, Full Load	17,600 LT (17,882 mt)
Draft, Full Load	27 ft-0 in. (8.3 m)
Rated Horsepower	22,000 (total)
Speed	18 knots
Endurance	12,000 n. mi
Accommodations	175
Propulsion Type	Geared turbine

Maury (Oceanographic Research and Survey Ship) (2 Ships). MSC oceanographic research and survey ships are used to study the ocean environment. They improve our Navy's capability in undersea warfare and enemy ship detection. The ships' multi-beam, wide-angle precision sonar systems permit continuous charting of a broad strip of ocean floor. Principal characteristics of the T-AGS 39 class ships are

Length Overall	499 ft-10 in. (152.4 m)
Maximum Beam	72 ft-0 in. (21.9 m)
Displacement, Full Load	15,821 LT (16,074 mt)

Fig. 8 Fleet stores ship USNS *Sirius* (T-AFS 8).

Table 2 Special mission support (23 total ships)

AGS (Surveying) — 9 total	AG (Navigation Test Support) — 1 total	AK (Undersea Cable) — 1 total
USNS *Silas Bent*	USNS *Vanguard*	USNS *Furman*
USNS *H. H. Hess*	AGM/AGDS (Missile Range Instrumentation) — 4 total	ARC (Cable Repairing) — 3 total
USNS *Chauvenet*	USNS *Range Sentinel*	USNS *Albert J. Myer*
USNS *Harkness*	USNS *Observation Island*	USNS *Neptune*
USNS *Kane*	USNS *Redstone*	USNS *Zeus*
USNS *Wilkes*	USNS *Point Loma* (AGDS)	AG/FF (Acoustic Research) — 2 total
USNS *Wyman*	AGOR (Oceanographic Research) — 3 total	USNS *Glover*
USNS *Maury*	USNS *Bartlett*	USNS *Hayes*
USNS *Tanner*	USNS *De Steiguer*	
	USNS *Lynch*	

Fig. 9 Missile range instrumentation ship USNS *Observation Island* (T-AGM 23).

Fig. 10 Oceanographic research and survey ship USNS *Maury* (T-AGS 39).

Fig. 11 Cable ship USNS *Zeus* (T-ARC 7).

Draft, Full Load	30 ft-6 in. (9.3 m)
Rated Horsepower	12,500 (per engine)
Speed	20 knots
Endurance	17,820 n. mi
Accommodations	108
Propulsion Type	Diesel, geared

Zeus (Cable Ship). MSC operates three cable ships. They transport, deploy, retrieve and repair submarine cables. The *Zeus* was the first cable ship built specifically for the Navy, and she can lay up to 1000 mi (1609 km) of cable in depths to 9000 ft (2745 m). Principal characteristics of T-ARC 7 class ships are

Length Overall	513 ft-5 in. (156.5 m)
Maximum Beam	73 ft-0 in. (22.3 m)
Displacement, Full Load	15,986 LT (16,242 mt)
Draft, Full Load	27 ft-9⁹⁄₁₆ in. (8.5 m)
Rated Horsepower	5000 (each shaft)
Speed	15 knots
Endurance	10,000 n. mi
Accommodations	126
Propulsion Type	Diesel electric

Strategic Sealift Force. MSC's third and largest force of ships is the Strategic Sealift Force. The mission of these ships is to deploy and sustain U.S. military forces, wherever needed, through delivery of equipment, petroleum products and other supplies. In peacetime more than 95 percent of our military dry cargo is carried on regularly scheduled U.S.-flag ships.

Figure 12 presents the mix of strategic ships in full operational status (FOS) just before Iraq's invasion of Kuwait in August 1990.

Prepositioning ships were strategically stationed around the world, all laden with military equipment and supplies. Eleven commercial dry cargo ships and 26 tankers under long term charter also played major roles in peace time sealift. The chartered dry cargo ships carried ammunition and transported supplies to out of the way places not regularly served by U.S.-flag ships. The chartered tankers carried most of military's petroleum products. At that time several other sealift vessels remained at their piers in reduced operating status (ROS).

Soon after Iraq's invasion, President Bush ordered U.S. forces to the Middle East, and sail orders were issued for two of MSC's three maritime prepositioning squadrons and ten of the afloat prepositioning ships. All eight of MSC's fast sealift ships, two hospital ships and two aviation logistics support ships, all of which had been in reduced operating status, were activated. By the end of September 1990, the number of Strategic Sealift Ships had almost tripled. (See Fig. 13.)

By January 15, 1991, the active strategic sealift force had grown to 273 ships. The greatest increases were in the area of chartered ships and newly activated ready reserve force (RRF) ships. (See Fig. 14.)

Following are some of the ships that make up the Strategic Sealift Force at the time of this writing (1992).

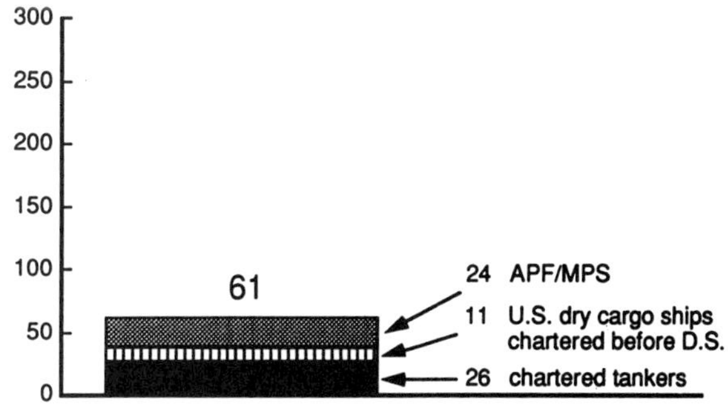

Fig. 12 Active strategic sealift, 1 August 1990.

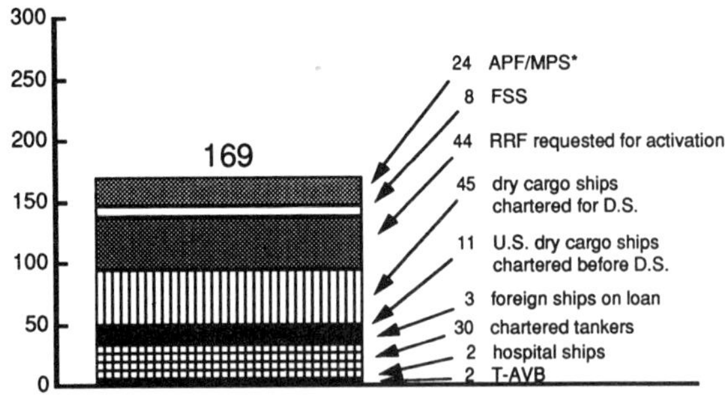

*Some of the APF/MPS converted to common user status after mid-August, but they are still counted as APF/MPS on the graph.

Fig. 13 Active strategic sealift, 30 September 1990.

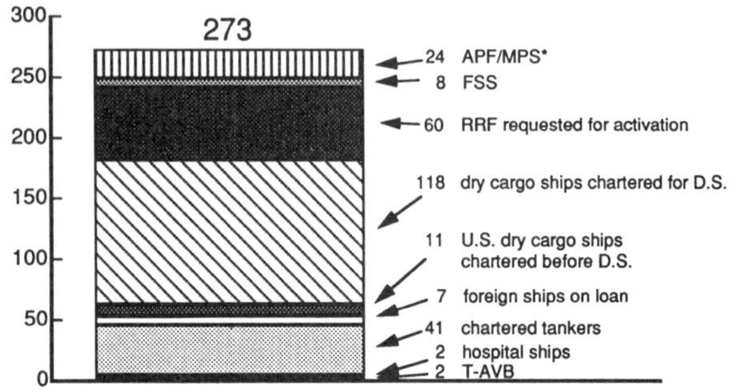

*Some of the APF/MPS converted to common user status after mid-August, but they are still counted as APF/MPS on the graph.

Fig. 14 Active strategic sealift, 15 January 1991.

Austral Rainbow (Afloat Prepositioning Ship). The *Austral Rainbow* is one of the Afloat Prepositioning Force (APF) ships. The successful deployment of U.S. military forces depends on the ability to act quickly. Recognizing that hostilities can erupt at any time, these ships give the United States rapid surge response. APF ships are loaded primarily with Army and Air Force equipment and supplies, but each is also equipped with a Navy fleet hospital. They are prepositioned in strategic locations around the world.

Sgt Matej Kocak Class (Maritime Prepositioning Ship Force) (13 Ships). MSC has 13 Maritime Prepositioning Ships (MPS) divided into 3 squadrons, each stationed at a potential trouble spot in the world. Each squadron can carry equipment and supplies, including fuel and water, to support a heavily mechanized 1600-man Marine expeditionary brigade for 30 days.

Bellatrix Class (Fast Sealift T-AKR) (8 Ships). Fast Sealift Ships (FSS) are the fastest cargo ships in the world, making them a key part of our rapid surge capability. Though they are almost the size of aircraft carriers, they can travel over 30 knots. The Navy purchased eight commercial containerships in the early 1980s and converted them for this service. Modifications provided roll on/roll off (RO/RO) capability, additional lifting capacity, and helicopter facilities. They are ideal for transporting Army wheeled and tracked unit equipment, such as tanks, helicopters, and other bulky military equipment. The eight ships, combined, can carry the equivalent of a full Army mechanized division. They are normally kept in reduced operating status, but can be activated and ready to sail to a loading berth in just 96 hours.

Mercy Class (Hospital Ships T-AH) (2 Ships). The *Mercy* and the *Comfort* are MSC's two hospital ships, designed to provide emergency, on-site medical care for our combat forces. Worldwide, there are only seven hospitals larger than either of these huge ships. Each has twelve operating rooms and a 1000-bed hospital. They are maintained in reduced operating status and can be fully activated, crewed, and ready for deployment with five days.

Keystone State Class (Crane Ships T-ACS) (8 Ships). The *Keystone State* is one of eight recently converted crane ships in the ready reserve fleet. They are specifically designed to unload military cargo from non-self-sustaining containerships at undeveloped or war-damaged ports.

Future Growth. The lessons learned from Iraq's invasion of Kuwait have identified a shortage of U.S. strategic sealift assets. Congress has directed the Department of Defense to accomplish a strategic sealift ship project to improve this condition. Additionally, as reductions in budgets and manpower authorizations continue to affect the Navy, the potential for transfer of Navy auxiliary ships (e.g., ARS, AO177, AOE, AOR's) to MSC will increase.

Military Sealift Command has a vital role in supporting our nation's defense. Its ships, both government owned and commercially chartered, directly contribute to the readiness and flexibility of the U.S. military services worldwide.

Additional Reading

Giblon, R. P., Elmer, W., and Nace, F. F., "Machinery Design of the *Barrett* Class MSTS Transports," *Trans.* SNAME, Vol. 61, 1953.

Miller, M. O., Hammett, J. W., and Murtphy, T. P., "The Development of the U.S. Navy Underway Replenishment Fleet," *Trans.* SNAME, Vol. 95, 1987.

Alan R. Reid is the Director of the Ship Introductory Division, Operating Department, Military Sealift Command, Department of the Navy.

Edward Barrett is the Director of New Ship Construction, Military Sealift Command, Department of the Navy.

Fighting Boats of the Navy

Richards T. Miller

Introduction

THE U.S. NAVY traditionally has been a blue water navy. Consequently for more than a century, each time it has become engaged in a conflict, it has found itself lacking some or all of the types of small craft needed for the inshore operations that have developed. In the years between wars most of the resources for new developments and construction have been spent on major combatants and auxiliaries. At the beginning of World War II, as in World War I, hundreds of yachts, fishing vessels, tugs, etc. were purchased and converted into patrol craft, gun boats, minesweepers, and local support vessels. Eventually special craft were designed to fulfill those functions, as well as for amphibious operations and inshore raiding in contested areas. The pattern of having to acquire specialized craft on a crash basis was repeated in the Korean War, and in Vietnam. From the start of World War II through Vietnam, the Navy's collective memory regarding these craft resided in an office under the able leadership of Sidney Peters. That office, the Bureau of Ships' Patrol, Service and Minecraft Type Desk, covered design, construction, procurement, and maintenance of these various craft.

Current usage of the word "boat" covers much more than the dictionary definition of "a small open vessel, or water craft, usually moved by oars or paddles..."; and it is not limited to its old nautical usage to cover craft that might be carried on a ship. There seems to be no precise definition of what is, or is not, a boat. While the largest patrol craft considered, *Cyclone* (PC-1), is 170 ft long, and therefore would appear to establish a reasonable upper limit for this section, the ocean minesweepers at 176 ft and mine countermeasure ships at 224 ft are also included in order to give complete coverage of mine warfare vessels.

While fighting boats served in World War II and in Korea, the operational geography of South Vietnam created the arch-typical theater for small combatant operations. There were some 40,000 square miles of coastal waters and 800 miles of coastline with smooth wide beaches. The rivers and canals of the Mekong Delta offered thousands of miles of aquatic highway protected from the sea. Some 40,000 junks plied those waters, many in pursuit of legitimate trading and fishing activities, but others carrying contraband of war from the north to, and between, enemy enclaves in the south. If the Navy's only mission on South Vietnam's rivers and in its coastal waters had been to police that traffic, it would have been task enough. But, in addition to that mission, the Navy had to conduct amphibious operations, and carry out general patrol duties and mine countermeasures. Those tasks resulted in the development of some unique adaptations of existing craft to perform special missions, as will be discussed.

Technological Advances

Hull Material. During World War II designers and builders of small craft made considerable use of marine-grade plywood and laminated timbers. This was made possible through the prior development of resorcinol and phenol-resorcinol glues and techniques for gluing the laminations together. These complemented standard wood construction materials and techniques so that a majority of fighting boats up to and including the 110-ft sub-chasers (SC/PC) and 136-ft mine sweepers (YMS) were of wooden construction. On the other hand, the 56-ft, or larger, landing craft (LCM), and larger patrol craft and mine countermeasures vessels, were built of all-welded steel. A few experimental craft were built of aluminum.

Since World War II the development of epoxy and polyester resins, together with fiberglass mat and woven roving, have given designers and builders a major new material for small craft construction. While not replacing wood it offers comparably light, strong structures. The development of weldable, marine-grade aluminum alloys made them equally important materials for small craft construction. Of 53 small combatant designs developed over the past 50 years, 12 were of wood, 9 of fiberglass reinforced plastic, 17 of steel, and 15 of aluminum construction. The wooden craft range in size from the 36-ft minesweeping launch (MSL) to the 224-ft mine countermeasures ship (MCM). The large wooden minesweepers were built with laminated oak frames, two inner layers of cross-diagonal planking, and one heavier fore-and-aft layer. The diagonal courses were fitted in lieu of the more conventional fore-and-aft planking with heavy diagonal bronze straps to provide shear strength. This was proposed and promoted by Meredith Johnson of the Bureau of Ships. Fiber reinforced plastic combatants range in size from the 31-ft river patrol boat (PBR) to the new 164-ft mine-hunting craft (MHC), adapted from the Italian LERICI-class design. Welded steel construction has been used for vessels as small as the 36-ft landing craft control boat (LCPL-MK4). Welded aluminum construction has been used in small combatants ranging in size from the 36-ft swimmer reconnaissance boat to the 165-ft patrol boats of the PG-84 class.

Machinery. Much of the history of small craft development concerns attempts to achieve higher and higher power in limited sizes and weights. Through World War II, with one notable exception, this was accomplished only with gasoline engines. In general, marine gasoline engines were available in weights of from three to five pounds per horsepower. Motor torpedo boats, however, required particularly high-power, compact engines. This was achieved using marine versions of water-cooled aircraft engines burning high-octane gasoline. Perhaps the most famous such engine was the Packard 4M2500. It had been developed by Packard for racing hydroplanes between World Wars I and II from the Liberty aircraft engine. Rated at 1350 hp originally, it weighed about one pound per horsepower. Ultimately Packard produced its W-100 model, which developed 2500 hp. However, because the Packard was a very specialized engine, it went out of production soon after the war.

Because of the risk of gasoline fires, the Navy actively sought suitable lightweight diesels to power its small craft. Initially it had to use General Motors GM8-268-A engines developing 600 hp at a weight of approximately ten pounds per horsepower for the 110-ft subchasers. By 1942 General Motors had delivered the first production model of its GM16-184-A "pancake" diesel, which delivered 1200 hp at a weight just over four pounds per horsepower. The firm also developed its Detroit Diesel-71 Series engines, which delivered 225 to 700 hp in 6-, 12-, and 16-cylinder models at weights of approximately ten pounds per horsepower. Because it was difficult to produce, and suffered serious maintenance problems, the pancake diesel never achieved strong acceptance, and was discontinued after the war. In its search for higher powers at reasonable weights for subchasers and larger landing craft, the Navy and General Motors developed the ingenious "quad," four GM6-71 diesels linked to a common bull gear by individual friction clutches. It developed 893 hp at 14.39 pounds per horsepower. It was used in many wartime and postwar landing craft and also in the MSI-1 class inshore minesweepers of the late 1950s. It was also used into the late 1960s in standard motor gunboats built for foreign navies. The other Series 71 diesels have also been used extensively over the years along with comparable engines by Gray Marine, Chrysler, Cummins, Hercules, Fairbanks Morse, Westerbeke and Hall Scott.

Shortly after World War II, the Navy recognized that high-performance, low-magnetic-signature diesel engines would be required for the next generation of minesweepers. The Bureau of Ships (BuShips) accordingly supported a program of engine research at the Packard Motor Company. It also planned a small, but carefully thought-out program of ship development. In October 1950, 2000 mines laid by an unsophisticated enemy kept some 50,000 U.S. troops embarked in 250 ships from landing at the otherwise unprotected port of Wonsan, North Korea. For lack of sufficient minesweepers an operation that should have taken no more than five days took fifteen. This reality of war forced a crash program of minesweeper design and production, and rushed the Packard engine from an orderly development program to war-scale production with disastrous effects. Built more to aircraft than to heavy industrial standards, it required aircraft-type maintenance. It took some eight years to correct the many mechanical ills with which it was plagued; thus it acquired an unsavory reputation. Whether that reputation was responsible or not, hoped-for commercial markets never materialized, so production of the engine ceased with completion of the minesweeper program.

When the Navy had a need for fast patrol boats during the Vietnam war it found itself with neither the hull design nor the high-performance diesels required to power such combatants. For the hull it was forced to go to the Norwegian *Nasty*-class. For power it purchased British Napier Deltic engines. These were developed after World War II to produce 2400 continuous horsepower (3100 maximum) at 5.33 pounds per continuous horsepower. For current high-performance requirements there is still no U.S.-built diesel. Instead, U.S. designers and builders must rely on U.S.-built gas turbines or on European-built diesels, such as the MTU (Motoren-und-Turbinen-Union) 12V, 16V, and 20V models. These diesels deliver 500 to 7400 hp at 5.5 to 6.0 pounds per horsepower.

After World War II gas turbines were first applied to warships by the British. Their Proteus marine adaptation of an aircraft gas turbine developed 3600 hp at the attractively low weight of 0.88 pounds per horsepower. The first U.S. watercraft to be powered solely by a gas turbine was a 24-ft personnel boat fitted with a 160 hp Boeing gas turbine with a unit weight of 1.97 pounds per horsepower, developed under a BuShips contract. Trials began in May 1950. The first major BuShips gas turbine contract, let in September 1950, was for Solar generators to be used in the 36-ft minesweepers (MSC). The first higher-powered U.S. small combatant vessel fitted with gas turbines was the 52-ft landing craft, swimmer reconnaissance (LCSR). This had two 50 hp Solar gas turbines weighing two pounds per horsepower.

Starting in the 1950s, gas turbines suitable for powering small combatants were developed from aircraft jet and turbo-prop engines. AVCO Lycoming developed the TF12, rated at 920 hp, and the TF14, rated at 1250 hp. These were derived from the T53 Army/Air Force helicopter engine. The TF25, ultimately rated at 2500 hp, was a further development of the TF14. AVCO's TF35, ultimately rated at 3500 hp, was developed from the T55 aircraft engine. The TF40, a development of the TF35, powered the prototype air-cushion landing craft (LCAC) of the 1980s. Three TF25s powered the prototype coastal patrol and interdiction craft (CPIC) of the 1970s.

The General Electric (GE) LM100 was a 1900 hp marine version of the T58 helicopter engine. It powered the Bell SK-5 hovercraft used in Vietnam. The larger LM1500, rated at 13,300 hp, developed by GE from the J79 jet engine, powered *Plainview* (AGEH-1) and the *Asheville* (PG-84) class. GE's LM2500 gas turbine, developed from the TF59 fan-jet and rated at 20,000 to 25,000 hp, became the Navy's standard gas turbine in the 1980s. It powers the PHM-1 class patrol hydrofoil craft, as well as the DD-963 class destroyers.

The AVCO and GE gas turbines offered specific weights of 0.37 to 0.92 pounds per horsepower without reduction gears. They were accordingly ideal for high-performance combatants requiring high power in limited sizes and weights. Their specific fuel rates, however, were relatively high. As a result, in some applications diesel engines were also installed for cruising or loitering modes of operation.

Hull Form. Given compact sources of power, high performance in conventional monohulls is limited by the wavemaking component of drag. In both hydroplanes and hydrofoil craft, this component can be significantly reduced by lifting the hull partly or completely out of the water with forces generated by the speed of the boat through the water. An alternative, the air-cushion vehicle, rides on a low-friction cushion of air generated by a separate power source. The three alternatives are limited to relatively small combatants by the "square-cube law" in that each square foot of area can support only a limited weight. Therefore available lift increases as the square of the craft's dimensions, whereas weight increases generally as the cube.

Although the principles of hydroplanes, hydrofoils, and air-cushion vehicles were known before World War I, only hydroplanes were exploited through World War II. Then, in the late 1940s, interest developed in hydrofoils. A series of small craft tests was instituted by the Bureau of Ships to develop automatic seakeeping control systems for fully submerged foils. In 1953 Vannevar Bush's Hydrofoil Corporation, in Annapolis, MD demonstrated the first submerged hydrofoil controlled by autopilot. This was the 20,000-lb *Lantern* (HC-4). Building on that experience, Gibbs and Cox designed *Sea Legs*, a 25-ft cabin cruiser hull supported by a canard foil configuration (main foil aft, stabilizing foil forward). An autopilot, responding to height sensor and accelerometer signals, controlled flaps to maintain height and attitude successfully in four-foot seas with a 1.5- to 2-ft clearance between hull and water. The *Sea Legs* control system was applied to the first U.S. Navy hydrofoil combatant, *High Point* (PCH-1). Its preliminary and contract designs were done by the Bureau of Ships. Detailed design and construction were by Boeing at the J. M. Martinac Yard in Tacoma WA. The 120-ton PCH-1 was followed by the 320-ton *Plainview* (AGEH-1), designed by Grumman in a Navy-sponsored competition, and ultimately built by Lockheed's Seattle shipyard. Next, two competing prototype hydrofoil gunboats, *Flagstaff* (PGH-1), and *Tucumcari* (PGH-2), were developed by Grumman and Boeing respectively in the late 1960s. Successful operation of these two boats in Vietnam led to the PHM program of the 1970s, the most recent application of this technology to U.S. Navy small combatants.

Successful application of the air-cushion principle to watercraft started with the work of Britain's Christopher Cockerell in the 1950s. Early craft could achieve ground clearances only in the order of one-twentieth to one-thirtieth of their beam. However, with the invention of flexible skirts in the late 1950s, ground clearance to hard structure was increased by a factor of ten, and cushion air capture efficiency was greatly improved. In 1963 the Bell Aerosystems Company won a Bureau of Ships competition to design and build a 20-ton experimental air cushion vehicle (ACV), SKMR-1. After extensive tests to determine its operational capabilities as a landing craft it was placed in mothballs and eventually broken up. Subsequent operational experience in Vietnam with the British-designed, Bell Aerosystems-built SRN-5 gave the Navy further confidence in the ACV concept. This led to procurement in the late 1970s of two experimental ACV amphibious assault landing craft (AALC) known as *Jeff* craft. Their success led to contracts for operational landing craft, air cushion (LCAC), starting in 1982.

The Navy also has experimented with a variation of the flexible-skirted ACV in the form of a rigid sidewall configuration known as the SES. Extensive tests were conducted on three smaller craft. These were the 79-ft and 78-ft SES-100A (designed and built by Aerojet) and 100B (designed and built by Bell), and the 159-ft SES-200 (Bell-Halter's BH-110). Three of the last-named vehicles have been operated successfully by the U.S. Coast Guard since 1986. Nevertheless, no operational craft of the type has been procured, or even contemplated by the Navy as of this writing.

Commercial designers and operators have recently become greatly interested in variations of catamaran hull forms, which are reputed to give high-performance characteristics to ferries. The Navy's principal interest in catamarans (after less-than-happy experiences with two catamaran submarine rescue ships and one catamaran oceanographic research ship) now resides in the slim waterline, twin hull or *Swath* configuration. As this is written, some 15 years' experience with the 200-ton test *Swath* craft *Kaimalino* has demonstrated the exceptional seakeeping characteristics of the hull form. However, the principle currently is being incorporated only in a class of surveillance ships (TAGOS) rather than the small combatants included in this chapter.

Weapons. By the end of World War II the Navy had developed a series of standardized weapons for small combatants. These included the 0.30-caliber[1] Lewis machine gun, the 0.45-caliber Thompson submachine gun, the 0.50-caliber machine gun, the 20 mm Oerlicon gun, the 40 mm Bofors gun, and the 3-inch/50 gun. They were carried in a variety of single, twin and multiple barrel mounts. One of the more notable of the latter was Elco's *Thunderbolt*, which increased PT firepower by incorporating two 0.50-caliber machine guns and two 20 mm guns in a single mount. Torpedoes, which initially had been fired from tubes, were side-launched from racks similar to depth charge racks. An ongoing 1945 Bureau of Ordnance weapon-development program for PTs was curtailed after the war for more urgent ASW and AAW weapons programs.

As the riverine warfare of Vietnam escalated in the 1960s, attention once again was directed toward producing lighter, more effective weapons for the small combatants. New versions of the standard World War II guns were developed or adapted, as were new types of guns. In the former category were a 40 mm grenade launcher, and a rapid fire RF 3-inch/50 gun, which was installed on the *Asheville*-class PGs. The latter category includes the 81 mm mortar, first tested in 1951, and the OTO-Melara 76 mm/62 gun installed in the *Pegasus*-class PHMs. Weapon development included improvements in a number of different gun mounts, some stabilized to provide greater accuracy, others combining several barrels to provide greater firepower. The Navy's MK48 mount is typical of these. Mod 0 carries a 20 mm cannon and a 40 mm grenade launcher; Mod 1 carries twin 7.62 mm machine guns and a 40 mm grenade launcher, and Mod 2 carries twin 0.50-caliber machine guns and a 40 mm grenade launcher.

Toward the end of World War II launchers for 4.5-inch fin-stabilized rockets (nine per launcher), for 2.36-inch (bazooka) rockets, and the 5-inch spin-stabilized (eight per MK107 launcher) rocket were fitted on PTs and PGMs. Recognizing the "punch" that missiles could give to gunboats, the Navy in 1971 fitted two launch cells for modified General Dynamics-designed Tartar-Bullpup surface-to-surface missiles to *Antelope* (PG-86) and to *Ready* (PG-87). These two PG-84 class motor gunboats had been fitted with the MK87 track-while-scan radar fire control system, developed from a Dutch system. The MK87 FCS offered superior compensation for ship motions over the MK63 FCS installed in the other boats of the class. *Grand Rapids* (PG-98) and *Douglas* (PG-100) were fitted with an improved standard ARM missile. Finally, in the 1980s, the PHM-1 class hydrofoil gunboats were fitted with six Harpoon surface-to-surface

[1] "0.30-caliber" refers to gun bore diameter in inches; "20 mm" refers to gun bore diameter in millimeters; "3-inch/50" indicates a bore diameter of 3 inches and a bore length 50 times the diameter, or 150 inches.

Fig. 1 Experimental hydrofoil patrol craft *Plainview*, AGEH-1.

missiles controlled by a MK87 FCS, in addition to their RF 3-inch/50 guns.

Other weapons that were available—and were occasionally installed on small riverine combatants—included the Army's M132 flame thrower and 105 mm howitzer.

Coastal Fighting Boats

Patrol Torpedo Boats. In the late 1940s the Bureau of Ships initiated a competition to develop replacement PTs for the aging and hard-used wooden craft of World War II. Four experimental boats, differing in details of hull form and structure, resulted. PT-809, designed and built by the Electric Boat Company, had a convex V, hard chine hull. Its all-riveted aluminum hull followed principles of aircraft construction. PT-810, designed by Sparkman and Stephens and built by the Bath Iron Works, had a convex V, hard chine hull form of shorter, wider proportions. Construction was riveted aluminum along ship structural principles. PT-811, designed by the Annapolis Yacht Yard (and completed by John Trumpy and Sons after acquiring the Annapolis facilities), had a "bell bottom" convex/concave bottom. Her all-welded aluminum structure had scantlings similar to those of PT-810. PT-812, designed and built by the Philadelphia Naval Shipyard, had a long, narrow, round-bottom hull along the lines of the German E-boats. She, too, was of aluminum construction. All four boats were fitted with four Packard 2500 hp gasoline engines. In 1956, PT-812 was fitted with two 4000 hp gas turbines on the outboard shafts, and two 800 hp diesels and controllable pitch propellers on the inboard shafts.

The following table shows the principal characteristics of the four boats.

	PT-809	PT-810	PT-811	PT-812
LOA, ft	98	89	95	105
Beam, ft	26	24	25	22
Draft, max., ft	6	6	6	6
Displacement, lb	189,350	182,000	173,000	200,032
Complement	20	18	22	22
Power, shp	10,000	10,000	10,000	10,000
Speed, knots	46	46.5	46	44.75
Armament				
40 mm guns	2	2	2	1
20 mm twin guns	2	2	2	2
81 mm mortar	1	1	1	1
MK-107 RL	0	0	0	1

Note: Torpedo launch racks were never fitted.

Fast Patrol Boat, PTF-3 Class. As the Vietnam War escalated the Navy searched for an appropriate design to supplement PTF-1 and -2 (ex PTs 810 and 811). The Norwegian *Nasty*-class was selected, with the first two boats, PTF-3 and -4 delivered in 1962 from their builder, Boat Service, Ltd. A/S, of Mandal, Norway. PTF-5 to -16 were delivered from Norway in 1963 and 1964. The final six boats, PTF-17 to -22, were built by John Trumpy and Sons, Inc., Annapolis MD in 1964–1968.

Fig. 2 Fast patrol boat PTF-2 (ex PT-810).

LOA	80 ft-4 in.
Beam	24 ft-6 in.
Draft	6 ft-9 in.
Displacement	64 tons, light; 76 tons, full load
Complement	19
Power	Two 3100 hp (max) Napier Deltic diesels
Speed	40+ knots
Armament	Two 40 mm guns
	Two 20 mm guns

Motor Gunboat, *Asheville*, PG-84 Class. The PG-84 class boats were designed in the early 1960s. With the exception of the four experimental PTs, this was the first class of high-performance, small craft designed by the Navy since World War II. With their all-welded aluminum hulls, and combined diesel or gas turbine power plants, they represented the latest advances in naval engineering. They were designed to provide patrol, blockade, and surveillance in coastal waters, perimeter defense for amphibious landings, and support for unconventional and guerrilla warfare. Fifteen boats were delivered between 1966 and 1969.

LOA	165 ft
Beam	23 ft-6 in.
Draft	8 ft
Displacement	225 tons, full load
Complement	25
Power	One GE LM1500 gas turbine of 13,300 hp
	Two Cummins 725 hp cruise diesels
Speed	37–40 knots
Armament	One 3 inch/50 RF gun
	One 40 mm gun
	Four 0.50-caliber machine guns
	Two ship-to-ship missile launchers (four boats)

Fast Patrol Craft, PCF MK1. The PCF MK1 fighting boats, with welded-aluminum hulls, were developed in 1965 from crew boats of the offshore oil industry. They proved useful in the near-shore, river, and bay war of South Vietnam. One hundred and four were built. In addition, eight boats of a slightly larger MK2 design were built in 1967 for the Military Assistance Program (MAP). The following table shows the principal characteristics of the MK1 boats.

LOA	50 ft-1 in.
Beam	13 ft-1 in.
Draft	3 ft-6 in.
Displacement	19 tons , full load
Complement	5
Power	Two 430 hp diesels
Speed	About 25 knots
Armament	One twin 0.50-caliber machine gun mount
	One single 0.50-caliber machine gun mount
	One 81 mm mortar

Coastal Patrol and Interdiction Craft (CPIC). In the early 1970s the Naval Inshore Warfare Craft Program was initiated to develop a heavy weather coastal patrol and interdiction craft (CPIC), and a smaller coastal patrol craft (CPC). Only CPIC itself was built (and that in only two copies) and tested. That was in 1974. Its double-chine planing hull form proved particularly capable in rough-water trials. This offered the possibility of a relatively simple, robust hull that could compete effectively with the more expensive and complex hydrofoil and surface-effect craft. Of longitudinally framed, welded aluminum construction, its slim and relatively shallow hull was much lighter than those of other boats of its type. Consequently one of the two hulls was subjected to extensive seakeeping and strength tests by the David Taylor Model Basin, Carderock MD.

LOA	99 ft-10.5 in.
Beam	18 ft-6.4 in.
Draft	6 ft-6 in.
Displacement	81.6 tons, full load
Power	Three AVCO Lycoming TF25A 2000 hp gt
	Two diesel out-drive units for low-speed loitering
Speed	49.5 knots in calm water
	38 knots in 4- to 6-ft waves
	35 knots in 5- to 9 -ft swells
Armament	One Emerson Electric twin 30 mm gun turret

Patrol Boat Mark III. In the late 1970s the Navy designed a new 65-ft patrol boat. Its most innovative characteristic was an asymmetric topside arrangement.

Fig. 3 Motor gunboat *Ashville*, PGM-84.

Fig. 4 Fast patrol craft PCF Mk I.

Fig. 5 Patrol boat Mk III.

Its superstructure was on the starboard side, and a heavy gun was located on the port side. This was in addition to the usual guns fore and aft. Thirteen of these craft were in U.S. Navy service in the 1980s. Three Mark IV boats, 3 feet longer and fitted with 650 hp diesels, were ordered in FY85 for service in the Canal Zone.

LOA	64 ft-11 in.
Beam	18 ft
Displacement	78,000 lb, full load
Power	Three 600 hp diesels
Speed	27 knots
Armament choices	One single 33 mm gun or one twin 20 mm gun or one 106 mm recoilless rifle or one torpedo launcher or one rocket launcher to port just forward of amidships; One twin 0.50-caliber machine gun or one twin 20 mm gun or one twin 7.62 mm gun on superstructure to starboard, just aft of amidships; and Two 0.50-caliber machine guns or two 7.62 mm machine guns or two 40 mm grenade launchers or two 20 mm guns to port and starboard aft

***Cyclone* (PC-1) Class.** In the late 1980s the Navy requested proposals for the design and construction of a new class of coastal patrol craft for interdiction and special warfare missions. Bollinger Shipyard won with its design developed from Vosper-Thornycroft's *Ramadan*-class hull. *Cyclone* (PC-1), the first of thirteen vessels to be built under the contract, was launched on February 1, 1992, with delivery anticipated, as this is written, in June 1992.

LOA	170 ft
Beam	25 ft
Draft	7.8 ft
Displacement	315 tons
Power	Four 3350 hp Paxman diesels
Speed	35 knots maximum 25 knots cruise
Armament	Two 0.50-caliber machine guns Two 7.62mm guns Two 25mm guns Two 40mm grenade launchers Stinger surface to air missiles

River Fighting Boats

Prototype Command Junk. The prototype command junk was one of a number of junk types built under a special program in Vietnam. Designed by LCdr. W. E. Hanks, USNR, it had typical "bat-wing" sails and an American diesel engine. Its hull lines showed the deadrise characteristics of the workboats of Hanks' native Chesapeake Bay.

Fig. 6 Coastal patrol craft *Cyclone*, PC-1.

LOA	55 ft
Beam	16 ft-8 in.
Draft	2 ft-8 in. (centerboard up)
Complement	7
Power	One 225 hp diesel
Armament	Small arms

Monitor. The monitor type gunboat, converted from LCM-6 landing craft hulls, was given a spoon bow. Early versions had been converted for the South Vietnam Navy. Some fifteen were converted between 1965 and 1967 for the U.S. Navy.

LOA	60 ft
Beam	17 ft-6 in.
Displacement	75 tons, full load
Power	Two 225 hp diesels
Armament	One 81 mm mortar
	One 40 mm guns
	Two 20 mm guns
	Two 0.50-caliber machine guns
	Four 0.30-caliber machine guns

Assault Support Patrol Boat. In the mid 1960s the Navy designed a new assault support patrol boat (ASPB) for riverine warfare in Vietnam. Eighty-six of these boats were built by Gunderson Brothers Engineering Corp., of which fourteen were for the South Vietnam Navy. In the late 1960s two similar boats, ASPB MK2 and MK3, were ordered to their designs from Sewart Seacraft and Sikorsky respectively.

	ASPB MK1	ASPB MK2	ASPB MK3
Material	Steel	Alum	GRP
Length	50 ft-1 in.	50 ft-0 in.	49 ft-8 in.
Beam	15 ft-2 in.	21 ft-4 in.	19 ft-8 in.
Draft	4 ft-3 in.	2 ft-9 in.	3 ft-5 in.
Displ.	76,814 lb	90,310 lb	74,537 lb
Engine	Two GM dsl 12V71	Two AVCO gt TF1416	Two United Aircraft gt ST6-570
Propulsion	Propeller	Water jet	Water jet
Power	430 hp	950 hp	443 hp
Speed	16 knots	25 knots	16+ knots
Armament	Two MK48 mounts	Three 20 mm guns	One 1.05 mm howitzer
	One 81 mm mortar	One 40 mm gun	One twin 20 mm gun
		One 81 mm mortar	One MK48 mount

Command Communications Boat. The Command Communication Boat (CCB) was another conversion of the steel LCM-6 landing craft for Vietnam operations. It, too, was fitted with a spoon bow. Eleven were converted between 1964 and 1967. Armament consisted of one 40 mm gun, two 20 mm guns, two 0.50-caliber machine guns, and two 0.30-caliber machine guns.

Armored Troop Carrier. The Armored Troop Carrier (ATC) was the third conversion of the LCM-6 for Vietnam. It retained the ramp bow of the original, but was heavily armored and armed. Ninety-two were converted between 1965 and 1967. Armament consisted of one 20 mm gun, two 0.50-caliber machine guns, and four 0.30-caliber machine guns.

Fig. 7 Gunboat *Monitor* (LCM-6 conversion).

River Patrol Boat. The River Patrol Boats (PBR) were Navy conversions of a standard commercial hull of glass-reinforced plastic construction. Instead of propellers, they were fitted with water jets. They proved to be satisfactory river patrol craft in South Vietnam. One hundred and eight of the MK1 boats were built in 1966. Eighty-one of the slightly larger MK2 boats were built in 1967.

LOA	31 ft or 31 ft-11 in.
Beam	10 ft-10 in. or 11 ft-7 in.
Draft	2 ft-2 in.
Displacement	14,500 lb or 16,000 lb
Power	Two 216 hp diesels
Speed	About 25 knots
Armament	Two twinned 0.50-caliber machine guns
	One single 0.50-caliber machine gun
	One grenade launcher

River Patrol Craft. The River Patrol Craft (RPC) were of welded steel, V-bottom construction. Because they were slower than the PBRs only 34 were procured in 1965. Some served as minesweepers for the U.S. Navy in Vietnam, but most were employed by the Vietnam Navy as gunboats.

Fig. 8 Armored troop carrier (ATC).

Fig. 9 River patrol boat PBR Mk I

LOA	35 ft-10 in.
Beam	10 ft-10 in.
Draft	3 ft-6 in.
Displacement	26,000 lb, full load
Power	Two 216 hp diesels
Speed	14 knots
Armament	One 0.30- and one 0.50-caliber (or two 0.30-caliber) twinned machine guns

Landing Craft

Landing craft in general are designed to transfer personnel and vehicles from ocean transports to a hostile beach. As such they can properly be designated as combatants.

36-Foot Landing Craft. Designed during World War II, the 36-ft Landing Craft (LCVP) were still being built in large numbers during the 1950s and 1960s to serve the needs of the Korean and Vietnamese wars. Early versions of the craft were planked with marine-grade plywood over wood frames, and provided with one-quarter inch steel armor. Later versions were constructed of fiberglass reinforced plastic.

LOA	35 ft-9 in.
Beam	10 ft-6 in.
Draft	3 ft-5 in.
Displacement	26,600 lb, full load
Power	One 225 hp diesel
Speed	9 knots
Armament	None

Landing Craft, Mechanized. The Mechanized Landing Craft (LCM-6 class) was based on the World War II LCM-3 class vehicle and personnel landing craft, with six feet of middlebody added. Its design was so versatile that 355 of the Mod 1 design were built in the 1950s; and 271 of the Mod 2 design, developed by Lukens Steel to further facilitate mass production, were built in the 1960s.

Fig. 10 36 ft landing craft, LCVP.

LOA	56 ft-2 in.
Beam	14 ft
Draft	3 ft-10 in.
Displacement	55 tons
Power	Two 225 hp diesels
Speed	9 knots
Armament	None

As was true of the LCM-6 class, the LCM-8 class is a direct and logical development of World War II landing craft. The first 146 of the 217 craft built between 1949 and 1967 are of welded steel construction. The rest are of welded aluminum. Able to carry the M-48 and M-60 tanks, they can support a greater variety of operations than the LCM-6; but because of their greater size and weight they are limited in the number and types of ships that can support them.

LOA	73 ft-8 in.
Beam	21 ft
Draft	5 ft-2 in.
Displacement	113 tons
Power	Two 325 hp diesels
Speed	9 knots
Armament	None

Utility Landing Craft. The LCU-1466 class of landing craft is slightly longer and wider than its World War II ancestor. Fifty-four of the class were built in the 1950s. Two of the LCU-1625 class were built in the mid 1960s. These were essentially LCU-1466s fitted with vertical-axis propellers. Their major characteristics are shown below.

LOA	118 ft
Beam	35 ft
Draft	6 ft
Displacement	360 tons, full load
Complement	14

Fig. 11 Landing craft, mechanized, LCM-6 class

Fig. 12 Utility landing craft, LCU-1466 class.

Power	Three 225 hp diesels
Speed	8 knots
Armament	Two 20 mm guns

The LCU-1610 class craft were a larger version of the LCU-1466s with drive-through arrangements so that several could link up to form a causeway. With the exception of one aluminum-hulled boat in the last group to be built, all the LCU-1610 class boats were of welded-steel construction. As an experiment, several of the boats of the class were fitted with vertical axis propellers, several with Kort nozzles, and several with right angle drive units that could rotate through 360 degrees to give thrust in any direction. Thirty-four boats were delivered in the 1960s.

LOA	135 ft
Beam	29 ft
Draft	6 ft
Displacement	375 tons, full load
Complement	12
Power	Two 1000 hp diesels
Speed	11 knots
Armament	Two 20 mm guns

Landing Craft (Air Cushion). The LCAC-1 class air cushion landing craft is the first operational craft to come out of the Navy's air cushion vehicle research

Fig. 13 Utility landing craft, LCU-1610 class.

Fig. 14 Landing craft, air-cushion, LCAC-1 class.

program. The prototype craft was delivered in 1984, and the first operational unit was activated in 1987. As of this date the Navy anticipates procurement of approximately 100 LCACs, of which some 20 have been completed as of 1990 by Textron Marine Systems and Avondale Gulfport Marine.

LOA	88 ft
Beam	47 ft
Payload	60 tons
Total weight	200 tons
Complement	5
Power plant	Four 3070 hp AVCO-Lycoming gt driving two shrouded reversible-pitch propellers; Four double-entry lift fans
Speed	40 knots
Armament	None

Amphibious Support Craft

Landing Craft Personnel, Large. Because they have been fitted out as small gunboats, the large landing craft for personnel (LCPL MK4) are included in this section. Some eight of the boats were fitted with two 0.50-caliber and two 0.30-caliber machine guns, and a 60 mm mortar for use in river combat. Sixteen LCPL MK11 hulls, slightly larger, were fitted out with only two 0.50-caliber machine guns.

	MK4	MK11
LOA	35 ft-10 in.	36 ft
Beam	11 ft-3 in.	13 ft-6 in.
Draft	3 ft-8 in.	3 ft-11 in.
Displacement	22,500 lb	19,320 lb
Construction	Steel	FRP
Power	300 hp	300 hp
Speed	About 19 knots	About 19 knots

Landing Craft, Swimmer Reconnaissance. In 1964 thirty special landing craft, the LCSR MK5, were built to deploy and recover swimmer reconnaissance teams. Because the MK5 boats were too long to be carried in standard Welin-type davits a new class of smaller boats, the *Seafox*, was designed and built in the late 1970s and early 1980s. About 45 of the latter units were procured. All of the boats were of fiberglass-reinforced plastic construction.

	MK5	*Seafox* Class
LOA	52 ft-4 in.	35 ft-6 in.
Beam	14 ft-9 in.	9 ft-9 in.
Draft	5 ft-6 in.	2 ft-6 in.
Displacement	50,000 lb	25,000 lb
Capacity	22 (incl. crew)	22 (incl. crew)
Power	Two 500 hp gt	Two 450 hp diesels

Mine Countermeasures Craft

The mine forces of the U.S. Navy have a motto, "Where the fleet goes, we have been." Consequently their ships and craft are truly combatants.

Minesweeping Launch. The mine sweeping launches (MSLs) were designed to be carried in Welin-type davits. The first hulls were of wood, later hulls were of fiberglass reinforced plastic construction. Diesel engines replaced gas turbines for propulsion power in later boats. All were fitted with gas turbine-powered generators required for magnetic minesweeping. Some 56 boats were built between 1948 and 1966.

LOA	36 ft
Beam	11 ft-7 in.
Draft	3 ft-8 in.
Displacement	23,100 lb, full load
Power	200 hp gt or 160 hp diesel
Speed	12 knots with gt 10 knots with diesel
Armament	None

Minesweeping Boats. The minesweeping boats of the MSB-5 class were built of wood (aside from a single experimental unit built of fiberglass reinforced plastic). Forty-nine boats were built between 1952 and 1956. Because the class's final weight exceeded the capacity of generally available cargo booms in any case, it was decided to build one even larger boat to try for improved seakeeping ability. John Trumpy and Sons of Annapolis designed and built this special boat, the MSB-29. It carried the same minesweeping gear and gas turbine-powered minesweeping generators as the MSB-5 class.

Fig. 15 36 ft minesweeping launch, MSL.

Fig. 16 Minesweeping boat, MSB-5 class.

	MSB-5	MSB-29
LOA	57 ft-3 in.	82 ft
Beam	15 ft-6 in.	19 ft
Draft	4 ft	5 ft-6 in.
Displacement	42 tons	81 tons
Power	600 hp diesel	600 hp diesel
Speed	14 knots	12 knots
Armament	Machine guns	None

Inshore Minesweeper. The Inshore Minesweeping class (MSI-1) craft were designed and built in the late 1950s as possible replacements for the coastal minesweeper class (MSC). They carried MSC-sized moored and acoustic sweep gear, and MSB-sized magnetic sweep gear. A later MSI-15 class was proposed and developed by Peterson Builders for the Military Assistance Program. These were fitted with twin screws, 500 hp each, and a gas turbine-powered (rather than diesel-powered) minesweeping generator. Those changes allowed the boat to carry MSC-sized magnetic sweep gear with no substantial change in other characteristics.

LOA	111 ft-9 in.
Beam	23 ft-5 in.
Draft	19 ft
Displacement	249 tons, full load
Power	One 650 hp quad diesel
Speed	12 knots
Armament	None

Fig. 17 Inshore minesweeper, MSI-1 class.

Coastal Minesweeper. The Coastal Minesweeper (MSC-121) class was built as a result of the renaissance of concern for mine countermeasures generated by our experience in the Korean War. Twenty-two of the boats were built between 1953 and 1956. They were of wooden construction and fitted with Packard diesels for propulsion power and minesweeping generators.

LOA	144 ft
Beam	28 ft
Draft	8 ft-6 in.
Displacement	370 tons, full load
Complement	39
Power	Two 600 hp diesels
Speed	14 knots
Armament	One 20 mm gun

Coastal Minehunter. One prototype coastal minehunter, MHC-43, was built in 1957. With the basic characteristics of the MSC-121 class, it was fitted with special underwater ordnance locators for hunting mines, and an "active" rudder for improved maneuverability at low ship speeds.

Coastal Minesweeper. The Coastal Minesweeper (MSC-289) class was designed and built in the early 1960s as an improvement over the MSC-121. Two boats went to the U.S. Navy; the rest were built for the Military Assistance Program. They had improved arrangements for handling the minesweeping gear, higher-powered, medium-speed, commercial-type propulsion diesels, and gas turbine powered minesweeping generators.

LOA	145 ft-5 in.
Beam	26 ft-6 in.
Draft	7 ft
Displacement	400 tons, full load
Complement	39
Construction	Wood
Power	Two 500 hp diesels
Armament	One 20 mm gun

Ocean Minesweeper. Design and construction of the Ocean Minesweeper (MSO) classes was also precipitated by the Korean War experience. All three classes, MSO-421, MSO-508 and MSO-519 were designed for heavy towing and long passages. Their wooden hulls were constructed with bronze and stainless steel fittings for minimum possible magnetic signatures. The 519-class boats were designed to be mine division commanders' flagships. Fifty-eight MSO-421 class, four MSO-508 class, and three MSO-519 class boats were built in the years 1953 to 1957.

	MSO-421	MSO-508	MSO-519
LOA	171 ft-6 in.	173 ft	189 ft-9 in.
Beam	34 ft	35 ft	35 ft
Draft	9 ft-6 in.	10 ft	10 ft
Displ., full load	735 tons	780 tons	968 tons
Complement	74	74	83
Power	1520 hp GM or 2800 hp Packard	2800 hp Packard	2800 hp Packard
Speed, knots	15 or 15.5	14	15
Armament	One 40 mm gun (all classes)		

Mine Countermeasures Ship. After years of neglecting the mine countermeasures problem, in 1982

Fig. 18 Coastal minesweeper, MSC-289 class.

Fig. 19 Ocean minesweeper, MSO-519 class.

Fig. 20 Mine countermeasures ship *Avenger*, MCM-1.

the Navy finally started a new class of oceangoing mine countermeasures ships with the laying of the keel of the *Avenger* (MCM-1) at Peterson Builders. Funding for the last three of a 14-ship class is contained in the FY 1990 shipbuilding budget. They have wooden hulls of a sophisticated structural design, and are capable of all types of mine countermeasures.

LOA	224 ft
Beam	39 ft
Displacement	1040 tons, full load
Complement	73
Power	Four 600 hp diesels on 2 shafts
Speed	14 knots
Armament	Two 0.50-caliber machine guns

Mine Hunter. In 1986 the Navy embarked on a program to build 17 fiberglass reinforced plastic hulled minehunters (MHC) based on the Italian Navy's *Lerici*-class. The first vessel, *Osprey* (MHC-51) is scheduled to be delivered by Intermarine USA of Savannah GA in 1991.

LOA	188 ft
Beam	36 ft
Displacement	840 tons, full load
Complement	45
Power	Two 800 hp aluminum block diesels
Propellers	Two Voith-Schneider
Speed	15 knots
Armament	Two 0.50-caliber machine guns

Hydrofoil Gunboats

Patrol Hydrofoil Missile Ships. As the culmination of some thirty odd years of experimenting with hydrofoils, the Navy now has a six-ship squadron of operational Patrol Hydrofoil Missile Ships (PHMs). Two *Pegasus*-class PHMs were included in the FY73 research and development budget, and four more in the FY75 shipbuilding budget, although final approval for their construction (by Boeing Marine Systems) was not granted until 1977. They are the Navy's fastest ships, with good hull borne range on their diesels, excellent seakeeping qualities, fast response to requirements for speed, and a potent punch. In the more than four years they have operated as a unit (1990), they have encountered few problems, and have established a high availability rate while participating in a variety of missions.

Fig. 21 Patrol hydrofoil missile ship PHM-5, *Pegasus* class.

LOA	133 ft
Beam	28 ft
Draft	6 ft-2 in., foils retracted
	23 ft-2 in., hull borne, foils extended
	8 ft-8 in. (nominal), foil borne
Displacement	231 tons
Complement	21
Propulsion	Water jet
Power	Two 800 hp diesels for hull borne
	One 18,000 hp gt for foil borne operations
Speed	12 knots, hull borne
	48 knots, foil borne
Armament	Eight Harpoon missiles
	One 76 mm gun

Conclusion

From the foregoing discussion of the development of fighting boats we may conclude that our blue water navy has been well served by a highly competent cadre comprising designers and builders, and suppliers of engines and ordnance. These individuals have been employed within the Navy's own establishment as well as within private industry. When the occasion demands, they have been able rather quickly to produce the specialized boats required for "green" or "brown" water operations.

Bibliography

"Boats of the United States Navy," NAV SHIPS 250-452, Dept. of the Navy, Naval Ships Systems Command, 1967.

Friedman, N., *U.S. Small Combatants*, U.S. Naval Institute Press, 1987.

Miller, R. T., "Fighting Boats of the U.S.," *Naval Review*, U.S. Naval Institute, 1968.

Miller, R. T., "Minesweepers," *Naval Review*, U.S. Naval Institute, 1967.

Modern Ships and Craft, Naval Engineers Journal Special Edition, Feb. 1985.

Sea Power—Special Edition, Navy League of the United States, Jan. 1990.

Weyers Warships of the World 1988/89, G. Albrecht, Ed., The Nautical & Aviation Publishing Co. of America, Baltimore.

CAPT. RICHARDS T. MILLER, USN (ret.), is a Fellow and Honorary Vice President of SNAME. He graduated from Webb Institute hoping to become a yacht designer. Circumstances, however, led him into a career as a Naval Constructor, but with a continuing interest in small craft. During World War II he supervised the construction of a wide variety of small fighting boats. In the early 1960s he was head of the Preliminary Design Branch of the Bureau of Ships in which position he was involved in the design of the *Asheville* as well as of an air cushion vehicle (SKMR-1) and a hydrofoil craft (AGEH-1). He has taken a leading role in the design of many kinds of small commercial and recreational craft. He is the co-author of two papers on sailing yacht design, one now published as a book.

Part VIII

Guardians of the Coasts and Beyond

United States Coast Guard Cutter Design, 1941–1991

ROBERT ERWIN JOHNSON, with ROBERT E. WILLIAMS

IN 1940 THE PRINCIPAL DUTIES of the U.S. Coast Guard were listed as follows: providing safety and security at sea through enforcement of the nation's navigation laws, saving life and giving assistance to vessels in distress, maintaining aids to navigation, and icebreaking in domestic waters. The Coast Guard was also a military force, intended to serve under the Navy Department in time of national emergency. To perform these many duties, the service had a fleet of seagoing cutters, patrol boats, buoy tenders, lifeboats, and other small craft. The largest components of the fleet, all steam powered, were seven 20-knot 327-foot cutters built in 1935–1937, ten 16-knot 250-footers completed in 1928–1930, four 15-knot 240-footers commissioned in 1921 or 1922, and six 13.5-knot 165-footers launched in 1932 and 1934.

According to the chief constructor at the time, "... under a basic requirement for reliability the major technical characteristics of a Coast Guard cutter can well be given the following sequence of priority: seaworthiness, speed, cruising radius, deck equipment for assistance work, and accommodations." While we might question the second priority, considering the moderate speed of the last three classes, most of the cutters designed during the next generation would generally fit this pattern.

The ten 250-foot cutters were transferred to the Royal Navy in 1941, under the Lend-Lease program enacted to aid Britain when it was standing nearly alone against the Axis nations. To replace them and three older vessels, Congress authorized 13 new cutters "of the usual Coast Guard type... of heavy construction to permit of working in waters where ice may be encountered." These became the *Owasco*-class, strongly built 255-footers with pilot house controlled turbo-electric drive, the last steam-propelled cutters built for the Coast Guard. These ships were completed in 1945 and 1946, too late for service in World War II. Their most notable employment was on the ocean stations established after the war to provide meteorological information and other assistance to transoceanic air liners. All were retired after the ocean stations were discontinued in the mid-1970s.

The Lighthouse Service had been transferred to the Coast Guard in 1939. Most of its lighthouse and buoy tenders had been designed to work in specific areas, but the Coast Guard required cutters to be multi-mission vessels. Its first buoy tenders, the 39 Cactus-class ships built during World War II, had ice-breaking hulls—nine were strengthened for work in ice—and enhanced search and rescue capability. These vessels were so successful that a number of them

Fig. 1 The *Owasco*-class cutter *Mendota* as completed. Note heavy armament.

Fig. 2 Winter, North Atlantic—the *Mendota* on Ocean Station Delta in 1965. This was the final 255-footer configuration.

have been re-engined and otherwise renovated to extend their lives another ten to fifteen years. A larger, but otherwise similar ship, the *Storis*, was completed in 1942 for use as a supply vessel in Greenland waters; she proved successful in a variety of roles and will probably serve past her fiftieth birthday.

The Coast Guard's first ocean icebreakers, the Wind-class, were also built during the war for service off Greenland. Their stubby, rounded hulls were of robust construction, with closely spaced frames and high tensile steel shell plating with a maximum thickness of 1.6 inches. They had twin screws aft and a bow propeller. The latter proved to be ineffectual in breaking heavy ice and so vulnerable to damage that it was soon removed and the bow tube capped. Each propeller was driven by an electric motor delivering 5000 shp at maximum rating, with six 2000 hp diesel generator sets. Ballasting systems capable of inducing a roll of ten degrees in 90 seconds helped to free the ships when beset in heavy ice.

Seven of these ships were built, two for the Navy, although the Coast Guard had to send three to the Soviet Union under Lend-Lease. When these were

Fig. 3 The *Cactus*-class buoy tender *Citrus* in the 1970s.

returned in 1950 and 1951, the Navy retained two, allocating the *Westwind* to the Coast Guard.

After the war, most of the icebreakers' heavy armament was removed, and the seaplane they originally carried was replaced by two helicopters, much more useful for locating leads through the ice. The Wind-class ships were remarkably successful, supporting the country's extensive polar programs for thirty years and more. Their after screws, however, sustained frequent ice damage until replaced by newly developed steel alloy wheels during the 1950s. They suffered occasional hull damage as well. The *Westwind* and the *Northwind*, the last survivors of the class, were decommissioned in 1989.

A variant on the Wind-class, the *Mackinaw*, was built during the war to help extend the shipping season on the Great Lakes. Tailored to conditions on the Lakes, she is longer and wider but shallower than her seagoing near-sisters. Like those ships, she proved a valuable asset and is, indeed, still in service.

Following the war many of the older cutters were retired and replaced by surplus naval vessels, of which salvage ships and fleet tugs were perhaps the most useful; three units of each remain in service in 1990. Eighteen small seaplane tenders were transferred for service on the ocean stations.

The Korean war brought a need for additional patrol craft. This led, in 1953, to the production of the first of 36 "seagoing patrol cutters." Twenty-one of these 95-footers carried antisubmarine and surface weapons, which were replaced by search and rescue gear in later units. They were named for North American capes in 1964. Increasingly heavy maintenance costs led the Coast Guard to retire these ships, beginning in the 1980s; the last was decommissioned in 1990.

In 1959 construction of the 95-footers was stopped in favor of a class of 82-foot twin-screw patrol boats with steel hulls and aluminum superstructures. Their design emphasized seagoing qualities and economic crewing: eight in place of the fifteen required on the 95-footers. All were given geographical "Point" names in 1964. By the time the construction program ended in 1970, no fewer than 79 of them had been built. Twenty-six served in the "Market Time" naval interdiction campaign during the war in Vietnam; then all were turned over to South Vietnam in 1969–70. The remaining 53 patrol boats are still in Coast Guard service and have recently been re-engined. They have proven to have exceptional seakeeping ability; an early report concluded, "These boats are well built and could probably weather anything."

Fig. 4 The icebreaker *Eastwind* as completed. Note seaplane and heavy armament.

Fig. 5 The *Northwind* after modernization in the 1970s. Her helicopter hangar is fully extended.

Many of the Coast Guard's 36-foot motor lifeboats required replacement by 1960. A survey of veteran lifeboat station personnel indicated dissatisfaction with the 36-footer's low speed, lack of towing capability, and poor steering station visibility; so a new design was prepared. The prototype 44-foot motor lifeboat, the CG 44300. was completed in 1962. Self-bailing and self-righting, she had diesel engines driving twin screws and a top speed of 15 knots. She also had power-assisted steering, an elevated steering station amidships, and a better location for the towing bitts. Her hull was ice-strengthened, and her port main engine could drive a pump for firefighting and salvage operations.

The CG 44300 was subjected to rigorous operational testing on both Atlantic and Pacific coasts, and her performance was reported to be "outstanding in all types of seas ... [including] large extremely dangerous

Fig. 6 The Great Lakes icebreaker *Mackinaw*. Her appearance has changed little during her long career.

Fig. 7 The 95-foot "seagoing patrol cutter" *Cape Henlopen* in 1987.

Fig. 8 The 82-foot patrol boat *Point Turner* in 1987.

seas on the bar and reefs." The design was put into production, and by the time the program was completed in 1972, 106 of the boats had been built by the Coast Guard, four for Italy, and one each for Canada and the U.K., both of which acquired additional units later. In service these 44-footers proved beyond any doubt that they were among "the finest rescue craft of [their] type in the world."

In 1990, a faster, more capable prototype 47-foot moter lifeboat was delivered. By the end of 1993, the prototype and 5 preproduction boats will be in service.

Design of the first postwar cutters, the *Reliance*-class, proceeded simultaneously with that of the motor lifeboats. Intended to replace the 125-foot and 165-foot vessels that were nearing the end of their lives, the new ships were superior in most regards. They were designed to carry out search and rescue missions

Fig. 9 A 44-foot motor lifeboat surmounts a breaker.

Fig. 10 The 210-foot cutter *Reliance* as completed.

up to 500 miles offshore, with an endurance of 5000 miles at 15 knots. The inclusion of a helicopter deck required a waterline length of 200 feet, and resulted in an unusual design. A large, high superstructure incorporated the bridge, cabin, electronic spaces and a helicopter workshop aft. The engine exhaust ducts were led through the transom stern to avoid turbulent stack gas above the helicopter deck and to minimize corrosion of topside electronic gear.

The first 5 of these twin-screw cutters were equipped with combined diesel and gas turbine (CODAG) propulsion systems, the first American vessels to be so equipped. Each shaft was powered by a 1500 hp diesel and a 1000 hp gas turbine. Initial reports on the vessels were quite favorable. They reached their designed speed of 18 knots with ease; their habitability set a new standard for the Coast Guard and their twin rudders made them highly maneuverable. The CODAG system, however, experienced problems caused by the different speed/torque characteristics of the two kinds of machinery. Moreover, the engine spaces were overcrowded. Starting with the sixth cutter, the CODAG system was abandoned in favor of twin diesels with the same total power as before.

Sixteen *Reliance*-class cutters were built, the last in 1969. Beginning in 1984 they were extensively renovated. The original CODAG units were re-engined with 2500 hp diesels. (The turbines had been removed in the 1970s, so these ships had been operating at reduced power.) The exhaust ducts were removed in favor of vertical ducts leading into normal stacks positioned and configured with benefit of wind tunnel tests.

Construction of the first postwar coastal buoy tenders began in 1963. The five Red-class vessels were shallow-draft, low-freeboard ships with twin controllable-pitch propellers. They were strengthened for light icebreaking. The last was completed in 1971.

In 1966 the Coast Guard broke from the Navy's practice of classifying its cutters—a "W" for Coast

Fig. 11 The 157-foot coastal buoy tender *Red Wood* upon completion in 1964.

Guard followed by PG, PC, AGL, etc. A system was adopted to better describe the cutters' functions: WHEC, high endurance cutter; WMEC, medium endurance cutter; WLB, seagoing buoy tender, etc. Early vessels of a new fleet of high endurance cutters were under construction at the time, flush decked 378-footers known as the *Hamilton*-class. The design featured a long aluminum superstructure, the after 80 feet of which formed the helicopter deck. These ships were rather heavily armed and given good habitability features in keeping with their high endurance. Their 20-knot cruising speed was provided by two 3500 shp diesels, each driving a controllable-pitch propeller, and they could attain a "burst" speed of 29 knots by clutching out the diesels and clutching in a pair of gas turbines.

The 378-foot WHEC-class consisted of the *Hamilton*, eight additional "Secretaries," and three "Heroes." There were minor differences between the sub-classes. The last was completed in 1972. Since 1986 these cutters have been undergoing extensive modernization, including upgrade of armament and combat systems, relocation of CIC from superstructure to second deck, and helicopter hangar and Navy LAMPS helicopter qualification. These major changes provided an opportunity to arrange better accommodations for both sexes.

The Coast Guard was transferred from the Treasury Department to the new Department of Transportation in 1967. Soon afterward it adopted distinctive markings for its vessels and aircraft: diagonal orange and blue hull stripes, with the Coast Guard emblem superimposed on the former, followed by COAST GUARD in block letters. Most foreign coast guards soon painted somewhat similar "racing stripes" on the bows of their ships.

The *Polar Star*, first of a new class of two icebreakers, was begun in 1972. With a length of 399 feet, the new ship was a good deal larger than her predecessors and accommodated two helicopters as well as laboratories for a variety of scientific missions. Her bow was of the improved "White" design, which allowed more of the ship's weight to ride up on the ice. She had triple controllable-pitch propellers aft, each driven by a 5000 shp motor energized by two 3000 bhp diesels. When breaking heavy ice, surge power could be provided to each shaft by 20,000 hp gas turbines. The controllable-pitch screws, a feature then unique in icebreakers, proved troublesome and the problem was solved only by regularly replacing them with rebuilt units. Nonetheless, the superiority of the *Polar Star* and her sister, the *Polar Sea*, is evident. They have been the nation's only polar icebreakers since 1989.

The 110-foot tugs of prewar design, that had proven so useful for breaking ice in domestic waters, also required replacement. They, too, were succeeded by larger, more capable units, the 140-foot Bay-class icebreaking tugs. Nine of these highly successful vessels were completed between 1979 and 1988. Their icebreaking capability is enhanced by a hull air lubrication (bubbler) system housed in a portable van on the fantail. This seems to have been the system's first U.S. application.

Adoption of the 200-mile economic resources zone in 1977 made the construction of additional

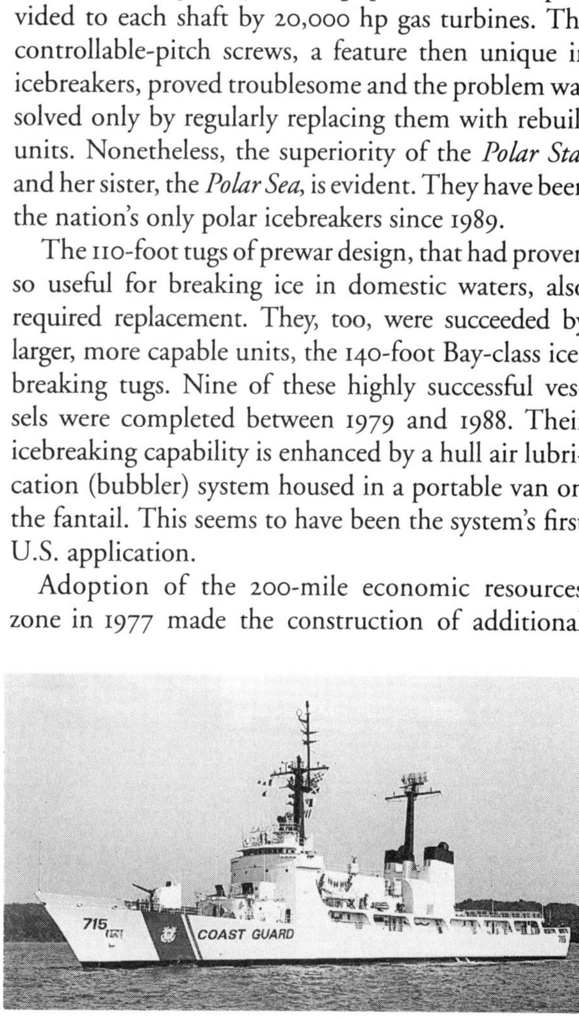

Fig. 12 The 378-foot high endurance cutter *Hamilton* as completed.

Fig. 13 The *Polar*-class icebreaker *Polar Sea.*

medium-endurance cutters necessary, especially since the venerable 327-footers were finally approaching the end of their service. A contract for four Famous-class 270-footers was let in 1977. The ability to operate a helicopter during a two-week patrol was a principal de- sign requirement; the ships have telescoping hangars and active fin stabilization to facilitate this.

When the first of these cutters, the *Bear,* was on her delivery voyage from Tacoma, she encountered a rogue wave that damaged the forward gun mount irreparably and the superstructure less seriously. Later units were delayed by lawsuits over construction contracts and modifications to the design, during which weight margins were consumed. These cutters were initially reputed to be poor sea boats. With time in service, however, this complaint has largely been overcome. Their sophisticated electronic command and control systems have made them useful in a number of roles. The thirteenth, and last, ship of the class, the *Mohawk,* was commissioned in 1991.

As drug interdiction came to be accorded greater importance among the service's responsibilities, the lack of adequate patrol craft was noted. To fill this need quickly, the Coast Guard chose the well-developed

Fig. 14 The 140-foot icebreaking tug *Bristol Bay.* Her "bubbler" system van is not embarked.

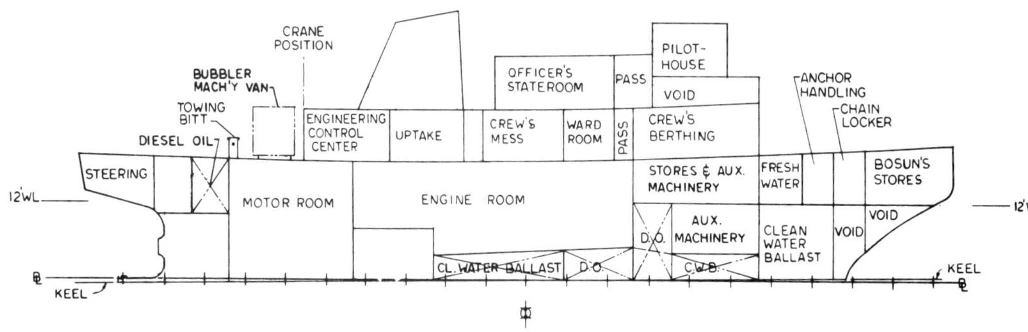

Fig. 15 Inboard profile of 140-foot *Bay*-class tugs.

Fig. 16 The *Famous*-class cutter *Northland*.

British Vosper Thornycroft 110-foot patrol boat. Forty-nine of these Island-class craft with planing hulls were built. The first 37 were powered with diesel engines that were ill-suited for search and rescue because they would stall well above safe towing speed. The last twelve units have trolling clutches, which permit a lower minimum speed. Meanwhile, a coastal patrol craft concept is planned to replace the 82-ft Point-class patrol boats.

The planned replacement of the buoy tender fleet began in January 1993, when a contract was awarded for the first of the 225-foot *Juniper*-class seagoing buoy tenders to replace the seemingly ageless 180-footers. The new buoy tenders will have a 15-knot speed, 6000-mile endurance, a dynamic positioning system, light icebreaking capability, a 20-ton buoy crane, and a spilled oil recovery system. In June, a contract was awarded for the 175-foot *Ida Lewis*, the first of the Keeper-class coastal buoy tender replacements. These tenders will have a 12-knot speed, 2000-mile endurance, 10-ton crane, dynamic positioning system, and ice strengthening.

In July 1993 a contract was awarded for the *Healy*, a 420-foot, 15,300-ton polar icebreaker with a diesel electric ac/ac cycloconverter propulsion system capable of breaking 4.5 feet of ice continuously at 3 knots and 8 feet by ramming. *Healy*'s principal mission is to conduct scientific research in the high latitudes.

Thus, the United States Coast Guard should enter the 21st Century with a fleet of modern and highly capable vessels. Can it be that the days of 50-year old museum candidates and Navy cast-offs are really in the past?

Additional Reading

Chiu, R., Haciski, E., and Hirsimaki, P., "Application of Plastic Analysis to U.S. Coast Guard Icebreaker Shell Plating," *Trans.* SNAME, Vol. 89, 1981.

Cowart, K. K., "Development of Vessel Servicing Aids to Navigation for the U.S. Coast Guard," *Trans.* SNAME, Vol. 66, 1958.

Daidola, J. C. and Sheinberg, R., "A Procedure for the Structural Design of Icebreakers and Other Ships Navigating in Ice," *Trans.* SNAME, Vol. 96, 1988.

Edwards Jr., R. Y., Lewis, J. W., Wheaton, J. W., and Coburn Jr., J., "Full Scale and Model Tests of a Great Lakes Icebreaker," *Trans.* SNAME, Vol. 80, 1972.

Ferris, L. W., "The Proportions and Form of Ice breakers," *Trans.* SNAME, Vol. 67, 1959.

Fischer, R. W., "Noise Reduction Program for the USCG 95-Ft Patrol Boats: A Case History," *Trans.* SNAME, Vol. 92, 1984.

Johnson, H. F., "Development of Ice-Breaking Vessels for the U.S. Coast Guard," *Trans.* SNAME, Vol. 54, 1946.

Lewis, J. W. and Edwards Jr., R. Y., "Methods for Predicting Icebreaking and Ice Resistance Characteristics of Icebreakers," *Trans.* SNAME, Vol. 78, 1970.

Purcell, E. S., Allen, S. J., and Walker, R. T., "Structural Analysis of the U.S. Coast Guard Island Class Patrol Boat," *Trans.* SNAME, Vol. 96, 1988.

Zahn, P. B., Humphreys, D., and Phillips, L., "Full-Scale Towed Resistance Trials of USCGC *Mobile Bay* in Uniform Ice, *Trans.* SNAME, Vol. 95, 1987.

Appendix—Ship Class Characteristics

Class	Ships in Class	Displacement (full load), tons	Length, ft-in.	Beam, ft-in.	Maximum Draft, ft-in.	Propulsion	Screws	SHP	Speed, knots	Endurance
Owasco	13	1978	254	43	17-3	Turbo-electric	1	4000	19	10,379 n.mi @ 10 knots
Cactus	39	1025	180	37	13	Diesel-electric	1	1000 1200	12.8 13.5	13,000 n.mi @ 10.5 knots
Storis	1	1715	230	43	15	Diesel-electric	1	1800	14	15,500 n.mi @ 10 knots
Wind	7	6515	269	63-6	29	Diesel-electric	3*	10,000	16	38,000 n.mi @ 11.5 knots
Mackinaw	1	5253	290	74	19	Diesel-electric	3	10,000	18.7	41,000 n.mi @ 9 knots
Cape	36	105	95	20	6-5	Diesel	2	2324	20	1700 n.mi @ 12 knots

Class	Ships in Class	Displacement (full load), tons	Length, ft-in.	Beam, ft-in.	Maximum Draft, ft-in.	Propulsion	Screws	SHP	Speed, knots	Endurance
Point	79	69	82-10	17-1	5-8	Diesel	2	1200 1600	16.8 22.9	1500 n.mi @ 9.4 knots
Motor Lifeboat	106	15.8	44-10	12-8	3	Diesel	2	400	15	200 n.mi @ 10 knots
Reliance (A)	5	930	210-6	34	10-6	CODAG	2 CP	5000	18	6100 n.mi @ 14 knots
Reliance (B)	11	1007	210-6	34	10-6	Diesel	2 CP	5000	18	2700 n.mi @ 16 knots
Red	5	512	157	33	6	Diesel	2 CP	1800	13	3000 n.mi @ 11.6 knots
Hamilton-Hero	12	3050	378	42-8	20	CODAG	2 CP	7000 Diesel 36,000 Gas	20 29	14,000 n.mi @ 11 knots
Polar	2	13,190	399	86	31	CODAG	3 CP	18,000 Dies-elec 60,000 Gas	18	28,275 n.mi @ 13 knots
Bay	9	662	140	37-6	12-5	Diesel-electric	1	2500	14.7	4000 n.mi @ 12 knots
Famous	13	1780	270	38	13-6	Diesel	2 CP	7000	19.5	10,250 n.mi @ 12 knots
Island	49	161	110	21	6-10	Diesel	2	5800	30	1350 n.mi @ 15 knots
Juniper	16	2000	225	46	13	Diesel	1	6200	15	6000 n.mi @ 12 knots
Keeper	14	845	175	36	8	Diesel	2	1710	12	2000 n.mi @ 10 knots
Healy	1	15,300	420	82	28	Diesel-electric	2	30,000	18	33,000 n.mi @ 12 knots
Motor Lifeboat	105	18.6	47-11	14	4-4	Diesel	2	850	25	200 n.mi @ 25 knots

* Includes bow screw, later removed from all.

Prof. Robert Erwin Johnson has been a member of the History faculty at the University of Alabama since 1956. He served in the Coast Guard 1941–46, and earned a PhD in history at Claremont Graduate School in 1956. He is the author of several works in naval and maritime history, including *Guardians of the Sea: History of the U.S. Coast Guard, 1915 to the Present* (Naval Institute Press, Annapolis, MD, 1987).

CDR **Robert E. Williams** served in the Coast Guard for over 39 years in both military and civilian capacities. His sea duty included assignments in both engineering and operational billets aboard nine different cutters. His seagoing service culminated as commanding officer of the cutter *Mendota.* He retired from military service in 1965, but in 1973 returned to the Coast Guard as Chief of the Marine Safety Branch at the R&D Center, Groton, CT, and then Chief of the Technical Branch at Headquarters before retiring in October 1989.

The Canadian Coast Guard

ROB SNIDER, coordinator

With contributions by individuals in the Canadian Coast Guard

THE CANADIAN COAST GUARD (CCG) as we know it today is a relatively new organization. But, in historical terms, its role has been evolving since the earliest days of French settlement in Canada. In 1665, Jean Talon, Intendant of New France, committed public funds for marine purposes, including shipbuilding and the training of mariners in what is now Canada. In an early writing, Talon forecast things to come:

> The Canadian youth are improving their knowledge. They take to schools for sciences, arts, handicrafts and especially navigation and if this movement is sustained, there is every reason to hope that the country will produce, mariners, fishermen, seamen and skilled workmen.

Canadian Coast Guard—Early Beginnings

At the time of Confederation, in 1867, the Department of Marine and Fisheries became the Dominion body responsible for the affairs of shipping. When relations between Russia and Britain became strained in the late 1870s, the Canadian government began to think in terms of a great naval reserve to protect shipping both at sea and on the Great Lakes. The question Canada faced with regard to participation in the naval expansion was whether it should have its own navy, have no navy at all, or simply contribute to the Royal Navy. Sir Wilfred Laurier, Prime Minister of Canada from 1896 to 1911, decided Canada was to establish its own naval force and on May 4, 1910, brought in a bill called the Naval Service Act.

When the Department of Transport Act was passed on November 2, 1936, all agencies of transportation were placed under one authority. The Marine Branch came under the jurisdiction of the Department of Transport, Marine Services. Twenty-six years later, in 1962, the official title of Canadian Coast Guard was introduced.

Today the Canadian Coast Guard, a civilian organization, is responsible for icebreaking, northern resupply, marine regulations, search and rescue, environmental and emergency response, public harbours and ports, and marine navigational aids. It is also responsible for a network of marine telecommunications stations and facilities on the east and west coasts, in the Arctic, and on the Great Lakes. Its headquarters group and Northern Region office are located in Ottawa. Five regional offices operate out of St John's, Newfoundland; Dartmouth, Nova Scotia; Quebec City, Toronto, and Vancouver.

The Coast Guard's legal powers and responsibilities are provided mainly through the Canada Shipping Act, supplemented by the Arctic Waters Pollution Prevention Act, the Navigable Waters Protection Act, and regulations made under them.

The present fleet includes 66 major ships, 18 search and rescue lifeboats, 27 inshore rescue boats, four air cushion vehicles, 35 helicopters, one fixed-wing aircraft, and some 308 small craft.

Canadian Coast Guard Fleet

The development, operations and exploits of the Canadian Coast Guard fleet tell a fascinating story.

Icebreakers. The Coast Guard's powerful fleet of icebreakers assists with navigation along the East Coast, across the North and through the Great Lakes–St. Lawrence system. As well, the fleet contributes to flood control and protection against shoreline erosion.

Among the more notable ships of the Coast Guard's early years was the powerful icebreaker *N. B. McLean*, named after the leader of an expedition to Hudson Strait in 1927. The *McLean* was built at Halifax Shipyard early in 1930. This remarkably tough and reliable ship made 40 Arctic voyages escorting commercial ships and taking part in the yearly Arctic sealift from 1930 until 1970.

The only other icebreaker added to the fleet during the years of the Depression and the Second World War was the *Ernest Lapointe*, which came into service in 1941. Built by Davie Shipbuilding & Repair Company, Lauzon, Quebec, she was a twin screw vessel of modest size and power.

The hull shape of icebreaking ships is driven by one primary consideration: the lowest power required to make progress in ice. While traditionally shaped icebreakers are still being built, a number have been constructed with non-traditional forms. The selection of bow shape is greatly influenced by the mission profile, and this is reflected in the shapes adopted over the years.

Owing to increasing demands on Arctic sealift services in the summer months, and St. Lawrence operations in winter, it became necessary in the early 1950s to build a full-sized icebreaker. As a result, the *d'Iberville*, the largest icebreaker in the world at the time in terms of power and tonnage, joined the fleet in June 1953. Built by Davie Shipbuilding the previous year, the ship's hull form, size and propulsion power were a major departure from the pre-war design of icebreakers. After a long and distinguished career in the Arctic, this ship was taken out of service at the end of 1981.

A major innovation in icebreaker technology came in 1953 with the design and construction of the Canadian Navy Arctic patrol vessel *Labrador*. Built by Marine Industries Ltd., and commissioned in July 1954, she was the first Canadian-built diesel-electric propelled, bridge-controlled icebreaker. She joined the Coast Guard fleet in 1958.

In 1959, six medium-sized icebreaking tenders were added to the fleet. They were needed to meet increasing demands of northern weather stations, the Distant Early Warning Line, and increased commercial traffic in the North and through the Gulf of St. Lawrence. These vessels were:

- the *Sir Humphrey Gilbert*, built by Davie Shipbuilding, Lauzon, Que., employed in Hudson Strait and eastern Arctic;
- the *Sir William Alexander*, a diesel-electric twin-screw icebreaker, built at the Halifax Shipyard, employed in Hudson Strait and eastern Arctic;
- the *Wolfe*, built at Canadian Vickers, Montreal, and employed in eastern and northern waters;
- the *Camsell*, built at Burrard Drydock, Vancouver, which circumnavigated King William Island in a historic 1962 voyage, and employed in the western Arctic and on the West Coast;
- the *Alexander Henry*, completed at Port Arthur, Ontario, employed in the Great Lakes; and
- the *Tupper*, built by Marine Industries Ltd. at Sorel, Quebec, employed on the East Coast.

In the 1960s, with continued and rigorous demand for icebreakers in the north, three more were added to the fleet. The largest of them was the *John A. Macdonald*, built in 1960 by Davie Shipbuilding. She was a state-of-the-art vessel with a triple-screw, diesel-electric propulsion system. This was Canada's most powerful icebreaker at the time. She helped rescue the U.S. Coast Guard icebreaker *Northwind* from deep in the Polar pack in 1967, and escorted the tanker *Manhattan* through the Northwest Passage in 1969.

The other two ships were the *Simon Fraser*, built in 1960 by Burrard Drydock, and the *Thomas Carleton*, built in 1960 by Saint John Dry Dock, both light icebreakers with twin-screws and diesel-electric propulsion.

The *John Cabot* is a heavy icebreaker with capacity for cable-laying and ship repair. She was built in 1965 by Canadian Vickers, Montreal. She is equipped with twin-screw diesel electric drive. Three thrusters enable her to move sideways. She supports cable communications and is one of the few ships in the world capable of burying cable in the ocean floor. Equipped with state-of-the-art technology, including the capability to operate the Submersible Cable Assisted Repair And Burial units (generally known as Scarab robot

Fig. 1 *N. B. McLean*—heavy icebreaker (October 1963). (Courtesy Coast Guard News.)

Fig. 2 *D'Iberville* joined the Coast Guard fleet in 1953. (Courtesy Coast Guard News.)

submarines), she has been at the centre of international drama on more than one occasion. On June 27, 1985, for example, four days after the crash of an Air India aircraft off the coast of Ireland, the ship recovered the flight data and voice recorders, as well as other parts of the airliner, from a record depth of 2010 metres.

The *Louis S. St. Laurent*, the largest and most powerful Canadian icebreaker to date, was delivered by Canadian Vickers, Montreal, in 1969. She is currently being modernized at Halifax Dartmouth Industries, which has scheduled her for completion late in 1992.

The ship's hull form was developed on the basis of operational requirements and past experience. The ice belt was two inches thick and the frames, spaced at 16 inches, were combined with transverse webs and horizontal stringers to withstand the stresses inherent in icebreaking. Other features included a bow thruster, trimming and heeling systems, a flume stabilizing system and a helicopter hangar.

The modernization of the ship will involve a new power plant having five main generators, which will enable the ship to develop 20,985 kW of power. Icebreaking efficiency will be improved with the fitting of a new, modified R-class icebreaking bow. The bow thruster will be replaced by an air bubbler/hull lubrication system that reduces hull resistance while doubling as a bow thruster. The original helicopter lift system in the hangar will be replaced by a fixed helicopter hangar. This arrangement will also include a helicopter refueling station.

The *Pierre Radisson*, built in 1978 by Burrard Dry Dock, joined the fleet that same year and was the first of three R-class heavy icebreakers. The other two were the *Sir John Franklin*, built in 1979 by Burrard Dry Dock, and the *Des Groseilliers*, built by Port Weller Dry Dock. They were commissioned in 1979 and 1982, respectively.

In the mid-1970s, with more than three-quarters of the fleet past mid-life, the Coast Guard developed its Fleet Capital Investment Plan (FCIP) to systematically review and modernize its fleet. FCIP established a 30-year life-cycle replacement program for all Coast Guard ships except small search and rescue craft, which are to be replaced every 20 years.

Formally approved by the Cabinet in November 1979, FCIP was the first such structured capital investment plan undertaken by the federal government. During the initial phase, 15 new ships were built and three modernized.

As part of FCIP, six Type-1100 icebreaking buoy tenders were commissioned in 1986 and 1987. They are the *Martha L. Black* and the *George R. Pearkes*, both built in 1986 at Versatile Pacific Shipyards Inc. in Vancouver and Victoria; the *Sir Wilfred Laurier* (1986), built by Collingwood Shipyard Ltd., at Collingwood, Ont; the *Edward Cornwallis* (1986) and the *Sir William Alexander* (1987), both built by Marine Industries Ltd., at Sorel, Quebec and the *Ann Harvey* (1987), built by Halifax Industries Ltd.

The propulsion plant and machinery concept of these vessels is a technical breakthrough and a first for the North American marine industry. Although

Fig. 3 Giant sheaves on the *John Cabot* over which submarine cable is laid (April 1964). (Courtesy Transport Canada.)

Fig. 4 *C. D. Howe*—first supply vessel specifically designed for arctic waters. (Courtesy Coast Guard News.)

In 1990, however, in a period of continuing budgetary restraint and program cutbacks, the *Polar 8* was cancelled. A wealth of important and useful data and insight from the work already done is now available should a similar project ever be undertaken.

Buoy Service and Supply Vessels. In 1949, the *Edward Cornwallis* joined the fleet as a lighthouse supply and buoy vessel. Built at Canadian Vickers in Montreal, she was a twin screw vessel with a length of 78 metres, a displacement of 1965 tonnes, and Skinner Uniflow steam engines providing 2800 hp.

diesel-electric, the output is ac/ac rather than ac/dc, employing cyclo-converter technology to control shaft requirements for operating in ice. A bow thruster aids in precise ship-handling.

The *Henry Larsen* is a Type-1200 twin-screw heavy icebreaker, built in 1987 by Versatile Pacific Shipyards.

Self-propelled air cushion vehicles represent the latest in icebreaking technology. They have been particularly useful in breaking ice in shallow waters where conventional icebreakers are unable to operate.

The Coast Guard was also involved in testing an air cushion icebreaking bow (ACIB). When attached to the bow of a ship and pushed across the ice, the platform generates an air pocket below the level of ice that, as it expands, causes ice to break under its own weight. In early tests using a light icebreaker, a small ACIB substantially improved the ship's icebreaking capability.

A notable event occurred in 1985 when the Right Honourable Joe Clark announced plans in the House of Commons to build the *Polar 8*, projected to be the world's largest icebreaker. With its great power plant, it was to be capable of maintaining a steady speed of three knots through 2.5 metres of ice.

Polar 8 was designed as an escort icebreaker with considerable importance being placed on manoeuvrability and astern performance. One of the important features was a double hull throughout, with a capacity for 13,000 tonnes of fuel, 400 tonnes of lubricating oil, 400 tonnes of fresh water and, 875 tonnes of helicopter fuel.

The *C.D. Howe*, commissioned in 1950, was the first supply ship specifically designed for Arctic waters. Although not an icebreaker, she was strengthened for ice. She served faithfully for 19 years in the Eastern Arctic Patrol and later provided health services to the native population and isolated outposts.

In 1957, the *Montcalm* was built by Davie Shipbuilding. She was the second ship to bear that name, and was added to the fleet to increase lighthouse supply and buoy tending capacity, as well as icebreaking capability.

After the Second World War, the Department of Transport, with capital funds in short supply, began looking at the possibility of converting naval vessels for use in various Arctic operations and other marine and meteorological duties entrusted to the Department. During the 1950s, the Coast Guard acquired seven converted landing craft known as "bird-class" supply vessels. These had been built between 1944 and 1946 at various shipyards in the U.K. as landing craft barges known as LCT-8s. The conversion of landing

Fig. 5 The *Gannet* (July 1966) was one of the converted landing craft acquired by the Coast Guard after World War II. (Courtesy Transport Canada.)

Fig. 6 The weather ships *Vancouver* (foreground) and *Quadra*—built in 1966. (Courtesy Coast Guard News.)

vice. The *Spray, Spindrift*, and *Spume* have operated out of various ports on Lakes Ontario, Erie, and Huron throughout their long careers in search and rescue operations. They remain unique in the Coast Guard as the only wooden vessels of significant size in the fleet.

The *Alert*, built in 1969 by Davie Shipbuilding, was the first special purpose, offshore search and rescue ship to join the Coast Guard.

The most recent additions to the fleet are two Type-500 search and rescue ships, the *Gordon Reid* and the *John Jacobson*, built in 1990 by Versatile Pacific Shipyards, Esquimalt, B.C. The Type-500s were specifically designed for their task, having exceptional sea worthiness and manoeuvrability. A transom stern ramp enables them to launch and recover rescue craft in severe weather.

craft, such as the *Eider, Raven, Auk, Puffin, Gannet, Skua*, and *Nanook* to Northern Arctic Supply ships represented one successful innovative approach.

Weather Ships. Other conversions became the first three Canadian Coast Guard weather ships, namely the *St Stephen*, the *St Catharines*, and the *Stonetown*. These highly-equipped ships were converted in 1950 from the last of the obsolete steam-propelled naval frigates built in the 1943-44 period as anti-submarine vessels.

The *St. Catharines* and *Stonetown* served for more than 16 years in the harsh environment of the North Pacific until they were replaced by two sophisticated, purpose-built weather ships, the *Vancouver* and the *Quadra* in 1966 and 1967, respectively. The *St Stephen* stayed in Victoria on standby, but was never called to active duty, a tribute to the quality and upkeep of the wartime frigates built at Esquimalt, near Victoria, B.C.

The *Quadra*, the second ship to bear the name, and her sister ship, the *Vancouver*, were both built in 1966 by Burrard Drydock in North Vancouver. In addition to weather ship duties, they also served as search and rescue ships, and provided radio beacon service to trans-Pacific shipping and aircraft. Both were eventually made redundant with the advent of satellite navigation and communication technology.

Search and Rescue Vessels. In the 1960s, the Coast Guard began to modernize its search and rescue fleet with the acquisition of 13-metre self-righting steel hulled lifeboats, 28.5-metre R-class cutters, and wooden hulled S-class cutters. While the R-class cutters have since been replaced, many of the original 13-metre lifeboats and the S-class cutters are still in ser-

Air Cushion Vehicles. Air cushion vehicles (ACV) became part of the Department of Transport in 1968, when the SRN-5, built by British Hovercraft Corp. in Cowes, England, entered Canadian service.

In 1972 the Transport Development Agency accepted delivery of *Voyageur*, a prototype ACV, for evaluation. *Voyageur* was developed by Bell Aerospace Canada in a program jointly funded by that company and the Department of Industry, Trade and Commerce. The *Voyageur* was a general purpose, amphibious ACV with a large work deck area for cargo. Powered by two 635 kW Pratt & Whitney turbo shaft engines, she was able to operate at speeds of up to 50 knots.

Fig. 7 The wooden hull of the search and rescue cutter *Spume* during construction (ca. 1963). (Courtesy Transport Canada.)

Fig. 8 The 95-foot wooden hulled search and rescue cutter *Spume* in service on the Great Lakes (January 1964). (Courtesy Coast Guard News.)

The Coast Guard evaluated the *Voyageur*, first in 1972 on the St Lawrence and the Great Lakes, then in 1973 on the Mackenzie River, N.W.T. During the evaluations, it successfully demonstrated its buoy tending capabilities. Modifications were made, and the *Voyageur* was accepted by the CCG Evaluation and Development unit in January 1974. She was first based at Parry Sound, Ont., where she was evaluated for icebreaking capabilities.

In December 1974 the *Voyageur* was moved to a new base in Montreal where she was used extensively for navigational aids work, construction support, search and rescue, and icebreaking and flood control on the St Lawrence River and its tributaries.

The first SRN-6, three metres longer than the SRN-5, was bought in 1977. Later, two scrap SRN-6 hovercraft were re-built by Coast Guard ACV mechanics. They used spare parts from the SRN-5, which had to be retired, as well as various spares from British Hovercraft Corporation. These entered service in 1982 and 1986, respectively.

The *Voyageur* was replaced in August 1987 by the *Waban-aki*, built by British Hovercraft Corp. The *Waban-aki* is powered by four lightweight high-speed diesel engines producing 1760 kW of power, allowing operating speeds of up to 50 knots.

At present, there are four ACVs in the Coast Guard fleet: The *Waban-aki* based in Montreal, one SRN-6 at Parksville on Vancouver Island, and two SRN-6s at Vancouver International Airport.

Special Purpose Vessels. The Canadian Coast Guard assumed responsibility for maintaining the Mackenzie–Athabaska waterway system in the Northwest Territories in 1955.

Special purpose ships were designed to operate in shallow waters with strong currents as is found in the MacKenzie River system. They also had to be capable of carrying out buoy tending, construction work, and resupply. The first such ships were the *Dumit* and the *Miskanaw*, built in 1958 by Allied Shipbuilders Ltd., North Vancouver.

The *Eckaloo* was built in 1961 by Allied Shipbuilders Ltd. A replacement ship built by Vancouver Shipyards and bearing the same name, was commissioned in 1988. The new *Eckaloo* has a flat bottom, propellers recessed flush with the hull to prevent damage when the ship touches bottom and, to increase manoeuvrability, four rudders.

The *Tembah* was built in 1963 by Allied Shipbuilders, while the original *Dumit* was replaced in 1979 by another ship from the same shipyard. In keeping with tradition, she was given the same name.

The Canadian Coast Guard in the North

Great Britain transferred her rights of Arctic sovereignty to Canada in 1880, a move that would have important ramifications for the Coast Guard. It was not until 1903, however, that Canada took steps to assert its sovereignty.

In 1903, the *Neptune* was the first ship sent to the North by the Department to establish permanent stations for collecting customs, administering justice, and enforcing law and order.

The following year, Capt. J.E. Bernier, took command of the *Arctic*, which had been bought from Germany and had been designed for Antarctic service. During an expedition in 1908–09, Bernier penetrated as far west as Cape Hay on Melville Island. It was here, on July 1, 1909, that he erected a tablet proclaiming the annexation by Canada of the entire Arctic archipelago. Bernier regularly continued his northern patrols in his beloved *Arctic* over the years 1922–1925. Those patrols formed the first series of what was later to become known as the Eastern Arctic Patrol.

Fig. 9 The air-cushion vehicle (ACV) *Waban-Aki*. (Courtesy Transport Canada.)

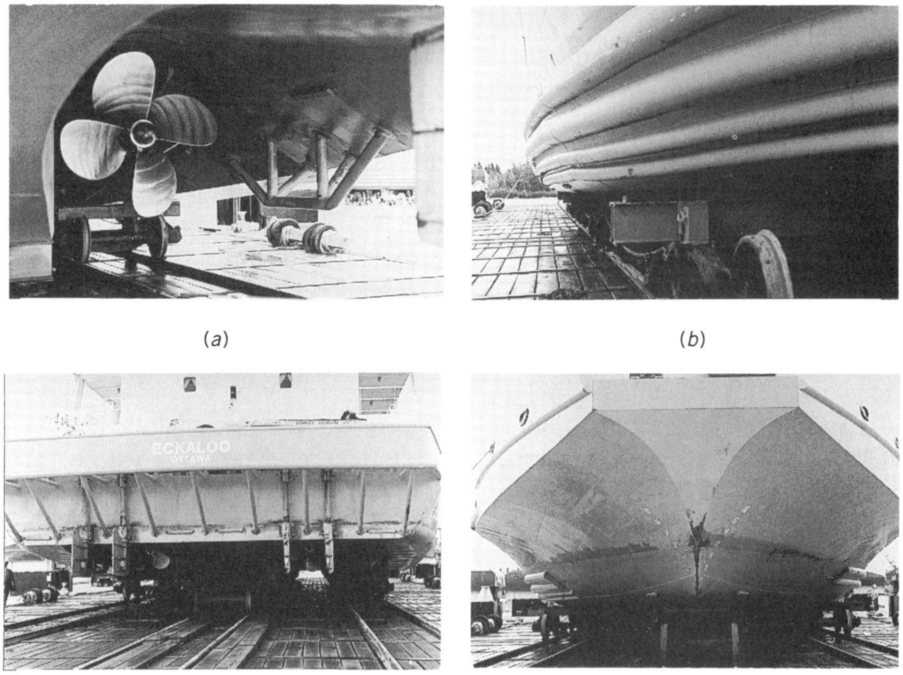

Fig. 10 *Eckaloo*: (*a*) propellers are recessed flush with the hull, tucked up in tunnels, so they will not be damaged when the ship touches bottom; (*b*) skin cooler pipes are welded along the sides, Instead of a salt-water heat exchanger, which would suck up mud when the ship run up on the beach; (*c*) four rudders are to increase manoeuvrability in shallow draft conditions; and (*d*) the bow. (Courtesy Coast Guard News.)

Coast Guard icebreakers sail to the Arctic each summer to help the government's annual Eastern Arctic resupply. The icebreakers escort a government-chartered fleet which delivers goods needed by residents of remote settlements and military installations until the next sealift—a year later.

In addition to its sealift activities the Coast Guard has spent approximately $20 million on valuable research into ship–ice interaction and Arctic ice conditions.

As experience has grown and research progressed, shortcomings with the existing Arctic marine regulations have been identified. They relate to the ship ice-class categorization system, the lack of flexibility in the Arctic waters zone–date control system, inadequate structural design requirements, lack of standards for structural steels, and crew training.

In 1985, the Commissioner of the Canadian Coast Guard formed a government–industry committee to review the regulations and to propose revisions to the regulations. The work proceeded from a basic concept that all aspects of Arctic navigation such as ship control, operating modes, ice information, and training, are interrelated. The outcome was a compact set of proposals for ship categories, a navigation control system, design ice loads, structural strength, material standards, and ship subdivision. These proposals are contained in the "Proposals for the Revision of the Arctic Shipping Pollution Prevention Regulations" report (TP9981). They were presented at the SNAME Ice-Tech '90 Symposium in Calgary, Alberta, in March 1990.

Navigation Systems

The Coast Guard provides services designed to facilitate the safe navigation of vessels in Canadian waters.

Fig. 11 *Tembah* on 19 kilometre portage en route to launching (October 1963). (Courtesy Transport Canada.)

These services include lighthouses, an extensive network of conventional and electronic aids to navigation, safety information broadcasts, a system of traffic lanes and traffic separation zones, and a series of ship traffic management stations to monitor and guide ships through areas of significant traffic.

Ever higher radio frequencies led to the evolution of radar in the 1930s and 1940s. For many years a shore-based marine radar station, established during the Second World War, guided ships in and out of Halifax Harbour.

During the 1940s and 1950s, experiments in different ways to modulate the radio signal led to the unfolding hyperbolic navigation systems. As a result, Loran-A, Decca Navigator, and later Loran-C, supplanted the direction finding stations in the Coast Guard's radio navigation service. Many ships were fitted with receivers for these signals, particularly for coastal and transoceanic voyages.

Radio telex, radio facsimile, and the improved voice transmissions provided by frequency modulated transmitters developed in the 1960s and 1970s came into widespread use.

The 1980s ushered in the era of satellites. The International Maritime Satellite organization now provides ships with voice and data services in all major ocean areas of the world. The Canadian Coast Guard was instrumental in its establishment.

A joint Canada/U.S./France/C.I.S satellite system, known as COSPAS/SARSAT, routinely locates distress alerting signals worldwide from ships within minutes, allowing rescue forces to be marshalled quickly. The new GPS Navstar satellite constellation is a global positioning system that provides a suitably equipped ship with its position within metres of its true location. The new Canadian satellite system, MSAT, designed to improve all mobile communications, shows promise for shipping. It can provide highly flexible communications using portable equipment from anywhere in Canada except the high Arctic.

The advent of the microprocessor offers the key to integration on a ship's bridge of several electronic systems for simultaneously extracting the needed information in visual form.

Canadian Coast Guard and the Environment

The Coast Guard also plays an active role in environmental protection and oil pollution clean-up.

Following the grounding of the tanker *Arrow* in Chedabucto Bay in 1970, the Parliament of Canada amended the Canada Shipping Act to strengthen its capability to deal with problems associated with ship pollution and its aftermath of liability and compensation.

Not long after, the Government of Canada recognized its responsibility to protect the fragile Arctic waters, and passed the Arctic Waters Pollution Prevention Act, which became law in 1972. Specific requirements for ship construction, machinery, equipment, personnel, and certification were set out in the Arctic Shipping Pollution Prevention Regulations made under the Act.

A national response organization, known as the Canadian Coast Guard Emergencies Organization, was set up in 1970 to deal with oil spill emergencies anywhere in Canada's navigable waters.

The Coast Guard actively supports the development of new technology and research into better ways to respond to oil spills. A training program for on-scene commanders, those responsible for decisions at the spill site, boasts attendees from around the world.

Regardless of how well prepared any organization may be, incidents continue to occur. The *Exxon Valdez* spill in Prince William Sound in March 1989 happened while Canada was still cleaning up a December 1988 oil spill from the barge *Nestucca* off the coast of Washington State. The *Nestucca* spilled about 875,000 litres of bunker oil, some of which drifted north and washed up along the shores of Vancouver Island.

Largely as a result of those spills, on June 9, 1989, the Prime Minister of Canada appointed the Public Review Panel on Tanker Safety and Marine Spills Response Capability to assess current response capabilities and set out a comprehensive set of recommendations to address existing shortcomings.

A review and evaluation of the Public Review Panel's 107 recommendations was undertaken by the Canadian Coast Guard. In addition, a Tanker Safety Implementation Team was established to examine the feasibility, cost and implications of these recommendations. Some were immediately implemented, while others, such as the enhancement of coastal aerial surveillance and the purchase of new oil spill response equipment, will be implemented over time as funding permits. Still others, such as increasing polluter fines, will require legislative changes to the Canada Shipping Act.

The nature of emergency operations has broadened over time to include not only the important function of emergency response operations, but also environmental protection, environmental review and assessment, and disaster management. As a result, the original Emergencies Organization has evolved to meet these needs and to undertake added responsibilities. The branch, centred in Ottawa, has been reorganized and renamed the Environmental Response and Emergency Planning Branch, and employs experienced response personnel in each of the six Coast Guard regions.

It is a national commitment that has in turn led to an international dedication to better protect all waters and shores. The response to the Persian Gulf oil spill in March 1991 demonstrated that the Canadian Coast Guard is indeed part of the international community that came together to respond to one of the world's largest spills.

Summary

The Canadian Coast Guard has a long and distinguished lineage. It provides essential services to help ensure safe and effective marine transport. Its services are available not only to commercial shipping, but also to the fishing industry, and the boating public. It provides public commercial ports, and it zealously guards the marine environment. The main activities of its fleet include providing and maintaining marine navigational aids, clearing ice from shipping lanes and harbours, and escorting commercial ships. It also provides a dedicated search and rescue capability coordinated with the Department of National Defence.

The 1990s promise to be a period of transition for the Coast Guard; increasing use of automated controls, a reduction of human resources because of electronic systems, and tighter controls of all shipping activities will have a major impact. The Coast Guard fleet is currently being restructured to provide the best available mix to meet the specific program needs of each region as efficiently as possible. The restructuring includes redeploying ten ships between regions, the addition of two catamaran survey vessels and another air cushion vehicle, and decommissioning ten ships, all by the mid-1990s.

Lastly, and most importantly, the Coast Guard does not forget the dedicated men and women who make up the organization. In this age of computers and electronics, it is their special skills that enable the Coast Guard to accomplish its assigned service and safety roles.

Bibliography

Appleton, T. E., *Usque Ad Mare: A History of the Canadian Coast Guard and Marine Services*, The Queen's Printer, Ottawa, 1969.

Debates, Canadian Parliament House of Commons, June 6, 1989.

Canadian Encyclopedia, Canadian Coast Guard, 2nd ed., Vol. 1, Hurtig Publishers, Edmonton, 1988, p. 336.

Dick, R. A. and Laframboise, J. E., "An Empirical Review of the Design and Performance of Icebreakers," *Marine Technology*, Vol. 26, No. 2, April 1989, pp. 145–159.

Mantle, P. J., *Air Cushion Craft Development*, 1st rev., U.S. Government Printing Office, Washington, D.C., 1980.

Transit Management in the Northwest Passage, D. L. Vanderzwaag and C. L. Lamson, Eds., Cambridge University Press, Cambridge, U.K., 1986.

Additional Reading

German, J. G., "Design and Construction of Icebreakers," *Trans.* SNAME, Vol. 67, 1959.

Matthews, F. W., "Stability and Control of HMCS *Labrador*," *Trans.* SNAME, Vol. 67, 1959.

Noble, P. G., Allen, R. J., Dunne, M. S., and Johnson, B., "Ice Effect Trials in Arctic Waters on CCGS *Louis S. St. Laurent*," *Trans.* SNAME, Vol. 86, 1978.

Tatinchaux, J.-C., Alekseyev, J., Enkvist, E., Kitagawa, H., Narita, S., Schwarz, J., Takekuma, K., and Williams, F. M., "Comparative Model Tests in Ice of a Canadian Coast Guard R-Class Icebreaker," *Trans.* SNAME, Vol. 97, 1989.

Watson, A., "Operation of Department of Transport Icebreakers in Canada," *Trans.* SNAME, Vol. 67, 1959.

The Canadian Coast Guard is dedicated to the safety of the Canadian public and the protection of the marine environment. The motto by which the Coast Guard operates and conducts its business is "Safety First, Service Always." The Canadian Coast Guard, a civilian organization, maintains a presence in all three coastal areas (Atlantic, Pacific, and Arctic), and major inland waterways.

Part IX

Craft for Work and Pleasure

Canadian Ferry Services

George S. Gibb

A Mari Usque ad Mare

"From sea to sea." If it were possible to condense a description of Canada into four words, this translation of the motto under her beautiful coat of arms, while inadequate, would have to suffice. From sea to sea, Canada encompasses the world's longest coastline and its greatest concentration of fresh water. Coasts and lake shores are generously endowed with inlets, estuaries, coves, and rivers. Human curiosity compels us to discover what lies on the opposite side of a body of water; subsequent social necessity requires that, once curiosity has been satisfied, repeated communication usually ensues. It is thus that ferry systems evolve.

The geography of Canada virtually demands that ferries be used for communication in coastal regions and lakelands. The First People used canoes to cross to regions that were otherwise inaccessible, or to avoid long roundabout land trips. Different lifestyles and urgencies from those widely accepted today would not have required regularly scheduled crossings in those days, but repeated use of the same departure and arrival points was practiced by the Native people long before the arrival of Europeans. Even today, scheduled times are not inflexible; repetition over the same route is the essential constant for a ferry service.

Having acknowledged that ferry operations predate formal naval architecture and marine engineering, we can now look at the state of ferry services across Canada today with some emphasis on the technical aspects. Our study will cover three regions—Western Canada; Great Lakes and St. Lawrence; and Atlantic Canada, with special emphasis on the period 1940–1990.

Owing to space constraints, a complete catalogue of Canadian ferry systems is not presented. What follows, then, is a sampling of representative services. These have been chosen with an eye to presenting a balanced picture, coast to coast. Also, the principal technical developments of the past 50 years are traced.

Western Canada: The Fleet of the British Columbia Ferry Corporation

Origin and Development. The British Columbia Ferry Corp. came into being in 1958 after a period of disruption of existing services operated by private lines. A combination of industrial disputes and management decisions to discontinue certain services left a gap in the transportation of passengers and vehicles between Vancouver Island and the lower mainland of British Columbia.

The provincial government of the day was not convinced that this essential service could be safely left in the hands of private enterprise, and so took the initiative of introducing a government-run ferry service under the name British Columbia Ferry Corp.

Fig. 1 Crossing the strait.

Fig. 2 The *Sidney* on trials.

This undertaking was not regarded by the industry as having any hope of success; now, 32 years later, the government's decision has been vindicated in the existence of a popular, efficient service that, starting with only two ships, developed into a fleet of 38 ships by 1990/91 and is still expanding. To be self-sufficient, a fleet of this size requires supporting services ashore and these have developed as the Corporation grew. Access for loading and disembarking efficiently and swiftly is provided at the terminals, including two-level ramps where traffic density so requires; fleet maintenance is undertaken at the Corporation's own dock, leaving only major repair and refit work to be done by private shipyards.

The Original Vessels. Service was started in June 1960 with the entry into service of the *Sidney* and *Tsawwassen* (now *Queen of Sidney* and *Queen of Tsawwassen*). These two ships enjoyed instant success, their only deficiency being inadequate dining facilities. This was quickly remedied by adding a dining room at the aft end of the Promenade Deck. Today, more than 30 years later, these two ships retain their original appearance, unlike their younger sisters, which have undergone considerable modification. To some eyes, these are still the two most handsome ships in the fleet. The nice balance of the outboard profile with the bow and stern loading doors required for straight-through roll-on, roll-off operation can be seen in the photo of the *Sidney* on trials (Fig. 2).

Design	T. A. McLaren, P. Eng., Vancouver, B.C. in association with P. Spaulding, Seattle WA
Construction	*Queen of Sidney*—Victoria Machinery Depot *Queen of Tsawwassen*—Burrard Drydock Co.
LOA	342 ft-4½ in.
LBP	308 ft-0 in.
Beam	76 ft-4 in.
Service speed	17 knots
Propulsion	Twin screw, each 3000 bhp
Passengers	969
Cars	106

The addition of platform decks attached to the center casing port and starboard in 1966 increased car capacity to 138.

Fig. 3 "C" class *Queen of Oak Bay.*

The "V" Class. Two new ships were added to the fleet in 1963: *City of Victoria* and *City of Vancouver* (now *Queen of Victoria* and *Queen of Vancouver*). The design of these ships was developed from the *Sidney* and *Tsawwassen* with improvements arising from experience with the earlier vessels, in particular, an increase of two feet in beam and an eleven percent increase in horsepower.

Construction	*Queen of Victoria*—Victoria Machinery Depot *Queen of Vancouver*—Burrard Drydock Co.
LOA	342 ft-4½ in.
LBP	308 ft-0 in.
Beam	78 ft-4 in.
Service speed	17 knots
Propulsion	4 diesels driving twin screws, total 6664 bhp
Passengers	969
Cars	106

The fleet was further augmented in 1963 by the delivery of two more ferries of the same class, the *Queen of Saanich* and *Queen of Esquimalt*, and in 1964, by the *Queens of Nanaimo, Burnaby,* and *New Westminster*, all built by Victoria Machinery Depot.

The capacities of the "V" class have been progressively increased by adding stowable platform decks and by lengthening. A further increase in capacity was achieved by raising the Promenade Deck and superstructure of the *Queens of Victoria, Vancouver, Saanich,* and *Esquimalt* "en bloc" and inserting a new car deck. This process (illustrated in Fig. 5) has been completed for at least one other vessel of the "V" class (*Queen of New Westminster*).

The "C" Class. The success of the ferry service generated a demand for additional capacity; so in 1976 three new super-ferries were added to the fleet (see Fig. 3). In a departure from the more conventional configuration of the existing *Queens*, these were designed as double-enders, similar to those operated by Washington State Ferries. While they are sister ships, one of the trio, the *Queen of Alberni*, has a much shorter superstructure. This ship is designed to accommodate over-height vehicles on the vehicle deck, which, because of the shorter superstructure, is open at each end. Otherwise, she has the same hull and machinery as her sister ships, the *Queens of Cowichan* and *Coquitlam*. The double-ended configuration eliminates the need for turnaround at the terminals.

Design	Nickum, Spaulding Associates, Seattle WA
Construction	*Queen of Alberni*—Vancouver Shipyards *Queen of Cowichan*—Yarrows Ltd. *Queen of Coquitlam*—Burrard Drydock Co.
LOA	457 ft-0 in.
LBP	417 ft-0 in.
Beam	89 ft-0 in.
Service speed	19 knots
Propulsion	See Fig. 4
Passengers	1100 (on *Alberni*: 450)
Cars	362 (on *Alberni*: 145)

Inter-Island Fleet. In the latter half of 1961, the Corporation purchased the privately run Gulf Island Service, thereby acquiring a fleet of smaller double-ended ferries of wooden construction to service the many "Gulf" (Gulf of Georgia) Islands off the southern end

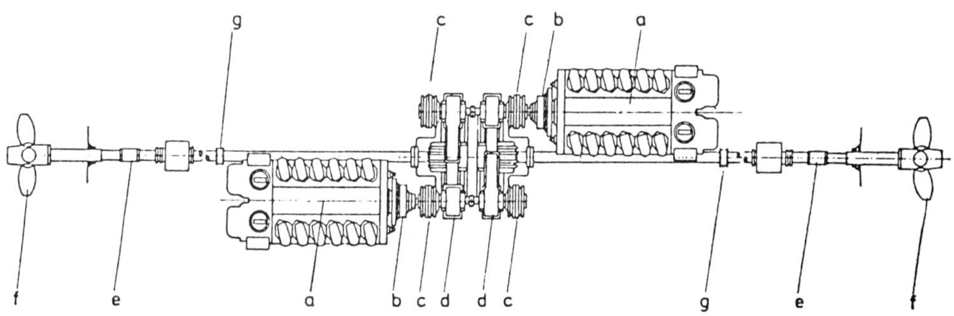

Key

A. MaK 12M551AK propulsion diesel engines
B. Lohmann and Stolterfoht Spiroflex KJS350 highly-elastic shaft couplings
C. Lohmann and Stolterfoht Pneumastar KUZ1153 torsionally stiff clutches
D. Lohmann and Stolterfoht Navilus GVQ1575 twin-input single-output gears
E. Propeller shafts
F. KaMeWa C.P. propellers
G. Shaft brakes

Fig. 4 Diagram showing the unconventional layout of the machinery. A total of nine alternative manoeuvres are possible. All three ferries are identical below the main deck. The engine rooms are spacious, well ventilated, and very well illuminated. The propulsion machinery consists of two MaK 12M551AK diesel engines, each developing 5825 hp @ 429 rpm. The engines are installed fore and aft of midships, off the centreline.

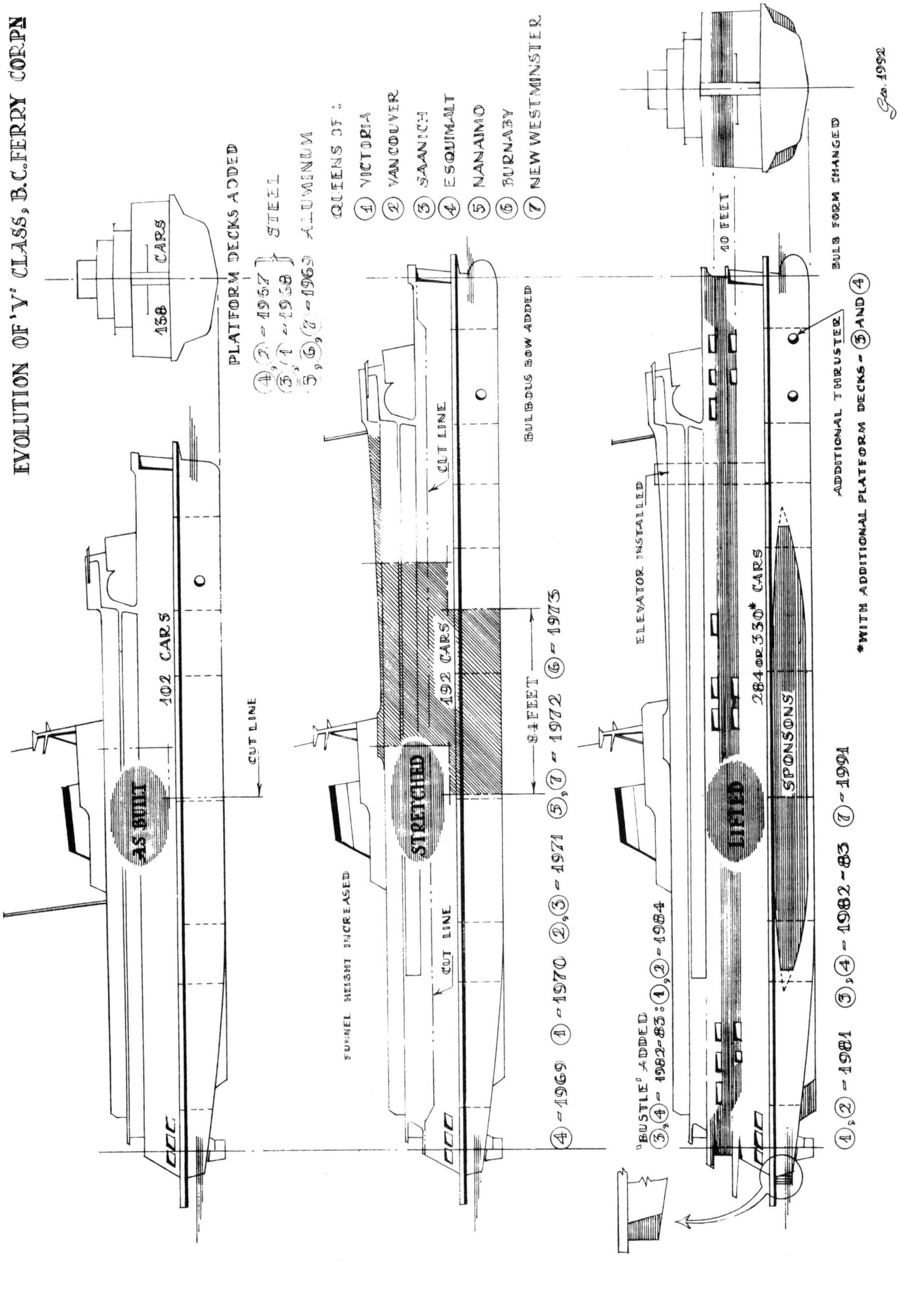

Fig. 5 Stretching and raising the "V"-class ferries.

Fig. 6 Raising procedure for "V" class. Design: K. E. Hanson, Copenhagen. Construction: Burrard-Yarrows Corp., Victoria, B.C.

of Vancouver Island. These vessels have been replaced by three vessels of the Bowen class, which are conventional double-enders with open decks fore and aft, and have since been lengthened and re-engined.

Conversions. Ever since its inception, the Corporation has been challenged by the necessity to expand due to the demand generated by public acceptance of the service. As well as introducing new vessels, and in some cases acquiring existing ones, the Corporation has also increased capacity by enhancing ferries already in the fleet.

Fig. 7 *Mill Bay.*

This enhancement has been accomplished by adding platform decks, lengthening hulls, raising topsides to introduce additional full length car decks and, in one case, dividing a hull lengthwise, separating the two halves, and integrating them into a new catamaran structure.

Of the ships operating on the major routes, all but the original two, the *Queens of Sidney* and *Tsawwassen* and the four "C" class boats have undergone radical conversion.

The sequence of modifying the original "V" class is shown in Fig. 5. Internal and external views of the raising procedure in progress are shown in Fig. 6.

Engineering Details. *Casing Arrangements.* Only one ship in the fleet has side casings. For the rest there are two configurations. Some have centerline casings or twin casings disposed about the centerline with traffic lanes outboard and on the centerline. The one exception to this is the *Mill Bay*, which has a single off-center casing with outboard lanes (Fig. 7).

Deck Machinery. The original two ships and the seven "V" class ferries use capstans rather than windlasses for anchor handling.

Bridge Control and Machinery Control Rooms. While the Corporation did not pioneer bridge control of main machinery, they were among the earlier of the major fleet operators to adopt what is today common practice.

"C" Class Main Propulsion Machinery and Steering. Figure 4 is reproduced from a paper presented to the 1977 Annual Meeting of the Canadian Shipbuilding and Ship Repairing Association by George Baldwin, and shows the innovative machinery arrangement of the "C" class ferries. Unlike their older sisters, the ships in this class do not have transverse thrusters. There is a rudder at each end on the center line. These rudders swing to 45 degrees port and starboard rather than the more usual 34 or 35 degrees. With both rudders hard over to the same side, the propeller wash reacts on them in a transverse direction to assist in manoeuvering alongside.

Fig. 8 Artist's rendering of BC Ferries' "S"-class vessel.

New Construction—Current Programs. *Inter-Island Service.* Construction of two double-ended ferries of 85-car capacity was started in mid-1990 with delivery in June 1991 for the first ship and December of the same year for the second. These are double-ended with open decks at each end. Propulsion will be by two Z drives at each end.

Design	Wartsila Marine, Vancouver and S.H.M. Marine International, Victoria
Construction	Vancouver Shipyards
LOA	314 ft-10 in.
Beam	69 ft-6 in. (over guards)
Service speed	14 knots
Propulsion	5000 bhp, diesel electric
Passengers	330
Cars	85

Vancouver Island to Mainland. A major addition to the fleet will be the delivery to the Corporation of two new super ferries, identified as the "S" class (Fig. 8)—one in March 1993 and the other in March 1994. These two new ships, with a capacity of 470 cars, will be the largest in the fleet. Initial design shows them to be of conventional profile with bow and stern loading.

Design	Concept: K. E. Hansen, Copenhagen Development: Polar Design Associates, Vancouver
Construction	Integrated Ferry Constructors Ltd.[1]
LOA	549 ft-5 in.
LBP	511 ft-8 in.
Beam	87 ft-3 in.
Service speed	19.5 knots
Propulsion	15,000 bhp, twin screw, variable pitch, hi-lift propellers, 2 bow thrusters
Passengers	2000
Cars	470

The Effect of Conversion on Performance. The five "V"-class ships already in service in the raised configuration have been re-engined, their propellers being changed from fixed pitch reversible to unidirectional, variable pitch.

Considering only these five, the raising operation involved the most significant change to the vessels' characteristics. As seen in Fig. 5, the raising required the addition of sponsons on each side of the hull below the Main Car Deck.

When the first two ships so modified returned to service it was noted that the sponsons introduced secondary waves in the wake pattern. This was eliminated by fairing the forward and after ends more smoothly into the shell, as indicated by the broken lines on the profile sketch.

Two other appendages will be noted on this profile sketch. One is a bustle added to the transom, which is said to have produced a three percent increase in speed. The other is an elongation to the aft end of the skeg to eliminate interference between port and starboard propellers when the propulsion system was changed.

	As Built	After Conversion
Loaded draft	13 ft-0 in.	14 ft-5 in.
bhp	6664	8500
Service speed	16.3 knots	18.3 knots

Western Canada: Inland and Coastal Ferries of the Ministry of Transportation, British Columbia

This part of our study begins with an apparent contradiction in that there has been no coastal service operated by the Ministry since 1985, when this part of the fleet was transferred to the British Columbia Ferry Corporation. It is included under this heading rather than in the foregoing paragraphs as it was the Ministry's responsibility for the greater part of the period under discussion.

[1] A consortium of three B.C. shipbuilders: Yarrows Ltd., Pacific Rim Shipyards Ltd., and Allied Shipbuilders Ltd.

Fig. 9 Reaction ferry *Usk.*

Origin and Development. The interior waters of British Columbia, its lakes and rivers, are numerous and extensive, as are the inlets and sounds around its coast. In the earlier days, when there were no highways and few railroads in the interior, many bodies of water lay athwart established trading routes, so means had to be provided to cross them. And means of crossing were indeed provided, in an impressive array of ferries employing every conceivable means of propulsion. In the early days these ranged from such simple manual methods as rowing, poling, and pulling on overhead ropes, through mechanical means such as paddle wheels, outboard motors, tug and barge, conventional screw propulsion, and guided cable operation, to guided hovercraft and—of only academic interest to naval architects and marine engineers—aerial basket carriages.

One type not specifically mentioned in the preceding paragraph, but extensively used by the ministry, is the reaction ferry. This is a development of a system used in more primitive societies where river current is used to propel the vessel from one side to the other. This application of natural laws could not be simpler. A boat is permanently attached by a long rope to a post driven into the river bed upstream of the crossing, the whole affair having the character of a pendulum operating in a horizontal plane. The boatman, using a steering oar, alters the angle of attack of the hull with the current and this generates a reaction that propels the ferry across the current.

This simple expropriation of the powers of nature is still in use today at six ferry crossings operated by the Ministry, but in a rather more sophisticated fashion developed to overcome the danger of damage to rope work at the surface from waterborne timber and debris. The ferry is rigged to an overhead cable by means of a bridle. The bridle wire is spooled onto a hand operated winch that performs the same function as the steering oar in the basic version. Turning the winch drum changes the attitude from neutral, i.e. hull parallel to the current, to the angle of attack required to produce lateral motion. A typical reaction ferry has a platform deck spanning twin hulls, as seen in the photo of the *Usk* crossing the Skeena River (Fig. 9).

While it is unnecessary to deal at length with the history of the Ministry's ferry operations it is appropriate to look back to 1933, when a noteworthy transition took place on Kootenay Lake. The Canadian Pacific Railway had for many years provided vital services on the Kootenay, Arrow and Okangan Lakes and their connecting rivers; but, with railroad and highway development, these routes became progressively less used and consequently uneconomical. As a result CPR had to dispose of one of the last of their fleet of sternwheel steamers, the *Nasookin* (Fig. 10). Fortunately, because of highway development, this corresponded with a need for a vehicle ferry across Kootenay Lake and so she was purchased by the Ministry in 1933 for this service, which she provided until 1947 when she was replaced by a newly constructed ferry, the *Anscomb.*

Today the Ministry's inland fleet consists of two distinct elements One comprises units not requiring certification, including reaction ferries, cable ferries, and tug/barge combinations. The other includes self-propelled vessels subject to regulation by the Canadian Coast Guard. The inland fleet, not having access

Fig. 10 Sternwheeler *Nasookin.*

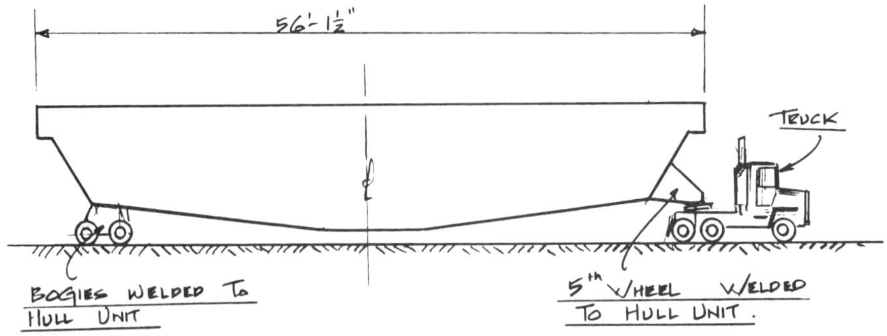

Fig. 11 Francois Lake ferry typical shipping arrangement.

to coastal repair yards, is provided with facilities for repair and maintenance locally, including underwater work. The two Kootenay Lake ferries are serviced by a floating dock capable of lifting 896 tons. The Francois Lake ferry also has its own floating dock, while the rest of the inland ferries are serviced by marine railways.

The Canadian shipbuilding industry cannot claim to be the originators of the process whereby a ship intended for service in a location far removed from the shipyard is shipped in sections to its final destination. We have, however, no need to be modest about our success with this method. It has been practised since before World War II, and used for a wide variety of vessels, such as tugs, barges, dredges, floating log sorters and, of course, ferries. In the earlier days ships destined for service in Western Canada came from as far away as Sorel, Quebec; but today, the industry in British Columbia has this line of business well in hand.

Galena. An example of breakdown for re-erection, the double-ended ferry *Galena* (Fig. 12) was built in Vancouver and transported in sub-erection units by road to Castlegar at the south end of the Lower Arrow Lake, more than 400 miles from the shipyard. There she was re-assembled and outfitted for delivery to the north end of the Upper Arrow Lake.

Design	Case Existological Laboratories, Victoria
Construction	Allied Shipbuilders Ltd.
LOA	241 ft-0 in.
LBP	165 ft-0 in.
Beam	61 ft-4 in.
Service speed	10 knots
Propulsion	Diesel electric, total bhp 700, Voith Schneider propellers each end
Passengers	200
Cars	50

Anscomb. As was noted earlier, the *Anscomb* (Fig. 13) replaced the ex-CPR sternwheeler *Nasookin.* Of riveted construction, she was built on-site in 1946 using the CPR marine ways at Nelson on the Kootenay river. She was modernized twice, once in 1960 when her superstructure was raised to provide increased overhead clearance, and again in 1972 when all structure above the car deck was removed, the entire superstructure renewed and modernized, and the vessel re-engined.

Design	T. Halliday
Construction	Western Bridge and Steel Fabricators
LOA	182 ft-6 in.
LBP	169 ft-6 in.
Beam	46 ft-3 in.
Service speed	12 knots
Propulsion	Twin screw diesel, total 800 bhp
Passengers	150
Cars	40

Cable Ferries. Cable ferries are fitted with power operated winches rigged to cables fixed to each bank of a river. The winch pulls the ferry across the river by spooling on and spooling off as it progresses from one side to the other. They are most suitable for short crossings on slow moving rivers and do not require certificated crews. The Ministry has operated many cable ferries over the years but most have been taken

Fig. 12 *Galena.*

Fig. 13 *Anscomb.*

Fig. 15 *Tenaka.*

out of service, having given way to bridges. There are still four cable ferries in operation today.

Probably the most noteworthy cable ferry of all, the *Tsimeh,* had only a brief existence. She provided a crossing of the Fort Nelson River at a place where the presence of debris in spring and fall and pancake ice in winter made it unsuitable for a displacement hull. The solution was rather innovative—a hoverlift cable ferry. Entering service in 1979, she had a useful life of some three years before being destroyed by fire. As a bridge was erected at this crossing in 1984 it is not likely that this surface effect, guided cable style will be revived.

The Coastal Fleet. For some seventeen years before the inauguration of services by the British Columbia Ferry Corporation, the Ministry of Transportation and Highways provided inter-island and estuarial ferry crossings on the southwest coast of the province, and continued to do so until 1985 when the Corporation purchased the service from the Ministry.

Services were provided by a variety of ferries ranging in style from the simplest self-propelled deck barges with only the most basic deckhouse to conventional double-enders of the type found world wide today. Examples include the *Quinitsa* (Fig. 14), and Ro/Ro's of more shipshape configuration—or at least having identifiable forward and after ends, port and starboard sides—as exemplified by the *Tenaka* (Fig. 15). She was designed to serve the smaller islands in the sheltered waters between Vancouver Island and the mainland. She was built in 1964 and re-engined in 1977. *Tenaka*:

Construction	Victoria Machinery Depot, Victoria, B.C.
LOA	151 ft-6 in.
LBP	141 ft-0 in.
Beam	43 ft-5 in.
Service speed	12 knots
Propulsion	1650 bhp
Passengers	100
Cars	26

The People Movers. *The Deficient, Inefficient, Insufficient Bridge.* In an urban setting bridges are not always the best way of moving people over water and this is especially true when one vehicle transports one person. This is most apparent at rush hour.

A foot passenger ferry was operated between downtown Vancouver and North Vancouver on the opposite side of Burrard Inlet until 1958. This is a broad part of the inlet, bracketed between two highway bridges. On the jacket of the book *Ferry Across the Harbour* by Capt. James Barr (a history of this crossing), the publishers comment that on his retirement "Captain Barr decided to assemble the story of this ferry service that came to its end when great steel spans were flung across both east and west narrows of the Burrard Inlet basin, making water transport redundant." If we paraphrase Parkinson's Law, we can say that traffic expands so as to fill the lanes available for its occupancy. Unfortunately, the rate of expansion of pavement is linear and of traffic, exponential. The "great steel spans" may have made water transport redundant, but only temporarily. Traffic expanded.

In 1974 the Bureau of Transit Services, a provincial authority, conceived the idea of linking the bus services in Vancouver and North Vancouver by ferry across the inlet. Linking bus with bus meant passengers only.

The Seabus System. The concept encompassed a total system consisting of the ferries themselves plus all supporting services. The system would

Fig. 14 *Quinitsa.*

Fig. 16 Seabus approaching terminal.

receive passengers from street buses and deliver them to street buses on the opposite shore. Thus it had to include terminals providing an easy flow of foot traffic, shortest practical walking distances, and adaptability to the tide. Administration and maintenance facilities had to be supplied. The successful bidder for the design and supervision of construction of the total concept was Case Existological Laboratories Ltd (CELL) of Victoria, B.C. This organization, directed by John Case, was principally engaged in marine engineering, ship design, and ocean science; and so contracted civil engineering design out to Fenco Consultants Ltd.

Two all-aluminum, catamaran ferries of distinctive, innovative, and attractive appearance were designed. They are capable of carrying 400 foot passengers in comfort and safety. Comfort and safety dictated certain features such as 400 seats for 400 passengers, no standees; catamaran hulls for stability and capacity; and double-ended configuration for straight-in, straight-out docking. Six centrally controlled, pneumatically operated doors on each side permit a full complement of passenger to disembark in less than one minute. When all departing passengers are ashore, incoming passengers embark through the doors on the opposite side. To accomplish this, the terminals are designed to direct embarking and disembarking passengers in separate streams.

The floating terminals, E-shaped in plan, are made from pre-stressed, post tensioned concrete requiring minimum maintenance. They support enclosed, heated passenger areas connected to the shore by hinged ramps. At the Vancouver city end, escalators take passengers up to an enclosed overpass that spans railroad tracks and connects directly at street level to downtown.

This system has been operating since 1977, sharing the inlet with cargo liners, tugs, barges, and floatplanes without serious incident and enjoying an immense and well-deserved popularity with the travelling public. The ability to expand the service simply by increasing the number of ferries is built into the design.

The original ferries are the *Burrard Beaver* and the *Burrard Otter*.

Construction	Yarrows Ltd., Victoria, B.C.
LOA	112 ft-6 in.
Beam	41 ft-6 in.
Service speed	13.5 knots
Propulsion	4 diesels, each driving a right angle drive unit; 1600 total bhp
Passengers	400

Great Lakes and St Lawrence

Development of ferry services on the inland seas and the great natural waterway leading from the interior to the ocean was inevitable. On the Great Lakes and the St Lawrence there are many offshore islands and river crossings that today are served by bridges, but there still remain some important crossings served by automobile and passenger ferries, many of which manage to provide a service year-round, albeit with occasional interruptions during the severest of winter conditions.

Great Lakes Ferries. The longest ferry trip on the Canadian lakes is across Georgian Bay, the northeast bight of Lake Huron, between Tobermory on the northern tip of the Bruce Peninsula and South Baymouth on Manitoulin Island. This crossing provides direct north/south access to the Trans-Canada Highway from the populous southwestern Ontario region on the west end of Lake Ontario and the north shore of Lake Erie. Service is from late April to mid-October and is provided by the Ro/Ro *Chi-Cheemaun* (Ojibway for Big Canoe) (Fig. 17). Built in 1974, she has a stern door and bow visor, straight-through loading-unloading, and is fitted with active stabilizers. Originally fitted with mezzanine decks port and starboard, outboard of her twin casings, she was later retrofitted

Fig. 17 *Chi-Cheemaun.*

Fig. 18 *Quinte Loyalist.*

with stowable platform decks to increase her car capacity. *Chi-Cheemaun*:

Design	Case Existological Laboratories Ltd., Victoria, B.C.
Construction	Collingwood Shipbuilding and Engineering Co., Ont.
LOA	365 ft-5 in.
LBP	334 ft-5 in.
Beam	62 ft-4 in.
Service speed	17.25 knots
Propulsion	2 diesels, twin screw, bow thruster; 7000 total bhp
Passengers	530
Cars	113
High vehicles	17

The Ministry of Transportation and Communication provides two other crossings, without the services of an intermediary operating company. Since they are integral elements of the provincial highway system, passage is free for vehicles and passengers. The first of these, between Adulphustown and Glenora on the Bay of Quinte is a year-round, 24-hour service using two ferries, the *Quinte Loyalist* (Fig. 18) and the *Glenora.*

The other free crossing is between the city of Kingston at the extreme east end of Lake Ontario and Wolfe Island in mid channel. This is where the lake waters end and the St. Lawrence begins. This is also a year-round service with a reduced schedule at night and in the winter months. *Wolf Islander II* (Fig. 19):

Design	Case Existological Laboratories Ltd., Victoria, B.C.
Construction	Port Arthur Shipbuilding Co., Ltd., Thunder Bay, Ont.
LOA	205 ft-0 in.
LBP	193 ft-6 in.
Beam	65 ft-0 in.
Ice class	2
Service speed	10.5 knots
Propulsion	4 right angle drive units, 2100 total bhp
Passengers	338
Cars	50

Fig. 19 *Wolf Islander II.*

The St Lawrence Ferries. *Quebec–Levis: Early History.* From the earliest days of French settlement until the middle of the 19th Century, crossings between Quebec City and the south shore were undertaken by individuals who ferried passengers and goods in canoes. Service was provided twelve months per year.

Winter crossings, depending on the vagaries of climate and current, were accomplished partly on open water and partly by manhandling over the ice.

In mid-century steam side wheelers appeared, operated by enterprising local businessmen. Steam propulsion still experienced some competition from manpower before it eventually prevailed. But another form of propulsion the "bateau a manege," or "horseboat," also ran in opposition for a brief period. "Manege" can be translated as "equitation," but also as "carousel," thus providing a capsule description of the motive power for this scow type side wheeler (see endpiece figure, p. 473).

Two significant dates were 1856 when a winter ferry went into service and 1862 when screw propulsion was introduced.

Competition between rival operators was neither efficient nor well regulated. Recognizing this, the municipal authorities enacted progressively more stringent regulations, culminating in the award by auction of a contract—to the sole bidder—in 1863. From then until 1971 the service was provided by a variety of private companies, subject to regulation by a joint committee having equal representation from Quebec and the south shore municipalities.

Before leaving the private ferry service we should record the ironic events of Jan. 31/Feb. 1, 1955. Winter ferries were sometimes unable to make an uneventful crossing. Ice fields and currents could carry them downstream where they might remain for hours before a change in the conditions allowed them to break free and steam back to Quebec or Levis. In mid-afternoon of January 31, attempting to cross the ebb tide, the Clyde-built ferry *Cite de Levis* found herself ice-bound five miles downstream, with the temperature dropping to –36 °C and in an impenetrable fog. Unable to extricate himself from this unhappy situation, the captain called for assistance by the icebreaker *Lady Grey*, on station in Quebec harbour.

At dawn on Feb 1 *Lady Grey* set out to free the ferry, but in the meantime the *Cite de Levis* was succeeding in her own efforts to break loose. Unfortunately, success came exactly as the icebreaker appeared out of the fog, dead ahead. The prevailing notion on the Clydeside those days was that nothing could beat a Clyde-built ship; but ramming an icebreaker and sinking it was no way to prove the point. The ferry, having sustained no damage, returned to Quebec with the rescued crew of the icebreaker.

The Quebec-Levis crossing was only one of many across the St Lawrence both upstream and downstream of the old capital. Today nearly all the ferries upstream have been supplanted by bridges in both Quebec and Ontario. Downstream the river widens considerably and ferries have survived as important links between the south shore and the more sparsely populated north shore. This study, then, will confine itself to the crossings downstream of Quebec City.

LaSociete des Traversiers Quebec–Levis, 1971–1974. For many years the provincial government had been aware of the important role the St Lawrence ferries played in the highway system and had subsidized private enterprise. In 1969, when the need for modernization became evident, a study commissioned by the government resulted in the Quebec–Levis crossing becoming a provincial corporation: La Societie des Traversiers Quebec–Levis. That was in 1971.

Meanwhile, two all-season ferries had been commissioned and were delivered in 1971. Built in the principal local shipyard in the respectable time of thirteen months, the *Lomer-Gouin* and *Alphonse Desjardins* (Fig. 20) are still in service. They have center line casings, side loading ramps and a somewhat modest overhead clearance of just twelve feet. Passengers accommodation on the Promenade Deck provides shelter from inclement weather. There is also an open deck for a pleasant walk-around when the weather is fine. These ships do not have an ice class but are reinforced for ice navigation. *Alphonse Desjardins*:

Fig. 20 *Alphonse Desjardins.*

Construction	Davie Shipbuilding Ltd., Levis, P.Q.
LOA	214 ft-0 in.
LBP	200 ft-0 in.
Beam	71 ft-6 in.
Service speed	12 knots
Propulsion	Twin engine, single v.p. propeller; 3240 total bhp
Passengers	700
Cars	55

La Societe des Traversiers du Quebec—1974. "Le Quebec" as distinct from "Quebec" signifies the Province as distinct from the city, so "du Quebec" indicates the expansion of the provincial ferry corporation to areas outside the capital region. Today the Societe provides service to two islands downstream and to two river crossings forming links of major importance in the highway system of the province. Another, shorter crossing is provided at a narrow point in the river at Sorel, fifty miles northeast of Montreal. The fleet consists of eleven ferries capable of transporting a total of 5046 passengers and 665 vehicles.

Ile aux Coudres–St. Joseph de la Rive. Two ferries, *Trois Rivieres* and *Joseph Savard* (Fig. 21), provide the only highway link to the pretty little "island of hazelnuts." The service is year-round. *Joseph Savard*:

Built	1985
LOA	205 ft-8 in.
LBP	182 ft-8 in.
Beam	71 ft-10 in.
Service speed	13 knots
Ice class	1
Passengers	400
Cars	59

Tadoussac–Baie Ste Catherine. Provincial highway 138 follows the north shore from Quebec to Sept Iles, a driving distance of 420 miles. At a little less than half way, the south flowing Saguenay River joins the St Lawrence. Two ferries cross the Saguenay, a distance of one mile, in ten minutes, saving a detour inland of 160 miles. These sister ships, built in 1980, provide a year-round 24 hours per day service, reflecting the importance of this crossing as a vital link in the north shore highway system.

Fig. 21 *Joseph Savard.*

Fig. 22 *Saguenay.*

These ferries have configurations that are logical developments of the design of the previous units, which had been in operation for 25 years. The new units have retained the location of the uptake casing, service areas, and passenger accommodation, which are offset to one side of the vehicle deck. A bridge with windows fore and aft on the center line as well as on the sides, spans the entire vehicle deck at a height above the deck of 18 ft-1 in. A double control console is located outboard on the side opposite the uptake casings. The significant difference between this arrangement and that of the previous ferries is that the officer of the watch is not permanently confined to one side of the ship, and thus has a much better line of sight in the direction of travel. The photos show the configuration of the previous vessel, *Saguenay* (Fig. 22), in comparison with the improved version (Fig. 23). *Armand-Imbeau/Jos-Deschenes*:

Design	Developed by La Societe des Traversiers and the shipbuilder
Construction	La Marine Industrie Ltee, Sorel, Quebec
LOA	203 ft-7 in.
LBP	183 ft-9 in.
Beam	70 ft-6 in.
Ice class	1
Service speed	12 knots
Propulsion	2 diesels, 2480 bhp, controllable pitch propeller at each end
Passengers	400
Cars	60

Fig. 23 *Armand-Imbeau.*

Matane–Baie Comeau–Godbout. Matane is the south shore terminal of a ferry that crosses alternately to Godbout—34.5 miles and 2.25 hours and Baie Comeau—39 miles and a little over two and a half hours. This crossing is served year round by the *Camille Marcoux* (Fig. 24), the only vessel in the fleet having a bow visor and stern door and, consequently, totally enclosed car decks. She was built in 1974. *Camille Marcoux*:

Design	Vickers Engineering Services Inc.
Construction	La Marine Industrie Ltee, Sorel, Quebec
LOA	310 ft-8 in.
LBP	279 ft-11 in.
Beam	60 ft-6 in.
Ice class	1
Service speed	15 knots
Propulsion	Twin diesel, 9600 bhp, controllable pitch propellers, bow thruster
Passengers	600
Cars	126

Atlantic Canada

Origin and Development. Ferry service between the various Canadian maritime provinces developed with the expansion of railways in the late 19th Century. The service formed a necessary link in the transport chain joining the island provinces to mainland Canada. Rail, sea, and again rail involved some inconvenient transshipment in the early days, but it was better than isolation and did much to help the central government meet the terms of confederation.

The earlier ferries were conventional ships, but railcar ferries and Ro/Ro's were built, or suitable existing ships were acquired. The Canadian government had been directly involved in these services since early in this century, especially in building and operating the first railcar ferry. This was the reason that the

Fig. 24 *Camille-Marcoux.*

Fig. 25 *Sir Robert Bond.*

national railway system was given the task of operating the maritime ferry services. Under Canadian National Railways the service was not autonomous and the administration was somewhat fragmented, with responsibilities being spread over several different departments both in the government and in the railway. This situation was rectified in 1976 when CN Marine became a separate division of Canadian National Railways. Total autonomy was achieved in 1986 when the fleet, equipment and facilities were transferred to a new, independent Crown Corporation, Marine Atlantic Inc., which operates coastal passenger-cargo ships along the Labrador coast as well as the inter-provincial ferries. There is also a seasonal ferry operated on the Labrador coast. Like the British Columbia Ferry Corp., Marine Atlantic operates a complete system including modern terminal facilities, and a drydock and repair yard in St. John's, Newfoundland.

The Routes. On the West Coast the British Columbia Ferry Corp. operates many fairly short trips in relatively sheltered waters. This is reflected in the open nature of the structure above the main vehicle deck. This configuration is found in only two of Marine Atlantic ferries; and these operate only in the ice-free season. The open waters, severe winters and, in most cases, long distances involved, have resulted in a fleet much different in structure and appearance from their western counterparts. Even on the shortest route (a mere eight nautical miles between Cape Tormentine, New Brunswick and Borden, Prince Edward Island), the year-round ferry, *Abegweit,* has bow and stern doors; and the winter ferry, *John Hamilton Gray,* has vehicular access through a stern door.

The longest route, from Lewisporte, Newfoundland to Goose Bay, Labrador, about 520 nautical miles, is served during a short summer period by the *Sir Robert Bond* (Fig. 25) assisted when necessary by the *John Hamilton Gray* (Fig. 31).

From Nova Scotia to southern Newfoundland there are two crossings, both originating in North Sydney, Nova Scotia, one to Port aux Basques and the other to Argentia. The former is served by the Corporation's two newest and largest ships, *Caribou* and *Joseph and Clara Smallwood.* The latter is served by the *Ambrose Shea.*

The Atlantic Fleet. *Prince Edward Island to New Brunswick.* Prince Edward Island's connection to the mainland is provided by the *Abegweit.* It would be improper, however, to discuss today's *Abegweit* without first mentioning her predecessor of the same name. Built in 1947, she was, for her time, a most innovative and technically advanced icebreaking ferry, with eight diesels driving four electric motors operating two propellers forward and two aft. She was replaced in 1981 by a new ship of the same name on the same Borden to Cape Tormentine crossing.

The new *Abegweit* (Fig. 27), a rather unusual looking ship, is clearly custom-designed to the service she provides. With twin screws aft, twin thrusters forward and a full docking bridge at the stern, she is well suited to manoeuvering in the confined approaches to the terminal at Borden, Prince Edward Island. She has two vehicle decks with bow and stern door access. *Abegweit*:

Design	Concept: Warsila Shipyard, Finland Development: German and Milne Inc., Montreal, Quebec
Construction	Saint John Shipbuilding Co., Ltd., Saint John, N.B.
LOA	401 ft-6 in.
LBP	384 ft-9 in.
Beam	70 ft-6 in.
Ice class	1A Super

Fig. 26 *Ambrose Shea.*

Fig. 27 *Abegweit.*

Service speed	12 knots
Propulsion	6 diesels, 22,560 total bhp, twin screw, twin bow thrusters, single stern thruster added in 1990
Passengers	974
Cars	250
Tractor-Trailers	40

Nova Scotia to Newfoundland. The *Caribou*, delivered in March 1986, and the *Smallwood* (Fig. 28), delivered in October 1989, are the two newest and largest ferries operating between Nova Scotia and Newfoundland. In most respects they are sister ships but the *Smallwood* incorporates significant changes in victualing arrangements, machinery layout, and passenger accommodation arising from operational experience. A most noticeable change was the elimination on *Smallwood* of *Caribou*'s innovative stores replenishing system—a most important consideration for a ship offering full facilities for 1200 passengers, including overnight cabins for 196, and quick turnaround at each terminal.

Caribou's installation has port and starboard stores cranes on the topmost deck abaft the twin funnels. These are used to lift pre-stocked containers, including freezers, through a generous watertight hatch into the galley below. This was eliminated on the *Smallwood* and was replaced by fixed insulated chambers, and a freight elevator from the vehicle deck to the galley area.

These ships have two-level loading, stern doors, bow visor with interior door, passenger elevators, escalators, and stabilizers. These features plus comfortable overnight cabins, reclining seat lounges, and day lounges, make for a safe and pleasant voyage.

As the vehicle decks are totally enclosed, a high volume ventilation system consisting of reversible fans (forward, aft, and amidships) can be programmed to clear vehicle exhaust while loading and disembarking.

During construction of the *Smallwood* the shipbuilders introduced the highest level of pre-outfitting of sub-erection units, performed under cover in heated shops. In spite of some occasional difficulties, this operation must be considered successful. *Smallwood*:

Design	Concept: K. E. Hansen, Copenhagen Development: German and Milne, Montreal, Quebec
Construction	MIL Davie Ltd., Lauzon, Quebec
LOA	587 ft-1½ in.
LBP	541 ft-2½ in.
Beam	82 ft-0 in.
Ice class	Northern Baltic 1A Super
Service speed	15 knots
Propulsion	4 diesels, 28,000 total bhp, twin screws, 2 thrusters forward, 2 aft
Passengers	1200
Cars	350
Tractor-Trailers	91

Nova Scotia to Maine. The Crossing between Nova Scotia and Maine is Canada's only international

Fig. 28 *Smallwood.*

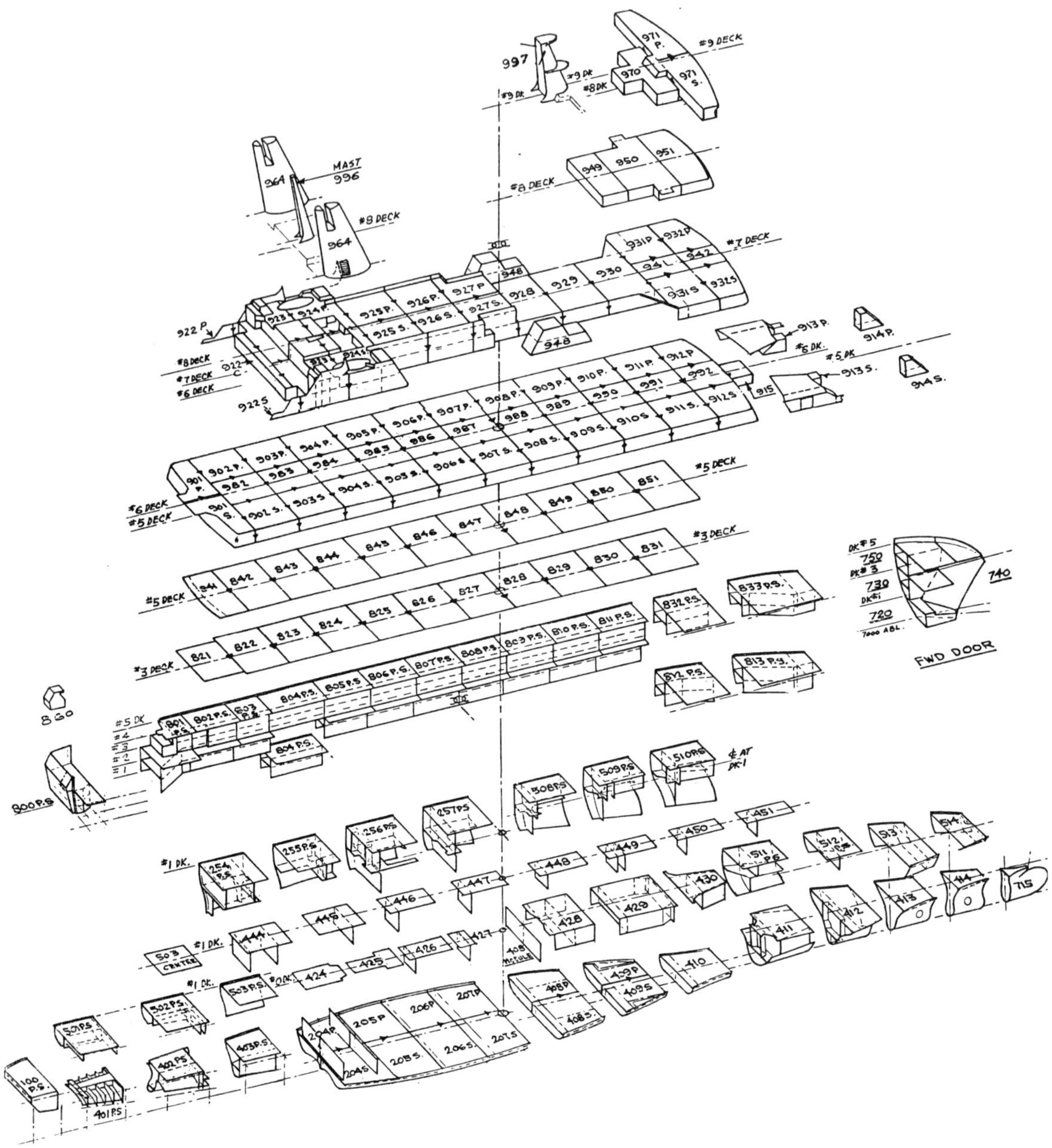

Fig. 29 *Smallwood*—Indentification of blocks.

ferry service. Today it is provided by a Yugoslavian-built ship acquired in 1982, but the route was inaugurated in January 1956 when the newly built ice-breaking ferry *Bluenose* was delivered. She provided a return-trip service three times per week for 21 years. During that time Ro/Ro technology developed far beyond the side-loading access of the *Bluenose*, and she was no longer able to provide a suitable service for the more modern tractor-trailers that were an important element of her payload. Today's *Bluenose* has two-level loading with stern doors and bow visor. *Bluenose* (1956):

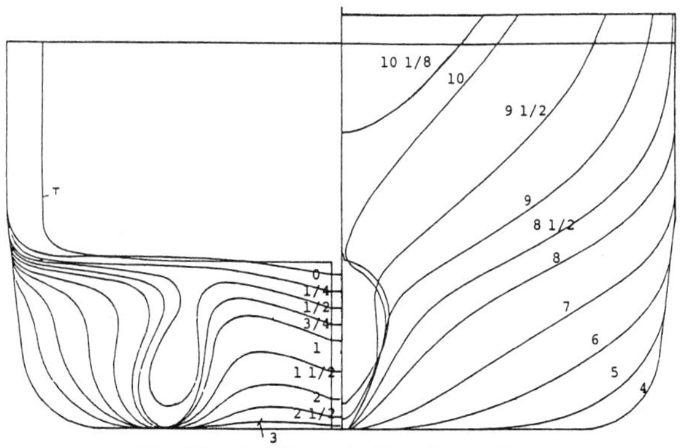

Fig. 30 *Smallwood*—final body plan.

Fig. 31 *John Hamilton Gray.*

Design	Milne, Gilmore & German, Montreal, Quebec
Construction	Davie Shipbuilding Ltd., Lauzon, Quebec
LOA	346 ft-0 in.
Beam	65 ft-0 in.
Service speed	18.5 knots
Propulsion	6 diesels, 12,000 total bhp, twin screws
Passengers	600
Cars	150

Closure

This chapter opened with a reference to Canada's original ferry systems, as exemplified by the native canoe.

Since then a variety of types has evolved to suit specific needs. Canadian designs to meet these needs are in many instances not much different from their counterparts in other parts of the world.

However, most of our designs are products of Canadian naval architects or adaptations of concepts originating elsewhere, modified to suit our needs, climate, and our interpretation of rules governing safety.

The history of ferry design in Canada shows that a small design community can develop innovative and elegant solutions to a wide range of challenging operational requirements.

George S. Gibb is a native of Clydeside, Glasgow, whose early training took place in the Fairfield shipyard. He started there in the mould loft and then entered apprenticeship in the hull drawing office. During World War II he spent four years in the Royal Air Force, after which he returned to Fairfield as an assistant manager of production. In 1954 he emigrated to Canada, which has been his home ever since. His Canadian maritime career has been marked by alternating periods of employment by shipyards and by ship owners, and on both coasts and the Great Lakes. Before he retired in 1989 he was a member of the National Association of Marine Surveyors (U.S.) and the Society of Consulting Marine Engineers & Ship Surveyors (U.K.).

St Lawrence horseboat ferry, mid-19th Century. Cattle forward, passengers aft. Note "four horsepower" amidships.

Inland Waterway Vessels

DONALD P. COURTSAL

Introduction

THE INLAND and coastal waterways of the United States, excluding the Great Lakes, consist of over 25,000 miles of channels that are commercially navigable. These channels extend along almost the entire length of the Gulf of Mexico and East Coast, the Mississippi River and its tributaries, the New York State Barge Canal and Hudson River System, the Columbia River System and several smaller river systems on all three coasts and the Great Lakes. The Mississippi River System, the Gulf Intracoastal Waterways and the other navigable river systems that empty into the Gulf of Mexico are commonly referred to as the Western Rivers System.

Based on statistical data contained in Ref. 1 for 1988, the latest year for which data are available, nearly 90 percent of the traffic (measured in tons) that was moved by barges on the inland waterways was moved on the Western Rivers System. The transport system that has developed to move this traffic is unique and is used now on other inland waterways, although usually with some modifications to suit local conditions. The remainder of the traffic is moved in a wide range of barges that are generally towed by hawser and there is little in common among them. In view of these factors, this study will be limited to the Western Rivers System.

The fifty-year period that ended in 1943 saw major changes in how cargo was transported on the western rivers. Among them was the completion of the system of locks and dams on the Ohio River and the Upper Mississippi, which made year round navigation possible, with operating drafts of nine feet. Reference 2 contains an excellent summary of the history of the development of these waterways.

By 1943, screw propellers and diesel engines had been introduced and the days of paddle wheels and steam propulsion were numbered. The Kort nozzle had been installed on a number of towboats and had provided significant improvements in performance. The construction of vessels had changed from wood to combinations of wood and steel, then to all-steel. Riveting had generally been replaced by welding and methods for mass production were being developed.

In a paper presented before the Waterways Division of the American Society of Civil Engineers in 1940 [3], Alex W. Dann noted that a century earlier the objective of river improvement was to provide 24 inches of water and that by 1940 we were operating at 9 feet. He concluded that there was every expectation that our rivers would someday carry barges with much greater draft and that wider and deeper channels, better barges and towboats, and more skillful crews would increase the amount of freight handled in each flotilla. The next fifty years did not lead to the deeper channels that Dann had forecast, but he was correct in the second part of his prediction. Better barges and towboats and more skillful crews have yielded great increases in the amount of freight handled by a single flotilla and that is one of the major accomplishment of the fifty-year period from 1943 to 1993.

Operations

Reference 3 noted that in 1940 the Western Rivers System consisted of approximately 6000 miles of navigable channels with depths exceeding 9 feet. There are now over 14,000 miles of channels in this system and 8700 miles of these have normal operating depths of 9 feet or more (see Fig. 1). Again from Ref. 1, we find that in 1947, 205 million short tons of cargo moved on the inland waterways. By 1988 this had increased to 672 million tons. Data compiled by the American Waterways Operators [4] show that this traffic is moved in a fleet of 27,046 dry cargo barges having a total capacity of 37.7 million short tons, and 4,043 liquid cargo barges with a total capacity of 10.8 million short tons. Propulsion of these barge fleets is accomplished by a fleet of 5188 towboats and tugs with cumulative installed power of 8,550,000 horsepower.

In their 1943 paper, V. B. Edwards and F. C. Cole [5] stated the following, "For the present, river transportation will be limited to moving its natural cargoes,

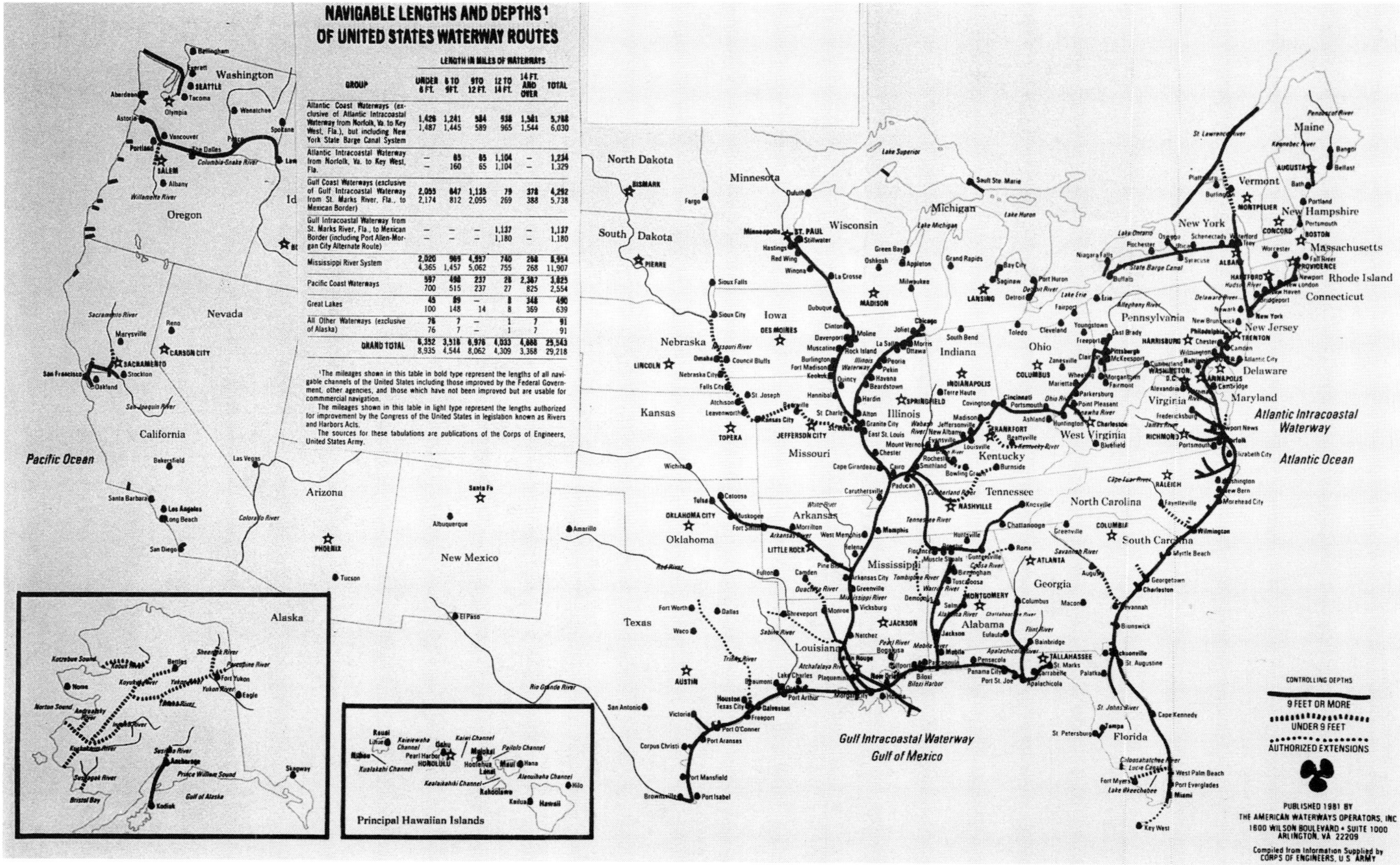

Fig. 1 Waterways of the United States.

Table 1 Principal products carried

Product	Percent of Total Tonnage Carried 1960	1988
Coal and coke	37.7	36.2
Petroleum and products	25.6	18.5
Sand, gravel, and stone	11.7	9.7
Grains	4.9	17.3
Iron and steel	3.2	2.6
Chemicals and related products	2.9	8.9
All others	14.0	6.8

bulky raw materials." Fifty years later, those are still the primary products moved on the inland waterways. The product mix, however, has changed following changes in technology and living standards in this country. As an example, Table 1 compares the percentage of total tonnage of various products carried on the Mississippi River System in 1960 [2] with those carried in 1988 [6].

The major changes were in petroleum products, grains and chemicals and related products. It is significant to note that, in absolute terms, the amount of coal and coke moved in 1988 was more than twice that moved in 1960 even though the percentage remained nearly constant. The amount of grain products, however, was about eight times greater in 1988 than in 1960.

Products moving on the Western Rivers System fall into one of three categories. The first, and by far the largest group, comprises regulated products moving under published tariff rates and conditions established by the Interstate Commerce Commission. Regulated products are best defined as those that can be counted readily. There are no restrictions on the number of different regulated products that can be carried in a barge or a fleet of barges. The second major category comprises unregulated products that are moved by carriers that are exempt for hire, i.e., contract carriers. These are almost exclusively bulk products and the only restriction imposed on their movement is that no more than three different products, exclusive of bulk liquids, may be moved in the same tow. This rather eccentric restriction is commonly referred to as the "Mixing Rule." The last category of products are those moved by private carriers. These products are owned by the carrier and have either been manufactured by the carrier or will be used by them in their manufacturing process.

Barges are moved on the Western Rivers System in groups that are usually referred to as "tows" or "fleets." (Earlier in our fifty-year period the term "flotilla" was commonly used.) Figure 2 shows two typical common-carrier fleets that have just completed a turning maneuver, which will be discussed later in this paper. The barges are secured to one another and to the propulsive unit with rope, wire and chain rigging. The propulsive units are called "towboats," although they invariably push and otherwise control the fleet from astern. The maximum size of the fleet is controlled by the channels in which they operate.

The channels are generally classified as either locking or open. On a locking river the channel depth and

Fig. 2 Typical common-carrier fleets.

flow of water are controlled by a series of dams, and traffic is moved past these dams through locks. In open-river channels the flow is largely unrestricted, but the depth is controlled by various means such as dredging, straightening bends, and constructing breakwaters (diagonal masonry deflectors). The Mississippi River below St. Louis, the Missouri River, and the Gulf Intracoastal Waterway east of New Orleans are examples of open channels. The Ohio River, the Mississippi River above St. Louis, the Illinois River, and parts of the Gulf Intracoastal Waterway west of New Orleans are examples of locking channels.

The size of lock chambers at the dams varies considerably and some dams have two chambers (in parallel) while others have only one. On the Ohio, the Upper Mississippi, and the Illinois Rivers the widths of all the chambers are 110 ft and the length of the long chamber is either 600 or 1200 ft. On the Monongahela and Allegheny Rivers the older chambers are generally 56 ft wide and 360 ft long, but new construction underway is providing the larger chambers that are standard on the Ohio. On the Gulf Intracoastal Waterway, chambers are 56 and 75 ft wide, and the lengths range from 260 to 1200 ft. Reference 2 contains a complete list of chamber sizes with the exception of those on the Gulf Intracoastal Waterway.

Operating drafts on these channels are nominally controlled by the limiting depth, as established by the Army Corps of Engineers, which means that under normal circumstances in a normal year the depth over the entire length of the channel will never be less than the limiting depth. Seasonal variations can have a major effect on this, particularly on an open river where the water depth is not controlled by dams. The Missouri River is a good example. The limiting depth, as established by the Corps, is 6½ ft from Omaha to Kansas City and 7 ft from Kansas City to the mouth of the river. During the spring and early summer the minimum channel depth is usually much greater than the limiting depth, owing to the spring floods, so barges can be loaded to deeper drafts to take advantage of the extra depth. From midsummer until the following spring, the minimum channel depth is usually the limiting depth and in some particularly dry years it may be less than the limiting depth. Barge operators keep a close check on the water depths in order to optimize their operations. Here again there is risk that must be weighed against the potential reward. If one loads to 9 ft in the face of falling water levels, the barge or barges could ground out and perhaps remain aground until the next high water season.

If we refer again to Fig. 2, a casual look might lead us to conclude that a barge fleet is a single rigid unit. Unfortunately, this is far from true. As noted earlier, the barges are secured to one another and to the towboat with a combination of rope, wire, and sometimes even chain rigging. Figure 3 is a photograph of typical rigging arrangements. If we consider the masses of the individual barges and the velocities at which they are moved, it becomes clear that the rigging must be relatively quite frail. When steering in nominally straight stretches of a channel, a fleet can be observed to whip noticeably if rudder angles of more than four or five degrees are used. The crews of river tows are constantly at work tightening rigging that has slipped or stretched. When maneuvering a tow, the pilot must bear in mind the amount of power and steering force that can be applied to the rigging without causing failure. Maneuvering tows successfully requires foresight and patience. When disaster strikes, the pilot may have to decide whether it is better to let the whole fleet get into trouble as a unit or take a chance on breaking it up and having individual units with which to cope. This requires judgement and experience and the ability to make decisions quickly. As with many such professions, hours of boredom are interrupted by moments of heart-stopping action. A more complete discussion of maneuvering tows on the Western Rivers is included in Ref. 7.

Channel conditions, lock sizes, the amounts of material to be moved, and the distribution of that material along the operating route all influence operating patterns. If a specific amount of material, such as coal or sand and gravel, is to be moved from one point to another in the same general area of a river, the development of an operating pattern is relatively

Fig. 3 Typical rigging arrangements.

simple. As the length of the move increases and expands over a wider range of operating conditions, the complexity of the operating pattern also grows. Fleet characteristics, size and number of towboats, and operating schedules are developed to maximize profit.

A regulated carrier operating from New Orleans to the upper reaches of the system has a more difficult situation. Consider the Pittsburgh to New Orleans route as an example. Fleets depart from Pittsburgh on a fixed sailing schedule and the number and sizes of the barges will vary from trip to trip. The largest fleet that will normally be moved on the Ohio (a locking river) consists of 15 jumbo barges (195 ft × 35 ft) or 22 standard barges (175 ft × 26 ft), or some combination of the two, and is limited by the 1200-ft × 110-ft lock chambers. The fleet passes through these long chambers as a unit in what is known as a single-locking. Many of the dams on the Ohio have chambers that are only 600 ft long, and fleets are passed through these chambers in two sections in a double-locking maneuver. This procedure is described in detail in Ref. 7.

As the fleet moves down the Ohio, barges are picked up and dropped off and usually the fleet grows as it moves downstream. In determining sailing schedules, the ideal frequency is one that matches traffic flow well enough to achieve a maximum size fleet by the time it reaches the confluence of the Ohio and Mississippi Rivers, at Cairo, Illinois. Since the Mississippi south of Cairo is an open river, the size of the fleet can be increased if barges are available and if the towboat is powerful enough to handle the larger fleet in the faster currents that are normally associated with the lower Mississippi. Barges from the Upper Mississippi, the Illinois and the Missouri Rivers are usually waiting to be moved south at Cairo so that the only factor that restricts the size of the fleet, other than the physical limitations of the channels, is the power of the towboat. Fleets of 30 to 35 jumbo barges are common on the Lower Mississippi and mixed fleets composed of standard and jumbo barges have involved more than forty barges. Fleets of this size, when fully loaded, carry more than 40,000 short tons of cargo as compared to the 20,000 tons carried by a maximum-size locking-river tow. The length of one of these large fleets often exceeds 1000 ft and the width may approach 250 ft, as great as any oceangoing ship in the world.

The amount of horsepower that can be used efficiently on a locking river is limited by a number of factors. First, fleet size is limited, therefore higher power does not permit bigger fleets but only higher speeds. Speed is limited, however, by relatively narrow channels, large numbers of landings, piers and other structures where wake may cause damage, and by water depths that are shallow enough to have a significant effect on fleet resistance, particularly loaded fleets. Finally, at least ten to fifteen percent of the passage time on a locking river is spent waiting to pass through locks or in actually locking through and during those periods engines are idling or operating at very low powers and higher installed power clearly has no value.

Towboats in the range of 3000 to 4000 hp are generally thought to be about the right size for locking rivers. Unfortunately a 3000 hp boat may have difficulty handling a maximum-size locking river fleet on the Lower Mississippi. The optimum plan then would be to use a 3000 to 4000 hp boat on the Ohio, and a 6000 to 8000 hp boat on the Lower Mississippi. That system is now common and is called a "turning operation." Figure 2 shows two boats that have just completed this maneuver. The one on the left is a 3200 hp boat while the one on the right is a 6400 hp boat. In cases where fleet size does not change, or changes very little, over the length of the trip, boats in the 4000 to 6000 hp range may be used over the entire length of the trip. Exempt carriers handling a common carrier tow, and fully-integrated unit tows (normally used to move petroleum products), are typical examples of this. The development of fleet integration and its advantages will be discussed in a later section on hull forms.

An important factor in building an efficient transport system on the Western Rivers was the development of the two-watch system of operation. Fifty years ago, larger boats had crews of from 20 to 25 people. Today, boats moving the larger common carrier tows will have crews of twelve or thirteen, and those moving the larger fully integrated unit tows will have crews of nine or ten.

The development of the two-watch system (six hours on, six off) was possible only because diesel-propelled towboats are treated by law as uninspected vessels and thus are not subject to Coast Guard crew requirements. This gave the operators the ability to develop watchstanding procedures that were consistent with safe and efficient operations in their particular environment. In addition, pilot houses and engine rooms were designed for one-person operation [7].

Vessel Types and Arrangements

In his 1950 paper, Albert J. Dawson [8] stated that the proportions of barges frequently reflect channel

Fig. 4 Standard open hopper barge fitted with two rake ends.

conditions, industry demands, and lock size. Things haven't changed much over the forty-year period since Mr. Dawson wrote those words. Dimensions have changed, but usually only as refinements to those that were already common in Mr. Dawson's day.

At the end of World War II the "standard" coal barge, which was 175 ft long and 26 ft wide, was the most common size. These carried about 900 short tons, had a depth of 10 ft, and were usually raked at each end. The barge shown in Figure 4 is an example. Barges of this size were originally built to fit the locks on the Monongahela and Allegheny Rivers, which were 360 ft × 56 ft. Barges of this size are still occasionally built today although the width has generally been increased to 27 ft and the depth to 11 ft. The increase in width was brought about by the realization that increasing beam is one of the cheapest ways to increase cargo capacity and that four feet of clearance in the lock chambers was more than was really required. The increased hull depth reflects the increased channel depth that became generally available as the years passed. Most of these later barges were built with one rake and one square end because it had become possible, in most cases, to turn the barges around before moving them any great distance and because this configuration increased both deadweight capacity and fleet speed, as explained later in the section on hull form.

The "jumbo" hopper barge, which now has become the standard of the industry, began to appear in significant numbers around 1950. These are 195 ft × 35 ft and 11 or 12 ft deep, with newer barges generally having the 12-ft depth. The dimensions of jumbo barges are based on lock chambers that are 110 ft wide and have lengths ranging from 600 to 1200 ft. Each such barge has a deadweight of 1400 to 1500 short tons and they are usually built with one square end and one raked end (semi-integrated). In recent years some 200-ft barges have been built. This was the result of the discovery that 600-ft lock chambers are actually 605- or 606-ft long and even a tow consisting of all-200-ft barges will fit, provided that the pilot brings it in gently. In an effort to take advantage of the positive effects of fleet integration, which will be discussed in a later section, some barges have been built with two square ends (box-type). This is more common with contract carriers where there is good control of the mix of barges in a fleet. The losses caused by having a loaded box barge on the head of a tow, will quickly eliminate any gains that may have been achieved through fleet integration.

Both the standard and jumbo units are used as open or covered barges, depending on what they are moving. The hatch covers that are fitted are weathertight, which means they will protect the cargo from precipitation, and are furnished in either a lift-off or rolling configuration. Lift-off covers are usually used for cargoes such as grain where cost is a major factor and where the operator can be certain that there will be a crane available at the unloading dock to handle the covers. Rolling covers telescope and can be positioned to expose half of the cargo compartment. If the barge has little trim, and if the covers have been reasonably well maintained, they can be moved by one or two people. Figure 5 shows a typical semi-integrated, jumbo hopper barge fitted with rolling covers.

Jumbo barges and standard barges built in the recent past, are usually fitted with watertight bulkheads

Fig. 5 Semi-integrated jumbo barge with rolling covers.

every forty feet between the collision bulkhead and the square end. These barges satisfy a one-compartment standard of subdivision. In the early years of the period being covered by this chapter, when standard barges were still being built in great numbers, many had only two watertight bulkheads, and these were the collision bulkheads at the rake ends. The entire midbody was one large compartment, and an underwater hole anywhere in that compartment would cause the barge to sink.

Tank barges can be divided into two general categories. Those that move relatively small lots of chemicals, or petroleum products, are similar in size to hopper barges and may move in fleets with hopper barges. The second category contains barges that move large quantities of liquid products point-to-point, usually on a contract basis. These barges operate in fleets known as unit tows, and have lengths ranging from 140 to 297 ft, and widths that vary from 50 to 54 ft. In a unit tow operation the barges remain together for the entire trip and the towboat waits with the barges while they are loaded and unloaded. This is in contrast to a common carrier tow, where barges are dropped off and picked up along the way, and the towboat continues on its way with a fleet that is constantly changing. The width of a unit tow barge is determined by the 110-ft wide lock chambers. Designed values vary between 50 and 54 ft as the result of each operator's analysis of the value of increased cargo capacity versus the cost of increased barge maintenance, and damage claims from the Corps of Engineers.

Unit tows operate in single- or double-width configurations depending on the requirements of the move in which they are employed. Most of these tows will use the Gulf Intracoastal Waterway west of New Orleans because that is the source of much of their cargo. Lock size on that waterway allows only single-width tows, so fleets that operate double-width elsewhere must be split up here. A second, chartered boat will then take charge of half the tow.

The length of unit-tow barges is set by the length of lock chambers and structural considerations. The length-to-depth ratio for river barges is limited by ABS to about 25 unless special arrangements are made, and a ratio of 30 is the absolute maximum. Since hull depths of 12 ft are normal, barge lengths are limited to about 300 ft if we wish to avoid increased depth or heavier structure. The maximum tow length that can be accommodated on the Gulf Intracoastal Waterway is 1180 ft. As noted earlier, fleets operating on the other major rivers must accommodate a 600-ft locking module. The leading section of a barge string will usually consist of a lead unit and a long box, each having a length of around 295 ft. This can vary from 290 ft all the way up to 297½ ft. The second section of the string is made up of the towboat, a trail unit, having the same length as the lead unit, and a short box. If we assume that the towboat is 140 ft long, which is common for the boats in this service, the maximum length of the short box would be about 150 ft. Normally the length of these barges ranges from 140 to 150 ft.

At the beginning of the period under review, most of the tank barges in operation were of single-skin configuration. As higher-value chemical cargoes began to be moved in quantity, double-skin barges began to appear. Leaks in barges are not uncommon and are caused by such things as groundings, collisions with other barges, striking locks or docks, and damage from floating trees and other objects carried along with the current. The double-skin barge protects the cargo from loss and contamination and is not much more expensive to construct. In addition, these barges are much easier to clean when changing cargoes, since all of the framing members are outside of the tanks. These arrangements were first seen in the 195-ft barges because they handled much of the chemical cargoes. In 1970 the U.S. Coast Guard published its regulations for handling hazardous cargoes [9]. Two of the three categories of barges listed there were required to be of double-skin construction. Virtually all of the 195-ft barges built after that were of that construction because the operators needed the versatility associated with these barges and, as noted earlier, the construction costs were not much greater when the barges were built in quantity. Double-skin construction did not move as quickly into the unit tows because they were involved primarily in moving fuels and other lower-value petroleum products, so product contamination or loss were not as important. As environmental regulations became stricter, and enforcement more common, leaky barges became a more serious consideration, and the move to double-skin construction in unit tows became more rapid.

The introduction of the Coast Guard rules for carrying hazardous cargoes brought about a greater awareness of the stability characteristics of barges. The intact stability requirements for all three categories of barges is the same and is a function of the area in which the barge is to be operated. Type II barges have to meet a one-compartment standard in the damaged condition and Type I barges a two-compartment standard. The one-compartment standard is easy to meet in the standard 195-ft double-skin barge, and most have been certified as Type II hulls as a matter of course. The two-compartment standard is much more difficult to meet and only those barges having a particular need for this certification have obtained it.

The most common cargo-handling system on a tank barge is either a deck-mounted centrifugal pump or a deep-well pump driven by a diesel engine. There will usually be a single transverse loading and unloading header connected to a common line connecting to the cargo tanks. Much more complicated systems have been installed, but these are designed for a specific product, or products, and are as numerous and varied as there are products to be handled. Many barges are fitted with heating coils with steam usually being supplied by the unloading facility.

There is a wide range of special purpose barges for moving things such as cement, flour, sulphur, and sand and gravel. The last of these is the only one where significant numbers of barges are involved. Deck barges with 3- or 4-ft coamings are common in this service because the cargoes are frequently loaded with a lot of water still in the product and this permits the water to drain off through freeing ports in the coamings. In cases where beam is limited, as in the Pittsburgh area, this extremely wet cargo can pose a stability problem in design owing to the difficulty of determining its center of gravity. These barges range in length from 100 to 195 ft, and are frequently fitted with two rake ends.

With decreases in the size of both propulsion machinery and crews, the size of towboats has generally been reduced over the period under study here. This has occurred in spite of the dramatic increase in the power of these vessels. The dimensions of new boats today are normally a direct function of the power of the engines to be installed. The length of the stern is a function of the propeller diameter which in turn is directly related to the power to be absorbed and to draft limitations. The overall length of the boat is usually a function of the amount of fuel to be carried, which again is directly related to engine power. In cases where only limited fuel capacity is required, the need to be able to control the trim of the boat may establish the minimum acceptable length. The beam of the boat is controlled by propeller diameter, where two or more propellers are to be fitted, and here again this relates directly to engine power. Most river towboats are fitted with at least two propellers in order to achieve the required maneuverability. Channel limitations will also affect dimensions but are usually of primary importance only in selecting hull depth and propeller diameter, both of which are related to operating draft.

Twin-screw towboats seem to be the optimum compromise between the requirements for maneuverability and the desire for propulsive efficiency. The limitations of channel depth will sometimes force the fitting of additional propellers if higher power is needed. As an example, the largest propeller diameter used on boats for 9-ft channels is about 10 ft. This size can successfully absorb 3500 shp, thus a 7000 shp boat is about the largest twin-screw boat that can operate efficiently in these channels. Triple-screw boats of about 10,500 shp have, however, been used with success in 9-ft channels.

The weight of the hull and the weight of fuel, or of fuel and compensating ballast, are the two major weights to be considered in establishing towboat arrangements that will provide satisfactory control over draft and trim in all operating conditions. The weight of engines and reduction gears is normally less than five percent of the total fully-loaded displacement, and their location should be determined by such things as length of shafting, foundation design, and quarters arrangements rather than any effect they may have on trim.

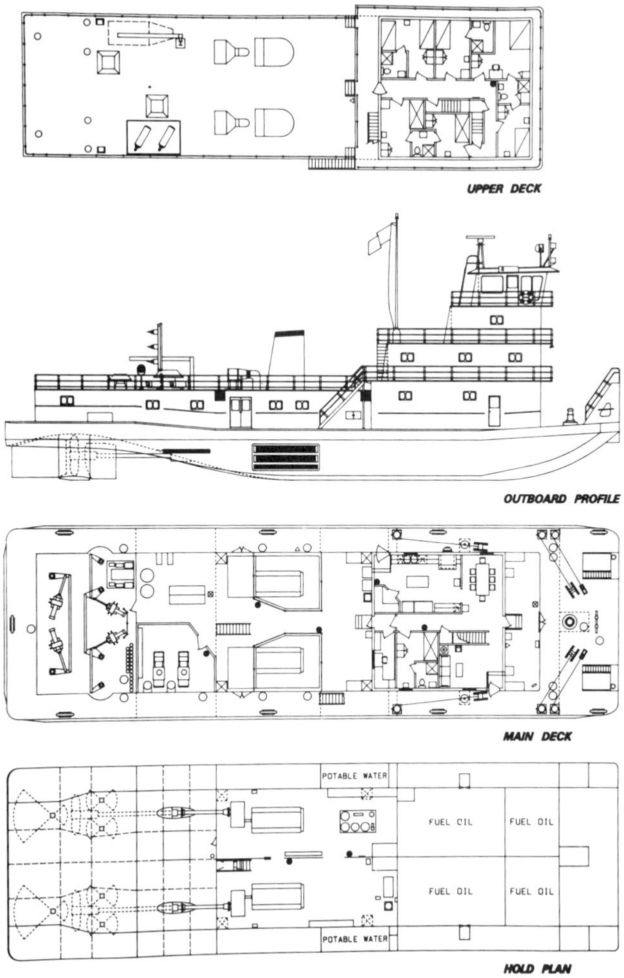

Fig. 6 General arrangement—5000 shp towboat.

Figure 6 shows the general arrangement of a recent 5000 shp twin-screw towboat and is included here to help you visualize some of the points that will be discussed. The engines have been located far enough aft to permit all of the quarters to be located forward of the engine room, yet far enough forward to provide adequate access to the space under the reduction gears. Owing to the tunnel stern configuration, there is a large excess of weight over buoyancy in the aft portion of the boat. In order to compensate for this, the fuel and ballast tanks are located well forward of the center of flotation. As fuel is used, ballast must be added to control trim. If this is not done, the boat will trim by the stern and the crew will have to slack the face wires if they are to avoid breaking them as the bow rises. The boat's performance will also be affected if the trim is allowed to become extreme. When the boat refuels, it is important to remove the ballast in order to keep the stern submerged. If the stern is allowed to rise too much, the propellers will suck air when backing and the boat will lose a major part of its backing power.

As noted earlier, an effort is made to locate crew quarters forward of the engine room. This has become more important as the power of boats has increased. No matter how vibration-free boats are when going ahead, they will inevitably vibrate badly in the stern when backing hard, particularly when the flanking rudders are in use. The propellers will also generate a lot of cavitation noise under these conditions. When a fleet is downbound on the Lower Mississippi, the towboat may spend hours backing and flanking around the bends. The crew has a much better chance of getting some rest if their quarters are located forward. An effort is also made to place public spaces between the engine room and the sleeping spaces. The noise generated by modern, turbo-charged diesel engines is piercing, and sound insulation is more effective if augmented by some intervening space.

Stability characteristics of larger towboats have usually been more than adequate owing to the beam required for propeller separation. In smaller vessels, those used around landings and fleeting areas, stability can be a problem. These vessels must have high pilot houses to provide visibility over empty barges and they tend to be rather narrow in order to fit into tight spaces. The result is a vessel that is tender and can be dangerous when fuel is low.

In the earlier years of the period being covered here, the size and arrangement of machinery in towboats was such that most of the boats could not survive an underwater puncture in the engine room. This was not thought to be a major problem as it was usually possible to beach a boat before it sank. Technological developments led to smaller and more compact engines and auxiliaries and it became possible to shorten engine rooms enough to permit boats to stay afloat even after the engine room was holed. This became more important as channels were deepened and current velocities increased, particularly on the Lower Mississippi, and it became increasingly difficult to beach a boat after trouble developed. Referring to Figure 6, note that the shaft alley and the lower engine room are separated by a bulkhead. This is typical of many of the boats in service today. Most of them have a watertight door that connects these spaces. The problem here is one of trying to keep the door closed during normal operations. The engineers usually check the shaft bearings at least once during a watch and many of them can't be bothered to close the door after making their inspection. The chance of getting this door closed after flooding begins is small and these boats cannot survive flooding both spaces. Some owners have eliminated this door, but provide access to the shaft alley from the main deck.

Hull Form

As noted earlier, this fifty-year period has been one of refinement and development rather than new discovery. This has been particularly true in the area of hull form. The material included in Dawson's 1950 paper [8] is still the basis for much of what is done in barge design today. Hopper barges and 195-ft tank barges usually have a six-inch bilge radius rather than the one-foot minimum advocated in the reference, but this is based on economic considerations and more concentration on fleet integration rather than any disagreement with the basic data.

Reference 8 contains some excellent data on the effects of fleet integration. In common-carrier tows partial integration can be accomplished by aligning semi-integrated barges (barges with one rake end and one square end) head to head and stern to stern. This assumes that all of the barges in the tow are either loaded or empty. Under these circumstances, an improvement of about three to five percent in operating speed can be realized.

Much bigger improvements can be realized in fully-integrated unit tows (rakes fitted at ends of the fleet only) where the barges always operate together and are normally loaded to the same drafts. In some cases pins are fitted in order to eliminate any small misalignments. In this case speed improvements in the order of eight to ten percent are possible when compared to barges fitted with two rake ends. Substantial increases in deadweight capacity are also achieved with fleet integration, as noted earlier.

The most economic length of barge rake is a function of operating speed. If a barge is being moved only short distances and spends a great deal of time being loaded and unloaded, then there is little point in giving up carrying capacity in order to achieve a higher speed. Hopper barges and 195-ft tank barges are usually fitted with a rake having a 40-ft radius and a length of 25 to 30 ft between the headlog and point of tangency to the base line. The bilge radius along the rake is almost invariably 30 inches. These barges generally operate in the 7 to 8 mph speed range.

Rakes fitted on the lead units of fully-integrated unit tows are much longer and have a much larger radius. Speeds of these tows range from 10 to 14 mph. These tows may be fitted with scow bows (flat all the way across) or spoon bows, which have a molded shape typical of oceangoing vessels. Both forms perform well in this speed range with the scow bow having a small advantage in single-string towing at all speeds and in double-string towing at speeds above 12 mph.

The hull form of towboats has changed more than that of barges over the past fifty years. At the beginning of the period the amount of power provided by the available engines could be absorbed comfortably by a propeller having a diameter small enough to fit beneath the normal waterline. This allowed the use of moderate tunnels and exit slopes that were horizontal or nearly so. Hull sides met the bottom in a gentle radius without any problem of air being drawn into the propeller. Bows were ship-shaped under water and flared out to a square shape at the deck. There was little parallel middlebody. Construction of these shapes was labor-intensive but this had not yet become a problem.

As engine development made greater power available, the size and loading of propellers increased. Since the draft of these vessels was limited to about 9 ft, the only way for the propeller to move was up. Maximum propeller emergence in modern boats is about 18 inches. This has led to much longer tunnels forward of the propellers in order to lead water uniformly into the propellers without turbulence. In addition, the lower edges of the sides of the hull forward of the propellers are raised as far as possible to promote uniform flow into the propellers while still maintaining enough depth at the side to prevent air from being sucked in when backing.

The tunnels aft of the propellers must be long enough to provide reasonably flat exit slopes while again maintaining sufficient depth at side to prevent air from being pulled in. The gentle radius at the point where the side meets the bottom has given way to a hard corner because it is more effective in preventing air inflow and does not have a major effect on ahead performance.

The ship-shaped bow was long thought to provide better performance. Recent model tests have shown that the scow-type bow performs somewhat better behind a fleet and this is probably because it fits into the flow patterns of the fleet without generating as much turbulence. The length of the rake radius is unimportant as long as the boat is operating behind a fleet. While boats don't earn money when running free, some consideration must be given to that eventuality and some reasonable rake radius and length should be selected to accommodate this.

Model tests have been an important tool in the development of towboat designs, particularly in the design of appendages. In 1943, at the beginning of this fifty-year period, Dawson [10] reported on an extensive testing program, both models and full scale, that was devoted to a study of Kort nozzle propulsion systems. During the middle years of the period under review, a lot of fleet model tests were made to compare various hull forms and fleeting arrangements and to try to determine speed and power relationships. Techniques and facilities were developed for testing these fleets in shallow water. These tests were expensive and, for reasons described in detail in Ref. 7, the speed and power relationships developed were not directly applicable to full-scale fleets. Testing in the later years of this period returned to more limited investigations of appendages, including modifications to the Kort nozzle.

As a result of these later tests, nozzle lengths were reduced, the apron at the bottom of the leading edge of the nozzle was removed, propeller tip clearances were reduced to ½ inch or less, and it was learned that locating the shaft strut aft of the propeller improved ahead performance by a small amount. The strut-aft configuration has never been adopted generally, however, because of factors that tend to offset the improved performance [7]. One of the more important developments from this period was a Kort nozzle with significant improvements in backing performance and only relatively minor losses when going ahead. Reference 11 contains a more detailed discussion of this development with drawings of the shape of the new nozzle.

Almost all river towboats are fitted with a steering rudder aft of each propeller and two backing rudders forward of each. Reference 10 contains excellent photographs of a typical rudder arrangement and Figure 6 of this chapter also shows the rudder arrangement. The backing rudders have a strong negative effect on ahead performance but are absolutely necessary for satisfactory backing ability. The steering linkage for the backing rudders is such that they will have a toe angle of about 15 degrees away from centerline when the rudders are amidships and will be parallel to one another when hard over. This arrangement reduces the losses in ahead performance.

Hull Structure

At the close of World War II virtually all barges were built with transverse framing. Gradually this has changed to nearly all-longitudinal framing or to a combination of the two systems. This change has been made to accommodate new fabrication equipment and techniques and in a response to a need for greater longitudinal strength. References 2 and 7 include typical midship sections and structural plans of some recent barges.

While wave loadings are not a factor in the design of river equipment, still-water bending moments can be much larger than would ever be permitted in oceangoing vessels. Cargo handling is usually done at terminals where the only objective is to load or unload the barges in the shortest time. If a barge buckles owing to poor partial cargo distribution, it is simply moved back to the holding fleet and the owner is notified. The development of high-capacity material-handling equipment has made this problem even worse. Before its introduction, barges were loaded and unloaded in at least two passes, and usually three. With this new equipment, however, barges are unloaded in a single pass, and the resulting hull girder stresses are much larger.

The U.S. Coast Guard regulations for carrying hazardous products introduced longitudinal strength criteria for tank barges that are to transport products that must be carried in Type I or Type II hulls. This standard assumes a pinnacle loading that might result if a loaded barge were grounded at high speed and driven up the bank. A complete description of these requirements is included in Ref. 7.

Since the loading of hopper barges is frequently volume-controlled, the naval architect makes every effort to enlarge the cargo holds. For this reason, the depth of the doublebottom and the width of the wing walls are kept small. There are no transverse bulkheads in the cargo space and this leads to transverse framing. The doublebottom depth on a typical hopper barge has been held to 14 or 15 inches and, bearing in mind that the unsupported span is about 27 ft, it can be seen that the transverse strength of these hulls is a challenge for the naval architect. Reference 7 includes a detailed discussion of this problem.

The subject of designing for impact and other local loads is one in which there are probably as many opinions as there are people involved. It is pretty universally accepted that if a barge is to be unloaded regularly with clamshell buckets, the wing walls should be transversely framed, particularly if they are sloped. I believe that the rules of the American Bureau of Shipping represent a good minimum only for barges in light service. If a barge is going to be doing a lot of locking, then corners and bulkheads need to be protected with heavy wear plates. If a barge is going to move corrosive materials, such as salt, then hopper plating should be much thicker.

The two most important considerations in towboat structures are local loadings and vibrations. Over the years, as power increased, the depth of the aft portion of the hull decreased, owing to the larger tunnels, and the local loads on nozzles and other appendages increased. It was important to increase the stiffness of the after section of the boats in order to deal with the vibratory loads. This was accomplished by raising the main deck aft, fitting trusses in the steering engine room, and using thicker plating and closer frame spacing. Continuous welding was generally used in attaching stiffeners to shell in this area and in all of the foundation structures for the appendages.

The structure of Kort nozzles evolved as propeller loading was increased. Ordinary steel could not withstand the cavitation in way of the propeller tips, so stainless steel bands were installed. The welding on the surface of these bands had to be ground smooth to avoid high rates of erosion. All of the welding on the back of the nozzle throat had to be continuous in order to keep the nozzles together. The structure that connected the nozzle to the hull had to run uninterrupted to the main deck in order to avoid cracking and other failures.

The nature of the service in which a towboat operates requires that it be capable of absorbing repeated impacts from almost every direction. The faces of the towing knees, headlogs, sternlogs, and the four corners of the boat are built of heavy plate that is strongly reinforced and often backed with concrete. Three-quarter-inch formed plate fenders filled with concrete are usually installed along the sides and across the stern. The use of concrete makes repairs more expensive, but the frequency of need is greatly reduced.

Main Propulsion

According to George Grunthaner [2], sternwheel towboats were superior to many propeller-driven towboats until the advent of the Kort nozzle. Even after the coming of the twin-screw, Kort-nozzle boat, operators found that the sternwheeler was turning in a better performance in downstream towing. Those were the days in which skiffs were used to run 1200-ft check lines to trees on the bank. Nevertheless, propeller-powered boats replaced paddlewheelers in the late 1940s and early 1950s. This came about because of the higher power that was available from diesel engines and the smaller machinery space required. In addition, fewer craftsmen were available to build good paddlewheels.

Because of the scarcity of diesel engines during World War II, a number of propeller-driven towboats were built with steam reciprocating engines. Despite the less expensive, lower grade fuels associated with these boats, their construction ceased as soon as diesel engines became available again. In addition to the reasons noted above, diesel-powered vessels were uninspected under the law and could operate with smaller crews than could steam-powered vessels, which were included as inspected vessels. Some of the steam-powered vessels built in the 1940s continued in operation until the late 1960s and, despite the handicaps noted above, turned in good performances for their owners.

During most of the period under review here, the diesel engines used in towboats were locomotive engine derivatives or types of engines used in trucks and construction equipment. Since both of those markets are large and international in scope, manufacturers spent a lot of money on research and development in order to remain competitive in their markets. The towing industry benefitted from this and a lot of progress was made over the years in fuel efficiency and in engine power-to-weight ratio. The power available per cylinder also rose steadily during this period.

Until the first fuel shortage in the mid-1970s, the price of diesel fuel was low and no real consideration was given to the use of alternative fuels. While many of the engines in operation at the end of the 1970s were two-cycle engines and not adaptable to heavier fuels, there were some four-cycle engines in service. In the early 1980s consideration was given to using blended fuels in these engines. The American Commercial Barge Line Company conducted an extensive test program on the use of blended fuels. R. Peter Spock reported the results in Ref. 12.

The same author later reported [13] that maintenance problems resulting from the use of blended fuels in four-cycle engines designed for diesel fuel had led to the abandonment of that program. This later paper reports on the installation of engines designed for the use of heavy fuels in the towboat *Bill Elmer*. Since that paper was written, a number of vessels have been powered with similar engines and it is my understanding that these installations have been successful. These engines are not substantially heavier than standard diesel-fueled engines although they do require a little more space for the required fuel-treating equipment. A drawback is that the fuel capacity and operating area of the boats must be related to the availability of fuel.

Towboats are normally fitted with reverse-reduction gears, with disk clutches being used on smaller high-speed engines and inflatable clutches on the larger engines. Propulsion shafting is usually designed to American Bureau of Shipping standards and vessels that operate on the Upper Mississippi or Illinois Rivers will often have shafting, rudder stocks, propellers and reduction gears strengthened for operating in ice. Both water- and oil-lubricated bearings are used in the struts. Reference 7 includes a discussion of the advantages of each.

Auxiliaries, Outfit, and Equipment

The electrical system on modern towboats generally consists of two diesel-driven generators either of which is capable of carrying the full load. Engines are either air-cooled or water-cooled through skin coolers or heat exchangers mounted on the hull. Power distribution and generator controls run the gamut from being basic (with generator controls mounted on the generators and power distributed through separate panels) to complex switchboard arrangements that may even include automatic generator-starting and load-transfer equipment.

Steering systems are usually electro-hydraulic, although there has been some use of air operated systems in the past. Because of the complex rudder arrangements that are normal on river towboats, steering systems are usually manufactured by the boat builders. Hydraulic packages and cylinders will generally be purchased, but all of the linkages and cylinder mountings will be designed and built by the yard. Mechanical linkages are commonly used for control owing to the relatively short distances between the steering engine and the pilothouse; but air systems are also used. A more detailed discussion of steering systems can be found in Ref. 2.

Fluid systems on towboats also vary widely. Even if we ignore the differences caused by the use of heavy fuels, we find a wide range of approaches even on the boats burning ordinary diesel fuel. Some owners insist on separate day tanks for fuel oil with service pumps, special filters and water separators. These boats will also have lube oil centrifuges and separate bilge, ballast and fire pumps, frequently with backup systems. On other boats the main engines will take fuel directly from the bunkers with only the manufacturer's filters and strainers in the line. Lube oil is not centrifuged, but simply changed when it is no longer within acceptable limits. Bilge and ballast pumps may be no more than portable electric units. The more complicated and redundant systems generally will provide more reliability, and each owner must decide what level of reliability is appropriate. The question of crew size and time available for maintenance will also influence this decision. Reference 2 includes some detailed descriptions of piping systems and some of the factors that should be considered in their design.

Air conditioning did not exist on towboats at the beginning of the period we are studying, but has become more common as the years have passed. Living quarters, mess rooms, pilothouses and sometimes the galley are usually air conditioned on most of the larger boats being built today. These systems can be simply electrically-powered window units or central chilled air or chilled water systems. Space heating on the smaller boats is normally electric and on the larger boats is accomplished by steam or hot water.

Conclusion

During the past fifty years, the amount of cargo moved on the Western Rivers has more than tripled. Moreover, the extent of the navigable channels has more than doubled. While the physical changes in the waterways were significant, technical improvementrs in equipment, and operating techniques have had an even greater beneficial impact. Edwards and Cole [5], in closing their paper stressed the growth in the use of the scientific approach to the solution of problems. As they put it, "Sound design is often the result of something akin to genius, but far more often it is the reward of painstaking and thorough investigation." The truth of that statement has been apparent during the past fifty years and I suspect will also be true during the next fifty.

Acknowledgments

I am indebted to those many long-time colleagues who worked with me in learning about, and developing many of the concepts and ideas described above. James Nivin of American Commercial Barge Line was particularly helpful in furnishing information on the design of backing nozzles and on the installation of heavy-fuel engines.

I appreciate the cooperation of Viking Maritec, Inc. for providing access to photographs acquired from the Dravo Corporation. I am also grateful to the American Waterways Operators for providing a copy of their map of the waterways of the United States.

Finally, I am grateful to my wife, Frances Chase Courtsal, for her able assistance in sorting out my exposition. Without her contribution, one would have great difficulty in understanding the points I have tried to make.

References

1 "Waterborne Commerce of the United States," Calendar Year 1988, Part 5, National Summaries, Water Resources Support Center, Department of the Army Corps of Engineers.
2 Grunthaner, G. L., "Commercial Transportation on the Inland Waterways," SNAME, 1962.
3 Dann, A. W., "Modern Developments in Towboats and Barges for Inland Waterways," Waterways Division Program, Spring Meeting of the American Society of Civil Engineers, 1940.
4 "Some Facts You Should Know . . . From America's Inland and Coastal Barge and Towing Industry," The American Waterways Operators, 1989.
5 Edwards, V. B., and Cole, F. C., "Water Transportation on Inland Rivers," *Historical Transactions, 1893–1943*, SNAME.
6 "Waterborne Commerce of the United States," Calendar Year 1988, Part 2, Waterways and Harbors—Gulf Coast, Mississippi River System and Antilles, Water Resources Support Center, Department of the Army Corps of Engineers.
7 Courtsal, D. P., "The Marine Business in the Central United States," SNAME, 1971.
8 Dawson, A. J., "The Design of Inland Waterway Barges," SNAME, 1950.
9 "Certain Bulk Dangerous Cargoes on Unmanned Tank Barges," Title 46, Code of Federal Regulations, Subchapter O, *Federal Register*, Vol. 35, No. 38, Part II, Feb. 25, 1970.
10 Dawson, A. J., "Recent Towboat Developments with Particular Reference to Kort Nozzle Propulsion," SNAME, 1942.
11 Oosterveld, M. W. C., "Ducted Propeller Systems Suitable for Tugs and Pushboats," *International Shipbuilding Progress*, Rotterdam, Vol. 19, No. 219, Nov. 1972.
12 Spock, R. P., "The ACBL Blended Fuel Program—An Overview," Winter Meeting, Great Lakes and Great Rivers Section, SNAME, Jan. 28, 1982.
13 Spock, R. P., "Repowering the M/V *Bill Elmer*," Work Boat Show, Jan. 19–22, 1984.

Donald P. Courtsal is a Vice President of SNAME. Following graduation from the U.S. Coast Guard Academy, he went on to MIT, where he earned a Master's Degree. He spent seventeen years in the Engineering Department of the Dravo Corporation designing the full range of equipment for the Western Rivers, and rising to the rank of Chief Marine Engineer. In 1975 he was appointed Vice President of Dravo and General Manager of the Engineering Works Division. In this position he was responsible for all of the construction of marine equipment produced by Dravo during the next seven years. He then served two years as Treasurer and one as Senior Vice President responsible for manufacturing activities.

Tugs

Donald C. Hankin

Introduction

It is doubtful that there is anything afloat more uniquely interesting than those little tugboats. They come alongside a giant ship and with little apparent effort gracefully guide them into or out of their berthing places. They are seen alongside a huge barge or on the end of a lengthy hawser towing a barge far out at sea. They attract the interest and imagination of seamen and landlubbers alike.

In years gone by, all tugs seemed to be alike and, for the most part (aside from size), they were. Crewing arrangements depended on the tug's service (harbor, coastwise, or deep sea towing). There were "lunch box" tugs whose crews reported daily at a scheduled hour and worked until their assignments were completed. There were the double crew boats manned for a seven-day period and then rotated; and there were the boats that would keep a crew aboard for thirty and sometimes sixty days. In all cases the men who manned them were seasoned sailors, excellent boatmen who could handle their craft with expertise, and loved their boats and maintained them with pride. Most tugs were a joy to behold, and were maintained by the crews themselves. Deck hands could be seen continually polishing brass, cleaning and painting, and above all keeping the captain's coffee cup full. Engine rooms were the private domain of the engineers and oilers, and no deck hand would enter without permission. Engines were maintained and repaired by the engine crew, who were proud that their tug worked the greatest number of days without the need to go into a shipyard for repairs. Engine rooms were maintained and painted, with all auxiliary machinery kept in good working order.

The accommodations were small but adequate. Except for the "lunch box" boats, there was usually one room on the upper deck for the captain and mate to share, and four rooms on the main deck for deck hands, cook, engineer and oiler. The galley was usually at the after end of the deck house; it was fitted with an oil burning stove, a sink with a hand pump, a table with stools and a small refrigerator. Most cooks were more than competent and the best meals anywhere were prepared and served on a tug boat.

Until after World War II there were working in the harbors many tugs that would fit this description. They had wooden hulls, steam reciprocating engines, and coal burning Scotch boilers. Diesel powered tugs began to appear in the early thirties, though struggling for acceptance. In both cases, engines were direct reversing at the engine control. Signals were passed from the pilot house by hand-pulled levers that operated a gong and bell in the engine room. The harmony of control was always more than fascinating. On many occasions the engineer would maneuver the engines before the bell signal arrived because he knew what the pilot wanted. When making up to or alongside a barge, the engineer would maneuver the engine to allow the deck hands to take in the slack in the lines without any bells from the pilot. I have seen occasions when entering a slip that the engineer would maneuver and execute a perfect landing at the proper spot on the pier for the deck hand to heave his line without any bells from the pilot. Such was life aboard the "bell boats."

Technical Developments

There were major changes in the tug boat industry during the post World War II years and these changes were brought on by developments during the war. The national economy was growing, and many of the industrial tools needed replacement. Further, there was now a new view of machinery and manufacturing methods. Where in the pre-war years, wood hulls were spiked and steel tugs were riveted, now new and better welding methods were available. Layout, lofting and fabrication procedures rapidly changed. No longer were templates coming from the loft with rivet hole layouts; and no more were plates sent to the fabricating shop to punch those holes. Now the plates were laid out from paper templates, and burned to shape with trim margins allowed for erection. Erection was by

Fig. 1 Defense Plant Corp. (DPC) tug DPC56—built October 1943 by Ira S. Bushey & Sons, Inc., 85 ft × 24 ft × 10 ft-6 in., 700 bhp, 8 cylinder General Motors 567A main engine.

modular section consisting of bow, fuel tanks, engine room and stern sections. Modules were turned during assembly to allow downhand welding and sand blasting before aligning the sections on the building ways.

Hull forms also changed. Where previously tugs were designed to be well rounded, sleek, full molded vessels, they now became of triple chine or double chine and in many cases single chine construction. No more rolled or furnaced plates. Sheer strakes became straight sided and hulls were constructed without attached steel fender systems. Sharp stem and rounded sterns were gone. Economy of construction became the rule, and thus we entered the fifties.

Shipyards, large and small, were searching everywhere for contracts. Some turned to scrapping and salvage and were placing surplus machinery on the market that made ship owners think of new boats. Then began the rush to new construction.

It was possible in the late forties to buy an entire U.S. Navy PC propulsion system for $2500. This included an engine, a reverse and reduction gear, heat exchangers, oil coolers, auxiliary pumps, silencers, diesel generator set, switchgear, and any or all piping.

Because there were many submarines being sold as surplus, owners and architects as well as engine builders turned their thinking to diesel electric propulsion. The competition that followed was largely between the sale of reverse, reduction gear and diesel electric propulsion, and the race for power was on. Where pre-war tugs were satisfactory at 400 to 600 bhp, we were now building tugs up to 1600 bhp. Previously important design standards were gradually eroded. In earlier years much care was taken to produce a hull form combining low wave-making resistance and high propulsive coefficient. Now our interest turned to putting the maximum possible horsepower into any given space. We also tried to increase bunker capacity to allow the tug to stay at work for longer periods.

These radical changes required a retraining of crew personnel. Railroads were big buyers of new boats and were moving from steam to diesel so rapidly that operators and maintenance personnel were not able, at first, to keep pace. Captains had to learn the difference between a slow bell to the engineer and the surge of clutch engagement from the new pilot house control. Manila lines parted, bitts were torn from the decks of barges, and deck personnel were injured. The engine crew complained about the noise of the engines; no more the pleasant hiss of steam blow-by or the quiet thump of a condensate pump. Ear muffs became standard gear and ear damage complaints were increasing. Engineers, with minimal training were at a complete loss until they gained sufficient knowledge of the new machinery. More upsetting were the problems of broken oil rings or cracked pistons and cylinder liners. Maintenance personnel, who for years, were hand lapping brass bearing shells with files and scrapers, were now installing manufacturers' precision bearing shells, and setting up bolts and nuts to precise torque wrench specifications. Despite all, in short order, training and education prevailed and we were on our way into a new era of tugboating.

Because of the increased power, manila lines were found to be inadequate. New products soon appeared on the market. Nylon, Dacron and polypropylene lines were lighter and stronger; they were smaller in circumference and easier to handle. They soon replaced manila; and deck hands rapidly learned the new methods of splicing. Soon a new problem arose. Because there was no more scrap manila, the deck hands had no material from which to weave side fenders and bow fenders. This gave birth to a new industry. Seeing the need, some companies soon were manufacturing laminated fenders from scrap truck tires, and the rubber industry began extruding preformed, hollow, square section fenders that were held

Fig. 2 Tug *McAllister Brothers*—built December 1958 by Jacobson Shipyard, 95 ft × 25 ft-10 in. × 13 ft-6 in., 1600 bhp General Motors 16-567-C engine.

Fig. 3 Tug *Jabbar*—built 1978 by Main Iron Works, 126 ft × 36 ft × 18 ft-7 in., 5750 bhp, 900 rpm model 16-645-E7 Electromotive Marine Diesel engine.

in place around the stem with chain attached with shackles to pad eyes welded to the hull. The square oak timbers that in earlier days were secured to the bulwark rail were soon replaced with laminated or molded fenders. "Bum boats" were a well established business in the major harbors. They would buy, and sometimes borrow without the owner's consent, used manila to weave bow fenders and it was not uncommon to see a scow adrift when someone had "borrowed" the mooring lines. Gradually the bum boats disappeared as the industry changed.

Tug boat navigation until the early fifties was only by magnetic compass. In 1952 I made a trip delivering a new tug from Camden to Jersey City, N.J. Before departing the Delaware River we had the compass adjusted, and while heading north on the coast we found the compass to have an eight degree error on that heading. Our radio was a single frequency company unit that could be used over only a one- or two-mile range. When approaching the narrows a heavy fog set in. From there to the pier it was dead slow, foghorn and bell. A lookout was on the bow and all eyes and ears strained for sight and sound as we navigated through ship anchorages, between ferries, and ships underway. To me, a miracle of seamanship, to the captain, another day at sea.

Beginning with the late forties and early fifties developments for marine navigation were rapidly taking place. Radar sets were on the market as well as Loran. Gyro compasses and radios were in use as never before. Tugboats today are equipped with all of the latest electronics. Most tugs have two radars, magnetic compass, gyro compass, auto pilots, Loran C, depth recorders, single side band radio, company radios, self monitoring radio, tachometer, and rudder angle indicators. Some tugs, especially those in ocean service, are equipped with Comsat (satellite navigation).

With the advanced technology and new labor agreements, tugs were fitted with engine room automation. Control panels were installed in the pilot house and captains were now able to start and stop the main engines as well as being able to engage the clutches and maneuver from control stations in the pilot house or from a remote location on deck. Steering engines, generators, electric load and transfer are now also included in the automation system. Alarm panels in the pilot house, engine room and engineer's room alert the officers to any incipient problems, whether with machinery or with various hull systems. Today's tug generally carries but one engineer. He works eight hours a day and sets his own work schedule. The maintenance work is now done largely by

Fig. 4 Tug *Marjorie B. McAllister* in pushing notch of tank barge *Cibro Philadelphia*. Tug pilot house in elevated position. Tug built in October 1974 by Main Iron Works: 106 ft-8 in. × 30 ft × 14 ft-8 in., 3900 bhp, two 16 cylinder 16-645-E2 Electromotive Marine Diesel engines. Tank barge built in September 1978 by Todd Shipyards, 425 ft × 74 ft-6 in. × 36 ft-5 in.

work crews ashore and many companies have employed their own electronics service technician to keep the equipment in repair.

As the use of electronics was changing, so too was life aboard. The crew rooms became large, with more concern given to creature comforts. Bunks were made more comfortable with springs and mattresses; desks and chairs were installed, closets replaced lockers, and wash stands were added. Usually there were only two men to a room. The captain, mate, cook, and engineer all rated single rooms. Galleys and mess rooms are now as well equipped as the best kitchen ashore. Refrigerators for daily use and a freezer for storage are installed. Electric ranges, ovens, hot plates, toasters, and an automatic dishwasher are installed. Hot water is readily available. To ensure crew comfort, tugs are fitted with thermostatically controlled heating and air conditioning in all living and working spaces. Rooms are well painted, and deck coverings (some tugs have carpeting) are selected to please the eye. Life aboard today's tug (with radio, television, and VCR) is as comfortable as life ashore.

Painting has changed with the growing sophistication of the petrochemical industry. Owners in the past were content to have the interior of the hull wire brushed and coated with red lead. Since red lead has been removed from the market by federal regulation, vinyl and epoxy systems are now in use. All steel interior and exterior surfaces are sandblasted to white metal before those protective coatings are applied. Where bottom anti-fouling paints in the past contained copper and arsenic, there are now in use less environmentally damaging coatings that effectively serve that purpose.

With the increased use of electronics and electric gear, the electric plant was modernized as well. As late as the forties and fifties, 115 V dc was still in use. The tugs were fitted with a 56-cell storage battery mounted in a lead lined box in the engine room and attended daily by the engineer with his hydrometer and distilled water. The batteries were maintained by a 20 kW diesel generator and a 10 kW generator driven by belts from the propeller shaft or engine. The switchboard was a live front, knife switch panel. Its primary use was for lighting, with a regulator to maintain constant voltage. Some tugs were fitted with a capstan, which (because of its necessary variable speed characteristic) made direct current a necessity. One advantage of direct current was that the tug could be left unattended overnight on battery power. If there were other advantages they were hard to find. For the most part direct current was troublesome and required constant care.

As we moved into the fifties and sixties, the use of alternating current was on the increase. Electric loads were growing and electronic equipment was becoming more in demand. Generators moved from 120 V single phase to 440 V three-phase; and generator manufacturers were producing reliable self regulating machines. All manner of motor-driven equipment, from capstans to air conditioning, was being added. Where dc systems were designed and operated at 25 kW, the alternating system with two diesel driven machines at 100 kVA were installed and were capable of parallel operation. The live front switchboard was replaced with a modern dead front, circuit breaker, multiple panel, unit capable of complete generator control.

Distributing direct current required large diameter armored cable, which was expensive and difficult to handle. The alternating current three-phase wiring systems uses plastic insulated, armored cable that is smaller, easier to handle, less costly, and more readily available. Further, because direct current is no longer in use ashore, it has become more difficult to find parts or to employ electricians who can handle it. On the other hand, alternating current equipment and electricians are readily available and can be found in all ports. Shore power for tug lay up is also easier to locate at most piers.

Because there is still a need for variable speed machines, hydraulics have found a place on the modern tug. Capstans are powered by low speed piston and roller hydraulic motors controlled by lever valves and constant speed motor-driven pumps. Steering is done using hydraulic cylinders and solenoid-operated four-

Fig. 5 150-ft oil field tug (Trinity Marine).

Fig. 6 Cut-away of the 225-ft oil field supply vessel *Sea Titan* (Halter Marine).

way valves mounted on a storage tank together with two hydraulic pumps. Likewise, windlasses and towing machines are hydraulic powered. Pilot houses are raised and lowered by hydraulic cylinders. With the advent of hydraulics, quick-coupled high-pressure hoses have become popular. They are easy to install, with no need for welded fittings, seamless pipe, or costly pipe line flushing. The hose and couplings are easily carried on board, and emergency repairs are quickly made.

During the seventies the world awoke to the dangers of ocean pollution. The U.S. Coast Guard required all vessels to stop discharging waste water overboard. Marine sanitation systems were therefore required. Today all tugs are so equipped.

Many and varied tug designs appeared during the seventies and eighties. Innovations were aimed at obtaining maximum towing speed, higher maneuverability, or economy of operation. While most tugs today are fitted with open propellers, many were fitted with Kort nozzles in order to gain greater thrust at lower towing speeds. Tugs fitted with Z propeller drives proved to be highly maneuverable, but more costly to install. Some tugs were fitted with controllable, reversing pitch propellers. All these have found a place for their own particular benefits; but, for the most part, well designed, open propellers remain most popular. Stainless steel has replaced bronze in this application.

In past years tugs were hired based on engine horsepower. This led some owners to claim exaggerated ratings. Charterers, in turn, began to write bollard pull requirements into their contracts, and tug owners were required to obtain bollard pull certificates as part of their tug's documentation. In the era of the 1600 bhp open-propeller tug, a bollard pull of 30,000 lb could be expected. Today, a 6000 bhp tug fitted with a Kort nozzle is able to produce 180,000 lb.

Following the oil crisis in the seventies, operators began to investigate engines that could operate on heavy, No. 4 fuel oil. This was of special interest to several new tugs intended for long ocean towing, where fuel bills were of major concern. Such heavy fuel, however, proved difficult to use. The bunker tanks had to be heated, and the oil had to be run through a centrifuge before going to the engine. Engineers had to be sure of the quality of the fuel delivered to ensure a minimum of service problems. Consequently, because of these problems, and in spite of the high cost, most tugs today continue to use No. 2 diesel oil.

The preceding paragraphs were written to give a broad picture of the evolutionary development of the tug boat. There were, however, certain external forces affecting the industry, and these are taken up below.

External Influences

Most tank barges of the forties were as small as 4000-bbl capacity. A large barge might carry 20,000 bbls. There were many of them, and distribution was generally from the local refineries to oil dealers at scattered locations. Cargo ships were still carrying break cargo, as containers were not yet in use, and common sizes ranged from 6000 to 8000 dwt. A large ship of that era would be a C-3 cargo liner, or a T-2 tanker, of World War II vintage. The barges were easily handled by one 600 bhp tug, while the ships were handled by one 1000-bhp, or two 800-bhp tugs, either towing or docking. Land transport was largely by railroad, the trucking industry being new and growing. Coal was the big item by rail, and freight yards with coal car

Fig. 7 Catamaran tug and barge.

dumpers were found in many major ports. Coal scows of 1000-ton capacity and carfloats could be seen in large fleets being distributed to power plants in the harbor. These, too, were handled by 1000-hp tugs.

The national economy was growing rapidly. Oil pipelines were being installed from refineries in the south and soon reached all the large cities. Power companies were converting to oil, so the demand for coal was waning. Harbor scows and carfloats were gradually disappearing. The local demand for oil increased and the need for larger barges was evident. Soon barges were being built as large as 60,000 bbl capacity, and it became more practical to move petroleum products by sea.

As break bulk cargo ships gave way to larger containerships, tugs of 1600 bhp were less than adequate. By the late fifties and early sixties new tugs were likely to be in the 2400 to 3000 bhp range.

Registered tonnage had never been a problem for tug boats, although all steam vessels were required to be designed, built, approved and inspected by the Coast Guard, regardless of gross tonnage. Diesel tugs rarely exceeded 300 gross tons and were therefore uninspected vessels. Owners were careful that any new building did not exceed that 300-ton limit. Problems sometimes arose, however, because another rule applied to vessels registering over 200 gross tons. When engaged on ocean voyages, such vessels were required to be manned by licensed personnel.

Because many of the steam boat personnel had either retired or were not in the local unions, licensed people were at a premium. Men who were deck hands for years and held union seniority were not anxious to sit for license and lose what seniority they had. The maritime schools were not able to turn out enough people to fill the available billets. The only answer was to convert and re-admeasure existing tugs; and to design all new tugs so as keep the gross tonnage below 200. Naval architects employed all manner of rule-beating schemes (often to the detriment of seaworthiness) to attain that aim. As a result, 120-ft tugs that in prior years would have exceeded 300 gross tons were now being built to measure less than 200 tons.

Tug boats up to the forties and early fifties were, for the most part, under 100 ft in length, and rarely exceeded 28 ft in beam. Many tugs of that era were built with a beam of only 25 ft. As the power of the tugboats increased up to 2000 bhp, insufficient stability became apparent after several fishing vessels and tugs capsized. There were tugs operating during the fifties with less than one ft of GM, and boats that were known to be cranky were treated with great care; only the skill of the operators prevented serious problems. Transverse stability was estimated by the naval architect during design, and proved by an inclining experiment on completion. Stability standards were for the most part unknown and acceptability was determined by the architect. During the sixties the Coast Guard began to take an interest and solicited the industry for suggested rule making criteria for static stability, as well as towline and weather criteria. By 1970 both the Coast Guard and ABS had agreed on acceptable

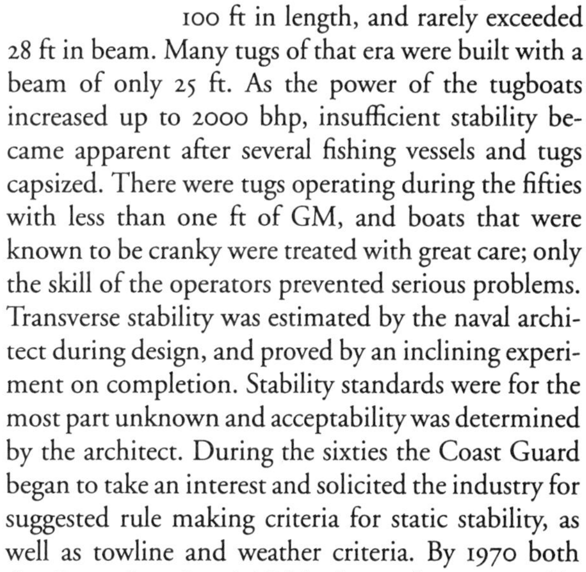

Fig. 8 Ship docking tug *Lindsey*: 70 ft × 30 ft × 12 ft-6 in., twin screw, 2400 hp, 360° rotating nozzle.

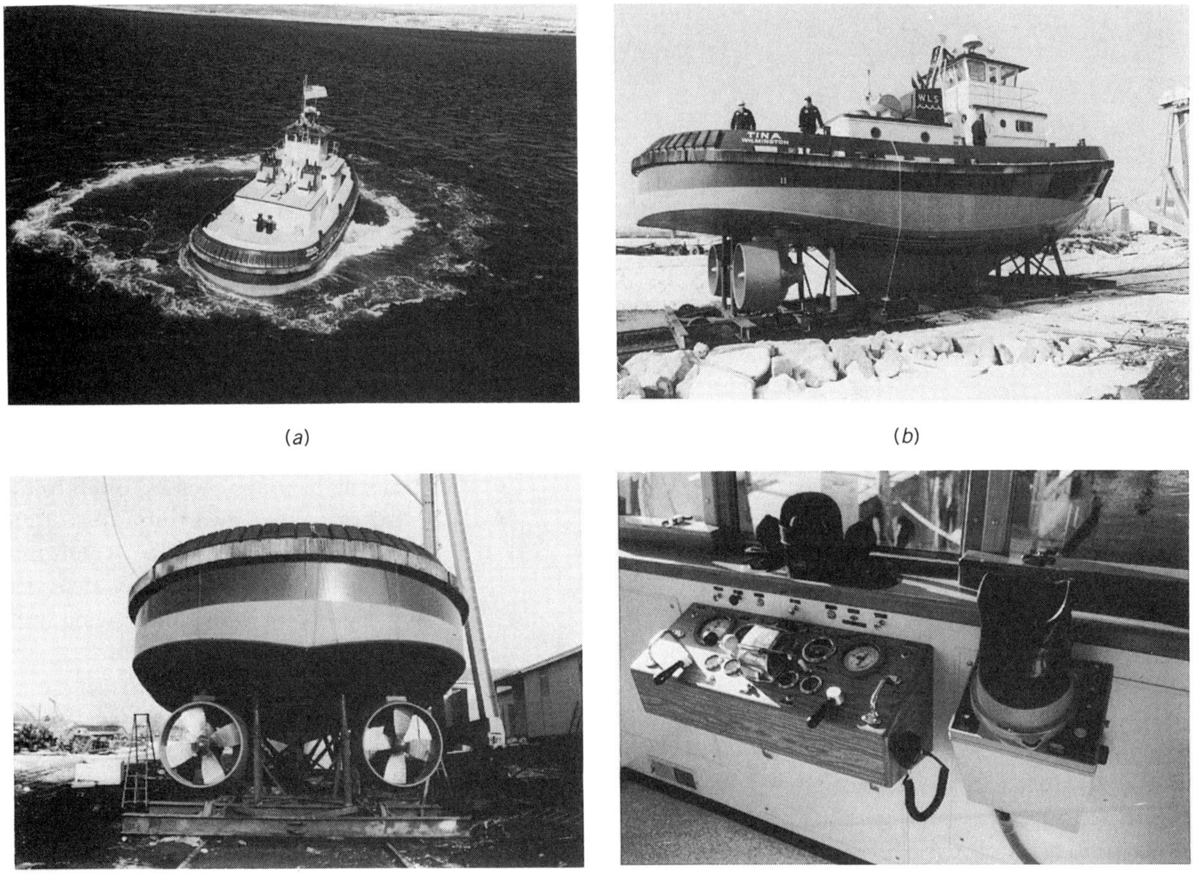

(a) (b) (c) (d)

Fig. 9 The tug *Tina* (1977), built by Gladding-Hearn Shipbuilding: (*a*) maneuvering through 360° rotation in 20 seconds; (*b*) underbody, showing fin keel and propulsion units; (*c*) 360° rotating Kort nozzle; and (*d*) forward-facing control console (the after end of the pilothouse is fitted with a duplicate).

standards. After some adjustment, the new rules, were entered into the Federal Register. A period of adjustment followed. Changes were made to allow existing tugs to comply. Some had to raise door coamings, some had to remove doors from the house sides entirely, and others had to raise a section of deck.

As we approached the seventies and eighties barges were reaching and exceeding the 100,000-bbl capacity, and the ocean trade for tugs and barges was greatly increased. Tugs of 2400 bhp were no longer the big work horses. Some that were now being built were 120 ft long and 34 ft in beam. Brake horsepower was now up to 6000. Tug crews were not eager to go to sea with single screw boats and so many owners turned to twin screw arrangements.

These higher powers made fiber hawsers obsolete. They were replaced by steel wire rope carried on the drum of a towing engine. Wire ropes 2¼ in. in diameter and 2400 ft long were in common use. For towing at sea the barges were fitted with chain bridles attached to a nylon stretch line about 150 ft long and shackled to the steel hawser. Barges continued to grow, and by the eighties some were in excess of 100,000 bbl capacity. New ideas were being tried to increase towing speed and decrease turnaround time. Barges were being fitted with deep notches shaped to fit the bow of the tug. The tug could then be securely lashed into the notch. So arranged, with the tug in the pushing position, the barge was easier to control, and yawing was no longer a problem. This worked well in seas of up to 4 ft; but if the sea state worsened, the tug would drop out of the notch, put out a hawser, and tow the barge in the traditional way. This system was used for short coastwise trading but was troublesome between the light and loaded conditions of the barge. Many clever new notch attachments were tried and patented around the world. As just one example, some deep notches were tailored to the tugs, which were so attached that the tug became an integral part of the barge.

An early example of the integrated tug/barge unit was built to carry dry cargo as well as liquid cargo.

Fig. 10 McDermott Shipyards' 105-ft standard tug *Dorothy Moran*: 105 ft × 32 ft × 15 ft-9 in., 3000 bhp, two General Motors 12-645-E6 engines.

After completion the owners were required to demonstrate the tug's capability to enter and leave the notch, pick up the barge on a hawser, and also prove the stability of the tug. There were trying times to prove all of this and, after much testing, approval was eventually granted. This beginning in the early sixties started the industry off into a world of new problems. Not only pitching and rolling had to be considered, but the ability to see over the barge when in the light condition. Tugs were fitted with second pilot houses set atop the existing house. Some tugs were fitted with elevating pilot houses. This gave the tug an advantage in docking ships. Tug crews preferred the pushing mode, for there was increased safety and comfort. The detrimental factor was knowing what the sea state was and when the tug should get out of the notch. There are reports of units today that can successfully survive seas of up to 10 ft. On the other hand, there have been cases when the captain waited too long and caused severe damage to both the tug and the barge when making his exit.

One major reason for integration was to gain sea speed, but it is doubtful that this has been much of a benefit. Today's bigger, more powerful tugs and well designed barges seem to maintain sea speeds as well on a hawser as in the notch. Still, there are other benefits, and integrated tug/barges have a definite place in the industry.

Conclusion

In the past 50 years, tugboats have been transformed. Displacement of the early tugs rarely exceeded 250 tons, whereas today's tugs can approach 700 tons. Horsepower has increased from 600 to 6000. Machinery, electric and electronic equipment, outfitting, and livability comforts today are as fine as on any ship afloat. Those "little tugs" have evolved into powerful giants that are awesome to behold and exciting to operate. Visitors on today's tugs are utterly amazed that their exterior appearances are so completely contrary to the actual internal size and accommodations on board.

To speculate as to which vessel of all those afloat is the most interesting and lovable would, no doubt, evoke much discussion among owners, naval architects, and builders. I, for one, as you may have guessed, would loudly vote tugboats number one.

Donald C. Hankin spent most of his 44-year professional career designing, building, or operating tug boats. He started as a quarterman in the old Cramp Shipbuilding Company's Outfitting Department, then moved on to the RTC Shipbuilding Corp. as Chief Engineer. He later took over as Vice President and General Manager of the John H. Mathis Company. The final eighteen years of his career were spent as Vice President for Engineering with McAllister Brothers, in New York. Mr Hankin retired in 1986. He has been an Associate Member of SNAME since 1951.

Fireboats

Donald C. Hankin

High speed containerships made their appearance in the marine industry during the post World War II decade, and their mark on the shipping world was well established by 1960. As containers replaced break bulk cargo, the smaller ships soon disappeared and the waterfronts of the world underwent massive changes. The warehouse storage of cargoes on piers rapidly declined and was replaced by huge container ports located well away from large city commercial areas. Those piers, which had been notorious fire hazards, were soon vacant. Many were demolished to make room for new commercial and residential growth. The need for numerous fireboats and fireboat stations, thereby decreased. Many cities that had planned powerful new diesel fireboats soon realized far fewer would be required. Further, the older large steam powered boats, with their high costs for crew and maintenance, were better replaced with smaller diesel-powered boats.

One City of New York fireboat, the *John J. Harvey* (130 ft × 28 ft) had been built in 1931 and was fitted with Sterling Viking gasoline engines. In 1957 the hull was found to be in sound condition and was refitted with five Fairbanks Morse engines, each of 600 bhp. These were connected to five 340 kW generators furnishing power to two 1065 hp propelling motors. Electric motors were connected to four pumps having a total capacity of 16,000 gpm at 150 psi. The conversion resulted in a substantial increase in speed and pumping capacity.

Between 1958 and 1961 NewYork City built four fireboats at the John H. Mathis Shipyard in Camden NJ. These had twin controllable pitch propellers.

These four boats were 105 ft-6 in. long, with a 27 ft beam. One has been retired from service.

Philadelphia also inaugurated an expansion program for increased fire protection. Their problem was the same as other cities. There were new container ports and fewer break bulk cargo piers. Four boats, designed by Thomas D. Bowes, were built by the RTC Shipbuilding Corp. in Camden NJ.

These boats were 79 ft long by 19 ft beam and were powered by GM Detroit 6-71 quad engines rated 800

Fig. 1 *John J. Harvey*, built 1931 by Todd Shipyard. Plans by Henry Gislow. Since conversion in 1957, she remains the fastest of New York City fireboats. (Top photo courtesy University of Baltimore Library; bottom photo courtesy Al Trojanowicz.)

Fig. 2 *Governor Alfred E. Smith*: two engines, 500 bhp each, for propulsion and two engines, 500 bhp each, for centrifugal pumps having a total capacity of 8000 gpm at 150 psi. (Courtesy Al Trojanowicz.)

Fig. 3 *Delaware.* (Courtesy Philadelphia Fire Department.)

Fig. 4 The City of Baltimore fireboat *Mayor Thomas D. Alesandro, Jr.*, designed by Thomas D. Bowes and built by the RTC Shipbuilding Corp. in Camden NJ, continues in service.

Fig. 5 Baltimore has found more practical use for smaller, faster boats such as (*top*) *Fire Rescue 1* and (*bottom*) *Fire Rescue.*

bhp. Each of the single engines was connected to a centrifugal pump rated at 600 gpm and 150 psi. The engines could of course be used for either propulsion or pumping. Several of these boats have been retired from service.

The day of the massive, high powered, large pumping capacity of the twenties and thirties climaxed with the close of World War II. There were several reasons for these changes. Material handling methods were drastically changed by containerization, piers disappeared from the waterfronts and the cities were faced with economic pressures. Further, the population of motor boats was increasing and fires and other accidents were numerous. The work of the fire departments has evolved into sea rescue as well as fire protection. The boats needed today are small, fast, and able to work in shallow water.

Donald C. Hankin spent most of his 44-year professional career designing, building, or operating tug boats. He started as a quarterman in the old Cramp Shipbuilding Company's Outfitting Department, then moved on to the RTC Shipbuilding Corp. as Chief Engineer. He later took over as Vice President and General Manager of the John H. Mathis Company. The final eighteen years of his career were spent as Vice President for Engineering with McAllister Brothers, in New York. Mr Hankin retired in 1986. He has been an Associate Member of SNAME since 1951.

Oceanographic Ships

John Daidola

Introduction

The revelation that most of the progress in developing our extensive knowledge of the oceans has been made in the approximately 50 years since World War II may be surprising [1]. This has been accomplished by a relatively small but increasing number of specialized ships that have carried oceanographers and their equipment to the far reaches of the globe. There they conduct research on the behavior of the sea surface and currents, temperature gradients, bottom topography, biological activity, composition beneath the earth's crust, and the sea's chemical composition, to name just a few. Yet even today the ocean is sometimes referred to as the last frontier, even more mystifying than outer space. Oceanographic investigations continue to attract worldwide interest.

The span of postwar years up through 1967 has been divided into periods by Nelson [2]. Extending these periods to 1993 could result in the following:

1941–1945	The War Years
1946–1956	The Postwar Years
1957–1958	The Years of Impetus
1960–1967	The New Emergence
1967–1993	The Years of Expansion

During World War II, the Navy vigorously pursued oceanographic research to meet the vital needs of defense, turning to private and academic institutions for the scientific expertise required. The urgent need for extensive hydrographic surveys, especially in the Pacific, saw the enlistment of a varied lot of converted ships. The Coast and Geodetic Survey (now the National Oceanic and Atmospheric Administration, or NOAA) gave a number of its vessels to the Navy and operated the rest to support the war effort.

Following World War II, the government, stimulated by the techniques and programs that evolved during the war, continued to support oceanographic research at universities and other institutions. This it did through such agencies as the Office of Naval Research (created in 1946), the Navy Hydrographic Office (now the Naval Oceanographic Office), the Bureau of Ships (now the Naval Ship Systems Command), the Atomic Energy Commission (established in 1946) and the National Science Foundation (established in 1950). Numerous ocean-oriented educational and research programs were launched at such universities as Columbia, Johns Hopkins, Rhode Island, New York, Oregon State, Miami, and Texas A&M. At the same time, applied ocean research directly related to defense problems was being expanded in the Navy's own laboratories such as the Naval Research Laboratory, Navy Underwater Sound Laboratory, Naval Electronics Laboratory, and Naval Ordnance Laboratory. During the postwar period the oceanographic research fleet and the Navy's hydrographic ships consisted wholly of converted ships with two exceptions, the *Atlantis* of Woods Hole Oceano-

Fig. 1 *Atlantis.*

Fig. 2 AGOR-3, MR&S.

graphic Institution (WHOI) and the *Velero IV* of the University of Southern California.

In the early 1950s the Office of Naval Research (ONR) became concerned that no specially designed American oceanographic research ship had been built since the *Atlantis*, in 1931. (See Fig. 1.) Nor was ONR particularly impressed with the then present hodgepodge of jury-rig vessels. This led ONR to establish a ship panel at the June 1952 symposium on Oceanographic Instrumentation in Rancho Santa Fe CA. This panel stimulated the oceanographic community's interest in ship design, and many concepts and recommendations evolved. Consequently, in that same year, ONR provided funds to WHOI to conduct a pre-design engineering study of oceanographic vessels reflecting existing ideas and exploring further the most logical choices of power plant, ship size, and outfitting [3].

Shortly after that, ONR sponsored a conference in Washington on oceanographic research vessels, which resulted in the appointment of a steering committee for the design of an oceanographic research ship. This committee prepared a set of tentative characteristics. In 1955 the Navy's Bureau of Ships undertook the preliminary design of an oceanographic ship. Their intent was to investigate the feasibility of those tentative characteristics. An additional intent was to let the steering committee members look at something concrete during their discussions. That design was aired as a paper presented to SNAME in 1956 [4]. The project, which eventually resulted in the basic design for the AGOR 3-class research vessels (Fig. 2) was to remain dormant, however, for more than three years while awaiting the appropriation of construction funds [5].

The concepts and guidelines developed by the National Academy of Sciences' Committee on Oceanography (NASCO) report of 1959 [6] provided the basis for long-range plans generated by the various Federal agencies. Of particular significance was the Navy's Ten Year Program in Oceanography (TENOC). The 10-year ship construction program recommended by TENOC called for 20 research ships (including 12 for institutions), 9 oceanographic surveying ships, and 11 hydrographic surveying ships. That construction program began in 1960 and continued throughout the decade. The ships incorporated new concepts such as cycloidal propellers in the AGOR-14 class (Fig. 3) and a catamaran hull in the T-AGOR 16-class (Fig. 4).

The SNAME *Transactions* contain plentiful information on oceanographic ships. In 1959 Minot [5] discussed the significance of oceanography to the future of marine transport. In 1960, Rosenblatt [14] presented significant design information on oceanographic ships and noted that Isaacs and Iselin [15] had published the earliest thoughts on oceanographic ships in the United States. Rosenblatt, et al [16] extended relevant information up to 1970. In 1985 Daidola and Griffin [7] presented updated design information and appended a definitive compendium of pertinent references dating back to 1970.

Minot [3] gave a synopsis of the status of the world's oceanographic fleet in 1970. In 1960 the U.S. fleet consisted of approximately 40 ships of all sizes, most of which were conversions. By 1970 the number of oceanographic ships operating under the U.S. flag had expanded to over 290, while more than 260 were

Fig. 3 AGOR-14, MR&S.

Fig. 4 AGOR-16, MR&S.

under foreign flags. A listing of oceanographic ships built since 1970 is given by Daidola and Griffin [7].

The number of nations and organizations employing oceanographic ships is diverse. As an example, an outline for the United States, developed from data in Ref. 8, is given in Table 1.

The U.S. federally supported portion of the oceanographic fleet consists of about 60 ships, of which about 45 are capable of deep-ocean, "blue water," operations. The nominal 60-vessel fleet is coordinated by the Federal Oceanographic Fleet Coordinating Committee (FOFCC). It now includes 43 research ships owned or operated directly by federal agencies. The majority of these are specialized vessels dedicated to specific assignments such as mapping, charting, fisheries and other mission-oriented surveys. These vessels are generally not available for academic research. Also included under FOFCC is the 20-ship fleet of the University-National Oceanographic Laboratory System (UNOLS). This fleet comprises university operated ships that carry out most of the basic research of the national oceanographic program.

The current UNOLS fleet's 20 vessels are operated by member institutions and are between 100 and 250 ft in length. Support for their operation comes from the federal agencies that use the fleet, coordinated through the National Science Foundation. Eighteen universities and research institutions are members of UNOLS, and another 31 are associated members. Their ships are primarily general purpose types with special capabilities in the basic sciences. The ship-operating institutions are autonomous, but UNOLS members acting jointly schedule the ships and coordinate performance standards [10].

Dinsmore [11] indicates that in 1982 some 72 nations were operating something like a total of 720 research ships. Nations then operating ten or more vessels are listed in Table 2. In 1986 Canada, France, the U.K., Japan and the Soviet Union, along with the United States, all had oceanographic fleets in excess of 20 ships [12]. A principal difference between the composition of the research fleets of the United States and other nations is the foreign emphasis on fisheries

Table 1 Groups involved in oceanographic vessel operation in the United States

		Federal Government		
Purpose	Academic and Research Labs	Direct Operators of Vessels	Fund Research or Survey at Academic Labs and Through Industrial Firms	Industrial Firms
Basic research	Independent labs, Private universities, State universities	U.S. Navy, U.S. Coast Guard, Evironmental Protection Agency	National Science Foundation, Nuclear Regulatory Commission, National Institute of Health, U.S. Navy, NOAA[a] Energy Dept.	Limited support
Military R&D	Some by academic labs	U.S. Navy	U.S. Navy	Under contract to Federal Government
Hydrographic survey	Some by academic labs	NOAA,[a] U.S. Coast Guard,[b] U.S. Naval Oceanographic Office	U.S. Navy, Marine Minerals Service (Interior Department)	Under contract to Federal Government
Applied fisheries research	Some by academic labs	NOAA[a]	NOAA[a]	Limited number of firms
Resources development	Limited	NOAA[a]	NOAA[a]	Offshore oil exploration with seismic and deep drilling vessels
Service	Limited	None	None	Many firms maintain research vessel fleets primarily for charter

[a] National Oceanic and Atmospheric Administration.
[b] Ships either manned by Navy Operations or the Military Sealift Command (MSC).

Table 2 Countries operating ten or more seagoing oceanographic ships (1982)

Country	Number of Ships
Argentina	10
Australia	10
Brazil	12
Canada	25
France	27
Germany	15
Italy	10
Japan	94
Sweden	11
Great Britain	39
United States	115
Soviet Union	194
All others (60)	158
Total (72)	720

research (60 such ships for Russia as compared to 9 for the United States). The Russian deep-ocean-capable fleet is about three times as large as our own.

Scientific Requirements

The scientific requirements for an oceanographic ship are driven by the user and the type of mission. Nevertheless, while there are differences between military, academic and commercial users, the survey or research has a common scientific technological base. Academic interests and technological possibilities for new ship designs have recently been reviewed at a number of widely attended government sponsored symposia through the UNOLS mechanism. Scientific requirements thus derived by UNOLS [13] have been taken as representative of what might be required by any organization and its oceanographers. These requirements dictate that a list of priorities for oceanographic ship designs be developed. The factors in Table 3 have been ranked by groups of practicing scientists from all disciplines within UNOLS [12].

The overall characteristics of an oceanographic ship will be determined by its intended use, which will identify the necessary equipment and scientific complement, as well as minimum range and endurance.

The type of equipment will depend on the vessel's function. A possible way of identifying some bounds on range, endurance and vessel size can be gleaned by considering the water depth ranges given in Ref. 8, Table 4. Some organizations prefer to identify vessels only by range and endurance.

Different oceanographic ships have evolved from the classical general-purpose vessel with laboratories and equipment suitable for two or more studies in the areas of physical, chemical, biological, geological and geophysical or ocean science. Marine geology and geophysical vessels for the purpose of investigating the sea floor, oceanic crust, margins and lithosphere have made their presence in the last 25 years, particularly in oil exploration. Specialized fishery research ships have evolved, although biological work carried out as part of the studies on a general-purpose oceanographic ship can relate to that function. Fishery ships generally fall into the categories of development work in fisheries, or applied research in fishing methods and gear, fish handling, processing and preservation of fish quality on board, as well as for exploratory fishing.

Design of Oceanographic Ships

The fundamental principles underlying the design of oceanographic ships are, of course, those applicable

Table 3 Priorities of oceanographic ship requirements

1. Seakeeping
   - seakindliness
   - station-keeping
2. Work Environment
   - lab spaces and arrangements
   - deck working area, overside handling
   - winches and wire flexibility
3. Endurance
   - range
   - days at sea
4. Science Complement
5. Operating Economy
6. Acoustical Characteristics
7. Speed
   - ship control
8. Payload
   - science storage
   - weight handling

Table 4 Classification of vessel size by water depth

200 m	Research addressing the continental shelf and shallow fisheries studies. Small coastal research ships or boats under 30.5 m (100 ft) in length are suitable.
1000–2000 m	Research addressing most marine life, the sound channel axis and the upper sides of islands and continents. Smaller coastal vessels under 45.7 m (150 ft) in length are suitable.
6000 m	Research addressing the ocean bottom and full depth of the oceans over 98 percent of their area. Oceangoing ships over 45.7 m (150 ft) in length generally required.
12,000 m	Research addressing the world's great trench systems. Oceangoing ships over 45.7 m (150 ft) in length with capacity for equipment to work at great depths are generally required.

Fig. 5 T-AGS 39, MR&S.

Fig. 6 T-AGS 51, Halter Marine.

Fig. 7 AGOR 23, Halter Marine.

in the design of any ship. However, in a number of specific areas there are special considerations to be taken into account. The technical literature on oceanographic ships is extensive, particularly as presented in the SNAME *Transactions* previously discussed.

In more recent times, the displacement type monohull has given way to other hull forms in some services. Small waterplane area twin-hull (SWATH) vessels offer certain advantages. They are less susceptible to wave induced motions; they have less wave making resistance at high speeds, and have longer natural periods of motion than a conventional monohull resulting in enhanced seakeeping characteristics. Reference 7 presents a summary of oceanographic SWATH ships constructed since 1977. It also projects several future designs, and takes up the pros and cons of generic SWATH and monohull designs. Advanced vehicles have been proposed for various oceanographic missions where high speed is useful, and hovercraft have been used in hydrographic surveys because of their unique ability to cope routinely with a grounding [17].

Fig. 8 T-AGOS 19, McDermott Shipyard.

Examples of Recent Construction

Recently a number of vessels have been delivered that exemplify the development over the last 50 years.

The U.S. Navy has undertaken construction of several classes of ships designed and built predominantly to commercial standards. The T-AGS 39-class, Fig. 5, delivered in the late 1980s, are large ocean survey ships of 500 ft LOA fitted with bottom topography survey systems. The T-AGS 51-class ships, Fig. 6, of the early 1990s are 205 ft LOA survey vessels, which carry launches for data collection in coastal regions with depths between 10 and 600 m and in deep water to 4000 m. The AGOR 23, Fig. 7, is a new class of 274 ft LOA oceanographic ships capable of operating worldwide in all seasons. They are suitable for use by navy laboratories, contractors and academic institutions. The first was delivered in 1991 and loaned to the University of Washington. The novel T-AGOS 19-class, Fig. 8, are 235 ft LOA SWATH ships.

The National Science Foundation has contracted for the charter of the *Nathaniel B. Palmer*, a 308 ft-6 in. LOA research icebreaker that was completed and entered the U.S. Antarctic Program service in 1992. See Fig. 9.

Other countries have been busy building oceanographic ships. For example, the German *Meteor*-class, Fig. 10, began operations in 1986.

The British commissioned the 227-ft LOA general purpose research ship *Charles Darwin*, Fig. 11, in 1984

Fig. 9 *Nathaniel B. Palmer*, Chouest Offshore.

Fig. 10 German *Meteor*, Harmster-Werften Shipbuilding.

Fig. 11 *Charles Darwin*, Appledore Shipbuilders.

Fig. 12 *James Clark Ross*, Swan Hunter Shipbuilders.

Fig. 13 *R. B. Young*, Allied Shipbuilders.

Fig. 14 R/V *Akademik Sergey Vavilov*, Hollming Shipyards, Finland.

for the National Environment Research Council. More recently, the 328-ft LOA *James Clark Ross*, Fig. 12, has been delivered to the British Antarctic Survey for oceanography and marine geoscience.

The Department of Fisheries and Oceans of Canada has recently obtained a 105 ft-7 in. LOA oceanographic and hydrographic survey vessel, the *R. B. Young*, shown in Fig. 13.

Finland delivered the *Akademik Sergey Vavilov*-class ships to the Soviet Union in 1989 and thereafter. These general purpose oceanographic vessels are 384 ft LOA. See Fig. 14.

Large geophysical survey ships for exploring such marine resources as oil and natural gas have also been

Fig. 15 *Mobile Search*, Mobil Oil.

Fig. 16 *Shell America*, Marinette Marine.

constructed during the latter years of this period. In 1982, the Japanese built the 323-ft LOA *Mobil Search* for a U.S. owner as the world's largest geophysical vessel, Fig. 15. In the same time frame, the 300-ft LOA U.S.-built seismographic research ship *Shell America* was built for another U.S. owner. See Fig. 16.

Conclusion

The past fifty years have seen a most impressive and explosive growth in the development of oceanographic research ships. Because of their multiplicity of purposes, they vary widely in size, configuration, and scientific gear. The diversity of organizations involved in their operation has additional impact in all this.

Although technical change has been rapid, we see no reason to believe that it will now slow down. On the contrary, ever-evolving needs, and the increasing numbers of interested users, point to continuing evolution of these versatile craft.

References

1 Spencer, D. W., "Introduction: Ocean Science and Ships," *Oceanus*, Vol. 25, No. 1, Spring 1982.

2 Nelson, S. B., *Oceanographic Ships Fore and Aft*, Office of the Oceanographer of the Navy, U.S. Gov't. Printing Office, 1971.

3 Minot, F., "Report on a Pre-design Engineering Study of the Development of Superior Ships for Oceanographic Research," Ref. No. 58-26, Woods Hole Oceanographic Institution, 1953.

4 Kielhorn, W. V., and Miller, R. T., "Oceanography and a Proposed Oceanographic Research Ship," Chesapeake Section, SNAME, Oct. 15, 1956.

5 Minot, F., "The Significance of Oceanography to the Future of Marine Transportation," *Trans.* SNAME, Vol. 67, 1959.

6 *Oceanography 1960 to 1970*, Committee on Oceanography of the National Academy of Sciences National Research Council, 1959, Chapt. 6.

7 Daidola, J. C., and Griffin, J. J., "Developments in the Design of Oceanographic Ships," *Trans.* SNAME, Vol. 94, 1986, pp. 121–158.

8 Vine, A. C., and Leiby, J., "Status of Oceanographic Ships in the United States with a View Into Future Developments," Ref. No. 67-11, Woods Hole Oceanographic Institution, 1967.

9 "Buyers Guide Directory 1985–1986," *Sea Technology*, 1985.

10 Webster, F. (College of Marine Studies, University of Delaware), Hearing of U.S. House of Representatives, Merchant Marine and Fisheries Committee, June 24, 1986.

11 Dinsmore, R., "The University Fleet," *Oceanus*, Vol. 25, No. 1, Spring 1982.

12 Winokur, R. S. (Technical Director, Office of the Oceanographer of the Navy, U.S. Navy), Hearing of U.S. House of Representatives, Merchant Marine and Fisheries Committee, June 24, 1986.

13 "A Plan for Improved Capability of the University Oceanographic Research Fleet," UNOLS, April 1986.

14 Rosenblatt, L., "Design of Modern Oceanographic Research Ships," *Trans.* SNAME, Vol. 68, 1960.

15 Isaacs, J. D., and Iselin, C. O'D., "Oceanographic Instrumentation," Division of Physical Sciences, National Academy of Science-National Research Council, Publication No. 309.

16 Rosenblatt, L., Serim, F. K., and Goode, R. W., "Ten Years of Progress in Research Ships Design: WHEO 701 Class Oceanographic Cutter," *Trans.* SNAME, Vol. 78, 1970.

17 *Ship & Boat International*, May 1986, p. 9.

Dr. John Daidola earned his degree in naval architecture and marine engineering at the University of Michigan. A long time employee of M. Rosenblatt and Son, he now holds the position of Assistant Vice President, Basic Ship Design Division in that firm. He is well known for his creative work in designing oceanographic research ships.

Submersible Service Craft

Eugene Allmendinger

The last fifty years have seen the development of a great variety of small to relatively large underwater vehicles designed and built for commercial, scientific, military and recreation related missions. These vehicles may be categorized as shown in Table 1 and pictured in Fig. 1, but realizing that certain hybrid systems exist that do not fit neatly into any category. As seen, the two major categories are manned and unmanned vehicles depending, simply, on whether humans are on board or at remotely located stations attending to their operations. The term "submersible" is often reserved for any manned underwater vehicle and is so used in defining the scope of this presentation. It is also used to distinguish between the small, limited endurance submersible and the large, self-sufficient naval "submarine." Terms such as "towed," "remote operated vehicle" (ROV) and "autonomous underwater vehicle" (AUV) are usually associated with unmanned vehicles—ROVs and AUVs being distinguished by whether or not they are tethered to another element of the mission system, such as a surface support ship. Both submersibles and unmanned vehicles may be used together in certain missions. A prime example is the series of dives to the *Titanic* in 1986 by the submersible *Alvin* with the small ROV *Jason Junior* tethered to it.

Table 1 Types of underwater vehicles

- A. Manned vehicles—submersibles
  - 1. One-atmosphere submersibles
    - a. Tethered—observation/work bells
    - b. Untethered—free-swimming submersibles
  - 2. Ambient-pressure submersibles
    - a. Tethered—sealabs/habitats
    - b. Untethered—diver assist/delivery submersibles
  - 3. One-atmosphere/ambient-pressure submersibles
    - a. Tethered—mobile diving unit
    - b. Untethered—diver-lockout submersibles
- B. Unmanned vehicles
  - 1. Tethered vehicles
    - a. Towed—all non-self-propelled vehicles
    - b. Self-propelled—remotely operated vehicles (ROV)
  - 2. Untethered vehicles
    - a. Preprogrammed—torpedoes
    - b. Autonomous—autonomous underwater vehicles (AUV)

Submersible types include one-atmosphere, ambient-pressure, and combination vehicles. In the first instance, the pressure exerted on the occupants remains essentially at one-atmosphere during the dive by virtue of their being encapsulated in pressure-resisting shells called pressure hulls. In the second instance, the occupants (divers) are subjected to full ambient sea pressure although in certain systems, such as sea labs, shell structures protect them from direct contact with the sea. Combination vehicles, usually

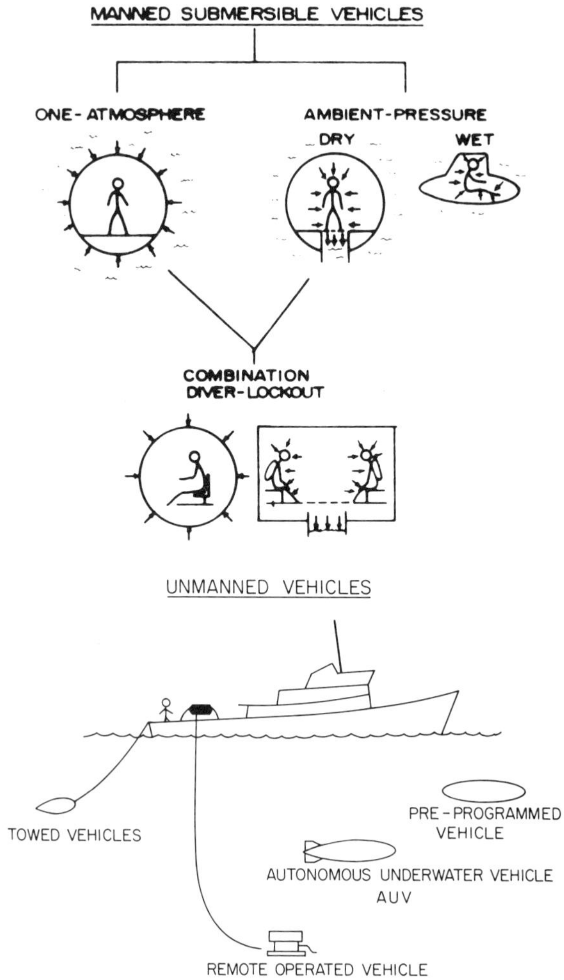

Fig. 1 Categories of underwater vehicles.

Table 2 Representative submersibles from various periods of submersible development

Period	Submersible	Fig. No.	Type[a]	Depth, m (ft)	Crew/Other	Builder
Prelude	Trieste	3	1A-UT	10,912 (35,800)	1/2 other	A. Piccard/USN
1	Alvin	4	1A-UT	1830 (6000)	1/2 sci	Gen. Mills/Litton
1	Aluminaut	5	1A-UT	1830 (6000)	2/4 other	E. B. Div./Gen. Dy. Corp.
1	Deep Quest	6	1A-UT	2440 (8000)	1/3 other	Lockheed Sp. & Mi.
1	Deepstar 4000	7	1A-UT	1220 (4000)	1/3 other	Westinghouse Corp.
1	Sea Lab II	8	AP-T	62 (205)	0/10 divers	U.S. Navy
2	Alvin II	9	1A-UT	4000 (13,125)	1/2 sci	Rebuilt by W.H.O.I.
2	PC-15	10	1A/AP-UT	366 (1200)	2/3 divers	Perry Offshore, Inc.
2	Pisces 8[b]	11	1A-UT	1000 (3280)	1/1-2 other	HYCO
2	DSRV Mystic	12	1A-UT	1525 (5000)	3/24 rescues	U.S. Navy
2	Arms 1	13	1A-T	914 (3000)	1/1 other	Perry Offshore, Inc.
3	Sea Cliff II	14	1A-UT	6095 (20,000)	1/2 other	U.S. Navy
3	Seeview (ex PC-1201)	15	1A-UT	365 (1200)	1/8 pass	Perry Offshore, Inc.
3	Atlantis III[b]	16	1A-UT	43 (150)	2/48 pass	Sub Aquatics Dev. Corp.

[a] 1A—one atmosphere; AP—ambient pressure; UT—untethered; T—tethered.
[b] Canadian-built submersibles.

called diver lock-out submersibles, combine the features of both types with crew and divers being in one-atmosphere and ambient-pressure hulls respectively. Sub-types of submersibles include tethered and untethered, or free-swimming, vehicles depending on whether or not they are physically linked to another object. Linkages include small fiber-optic cables, larger tension-bearing cables, and drill pipe strings. These tethers may transmit lifting/lowering tension forces, power, communications and breathing gases.

The following review of fifty years of submersible development (1943–93) is necessarily brief. It focuses on events and forces influencing this development, with Table 2 and Figs. 3 to 16 providing examples of a few of the submersibles associated with the various periods of a half-century of undersea activity in the United States.

Submersible Development

The past fifty years of submersible development may be divided into four periods whose end-points are marked both by well-defined, milestone events and by more arbitrarily assigned dates. They are periods of "Prelude" to submersible development (1943–63), "First Generation" (1963–69), "Second Generation" (1969–83) and "Third Generation" (1983–93+).

As an introduction to these periods, it is appropriate to pay tribute to the beginning of modern submersible development. This was the Beebe–Barton bathysphere built in 1930. The bathysphere, Fig. 2, is an example of a one-atmosphere, tethered submersible whose 1.37 m (4.5-ft) diameter, 0.04 m (1.5-in.) thick steel sphere permitted its occupants to dive to about 914 m (3000 ft) in 1934.

"Prelude" years, particularly those of World War II (when we were much involved with intense submarine and anti-submarine warfare), saw dramatic advances in undersea science and technology, notably in areas such as acoustics, electronics, and metallurgy, which were to be prime contributors to the development of future submersible systems. Perhaps a key event motivating this development, at least for science-related missions, was a high-level meeting of scientists in February 1956. That gathering culminated in a resolution stressing the need for a national program providing facilities for man to penetrate the depths of the oceans. The resolution led, eventually, to the development of the research submersible *Alvin*, launched in 1964. In 1958, however, the United States entered the submersible field when the Navy purchased Auguste Piccard's bathyscaphe *Trieste*. In 1960 that submersible made a historic dive to the bottom of the Mariana Trench at 10,912 m (35,800 ft), a record that has never been equalled. *Trieste I*, shown in Fig. 3, is an example

Fig. 2 Beebe–Barton bathysphere. (Courtesy National Geographic Society.)

Fig. 3 *Trieste I.*

Fig. 5 *Aluminaut.*

of a one-atmosphere, untethered submersible. Aviation gasoline was used to provide buoyancy required to support the 2.19 m (7.2-ft) diameter, 0.12 m (4.7-in.) thick, 13.0 tonne (12.8-long ton) forged steel sphere used for that dive. Modified by the Navy to become *Trieste II* with enhanced operating characteristics, this submersible served well into the "First Generation" period before her retirement.

The "First Generation" period was introduced tragically on April 10, 1963 when the nuclear submarine *Thresher* was lost in 2560 m (8400 ft) of water 200 miles off Cape Cod. *Thresher*'s legacy was a sharply focused national attention on the ocean environment, highlighting the frailty of humans attempting to work in its depths. *Trieste II* was uniquely capable of diving to that depth, but her abilities were limited to visual observations plus photographing and retrieving small objects from the wreckage. The following years saw a great flurry of government and private activity. Acting quickly, the Navy formed the Deep Submergence Systems Review Group (DSSRG). Resulting action led to a call for the development of a broad array of underwater systems. The Congress reacted with the passage of P.L.-454, the Marine Resources and Development Act of 1966, and the formation of high-level, marine related councils. One of the most prestigious and comprehensive documents written during this period was the Stratton Commission Report, "Our Nation and the Sea." The report advocated developing underwater work systems with capabilities down to 6096 m (20,000 ft). It is little wonder, then, that all this potential market stimuli created a great upsurge of activity with companies large and small, many in the aerospace industry, racing to establish themselves in some area of submersible technology. Indeed, this period might well be called one of "Great Expectations." The thought prevailed that the federal government would support "inner space" R&D just as it was supporting outer space R&D, and that a "wet NASA" would result. Submersibles of all types were built. Among the examples of First Generation submersibles listed in Table 2, *Alvin* and *Aluminaut* participated in the successful search for and recovery of the hydrogen bomb dropped into the Mediterranean

Fig. 4 *Alvin I.*

Fig. 6 *Deep Quest.*

Fig. 7 *Deepstar 4000.*

Fig. 9 *Alvin II.*

when two U.S. Air Force planes collided near Palomares, Spain in 1966.

The spring of 1969 may be cited as the beginning of the "Second Generation" period. That spring, in an address to a professional society, then Vice President Agnew made it clear that there would be no "wet NASA." In addition, competing pressures from the Vietnam War caused the Navy to reassess its interests and significantly reduce its funding for submersible technology. Thus it was, with few customers in sight, that the great expectations faded, with many companies deserting the field. Problems were compounded in that many submersibles of this period were built on speculation, or to demonstrate company capabilities so as to to improve standings on government bidder lists. Designs were often based on mission requirements unrelated to well-defined markets. In effect, submersibles were built and then went looking for work. It is little wonder, then, that this period was initiated by a few years in the "Doldrums" when many submersibles were laid up or scrapped, with the industry reaching a low ebb of activity. For those few organizations maintaining a continuing interest, those were years for taking stock of the lessons learned in the previous period, adapting existing submersibles, and creating new designs for specific markets.

By far, the largest market for submersibles during a major portion of this period was the offshore oil and gas industry, which required a great variety of underwater capabilities. Submersible systems of many types were used in meeting those requirements. The zenith of this market was reached in about 1975. The following years, however, saw a precipitous decline in the use of submersibles owing largely to rapid advances in unmanned vehicle technology, and the increasing use of their much less costly capabilities.

Submersibles nevertheless remained viable for markets with missions requiring capabilities for which man still enjoyed superiority over machines. Examples included man's superior stereo and peripheral vision, and brain power, which far surpassed that of artificial intelligence software then in the early stages

Fig. 8 *Sea Lab II.*

Fig. 10 *PC-15.*

Fig. 11 *Pisces 8.*

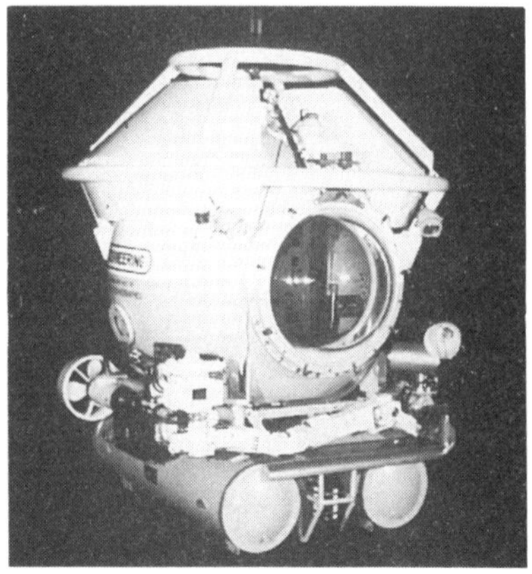

Fig. 13 *Arms I.*

of development. Thus, missions with complex tasks were still well-served by submersibles. Included here were underwater science, and Navy search and rescue missions. Some notable scientific missions of this period using submersibles included the 1974 French-American Mid-Ocean Undersea Study (FAMOUS) and the 1975 East Pacific Rise Study, which first discovered thermal vents on the deep ocean floor. *Alvin* (see Table 2), as rebuilt by the Woods Hole Oceanographic Institution, participated in both missions. The Navy mission was served by two Deep Submergence Rescue (DSRV) submersibles: *Avalon* and *Mystic* (see Table 2), whose mission was to locate and rescue crews of bottomed submarines. This was one of the five missions identified after *Thresher*'s loss. Other representative submersibles of this period appear in Table 2 and in figures referenced therein.

The year 1983 is arbitrarily chosen as the beginning of the final stage of submersible development, which we have called the "Third Generation." This was a period of focus on underwater science and recreaction markets. From the perspective of science, 1983 saw a major recommendation of the Stratton Commission Report (which was to "establish underwater work systems with capabilities down to 6096 m [20,000 ft]") finally realized with the Navy's *Sea Cliff*, with a titanium pressure hull, being certified to that depth. *Sea Cliff*, replacing the much less capable and long since retired *Trieste*, was made available to civilian scientists on an "as convenient" basis, as were other Navy submersibles such as *Turtle* and the nuclear powered NR-1. This arrangement had its obvious shortcomings and the civilian oceanographic community, primarily through the University National Oceanographic Laboratory System (UNOLS), voiced its need for suitable dedicated submersibles. Those needs were related to four depth regimes: (1) shallow—to 914 m (3000 ft); (2) intermediate—to 4000 m (13,125 ft); (3) deep—to 6096 m (20,000 ft); and (4) ultra deep—to 11,000 m (36,000 ft), which is approximately the maximum known depth. Figure 17 indicates the percentage of sea floor area of the world that can be reached by submersibles with various depth capabilities.

Shallow and intermediate depth scientific needs for the most part have been met using existing submersibles of this and previous periods, including the ven-

Fig. 12 DSRV *Mystic.*

Fig. 14 *Sea Cliff.* (Courtesy DS Oceans, Ltd.)

Fig. 15 *Seeview.* (Courtesy DS Oceans, Ltd.)

erable *Alvin*, now fitted with a titanium pressure hull, and certified to a depth of 4000 m (13,125 ft). Developments of this period have occurred in submersible systems and equipment rather than in new submersible designs. These developments include robotics, fiber optics, advanced imaging and sensors, precision underwater navigation, and advanced data management systems. A UNOLS committee has also called for a large research submersible with extended underwater endurance to serve scientific needs in the shallow depth regime. Deep water needs have been met, to a limited extent, by using *Sea Cliff*, one of only five submersibles in the world capable of diving to 5096 m (20,000 ft). UNOLS strongly recommended the development of a civilian research submersible with this depth capability by the mid-1990s. The ultra deep regime, accounting for about three percent of the world's sea floor area that lies below the 6096 m (20,000 ft) depth (see Fig. 17), has remained inaccessible to scientists since *Trieste*'s historic dive in 1960. UNOLS proposed that a submersible be developed to meet scientific needs in this regime by the end of this Century. A group of highly placed individuals associated with the oceanographic community has recently spearheaded a Deepest Ocean Presence initiative vigorously advocating that development.

The underwater recreation market, involving tourist-carrying submersibles, has grown dramatically on

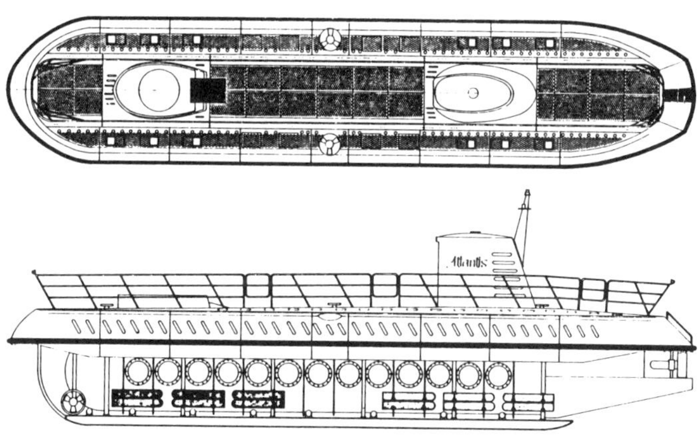

Fig. 16 *Atlantis III.* (Courtesy DS Oceans, Ltd.)

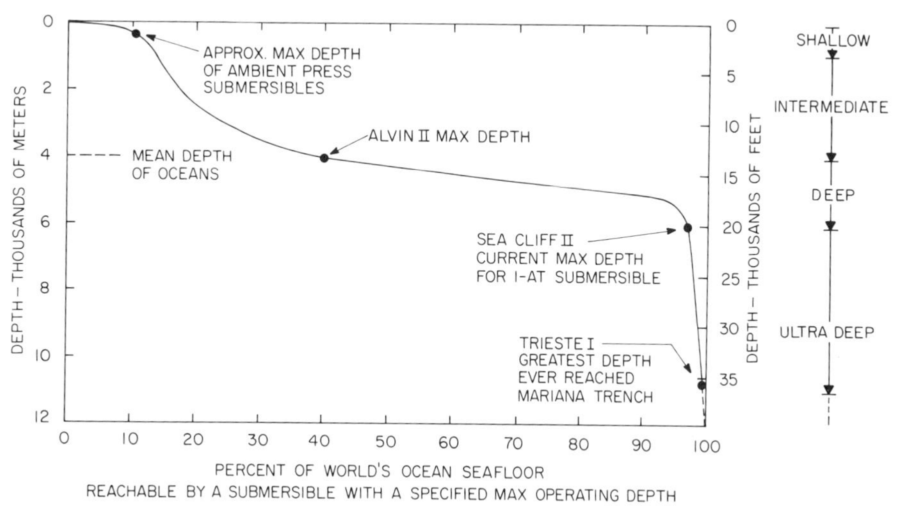

Fig. 17

a world-wide basis during this period. While U.S. companies have had relatively little to do with the design and construction of these unique submersibles, the American Bureau of Shipping and the U.S. Coast Guard have played significant roles in their development through their classification and certification activities.

Tourist submersibles date back to 1964, when the *Auguste Piccard* carried passengers into the depths of Lake Geneva during the Swiss National Exposition. Nearly twenty years elapsed before development resumed in 1983 with the conversion of several U.S. built, diver-lockout work submersibles for the tourist market on Grand Cayman Island, B.W.I. These carried a few passengers (2–4) to relatively great depths (250 m [820 ft]). The first contemporary submersible specifically designed for the tourist market began operations in 1986, and about thirty have been built to date. Most are relatively large, carrying a greater number of passengers (28–48) to shallower depths (30–49 m [100–160 ft]). One U.S. company has designed and built a submersible with an innovative hull of acrylic. This material, having the same index of refraction as seawater, should enable passengers to feel almost a part of their environment.

This chapter has focused on submersible development in the United States based on events and initiatives originating in this country as well as worldwide. Owing, however, to our close technological ties with Canada, it is appropriate to take note of the significant Canadian accomplishments in the field. This activity has been concentrated essentially in Vancouver, B.C. One company there was the second largest manufacturer of one-atmosphere, untethered submersibles in the world (the *Pisces*-class), while a second is currently the world's leader in producing tourist submersibles (the *Atlantis*-class). Representative Canadian submersibles appear in Table 2.

A graduate of the University of Michigan, **Prof. Eugene Allmendinger** has been involved in submersible/submarine design since 1943. In 1968 he instituted our country's first course in submersible design (at the University of Hawaii). He has taught the subject at the University of New Hampshire as well as at technical institutes in Norway and Sweden. He has served on submersible technical panels of SNAME and MTS, and was a member of the Submersible Science Study group sponsored by NSF, ONR, and NOAA. He was the editor and principal author of the SNAME book *Submersible Vehicle Systems Design.*

Yachts and Other Pleasure Craft

Olin J. Stephens II

Intent

My intent here is to trace the principal trends that we have seen in the design of pleasure craft during the past fifty years. By way of introduction, however, a few paragraphs will be devoted to developments that occurred between the two Great Wars.

Although they may have some things in common, I have chosen for the most part to give sailing craft and power craft separate treatment.

Wherever possible, I draw attention to external factors that have influenced the trends that we have seen in these designs.

Between the Wars

Sailing Craft. In the decades leading up to World War II, a group of rating rules formed the single most important influence on sailing yacht design. The International Rules led to the dominance of the "meter" in racing boats ranging in size up to the twelve meter class. Larger racing yachts, such as the J-class, were governed by rules set up by the New York Yacht Club. The Cruising Club of America, founded in 1925, was instrumental in fostering interest in offshore cruising and racing in this country. It, too, published rules that strongly influenced designs.

In addition to the above, there came into being numerous one-design rules applicable to small racing craft, particularly on lakes and other protected waters.

In those days sailing yacht design tended to be highly individualistic one-man efforts. In this country the field was dominated by such names as Nathanael and Francis Herreshoff, Starling Burgess, William Gardner, B. B. Crowninshield, George Owen, John G. Alden, William Hand, Philip Rhodes, and the partnership of Sparkman and Stephens.

Design methods were gradually evolving, with ever-increasing weight given to science, and less to art. Hull structural details initially tended to be whatever the builder considered proper. Spars were usually of spruce, perhaps hollow; and relatively simple engineering analyses were used to estimate rigging loads. Hull form design was largely a black art until Prof. K. S. M. Davidson of Stevens Institute came along. He developed reasonably reliable ways to predict hull form resistance using small models that were within the economic reach of the typical yacht owner. His techniques were proven in 1937 with the success of the J-class *Ranger* in her defense of the America's Cup.

Just a word about costs. At the lowest point of the Depression, say 1931, the cost of a well-built auxiliary sailing boat, without sails or design fee, came to no more than fifty cents per pound of displacement. Today, owing largely to inflation, but also to generally higher standards, the cost would be at least eight dollars per pound.

Power Boats. In the 1920s power boats were slow by today's standards. Some were large and luxurious (just as most of us visualize a true yacht to be), but the majority (cruisers) tended to be pretty small. The yachts were usually steel, the cruisers wood. The larger craft may have used diesel engines, but the smaller ones usually had gasoline engines. A few of the largest yachts, such as J. P. Morgan's *Corsair*, used steam.

The driving force for change in power yachts was the change in available engines. Owing largely to developments in the automotive field, marine-adapted engines continued to grow lighter and cheaper. That trend led steadily to higher powers and higher speeds.

Speed boat racing became a popular sport, and speed boat designers such as Gar Wood applied their experience and skills to turn out standard design high speed runabouts for the upper middle class. Then, in the 1930s, the advent of the outboard motor (together with mass produced hulls), made motor boating available to the general public.

Well known names in power yacht or racing boat design included, in addition to Gar Wood, Clinton Crane, Cox and Stevens, Henry J. Gielow, Walter J. McInness, William Hand, Purdy of Port Washington, Crouch/Nevins of City Island, and John L. Hacker. Well known builders of production boats included the Electric Boat Co., Dawn, ACF, and Wheeler.

Post-War Developments in Sailing Craft

First a disclaimer. My professional experience has been heavily centered on the East Coast. I am under the impression that before World War II that area was this country's center of yachting activity. Since the war, however, that activity has become widely diffused. Recreational boats are now designed, built and operated along every coast of North America, on the Great Lakes, and countless inland lakes and rivers. How did this come about? War time experience brought many people to the sea so that interest in sail and power boats, and the ability to handle them became widespread. Lakes, great and small were gradually covered with boats, and the trend has continued. Small boats on trailers are seen in yards and garages all over the country. Industry records show the sale of 143,000 small boats in 1950, increasing to 310,000 in 1970.

As one example of geographic diffusion, let me call attention to California and Washington State. Those regions have produced some of our most successful designers, builders and sailors. The weather is generally favorable for sailing and the aircraft industry is a source of technical support that has led to stronger yet lighter hulls. Airframe production techniques have also led to standard boats being produced in quantity.

This history is as seen from this country and aims to report American activities. Nevertheless, yachting has developed in a world-wide sense. Some important developments have come to us from abroad and they will be noted.

Sailing Craft. Table 1 shows the major dimensions of successful offshore sail boats that may be taken as typical representatives of the indicated years. Figure 1 plots beam/cube root of displacement vs length and date. As you may note, the trend is from heavy to light, with increasing beam. This has been made possible by new structural materials and encouraged by the economies of light displacement. Though a large segment of the offshore fleet did more and more inshore racing, the fact that the longest transoceanic racing required offwind ability called attention to the speed of lighter boats. Often, without apparent reason, the many followed the few; and racing boats led the way.

Offshore. Boats built for offshore racing have been influential in setting patterns followed by a large proportion of the stock boat fleet. Major changes in the boats took place under a succession of rating rules. The standards of the CCA rule demanded rather heavy proportions, which prevailed in the United States through the 1960s until the International Offshore Rule (IOR) [32] was adopted. Among the smaller boats beamy centerboarders became winners. The moderate proportions of the best larger boats will be found in Table 1. In most other parts of the world the less restrictive RORC rule promoted greater experimentation. Designed to that rule right after the war, *Myth of Malham* (Fig. 2), a radically light, narrow boat with a deep keel and short ends, achieved great success. Except for the contrasting beam, *Myth*

Table 1 Selected seagoing yachts, 1930–1990

Nr	NAME	YEAR	DWL	DISP	BEAM	BM/DI	DLR
1	DORADE	30	37.25	40000	10.50	.308	345
2	BRILLIANT	32	49.00	85000	14.17	.323	323
3	STORMY W	34	39.75	44800	12.52	.354	318
4	ZEEAREND	36	40.00	45260	12.52	.353	316
5	NYYC 32	36	32.00	24500	10.58	.366	334
6	DJINN	38	44.75	59950	13.33	.342	299
7	BOLERO	49	51.00	93800	15.08	.333	316
8	FINISTERRE	54	27.50	18646	11.25	.426	400
9	WINDBOURNE	56	27.83	18230	10.87	.414	378
10	HESTIA	59	25.50	16230	9.67	.383	437
11	ANITRA	59	33.50	29762	12.33	.399	353
12	MALAY	60	31.98	17250	10.00	.388	235
13	PALYNODIE	62	28.75	16950	10.08	.394	318
14	CAL 40	63	31.18	16196	10.92	.433	239
15	CLARION W	63	30.00	22175	10.87	.388	367
16	CHIMAERA	67	36.00	22000	11.17	.400	211
17	EQUATION	67	60.91	57727	15.58	.405	114
18	PALAWAN 3	68	42.07	49215	12.49	.342	295
19	HOLGER DAN	68	34.94	34865	13.41	.412	365
20	ELIXIR	69	27.50	15810	10.10	.404	339
21	YANKEE G	70	45.25	37500	14.07	.422	181
22	RUNNING T	70	49.29	46000	14.02	.393	171
23	STORM	71	27.34	17524	10.03	.387	383
24	SWAN 44	73	35.75	30115	12.55	.405	294
25	SCARAM'CHE	73	43.97	32000	14.60	.461	168
26	DORA 1V	73	49.66	49000	15.39	.422	179
27	KIALOA	74	67.75	73500	17.08	.409	106
28	TENACIOUS	75	40.06	27500	13.78	.458	191
29	PROSPECT W	75	32.08	21414	12.20	.441	290
30	BLAZE	75	31.78	20340	12.72	.468	283
31	JIMINY C	75	31.56	7889	10.82	.545	112
32	TAR 10	76	28.51	8527	9.45	.464	164
33	EMILY	76	34.81	29580	12.56	.408	313
34	SCARM'CH 2	77	40.44	27500	13.84	.460	186
35	CAL 329	78	33.81	14250	12.01	.497	165
36	ARCADIA	78	47.98	59407	14.14	.364	240
37	KIALOA+	80	70.42	80000	18.54	.432	102
38	ROBIN	82	29.47	19434	11.42	.426	339
39	KARIATIS	86	57.38	86126	17.92	.407	204
40	RAMBO	86	28.37	5742	11.02	.617	112
41	CANNONB'L	88	60.31	73605	18.95	.454	150
42	PIED PIP'R	88	61.67	27111	15.40	.514	52
43	DRAGON F'R	88	33.03	21444	12.50	.451	266
44	COLLABOR'N	90	38.51	19600	13.89	.517	153
45	IONA	90	40.27	22051	13.80	.494	151

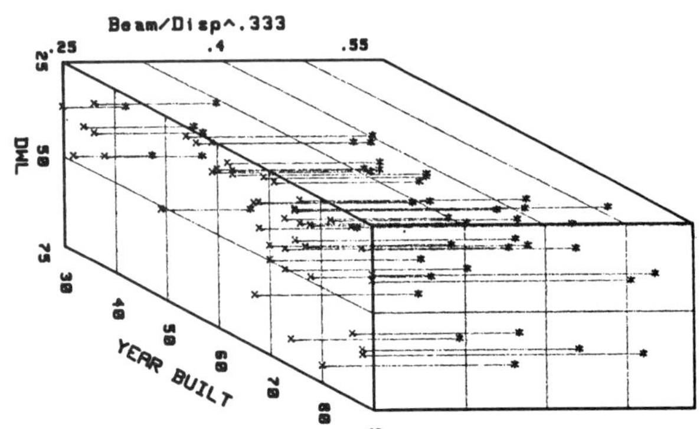

Fig. 1 Plot of beam/cube root of displacement versus length, 1930–1990 (Jan. 4, 1992).

foreshadowed the later trend toward short ends and light weight. She was designed by Laurent Giles with input from her owner, Capt. John Illingworth. Although virtually a breakthrough, her followers did not seem to improve on her performance and one surmises that she may have been hard to sail well.

My impression is that design advances were slow to come, and that the most successful yachts here and abroad were those that best fitted the different rating rules. After 1970, when the (IOR) had reconciled the great differences between the American and British rules, the character of offshore design changed rapidly and building was active.

At first the best boats had moderate proportions, but it gradually became apparent that light, beamy boats had become the usual winners. International racing was active and proponents of the rule found the boats fast and fun to sail; but the loss of fifteen lives in the Fastnet Race of 1979 resulted in serious concern regarding the common design, which had a limited range of positive stability and required the weight of the crew on the weather rail in order to do well.

In this country such concerns led to support for the Pratt Project (see below) and the development of an alternative to the IOR that would offer better safety and better cruising accommodations. A fair chance for boats of varied character was the objective, and the IMS, outcome of the Pratt Project, has gained in popularity and appears as a likely answer to that demand.

In racing under both rules, time has brought specialization and more extreme designs. The last decade has seen less offshore activity as many of the contestants have seemed to prefer shorter course racing while costs have climbed. Within level rating groups, however, competition has remained active.

Professional sailors have become active in specialized racing worldwide. Multihulls and water ballasted single-hulls have been widely used, mainly by professionals. French designers and sailors have had great success in this field. Speed records are broken every year.

Inshore Classes and Day Sailers. Some avid small boat sailors managed to maintain their competitive sailing throughout the war. Then, with peace came a period of rapid growth. Initially the emphasis was with pre-war one-designs such as Stars, Lightnings, and Internationals. There was, however, a demand for smaller, less expensive boats. (In 1937 the first Lightnings cost just under $500. This jumped to almost $3000 after the war.)

At this time fibreglass and its derivatives became available, and the beneficial impact was felt through-

Fig. 2 English offshore racer *Myth of Malham*, 1949. (Photo by Beken of Cowes Ltd.)

out the small boat industry. The new material produced lighter, less expensive hulls capable of high sailing performance while operated by minimum crews. Today there are countless such classes, but Bruce Kirby's Laser design is an outstanding example, and one with many thousands of units spread over the world.

Small Auxiliaries. Similarly small auxiliaries have multiplied. These generally have cruising accommodations ranging from Spartan to deluxe depending on size, price and the degree of racing emphasis. Today an owner has a wide choice; although, as this is written financial stringency, coupled with a lack of space in slips, has curtailed production and closed many shops.

Open Classes. Owing to the price advantages of production methods the majority of new boats have been in the one-design category. Thus open class racing in the conventional sense has grown but slowly. The six, eight, and twelve meter classes have had their active supporters, notably in the twelves as the vehicle of the America's Cup from 1958 to 1986. Applied only

Although many of the developments in design and construction centered on the Coasts, particularly the East Coast, the Midwest, particularly the Great Lakes area, represented one of the greatest areas of interest in recreational boating.

Unique to the inland lakes of Minnesota, Wisconsin, and Michigan are the Inland Lakes Scows. With a history that goes back to the early years of the century, developments in these high performance racing yachts reflected advances in sails, rigging, and construction materials that were occurring on the coasts. A paper presented at the 1989 Yachting History Symposium, sponsored by Mystic Seaport Museum, provided an excellent history and update of this class.

Another popular class, the Thistle, *had its origins in the Midwest. Sandy Douglas, an enthusiastic sailor of 14-foot dinghies, was employed in the mold loft of the American Ship Building Company during the war years. As the war drew to its conclusion and the demand for naval vessels was tapering off, Sandy apparently was finding some spare time and laid down the lines for the* Thistle *on the floor of the shipyard's mold loft. With the return to civilian enterprises, Sandy established a company to build not only the fourteens, but also the* Thistles, *an enlarged, one-design version of the fourteens. This must have been one of the first enterprises to produce laminated wood hulls on a production basis.* Wooden Boat *magazine featured the history and development of the* Thistle *in issue 29.*

Another of the world's popular classes, the vintage Snipe, *designed in the 1920s as one of the first of the poor man's yachts, has enjoyed continued popularity on the inland lakes of the Midwest as well as other areas of the country. The Snipe organization, like others of the older one-design classes, has modified its rules to permit the use of newer materials without diminishing the competitiveness of the older boats.*

The 1970s and '80s saw a resurgence of interest in the classic wooden boats of the 1920s and '30s. In some areas, such as the le Chenaux Islands of Michigan, the lakes of Ontario and the Lake George area, some of these boats have been maintained and operated by the same families through generations. For the most part, however, people who had acquired an appreciation for these classic boats had only two or three choices. If lucky, they could locate an old hull in reasonably good shape that could be restored with only a moderate amount of material replacement and perhaps a re-engining. More often, they would have to start with something little more than a pile of rotted wood and rusty iron. By diligent personal effort, or very expensive professional labor, a "reconstruction" could be produced that retained little more than the shape of the original. The final alternative, a replica, was becoming more readily, if very expensively, available. Several firms, Grand Craft in Michigan and others in both the Great Lakes and Lake George area, were producing replicas of Hacker, Chris Craft, and Gar Wood boats, as well as contemporary designs that still reflected the character of the earlier boats. The replicas, although resembling the earlier boats in appearance and finish, utilize contemporary materials and methods as well as current high-powered lightweight engines. Thus, both performance and life expectancy are much greater than those of the prototypes. Organizations such as the Antique and Classic Boat Society and several of the so-called Marquee organizations representing owners of Chris Craft, Gar Woods, etc. provided both encouragement and guidance for the acquisition, restoration, and maintenance of these boats.

The last stand of the Universal-class sailing yachts was probably in the Great Lakes area. R-class sloops raced actively in the Cleveland, Chicago, and Toronto areas through the fifties. A few also remained in Detroit, although the 6 Meters were the more popular racing class in those waters.—W. C. Cowles

to the largest racing boats, the Universal Rule died. Locally popular types such as dinghies and inland lake scows have spread widely. One class or, perhaps type, has led all others in popularity and growth, namely the sailboard. It has the advantages of low cost, ease of transport and high speed owing to light weight coupled with crew-provided stability, and it can be sailed anywhere, including very shallow water. Although handling requirements differ from those of conventional sailing, its use of the forces of the air and water interface demands the consideration of sailing fundamentals. The sailboard's appearance during the sixties brought thousands to the sport. Records for speed under sail of better than forty knots have been held by sailboards [11].

America's Cup. The pre-war America's Cup activity was in the large J-class yachts. This series climaxed in 1937 with *Ranger*'s decisive win. Inflated post war costs discouraged America's Cup activity. This led the New York Yacht Club to petition the courts to change the deed of gift to reduce the minimum waterline length from 65 to 44 ft. This made a match possible in the twelve meter class, which was still popular in England. The Royal Yacht Squadron (Great Britain) thereupon challenged the New York Yacht Club for a match in the summer of 1958. Reference 12 offers a good account of Cup racing through that year.

Three new twelves were built to compete for the right to defend, and the 1939 *Vim* was returned to good racing trim. In the trials the new *Columbia*, a near sister of *Vim*'s, just edged the older boat and beat the British *Sceptre* in four straight races. From then until 1986 the America's Cup was confined to twelve meter boats, with races being held about every third year.

The American representatives were consistent winners until 1983, when the Australian yacht *Australia* with a radical wing keel (see Fig. 3) succeeded in beating the conventional defender, *Victory*, off Newport RI, taking the cup to Western Australia.

In 1986 off Freemantle, Western Australia, one of five American challengers, *Stars and Stripes,* took the cup for the San Diego Yacht Club. Lack of an announced venue or class for the defense left an opening for a legal challenge from New Zealand in a 90-ft waterline yacht and at an early date. Unsuccessful in legal steps to invalidate the challenge, San Diego built a 60-ft catamaran to defend. This precipitated further legal action both before and after the race, but the catamaran easily beat the big monohull challenger. In the final court action the San Diego victory was confirmed. If I may editorialize, this mismatch, seemingly initiated with no sporting intent, seems a sad turn in the history of the competition.

While the competitors were still in the courts, an independent agreement was reached between the countries that had raced in Freemantle to work out a new America's Cup rating rule to apply in future contests. This was agreed upon and the first match under the new rule was held off San Diego early in 1992. This followed an extended series of trials, in which eight foreign challenger candidates took part separately from three defense candidates. The competition ended in May with *Il Moro*, the Italian challenger, losing the Cup races to the defender, *America 3* in five races. Application of the America's Cup rule resulted in boats well suited to the light Pacific weather. Some features of this rule will be discussed below.

Standardized Boats. A major development that gained momentum from the forties through the eighties was the wide acceptance of production boats. FRP materials offered relatively low first cost and easy maintenance. First used for dinghies and other small types, they are now found in sizes up to 80 footers, and in both heavy cruising boats and light racers. Pierson Yachts of Bristol RI was possibly first on the scene with the Ensign and Alberg-30 classes. Virtually all boats over about 30 ft long were fitted with inboard auxiliary power. Gasoline engines were initially most

Fig. 3 *Australia II* keel, September 1983. (© Stanley Rosenfeld photographer.)

popular; but as small diesels were developed, they became universal, offering both safety from fire and extended cruising radius. Smaller cruisers usually carried outboard power. Overall, the growth of the fleet of small boats for all purposes has been spectacular, though the most recent years have been slow.

Ocean Cruising Yachts. Sailing yachts for ocean cruising seem to fall into two categories. Some, designed for racing substitute speed for comfort. True cruising boats are more stoutly built, place no reliance on human ballast, and have generous ranges of stability. Some boast such amenities as electric generators to serve large freezers, cabin heating and air conditioning, and fresh water making.

During the later years of our period a number of yachts, of 90 ft and up, both sail and power, were built for owners who could send them to foreign cruising grounds and fly over to enjoy new cruising conditions. These yachts often carried complete and complex communication equipment, although the use of advanced electronics was by no means limited to the bigger vessels.

Maxi Yachts. Most ocean races had upper and lower limits on the size of the contestants. Owners wishing to finish first often built to the upper limit. For years it was 72 ft overall, but recently some maxi owners have formed an association setting some rules of their own and establishing a schedule of worldwide racing. These maxi yachts (Fig. 4), about 80 ft overall, are, in contrast to the cruisers, lightly built and carry minimal accommodations. They are out and out racing yachts. Usually they have been sailed from series to series by delivery crews, though some have been shipped in special containers.

Although traditional rating rules prohibit shifting ballast, certain recently developed classes permit pumping water from side to side.

Multihulls. Catamarans and other multihull boats (Figs. 5 and 6) combine great stability with light weight to offer high speed under sail. Although the idea is old, there were obstacles to wide acceptance. These included a limited range of stability and structural complications. Recently the urge for speed (and some new quick release controls) have offset stability concerns, while new structural techniques have been used to join the hulls, with the result that popularity has been explosive, especially in professional long distance racing. Since the middle sixties small day sailers and racers, such as the Hobey Cats, have grown into the thousands, while larger oceangoing catamarans and trimarans have broken sailing ship records between ports all over the world. The 1989 America's

Fig. 4 Maxi yacht *Timoneer.*

Fig. 5 Multihull *Fleury Michon.*

Cup series, in which the catamaran walked away from the large monohull, simply confirmed the well known speed advantage of multihulls.

Motor Sailers. Yachtsmen who want to enjoy pleasant cruising under sail, but who also want to make progress when the wind is not right, are likely customers for motor sailers. Such boats, particularly in the larger sizes (up to 80 ft), may hope to maintain reasonable speed under either power or sail. In the early years of our time frame engines were relatively large and heavy, thus precluding their use in all but the larger boats. Now, however, smaller, lighter engines extend the practical range downward. Nevertheless, at the small end of the scale good all around performance is unlikely.

William Hand pioneered the motor sailer as a seagoing power boat. An early Luders design for Clifford Mallory broadened the concept toward sail. Philip Rhodes, S&S and others later developed types with still more sailing ability. As engine power grew and hulls could better combine lightness with strength, performance improved. A trend setting, light displacement motor sailer of the sixties was the *Blue Leopard* (Fig. 8), designed by the Englishman Laurent Giles. Somewhat more than 80 ft long, she could do as much as 12 knots under either sail or power.

Fig. 6 Multihull *Primagaz.*

Fig. 7 The sailboard has the advantages of low cost, ease of transport and high speed owing to light weight coupled with crew-provided stability, and it can be sailed anywhere, including very shallow water.

Cruise and Charter Vessels. The popularity of sailing has led to the building of sailing boats for both cruise and charter purposes. Types range from standard or modified production models for bare boat charters to larger yachts with paid crews of one or more. Those for bare boat use must be simple and rugged in their rig and equipment and are apt to be provided with more berths than appropriate to their dimensions, which run from 30 to 50 ft in length. Commercially operated fleets of these boats are now found in many popular cruising areas such as the Caribbean. Boats chartered with crew are more apt to be privately owned larger yachts, offered to others through either charter organizations or yacht brokers.

Cruise vessels range in size from large commercial cruise ships, some of which have been fitted with sails, to small coastal types, which may be crewed by their paying passengers. They must meet Coast Guard requirements that have recently been tightened so that rigs, usually schooner or ketch, are moderate in area. These boats are generally beamy to provide space for accommodations, and shoal in draft for ease in docking. These proportions do not lend themselves to excellent performance under sail unless weights are kept light so that crew weight provides sail carrying power. Virtually all cruise vessels have some form of auxiliary power.

Design Technology: Sail. *General.* During the past half century the trend in yacht design technique has been toward less art and more science. Nathanael Herreshoff derived hull lines from his hand-carved models, and used those lines largely to enlarge the offsets to full scale. Today, lines have become the foundation of detailed analysis. Perhaps the most significant development has been the quantification of performance through vector analysis of the individual elements constituting the drive and drag forces.

I surmise that before Davidson, few sailing yacht designers thought of either hull drag or sail drive in precise quantitative terms, though it is recorded that the Scottish naval architect G. L. Watson tested a model of *Valkyrie III* before the turn of the century and that Starling Burgess tested a model of *Rainbow* in Washington in 1933. He is also known to have tested *Rainbow* at the University of Michigan.

Davidson. Davidson's work of the mid-thirties marked a turning point in the numerical analysis of yacht design. He applied sail forces derived from driving and heeling coefficients found from measurements of a single rig under sail, though confirmed by comparison with aerodynamic theory. These were used to predict windward sailing performance. Later Paul Spens [13] extended these for reaching and running angles so that speed could be computed on any course.

Davidson concentrated on windward hull performance by testing the models with side forces and heel angles appropriate to wind strength and stability. Much more recently, high speed tests have simulated downwind sailing up to speeds such as Froude Number 0.65.

Pratt Project. The next big step in design for performance resulted from the Pratt Project at MIT,

Fig. 8 Motor sailer *Blue Leopard*, England, 1963. (Photo by Beken of Cowes.)

Fig. 9 America's Cup *Ranger.*

the seeds for which were planted by Prof. J. N. Newman's plea for a rational rating rule based on the mechanics of yacht performance {14]. A group of sailors who were disenchanted with the IOR rules collected a fund to support his proposed work. The project included a series of coordinated model tests, and an analytical study of sail forces. From this was derived a computer-assisted method to predict sailing performance under various conditions, as described in Refs. 14 and 15. The success of the project is shown by its use in the Measurement Handicap System, (MHS) now the IMS ("I" for International). Although the project was supported for use as a rating rule, the fall-out in the form of organized model test data and the formulation of the computer program for performance prediction (VPP) has been invaluable to yacht designers well beyond the racing field.

Computers. Since Davidson's contributions, most of our advances in yacht design have been made possible by the computer. Real progress dates back to the early 1970s with pioneering work by Dr John Letcher [16], Dr Hugo Meyers [17], George Hazen [18], Prof. Jerome Milgram, and others, including those involved in the Pratt project mentioned above. Their analytical tools include velocity prediction programs, now known as VPPs. These are now bringing reasonably close agreement with model tests, which I think is good news because model testing is expensive, and consistent results are often difficult to achieve.

Other computer programs are now available for predicting such performance factors as induced drag, effect of heel, and effect of wave encounters. Different combinations of sail shape and sail combinations can also be analyzed.

An example of advancing hull design was given by the Australian America's Cup success in 1983, when a triad consisting of an able yacht designer, a hydrodynamicist skilled in tank work, and an aerodynamacist combined their talents to design a breakthrough wing keel. In the next cup match similar studies supported by simulations using the match conditions in conjunction with weather records and a good VPP made it possible to bring back the cup.

Aerodynamic progress is exemplified in the use of rigid wing sails on various high performance catamarans, capped by the sail used in the 1989 America's Cup defense.

As in other branches of the marine industry, computers are widely used in many facets of yacht design in addition to those mentioned above. Of particular interest has been the work of Letcher [20] and Hazen in generating and fairing lines.

Capsize Study. During the 1979 Fastnet Race (England) a sudden blow produced steep breaking waves, leading to many mishaps, particularly among the smaller boats, some of which capsized. Fifteen lives were lost. This focused attention on the need to determine what design characteristics were central to the tragedy. Studies [39] and [40] were made in the United States and England, which led to the formation of a joint committee of the USYRU and SNAME, to study yacht capsizing in terms of its mechanisms, using both model tests and analytical methods. The work of this committee, combined with the British study, threw valuable new light on the subject of yacht safety [41,42]. Attention was drawn to the limited range of positive stability occurring when overly generous beam is combined with light displacement and a high center of gravity. The favorable effect of a large transverse moment of inertia was also confirmed by model tests.

New Studies. Following the technical initiatives illustrated by the Pratt Project, and as a response to the success of Australia in the 1983 America's Cup races, many possibilities were explored in the process of designing succeeding contenders. Keel shape and other hull form factors have been explored in depth by means of both model work and computation. Keel studies have been directed toward both lower drag and lower center of gravity. New insights have been gained relative to flow fields affecting lift, drag, and interference effects at the juncture of keel and hull. The currently popular clear demarkation between keel and hull suggests the application of appropriate Reynold's Numbers to different area, thus improving predictions of both viscous and wave drag. Some experts point to the possibility of maintaining laminar flow over full scale keels.

Other current studies are considering the effects of midsection shape, waterplane loading, and the relationship between ship moment of inertia, displacement, and rough water performance. Recent model tests and analytical work at the University of Michigan, for example, brought a better understanding of the relationship between seakeeping and moment of inertia. This was done for a relatively new organization called PACT (Partnership for America's Cup Technology), a consortium made up largely of American contenders for defense of the Cup. The overall work (at Michigan and elsewhere) has included thorough comparisons of model tests and computed forces. Although still proprietary, in time the results will be made public.

Other Model Studies. Since the mid-thirties model testing has been an accepted part of the design of racing yachts and others calling for good performance. Most tests are made with the models both upright and at several heel angles. Tests are run in waves as well as in smooth water [22]. Side force are usually applied, either an amount appropriate to the heel angle and stability, or at several values to permit cross fairing of the data. Flow is visualized using tufts, colored liquids, or laser methods. Models weighing as little as 27 pounds (1/13 scale for 12 meters) were generally accepted up to the late sixties. In 1970 disappointment with two new twelves and one that had been altered led to a general change to much larger models; a scale of three to one being widely used for twelves. These gave somewhat better results, though not without some disturbing scatter in repeated runs. Scale questions are discussed by Kirkman and Pedrick [21].

Fig. 10 *Ranger* model under test in the Davidson Laboratory towing tank.

Outfitting. In what comes to be as regarded as standard equipment, yachts, whether sail or power, have become increasingly complicated over the years. Such complexity has caused the cost of yachts to increase far faster than the general cost of living. In the mechanical and electrical areas, the installations have become extremely complex. This brings a need to coordinate the installation and operation of many different pieces of auxiliary machinery and equipment. In preparing the plans of larger yachts CAD methods have become invaluable in laying out the complex piping and electrical systems.

Not only must the equipment operate properly, it must do so quietly. Then in addition the accommodations must be not only comfortable and quiet, but also beautiful. For such reasons it is frequently advisable to employ consultants in one or more of the specialized areas, ranging from mechanical and electrical installations to interior decoration, noise suppression and superstructure design.

Advanced designs. When greater speed is the object, as in most advanced designs, attention focuses on the combination of reduced drag (usually weight) and increased drive (usually stability). In a series of high speed sailing trials in England, speeds have reached 40 knots by sailing craft with those characteristics in the extreme. Catamarans, trimarans, long single hulls with outriggers or trapezes, and hydrofoils have been used, sometimes in combination. In 1990 a sailboard first reached 40 knots, probably with the displacement advantage of the lifting force given by a rig heeled to weather.

Spars. During our period aluminum alloy spars have replaced wood in the great majority of sailing yachts. In the critical relationship of Young's modulus to density, the materials are nearly equal, but with aluminum the thinner walls provide a greater moment of inertia for the same weight and outside diameter. Standard sections and large production have moderated costs and the attachment of rigging is easy with the metal spars. Steel also has similar properties, but an otherwise well proportioned small steel mast would have delicately thin walls, so its use is restricted to large yachts. High modulus composites (usually employing carbon fibres) are fast coming into use, though often restricted by rules in the interest of cost.

Most spars are analyzed by Euler column theory, with loads determined from stability calculations, which often have been confirmed by tension dynamometers in the rigging. Assumptions on end fixity have been based on experience. Computer codes have been used to estimate load distribution. Finite element studies have been made but, so far, do not seem to have taken the place of the simpler Euler method. Loads vary greatly with wind and sea conditions, as well as loadings resulting from crew decisions, so that factors based on experience are usually accepted as the best guides. A complicating factor results from the advantages of mast bending under sail to vary sail shape, while most rules prohibit permanently bent masts.

Sails. Over the past forty years new materials and a better understanding of aerodynamics have combined to produce greatly improved sails.

The change from cotton to synthetics marked relief from the constant drying of sails to avoid mildew. Synthetic fibers began to appear before the war, but it was not until the 1958 America's Cup races that polyester fibers (Dacron in the U.S.) proved their worth. Fibers made from many different resins have been tested. Now Kevlar is also widely used in place of Dacron for some racing sails. It is expensive and tears easily but is highly resistant to stretch. Light sheet Mylar has also found a place. A recent development applies materials of differing weight, composition and direction of weave to specific parts of a sail depending on expected loading. Over the years sails have grown much lighter, while they have been cut so that the crew have better and more complete control of their shape and loading. New fabrics hold their shape under pressure better than older ones, but the pressure to save weight aloft places a premium on short-lived light fabrics. The lighter weight of modern sails is an advantage in both sailing and handling.

The shape of modern mainsails can be controlled by the adjustment of tension on the luff, foot, or vang in a way that was impossible earlier, thus making fewer sails necessary to meet varying conditions.

Computer programs have been widely used in cutting the fabric to give the desired shape, but the analytical determination of the best aerodynamic shape seems to be dependent on studies and computer codes not yet well developed. Once we reach a better understanding of the interaction between sail forces and sail shape, the way should be clear to match accurately hull and sail characteristics, thus leading to improved performance.

Milgram [23,24], Greeley [25], and others have analyzed sail shape with a view to improving effectiveness. Hazen's work [26] has also been significant. Rule requirements and the characteristics of materials make such studies difficult and limit their applicability; but progress is being made with the help of aircraft experience.

At this writing the ratio of cost to certainty with regard to sail definition seems high, but we may hope that this sort of investigation will bring the time when the designer will have the same control over the sails that he now has over the hull.

Reefing. Various reefing systems have been used. In the fifties and sixties roller reefing booms were popular. Later, so-called jiffy reefing was widely used. More recently stowaway masts have become popular, particularly as they can be motor driven to save work. Roller jibs are also in wide use though generally for furling rather than reefing. Headsails have been made easier to shift by a change from hanks to one of many luff groove devices, despite the inconvenience, if not danger, in that the sail is no longer captive as it comes out of the groove.

Stow away masts allow the sail to be rolled inside as a method of reefing or stowing. Large units can be powered, making for ease of handling, as a result of which large sailing auxiliaries have had increased popularity. Although this practice is fundamental to much large sailing yacht design, I hesitate to endorse the dependence on motor driven reefing arrangements because of the danger inherent in such operating difficulties as loss of power or jamming.

Arrangements. One factor for change in yacht arrangements is the growing numbers of females aboard. We also note that both the females and their male counterparts have taken over housekeeping duties that were previously done by paid hands. Thus galleys have become more accessible and located closer to the center of deck activity, usually aft. Except in large boats, accommodations for the hands, usually in the fore peak, have been generally eliminated. Production boats, in particular, have been designed to appeal to the ladies. Considerations of seagoing comfort have often been sacrificed to gain comfort in harbor. Berth widths have increased and decorator colors have been introduced below.

Among racing types the trend has been the reverse. Comfort has been secondary if considered at all. Pipe berths have come into use, generally in tiers so that sleeping crew members can have their weight to weather where it helps the most, though the off watch is often kept on the weather rail. Deck plans have received great attention so that clear and direct leads to winches are provided. The winches must be accessible with room to swing the handles, whether of deck or pedestal type. In larger boats pedestals are linked for increased power.

Measurement Rules (Sail). Because a successful racing boat must be designed to minimize rating in relation to speed, rating rules will have a profound impact on shape. Thus throughout history the rules in force have determined the geometry of hull and rig; they have been "type forming." Whether the type has been approved or rejected has usually determined the lifetime of the rule.

The International Rule. With a tight link between length and displacement enforced by a stiff penalty on displacement below standard, the International Rule [28] has been highly type forming. The rule was used for the six, eight and twelve meter classes. It was little changed during the period under review, although new structural materials, such as aluminum alloy and FRP, were provided for by Lloyds Register. With a new rule to be used in the America's Cup, it seems unlikely that the International Rule will be further used except in the six and eight meter classes.

The New America's Cup Rule. A new rule [29] for use in the America's Cup has been formulated by an international committee of designers with America's Cup experience. In form this rule resembles the older International Rule [29], but the parameters and limits are such as to produce a boat about 60 ft on the waterline with light displacement, a large rig and a deep keel. Scantlings are controlled by the rule, but the hulls have turned out to be very light and easily punctured. Highly engineered composites have been used for both hull and rig.

Offshore Rules. In the United States in the fifties and sixties the typical offshore racing owner was a cruising man who liked to race. In Europe there was more differentiation between the cruiser and the racer. The Cruising Club of America was the *de facto* leader of offshore racing here, and before the adoption of IOR and IMS its rule [30], was used by other influential groups such as the New York Yacht Club. It was also used on the Southern Circuit, a series of winter races around Florida that had great popularity. This rule had been adopted in 1934. Using base proportions that were desirable for cruising as well as for racing, it applied penalties and bonuses to variations that affected speed. The resulting type was good for cruising, but less so for racing. By the sixties, measurement problems were showing up, the worst being the difference that often occurred between measurement and sailing trim and weighed displacement. Additionally a simplistic length measurement was opening the way to exploitation by means of U-shaped ends.

In England, the Royal Ocean Racing Club employed a less confining rule [32], which gave good racing, although to American eyes a boat's ends tended to be unduly fine and the scantling provisions offered an obvious loophole. In other countries there were other rules. As international racing became increasingly popular, the need for a uniform rule was recognized.

IOR. In June 1961 an *ad hoc* group from Germany, Great Britain, Sweden and the United States met in Bremen to discuss possibilities. This led to forming an expanded association called the Offshore Rules Coordinating Committee, which later met in London, and in 1967 appointed the International Technical Committee. The result was the formulation and acceptance of the International Offshore Rule (IOR) [32], which became operative in 1969.

It may be said that the IOR led to much good international racing in fast boats, but that over time and in the minds of many the type that developed had serious drawbacks at sea. This was driven home in the Fastnet Race of 1979. Studies that followed made it apparent that pressure for favorable ratings had led to boats having a dangerously high ratio of beam to depth combined with a high center of gravity, resulting in hulls with a limited range of positive stability.

IMS. In the United States some had been unhappy about the IOR rules from their beginning. The Pratt Project, and the ability to predict speeds, led directly to the Measurement Handicap System. The aim was to write a measurement rule that would encourage boats that were both safer and more comfortable than the IOR type. It was also recognized that, except in a near-identical fleet, the winner was determined by weather conditions sure to suit one boat better than another. This led to a rule that adjusted time allowances to weather and course. The IMS was exactly that as the time allowances were based on the VPP applied in accordance with either assumed or experienced conditions.

An MHS-class was first raced in 1978 and has been continued to the present. In 1986 the Offshore Racing Council adopted the rule as an alternative for offshore competition under the name International Measurement System (IMS) [33]. Recently it has appeared that some new winning boats, specifically designed to the rule, have exposed vulnerable points in the rule. The technical committee recognizes the need to keep ahead of designers and intends to adjust the rules so as to keep the way open to the dual-purpose boat. One concern, for example, is the matter of the adjustment for weather conditions.

Other Rules. Further information on these and other rules may be found in Ref. 34.

A Personal Observation. In my opinion, open class rating rules have provided needed alternatives to the one-design classes by giving the owner and designer the opportunity to exercise individual preferences; yet it seems a fact that the best racing has happened when type forming within a rule has resulted in a close approach to the one-design condition.

Power Boats

Power boat technology has always been driven by the availability of lighter, more powerful engines and lighter hull materials. A resulting general trend has been toward lighter, faster boats, and—until the past two decades—a growing popularity for boats in the smaller sizes. Demographic influences have included growing prosperity in the upper income range and the attractions of wintering in warm climates.

After the war boats of all types came into great demand, the smallest, such as outboards, leading the way. As the economy grew the standardized cruiser became popular. Chris Craft came out right after the war with a mass produced line of wooden power cruisers. Hattaras Yachts seems to have been first with fibreglass, offering a popular 41-ft hull designed by Jack Hargrave. That segment of the industry grew as other builders saw the possibilities of standardization and FRP construction.

Outboard powered boats were built in growing volume. FRP and aluminum became the most popular materials with wood a distant third. Types multiplied to suit special purposes such as fishing, racing, water skiing and cruising. With increasing power, cruisers up to 40 ft in length have been outboard powered. Rubber inflatables found their own niche after becoming popular abroad.

Some inboard powered boats were built for speed, others for comfort. Light weight gasoline engines of up to 1000 hp became available, encouraging offshore racing, while the more sedentary element enjoyed luxuriously outfitted cruisers in the 50- to 100-ft range. As the present date approached, the larger boats especially, grew faster until speeds of 25 to 35 knots were commonly achieved with lightweight diesel power. In a counter-trend a number of slower trawler-like yachts were built by owners requiring great cruising radius and comfort at sea.

Since the late seventies a group of the largest new yachts have become known as mega-yachts. Although they may be driven by either power or sail they share many characteristics in addition to size. One of the most important is the ability to communicate with all parts of the world. By virtue of the time an owner may spend on board, the mega-yacht typically has luxurious accommodations.

Costs have risen with inflation and complexity. Between the late thirties and late eighties costs have increased about ten fold (or more if higher powers and more electronics are taken into account). Recently advertised costs of stock cruisers have been between five and six dollars per pound. Early power boat prices seem to have been about fifty cents a pound. Custom design and construction could add 50 to 100 percent to such figures.

Design Technology: Power Boats. *Engines.* The trend in propulsion systems has combined rapidly growing power with lighter weight. Outboard motors have come to dominate the gasoline powered group, while diesels have taken over the high power range.

Outboard motors are used in far greater numbers than inboard engines. Their numbers have grown rapidly through most of our period, more than doubling between 1945 and 1970, though leveling out more recently. A pre-war range of available power of up to 22 hp has now gone as high as 300. Average weights dropped from five pounds per hp to about three in 1960, and are now down to less than two in the top power range.

Inboard gasoline engines built for marine use peaked at the time of the war and rapidly declined after that, as suitable automotive engines became available in quantity. Now there are few special marine engines on the market, but many automotive conversions.

Inboard diesels now cover a wide power range so they are used for auxiliaries such as generators and pumps as well as main engines. Following the war marine diesels were built by such firms as Caterpillar,

Fig. 11 Ninety-five foot diesel-powered motor yacht *Kalamoun.* (Sparkman & Stephens, Inc., New York.)

Fig. 12 Ninety-foot offshore motor yacht built by Palmer Johnson, Inc., Sturgeon Bay, Wisconsin.

Cummins, General Motors, Gray, and Volvo (Sweden). Ranging up to 400 hp, such engines ran at around three times the weight and cost of gasoline engines, while offering greater operating economy, life and safety. Among the most widely used were the General Motors two-cycle 71 and 110 series. Thousands of GM 6-71s powered landing craft during the war, making them widely known. In the late fifties turbo-charging added close to 20 percent to their power and lengthened the life of the series. Ultimately more efficient four-cycle engines, also turbo-charged (of which Caterpillar is an example) became the preferred type in the 400–500 hp range. Today diesels have taken over the market except in the racing and outboard fields.

Propulsion Systems. The availability of more and more power over the years, coupled with the wish for speed, has led to an intense study of propulsion methods. The most common problem has been that of propeller cavitation. In the past the answer has been to use larger diameters while reducing rpm until reaching limits imposed by space, angularity and appendage drag. We shall next look at some of the many ways to apply power.

The outboard motor with its conventional propeller, angle drive and directional control has been the most widely used method of driving small boats, followed by its near relative the inboard/outboard, or Z drive, as size has grown. Both were useful at high speeds in their ability to direct the thrust in the vertical and horizontal planes so as to eliminate the eccentric propeller loading caused by the shaft angle in an inboard installation. Outboard drives are easily installed in a wide variety of hulls suited to high speeds in which it was advantageous to carry weight well aft.

When arrangements require engine weight amidships, the conventional shaft and strut carry the power to a conventional propeller, usually three bladed. Reduction gears are frequently used in the interest of propeller efficiency.

Another option that has recently become popular is the water jet. Limitations on pass-through and discharge areas make this essentially a constant thrust device with acceptable efficiency depending on high speed. Jets have the merit of reduced appendage drag as well as reduced noise and vibration. A serious deficiency is in extended cruising at low speeds. Thus the jet is sometimes combined with a low power submerged propeller for cruising, while the jet is used for running at speed.

In some racing applications step-up gears are used with super-cavitating propellers, which are held to a small diameter to reduce torque on the hull. The blades are designed to accept the sacrifice of low pressure on the back while maximizing thrust on the face. Recent practice, however, has favored the water jet for high speeds.

The recent past has also seen some interest in the surface piercing propeller. This has much in common with the super cavitating type in that ventilating to the atmosphere also destroys the usefulness of the back surface, while relying on the face to accelerate the water. The surface propeller is fitted to a shaft as close as possible to the bottom of the hull so that at speed only the lower half diameter is in contact with the water.

Air drive has also been used for the advantages it offers in shallow water. Recently, however, the jet has been favored in that service.

It seems worth noting that, at low and moderate speeds, or until cavitation is unavoidable, the conventional screw propeller remains the most efficient method of propulsion. The alternatives seldom approach the efficiency of the submerged propeller until that efficiency is degraded by cavitation. On the other hand, other options may offer the advantage of reduced appendage drag and drive angularity.

Hulls. The demand for speed and the availability of ever greater power has influenced hull design. The trend over our period has been away from round bottoms and toward vee shapes. Again, there has been a wide spread between small and large against slow and fast; and hull form has been varied to suit the circumstances. Catamarans dominate the racing hydroplanes except those for offshore competition.

A Froude number of something over 0.6 usually signals a change from round to vee bottom. Post war fishing boats and cruisers were often modest in speed, so round bottoms were appropriate. But, with increasing power and speed, the vee bottom became appropriate. With that form it was possible to carry the abovewater beam well forward, benefitting accommodations. Not infrequently, at medium speeds, a vee bow is blended into a rounded stern.

Deadrise angles have tended to increase over the years and now range up to 30 degrees. Older practice was to use a warped bottom, with deadrise decreasing from bow to stern. In offshore power boat racing hulls the uniform deep vee is necessary to moderate accelerations in rough water. Experience shows that, although in theory the great deadrise increases resistance, it is offset by the better ride. Moreover, increases in installed power have made small differences in hull drag unimportant.

Power boat racing has changed rapidly since the forties and has branched into several different classes. The Gold Cup class grew from a gentleman's fast runabout with narrow beam, flat deadrise, and a straight bottom, often with multiple steps. Speeds increased to the 100 mph range with the use of surplus airplane engines after the war. Hull forms tended toward greater beams. In one type, represented by the 1953 record holder, *Slo-Mo-Shun*, the bow was supported by widely spaced sponsons, while the stern was supported on the propeller alone; this was the "proprider type." From this, and following success among outboards, the related catamaran form took over.

Offshore racing saw rapid growth starting in the late sixties, contributing to the spread of the deep vee hull and the stern, or Z, drive. The Hunt designed Moppie type led the way followed by the Cigarette and related high deadrise designs.

Materials. As in the case of sailing craft, the availability of improved materials has had its impact on power boat design. The materials that were common in the forties, namely wood and steel, have been widely replaced by aluminum alloy and FRP, although wood retains a place in one-off construction among many who value its aesthetic appeal.

Right after the war, wood was the material of choice for both sail and power yachts of less than 80 ft in length. Classic wooden boats were built by setting up molds to the inside of the planking. Longitudinal ribbands were fitted on the molds and steam bent transverse frames fitted inside the ribbands, which were then replaced by the planking. Planking was ordinarily applied from both deck and keel, meeting at a shutter strake. Clamps and shelves to support the beams were continuous longitudinals fitted at the frame heads. The heels of the frames were joined by transverse floors bolted through the keel. Consideration of the number and variety of these steps will suggest reasons for the relative economy of FRP construction as discussed below.

Central American mahogany was used for top quality planking, though replaced by cedar in smaller craft, and yellow pine or fir on larger, but less expensive boats. Because of its long life and hard finish, teak was another good choice when its weight was acceptable. To save money, double sawn frames with overlapping joints were often used instead of bent frames. White oak was the preferred material for either type of frame, while light weight bent frames of elm were used on some racing boats.

Silicon bronze screws and bolts, in the proprietary alloy Everdur, were considered the best wooden hull fastenings. Copper rivets were also popular. In larger, heavier boats galvanized steel spikes, bolts, and drifts were often used for the sake of economy.

The forces that were to bring in FRP introduced builders to water resistant resins for finishes and adhesives. These made it practical to use laminated wood in many applications such as bent frames and heavy sections, for which good wood was becoming scarce. The Gougeon brothers developed a technique called "cold moulded" wood construction, using water resistant epoxy resins as glue. They used it to replace point metal fastenings with areas of adhesive so that the grain of thin wood laminates could be directed to best advantage for hull strength. Considerations of tensile strength and elastic modulus relative to density show laminated wood to be an excellent structural material when bonded by such means.

Few large (over 85-ft) yachts were built right after the war, but, for those few, steel remained the preferred material. The choice was between riveting and welding with the latter gaining acceptance.

The development of corrosion resistant alloys during and after the war made aluminum alloy an attractive material for light hulls. When compared to wood, the smaller dimensions of aluminum offer greater interior space, while the better continuity of end joints over point fastenings, and the ability to use hull tanks, are further advantages. The use of aluminum alloy spread from the fifties on, although now, except in the largest yachts, various glass related composites are taking its place.

During the war, government supported research developed fibre reinforced plastics as desirable hull materials. Several pioneers brought out small standardized outboard boats and sailing auxiliaries. Initially the hulls were not more than about 30 ft long; but, with experience, acceptance grew so that some FRP hulls are now up to 100 ft or more in length. Once tooling is in place labor costs are low and quantity production is relatively inexpensive, while tough finishes help to minimize maintenance cost. FRP in one of its many forms has become the way of quantity production as the cost of one-off work has soared.

Standard references such as [35] may be consulted for information about the many ways in which these materials are being applied in boat building; but be aware that the techniques are rapidly changing.

Ferro-cement started to gain acceptance in the 1950s. It is considered easy for amateur builders to use and some successful boats have been turned out. To some it lacks appeal because it is heavy in relation to its strength and because it is difficult to follow lines drawings accurately.

Outfitting. Marine hardware has changed with the hulls and power plants. Up to the start of our period most deck hardware was bronze, generally cast, with some forged steel for heavier or cruder parts. Development has been toward stainless steel, generally stamped or formed from sheet. In high performance applications a good deal of titanium is now used. Under some rules the so-called exotic materials, such as carbon fibre and titanium, are prohibited because of their cost.

Essential in the world of yacht racing, winches are at least a convenience on any sailing vessel. Like so much that is useful to sailors, those in use in the forties were based on Herreshoff patterns. Even those that seem most sophisticated are old in principle. The most elementary and most usual winch in early use was the direct acting sheet or halyard type, with a removable handle at the top. This type is useful for wire halyards and may be used for sheets in smaller boats, although the top handle slows the taking of turns. When located on a side deck with the underside accessible, a simple expedient was to run the handle down through the deck. This arrangement was used for Genoa jib sheets on six and eight meters, usually with added gearing providing two speeds by reversing handle direction.

Pedestal winches came from the J boats to ever smaller racing yachts looking for any possible advantage. Such makers as Barient and Lewmar went to the use of titanium drums to save weight, and roller bearings to reduce friction. They featured up to four gear ratios, with clever provisions for quick gear change. Standard units are made today in a range of size and power. Frequently pedestal and drum connections can be directed to apply all available power where most needed.

Fig. 13 *Vamp X*, 1966. (© Rosenfeld Collection, Mystic Seaport Museum, Inc.)

Among the most radical changes occurring in the seventies and eighties was the growth and wide use of powerful communication and navigating equipment. Earlier we had long range radio telephone, then radar, fresh from its wartime use and widely used as costs came down. Loran closely followed as well as facsimile transmission of weather maps, and then the now ubiquitous fax machine. Satellites for both navigation and communication are now important in offering the yacht owner the facilities of his office in any part of the world. It is remarkable how often such equipment is found on 40 or 50 footers.

Designers and Builders

Following World War II, with growing prosperity and easy air travel, yacht design became an international pursuit. Americans had the initial advantage of lesser wartime disruption but soon received competition from the far side of the world. Bruce Farr of New Zealand is a prominent example.

Several yacht designers who were active in the pre-war period have already been mentioned. Among these, John Alden was still active after the war, but at a reduced level. He sold his firm in 1955. Philip Rhodes and S&S came out of the war best prepared to go back to yacht work. They were gradually joined by others. Space does not permit much more than a listing of the most active. Ray Hunt worked imaginatively on

Fig. 14 *Noryema*, 1972. (© Rosenfeld Collection, Mystic Seaport Museum, Inc.)

both sail and power boats. His twelve meter *Easterner* of 1958 was a fast boat, if not an outright winner. His 110 and 210 one-designs were original, fast, inexpensive and popular. In power his major successes were with the deep vee and with the inverted vee of the Boston whaler. K. Aage Nielsen was known for his handsome and thorough work. Charles Morgan of Florida was active in production boats, offshore competitors and later the twelve meter *Heritage*. Jack Hargrave, also of Florida, did fine large power boats. The name of Fred Lord should be mentioned for his work with racing power boats and other carefully developed designs. Arno Apel of Atlantic City developed the three point hydroplane. Combined designers and builders of many power cruisers were Broward of Florida and Burger of Manitowoc WI.

As post-war racing grew, Dick Carter was successful. With him for a short time, and, later, on his own was Doug Peterson of California. German Frers, Jr., came from Argentina to work for S&S and went on to active and successful independence. Bill Lee of California developed the ULDB type, while Bill Lapworth pioneered the separated keel and rudder, now universal in racing yachts.

Meanwhile, other outstanding designers were appearing overseas, especially in England and New Zealand, while well built yachts were coming out of European yards. Unfortunately, however, we have no room for details here.

Among American builders, Palmer Johnson of Sturgeon Bay WI has probably led in taking the place held by Herreshoff up to the twenties and then by Henry B. Nevins of City Island NY. Stephens Brothers have been leaders in California. During the fifties and sixties Chris Craft became the largest builder of small and medium sized power boats (22 to 55 ft) in our country, while Hattaras turned out many power cruisers in a slightly larger range. More recently, Pearson Yachts and Tartan Marine have attained prominence. Ted Hood, who was a leading designer and sail maker, has become highly successful with semi-custom and custom yachts built in Marblehead and Taiwan.

Such lists must be incomplete and apologies are offered to those omitted by inadvertence.

Regulation and Classification

Aside from racing rules, in this country yachts are fairly free of restrictions on design and construction. individual states may require that boats above a given size be licensed and numbered. Larger boats may be licensed and documented by the U.S. Coast Guard, particularly if intended for operation in coastal or open waters. All must meet minimal requirements for life preservers and other safety devices. It is only on vessels carrying passengers for hire that there are requirements in regard to stability, structure and arrangement [36].

The American Yacht and Boat Council (ABYC) is an industry-supported organization that has established accepted standards [37] for important elements of structure, arrangement, material and equipment. This provides valuable guidance to designers and builders, especially useful in the production boat field.

The classification societies also play a part in yacht construction. Larger yachts are classed by ABS or Lloyds, and the rules of one or the other are frequently followed by the builders of standardized boats, either with or without full inspection and certification.

At the request of the ITC (the technical committee of the Offshore Rating Council) the American Bureau of Shipping in the late seventies prepared scantling rules for offshore racing yachts. Those rules establish loading criteria for various hull locations according to the size and character of the boat. To receive approval, the proposed scantlings must be shown to be sufficient to withstand those loads. The need for such a system must be appreciated when one realizes how hard racing yachts are driven by the crew's urge to win.

Over many years Lloyds Register has collaborated with the IYRU to provide rules for the open rating classes under the International Rule. Before the seventies, either wood, steel or composite construction was required; but, in preparation for the 1973 America's Cup series, Lloyd's prepared rules for aluminum alloy; and, for 1986 the rules were made applicable to FRP construction.

Virtually all of the classification societies have rules for power and sailing yachts, and they are widely used in the larger units, but more often than not the scantlings are too heavy for racing yachts.

Bibliographical Notes

In addition to references in the text, a number of more general texts are worth mention.

John Parkinson, Jr. has written histories of the Cruising Club of America to 1960 and of the New York Yacht Club to 1973. Philip H. Tobey has done a history of the Eastern Yacht Club of Marblehead. Together these three cover a great deal of East coast sailing. Since the Stone/Taylor America's Cup history [12], there have been several books on Cup racing in specific seasons. A general account ending with the 1977 series is Ian Dear's *The America's Cup, an Informal History* [38].

Personal views of the earlier period of this study can be found in the *Memoirs* of Clinton Crane and in *Sherman Hoyt's Memoirs.*

Two excellent papers, [27] and [19], are found in the SNAME *Transactions* for 1963 and 1990. The first is by Henry and Miller; the second by Miller and Kirkman. Those papers provide a good perspective of design practices as they have developed over time.

References

1 *International Rule*, Year Book, North American Yacht Racing Union, ca. 1946.

2 *Universal Rule*, Year Book, North American Yacht Racing Union, ca. 1934.

3 Loomis, A. L., *Ocean Racing*, Van Nordstrand, New York, 1936.

4 *R Class Scantling Rules*, Lloyds Register of Shipping, ca. 1928.

5 Herreshoff, N. G., *Rules for the Construction, the Scantlings, and Other Proportions of Wooden Yachts*, New York Yacht Club, Jan. 1928.

6 Nevins, H. B., "Proposed Scantling Rules for Wooden Yachts," Henry B. Nevins, Inc, City Island, NY, ca. 1928.

7 Froude, W., "Greyhound Experiment" (described by A. Robb), *History of Technology*, Vol. V, Singer et al, 1958.

8 Taylor, D. W., *The Speed and Power of Ships*, Washington, 1933.

9 Davidson, K. S. M., "Some Experimental Studies of the Sailing Yacht," *Trans.* SNAME, 1936.

10 Fostle, D. W., *Speedboat*, Mystic, 1988.

11 Johnson, P., "Sailing Records," compiled yearly by P. Johnson for Offshore Racing Council, London.

12 Stone, H., and Taylor, W. H., *The America's Cup Races*, New York, 1958.

13 Spens, P., et al, "Some Further Studies of the Sailing Yacht," *Trans.* SNAME, 1967.

14 Kerwin, J., "A Velocity Prediction for Sailing Yachts," New England Sailing Symposium, New London, CT, 1976.

15 Kerwin, J., "A Velocity Prediction Program for Ocean Racing Yachts," rev. to Feb. 1978, MIT Rept. 78-11.

16 Letcher, J., "Handicapping Rules and Performance of Sailing Yachts," Chesapeake Sailing Yacht Symposium, Annapolis, 1974.

17 Myers, H., "Theory of Sailing Applied to Ocean Racing Yachts," Chesapeake Sailing Yacht Symposium, 1975.

18 Hazen, G., "Yacht Design with Computers," Chesapeake Sailing Yacht Symposium, Annapolis, 1983.

19 Miller, R., and Kirkman, K., "Sailing Yacht Design: A New Appreciation of a Fine Art," SNAME, San Francisco, 1990.

20 Letcher, J., "Mathematical Hull Design for Sailing Yachts," Chesapeake Sailing Yacht Symposium, Annapolis, 1981.

21 Kirkman, K., and Pedrick, D., "Scale Effects in Sailing Yacht Hydrodynamic Testing," SNAME, New York, 1974.

22 Kapsenberg, G., "A New Technique for Testing a Yacht in Waves," Chesapeake Sailing Yacht Symposium, 1991.

23 Milgram, J., "Analytical Design of Yacht Sails," *Trans.* SNAME, 1968.

24 Milgram, J., "Sail Force Coefficients for Systematic Rig Variations," SNAME TR Report N; HR-10, 1971.

25 Greeley, D., et al, "Scientific Sail Shape Design," The Ninth Chesapeake Sailing Yacht Symposium, 1989.

26 Hazen, G., "A Model of Sail Aerodynamics for Diverse Rig Types," New England Yacht Symposium, New London, CT, 1980.

27 Henry, R. G., and Miller, R. T., "Sailing Yacht Design—An Appreciation of a Fine Art," *Trans.* SNAME, 1963.

28 See Ref. [1] above or IYRU, London.

29 "International America's Cup Class Rule for America's Cup XXVIII," CORD, 1989.

30 Strohmeier, D., "Yacht Rating," Chesapeake Sailing Yacht Symposium, Annapolis, 1974.

31 Larsson, L., "Scientific Methods in Yacht Design," Stockholm, 1990.

32 "International Offshore Rule," Offshore Racing Council, London.

33 "International Measurement System," Offshore Racing Council, London.
34 "Midget Ocean Class Rule," *US Sailing*, Newport.
35 "Marine Survey Manual for Fibreglass Reinforced Plastics," Gibbs & Cox, Inc., New York, 1962.
36 Various U.S. Coast Guard regulations.
37 "American Boat and Yacht Council, Recommended Practices." American Boat and Yacht Council, Marysville, Maryland.
38 Dear, I., *The America's Cup, an Informal History*, New York, 1980.
39 Stephens, Kirkman, and Peterson, "Sailing Yacht Capsizing," Chesapeake Sailing Yacht Symposium, Annapolis, 1981.
40 Forbes, H., Laing, M., and Myatt, J., "1979 Fastnet Race Inquiry."
41 McCurdy, R. C. "Final Report of the Joint Committee on Safety from Capsizing of USYRU and SNAME," June 1985.
42 Claughton and Handley, "An Investigation Into the Stability of Sailing Yachts in Large Breaking Waves," Wolfson Unit, University of Southampton, Jan. 1984.

Olin J. Stephens II is the doyen of American yacht designers. He was actively engaged in the practice from 1928 until 1978 as chief designer of the Sparkman and Stephens office. He was a co-designer of the successful America's Cup defender *Ranger* (1937). In 1958, an S&S design, the 12 Meter *Columbia*, successfully defended the America's Cup, followed by other S&S defenders, *Constellation*, *Intrepid* (twice), *Courageous* (twice), and *Freedom* (1980). He maintains his interest in racing rules and has served as a member of some four technical committees involved in writing such rules. A member of SNAME since 1934, his outstanding achievements have been recognized in his being named a Fellow and Honorary Member. He is also a recipient of the Society's Taylor Medal, the Nathanael W. Herreschoff Trophy and the Richard Nye Trophy of US Sailing, and the Beppe Croce Award of the IYRU.

Part x

Fishing Vessels

Canadian Fishing and Ancillary Craft

INAM MAJID

Introduction

IN THE EARLY 18TH CENTURY, Canada's Atlantic area settlers fished in small, open boats using single lines with hooks. In the second half of that century schooners, such as the *Bluenose*, carried dories, which the fishermen would row out, bringing back fish for splitting and salting. In the 19th Century the seining method and longlines led to the development of seiners and longliners. The industrial advancements of the 19th century changed the construction material from wood to steel. This chapter gives a brief history of the changes in fishing vessel design, construction and construction materials, improvements in aids to fishing and advances in shipboard fish handling and storage facilities over the last fifty years. It discusses one of the more popular designs, namely the Cape Islander.

From the research and scientific viewpoint, the paper relates the establishment of marine biological stations in Canada, the first "moveable" wooden station of June 1899, and the 50-ft *Edward E. Prince*. It describes the diesel ketch *Calanus*, designed for fisheries research in Hudson Bay and her successor, the "*Calanus Replacement*." It describes the 243-ft chartered vessel *Gaddus Atlantica* and her replacement the *FARV-A*. It also describes the fisheries surveillance vessels *Chebucto* and *Cobequid Bay*, the 57-ft GRP Reef-class vessels and the SWATH *Frederick G. Creed*.

Historical Background

Canada's Atlantic fishery was among the attractions that drew the early Europeans to the northern half of North America. In the 16th Century, cod were plentiful off the coast of Newfoundland and fishing was carried out close to shore, directly from small boats using hooks and lines (Fig. 1). The fish were cleaned, split, lightly salted, and dried. In this state the fish could be transported great distances.

The second half of the 18th Century saw the advent of schooners for fishing. These fore-and-aft rigged vessels, like the *Bluenose*, fished for cod, halibut, and haddock. Fishermen rowed out from these ships in dories and brought fish back for splitting and salting on board the schooners. In the 19th Century, American fishermen improved on the "beach seine" method of fishing (which required points on land to help encircle fish within a net) when they developed the "purse seine" method. Operating in open water, a net hanging from a line of corks was used to surround surface-schooling fish. A purse line at the bottom of the net was then tightened to enclose the fish in a bowl. The fish were then "dipped" out. During the same period French fishermen introduced what are now known as "longlines." These were long "ground lines" anchored on or near the bottom with many short branches with hooks, so that many more hooks could be used on a single line.

The industrial advances of the 19th Century also brought changes to materials used in ship construction. More reliable, safer and larger steel ships began replacing the traditional wooden vessels, and steam replaced sail. The greater speed of these ships allowed fresh fish from the Grand Banks to be taken to market before the fish spoiled.

After World War I the Europeans introduced the much more reliable trawlers to the Western Atlantic.

Fig. 1 Drawing of a 16th-century Newfoundland fishing ship. The cod were so plentiful that they could be hauled in with baskets. (Courtesy National Archives of Canada.)

These large vessels, which dragged conical nets along the bottom, increased the catch size many times over.

World War II brought in new technologies that greatly increased productivity. Radio, radar and sonar were rapidly adopted by the world's fishing fleets. To ensure continued supply of fish through fisheries assessment and research, the Canadian Government approved the establishment of the first "moveable" marine biological research station in 1889. The station was designed as a rectangular wooden building that could be placed on a scow and towed from place to place along the coast. It was sited at St Andrews, N.B. where marine life is swept in with the large tides (often greater than 28 ft) and left behind on the beach and in pools at ebb tide. The station was 50-ft long and had a 30 × 15-ft deckhouse containing offices and laboratories.

Over the next 25 years various types of small motorboats were used for local fisheries research. In 1913 a 50-ft wooden motorboat was acquired by the Department of Marine and Fisheries. This boat, the *Edward E. Prince*, was able to sleep six, thus allowing the scientists to go away on longer trips.

Fishing Vessels

The type of vessel used for fishing is dictated by the kind of fish sought. Operations on the offshore fishing grounds are based on schools (where fish feed and spawn) of demersal (those feeding near the floor, mainly on offshore banks) and pelagic stock (occupying the surface of the open sea). Such fishing involves large ships. The presence inshore of littoral stocks (found within the 50–60-fathom contour adjacent to the coastline) and some demersal and pelagic species provides the basis for various small boat fisheries. Inshore craft, generally defined as those under 25 gross tonnes, account for 95 percent of Canadian fishing vessels. One such kind of craft has therefore been chosen to illustrate typical developments in boat design over the past fifty years.

The Cape Island Boat. The development of the Cape Islander began around the turn of the 20th Century on Cape Sable Island, NS, when Ephraim Atkinson built a gasoline-powered boat for the local fishermen [1]. These boats were probably patterned after a pleasure craft of the period and were powered by engines of about 5 hp. They were about 25 ft long, of fairly narrow beam, with only a canvas dodger for protection of the crew. The boats were initially used to fish for lobster, but were later adapted for longlining and gillnetting and later still for dragging.

As larger and more powerful engines became available over the next 30 years, the boats gradually increased in size to accommodate more lobster traps and to allow for larger working platforms (Fig. 2). The boats were carvel planked with single or double bent frames. They were built of local wood with poor rot resistance and seldom lasted more than ten years. The length, at this time, was about 35 ft for ease of maneuvering around the buoys when lobster fishing.

By the 1950s a stand-up windshield and then a cabin were fitted at the forward end of the cockpit, which provided better protection from the elements. This was later closed in to form a true wheelhouse. There was little change in the size or shape during this period. The motive power was, typically, high-speed car engines.

In the mid-1960s, the boats grew somewhat in size, and a watertight cockpit floor was added, which greatly improved their seaworthiness by obviating the necessity of continuous pumping in rough weather, which the earlier boats had to do.

In the mid-1970s, fiberglass hulls were tried experimentally, but being prohibitive in cost, gained little popularity. The boats, however, continued to grow in size. This was done to increase lobster trap carrying

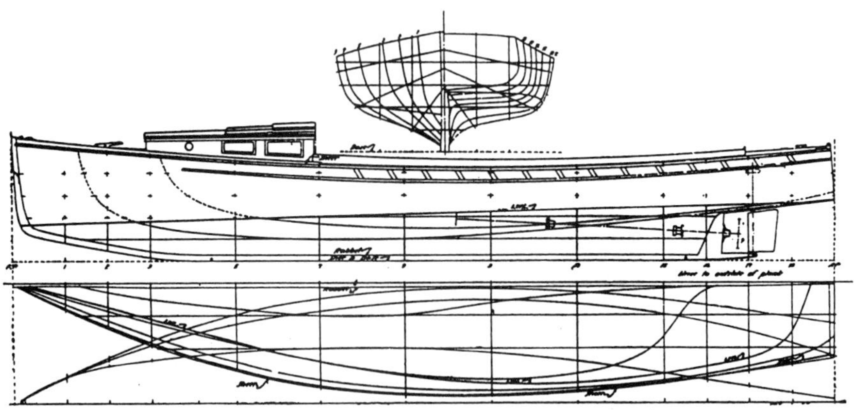

Fig. 2 A typical early Cape Island boat, showing the long low hull.

capacity for "opening day," as well as increased operating range, and improved seakeeping. New fishing techniques such as dragging also required stronger and more powerful boats.

The building boom of the early 1980s signaled the return of the glass boat, all but eliminating the wooden ones. The length of these boats was restricted to 42–45 ft due mainly to the "45-ft maximum length" rule for licensing. With growing volumetric needs, some boats grew to 22-ft beams with proportionately larger draughts and topsides (Fig. 3).

In the mid-1970s marine diesel engines began to replace gasoline engines, and the replacement has been so complete that today there are few Cape Islanders with gasoline engines.

After World War II the fishermen began using echo sounders to locate fish. In the late-1960s echo sounders were replaced by sonar. The most notable changes have been in the fields of ship navigation and positioning. This has saved fishermen valuable time by allowing them to return directly to the exact locations where fish had previously been harvested.

The "ice" boats, which could hold a catch for no more than ten days, have been replaced by "freezer boats," which now stay at sea as long as necessary to fill their holds before returning to port.

Because of the improvement in fishing aids and equipment, fishing vessels are becoming multifaceted, and a single ship is now capable of fishing for herring, salmon and cod as well as using longlines and pots.

Fishery Ancillary Vessels

Fisheries Surveillance Vessels. To prevent illegal fishing or over-fishing, many national governments undertake patrolling and surveillance of the fishing waters within their jurisdiction. In the early days of the 20th Century, such patrolling was carried out by the country's navy. Indeed, Iceland's navy to this day serves as its fisheries protection service. Today, however, Canada's fishery enforcement is carried out by special vessels owned and operated by the Department of Fisheries and Oceans.

Back in 1966, among the first fisheries patrol craft to be built of steel was the still-extant *Chebucto*. Her main hull is of steel, but her superstructure above the boat deck is of aluminum alloy (Fig. 4). She is 179 ft long, with a beam of 31 ft and a maximum draught of 11 ft-4 in. She is fitted with two Fairbanks-Morse 38D diesel engines developing a total of 3400 bhp. Her one controllable-pitch four-bladed propeller gives her a maximum speed of 14.75 knots. Her range varies from 4600 nautical miles at full speed to 7320 at economical speed. She has two electric generators, Westinghouse CVG 3 and CVG 1, delivering a total of 153.2 kVA. Her drinking water supply of 23 tonnes can last over 15 days. Deck equipment includes a 1-ton crane and an 8-cwt boom. She has an impressive suite of navigation and communication equipment, which includes: two Decca radars; Kelvin-Hughes and Simrad Scientific sounders; Decca Navigator, Loran A and Loran C positioning equipment; two magnetic compasses and a gyrocompass, and various radio and direction finding sets. For boarding and enforcement she has a 25 ft-6 in. fiberglass, specially designed U.S. Coast Guard boat and a four-person Avon Sea Rider.

A smaller patrol vessel, the *Cobequid Bay*, was built the next year, 1967, for surveillance of the closer inshore regions [2]. This boat, shown in Fig. 5, is of carvel planked wooden construction, 70 ft long, with

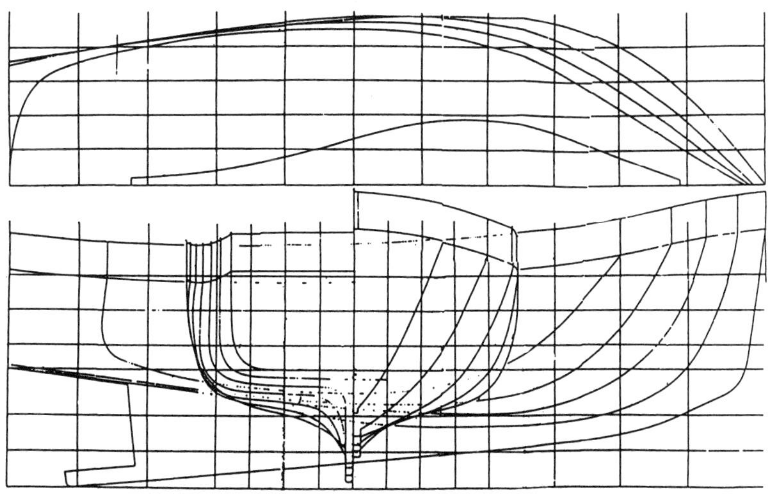

Fig. 3 An extreme example of a modern Cape boat, LOL 44 ft.

Fig. 4 *Chebucto.*

set. It was at this time that anti-pollution regulations for disposal of sewage and grey water were coming into force, so this ship is fitted with vacuum toilets. For enforcement and boarding purposes she carries a 16-ft speed launch.

In the mid-1980s a new class of surveillance vessel was developed for the inshore fisheries of the Canadian West Coast. The prototype was christened *Bajo Reef.* The lessons learned from her were incorporated into the *Lewis Reef* (Fig. 6) and then into the latest, the *Robson Reef,* completed in June 1990. These depart radically from previous patrol and surveillance vessels. Each is 56

a beam of 18 ft-3 in. and a draught of 7 ft-6 in. Whereas the *Chebucto* has a total complement of eighteen, the *Cobequid Bay* carries only four. Her main engine is a Cummins V12, 340 bhp diesel driving one fixed-pitch propeller. She has a range of 1000 nautical miles at a maximum speed of 10.5 knots and 1300 at 8 knots. A total of 80 kVA electric power is generated by two identical Lister HW12 generators. Her deck machinery, for surveillance purposes, consists of one 700-lb winch (for hauling in illegal fishing gear and, at times, offering assistance to disabled fishing craft). For her size, the *Cobequid Bay* carries an even more impressive suite of navigation and communication equipment: a Kelvin and White compass; two radars, one with 64-mile range and the other with 25-mile range; a sounder with 110-fathom range; a Decca navigator and a Loran A for positioning. Her communication equipment consists of a 50-channel VHF, a 40-channel CB Messenger, and a 25-channel Single Side Band

ft-9 in. long, with a beam of 17 ft-4 in. and a draught of 6 ft-10.5 in. The hull is of semi-displacement form, made of closed cell, cored, glass reinforced plastic. Each is powered by a Deutz 816, 6-cylinder diesel engine developing 440 bhp. Propulsion is through a four-bladed controllable-pitch propeller. This class is also fitted with an 18-in. bow thruster for station keeping. Two 35 bhp Nissan diesel engines generate a total of 36 kVA. These ships have a maximum speed of 12 knots and can work continuously at speeds as low as 1.5 knots. They carry 1200 imp gal of diesel fuel and 350 gal of fresh water. Even though they are coastal vessels, they have an endurance of ten days. Each vessel is fitted with a crane having a capacity of 1400 lbs at 15 ft. For boarding and enforcement purposes each vessel carries a 14-ft rigid hull inflatable boat. For assessment of the herring fish the vessels are fitted with a side scan sonar. In addition to the mandatory radio, radar and communications equipment, these vessels

Fig. 5 *Cobequid Bay.*

Fig. 6 *Lewis Reef.*

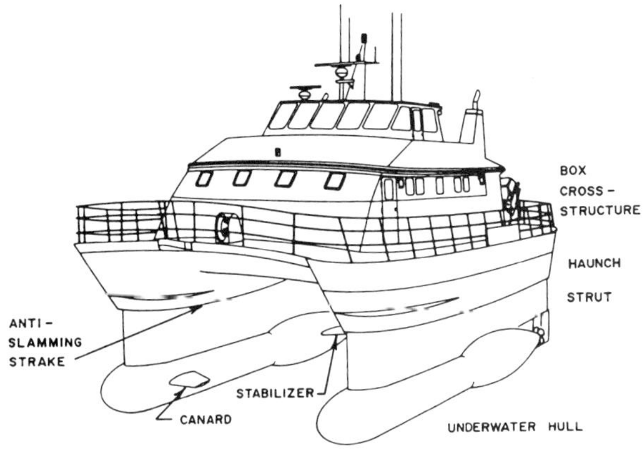

Fig. 7 *Frederick G. Creed.*

carry 10-color echo sounders, cellular telephones, DATAHAIL systems for communications over the secondary Single Side Band radio, and Loran C with all the RAMcards for the Canadian West Coast. These modern vessels have a crew of only three and often carry two fisheries enforcement officers as passengers.

The latest vessel to join the fisheries surveillance group is the *Frederick G. Creed* (Figs. 7 and 8) She is a fourth generation Small Waterplane Area Twin Hull (SWATH) vessel that has undergone extensive operational trials with the Canadian Department of Fisheries and Oceans and the U.S. National Oceanic and Atmospheric Administration. The *Creed* has a marine-grade aluminum alloy hull. She is 67 ft long, with a beam of 32 ft and a full load draught of 8 ft-6 in. at a displacement of 81.3 tonnes. She has a maximum continuous speed of 24 knots and a range of 2000 nautical miles at 10 knots. Each of her two 48-in. fixed-pitch propellers is powered by a 1080 bhp Detroit Diesel 12V92TA, 2300 rpm engine, and the electrical power is provided by two 33 kW marine diesels. Being a SWATH she demonstrates far superior seakeeping qualities while under way than does a monohull. These qualities owe, in part, to her Ride Control System [3], which consists of Relative Bow Motion (RBM) sensor, the DTH 2000 sensing/control unit and four stabilizing fins. The RBM sensor is a transducer fitted in the box cross structure. The canards and stabilizers are fitted to the inboard sides of the underwater hulls. The RBM transducer provides a rate of change of the bow position, which is fed into and added to the attitude acceleration component of the DTH 2000. In effect the system's operation is similar to that of an autopilot in that it determines a rate of change as well as an anticipated change and transmits them to the stabilizers and canards to control the motion of the ship in a seaway. This system combines so well with the small waterplane area that during trials the *Creed* outperformed the *Louisburg*, a ship nearly twice her length at 125 ft, and more than twice her displacement at 250 tonnes. There are, at present, no intact stability standards promulgated by the Canadian Coast Guard Ship Safety Branch (SSB). They have, however, agreed to assess the *Creed* under the same criteria as monohulls. It goes without saying that the *Creed* passes the ice-free and iced-up stability conditions with flying colors. The navigation suite of the *Creed* is up-to-date, commercially available equipment, which includes multiple radars and echo sounders, a Loran C navigator and a satellite navigation (SATNAV) receiver. She is also fitted with a highly sophisticated echo sounder for hydrographic and oceanographic observations.

Fisheries Research Vessels. Fisheries protection and enforcement vessels endeavor to regulate the amount and type of fish harvested from the various fishing grounds within their surveillance. The allowable catches, however, depend upon the fish reserves available. The availability of the fish resources is assessed by the various nations' fisheries research agencies.

Research in the world's fish resources is relatively new. Few research vessels existed before the late 1940s. Any research that was done was carried out on an opportunity basis and on board any fishing (or indeed, non-fishing) vessel that may have been available. One of the earliest examples is the *Beacon Hill*, a steel hulled, *Prestonian*-class, Royal Canadian Navy Frigate: FFE 303 [4]. She was built in 1944, was 286 ft long, with a beam of 36 ft-7 in., and a mean draught of about 15 ft-6 in. She had two fixed-blade propellers each driven by a four-cylinder 5500 hp triple expansion reciprocating engine. She carried the normal naval electronic suite of the day: Sperry radar, Sperry Loran, gyrocompass, Chernikieff Log, sonar and echo sounders. During fisheries research deployments she was temporarily fitted with gasoline powered winches, carried about six scientists and conducted

Fig. 8 *Creed* underway.

Fig. 9 *Calanus.*

light mid-water trawls with some hydrographic casts and bathythermographic observations.

Another example is the *Investigator I* [5]. This was a wooden fishing vessel built in 1944. In 1947 she was converted into a fisheries research vessel. She was 53 ft-2 in. long, with a beam of 15 ft-8 in., and had a draught of 8 ft-10 in. She had one fixed-pitch propeller driven by a 225 hp Rolls-Royce diesel engine, which gave her a cruising speed of 8 knots and a range of only 900 nautical miles. By today's standards she was very sparsely fitted out for fisheries research: two winches and an A-frame. She was used primarily for "tagging" experiments.

Among the first purpose-built fisheries research vessels were the *Harengus* and the *Investigator II* [4]. The *Harengus*, built in 1946, was a wooden trawler, 84 ft long, with a beam of 20 ft and drew 12 ft-6in. Diesel engines had by now replaced steam, and this ship was fitted with a single propeller driven by two 135 hp Caterpillar engines. She had radar and Loran and two echo sounders. She had two winches: a two-drum winch with power taken off the main engine and an electric winch. She had no laboratories and little or no acoustic isolation. The *Investigator II*, also built in 1946, was a "dragger." She, too, was wooden hulled, 85 ft long with a beam of 22 ft and draught of 10 ft. She had an Atlas Imperial 250 hp engine and a single fixed-pitch propeller. She carried navigation and echo sounding equipment similar to that in the *Harengus*. She also carried two winches, of which one was powered from the engine and one was electrically driven. Like the *Harengus*, she, too, had no laboratories and no acoustic isolation. She was mainly employed on ground fish research on the Grand Banks. In 1948 a new vessel, the *Calanus*, was specially built for fisheries and biological research in the Eastern Canadian Arctic. Research was conducted only during the summer months, and for the rest of the year she was beached. (See Fig. 9.) She was of wood, 49 ft-6 in. long with a beam of 15 ft and a draught of 7 ft-6 in. She was designed after Nansen's *Fram*, and was built in Mahone Bay NS. She had a rounded bottom, shaped to allow her to rise when squeezed by ice, rather than be crushed. This was successfully demonstrated on one occasion when the *Calanus* was caught in heavy ice in the Hudson Strait. Ice converged from both sides, she rose, rolled to starboard, lay momentarily on the ice then slid slowly into a lead which had opened astern! Her ribs were of white oak, while her underwater planking was of yellow birch, sheathed with oak for protection from the ice. Her above-water planking was oak. She was fitted with one Rolls-Royce SF 65M, 136 bhp diesel engine driving a single fixed-pitch propeller. She was ketch rigged for emergencies in remote Arctic regions. Her main mast was fully 40 ft long! She had a total complement of five, and the master was the only official crew member! The scientific staff served variously as part-time crew. For navigation she carried a magnetic compass, a Sperry gyrocompass, a Marconi Type 25 radio, a Kelvin Hughes Type 17 radar and a Simrad echo sounder. Even for operations in such a remote area she had no watermaking facilities. Instead she carried 1.1 tons of drinking water. Being the only research ship in the far north the *Calanus* had to be a jack-of-all-trades. As such, her duties included oceanographic stations (standard casts and vertically collected biological samples) to 3300 ft; trawling, dredging and horizontal hauling to 605 ft; general near-shore biology, chemistry and physical surveys, and marine manual observations. To accomplish these tasks she was fitted with the most up-to-date equipment of the time: an oceanographic winch with 3300 ft of 3/32-in. wire; a double-drum trawl winch with 1500 ft of wire; two A-frames with meter blocks; various water samplers and nets, and an echo sounder. She last sailed in the summer of 1979.

In 1959 another vessel was built. This, the *Salvelinus*, was designed to serve in the Western Canadian Arctic (Fig. 10) [2]. She had a wooden hull with red oak for planks and frames, was strengthened for Arctic conditions with greenheart sheathing over the hull and steel on the bow. She was 35 ft-5 in. long, with a beam of 12 ft and drew 3 ft-2 in. She had a single fixed-pitch propeller and was powered by a GM model 4017C 95 bhp diesel engine. She had no generator

Fig. 10 *Salvelinus.*

and used 32 V dc lead acid batteries, which were charged through an alternator. She had a maximum speed of 8.6 knots and a range of only 500 nautical miles. Her navigation equipment consisted of a Sperry gyrocompass, an echo sounder, a radar and a radio transceiver. Her fisheries and research equipment consisted of one trawl winch, a hydrographic winch capable of 1-tonne lift, with a wire capacity of 1650 ft, a towing boom and an overhead boom. She was meant for inshore fisheries and hydrographic research, but even for these limited roles the ship's speed and that of her winches was too slow.

Among the more recent vessels is the *E. E. Prince* (Fig. 11) [2]. She was specially built in 1966 for fisheries research. She is steel hulled, 130 ft long, with a 27-ft beam and a draught of 12 ft. She has a cruising range of 3000 nautical miles at 8 knots and an endurance of 14 days. She carries nearly 7000 imp gal of fresh water. Her fisheries research-related spaces include a 75 sq ft dry laboratory, a 160 sq ft wet laboratory, 175 cu ft of cold storage for fish samples, and over 540 sq ft of clear working deck. Her main engine is a Burmeister type 49-VO 600 bhp diesel, which drives a single 6 ft-10.5 in. controllable-pitch propeller. Her auxiliary power is provided by a Dorman 223 bhp generator, type 6QTV. Electrical power is generated by two 192 kW, 440/220/115 V ac 60 Hz generators. Her navigation equipment consists of X/S band radar, single-side band, Loran C, VHF radios and a doppler speed log. For fish-finding and location she has a SIMRAD bottom sounder and an ELAC mid-water sounder and a 3.5m sonar for towing. She carries two trawl winches: a Yankee Bottom trawl IGYPT, and a Brattvaag two-drum winch with 3000 ft of wire on each drum. She also has a HIAB 1-ton crane. On special occasions she can be fitted with portable winches to carry out specific tasks. Her accommodation is built to modern standards capable of carrying a mixed complement of men and women. She has a crew of thirteen and accommodates six scientists. She is employed basically for fisheries research, which includes bottom, mid-water, and shrimp trawling, as well as bottom and drifting longlining.

For fisheries research off the coasts of Newfoundland and Labrador the Canadian Department of Fisheries and Oceans has chartered a conventional stern trawler and converted it for research purposes. This vessel, the *Gaddus Atlantica*, was built in 1969. She is 243 ft long, with a beam of 44 ft and draft of 15.2 ft. She has a cruising range of 20,000 nautical miles at 12 knots and an endurance of 90 days. Her continuous minimum working speed is as low as one knot. She has a 4200 hp diesel engine driving a single fixed-pitch propeller. For deck and scientific purposes she has two five-ton cranes, an A-frame aft, one trawl winch and five scientific winches. She uses the standard commercial type Engels 145, Diamond XI trawl nets. Her navigation equipment consists of Satellite navigation, X/S band radar, Inmarsat, single side band, VHF, weather facsimile, and direction finder. Her research related spaces include: one 650 sq ft and one 110 sq ft laboratory, over 700 cu ft of cold storage, 160 sq ft of space for standard ISO containers, and a large wet laboratory with a conveyor belt from the fish area. She

Fig. 11 *E. E. Prince.*

has an ice classed hull and is manned by a crew of 28. Although she can accommodate 12 scientists, the standard of the accommodation leaves much to be desired.

The old *Calanus* (the Grand Old Lady!) has now been replaced by a new vessel. The *Calanus II* is steel hulled with an aluminum superstructure. She bears little resemblance to her predecessor. (See Fig. 12.) She is 65 ft-4 in. long, with a beam of 22 ft-7 in. and a draught of 9 ft-2 in. She has a service speed of 9.5 knots and a minimum working speed as low as 1.5 knots. She has a cruising range of 2000 nautical miles and an endurance of 30 days. She carries 1100 imp gal of fresh water and is fitted with a reverse osmosis plant, which can produce about 20 gal per hr. Like her predecessor she will spend the Arctic winters beached somewhere near Frobisher Bay. She has therefore been specially designed with a flat bottom. She will be pulled out of the water on a specially designed cradle, by a tractor. All her systems have been designed with winterizing in mind: all attractive electronic equipment will be removed and taken south; all pipelines will be drained; the ship will be closed up and locked. Provision has been made to fit heaters for the duration of the winter, if electric power can be tapped where she is beached.

She is fitted with one Caterpillar 3412, 500 bhp diesel engine which drives a controllable-pitch propeller inside a nozzle. She also has a bow thruster. Electric power (a total of 90 kW) is generated by two other Caterpillar engines. For her fisheries research role she is fitted with a crane having a capacity of 2800 lb at 20 ft, two trawl winches, one sounding/collecting winch, one gillnet roller and a linehauler for longlining. She also has a two-tonne A-frame. Her dry laboratory has an area of 120 sq ft, and the wet laboratory has 2800 sq ft. She has an open after working deck of 390 sq ft.

Her scientific and navigational equipment is the latest available: two radar sets, both with ten ranges each, one from 0.25 to 120 nautical miles and the other 0.25 to 60; a doppler speed log; a scientific dual frequency color echo sounder; a navigational echo sounder; one Loran C and one GPS receiver; a weather facsimile receiver, and other mandatory equipment. She is also fitted with an autopilot. She can accommodate a maximum of nine persons; the master in a single cabin and the others in double.

The *Gaddus Atlantica* is being replaced by a purpose-built vessel. To this end a new Fisheries Assessment Research Vessel, *FARV-A*, has been designed. The *FARV-A* is 236 ft-6 in. long, has a beam of 45 ft-11 in., and a draught of 16 ft-5 in. She has a steel hull

Fig. 12 *Calanus II.*

reinforced for operation in ice and an aluminum superstructure. She is powered by two identical marine diesel engines driving a single controllable-pitch propeller inside a nozzle. She has numerous innovative features.

Her two main trawl winches are fitted in a winch room on the lower deck, below the fishing deck. They are fitted below to provide shelter to the fishing crew and the scientists when operating in the northern winters, as well as to shorten the length of the fishing deck. The trawl winch control system is fully automated and is itself electronically controlled. She also carries all the other winches, net drums, power blocks, booms, cranes, A-frames, etc. that are necessary for successful fishing operations. Because she is also expected to fish in broken ice, she is fitted with ice davits to allow her to set and retrieve nets below the ice abaft her stern.

Her hull form was extensively model tested in water and ice tanks for resistance and powering and in open water for seakeeping. As a result of these trials she has been fitted with a stern bulb and a Schilling rudder. For position and station keeping she has an azimuthing bow thruster, which can be withdrawn into the ship when working in ice. Although her length-to-beam ratio of nearly five is rather high for a fishing vessel, it is not considered excessive in a research vessel. Her fishing deck is a multipurpose work area. She has the usual "arena" for fishing purposes on the centerline; while on the starboard side, for the whole length of the after deck, is a space reserved for oceanographic work. Between these two work areas is a removable longitudinal bulkhead which provides shelter, as well as separation, for the fishing deck. The "fish hopper," which is 430 cu ft (12 cu m) in volume is made of stainless steel and is removable to allow for structural maintenance and changing requirements.

In addition to the fishing and oceanographic work areas, she is fitted with a dark room and six laboratories for wet, dry, oceanographic, control, biology and fluorometry studies. She also has designated electronics, instruments and mechanical workshops. All laboratories have been carefully sound and vibration isolated by fitting vibration attenuating bulkheads and floating floors.

She has a crew of twenty and a scientific staff of twelve men and women, all accommodated in single cabins with either *en suite* or shared-between-two washrooms. Her propulsion machinery is fully automated with programmed logic controls so that equipment is automatically brought on line or shut down as required to meet the ship and scientific demands. Her propulsion is controlled directly from the wheelhouse consoles in such a manner that under normal conditions the most efficient propulsive combinations of pitch and rpm are automatically provided. When underwater noise control is required, the pitch and rpm can be separately manipulated so as to avoid unwanted frequencies.

The *FARV-A* is fitted with the most modern and sophisticated navigation and communication equipment: X-band and S-band radars, navigation echo sounders, fish finding echo sounder, net sounding equipment, cable link net sounding equipment, acoustic net sounding equipment and scientific sonar, Loran-C, SATNAV, Precision Navigation System (PNS), direction finders, weather facsimile, SATCOM, etc. She also has an elaborate internal voice communication system and a modern Navigation and Environmental Data Logging (NEDAL) System.

Conclusions

This chapter has briefly described the changes that have taken place in Canadian fishing craft, fisheries research ships, and fisheries patrol vessels over the past fifty years. It has shown how the shape and size of fishing craft have changed with the type of materials used, the available fish that were sought, and the pertinent laws and regulations. It has described the fisheries protection vessels used in regulating the catch, and the research ships used to assess the stocks of fish in Canadian waters. The general trend in all cases has been for increasingly sophisticated electronics to locate the fish, larger holds and higher speeds, and improvements in shipboard processing and storage to enable the ships to stay at sea longer and return with bigger catches. At the same time improvements in electronic detection and navigation equipment have enhanced the capabilities of patrol vessels and research ships. Each suceeding vessel has been fitted with the latest in radios, radars, echo sounders, and sonars. We are now at a stage where more attention is likely to be paid to the hull form. Conventional monohull displacement vessels seem to be ready for replacement by less conservative shapes such as catamarans and SWATHs; single purpose craft are already being replaced with vessels capable of carrying out multiple tasks.

References

1 Lacey, G. E. and Dalziel, J. W., "Synopsis of the Development of the Cape Island Boat," *CMVDR*, Sept. 1990.

2 Department of Fisheries and Oceans internal documents.

3 Goodyear, J. E., Hussey, D. J., et al, "An Evaluation of the SWATH Vessel *Frederick G. Creed* in the Canadian North Atlantic," U.S. Hydrographic Conference 90, Norfolk, VA, May 1–4, 1990.

4 "Oceanographic Vessels of the World: IGY World Data Centre A for Oceanography and the National Oceanographic Centre."

5 Johnstone, K., *The Aquatic Explorers, A History of the Fisheries Research Board of Canada*, University Press, Toronto, 1977.

6 Grainger, E. H., "*Calanus*: Grand Old Lady of Arctic Research," Canadian Geographic, April/May 1982.

CDR Inam Majid studied naval architecture with the Royal Navy at the Royal Naval Engineering College, Manadon, and the Royal Naval College, Greenwich. He retired as a Commander from the Pakistani Navy in 1971. Before joining the Department of Fisheries and Oceans, he served in various capacities in the Canadian Armed Forces and the Canadian Coast Guard. He is now the Chief of Naval Architecture, Ship Branch, Department of Fisheries and Oceans.

Development of Stern Trawlers

Luther Blount

Introduction

The concept for the modern American stern trawler was first given form with the launching of the *Narragansett* on January 9, 1963. Designed and constructed by my own company (Blount Marine Corporation in Warren, RI), the vessel was intended to be my answer to the massive foreign fishing fleets working right off the coast of New England.

At that time, the stern trawler was just coming into use overseas, but only in sizes greater than 600 tons. Indeed, many observers thought the concept was appropriate only for the large factory trawlers, working with large crews, and not for the American fishing industry, which was geared for catching and marketing fresh (unfrozen and unsalted) fish.

Side Trawling

In 1963 American trawling had changed but little since 1905, when the first steam propelled beam trawler, the *Spray*, was launched at the Fore River Shipyard in Quincy, MA. That vessel replaced the then conventional fishing schooner that fished with hand lining techniques. The power of the ship's steam engine allowed it to drag a trawl net across the ocean floor. The net was held open by a massive beam, giving rise to the term "beam trawling."

The beam technique for spreading the mouth of the net gradually gave way to the current method, that of using paravanes, then called "otter boards" and now termed "doors," fastened to the wings of the net. The forward motion of the towing vessel provided the transverse component of force to spread the net. The otter boards were a great advance, as they were significantly easier to stow on board.

With the advent of beam trawling, net and beam handling was done over the side of the boat, as the great length of the net spreading beam could not be handled elsewhere. This tradition was carried on in the use of the otter boards in lieu of the beam, and so "otter trawlers" (as they were called) also handled the net and the otter doors over the side.

Economic forces caused the evolution of the New England fishing fleet to gravitate away from the large steel trawlers such as the *Spray* to smaller, wooden boats. The typical New England trawler of the early 1960s was around 80 ft in length, characterized by low freeboard to facilitate fish handling over the side, and with an aft house incorporating the upper engine room, the pilothouse, and a head in the turtleback. The trawl winch, located just forward of the house was mechanically driven off the main engine. The fish hold, laden with crushed freshwater ice on the trip out and iced fish on the return trip was located just forward of the engine room. The crew accommodations were within a forecastle.

Mechanically, the boats were simple, with a diesel engine connected to a fixed pitch propeller through a mechanical clutch and a reverse and reduction gear. The deck mounted trawl winch was controlled locally through simple mechanical clutches and handwheel controlled band brakes. Direct current production off the main engine, which was running at variable speeds, provided basic lighting and power for elementary electronic gear. There was no mechanical refrigeration for either the preservation of the fish or comfort of the crew. Heat was provided by massive oil stoves located in the forecastle.

Handling and towing the nets and otter boards from the side of the boat presented several problems. First, the boat had to be carefully maneuvered to bring the net alongside without fouling the propeller. Fishing in wind, tidal currents and rough sea conditions further complicated these efforts. The standard practice in deploying or retrieving the net was to bring the vessel abeam of the seas and to leeward of the net and otter boards. The engine was then stopped, leaving the boat dead in the water and wallowing in the beam seas. Deck personnel were then deployed at the weather rail amidships to handle the net and fish. Fishing broadside to the advancing waves severely

Fig. 1 *Narragansett.*

limited the maximum sea conditions that were acceptable for working the vessel.

Secondly, with the net trawl warps affixed to but one side of the boat, it invariable circled to one side in towing the net unless significant rudder angles were applied, thereby wasting precious engine power that could have gone into towing the net. Maneuverability was obviously restricted, owing to the ability to circle in only one direction. As fishing became more and more a matter of avoiding bottom obstructions and following courses along bottom contours, the lack of maneuverability became of considerable importance.

Probably the most significant problem in using the otter trawl, however, was the retrieval of the net and fish over the side of the vessel. The cod end of the net, where the trapped fish collect during the tow, was hoisted aboard by overhead lifting gear using brails or straps manually placed around the net in the water. Hoisting the bag, as this process was termed, became particularly dangerous as the sea conditions worsened.

Stern Trawling

I happened to hit on the concept of the *Narragansett* during a swordfishing trip, during which I observed the large European factory trawlers catching great quantities of fish just off the New England coast on George's Bank. (At that time, the Soviet Union alone had 219 large fishing trawlers working there.) These were stern trawling ships, with large crews, and were able to operate far more efficiently and to catch far more fish than the American side trawlers operating alongside.

These foreign ships overcame the aforementioned limitations of the side handled otter trawl. As a result, the time to set and haul a net was cut about in two. Additionally, the fishing process was less limited by sea conditions, allowing continued operation even in rough seas. Nonproductive maneuvering to come alongside the net and bring the fish aboard was reduced, further saving both time and fuel costs. Perhaps most importantly, the increased maneuverability afforded by towing the trawl from both the port and starboard sides ensured larger catches in each trawling operation. Because American fishermen were still oriented to the delivery of fresh fish to their relatively close home ports, time per round trip was limited. Thus, economic success was closely linked to fish catching efficiency. This led me to develop a smaller, automated version of the stern trawler concept appropriate to the needs of the New England fishermen.

The Pioneer

The *Narragansett* was 83 ft LOA, 21 ft in beam, and had a designed light draft of 9 ft. The hull had a great deal of flare, both forward and at the stern. This reduced the light ship metacentric height, but added stability at large angles of heel by immersion of additional waterplane area.

The concept of handling the nets and gear over the stern required that the pilothouse be located forward, a major departure from tradition. The crew's quarters, with berths for eight men, were located below this deck house, along with an all-electric galley. The accommodation was a significant advance in

Fig. 2 Emptying the trawl net.

crew comfort over the traditional "bunkhouse" design, as two double staterooms and a four-berth stateroom were included, affording a degree of privacy not previously found on board a fishing boat of this size.

The fish hold was located aft of the forecastle. It had a capacity of 4800 cubic ft, and was foam insulated with fiberglass. The bottom was both insulated and covered with concrete. The hold was divided traditionally into pens, separated by removable pen boards. A centrally located lobster tank, with mechanically refrigerated, recirculated sea water was incorporated into the hold.

This left the very stern of the vessel for the engine room, which was uncharacteristically short. To adapt to the limited space, and to ensure good maneuverability, I adopted a unique propulsion arrangement. The main engine, a 380 hp Caterpillar D353 diesel, was located over the propeller, high in the boat and facing forward. An 18-in. wide gear belt drive to the stub shaft transmitted the power at a 3.5:1 reduction ratio. A 60-in. diameter Hustad three bladed controllable pitch propeller was controlled from the pilothouse by a single lever Morse cable device. The propulsion system developed a thrust of 11,000 pounds.

With the engine running at a constant speed, a power takeoff drive to a 5-kW, 220/110 V ac generator was included, as well as the traditional dc alternators. The power takeoff additionally drove the refrigeration compressor.

The requirement for an unobstructed aft work deck forced the outboard location of the engine room access. Engine combustion air and ventilation requirements were met with air intakes in this access house. The intakes were fitted with self closing louvers to prevent downflooding of seawater in the event of a severe roll.

Five thousand gallons of fuel, carried between the fish hold and the engine room, as well as under the hold, provided a cruising range of 2500 miles.

An auxiliary engine was located in the lower deckhouse to drive the deck mounted trawl winches. This engine was a 130 hp Caterpillar D330 diesel, driving the trawl winches through a torque converter. One 28 hp hydraulic pump was fitted to each of the two engines, permitting operation of the amidships mounted net reel from either of the engines.

The hydraulic pumps also provided power for two Gearmatic planetary reduction wire rope hoists mounted on the boom for lifting the cod end. These winches were each capable of a 5000-lb lift.

The trawl winches, designed and fabricated at my yard, had a capacity of 1000 fathoms of ⅝-in. diameter wire rope trawl warps. The winches were arranged for both local and remote control, and were located outboard of the net drum, which was on the centerline just aft of the deckhouse. The trawl warps were led aft through stern mounted towing blocks attached to port and starboard gallow frames. These gallows were mounted far enough forward of the rudder post to permit adequate maneuverability while towing under full power.

Between the gallows, and at the transom of the boat, an inclined ramp was cut into the main deck, to serve as a chute for setting and hauling the net.

Both the trawl winches and the net reel were controlled remotely from the aft-facing pilothouse control station. This station was designed to overlook the entire working deck, and included pneumatic controls for propeller pitch, steering, deck equipment (main and hoisting winches, net drum) and the various automated fishing controls.

To reduce crew requirements and ensure successful operation of the stern trawling system, a trawl warp measuring system was fitted to each winch. These interfaced with automatic interlock systems in the fishing gear controls to allow pushbutton operation.

In setting the trawl, the operator had the option of engaging a mechanical locking system from the control system that forced the port and starboard trawl warps to pay out identically, thus keeping the mouth of the net open to it's maximum extent. The length of trawl warp payed out could be dialed into the automatic pay-out device, stopping both winches automatically at the desired length of wire. Haulback was

Fig. 3 *Narragansett.*

accomplished by pressing a button that caused both winches to haul simultaneously until the doors reached the trawl blocks at the gallows, at which point the winches stopped automatically.

With the doors at the gallows, some deckhand intervention was required to bring the ground lines to the net reel. Rotation of the net reel, again controlled from the pilothouse, caused the wings and body of the net and its rollers to be reeled in through the stern ramp. With the belly of the net stretched out on the working deck between the net reel and the stern ramp, the cod end was hoisted and dumped on deck for sorting and processing the catch. At this point, repairs to the cod end were easily made, being accessible on the working deck.

In setting the trawl, the procedure was simply reversed.

In the best of the side trawlers, it had taken seven minutes to release the cod end after securing the otter boards on deck. In the *Narragansett* the operation took only about two and a half minutes. Similarly, the time to reset the net was reduced from seven to less than two minutes, and the net could be replaced in only three minutes.

If, during the course of trawling on the bottom, the net became fouled on an obstruction, a trawl wire shifter was automatically engaged, allowing the vessel to turn into the wind and haul the net, in a manner similar to that employed by a side trawler.

At any time during the fishing operation that the operator desired remote manual control of any of the fishing gear, this was available with fingertip control at the pilothouse console.

The ease of operation and minimal crew requirements of the system were demonstrated during the summer of 1963, as, assisted only by my twelve-year-old daughter as crew, I toured the East Coast aboard the *Narragansett* and performed all the fishing operations before many interested parties.

The *Narragansett* quickly proved her efficiency with consistent record breaking and highly profitable trips. Operating with a crew of only four (captain, engineer and two deckhands) the vessel exceeded the catches of the traditional side trawlers, fishing with crews of six to eight men. The boat worked consistently in winds greater than 45 knots, with no deck wetness or loss of fish from the pen boards on deck.

The American fishing industry, a tradition oriented branch of the historically conservative marine community, was slow to adopt the stern trawler, just as it had been resistant in the adoption of steel as the favored construction material from the traditional wood construction. Nevertheless, the productivity of the small stern trawler was just too obvious to ignore. Within ten years all new fishing vessel built in the United States were of this type. Today, few of the old side trawlers are still fishing, most having been phased out owing to age, or converted to scalloping, a fishery that still handles dredges over the sides.

Stern ramps and net reels have become the accepted practice, as has pilothouse controls of the trawl winches and other fishing gear. The automated controls first used on the *Narragansett* have just recently found their way into widespread international use, with the European trawlers leading the way, and large American processing vessels following the trend.

Epilogue

I have been granted several U.S. patents on my inventions. One was on the method used to disconnect the doors from the ground line, and another was for the double net drum configuration.

As to use of the patents, I have collected but one fee. I have allowed fishermen, naval architects and ship gear manufacturers to use my inventions indiscriminately and for free. In addition, several companies throughout the world were formed to build the net drums, remote controlled winches and associated gear involved with the fabrication of this special equipment that I designed for this stern trawler.

My original objective was to develop a trawler that would give my shipyard an edge in constructing fishing boats. Ironically, it took so long for the fishermen to accept the idea of a small stern trawler that, to keep my shipyard going, I had to concentrate on design and construction of passenger vessels. When the burst of trawler business finally came, my yard was so entrenched in passenger boat orders that I managed to work only a few of the stern trawlers into our production.

Additional Reading

Ridgely-Nevitt, C., "The Resistance of a High Displacement–Length Ratio Trawler Series," *Trans.* SNAME, Vol. 75, 1967.

Ridgely-Nevitt, C., "The Resistance of Trawler Hull Forms of 0.65 Prismatic Coefficient," *Trans.* SNAME, Vol. 64, 1956.

Symonds, R. F., and Trowbridge, H. O., "The Development of Beam Trawling in the North Atlantic," *Trans.* SNAME, Vol. 55, 1947.

LUTHER BLOUNT earned a degree in mechanical engineering at the Wentworth Institute of Technology in 1937. After twelve years in his family's seafood business, he started the Blount Marine Corporation and started building steel oyster boats. By 1990 his company had turned out nearly three hundred vessels of many types. Included among these were tugs, cruise boats, small tankers, offshore supply boats and many varieties of commercial fishing craft. A holder of twenty patents and three honorary degrees, he was inducted into the Rhode Island Hall of Fame in 1981 and granted the American Merchant Marine Achievement Award in 1989. Mr Blount is a member of SNAME.

Fisheries of the Pacific Coast

ARTHUR DE FEVER

Introduction

THE PACIFIC COAST of North America has for many years been a great producer of fresh, canned and frozen fish of many species. In the northwest region trolling boats are most common, with some operating out of San Francisco as well. Many of these have what is called the Monterey style hull, a double-ended, low profile, single-screw configuration with a small combined pilot house and galley. They are principally day fishing boats for the fresh fish markets and restaurants. The west coast seiner hull is popular for sardine fishing in the San Pedro and Los Angeles areas, where blue fin tuna were once plentiful. San Diego has long been identified with the tuna clipper, now the tuna seiner. San Diego was at one time known as the tuna capital of the world. For years there were ideal fishing waters off the Southern California coast and adjacent to the Channel Islands, Mexico and Central America. In recent years, however, major tuna fishing has shifted to American Samoa and adjacent waters as well as other parts of the world.

Commercial fishing and what has become the tuna industry began in the early 1900s, and by 1917 it began to flourish; packing plants quadrupled in number and Southern California was of major importance and destined to become, in a few years, one of the leading industries of California. The long-range tuna fleet was initiated in 1924 with a 68-ft flush deck boat, and the following year with a 100-ft boat, both by Campbell Shipyard of San Diego. In 1928 the first raised deck tuna clipper was built and this set the pattern of what was to follow. In 1938 Campbell Shipyard built 149-ft and 150-ft boats, each of 400-short ton capacity.

The evolution of the fishing vessel design over the past fifty years has advanced tremendously, particularly in the case of the tuna fleet. The early boats kept their catches with crushed ice in the fish holds. In later years (1937) the tuna clippers started to divide the underdeck hold into wells with ammonia refrigeration coils on all bulkheads, overhead and bottom, with circulating brine to cool down the catch.

In 1942 the industry had a setback, as many of the tuna clippers were taken over by the U.S. Navy for refrigerated food carriers to the South Pacific. Most of these were in the 120–150 ft size. In fact, the U.S. Navy at the start of the war had designed and built a number of YPs (Yard Patrols) 128 ft long, patterned after the tuna clipper. Many of these were sold into the tuna fleet after the war. This coincided with the fishing fleet building boom that started immediately after the war.

San Diego's fishing catch after World War II was the second largest in the nation, and San Diego boats brought in more tuna than those of any other port in the world. In attaining this premier position the tuna boats grew amazingly in size and complexity. Before World War II most fishing vessels were built of wood and the tuna fishing vessels formed the largest fleet both in numbers and in size. They were known as bait boats and featured heavy duty diesel engines and a three-compartment wooden bait tank on the after deck. Hulls were built of sawn frame construction and were insulated between the frames.

Types of Vessels

The Californian fishing fleet consists of gillnetters, trollers, trawlers, long-liners, small table or drum seiners, medium or large table purse seiners, bait boats, and the super tuna seiners of today (which emerged from the tuna bait boat). Some of the troller and small table seiners are equipped for all-purpose fishing. There are a few plank boats for swordfishing, together with small crab and lobster boats. (Plank boats take their name from the long plank-like bowsprit from which swordfish may be speared.)

Modern tuna seiners, among the largest and most complex of American fishing vessels, are the focus of this article. They are yacht-like in hull form and general appearance and have shown the most advances in design over the past half century. We have witnessed appreciable increases in speed and power, together with greater sophistication of machinery and equipment.

Fig. 1 The first stern engine room tuna seiner with fish wells forward and the first with a bulbous bow.

Up to the late 1950s and early 1960s most tuna clippers were of the bait boat design. The fishermen had learned over the years to attract schools of tuna by throwing out live bait (sardines or anchovies), found in abundance along the coasts of their fishing areas. Tuna would churn the water trying to get the bait and in so doing would also strike the feathered hooks held by the fishermen on lines at the end of bamboo poles. When tuna became too large to lift aboard by one man they would switch to teams of up to four men, each with his own pole, but all strung to a single hook. In the late 1950s and early 1960s the tuna fleet underwent a major transition, switching from the live bait hook-and-line method to the purse seine system. This proved to be more efficient and more productive. The transition to the present day tuna clipper design involved removing the bait tanks from the after deck and installing revolving net tables in their place. The modern design also features a stern ramp upon which to pull the skiff above the stacked net. The after section of the raised deck was also cut off and a large seine winch was installed on the main deck. This also required a large mast and boom to hoist a Puretic power block. These major revisions led to a new building program in the late 1960s and early 1970s. The new boats were larger; lengths increased to 260 ft and capacity to 2000 tons.

Developments in Design and Hull Form

After the war, starting in 1944 and 1945, there was a tuna clipper building boom. Most boat building yards capable and experienced in fish boat construction were filled with orders. Initially the average length was 95–100 ft and capacities ranged from 150 to 160 tons. But these rather quickly grew to 150 ft and 600-ton capacity. The first steel hulls appeared in the mid-1940s, but steel did not become popular until the mid-1960s, with the initial construction of the tuna seiners. This led to a vigorous new building boom that continued into the 1970s. Boats grew larger and finer hull forms became popular. Some were tank tested with various hull shapes to obtain optimum speed and power, and bulbous bows made their appearance.

The early steel live bait tuna clippers encountered stability complications. There were topside weight problems with the steel superstructure adding to the effects of free surface in the bait tanks. Some redesign was in order. The switch to purse seining solved the free surface problem, while the newer boats reduced topside weights further by switching to aluminum alloy superstructures, often with trimmed down profiles. Beams were increased as well. Some went to longitudinal framing. With these and other changes in design the tuna boats are now both more seaworthy and more productive. The discomforts and dangers of abrupt rolling brought on by excess stability are in some cases reduced through the use of flume stabilizers.

With this building boom of tuna seiners, new designs were developed. This brought about a total change in the configuration of the stern, and a large stern ramp for the seine skiff was added. The skiffs now are also built of steel, and the larger ones are over 35 ft long and 18 ft in beam, with diesel engines in excess of 300 hp. The early tuna clippers had a forward engine room and long shaft alley extending through fish wells aft. As vessels increased in size, the arrangement went to a stern engine room with fish wells forward. A piping tunnel leading forward from the engine room between the forward wells housed the brine circulating pumps and ammonia system valve manifolds. The trend of design has been to hulls with more beam and high rakish bows with considerable flare to keep the deck dry. Bulbous bows are now standard, too. The first of these modern tuna clippers, which I happened to design, was built in 1968 by PACECO. (See Fig. 1.)

Old wooden hull designs had considerable deadrise and deep fuel tanks; modern steel hulls have little deadrise and fuel is carried in the doublebottom.

The traditional problem of joining the aluminum superstructure to the steel hull was neatly solved by the Dupont Datacouple, a sandwich of steel and aluminum flat bars fused together with an insulating material. Now the superstructure can be preassembled on a jig simultaneously with the hull construction and then set on deck and welded in place.

Fig. 2 The modern tuna seiner of today. Designed and built by Campbell Industries.

Tuna boats today are capable of carrying great loads and can remain on the fishing grounds for long periods and navigate in the open seas in heavy weather. With the demand for greater speed (up to 16 and 17 knots) came the need for more powerful engines. This, in turn, required greater fuel capacity, not only for the more powerful engines, but to increase the range, for fishing in more distant waters. Some have both doublebottom tanks and deep tanks, carrying close to 200,000 gallons. In addition some fish wells are piped to the fuel system so they can carry fuel on the trip to the fishing grounds. With the longer ranges and longer times at sea, greater quantities of potable water are also needed. Consequently, larger water tanks are incorporated and, in addition, high-capacity distillation systems are standard.

Bow thrusters are commonly installed for good maneuverability and position keeping. Free standing masts with radio equipment and ship controls in the large crow's nest are now incorporated. There are also controls on the port side of the bridge and at the hydraulic control station on the after end of the bridge deck. Development and improvement of the Puretic power block and the seine winches over the years have added greatly to the efficiency of this style of fishing.

As in every part of the modern marine industry, computers are now widely used in designing tuna boats.

The tuna seiners of today provide comfortable accommodations for the crew, featuring many conveniences experienced at home and ashore, such as game rooms, lounge and reading rooms, and video motion pictures. The modern tuna seiner is depicted in Fig. 2.

Power and Power Systems

The tuna clippers of the 1940s were powered with heavy duty, slow-turning diesel engines in the 300–400 rpm range, such as the Atlas and Union engines, which were direct reversing. These were followed by the Superior, Washington and Enterprise diesel engines in the 600–1000 rpm range. In modern day tuna boats, Caterpillar, General Motors, Alco, Fairbanks-Morse and Electro-Motive diesel engines are used. These weigh considerably less per unit of power.

Diesel propulsion through reduction gears has been the practice since the purse seine building boom in the 1960s and 1970s. Power ratings have gone to 4000 hp and even above 5000 hp, where two engines have been geared to a common shaft and controllable-pitch propeller. A few twin-screw tuna purse seiners have been built and have proven successful; however, the single screw seems to be preferred because of less hazard of the seine net fouling the propeller.

Generators are of large capacity, two and three in number on large vessels, to handle the motor driven ammonia compressors, brine circulating pumps and other electrical equipment. Diesel engines are used to drive hydraulic systems for the large seine winch, Puretic power block, vang winches, boom topping winch, and other hydraulic gear. The bow thrusters are also commonly driven by diesel engines.

Engine room arrangements have improved significantly over the years. We now commonly find an air conditioned, soundproofed control station with heavy glass windows overlooking the machinery space. From here the engineer can operate the ship's switchboard and can monitor temperatures and oil pressures of all engines, as well as the temperature in each fish well.

Steering and Control

The steering systems are predominantly hydraulic with steering positions in the pilothouse, port side of bridge, and in the crow's nest. The seine winch, power block, boom vang winches and topping winch are all operated from one hydraulic control station at the aft end of the bridge deck, where visibility is good.

Arrangement and Outfitting

As previously mentioned, the modern tuna seiner has its engine room aft and brine refrigerated holds forward, with a piping tunnel between the wells. Fuel is carried in doublebottoms and deep tanks in the bow and stern. A compartment low and near the bow houses the bow thruster and its diesel engine. The main deck is totally enclosed and contains the following compartments listed in order starting in the bow: ship's stores, loading hatches to each of the fish wells (and the fish loading conveyers together with the fish loading hopper), upper engine room and, finally, the lazaret, or steering compartment. The raised deck above the main deck carries the anchor windlass at the extreme forward end. The superstructure extends from there to about mid-length. Aft of that is the mast and fish-catching gear, with the seine winch, purse davit, and hydraulic ring stripper on the port side. On the starboard side are the chase boats with hydraulic davit or hydraulic crane. On the center of this deck is the loading hatch with hopper. The exhaust stack is on the starboard side, and on the after deck is a huge net platform and skiff ramp.

The superstructure on this deck contains a large galley and mess room at the aft end, opposite which are refrigerated and freezer store rooms for provisions for trips of 90 days or more. Forward of the galley and provision rooms are the crew's quarters. Above the galley and crew's quarters is located the bridge deck, housing the pilothouse, radio room, chartroom and the captain's luxurious quarters. At the aft end of the bridge deck is the hydraulic control station for the seine winch, brailer and boom control winches. The top of the pilothouse is usually fitted with a helicopter landing pad. Alternatively, in some cases helicopters are landed on a platform on the foredeck. (The helicopters are used to spot and locate tuna schools.)

Along with the deck outfitting, the seine net is an all-important item. Some nets are now up to 1000 fathoms in length. The great improvement in nets has been the change from cotton to nylon webbing. Although nylon has a higher first cost, its greater strength and longevity make it more economical.

Navigation, Electronic, and Safety Equipment

Navigation and electronic equipment are items that are improved almost annually. Satellite navigation, radar, gyro compasses, weather fax receivers, Loran, automatic pilot, automatic direction finder and now GPS (short for Geographical Position System) have all added to the safety of fishing vessels. Bow thrusters can be tied into the automatic pilot to keep the boat on a steady heading. Fathometers, sonar and fish finders on colored display screens add much to the success of these vessels. Single side band radios, VHF sets, together with the amateur radio sets, leave little to be desired for communication. EPIRB (Emergency Position Indicating Radio Beacon) equipment and inflatable life rafts also add to the safety and life at sea.

Designers and Builders

Prominent designers over the past fifty years on the West Coast in the fishing industry and particularly in tuna boats have been H.C. Hanson, Bruce Newby, James E. Petrich and Robert Rados. I, too, have designed a number of the tuna vessels. Many builders have, or had, in-house designs, which have been successful; they include Campbell Industries, Martinac Shipbuilding, Tacoma Boatbuilding Company, Martinolich Shipbuilding Company, and Marco.

Other builders of the tuna vessel class during the past fifty years include Petrich Shipbuilding Company, Hodgson Greene Haldeman, United Concrete Pipe Corporation Steel Shipbuilding Division, Peterson Boatbuilding of Tacoma WA, Harbor Boat Building Company, San Diego Marine Construction, Lynch Shipbuilding, National Steel and Shipbuilding Company, Peterson Builders Inc. of Sturgeon Bay WI, and Pacific Coast Engineering. A few of these yards that were once prominent in the industry no longer exist.

Bibliography

Cole, J. A., "Fishing Vessels Of The Pacific Northwest: The Past 30 Years," *Marine Technology*, 1988.

Fishing Boats Of The World, J.-O. Traung, Ed., FAO, Rome, 1955.

Arthur De Fever has been a member of SNAME for 49 years. As an independent naval architect he has specialized in the design of tuna boats since before the start of World War II. He designed the first conversion from bait fishing tuna clipper to tuna seiner. He also designed the first tuna seiners with engines aft, bulbous bows, and twin-screw propulsion.

Fishing Vessel Design in the Pacific Northwest and Alaska

James A. Cole

Introduction

The fishery resources of Oregon, Washington, and Alaska consist of a variety of surface, midwater, and bottom species, as well as shellfish. They are found in the rivers, estuaries, coastal waters, and open ocean throughout the seasons of the year [1].

Consequently, the vessels of this region cover a variety of types. In the beginning of commercial fishing here, however, there were only a few types but they proved to be versatile and were adapted to suit several needs.

It is probable that the first fishing boats of the Pacific Northwest were the sailing gillnetters of the Columbia River in the 1880s [2]. Gillnet fishing is done by setting and retrieving a net that has a corkline, which allows the net to hang in the water like a fence. The fishing is done at night or in silt-laden estuaries where the fish will swim into the invisible net and become entangled by their gills. The first of these gillnet boats were built in San Francisco and were patterned after the Greek boats of the Sacremento River and the Italian boats of the Bay area, but had shallow draft hulls [3]. They were double ended, about 26 ft long, and were sail or oar powered. This kind of boat was the progenitor of several types that came a short time later. As gillnetters, they evolved into several configurations over the next fifty years. The sailing double enders operated on Bristol Bay in Alaska until 1951, when power boats became legal. The Columbia River gillnetters were sail powered until 1912, when the switch to gasoline power began. They then evolved into the bow picker gillnetters, where the net was set and retrieved over the bow to keep it out of the propeller. They remained about 26 to 30 ft long. The gillnetters on Puget Sound and Southeast Alaska evolved through the same process but their culmination as powered boats had them fishing off the stern.

The first descendent of the Columbia River gillnetter was the trolling boat, which was at first sail or oar powered. Gasoline power was installed in some of these in 1900 and they began to fish at sea [2]. Trolling is done with lines that trail from poles that are extended out over the sides of the boat. The lines are weighted and have hooks with lures at their lower ends [1]. Trolling had also started in Alaska in about 1885, so the use of gasoline engines can be assumed to have started there in the early 1900s [4]. The subsequent changes in the trollers made them deeper, decked over with a hatch, and a small forward deck house. The double ended hull remained popular although counter sterns were also adopted owing to their seakindliness. The size of these boats increased so that some were up to 60 ft long by the 1940s.

The West Coast seiner was originated on Puget Sound shortly after 1900 by Yugoslav fishermen in quest of migrating salmon. The original boats were oar powered skiffs, but the style of the seiners was established in about 1903 with boats about 30 ft long with pilot houses, foc'sles, fish holds, and gasoline engines. A hand-powered deck winch was used in conjunction with a crew of about eight to handle the net. Purse seining is done by encircling a school of fish with a net hanging from a cork-line. The purse line (which runs around the bottom of the net) is pulled in, closing the bottom in the same manner as a purse. The seiner type evolved into the signature fishing vessel of the West Coast, and variations of it are seen worldwide. Their registered length was limited to 50 ft for salmon seiners in Alaska, while their counterparts on Puget Sound grew to 70 ft LOA. The West Coast seiner was also converted to trawling and was a forerunner to stern trawling in the Pacific Northwest.

The commercial halibut fishery in the Pacific Northwest and Alaska was successfully started with steam powered boats that came from the Atlantic Coast in the 1890s. These were eventually replaced by the halibut schooners, which started fishing from Puget Sound in 1912 and 1913. The heavily built schooners averaged 70 ft in length, with fine bows

Fig. 1 Halibut schooner Polaris, a typical longliner as they appear today. She was built in 1913 and has been modified substantially over the years. Her current is Arne Einmo.

and counter sterns, making them excellent sea boats. I. Heggum of Seattle and John Strand of Tacoma built schooners decidedly Scandinavian in character, with their pilot house and deck house aft, and two masts carrying sails for steadying purposes and some propulsion. They originally fished with dories, but by the early 1920s were rigged for longlining, which they do to this day. Many of the halibut schooners built before 1920 are still fishing.

The materials used during this time were wood, of course. The boats up to about 60 ft in length had steamed and bent oak frames, oak or fir keels, fir planking, (the gillnetters and trollers might have cedar planking), fir or oak decks, and wear areas were trimmed in iron bark (the colloquial term for iron wood, a kind of eucalyptus from Australia). The larger or heavier hulls had double sawn frames or a combination of them with steamed and bent oak. Deck houses were of fir, red cedar or yellow cedar; and trim inside and out was of mahogany [5].

Propulsion machinery for the smaller boats was usually with a marine gasoline engine, and by the 1930s automotive engine conversions were available. The larger seiners, tuna clippers, and halibut schooners were powered with direct reversing marine diesel engines.

Deck machinery was simple and mechanically driven by line shafting from the propulsion engine. The boom-mounted and other small winches usually were driven by dc motors. In the 1920s the dungeness crab fishery started in the bays and estuaries. Those boats were, once again, adaptations of the Columbia River gillnetter, but they were more heavily built and had single cylinder gasoline engines of about eight horsepower.

The seiners were getting larger at this time, ranging farther from home; and naval architects were starting to be involved in their designs. The sardine seiners designed and built in Northwest yards for the Monterey sardine fishery were up to 90 ft in length.

By the late 1920s, the West Coast tuna clippers of about 100 ft LOA had evolved from the West Coast seiner type [6]. These were sophisticated boats for their time, with main engines of about 300 hp, refrigerated fish holds, radio direction finders, and two way radios. The quality of their construction matched their sophistication, and their customers were the tuna fishermen of southern California. The first local welded steel fishing boats appeared; they were seiners 40–50 ft in length. World War II intervened at this time and fishing vessels under construction were commandeered by the Armed Forces.

The boat builders of the Pacific Northwest quickly turned to the construction of wood and steel vessels for the war effort. Parallel efforts for military needs were bringing about the development of lighter weight diesel engines, synthetic fabrics and construction materials, welding technology, hydraulic powered machinery, and electronic equipment. All of these technological advances greatly influenced the design of all types of commercial vessels after the end of the war.

When World War II ended, and the adjustment to peace time business was complete, or nearly so, fishing craft construction resumed. The technical advancements that had come from all the military effort, and perhaps in some quarters the pleasure of building fishing boats again, helped the industry to more than pick up where it had left off. A period of steady change began.

Fishing Vessels of the Pacific Northwest and Alaska—1943 to 1993

Although this half century was a period of change or evolution, some facets of fishing vessel design and construction did not immediately change. Wood construction remained popular through the 1950s, owing in part to the woodworking skills nurtured during the war, when many wood minesweepers, tugs, work boats, patrol boats, and barges were built for the Navy or Army. (Minesweepers were also built during the Korean War.)

The war had awakened us and made us aware of our abilities; but the worldliness that resulted did not negate the influence of the ethnic groups on fishing craft design, principally in the design and construction of seiners of all sizes.

Technical advances such as fiberglass construction, wider use of hydraulic powered deck machinery, synthetic fibers for nets, and electronics brought about enhancements that the designers, builders, and fishermen were quite capable of exploiting. The development of the king crab industry in the 1960s brought another variation of the West Coast seiner with the raised foc'sle Alaskan king crab boat of about 100 to 120 ft in length. Later we began to stern trawl with these boats, and then went on to factory stern trawlers until our ports began to look more foreign than domestic. Changes in fishing regulations, though infrequent, would bring significant changes, as the following sections on gillnet boats and stern trawlers will tell.

Gillnet Boats

Gillnet boats have remained small, ranging from about 24 to 36 ft on Puget Sound, the Columbia River, and in Southeastern Alaska. They have remained this size because they are used by people who look upon commercial fishing as supplementary to their main line of work. Also, the Bristol Bay salmon fishery limits the length of the gillnetters to 32 ft. These boats were restricted to sail power as well through the 1950 salmon season. The boats were sprit rigged with a single mast and Spartan accommodations for two hardy fishermen. They were towed to the fishing grounds by small vessels called monkey boats. In spite of bad weather and strong tidal currents, each such boat could land 12,000 to 18,000 pounds of fish per day while fishing in shallow water [5].

During the 1950s and '60s, the gillnet boats evolved with transom sterns becoming the norm, and using gasoline or diesel power. A leading designer and builder of this type was Joe Dyer of Astoria OR. Hulls were usually of Douglas fir planking on oak frames. By the late 1950s, fiberglass construction was common, and welded aluminum boats, with chine hulls, were being produced in the 1960s [1]. The capacity during this time went from 13,000 pounds of fish for the wood boats to 16,000 pounds for the aluminum boats [5].

Smaller gillnetters became popular in Puget Sound during that time. The boats were owned and operated by part time fishermen with another primary source of income. The boats were trailerable with out-drive propulsion systems, hydraulic powered net reels and rollers, sleeping quarters for two, and a small galley. They had a moderate amount of electronics and could carry about 5000 pounds of salmon or other fish.

Where Bristol Bay was the last hold out of the sailing gillnetter, as mentioned earlier, it is now the scene of the most far-reaching development in high-powered gillnet boats. Their overall length is still limited to 32 ft, but the beam recently has gone up to 17 ft, with twin-screw propulsion totalling 1500 hp. The aim is to get to the better fishing spots and back in the face of increased competition. During good weather, these boats will carry 27,000 pounds of fish. Single chine, planing hulls are common, even with fiberglass hulls; but the larger, more powerful boats are of welded aluminum.

Trolling Boats

Nearly everyone who fishes commercially or has an appreciation for fishing boats has a high regard for trollers. They range from 30 to 60 ft in length, and travel the length of the Pacific Coast to the Gulf of Alaska, fishing for tuna or salmon in the summer and dungeness crab in the winter. Seaworthiness has always been their hallmark.

The powered trollers of the 1920s were deep, narrow boats. They were seaworthy, but were unable to plane and so sacrificed speed. During the post-war era their beam and depth increased by as much as 25 percent [10]. This increased their carrying capacity and, with more powerful engines available, their speed as well. A leading designer and builder of trolling boats was Frank Lowe of Coos Bay OR.

Both double-ended and counter stern hulls, with steam bent or double sawn oak frames and fir planking, were common in the trolling industry after World War II. Small seiners also worked successfully in this fishery. Hold capacities were moderate to permit more fuel to be carried for an extensive cruising range. A 42-ft troller of the post-war era would carry about 30,000 pounds of salmon and would have 800 gallons of fuel aboard, which was usually for a gasoline powered Chrysler marine engine. Electronic equipment was confined to a World War II style radio direction finder and a fathometer. The trolling gurdies were usually mechanically driven, but hydraulically powered units were becoming common.

Building trollers of fiberglass did not get started as soon as building fiberglass gillnet boats. That material is, however, well suited for troller construction, where a flared bow and counter stern combine for a sea-

Fig. 2 Trollers new and old. (Pen and ink drawing by author.)

worthy and durable hull. Steel and aluminum construction are also used although not as often.

Trollers have been fished traditionally by one or two people, and most of them by only one, who pilots the boat and handles the trolling gear from a cockpit at the stern. This practice has remained through the decades that these boats have fished our waters; but they are slightly larger now, and with better accommodations. Husband and wife duos are now more common.

The cool temperatures of the Pacific Northwest made it possible to preserve the salmon in iced fish holds, although holding coils to preserve the ice came into limited use in the late 1940s. The beamier trollers of today carry deck mounted ice makers that enable them to preserve their catch for the longer trips they take. They may also journey southward in the summer trolling for tuna.

Distant journeys have become necessary for these appealing craft, as the coastal salmon resources have dwindled because of over-fishing by commercial and sport salmon fishermen, and the reduction of the salmon runs from our coastal rivers owing to population growth.

West Coast Seiners

While local and Alaskan designers and builders were producing the 58-ft Alaska limit seiner in a steady and developing stream, larger variations of the West Coast seiner type were being developed that would go worldwide. In 1950 and '51, the Tacoma Boat Building Co. delivered two tuna bait boats to a Breton fisherman, which would change the way the French went after Atlantic tuna. When the Puretic power block was invented several years later, the French would convert their bait boat fleet to seiners, just as the California tuna fishermen were doing.

At about the same time the burgeoning anchovy fishery in Chile and Peru provided an opportunity for Marco, a Seattle based boat builder and machinery manufacturer. Marco brought to those countries the capability to design, build, and equip their own boats. A sister company in Chile operated a shipyard employing Chilean craftsmen and built longitudinally framed steel boats. They started with 54-footers in the mid 1950s, and advanced to 90-footers by the late 1960s. At the peak of construction, five 73-ft boats per month were produced.

Pacific Northwest builders had provided the highest quality wood seiners and tuna bait boats, principally from the Tacoma-based yards of J. M. Martinac, Tacoma Boatbuilding Co., and Western Boat Building Co. Those yards and others were owned and operated by families from either Yugoslavia or Norway who had an abundance of their countrymen as shipwrights and mechanics, and who were fishermen as well. This combination of capital, skill, and will power

Fig. 3 Shallow draft seiners. (Pen and ink drawing by author.)

produced top quality boats for fisheries the length of the Pacific Coast, and good repair capability for the local fleet.

The invention of the Puretic power block in the early 1950s greatly reduced the crew size requirements and increased productivity on local salmon seiners [6]. It also caused a number of California based tuna bait boats to be converted into seiners. This was followed by an almost continuous construction run of tuna seiners, designed and built in Puget Sound yards, that lasted for 30 years. The boats went from 130 × 30-ft wood tuna clipper conversions of 240 tons capacity through an evolution into steel boats of 165 × 34-ft and 650 tons; and then to 235 × 41-ft and 1200 tons capacity, and even larger.

I was the principal designer (under Mr. George Moore) of the last boat mentioned above. High speed was a major goal in its design. This led to twin propulsion engines, giving 5600 horsepower driving an 11-ft diameter controllable pitch propeller. The boat delivers a speed of over 19 knots while carrying 1200 tons of tuna.

King Crab Fishing Vessels

The vast majority of West Coast seiners are 55 to 100 ft long, and are true combination fishing boats. They have been used for bait fishing, seining, dragging, and longlining. The smaller ones can troll for salmon or tuna and pot fish for dungeness crab.

The Alaska king crab fishery started in the mid 1960s, and the boats that made up a good share of the fleet were the sardine seiners that had been idle for some years. They were converted to pot fishing and steel live-bait tanks were built into their aging wooden hulls. They served well for a few years, but were unsuited to the severe Alaskan winters. I recall one winter in the mid 1960s when one crab boat was lost per week for a two month period.

The local naval architects and builders responded with steel, house-forward boats with raised foc'sles to protect the deck area. Considerable attention was also paid to transverse stability, fuel oil, crab tank loading conditions, and the deck load of 700-pound crab pots, particularly during icing conditions. Even so, experience would reveal severe stability problems from icing, improper loading, and enough freeboard to assure a sufficient range of stability.

These boats ranged in length from about 90 to 125 ft during a construction period of 15 years. When the king crab fishery experienced a collapse in 1980, many of the boats were converted to stern trawlers. They had been designed with built-in stern ramps and other considerations for conversion to trawling, which was

Fig. 4 Combination fishing boat *Sea Wolf.*

required to obtain financing. This was the beginning of the bottom fishing industry here, and these would be the catcher boats that delivered to foreign factory ships in a joint venture fishery. When the 200-mile conservation zone was established due to the efforts of Senator Magnuson, the fishing community of the Pacific Northwest set to work designing and building factory trawlers that would replace the foreign processing ships and the joint venture fishery.

Factory Stern Trawlers

The longstanding tie with Scandinavia and especially Norway was to bring their naval architects and machinery manufacturers to Puget Sound. While many of our factory trawlers are converted rig-supply vessels 180–250 ft in length, they are equipped with Scandinavian (or other European) deck winches and processing and refrigeration machinery. One fine example of this cooperative effort is the *Northern Glacier*, a 201 × 44-ft factory stern trawler designed by Jensen Maritime Consultants of Seattle with Fiskerstrand & Eldöy of Ålesund, Norway. She was designed in Seattle, built in Tacoma by J. M. Martinac Shipbuilding, and has Norwegian propulsion machinery, including a controllable pitch propeller in a Kort nozzle, a bulbous bow, low pressure hydraulic deck machinery from Norway, German processing machinery, and refrigeration equipment from Denmark. The *Northern Glacier* will carry up to 1100 tons of frozen product that is caught and processed at sea by her crew of 40.

Conclusion

The fishing community of the Pacific Northwest and Alaska, like all others, has known how to respond to

Fig. 5 Factory stern trawler *Northern Glacier.*

opportunities. But the ability to adapt has only meant that more of our fishery resources have been exploited and over-fished. Safety has become an issue and economy of operation will have to be considered seriously when the next generation of boats is produced. Perhaps we will go back to boats of 100–120 ft in length because of their adaptability and lower cost. It will be interesting.

References

1 Cole, J. A., "Fishing Vessels of the Pacific Northwest, the Past Thirty Years," *Marine Technology*, April 1988.
2 Smith, C. L., *Salmon Fishers of the Columbia*, Oregon State University Press, Corvallis, OR, 1979.
3 Chapelle, H. I., *American Small Sailing Craft*, W. W. Norton & Company, New York.
4 Wick, C. I., *Ocean Harvest*, Superior Publishing Co., Seattle.
5 Faste, A. A. P., "Workboats of the Salmon Industry," Pacific Northwest Section, SNAME, 1952.
6 Gilbert, D., "Decade of Drama and Development," *Pacific Fisherman Magazine*, August 1952.
7 Hansen, H. C., "Pacific Gillnetters," in *Fishing Boats of the World*, Vol. 1, J.-O. Traung, Ed., FAO, Rome, 1955.
8 Leveau, C. W., "Aluminum and Its Use in Fishing Boats," in *Fishing Boats of the World*, Vol. 3, J.-O. Traung, Ed., FAO, Rome, 1967.
9 Carlton, J., "The Use of Reinforced Plastics Material in Small Vessel Construction, a 1987 Overview," Pacific Northwest Section, SNAME, 1987.
10 Hansen, H. C., "The Trolling Boat and Its Operations in the Pacific Coast," Pacific Northwest Section, SNAME, 1954.
11 Hanson, H. C., "Combination Fishing Vessels of the Pacific," Pacific Northwest Section, SNAME, 1954.
12 Nickum, G. C., "Proposed Stability Criteria," in *Fishing Boats of the World*, Vol. 1, J.-O. Traung, Ed., FAO, Rome, 1955.
13 Storch, R. L., "Small Boat Safety, The Alaskan King Crab Boat Experience," Pacific Northwest Section, SNAME, 1979.
14 Lerch, D. W., "The Application of Hydraulic Power to Fishing Gear," in *Modern Fishing Gear of the World*, Vol. 2, Fishing News (Books) Ltd., London, 1964.
15 Long, R. W., "The Evolution of Tuna Vessel Design with Emphasis on the Impact of the Federal Government Subsidy Program," Pacific Northwest Section, SNAME, 1969.
16 Schmidt, Jr., P. G., "Purse Seining, Deck Design and Equipment," in *Fishing Boats of the World*, Vol. 2, J.-O. Traung, Ed., FAO, Rome, 1960.
17 *Safety Study, Un-Inspected Commercial Fishing Vessel Safety*, National Transportation Safety Board, U.S. Government Printing Office, 1987.

JAMES A. COLE is a native of the Pacific Northwest and graduated from the University of Washington with a Bachelor's degree in industrial design in 1957. He has been employed in designing, building, and repairing vessels since then. As an avocation, he has been active in maritime historic research and preservation. He is known for his researched pen and ink drawings of fishing craft, some of which are included above.

Part XI

The Offshore Industry

Offshore Industry: Introduction

John J. Filson

Perspective

One of the truly new applications of naval architecture to evolve in the past fifty years is that of offshore technology. To offer a broad definition, offshore technology deals with those activities that attempt to accomplish something at sea at a specific place, usually dealing with the sea bottom. Petroleum exploration and production are the most common, but not the only, such activity. To understand the naval architect's role in the development of offshore technology, that industry itself must first be addressed. This offshore activity has presented a truly rich challenge to the engineering profession and has employed nearly all the creative expertise the profession has to offer. It has also offered a new definition of the profession: "ocean engineering."

The sea has been challenged and largely harnessed by the seafarer, shipbuilder, and naval architect. A great war was recently fought and won largely on the sea. Following that conflict, demand and confidence led to the sea for new challenges, and the first among these was the quest for oil. The Truman doctrine of 1947 laid out a legitimacy to exploit the outer continental shelf (OCS). The rest is an interplay of economics, technology, politics, and creativity.

Before World War II, oil exploration and production had already gone to sea in some modest ways, mostly approached as a practice of civil engineering, or "coastal engineering." Floating the working platforms had been more a means of transport and at best a tenuous way to perform the work. Following the war, looking at deeper water and larger financial exposure, it became necessary to find reliable ways to float and moor platforms on which to perform a variety of tasks. The war left a technical legacy that would help make these offshore activities possible. The opportunity for innovation and application of advanced technology was ripe. Operation afloat would become a preeminent means of performing almost any task required to exploit the resources of the sea.

So what has the offshore industry become, how did it get to be what it is, and what is its technology? In the next several chapters we shall try to answer these questions and to suggest where the industry may be headed. In doing so, we shall emphasize the role of the naval architect. Before we start, however, it must be appreciated that the development of the offshore industry has been driven by the quest for petroleum. This evokes an image of oil rigs and exploration and rightly so. By no means is it confined to that activity, but the overwhelming driving force and finance, its fuel, is indeed tied to petroleum: both up and down!

The history of the industry has been episodic. Our historic survey will follow these episodes. Following the present chapter, which attempts to define the industry, each of six major episodes are examined:

Early Years	before 1950
Invention	1950–1965
Development	1966–1973
Expansion	1974–1982
Consolidation	1983–1987
Regrowth	1988–1993

Each of these periods is discussed in its own chapter, sometimes tracing key events of that period beyond its arbitrary temporal confines. Each has its own peculiar history and driving forces. Each has its own important aspects. While there was hardly a coherent industry, or history for that matter, the early years set many precedents. The fifteen year period from 1950 into the mid 1960s was a period of invention in which most of the basic systems in use today were conceived. The period from the late 1960s up to OPEC oil crisis in 1973 was one of development in which much of the structure of the industry was established.

From 1973 and into 1982, along with the increasing demand for oil and its rise in price, the offshore industry, and the petroleum aspect in particular, went through an enormous expansion, only to dramatically reverse this trend afterwards. The following years were a period of consolidation, with radical restructuring of the industry. By 1988 signs of regrowth became evident, bringing the present question: Is it a mature

Fig. 1 *Super Servant III* with mine sweepers.

and stable industry or may we expect another episode of growth?

The Offshore Industry

General. The Offshore Industry includes the following diverse applications:

- Petroleum
- Mining
- Geotechnical research
- Sea energy conversion
- Dredging
- Space program
- Military facilities

The petroleum sector has dominated the industry. There are three major sectors associated with petroleum:

- Exploration
- Production
- Support

Equipment and technology from the petroleum sector is finding many applications in the non-petroleum sector. Behind the operational activity are many support activities employing marine technology. These include shipyards, equipment manufacturers, marine contractors, engineering companies, and consultants. The offshore industry has also had important impacts on regulatory bodies, classification societies, professional societies, and academic institutions. It has given a totally new dimension to ship model testing. These all touch on the profession of naval architecture in important ways.

The offshore industry has spawned new types of builders: shipyards specializing in mobile floating structures and the "offshore fabricators," an industrial entity altogether different from the traditional shipyard.

The offshore industry has developed primarily along the U.S. Gulf Coast, Texas, and Louisiana. While some parts developed in California, this declined during the 1980s. With the development of the North Sea as an oil province in the 1970s, much of the industry has developed around that location.

Looking at the petroleum side only, it should be pertinent to consider the players in the industry. The action centers on the oil companies. Here we find major, integrated entities, and the independents. Because they operate on an offshore lease or concession, these companies are frequently referred to as "operators." Supporting the operator is the drilling contractor, construction contractor, and a variety of service contractors. All rely on the talents of the naval architect.

Drilling contractors operate mobile, floating equipment that is designed, built, and operated largely in the manner of ships. The operators, however, are more prone to develop a fixed facility designed, procured, and built more in the manner of land facilities such as refineries. Both the engineering and construction practices differ from those employed by naval architects and shipbuilders. The typical drilling contractor will frequently rely on a general contractor to manage the project, while many fabricators and service contractors provide the needed components and services. The various contractors do, however, employ a variety of mobile marine equipment.

Fig. 2 MODU jackup *High Island* servicing a production facility. The production facility has three component features: the main production platform (upper right), a well protector (under the derrick), and a storage tank (underwater, upper left). All except the storage tank are fairly modern. The storage tank, over thirty years old, is a rectangular pontoon sitting on the bottom with four bottle columns. It was built by Bethlehem for Chevron, found to be in excellent condition, and refurbished and reinstalled in 1980 for Union Pacific Resources. A fitting component for tying in naval architecture is the supply boat (lower center), a vessel type originating offshore and now having found to have many diverse, non-offshore applications.

Drilling Companies. During the early period most offshore drilling was conducted by conventional land drilling companies. These were guaranteed business and financing by one or more oil companies. The more successful among them have become the established drilling companies of today:

- South East Drilling Company; now SEDCO/Forex
- Offshore Drilling and Exploration Company; now Diamond M
- ODECO
- The Offshore Drilling Company; now SONAT Offshore
- Penrod Drilling Company; now part of ENSCO
- Zapata Offshore
- Global Marine

Many names have been dropped from the list, many of which were acquired or merged over the years. Storm Drilling became part of ODECO in 1975 after it had become part of a leasing conglomerate, Dearborn Computer. Some like Western Offshore

Drilling Company (WODECO) disappeared within Fluor Drilling with obsolescence and scrapping of their drill barges.

Equipment And Systems

Exploration. Equipment for offshore exploration was originally performed by temporary land-type equipment deployed on fixed platforms. Soon, however, the need for mobile, self-contained equipment became evident. Looking back, imaginations must have run wild. Initial efforts produced two basic varieties of bottom founded units: "submersibles" and "jackups." Subsequent needs developed two types of floating units: "floaters" and "semisubmersibles." These four basic types of mobile offshore drilling units (MODU) have remained to this day.

Submersibles. The first of the bottom founded units, the submersible, floats and transits on a barge-like, buoyant mat. When positioned over its working site, it submerges by taking on ballast. To maintain stability during submerging, it generally features three or more large stability columns, on top of which is carried a separate deck structure high above the waves. Resting on the ocean floor, the submerged unit is ready to go to work. Some of the equipment, of course, is carried on the upper deck. Submersible units are generally limited to oceans depths of no more than 100 ft.

Usually built in a shipyard and like a ship, submersibles are particularly interesting to the naval architect. While the superstructure is essentially a flat deck supported by columns, the mat is a hull and the stability columns evoke one of the essential elements of naval architecture: static stability! For the most part, submersible design is a product of the naval architect.

Jackups. A jackup is essentially a barge with three or more legs that are lowered to the sea bottom during favorable conditions and used to elevate the barge above the sea. There are several varieties of jackups, some proprietary in nature. They can have legs with independent footings or all may be interconnected with a bottom mat. Methods of jacking include hydraulic cylinders and pins, as well as other varieties with rack and pinion systems. While the present trend is to employ three legs, four have been common, and some units have many more. Jackups are generally limited to water depths of 300 ft or less, but there are examples in the 400-ft range.

In the beginning the jackup was either a barge that had mechanical devices to lift it, or a large mechanical device that would float. Until jackups ventured into deep water and the legs became longer, stability was not a problem. Much later, with legs towering 300–450 ft above the hull when in transit, stability became critical. Other problems involved the strength of the legs and motions of the vessel. Soon model test tanks were discovered and stability criteria were developed.

Beginning with working depths comparable to those of the submersible units (i.e., 100 ft or less), the jackup quickly extended water depth capability to over 200 ft and later, with truss legs, to over 300 ft. Both types still work with surface wells and are used predominantly for development drilling.

Operations Afloat. As drilling moved to deeper waters, it became necessary to operate afloat. The earliest floating units were simply moored barges. One of the fundamental differences between floating and bottom founded units is the subsea system. With a bottom founded unit, the well is secured at the surface. That, however, is impractical when afloat. In the latter case, well control is achieved on the sea bottom with a blowout preventer (BOP), a system of protective valves hydraulically controlled from the surface. The well is accessed through a special jointed, tubular system attached to the BOP. This special device, called a "marine riser," requires active tension.

Another important aspect of operations afloat is station keeping. Although thruster systems were used in limited examples, the initial means of holding a floating drilling unit on station was spread moorings. This has become a technology in itself.

Floaters. The first floaters were simply barges, many of which were conversions. Soon, however, drill barges became drill ships. The main advantage of a drill ship is its mobility, followed by high payload. The principal drawback of a floater is its motions, which may at times require suspension of drilling. An important innovation that overcomes much of the problem is the turret mooring system. This allows the vessel to point to the weather, thereby reducing sea motions.

Semisubmersibles. Soon after the development of the drill barge, semisubmersible came into being. These allow a moored submersible unit to work afloat. These have now evolved into large structures, generally featuring a pair of fully submerged longitudinal pontoons with a number (typically six or eight) large-diameter stability columns and truss-work supporting a deck spanning the columns. The main advantage of a semisubmersible is minimum motions and the resultant improved drilling availability. Semisubmersibles tend to be payload limited and considerably less mobile than drill ships.

Fig. 3 First submersible, *Mr. Charlie*.

Fig. 4 Modern submersible, *Atwood Richmond*, under tow to Nigeria.

Fig. 5 Early jackup, *Scorpion*.

Fig. 6 Early submersible, *Transworld 54*.

Fig. 7 Bethlehem Steel mat-type jackup.

Fig. 8 Early jackup, *Mr. Gus I*.

Water depth limitation of drill ships and semisubmersibles depends only on the practicality of the station keeping system and drilling riser. An alternative station keeping system is dynamic positioning, which relies on computer controlled thrusters. This allows drilling in particularly deep water (over 3000 ft).

Arctic Technology. It has become evident that some of the greatest petroleum reserves lie in the Arctic. In the late 1970s several types of ship were developed, and these are of interest to the naval architect. Included here are not only drilling units, but all manner of support vessels. Canadian naval architects in particular produced several impressive ice breaking tug/supply boats.

Production

General. To the petroleum industry, the end objective of the offshore industry is production. Exploration is only an enabling step. The support sector provides the tools to develop a production facility. The function of such a facility is to raise the petroleum from the ground, to process it for shipment, and to deliver it to shore. In general they are fixed, permanent affairs and only in special circumstances are they mobile.

A variety of production activities are performed offshore. They generally include well control, separation of gas, oil and water, and the treatment of each as needed for disposition. In some cases injection of gas or water into the formation is included. Other possible functions include drilling, well servicing, storage, and offloading. These primary functions can demand a substantial industrial operation including accommodations, utilities and power, and logistics.

A wide variety of facility configurations has evolved to perform the above functions. The configuration depends on many factors including site characteristics, environment, the required suite of functions, performance capacity, length of service, logistics, regulation, economics, state of technology, and availability of support equipment.

Fixed Platforms. The earliest structures for offshore production were bottom founded, shallow water, pile supported platforms. They evolved into pile supported, tubular steel truss systems called "jackets," and continue to evolve as compliant towers, approaching water depths of more than 2000 ft. These systems include functional decks, which are lifted into place, and bear little resemblance to anything in shipbuilding.

A variant of this type, more prevalent in the North Sea, is the gravity base platform. Many of these include large subsea oil storage tanks, and the majority are of concrete. This variety of platform is frequently built afloat and requires significant application of the principles of naval architecture. Other sophisticated fixed platform variants were developed for arctic application, although never built.

The bottom founded platforms, as a group, are called "fixed platforms." Their design and construction have civil engineering roots that did not at first involve the principles of naval architecture. This changed, however, when the structures became quite large, requiring complex calculation for upending, launch, and the prediction and control of floating trim. The technology has ultimately developed into a branch of ocean engineering that is rich with applications of advanced hydrodynamics and structural mechanics.

Single Point Mooring. Another tool of production is the single point mooring (SPM) system. Early practice for transfer of petroleum to or from tankers, in other than sheltered waters, was the "sea berth." This consisted of a permanently set spread mooring system to which a tanker would temporarily moor. Such systems are weather limited. During World War II an alternative method was developed whereby a buoy was moored and the moored ship was allowed to weather vane about it. This was called "single point mooring," or SPM. Since the mid-1970s, the SPM has had a technically rich evolution, which now includes many variants, including a single anchored buoy with one or more articulations (SALM) and articulated columns, and a half million bbl storage buoy (SPAR).

Floating Storage. A direct outgrowth of the SPM is the permanent tanker mooring for a floating storage unit (FSU) and floating production, storage and offloading units (FPSO). Not to mention the vessels themselves, the mooring systems have produced good applications of both naval architecture and marine engineering. Early engineering work with moored tankers concentrated on model testing, but contemporary systems also offer opportunities for the most advanced application of hydrodynamic theory of ship motions in shallow water as well as the study of chaotic processes. Even the deep part of the structure of an articulated column uses design principles developed for submarines.

Floating Production. One important trend in the past twenty years has been the development of the floating production system (FPS). As suggested above, this began with storage tankers, which later began to perform various petroleum processing functions. This type of system began with the conversion of obsolete

Fig. 9 First semisubmersible, *Bluewater*.

Fig. 10 Early semisubmersible, *Sedco 135*.

Fig. 11 First-generation semisubmersible, *Df-95*.

Fig. 12 Old drill barge *Discoverer*.

Fig. 13 Reading & Bates Drilling Company's trendsetting semi-submersible *Zane Barnes* (designed by Friede & Goldman, New Orleans).

Fig. 14 Arctic drilling vessel *Kulluk*.

Fig. 15 Modern drill ship *Sedco 472*.

Fig. 16 Icebreaking tug/supply boat.

Fig. 17 Seismic research vessel *Shell America*.

Fig. 18 Multipurpose work vessel.

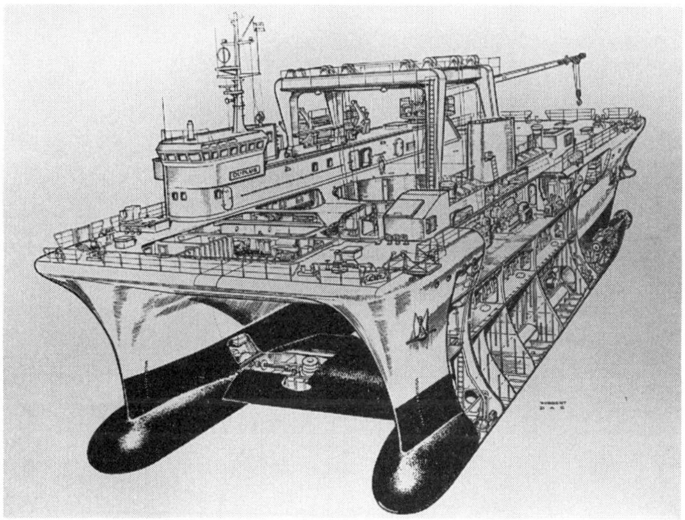

Fig. 19 Early SWATH offshore work vessel *Duplus.*

Fig. 20 Jacket-type fixed platform *Exxon Hondo.*

Fig. 21 *Ekofisk*—first gravity-based concrete platform in the North Sea.

Fig. 22 SPM: single-point mooring system.

Fig. 23 FPSO: floating production, storage and offloading system.

Fig. 24 *Conoco Hutton* TLP in the North Sea.

or otherwise surplus tankers and necessarily was inexorably tied to the practice of naval architecture and marine engineering. More recently, however, tightness in the availability of suitable tankers has led to purpose built units, with a trend toward specialized design.

Semisubmersible FPS. Another type of floating production system has evolved from the semisubmersible. It began in the mid 1970s with the conversion of the drilling semisubmersible *Transworld 58* by Hamilton Brothers to exploit the Argyle field in the North Sea. This was quickly followed by another North Sea conversion (by British Petroleum) for the Buchan A development. Since those beginnings there have been several conversions, some in Brazil, and one new building in the North Sea, *Balmoral*, by Sun Oil.

The semisubmersible FPS is an application of mobile drilling technology, although its specialized nature is rapidly changing it into a unique configuration. Its most notable distinctive feature is its mooring method, particularly in deepwater applications. These units are intended to remain fixed at one site for twenty years. This term of exposure requires expectation of environmental conditions much more extreme than used for mobile structures. Longevity and maintenance difficulties lead to some truly demanding design consideration.

TLPs. A parallel development to the FPS is the tension leg platform (TLP). While similar in appearance to the semisubmersible, the TLP is a fundamentally different development. It was conceived in the late 1960s and first built for the Hutton field in the North Sea in the late 1970s. The TLP is stationed by vertical tendons to restrain vertical motion as well as to provide lateral station keeping. While encompassing the same physical principles as the semisubmersible, the design application has many fundamental differences.

Support Vessels

General. Support is perhaps the oldest part of the offshore industry. As operations moved offshore, even in modest ways, floating equipment became essential. Even in their boldest activities, prudent men retain the familiar, the tried, and proven where they can, and innovate only where they need to. Land activities were taken to sea with the marine tools of the day, with a backdrop of experience in Louisiana's coastal swamps. The barges, tugs, and other equipment enlisted for this work were the seeds of the support fleet.

Specialization has evolved many new types of floating craft. To name but a few, these include: supply boats, anchor handling tugs, seismic vessels, heavy lift crane barges, launch barges, pipe laying barges, heavy transport barges, diving support vessels, and floating accommodations.

Today marine support equipment includes crane barges that can lift 7000 tons in a sea state 3, and moor in depths in excess of 2000 ft. Moreover, they can under their own power, outrun their anchor handling tugs. Gigantic tugs of 24,000 hp are not uncommon and hips that carry other ships are now available.

In many respects support vessels were the beginning of applied naval architecture in the offshore industry. Several new classes of floating craft have been put in front of the naval architect to design.

Much of the support fleet is used directly in constructing the permanent offshore facilities. Companies such a J. Ray McDermott, Brown and Root, Micopperi, and Herrema have become well known in this activity. Another, closely allied activity is towing and transport, with companies such as Smit, Wijsmuller, and Bugsier.

Vessels used for construction began as moored barges with a variety of equipment, which included cranes, pipe laying gear, and sometimes devices to launch large structures (jackets). Each application has evolved to yield some rather distinctive vessels, some displacement type ships, and some semisubmersibles. In addition, several kinds of small ships, some dynamically positioned, have evolved for a variety of tasks, many related to subsea installations. Another unique type is the "lift boat," a rather small jackup used for maintenance of platforms in shallow water. Unfortunately, for many years such craft were built while sorely in need of a naval architect.

Supply Ships. In the Gulf of Mexico, the small, towed transport barge was quickly replaced by the "boatruck," a 120–150 ft powered vessel with a short focsle and bridge far forward, and long open deck aft. Although little more than a self propelled barge, it quickly evolved into the more shiplike, offshore supply vessel.

After introduction to the demands of the North Sea, the Gulf of Mexico supply boat "grew up" into a handsome expression of ship design. Many are sophisticated vessels, in excess of 200 ft in length, high seakeeping standards, with power of 10,000 hp or more, and positioning thrusters. Some even have impressive ice breaking capabilities. Today the genre has a broad spectrum of sizes, outfitting, and functional capabilities. The offshore supply boat has become the root prototype for geophysical ships, diving support vessels, and anchor handling tugs, as well as naval and research vessels.

When built with higher power and suitable outfitting, these jacks-of-all-trades soon found work as tugs and anchor handling vessels. What the oceangoing tug is today owes much to the offshore industry. While one lineage of the oceangoing tug is quite clearly rooted in the Dutch ocean tugs, a more prevalent type has evolved from the offshore supply vessel.

Dry Transport. Transport of large and heavy cargoes continued to be on barges, but these grew into larger sizes and more specialization. While heavy transport for a period stabilized in a 400 × 100 ft format, with a variety of specialized features, an adequate supply of these led to the evolution of two rather special types, both larger than most ships! One is the huge, submersible, dry transport barge and ship. The other is the high capacity launch barge. The latter type has been built to sizes over 600 ft long, capable of transporting and launching at sea, structures weighing 40,000 tons. The dry transport ships are literally self propelled drydocks. The *Mighty Servant III* is capable of transporting cargoes of 30,000 tons or more. She made battlefield history during the Persian Gulf War in bringing the *Samuel B. Roberts* back to the United States for repairs after crippling mine damage.

Derrick Vessels. Derrick vessels have evolved rather dramatically. Beginning with essentially land-based cranes fitted on barges, they mounted ever larger lift capabilities suited to handling drilling rigs, some in the form of semisubmersibles and others self propelled ships. A recent example is the *Micoperi 7000.* It is an exceptionally large, sophisticated semisubmersible delivered in 1988. It is dynamically positioned for ultra deep water operation. It is not a singular achievement, for there are five similar ships not a great deal smaller.

Other. Some other sophisticated types of work vessels have developed for offshore application. These include accommodation barges, fire fighting ships, pipe laying ships, and diving support vessels. (It was a semisubmersible, the *Alexander Kijelland,* converted to an accommodations unit, that was responsible for the worst offshore disaster.) The pipe laying vessels have a similar, parallel evolution to the crane vessels discussed above, evolving from some fairly simple barges to large sophisticated semisubmersibles such as the *Viking Piper* and the *Saipem.*

Marine Technology

General. Until this point we have focused on the petroleum industry and its principal areas of support: drilling and offshore construction. Marine technology, allied industries, and more generic activities are, however, intertwined in a common background of technology, infrastructure, and regulation, all having deep involvement in offshore technology. The technology has become so rich and distinct that its professional practice has become known as "ocean engineering." Some look upon this as a richer, broader incarnation of naval architecture.

The venues of professional activities in marine technology include universities, major technical forums (OTC, OMAE, and SNAME); major research establishments of the major oil companies, industry associations and wider forums (e.g. API), and the U.S. Navy.

Its infrastructure is composed of the shipyards, fabricators, and equipment suppliers. World War II left us with an abundance of shipbuilding technology and resources, all left hungry for challenge and ready to adapt. The Bethlehem facility in Beaumont TX was most ready to meet this challenge and lead with appropriate technological development. Other builders, on the Gulf Coast in particular, were quick to follow, with many naval architects charting the course to meet the challenge. Many non-naval architects were to chart other paths and bring new ways of thinking into the profession. Likewise platform fabricators evolved, perhaps out of the inability of the shipbuilding industry to adapt to changing needs—or

Fig. 25 Semisubmersible crane barge *Balder*.

Fig. 26 Launch barge with jacket (top section) being towed out for installation in Gulf of Mexico. (Courtesy McDermott, Inc., 1981)

Fig. 27 FPSO: floating production, storage and offloading system.

Fig. 28 Heavy lift ship *Mighty Servant* transporting the articulated mooring column for floating storage unit "Challis" installation in the Timor Sea.

perhaps the designers' inability to appreciate the resource; probably both!

In the suppliers of equipment, the post war period left a legacy of experience, resources, ideas, and technology. Much of this capability found its way offshore. The traditional marine equipment suppliers were looked to for much of the equipment, but they also found new challenges. Demands long met by oil field equipment manufactures and expected by the industry provided new standards for marine equipment.

The Society of Naval Architects & Marine Engineers. The direct as well as indirect technical contribution from SNAME to offshore technology is much more significant that it would at first seem. Aside from the generic technology of naval architecture (such as stability, hull structure theory, and hydrodynamics) there has been considerable development of more specifically applicable technology, an area now becoming known as "ocean engineering." It is notable in this respect that some academic programs previously known as "naval architecture" now use this more generalized appellation.

One of the more significant contributions to ocean engineering was a 1957 SNAME *Transactions* paper: Rechtin, Steele, and Scales, "Engineering Problems Related to the Design of Offshore Mobile Platforms." This was the first paper on offshore petroleum activities to be published in the *Transactions*. Much of this material became a part of the ABS *Rules for Classing of MODUs* in 1967; and while the technology has improved and been extended in many ways since, what was published then is essentially valid today.

In 1965 Alan McClure presented a watershed of technology on semisubmersibles with his paper "Development of the Project Mohole Drilling Platform," in the SNAME *Transactions* of that year.

Many other SNAME papers have also made valuable contributions to offshore technology. Their rising numbers that appear in the annual *Transactions* reflect the profession's accelerating interest and involvement in ocean engineering. One offshore paper was presented in the 1950s, two in the 1960s, ten in the 1970s, thirteen in the 1980s, and an average of two to three per year in the 1990s.

Technical Forums. As any emerging technology would, the offshore industry has drawn from many other technologies. Naval architecture, marine engineering, civil engineering, and mechanical engineering have all had roles. In addition, of course, we must add petroleum engineering and more recently the rich offering of electronics. The early years were a period of take with little to give back. The years of invention were a period of competitive secrecy. But by the late 1960s came a desire and need to exchange ideas. The Offshore Technology Conference (OTC) is one of the best examples of this. Begun in 1968 by several sponsoring U.S. technical societies, including SNAME, it is a combined multi-disciplinary symposium and trade show. Other similar forums have also developed. Nevertheless, the symposium format has become the prevalent means of technical exchange.

Primarily on the section level, but with good examples on the national level, and through the SNAME Offshore Committee, SNAME has been a major participant in offshore technology. Many of the papers published in *Marine Technology* reflect this. Other forums include the American Petroleum Institute committees, other societies, and various university research programs. Another forum of exchange is the Joint Industry Project (JIP). This is a business format in which a collection of user and provider companies join in a funded study of interest. While closed to non-participants, it has become a major means of advancement of the technology. There are also university/government/industry research programs and consortia such as Sea Grant and OTRC.

Regulation. In many cases roles have developed for other agencies. This is reflected, for example, in the special regulatory history of offshore activities. It is also reflected in the insurance and financial institutions that serve the industry as well as classification, not to mention new areas of admiralty law.

The offshore industry developed at such a rate that regulation has barely been able to keep up. So long as it consisted of fairly conventional equipment, conventional classification and certification activities were the norm. Various technical changes in the 1960s, however, led to classing rules, USCG regulation, and the IMCO (now IMO) MODU Code.

Regulation and its evolution are discussed in a later chapter, but it also merits mention in this introductory chapter. Classing practice and USCG regulation have roles parallel to those in other maritime activities. Development of the outer continental shelf has, however, introduced another significant regulatory activity. In the United States the Department of the Interior, now through its Mineral Management Service (MMS), administers a permitting process and regulates safety and compliance with pollution regulation. As a practical matter floating equipment has been regulated primarily by the Coast Guard and classification societies. Conversely, fixed installations have been regulated through the MMS permitting process through their "OCS Orders." In addition, various design and manufacture codes have been developed in the petroleum industry through the American Petroleum Institute. What has occurred, then, is the development of design codes and regulations along two independent and rather different paths. With the introduction of floating production systems, the two paths come together and are now undergoing a reconciliation.

Design Codes. Driven by a need to provide self regulation, API developed and published recommended practices for planning, designing, constructing, and installing fixed offshore structures. The first such publication came out in 1967 and consisted of fewer than a dozen pages. It has since been revised perhaps a score of times and has now matured to include over several hundred documents. These consist of Recommended Practices, Bulletins, Specifications, and Standards. In 1988 the API committed to convert all its documents to international format, and has supported the process of conversion ever since. In some cases this was simply a matter of metrication. For the basic design methods used in fixed and floating structures, however, the conversion also meant translating from the deterministic "working stress design" method with essentially a single safety factor, to the probabilistic "reliability based design" method using partial safety factors. The conversion to the International Standards Organization format is bringing a much needed consistency to the industry during the internationalization stages of its growth.

Fig. 29 Semisubmersible accommodations and support vessel *Treasure Supporter*.

The SNAME Offshore Committee and its T&R committees have supported the API effort and have provided valuable professional resources in developing and documenting accepted design practices. Further, cooperative support of the American Bureau of

Shipping and Det norske Veritas have been instrumental indeed in producing a set of documents that can be used throughout the world to design safe offshore installations.

Driving Forces

One cannot complete an overview without considering the pertinent driving forces; and these in turn suggest the need for some historical perspective.

Because an industry must be fueled by a need and the money to meet that need, the oil industry has been the industry's principal driving force. Other than this, the driving forces have been political or economic, affected at times by unforeseen events.

The end of World War II with its booming economy marked the beginning of the industry as we see it today. The Truman doctrine of 1947 and subsequent acts gave legal legitimacy to exploring the coastal zone. The period was marked by continual discovery of oil farther offshore (principally in the the Gulf of Mexico), and a growing economy onshore. This development established much of the character and experience of the industry. In 1969 a second domestic province, the coast of California, was on the verge of development. A giant oil spill, however, produced the roots of political opposition that still stunt that development. With the special challenges of exceedingly deep water, open waters, and a seismic sea bottom, an altogether different set of challenges would surely have had a different impact on technology and the development of the industry.

The international component of the offshore industry had been fairly regional until the mid 1970s. Besides the Gulf of Mexico, Lake Maricaibo in Venezuela, and the Persian Gulf were the other active areas. Developments in Indonesian waters and the North Sea were still in their infancy. With the formation of OPEC in 1973, oil prices began a relentless rise which fueled an enormous expansion, perhaps explosion in the industry. Not only did it fuel the growth, it changed the industry from what was essentially an American dominated industry to one that is now decidedly international.

The great expansion of the 1970s led to high energy prices, which by the last of the decade saw inflation approaching 15 percent, interest rates exceeding 20 percent, business stagnation, and high unemployment. This and certain governmental policies caused major faltering in the industry. By 1983, however, demand had abated and development had produced far too much capacity. Oil prices fell dramatically. Only recently, in 1986, did they begin to rise again. With nearly ten years of low development activity, demand has again exceeded capacity, certainly on at least a local basis; and once again oil prices are stimulating activity.

John J. Filson earned a master's degree in naval architecture from the University of California in 1968 and has been employed in offshore engineering activities ever since. After graduation he was sent by the CBI Corp. to Dubai , where he worked for two years on various projects pertaining to oil storage and tanker loading. After that he spent 22 years with Omega Marine in a wide variety of offshore projects and in a wide variety of jobs, ranging from strictly technical to broad managerial responsibilities. He is currently employed by Alan C. McClure Associates. He is a member of SNAME and the Royal Institution of Naval Architects. He chaired the second SNAME Offshore Symposium, 1991.

Offshore Industry: Early Years

James E. Steele

Colonel Edwin Drake drilled the world's first oil well at Titusville PA in 1859. This was when the whaling industry had about reached its peak. In the 1850s the American whaling fleet consisted of some 700 ships manned by 20,000 seamen. An average voyage lasted 46 months and captured 40 whales with an output of less than 100 barrels of oil per whale [1]. Thus, the oil available to the national population of 30 million people was some half a million barrels per year. This comes to about one gallon per person per year.

Over the next forty years, petroleum became an important industry. Discovery and production spread from Pennsylvania to Ohio and West Virginia, to New York and Indiana, to California, and to Texas. It went overseas to the Russian fields at Baku on the Caspian Sea. Kerosene replaced candles and whale oil for lighting. The availability of petroleum based grease made the wheels of industry turn easier. In 1900, total world production was approximately 440 thousand barrels per day. Of this, some 225 thousand came from Russia, 180 thousand from the United States and the remaining 35 thousand from the rest of the world. In the United States annual consumption had grown close to a barrel per person per year.

By way of comparison, in 1990, with a population of 250 million, the United States uses 16 million barrels of oil per day, which amounts to 23 barrels per person per year. This is an increase in per capita consumption of oil by a factor of about one thousand since the height of the whaling days.

Early Drilling Tender Patent

During the last forty years of the nineteenth century, drilling for oil was spreading around the world, but the activity was confined to land. There is, however, evidence that people were thinking about drilling at sea, as shown by Figures 1 and 2, which are from patents issued in 1869. These concepts were evidently intended for use for civil engineering and for mining purposes; they were never built. Still, they indicate that the forty-year postponement in drilling over water may be explained by economics and by the ready availability of sites on land rather than by a lack of inventiveness. When the right time came, the technology developed rapidly.

Summerland

In the late 1890s a certain H. L. Williams was developing a shallow oil field at Summerland CA, close to Santa Barbara. The field appeared to extend south, under the Pacific Ocean. In 1897 Williams built a wooden pier out over the ocean and drilled the world's first offshore oil well. Several wells were drilled from this pier with the best production apparently coming from those farthest offshore. By 1902 several piers had been built, the longest extending out some 1230 feet from shore. A financial analysis of these wells is given in Ref. 2, which quotes from the *Summerland Advance-Courier* of February 1900:

> People outside of Summerland are continually inquiring: "What does it cost to sink an oil well?" As nearly as possible we will answer the question, taking the average depth of Summerland wells at 210 feet:

Cost of derrick	15.00
Drilling 210 ft at $0.80/ft	168.00
Casing (5⅝) at $1.20/ft	252.00
Tubing at $0.27/ft	56.70
Sucker rod at $0.045/ft	9.45
Triangle, pumping bbl., valves	20.00
Cable, 50 ft, at $0.03/ft	1.50
Tank, 50-bbl.	15.00
Incidental expenses (approx.)	15.00
Total	$552.65

The report goes on to estimate annual net earnings of $459 from the investment shown above.

Summerland wells were small producers and the field itself was a small one. Production gradually declined after 1902, resulting in a total production of some three million barrels.

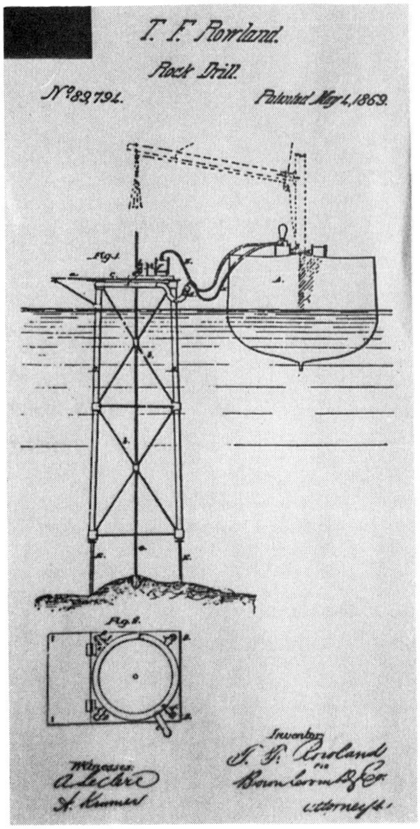

Fig. 1 Rowland patent, 1869.

Caddo Lake

Caddo Lake [2], partly in Texas and partly in Louisiana, is a few miles northwest of Shreveport. In the early 1900s the area around the lake was explored for oil. The resulting discovery of unwanted gas created problems for the driller and for the marketer. Enough oil was found, however, so that when the bed of the lake was offered for lease in 1910, the Gulf Oil Co. bid $30,000 for the right to drill 8000 acres of lake bottom. In the spring of 1911 Gulf drilled its first well from a platform supported on timber piles. This was the world's first well in inland waters.

In the next forty years Gulf drilled close to 300 holes in Caddo Lake and had produced over 13 million barrels of oil.

Lake Maracaibo

Marine drilling in Lake Maracaibo [2,3], Venezuela, developed after oil had first been found on the adjacent land, in a manner similar to that at Summerland and at Caddo Lake. In 1917 the Shell Oil Co. drilled the first small discovery well on the lake shore. This was followed by further modest lake shore development. Then in 1923 a 100,000 barrel per day gusher blew in, quickly attracting the attention of the world oil community. As the promising area on shore had already been leased, it was necessary to look to the lake itself for further opportunities. The lake is some 110 miles long by 70 miles wide. It has a maximum water depth of 120 feet, but a great deal is much shallower. The lake is slightly brackish. Hurricanes and currents are not a problem but gale force storm winds can raise seas of up to eight feet.

The gusher, Royal Dutch Shell's, *Barroso #2*, was two miles inland. The first well in the lake was drilled as close to this as possible while still remaining legally over water. The derrick sat on cast-in-place concrete footings a few feet out in the lake, while the remainder of the rig was on the beach. When oil was found, a drilling war started with Shell working the beach, and Lago and Gulf, the new operators, working in the shallows. After drilling some forty wells along the edge of the water, Lago decided to step offshore and to drill a well in 11 ft of water some 140 yards from the beach. The equipment was set up on platforms supported by wooden piles as at Summerland and Caddo Lake. In less than a year the piling had been destroyed by teredos. A period of experimentation followed during which many alternative forms of support were tried. Meanwhile the drilling progressed into ever deeper water, which required stronger piles and larger platforms. Soon they decided to combine the derrick and other equipment on a single platform, and then they found it economical to drill as many as four wells from the same platform. By this time, the wells were in truly deep water and the platforms were becoming expensive. The next step was to move some of the equipment from the platform to an adjacent barge. This transfer apparently started in 1934 and took place in several stages, starting with the boilers and then the

Fig. 2 *Samuel Lewis*, 1869.

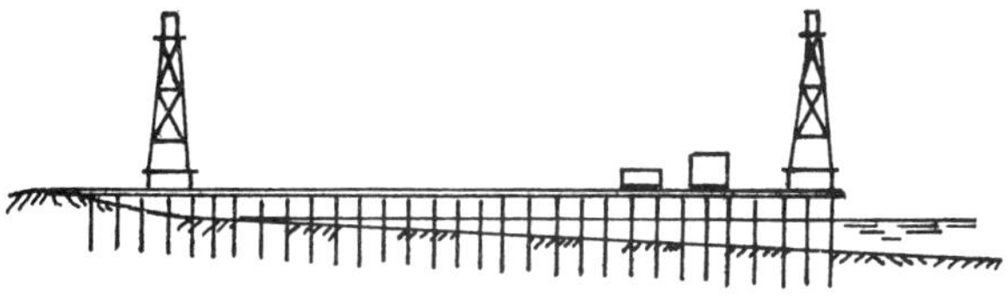

Fig. 3 Piers from shore: Summerland, CA, 1897; Caddo Lake, LA, 1911; Lake Maracaibo, Venezuela, 1924; Caspian Sea, Russia, 1925; and Creole, LA, 1933.

pipe rack and, finally, all of the equipment except the derrick was on the barge.

By the 1950s, these evolutions had led to an optimum design in 100 ft or more of water, in which a concrete platform, a steel substructure, and the derrick were supported on four prestressed concrete piles some 54 in. in diameter and 200 ft long. An accompanying tender barge (170 ft long by 70 ft wide by 12 ft deep) contained the pipe rack, mud system, and supplies. By this time, the cost of a 10,000-ft well was about half a million dollars, of which one fifth was for the concrete foundation structure and the rest for drilling. Lake Maracaibo had become the most productive field in the Western Hemisphere with close to 4000 wells yielding a cumulative total of over four billion barrels of oil.

Inland Drill Barge

The next innovation took place in the bays and swamps of Louisiana. This is salt dome country, and during the 1920s geologists had located a number of these inviting structures whose potential for oil production was spectacularly demonstrated by the discovery of the Spindletop gusher at Beaumont in 1901. Drilling in the Louisiana swamps was another problem. Much of the area was half land and half water; it had neither roads nor canals. Once a well site had been selected, two approaches were possible. If by land, it was necessary to build a corduroy log road through the swamp to haul in equipment, men, and supplies. Once on site, a mat of criss-crossed timbers was laid down to support the derrick and other equipment. If by water, a navigable canal had to be found or dredged to permit barges and boats to haul in the required material; in this case, it was usual to drive piling to support the operation. Whichever method was selected generally turned out to be both difficult and expensive. A better approach was called for and, not surprisingly, two separate individuals independently came up with the same solution. G. I. McBride worked for the Texas Co., which had acquired large holdings in southern Louisiana. McBride wrestled with the costs of building piling platforms in the swamp and felt it would be much simpler if the drilling rig, derrick and all, could be mounted on a barge that could be floated to the site and sunk to the bottom. The barge would be big enough to spread the load over the soft soil, and part of the load could be carried by the buoyancy of the barge. In addition to eliminating the need for the piling platform, this scheme would permit all the mechanical parts of the rig to be permanently pre-installed and piped up, thus saving set-up time.

When the Texas Co. agreed that this was the right way to go, they made a routine patent search and

Fig. 4 Summerland, CA, 1898.

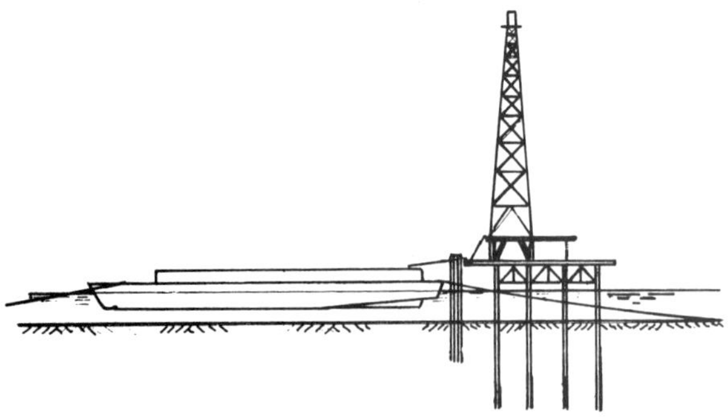

Fig. 5 Kerr & McGee converted YF tender in shallow water off Louisiana, 1947. Similar tender operations in Lake Maracaibo started in 1934.

found that Capt. Louis Giliasso had recently patented the identical idea. Giliasso had observed the activity in Lake Maracaibo and had wondered why it wouldn't be a good idea to mount the entire rig on a barge that could be easily floated from well to well. In his patent application he described a "practical apparatus for drilling oil wells in lake bottoms and other submerged lands, which will afford a firm, stable base for the drilling operations and which can be transported by floating from one location to another."

The Texas Company sought out Giliasso and obtained a license to use his invention. The first drill barge actually was made in three pieces. Two barges of identical size—120 ft × 24 ft × 10 ft—were permanently connected by transverse trusses above their decks, leaving an 8-ft slot to be used for the well location; it permitted the barge to be floated off after drilling was completed, leaving a small cluster of piles to protect the well head manifold. The derrick was mounted on the connecting trusses and the rest of the equipment could be conveniently located on the decks of the barges. It is said that this arrangement was selected, rather than a single large slotted barge as was later used, so that in the event the idea proved unworkable they could salvage the two barges and use them for general cargo transport. The third, smaller barge—generally 84 ft × 28 ft × 7 ft—was used to carry the boilers. In those days of steam powered drill rigs, they felt it prudent to keep the boiler fires away from the well head.

The first drill barge was built at the McClintic-Marshall plant at Leetsdale PA (later Bethlehem Steel Corp., Fabricating Division). It was floated down the Ohio and Mississippi Rivers and towed into an offshoot of the Gulf of Mexico called Lake Pelto, in Terrebonne Parish LA. On November 17, 1933 the seacocks were opened, the hull settled softly onto the bottom of the lake, and another new era in marine drilling began. The operation proceeded without difficulty. Drill barges quickly became the preferred tool wherever rivers, bayous, or canals existed for access or could be reasonably dredged. A paper by G. I. McBride in 1935 [4] describes the early operation of seven twin-hulled drill barges. He reports that such a barge had a shipyard cost of $27,000, while a boiler barge cost $9000; and the combination saved some $85,000 per year in operating expense when compared with the use of pile supported platforms. This is a pretty good return on your money.

The following table shows the production record of the Pennsylvania Shipyards (later Bethlehem Steel Corp., Shipbuilding Div.) in Beaumont TX, the most prolific producer of inland drill barges.

Year	Drill Barges	Boiler Barges
1934	2	5
1935	2	2
1936	4	1
1937	7	4
1938	11	8
1939	4	2
1940	3	3
1941	—	—
1942	—	—
1943	—	—
1944	1	1
1945	2	2
1946	—	—
1947	1	—
1948	2	—
1949	2	—
1950	2	—
1951	2	—
1952	6	—
1953	3	—
1954	1	—
1955	2	—
1956	9	—
1957	2	—
1958	—	—
1959	1	—
1960	—	—
1961	—	—
1962	1	—
1963	1	—
1964	1	—
Total	72	28

By the late 1940s the standard design consisted of a single barge that was about 170 ft long by 50 ft wide by 12 ft deep. An 8-ft slot extended half the length of the hull, and through it the well was drilled. Power was from a series of diesel engines mounted on the upper deck. They were linked by chain drives and were also linked to the drawworks (drilling winch) and to the mud pumps. The boiler barge had disappeared. The shipyard price ran to about $300,000; to which must be added the cost of the drill rig and other material furnished by the owner, which was in the neighborhood of $1,000,000.

These drill barges still operate in the inland waters of the Gulf Coast, mainly in Louisiana and Texas. They can navigate through the Intracoastal Canal and the various river systems. When the drilling site is in a swampy area it may be necessary to dredge a canal. When the site is in an open bay or lake, it may be necessary to lay down a mat of oyster shells to decrease the water depth. Operators have long been concerned that the barge will sink so far into the soft bottom that it may be difficult to raise. This was mentioned in McBride's 1935 paper [4]; he recounts that a system of 2-in. pipes was installed along the bottom through which water or air could be discharged to "break suction." In 112 moves made by these barges prior to the time of his paper, it was found necessary to use this system only once; in this instance the barge "was immediately freed by pumping water through the jets."

The inland drill barge was a successful innovation that finally combined the entire drill rig onto a single structure that served both as a floating hull during transport and as a spread footing when on site.

Fig. 6 Contemporary inland drill barge *Lewis Geer*, 1955 (Bethlehem Beaumont, TX, photo).

Gulf of Mexico

As the search for oil spread throughout the coastal plains and swamps of Louisiana and Texas, it became apparent that the productive geology continued under the adjacent waters of the Gulf of Mexico. By this time there had been experience with a number of methods that might be used to support the drill rig in shallow water. The first to be used here was a platform on wooden piles, similar to the first activity at Summerland, at Caddo Lake, and at Lake Maracaibo. In 1933, Pure Oil and Superior Oil jointly installed such a platform in twelve ft of water, some 3000 ft offshore near the town of Creole LA [2]. This proved to be a dry hole, but in 1937 the same operators struck oil from a similar wooden platform not far away.

Offshore exploration came to a halt during World War II, but after the war it started up again with a vigor. Several development techniques were pursued. In 1946, Magnolia built a self-contained platform some ten miles southeast of Eugene Island LA; this contained steel, as well as timber, piles. In 1947, Kerr and McGee installed a small steel-pile platform in 18 ft of water some 10.5 miles from shore in the Ship Shoal district of Louisiana. They followed the Lake Maracaibo practice of serving the platform with an adjacent tender containing the mud system, supplies, and some portions of the rig. The tender was converted from a war surplus YF supply barge. Pure Oil followed in 1948 with a similar platform and tender in 35 feet of water some nine miles off the coast of Vermillion Parish [5]. In this they used a converted LST for the tender.

Meanwhile, as the search for oil proceeded into deeper water along the coast of Louisiana, the standard inland drill barge design was modified to permit operating in deeper water. The obvious way to do this would be to deepen the hull. As the water gets deeper, however, it also usually gets rougher. To reduce the effect of waves, it was thought desirable to eliminate the portion of the hull at the waterline. To do this the depth of the barge was kept small and the working deck was raised some distance above the barge and supported on columns. This helped the wave force problem but it left the question of how stability was to be maintained once the deck of the barge was under water. Several going-on-bottom schemes were tried. In some schemes the water plane was maintained by the use of movable pontoons attached to the sides of the hull such that they would stay at the surface of the water as the main hull was sunk. After the main hull was on bottom, the pontoons were folded or slid down out of the wave action. The first such unit was the Hayward-Barnsdall barge *Breton Rig 20* [2]. It was designed by John Hayward and had stabilizing pontoons that slid up and down on vertical guide columns under the action of hydraulic cylinders. It was first used in 1949 in 18 ft of water. Its operation proved successful and the many worries about lack of stability, sinking unevenly in the mud, sticking to the bottom, scouring, and sliding were all found to be tolerable.

Service Vessels

In the early days of offshore operations in the Gulf of Mexico immediately after World War II, there was a need for service vessels to transport men and supplies and to tow drilling units. That need was filled by existing tugs and fishing boats, together with the surplus vessels from the war. The abstract of Ref. 6 states:

> Marine transportation equipment must be carefully selected so that it will be capable of operating with maximum safety and efficiency. To this end, twin screw steel-hull diesel engine driven vessels are preferred. In supplying 6 drilling rigs widely distributed off Grand Isle, LA, Humble Oil and Refining Company has 24 vessels in operation for personnel, standby and towing service, in addition to 17 cargo barges. Transportation expense, amounting from 20 to 30 per cent of total well costs, will vary according to the type and class of vessel selected and the number of rigs in operation. An efficient communications system is also essential to control vessel movements, and short-range radar has proved valuable as a navigational aid in reducing downtime to a minimum during periods of poor visibility. Weather conditions are of primary importance in offshore drilling and producing operations because they affect the continuity of work, they can cause serious damage to marine equipment, and may endanger the lives of personnel.
>
> An efficient weather forecasting service has been instrumental in helping to schedule transportation during periods when wave conditions of 4 ft or higher are considered to be critical for transfers between vessels and platforms. It has been estimated that these wave conditions are to be expected in 18 per cent of the time on the average; they have been forecasted during this interval with an accuracy of 79 per cent and predicted within 1.5 ft 91 percent of the time for the drilling sites in the immediate area of Grand Isle. When maximum wave heights reach 9 ft, practically all transfer of equipment and personnel ceases for reasons of safety.

Fig. 7 Hayward-Barnsdall barge *Breton Rig 20*, 1949.

Typical boats listed in Ref. 6 include:

- 125-ft tug, steel hull, twin screw, 1200 hp, 15 mph
- 83-ft crew boat, wood hull, twin screw, 300 hp, 12 mph
- 52-ft standby boat, wood hull, single screw, 150 hp, 12 mph
- 40-ft supervision boat, steel hull, twin screw, 300 hp, 17 mph
- 115-ft cargo boat, steel hull, triple screw, 450 hp, 7 mph

These were all originally built for other purposes. References 7 and 8 are also pertinent.

Summary

Looking back from 1990, we might divide the Age of Petroleum into three almost equal parts. The first, which lasted from Drake's 1859 well at Titusville PA until the Spindletop gusher of 1901 at Beaumont TX, took place almost entirely on land and mainly in the

United States or Russia. The second part took place during the first half of the 20th Century. During that period, drilling spread around the world and gradually started over water, first on fixed platforms and then on floating vessels of various types. The third part extends from shortly after World War II to the present date. Both onshore and offshore developments expanded at a rapid rate. The gradually accelerating pace of these three periods is illustrated in the following table:

	1859[a]	1900	1950	1990
United States				
Population, millions	30	75	150	250
Overall, billions bbl/year				
Oil production	0.0007	0.07	2.00	3.00
Oil consumption	0.0007	0.07	2.40	6.00
Per capita, gal/year				
Oil production	1	40	560	500
Oil consumption	1	40	670	1000
Offshore production, % of total	100	0	N/A	15
World				
Population, millions	1200	1600	2500	5200
Overall, billions bbl/year				
Oil production/consumption	—	0.2	7	23
Per capita, gal/year				
Oil production/consumption	—	4	117	183
Offshore production, % of total	—	0	N/A	25
Cumulative production, billions bbl	—	3	65	550

[a]1859 figures are for whale oil.

The worldwide cumulative production of 550 billion barrels through 1990 may be compared with an estimate of 3500 billion barrels originally in place, of which roughly half may be recoverable by presently known methods. This would mean that we have now used more than one-third of our supply of oil. It also means that, at our present rate, we will use it all in another fifty years.

References

1 *Log of Mystic Seaport*, Summer 1989.
2 Brantly, J. E., *History of Oil Well Drilling*, Gulf Publishing Co., 1971.
3 Donahue, J. and Halbouty, M. T., "In the Beginning," *Offshore*, May 1981.
4 McBride, G. I., "Drilling Barges," Spring Meeting, API, San Antonio, April 1935.
5 Alcorn, I. W., "The Pure Oil Company's Tideland Development," *World Oil*, May 1949.
6 Pennington, T. and Posgate, J. C., "Marine Transportation in Offshore Drilling and Production Operations," API, Chicago, Nov. 9, 1949.
7 Macy, R. H., "The Trend in Offshore Crew Boats and Supply Vessels," Gulf Coast Section, SNAME, New Orleans, May 3, 1957.
8 Henderson, W. H., "Operation of Marine Equipment for the Tidelands Oil Industry," Gulf Coast Section, SNAME, New Orleans, May 3, 1957.

JAMES E. STEELE graduated from Webb Institute in 1941 and set to work building destroyers at Consolidated Steel in Orange TX. In 1946 he shifted to the Pennsylvania Shipyard (later acquired by Bethlehem Steel Corp.) at Beaumont TX. There he was engaged in designing LPG barges, in developing LPG ship conversions, and in designing offshore drilling rigs. In 1958 he transferred to the Quincy shipyard, where he helped design the power plants for nuclear cruisers. He moved to Bethlehem Steel's home office in 1964; he is now retired.

Offshore Industry 1950–1965: Invention

Bruce G. Collipp

The era 1950 to 1965 was one of invention, innovation, and creation in the offshore industry. It would initiate many of the new ideas and kinds of equipment, establish most of the industry's precedents, and set the technological basis for the rapid expansion to follow. A few individuals and small groups would establish the foundations of an industry and adopt its uniform terminology. There were no textbooks, few regulations, and precious few naval architects. As this era would end, the industry would have moved from the shallow bays and protected waters to challenge the oceans of the world.

As it began, the offshore industry would continue to be driven by oil and gas. A long history on land and a successful progression through inland waters led to the grand frontier: the offshore. Success would require embracing new and unfamiliar technologies and a new professional: the naval architect. Exactly what he did and talked about were not understood by petroleum engineers. His acceptance was lukewarm at best and came about only because he might know something about things in the water. While the naval architect was at home in the shipyards, which would grow to serve the industry, the shipyards were still trying to relate to the oil business. A few enterprising naval architects who sensed the problem became consultants.

One naval architect recalls being informed, "We just paid a New York naval architecture firm $82,000 and they don't even understand our business. We will teach you the oil business, pay you $300 per month, and still not pay you that in 20 years." At another oil company meeting, the conversation ran: "I heard we hired a naval architect. . . . Yes, for our four-man offshore group. . . . We shouldn't be hiring people who don't know our business. When we hire people they should have an opportunity to career here." A few were hired because "Our competitors hired one."

Naval architects did play major roles in making the pursuit of offshore oil a reality. Richard Knapp with Exxon Production Research provided technical guidance for the large fleet of drilling tenders they used. Paul Wolff of Kerr McGee designed most of their pioneering submersibles. James Steele with Bethlehem Shipbuilding guided the design of tank battery barges and one of the first rigs to drill in 100 ft of water. As an employee of Shell Oil, I invented the semisubmersible, which permitted drilling in the harsh ocean environment, and Jerome Goldman with Friede and Goldman designed many of the shallow water submersibles and jackup rigs. Bob Macy with Ingalls Shipyard, John Graham with Global Marine, and Walter Michel with Friede and Goldman all played significant roles. In the 1950s graduates of Webb Institute, MIT, the University of Michigan, and others set the pattern for the industry's equipment.

More basic than this, however, was the ownership of the submerged lands. Following World War II and establishment of the United Nations, third world countries were attempting to claim UN ownership of the seabed and its mineral rights. In 1947, the Truman doctrine had proclaimed United States ownership out to the edge of the outer continental shelf, stating that the natural extension of the land out under the oceans belonged to the adjacent country. A state and federal ownership dispute persisted until the passing of the Tidelands Act in 1953. While the shallower water inside the twelve-mile limit would go to the adjacent state, the vast remainder would be administered "for the nation as a whole" by the Department of the Interior. Texas led off December 1, 1953 with its first offshore sale: 400,000 acres at an average price of $78.50 per acre.

There are three stages to development of offshore oil and gas: seismic exploration, exploration drilling, and production. Seismic exploration uses acoustic waves to probe buried sediments for clues to where the hydrocarbons may have been trapped. On land, rugged terrain can make this a slow and costly process. The most basic of our technology, the boat, would be found to offer a handy, inexpensive basic conveyance for seismic exploration and has found many other uses. The ease of deploying seismic equipment by boat vastly lowered the cost of seismic data acquisition

Fig. 1 A more modern version of drilling tender which began as Navy LST.

offshore and served to whet the appetite of the petroleum geologist (see Filson, Fig. 17).

Other than somewhat conventional marine equipment for construction and support, the production front would have to wait for significant contributions from the naval architect. Nevertheless, design and use of this equipment became an increasing challenge.

The main task for the naval architect was developing equipment for mobile exploration drilling. Operation afloat would present important technical problems and some costly lessons. How to maintain station, stability, and structural integrity in various operating and storm conditions was a central challenge. The first of these offerings, submersibles and jackups, would sit on bottom while operating, with temporary afloat operation only between work sites. Having been attracted to water too deep for such units, the industry would evolve drill barges and semisubmersibles as their replacements.

Equipment development progressed on two fronts: the Gulf of Mexico and coastal Southern California, both adjacent to important onshore oil fields. Rather different characteristics of the two venues, and perhaps personalities, would drive equipment choices.

The second ingredient required a basic change in how development drilling would occur. On land and in the swamps, the oil industry drilled a well and, if successful, completed and produced the fluids. Initially this philosophy continued. Large and small platforms were built to drill the exploration wells. Only about ten percent of exploration wells are successful. Although offshore the percentage was higher, large investments in platforms were being wasted. One of the first approaches was to reduce the size of the platform. Much of the equipment was placed on a moored ship. The portion on the ship could be reused at other locations. This was the drilling tender.

Near the end of the era, the expendable exploration well was conceived. Even if the exploration well were successful, it was abandoned. Additional expendable wells would be drilled until the magnitude of the discovery was known. Fewer permanent platforms were required and then only where they would be profitable. To abandon a well with pipe set in a producing zone was not an easy concept to embrace. But, statistically, it saved the large investment in unprofitable platforms. This change provided the opportunity for the rigs that would be called Mobile Offshore Drilling Units (MODUs). These rigs drilled wells with the platforms—a cost-effective exploration tool.

The ingredients were in place: the people, the laws, inexpensive seismic equipment, the development logic, and a vast new unexplored province. The stage was set. The Santa Barbara Channel and the Gulf of Mexico would be the test beds for the new equipment.

Gulf of Mexico

Drilling Tenders. Many in the industry had participated in World War II. For them the natural device to use at sea was a ship or barge. And the War had left a surplus of these.

Although the drilling tender had its origins in Lake Maracaibo and the swamps of Louisiana, operations were now going to be conducted in the unprotected

Fig. 2 Early submersible: *John Hayward* (ex-*Breton Rig 20*).

offshore waters. In 1946 Kerr McGee acquired two tracts in the Ship Shoal Area for $30,000. Block 32 was 10.5 miles offshore. Some means of reducing the cost of a fixed platform was desired. A small platform 38 × 71 ft held up by 24-in. pilings was constructed 20 ft above the water. Two war surplus YF barges and an LST were purchased. One YF barge was used for driving piles, the LST became a supply barge, and the other YF barge became a drilling tender moored at the platform. Mooring lines 1400 ft in length were used. The first offshore drilling tender went to work in 1947.

Exxon acquired 19 surplus LSTs. Originally built for $2 million, the surplus LST could be acquired for $75 thousand. About $600 thousand was required to strengthen and outfit the LST as a drilling tender. Four mooring lines forward and four aft consisting of $2\frac{1}{16}$-in. chain running out to spud piles held the tender in place. A tug stood by. In foul weather the tender was backed away and allowed to swing on a single storm mooring. The mooring lines had a conventional marine anchor just before the spud pile. Naval architect Richard Knapp explained, "that way if the civil engineers' piles fails, at least you will move off location, dragging an anchor."

New tenders were also being built at Gulf Coast shipyards. Typically, they were 270 ft long, 60 ft in beam and 27 ft deep; they displaced 6300 tons. By 1954, twenty-two tenders were in operation in the

Fig. 3 Early jackup: *Offshore No. 54*.

Gulf of Mexico and six were under construction. But the days of their dominance were numbered. Subject to considerable weather downtime, and saddled with the expense of a permanent platform, albeit smaller in size, exploration was still relatively expensive. Nevertheless, tenders are still used today in sheltered waters for development drilling.

Submersibles. John Hayward proclaimed that a barge fitted with two side pontoons could be built and sunk offshore. The side pontoons would provide the stability while the main hull was being submerged. By 1949 the Barnsdall-Hayward *Breton Rig 20* was built and successfully on location near the mouth of the Mississippi. War surplus LSMs were used as breakwaters. The first well to be drilled by a MODU was underway. Although successful, the industry did not rush to embrace the idea. It would be six years later, when a rig named *Mr. Charlie* (Filson chapter, Fig. 3) would be built by the Ocean Drilling and Exploration Co. It could operate in 40 ft of water. The same year, Kerr McGee, who had acquired the Barnsdall-Hayward rig, produced Kerr McGee *Rig 44*. The California Co. with Friede and Goldman as naval architects produced the *S-44*. It used recessed pontoons in the hull for stability while sinking. The pontoons were sent to the sea floor to provide a foundation. In 1956 the *S-55*, a larger version, was built by Avondale Shipyard. The rig capsized at the shipyard, providing the industry with one of its first tragedies.

All of the MODUs to this time had been barges. In 1956 Kerr McGee, with Paul Wolff as naval architect, built *Rig 46*. Cylindrical tubes formed a lower hull grid. Each corner had a cylindrical stabilizing column for submerging. The rig could operate in 70 ft of water. In 1962 the largest submersible ever built, Kerr McGee's *Rig 54*, was finished. The tubular hull was an equilateral triangle 388 ft on a side. Two hundred feet high, the rig could operate in 175 ft of water. It towered over New Orleans during its outfitting.

Jackups. What would become the most numerous type of MODU appeared on the scene in 1954. DeLong docks were a product of World War II technology. Pneumatic gripper jacks allowed the piling to be lowered to the sea floor and then the barge hull could be elevated above the seas. Offshore Company's *Barge No. 1* was built at American Bridge's yard in Orange TX. The piling were six ft in diameter and the barge was 160 ft long. *Barge No. 2* with 200 ft legs was built soon thereafter.

Close on the heels of *Barge No. 1*, Bethlehem Steel's Beaumont yard produced *Mr. Gus I* (Filson, Fig. 8). This rig, able to operate in 100 ft of water, was initially a product of the thinking of the time. Originally built in two pieces, the equipment was split much like a drilling tender. After it drilled its first well, the two pieces were joined. The original slip type grips were replaced by pins in the slots on the legs. The rig had the unique feature that its deck was permanently elevated. It floated on a lower hull, which was sunk to the sea floor when on location. Also in 1954, Royal Dutch Shell built an elevated deck rig. The Shell rig was the only rig ever built with square legs. Both the Shell rig and *Mr. Gus I* were lost in storms after a few years service.

Up in Vicksburg MS, R. G. LeTourneau, previously a maker of earth moving equipment, was building a three legged rig in principle similar to those described above. Zapata Offshore Co. was formed and the rig commissioned as the *Scorpion*. The open truss legs were triangular in section. With only three legs, there was great debate in the industry whether that few number was safe. Some believed that rigs with a larger number of legs had an advantage in case one leg failed. With time this prudent view was proven unnecessary, but it delayed the hiring of *Scorpion* for a period.

All of the above rigs would become known as *jackups* because they jacked up their deck after the legs were planted in the sea floor. As time progressed over half the rigs would be jackups.

Floating Rigs and Semisubmersibles. Things that float had not been forgotten. In 1959 Zapata Offshore converted a ship's hull into the *NOLA 1*. The drilling rig was mounted over the side. Equipped with a mooring system, the ship could operate in 200 ft of water. This was followed in 1962 by Zapata's *Sidewinder*. Equipped with outboard pontoons, the rig could drill over the side or through a slot in the hull. Reading and Bates Drilling Co. would take the process one step further. The *C.P. Baker* was a complete catamaran. Built in 1962, the rig drilled successfully until it encountered a gas pocket at a well depth of only 680 ft. The gas blew out, accumulating between the two hulls and exploded. Within 15 minutes the rig sank with a loss of life of 20 people. The explosion destroyed the intact buoyancy, which caused the sinking, but the event would be remembered. Considerable tests and research would examine the question of whether gas escaping from the sea floor could aerate the water so that a rig would sink. All tests and studies proved this thesis false, but it still persists.

Things that float really came into their own in 1962 when the first semisubmersible went to work. In 1954, my thesis at MIT, "Design of an Offshore Oil Drilling Rig," started my interest in stability and rig motions.

Fig. 4 First deepwater drill barge, *CUSS 1*.

When I went to work for Shell Oil, I found they had a top secret program for developing ideas and equipment for drilling afloat. By 1957 the University of California was testing the concept that a device could be built with natural periods outside the periods of the waves and so substantially reduce pitch and roll. With such thoughts in mind, I and others at Shell designed a floating drilling platform that would be almost immune to wave action. Howe and I published a paper that set the terminology for most of the kinds of rigs; but it would be another who would name this new rig. Meeting with Capt. McPhall of the U.S. Coast Guard for the purpose of having it certified, I explained that this new design had its hull submerged and therefore would be unaffected by waves. Capt. McPhall stated that if the hull were submerged it would sink. I countered that most of the hull, but not all, would be submerged. Capt. McPhall stated, "We will list it as a Super Manned Barge-Semisubmersible." The name stuck. In January 1962 the rig, *Bluewater No. 1*, spudded its first well in 297 ft of water, a record for the time. It worked well, but several years later a tornado spawned by a hurricane capsized the rig.

In 1963 the Ocean Drilling and Exploration Co. built the *Ocean Driller*, the first new built semisubmersible, and *Bluewater No. 2* was built on the West Coast. In 1965, the Southeastern Drilling Co. built a triangular shaped semisubmersible called the *Sedco 135*. That became the prototype for many following rigs. One third of the rigs operating by the end of this era would be of that configuration. Semisubmersibles would open the way for the industry to face all of the challenges of deep water and harsh weather.

Santa Barbara Channel

On the West Coast in the Santa Barbara Channel the challenge came from conditions that differed from those in the Gulf of Mexico. The State of California allowed drilling on unleased acreage only as long as the well stopped before it encountered the highest producing zone. A geologic zone in the Monterey sand was of particular interest to the oil companies. Offshore it was within view of land fields. The problem was that the water there was several hundreds of feet deep.

Continental, Union, Shell, and Superior oil companies formed a consortium called the "CUSS group." In 1953 a converted navy patrol boat, the *Submarex,* with the rig mounted over the side, was anchored on location and the first floating drilling started. In 1955 the CUSS group acquired a YFNB surplus hull, which they named *CUSS 1*. She was 260 ft long and had a hole in the center through which drilling was conducted. A revolutionary underwater drilling system using guidelines, underwater blowout preventers, and a marine conductor finished the package. The non-propelled *CUSS 1* was towed to position and anchored with a preset six point mooring system.

In 1955 the Richfield Oil Co. converted a YMS barge, which they named *LaCienca.* Standard Oil of California converted an LSM to the *Western Explorer*; and Humble, Exxon, converted another LSM to the

Fig. 5 Early semisubmersible: ODECO's *Ocean Queen.*

SM-1. With the open season on exploratory drilling, it was not unusual to see these rigs moored nearly side by side drilling away off the coast of California. This fleet also marked the high point of direct ownership of rigs by major oil companies.

In 1962, a new shipshaped rig called *Glomar II* was built. To be used on the West Coast, the contract called for Shell Oil to pay to tow the rig to the Gulf of Mexico if the contractor, Global Marine, wished. Analysis showed it would be cheaper to install propulsion machinery on the Gulf Coast, which was done. *Glomar II* found oil in the Santa Barbara Channel and later moved to Cook Inlet in Alaska to drill in the ice infested waters.

Unnoticed, because of secrecy, a small, 136-ft coring vessel, named *Eureka,* was built in 1961. Howard Shatto with Shell Oil reasoned that if the ship were fitted with two rotating outboard type propellers, given position data from a taut vertical wire line, and a computer to solve the mathematics, she could stay on location without a multi-cable mooring system. It worked. *Eureka* secretly drilled core holes in several thousand feet of water in the Gulf of Mexico and off the West Coast. The dynamically positioned drill ship was born.

Sidelights

In 1950 technology peculiar to the offshore oil drilling industry was nonexistent. Parts of naval architecture, coastal engineering, oceanography, and civil engineering were adopted. Perhaps oceanography was most lacking. Naval architects were taught the trochoidal wave theory. While perhaps correct, it gave little helpful quantified data. Naval architects were interested in reducing the force on a slender body moving through the water. The forces and motions of a body designed to remain at a fixed location had not been considered.

Certainly one of the breakthroughs came in the Morrisson equation. In this the force of the waves was separated into the part related to the velocity and the part related to acceleration of the water particles. Reid and Bretschneider at Texas A&M published a paper explaining this practical way of calculating waves forces. Meteorologist Al Glenn set about studying hurricanes: their intensity, frequency, and predictability. The concept of return frequency was born. The one hundred-year return frequency of a hurricane became, and still is, an industry benchmark. The computer as an analytical tool came into being. By the mid 1950s the slow (by today's standards) analogue computer was analyzing the plethora of data on ocean waves.

The landmark paper by Rechtin, Steele, and Scales was published in the SNAME *Transactions* (1957) during this period.

Of all the major oil companies, Shell Oil was perhaps the most dedicated to offshore operations, but their work was shrouded in secrecy to gain a competitive advantage. Shell engineers developed remote underwater vehicles, dynamic positioning, semisubmersibles, subsea drilling and producing equipment, and unguided reentry. In 1963 these

Fig. 6 Project Mohole: Semisubmersible ocean research vessel proposed for National Science Foundation.

findings were offered to the industry in a school with the unheard of tuition of $100 thousand. Seven competitor oil companies enrolled. Floating drilling and subsea completions thereby became a science known throughout the world.

Drilling Contractors. Many of the initial rigs were owned by major oil companies that had both the desire and the capital. For various reasons, however, oil companies didn't want to own the rigs. Historically they had had a few rigs, but generally relied on drilling contractors to furnish the equipment. But banks were reluctant to underwrite the financing of these new funny looking rigs. All early rigs were funded by co-ownership, oil company collateral, or long term leases. Five-year leases were common. By 1965 the offshore drilling contractor was established and several had developed sufficient financial stability to borrow their own money for new rig construction. A partial list of contractors follows:

Coral Drilling Co.	4 rigs
Global Marine Drilling Co.	11 rigs
Ocean Drilling and Exploration Co.	9 rigs
The Offshore Co.	16 rigs
Penrod Drilling Co.	5 rigs
Reading and Bates Drilling Co.	5 rigs
Southeastern Drilling Co.	5 rigs
Transworld Drilling Co.	7 rigs
Western Offshore Drilling and Exploration Co.	6 rigs
Zapata Offshore Co.	7 rigs

These are the companies that remain the backbone of the industry today. Starting slowly in 1966, an explosion of building would follow and eventually lead to a fleet of over 500 rigs.

The Mohole

At 2:06 AM on March 9, 1961, the drill ship *CUSS 1*, floating in 3111 ft of water, 20 miles off LaJolla CA, started coring in the sea floor. In 1952 a group of scientists had formed the American Miscellaneous Society (AMSOC). Eventually this group zeroed in on the objective of drilling a hole through the earth's crust into the mantle. In 1908 Andrija Mohorovicic of the University of Zagreb, Yugoslavia investigating the great earthquake of 1909 had ascertained that the earth had a crust that was separate from the next inner layer, the mantle. This discontinuity, dubbed the Moho, would have to be penetrated if the mantle was to be examined. The proposed penetration was named the Mohole. AMSOC under the leadership of Willard Bascom proposed to drill it. They chose to make the attempt offshore because the earth's crust is thinner there than under land. Eventually, with U.S. Government funding through the National Research Council, *CUSS 1* was outfitted with four outboard thrusters. The ship's officers navigating from buoys kept the ship on location. Tests off La Jolla were followed by five core holes in 11,700 ft of water off the Guadalupe Islands. The Mohole itself would have to be drilled in even deeper water and instead of a few hundred feet of penetration, a hole up to 30,000 ft deep would be required. A new rig would be needed. Brown and Root was selected as prime contractor.

In April 1963 the Brown and Root plans were unveiled. What they proposed was a giant for its time. It was to be a twin hulled, semisubmersible 279 ft long and 234 ft wide. The self propelled vessel would be held on station by six computer controlled thrusters. As it turned out, the project was killed by Washington politics and not owing to any lack of technical feasibility. The enormous engineering effort was not totally wasted, however, because the design served as the prototype for many following drill ships.

Alan McClure who would later found McClure and Associates was the chief naval architect. He published complete information on his design in the 1965 SNAME *Transactions.*

The project did give birth to the *Glomar Challenger,* a dynamically positioned drill ship, which would core the oceans depths and eventually provide the data from which today's plate tectonics would evolve. It also served to create public interest in "inner space." While the United States and the Soviets were vying for outer space, Mohole raised interest in the ocean's depths. This brought companies not oil related into offshore activities. Aquaculture, mineral mining, thermal energy conversion, and man in the sea all followed.

Other Marine Vessels

In support of the search for oil, other equipment had to be built. Seismic vessels, initially conversions, were built from the keel up. Massive derrick barges that could lift 700 tons were built to install self contained platforms. Pipe laying barges that could lay large diameter pipe in deep water to transport the oil and gas were constructed. Brown and Root and J. Ray McDermott became the leaders in the offshore construction industry.

Crew boats to transport men to the offshore rigs were specially designed. Some were hydrofoil craft. Supply boats to transport equipment were especially constructed. By 1965 the larger supply boats were 190 ft long and had 3600 hp engines. These were true oceangoing craft.

Summary

By 1965 most of the basics still used today had evolved. The computer and the microchip, better steels, and much other new technology would follow. But the oceans depths, the arctic and the harsh seas had been met. The infrastructure of the industry was in place. The role of the drilling contractor, the boat owner, and the major oil company had been defined. The basic strategy for developing the offshore potential had evolved. Numerically the industry was still small. The construction boom would follow. Even the universities were starting to recognize this new profession. Although it would encompass naval architecture and marine engineering, it was absorbing many other disciplines. A new term, ocean engineer, would start to emerge.

In 1965, George Bush, Chairman of Zapata Offshore Company and later President of the United States, in *Drilling Magazine* stated:

> . . . I am optimistic. . . . Today 10,000 square miles and internationally 260,000 square miles are under lease. . . . There are 17 rigs capable of drilling in 35 ft. of water, twelve for 35–80 ft., and 28 for over 100 ft. One immediate factor of concern is the North Sea. If it hits there won't be enough rigs. It is positively exhilarating to take a look at a world map and see the places where offshore drilling is underway or planned . . . and so I contemplate what we should be doing next to keep abreast of the rapid technological demand in our business

In 1963 L.C.S. Kobus, a naval architect in the industry wrote:

The Wave

A wave is a spirit
  That's restless and free
Sired by the wind
  And born on the sea.

Dashing and splashing
  And battering the shore
Dashing and splashing,
  And then it's no more.

The wave of the offshore industry was building.

Bruce G. Collipp earned a bachelor's degree in naval architecture from MIT in 1952. After two years as a ship's officer he accepted employment with Shell Oil, where he worked in a wide variety of offshore engineering responsibilities until 1987, when he retired from Shell to start his own consulting business. He is the author of more than fifty technical papers and holds five patents, all pertaining to offshore engineering. He is the recipient of the NAS Gibbs Medal and is a member of the National Academy of Engineering.

Offshore Industry 1966–1973: Development

Alan C. McClure

The creative trial and error activity of the 1950–1965 period finally began to give way to more sophisticated approaches to design and engineering of offshore structures. This resulted in a small number of types of floating craft for the needs of the offshore industry:

Type	Water Depth, ft
Drilling: bottom-founded rigs	
swamp barge	12
posted barge	10–20
submersibles	20–100
jackups	50–300
Drilling: floating	
platform tenders	less than 300
ship hull	greater than 200
semisubmersible	greater than 200
Construction	
construction barges	10–600

The naval architect still had to compete with the other engineering disciplines in this period. Drill ships were the province of the naval architect, but the various barges and jackups were influenced more by the civil engineer than the naval architect. Semisubmersibles (semis) were mixed, with naval architects involved in such complicated matters as stability and motions, but leaving structural design to the civil engineers. Naval architects were also involved to a minor extent in transportation and placing of offshore structures. One of the outstanding developments during this period was the B.P. 40s structure designed by Brown & Root Engineering with the aid of its naval architecture department. It was designed to be floated on its own oversized columns in a horizontal position. Once on station, it was to be flooded to rotate it to a vertical position and then settled onto the seabed. Extensive design and model tests were carried out, marking a comprehensive effort between the naval architects and other disciplines. The operation went smoothly, exactly as predicted from Rice University model tests.

Although in the early days we naval architects were called upon to work in multidisciplinary teams, we were seldom placed in charge. We were merely technicians, useful for solving certain types of problems that had somehow escaped the civil or mechanical engineering curricula. The obvious examples include stability of floating objects, wave forces, and resulting motions.

The method of determining motions of floating vessels using sea spectra and response amplitude operators (RAOs) of St Denis and Pierson (1953 SNAME *Transactions*), came into common use during this

Fig. 1 Jacket assembly.

period. In fact, naval architects were beginning to develop methods of design and analysis specifically for offshore problems, such as:

- Deep water permanent moorings,
- Wave forces on vessels moored in open water,
- Finite element structural analysis,
- Design of efficient supply boats, essentially self-propelled barges,
- Response of non-ship floating objects to the sea,
- Dynamic positioning, and
- Launching and righting of large subsea structures.

The Society of Naval Architects and Marine Engineers began to recognize a new branch of naval architecture. The 1944–1960 index lists one paper [1] related to offshore, and only one subject heading. The next index, 1961 to 1980, lists 15 papers under the headings:

- Ocean Engineering
- Ocean Mining
- Offshore Service Systems
- Offshore Thermal Energy Conversion
- Drilling Rigs
- Semi-Submerged Ships

Participation in the Offshore Technology Conference (OTC) first held in 1969 was a great step forward for SNAME in the offshore arena. Blakely Smith was instrumental in getting SNAME involved in the OTC from the beginning. He led the effort for several years as SNAME representative to the OTC executive committee. The conference was important as it was recognized universally among the offshore participants as the primary meeting place and technical exchange medium for engineers and scientists working in the offshore environment. It was vital for SNAME to be a part of this special fraternity. As the conference has evolved, there is little conflict between papers designed for OTC and those written for the SNAME *Transactions.* OTC papers tend to be fairly brief, six to twelve pages, and deal with a narrow subject or a narrow part of a larger subject. Approximately 20 of the 140 papers presented at the first OTC were by naval architects or were of direct interest to naval architects. These papers were on such subjects as wave motions of floating vessels, wave forces, deep sea mooring systems, and specialized offshore vessels. Since the OTC is composed of eleven technical societies, it appears that the naval architects held up their part of the bargain by presenting their relative share of papers.

It was during this period also that SNAME instituted a new committee as part of its technical and research arm. The new group was named the Marine Systems Committee and was intended to incorporate subjects in the marine field that were not already involved in existing committees. One such was the drilling and exploration fields. Panel MS-3 was formed in about 1970 to provide a forum for those of us with special interest in offshore activities. The first chairman of the Panel was Walter Michel of Friede & Goldman. He did an excellent job in starting the programs and in developing a meaningful approach. Under his leadership, a major program of analytical and experimental analysis of semisubmersible stability was undertaken. The program was funded by a group of about fourteen companies, including many major oil companies and some of the principal contracting companies and design firms. This work culminated in a paper presented to the Annual Meeting of the Society in 1976. By this time, I had taken over the chairmanship of the panel and the paper was written by Walter Michel, Edward Numata, and myself. The experimental work showed how difficult it is to cap-

Fig. 2 *BP 40*'s jacket afloat.

Fig. 3 Typical early R. G. LeTourneau slant-leg jackup rig—one of the more notable designs to come out of the 60's.

size a semisubmersible. In fact, it was found to be impossible to capsize an intact vessel owing to wave action alone, even with a small initial negative metacentric height, GM. Later, the American Bureau of Shipping extended its work and developed a new approach to semisubmersible stability requirements.

Outstanding vessels designed and built during this period are the SEDCO 700 series of semisubmersibles designed by SEDCO and their design arm, Earl & Wright, Inc. of San Francisco, and the Pacesetter design by Friede & Goldman. Sedco and Earl & Wright also designed an advanced drill ship called the *SEDCO 445*, which incorporated a dynamic positioning system, giving it almost unlimited water depth operational capability. A number of five-column Pentagone design semisubmersibles were built. These were designed by Forex Neptune of France, and proved highly successful as drilling platforms, although they were difficult to transport because of their rather large towing resistance. Both the SEDCO 700 series and the Friede & Goldman Pacesetters incorporate the twin hull multi-column design established earlier by Project Mohole. This gradually became the pattern for all later semisubmersibles. Ocean Drilling Co., known as ODECO, proceeded to build multi-hull designs with large numbers of columns. One of their most successful designs incorporated four parallel hulls for reduced towing resistance and about 16 columns on all four hulls. A large number of the successful SEDCO 135 designs, as well as three-column semisubmersibles designs designed by Friede & Goldman were also built. These proved satisfactory drilling vessels but, again, like the Pentagone, were difficult to transport because of their towing resistance.

References

1 Rechtin, Steele, and Scales, "Engineering Problems Related to the Design of Mobile Offshore Platforms," *Trans.* SNAME, 1957.

2 Honsinger, V. C., "Wave-Induced Vibrations of Offshore Structures," *Trans.* SNAME. 1963.

3 McClure, A. C., "Development of the Project Mohole Drilling Platform," *Trans.* SNAME, 1965.

4 Hooft, J. P., "A Mathematical Method for Determining Hydrodynamically Induced Forces on a Semi-Submersible," *Trans.* SNAME, 1971.

5 McTaggart, R. G., Gunderson, R.H., "On Damaged Stability of Drilling Vessels," *Trans.* SNAME, 1973.

ALAN C. MCCLURE* earned his BSE degree in naval architecture from the University of Michigan in 1949. Winning a SNAME scholarship, he went on to MIT for his master's degree, and then spent twelve years at Electric Boat. He was enticed away from there by Brown & Root to direct their Project Mohole. Five years later, upon termination of that project, he went to work for the Continental Oil Co. as a senior naval architect. In 1972 he went into business for himself. An author of many papers, he is a Life Fellow of SNAME holds the Blakely Smith Medal for outstanding accomplishments in ocean engineering.

*The editors note with deep regret the death of Mr. McClure on May 1, 1993.

Offshore Industry 1974–1982: The Age of Expansion

JOHN MERCIER

The Stage

THE PERIODS OF INVENTION and development set the stage for the growth of the offshore industry and foresaw practically all of the kinds of marine systems that have since been put into service. The expansion stage can be seen as largely a response to the impetus of the major shift in oil price after the 1973 OPEC embargo.

The dramatic expansion in size, scope, mission, and investment in offshore equipment entailed floating equipment of all kinds. While the offshore oil business originated and grew up mostly in the Gulf of Mexico, the innovations of the expansion era occurred largely overseas, especially in the North Sea. Exploration of the giant Dutch gas field at Groningen led to important gas discoveries in the southern reaches of the North Sea and, eventually, to the major oil finds of the central and northern North Sea. The weather and wave conditions there are more persistently severe than those in the American offshore, except for our occasional hurricanes. Among the important evolutions and innovations of this time period are:

(1) The third generation of mobile offshore drilling units, characterized by many vessels built to the same design, and with major technical contributions from naval architects and shipyards. The Sedco 700 series, the Aker H-3 class, and the Friede & Goldman Pacesetter design are examples. Figure 1 shows the *Alaska Star*, a Pacesetter-class semisubmersible.

(2) Fixed platforms in the new deep water fields (at this time water depths around 500 ft were considered "deep" in the North Sea) embodied new methods of floating from their fabrication sites to the offshore locations, and these were designed by naval architects. Conventional steel space frame structures were modified to be "self-floating" for the Thistle and other fields. A new kind of concrete gravity based structure was developed, which enabled topside drilling and production equipment to be installed inshore, before towing to such locations as, the Brent, Beryl, Cormorant, and Statfjord fields. An example is shown earlier in Fig. 21 of the Filson chapter.

(3) The equipment used to install offshore facilities, including crane vessels, pipe laying vessels, diving support vessels, and launch and cargo barges, increased in size and capability in response to requirements of this new and demanding arena. Heavy lift transport vessels were developed to support these challenging offshore projects. These enabled large modules and subassemblies to be built in remote locations, including the Far East, and then rapidly brought to site for integration with the rest of a project.

(4) Crude oil transport from these oil fields remote from land called for creative marine solutions. While

Fig. 1 Pacesetter-class semisubmersible *Alaska Star*.

Fig. 2 Brent "Spar": Single-point mooring buoy and oil storage and offloading facility at Shell's Brent Field in the North Sea.

many miles of large diameter pipelines were laid on the seabed for some of these requirements, the recently developed single-point mooring (SPM) concept was adopted for several large fields. In some cases, such as for Ekofisk, Brent, and Thistle, the SPM system was temporary, and was later supplanted by a pipeline system. For others, such as Beryl, Auk, and the giant Statfjord fields, all crude oil was (and still is) lifted by tankers.

(5) Oil and gas production to floating production platforms began in this period. Oil produced from the Argyll field went to the converted semisubmersible drilling vessel *Transworld 58* beginning in July 1975: Argyll produced the first North Sea crude oil to be landed in the United Kingdom. Soon after this breakthrough development came on stream, other forms of floating production came into operation. A popular adaptation entails converting a tanker, which is single-point moored, to process production from subsea wells and to store the crude oil for subsequent transfer to a trading vessel. These bold and novel configurations offer several substantial prospective benefits: reduced costs, reduced time to develop and deploy the production system, and capabilities for deep water applications.

(6) Technology to design, build, and operate these essentially new offshore systems was built on experience quickly learned: accidents occurred, theories were developed, regulators assumed their role, and competitive fabricators proposed alternative innovations. With the business growing in many parts of the world, various local considerations entered into choices of configurations; research was undertaken to support regulatory guidance and design practices (especially in the North Sea). Professionalism, encouraging the dissemination of ideas and data about the new systems, also afforded the growing ranks of engineers entering the offshore field a clear picture of the industry's main goals and directions. Schools of naval architecture and marine engineering, sensing the potential importance of this branch of maritime activity, introduced courses in offshore technology. Some even changed their names to reflect the new emphasis. In

Fig. 3 *Transworld 58*: First semisubmersible-based FPS, installed at Hamilton Brother's Argyll Field in the North Sea in 1977.

Alexander L. Kielland

In what was at that time the world's worst ever offshore industry disaster, 123 men lost their lives when the Norwegian semisubmersible accommodation platform Alexander L. Kielland *(Fig. 4) capsized and sank in stormy seas while positioned in the central section of the North Sea in the Ekofisk field at lat. 56°28′ N, long. 03°07′ E, on March 27, 1980.*

The disaster occurred when one of the five support legs, of some 122 feet in length, gave way and broke off during a force 10 storm. As a result, all production was halted in the Ekofisk field.

There were 212 men onboard at the time, including 172 Norwegians and 35 Britons, of whom only 89 were rescued, leaving 123 either dead or missing.

A task force of 11 helicopters, 42 merchant vessels, and 5 naval vessels raced to the area to search for survivors from the platform, which was chartered to Phillips Petroleum for use as an offshore hotel for oil workers resting from jobs on other rigs in the Ekofisk field. The Alexander L. Kielland *was connected by means of a gangway to the* Edda *production platform in the southwestern part of Ekofisk.*

The wreck was subsequently towed to Boknafjord, near Stavanger, in mid-April and into Gandsfjord, Stavanger harbor, on August 28. She was declared a constructive total loss, and several attempts to right her failed. It was as of then the largest oil rig insurance loss, with the Norwegian Oil Insurance Pool paying out the full insured value of $51.25 million (£23 million) to the owners in January 1981.

The wreck was blown up and sunk in a 2300-foot-deep fjord near Stavanger on November 18, 1983.

This was only the second major North Sea rig disaster involving serious loss of life, the previous worst being the December 1964 jackup rig Sea Gem *tragedy in which 13 men lost their lives as the rig was being prepared for a move to another location. There had only been one other North Sea total loss, the* Transocean III, *which sank about 100 miles east of Shetland on Jaunary 30, 1974.*

A number of regulatory changes occurred as a result of the Alexander L. Kielland *tragedy, including the requirement for better emergency evacuation equipment, improved inspections and quality control in construction, and redundancy in floating units. Redundancy arguments spawned many research studies, with a focus on risk analysis, quantified risk, and reliability studies in order to put offshore risks in perspective.*—Malcolm Sharples (Noble Denton, Houston)

the United States the Sea Grant Program sponsored many valuable research projects and educational activities to benefit the offshore industry.

Drilling Vessels

The semisubmersible drilling rig became the vessel of choice for drilling wells in areas like the North Sea, where rough sea environments called for its superior motion characteristics. During the interval from 1973 to 1982 the number of semisubmersibles in operation grew from about 30 to about 130. Aside from new buildings adding to the fleet, there were losses owing to accidents and scrappings. Many of the new semis were "sister ships," being built to a few basic designs, such as the Sedco 700 series (14 vessels), the Aker H-3 class (about 20 vessels) and the Friede & Goldman Pacesetter class. Designers became more conscious of the differences between requirements to work in the persistently harsh wind and waves of the North Sea or the Gulf of Alaska, and calmer areas of the world's oceans, such as the Gulf of Mexico, the South China Sea, offshore Australia, or the Persian Gulf. Designs were adapted to the local requirements. Competitive commercial and technical pressures were taken into account. These factors led to such diverse configurations as:

Cognac: In three pieces
Hondo: In two pieces, mate while afloat, then upend
Thistle: Self floating (also Magnus)
Concrete Gravity Structures: Float out with decks already fitted
Murchison: Big, barge-launched platform; compare launching a jacket with launching a ship.

The requirement for bigger offshore installations, especially in the North Sea, was the impetus to introduce new crane ships, both larger and more weather resistant. These enabled field installations to be done

Fig. 4 Single pontoon and column—all that remained afloat after capsizing of platform *Kielland*, March 1980, in the North Sea (from magazine *Noroil*, April 1980).

Ocean Ranger

The most expensive single marine loss ever at the time was the American semisubmersible drilling platform Ocean Ranger *(Fig. 5), which sank in atrocious weather conditions while in the Hibernia field off Newfoundland on February 15, 1982. She was valued at $86.5 million (£46 million). Tragically, all 84 men on board lost their lives.*

The Ocean Ranger *was positioned at lat. 46°43′33″ N, long. 48°50′13″ W, about 200 miles off St. John's, Newfoundland, when she reported on February 15 that she had developed a severe list and requested immediate Canadian Coast Guard assistance. Three lifeboats were launched from the platform but all on board perished when they capsized in the storm-tossed seas that reached heights of 50 feet in winds of 80 miles per hour. Due to the continuous battering it received, the* Ocean Ranger *completely sank in 260 of water. Only 22 bodies were recovered.*

The loss of the Ocean Ranger *was caused by a chain of events which resulted from a coincidence of severe storm conditions, design inadequacy and a lack of knowledgable human intervention. The design weaknesses included a failure to specify portlights of adequate strength and to provide a ballast control panel with components that were suitable for operation in an environment where there was a risk of exposure to sea water. The ballast control console was susceptible to common faults and presented confusing information as a result. Despite the portlight and malfunctioning of the ballast control panel, the loss could have been prevented by the intervention of a properly trained crew. The Royal Commission Report that followed concluded that had the crew only closed the deadlights, shut off the electrical and air supplies to the panel, cleaned up the water and glass and then retired for the evening, the* Ocean Ranger *and its crew would have survived the storm that night. In attempting to remedy the problems caused by ingress of water into the ballast control room, the crew, because of lack of understanding of the ballast system, reactivated the panel and unintentionally allowed water to enter the port pontoon. They failed to realize that the forward list and lack of head capability of the pumps made uprighting the rig impossible. Because of a misunderstanding about the manual control of the ballast control valves, they opened the valves instead of closing them, allowing even more ballast water to enter the tanks. The vulnerability of the chain lockers to flooding was also a contributing factor. Lack of survival suits available on board contributed to the loss of life, once the crew was in the water.*

Considerable regulatory activity followed, which led to increased training, improvement in regulatory standards with respect to ballasting systems, increased pressure to have knowledgeable ballast control operators on board other floating units, the development of more detailed emergency response plans, and a renewed initiative on the development of better emergency evacuation craft.— Malcolm Sharples (Noble Denton, Houston)

more efficiently and safely. The early offshore cranes were installed on ships or barges (and some were merely crawler cranes from shore-side operations placed temporarily on barges). The economic demands for bringing the North Sea fields on stream, however, promptly inspired the Heerema group, in particular, to develop huge, semisubmersible crane vessels. Two examples are *Balder* (see Filson's Fig. 25) and *Hermod.* With a single stroke, the standard for designing and building offshore processing equipment was transformed, and the reliability of offshore installation work upgraded. Now large topside modules could be lifted and placed on the jackets while the winds and waves were disturbing the work site. Moreover, setting and driving piles to secure the massive jackets to the seabed was also facilitated. The "weather window," during which projects could carry out the critical offshore operations, was considerably widened by these massive new tools.

Pipe laying, too, was being extended beyond previous service by the need for large diameter pipes to carry the crude from the wells to shoreside transfer facilities. These pipelines, some more than a hundred miles long, called for pipe laying vessels that could stay at sea doing their mission, even when the North Sea's daunting weather would have caused other, less capable, ships to cease work.

Diving support ships came into the market to support the growing role of divers in drilling, installation, and, increasingly, in inspection, maintenance and repair.

Fig. 5 *Ocean Ranger* just after delivery.

Offtake Systems

Direct loading of tankers from single point moorings in the near-vicinity of the oil production field originated in mild water areas like Dubai's Fateh field in the Persian Gulf. The long distances from shore of most of the North Sea oil fields encouraged operators to try this method of transportation from those fields. Ekofisk led the way with a temporary loading system, and operators in other fields, including Hamilton Brother's Argyll field, Mobil's Beryl (UK) and Statfjord (Norway) fields, successfully adapted this technique. Operators in other fields used tanker loading systems as temporary systems until the pipelines were ready. The effectiveness, economy, regularity, safety and environmental soundness of this method have been demonstrated by innumerable tanker loadings from those fields over more than two decades.

Glomar Java Sea

The worst maritime disaster of 1983 in terms of loss of life occurred during the night of October 25–26 when the American drilling vessel Glomar Java Sea *sank with her entire 81-man crew in the South China Sea at about lat. 17°49′ N, long. 108°53′ E, during the ferocity of tropical storm "Lex." She was last reported at 23.10 local time on October 25, 1983, when she transmitted a distress call advising that she was listing 15 degrees in winds of 75 miles per hour and 26-foot seas.*

She had been engaged in drilling exploratory oil and gas wells since January in an area south of Hainan. An empty lifeboat from the vessel was recovered on October 29, and assorted debris, including life jackets, was sighted later in the day in the general area where the Glomar Java Sea *was lost. The wreck was subsequently located lying upside-down on the seabed in 300 feet of water about 60 miles from Hainan Island, but no survivors were ever found despite an extensive air and sea search.*

With all the very expensive drilling equipment on board, the Glomar Java Sea *had a hull and machinery insurance value of $24.5 million.*—Malcolm Sharples (Noble Denton, Houston)

Floating Production

The world's first floating production system was brought on-stream by Hamilton Brothers Oil & Gas Co. at the Argyll field in the UK North Sea in June 1975. The semisubmersible drilling rig *Transworld 58* was converted to produce (initially) from four completed subsea wells. These were connected by seabed flowlines to a manifold located beneath the floating production facility. The owners achieved a number of important objectives by adapting this system. The field was discovered in early 1972 and the decision to use a floating production system was made in February 1973. Construction began in June 1974, and Argyll crude was the first UK North Sea oil to cross the beach at the beginning of the era. The time to start pumping oil was cut in half by this innovation, while the cost was cut by two-thirds.

The next floating production system to come on stream, about two years later, in August 1977, was for Shell at the Castellon field off Spain. This operation used a converted tanker to support a production system to produce a single subsea well. By the end of 1982 five more floating production systems were working. Four of them were based on tankers in the Mediterranean Sea, the South China Sea, and off California, with one additional semisubmersible production station at the Buchan field in the UK North Sea. The other kind of floating production system, the tension leg platform, did not come into service until August 1984, although the planning, design, and fabrication work for the first such unit were well underway by the beginning of 1983.

The advantages of floating production systems might seem to be confined to deep water, but the actual applications are by no means limited to this circumstance. Indeed, many of the tanker based units are in water less than 200 ft deep. Cost, schedule, and sometimes other commercial or political issues account for the growth in applications. It should be recognized, however, that the growth in use of these systems has not been as rapid as the prospective benefits revealed by Argyll, and some other projects, would indicate. Perhaps this is partly owing to management's natural caution and partly to the perception of relative complexity. In addition, not all subsequent projects have actually achieved the expected economic benefits, for a variety of reasons.

Bibliography

Allmendinger and St. Denis, "Problems of Ocean Platforms," Spring Meeting, SNAME, 1971.

Bernitsas, "A Three-Dimensional Nonlinear Large-Deflection Model for Dynamic Behavior of Risers, Pipelines and Cables," *Journal of Ship Research*, March 1982, p. 59.

Capanoglu, C., "Tension-Leg Platform Design: Interaction of Naval Architectural and Structural Design Considerations," *Marine Technology*, Oct. 1979, p. 343.

Carter, J. H. T. and Foolen, J., "Evolutionary Developments Advancing the Floating Production, Storage and Offloading Concept," *Proceedings*, OTC, Paper No. 4273, 1982.

Chou, "Conceptual Design Process of a Tension Leg Platform," *Trans.* SNAME, 1983.

Gerwick, Mansour, Price, and Thayamballi, "Feasibility and Comparative Studies for the Use of Prestressed Concrete in Large Storage/Processing Vessels," *Trans.* SNAME, 1978, p. 163.

Ghosh, Chou, and Huang, "A Rational Approach to the Design of a Pipelay/Derrick Semisubmersible Barge," *Trans.* SNAME, 1979, p. 40.

Gibbon and Minorsky, "Evolution of the Design of a Jack-up Type Offshore Structure," *Marine Technology,* Jan. 1975, p. 50.

Hammett, D. S., "Future Semi-Submersible Drilling Units," RINA International Symposium on Semi-Submersibles: the New Generation, London, 1983.

Hoffman and Fitzgerald, "Systems Approach to Offshore Crane Ship Operation," *Trans.* SNAME, 1978, p. 375.

Hooft, "A Mathematical Method of Determining Hydrodynamically Induced Forces on a Semi-Submersible," *Trans.* SNAME, 1971, p. 28.

Johnson and Cojeen, "An Investigation into the Loss of the Mobile Offshore Drilling Unit Ocean Ranger," *Marine Technology,* April 1985, p. 109.

Metcalf and Praught, "On the Installation Planning of a Self-Floating Offshore Platform," *Marine Technology,* July 1979, p. 279.

Numata, "Hydrodynamic Model Tests of Offshore Drilling Structures," *Marine Technology,* July 1966, p. 288.

Numata, Michel, and McClure, "Assessment of Stability Requirements for Semisubmersible Units," *Trans.* SNAME, 1976, p. 56.

Ochi, "Wave Statistics for the Design of Ships and Ocean Structures," *Trans.* SNAME, 1978, p. 47.

Sjanzberg, Greiner, Chen, and Rawstron, "Practical Design Approaches for the Analysis of Barge Performance in Offshore Transportation and Launching Operations," *Trans.* SNAME, 1980, p. 195.

Zumwalt, C., "Current Trends in Offshore Drilling Rigs," *Marine Technology,* April 1973, p. 172.

Zumwalt, C., "Optimal Preliminary Selection of Mobile Offshore Drilling Units," *Marine Technology,* Oct. 1982, p. 391.

John Mercier's college education started at Webb Institute, continued at MIT, and terminated (with a Ph.D.) at Stevens Institute in 1973. After some years of research at Hydronautics and Stevens, he joined Conoco, Inc., as a staff engineer, and is now a consultant to that company. He has been involved in all technical aspects of offshore exploration and production, especially in tension leg platforms. He coordinated the preliminary design of the world's first such facility, which was installed in the Hutton Field in the U.K. North Sea in 1984. A SNAME Fellow, Dr. Mercier's. pioneering work was recognized by his being awarded the Blakely Smith Medal in 1985.

Offshore Industry 1983–1987: Consolidation and Increased Cost Consciousness

Roger Mowell

The frenzied rig building that began in 1980 and 1981 continued; and during 1982 the competitive mobile offshore drilling unit (MODU) fleet was increased by 23 percent. There were 108 rigs delivered that year, bringing the total competitive fleet to 574 rigs. The additional rigs were fairly well employed until mid-1982, with about 96 percent of them in use. Day rates were off their record highs, but were still respectable.

With continued ordering of new rigs, it became obvious that an over-supply was developing. The main cause was the natural tendency of the typical drilling contractors to over-build, each believing that he alone must meet the anticipated demand. Another reason was that oil consumption flattened and, as a response, so did the price of crude oil. This resulted in reduced exploration budgets by the oil companies. The third factor in the rig oversupply was the U.S. tax law that provided tax shelters to limited partnerships. There was a joke going around the industry that every time a dentist, a doctor, and a tool pusher got together, a new drilling company was formed.

During 1982 there were 184 new rigs on order, including 104 jackups and 51 semisubmersibles. Fortunately, not all of these were built; some were options that were not exercised, and others were canceled. The competitive MODU fleet peaked at 603 in 1983–1984 and then was reduced by losses and retirements to 570 in 1987.

The demand for rigs stayed level at about 470 during 1982, dipped to 428 in mid-1983, and then recovered to 470 by year's end. During 1984 and early 1985 there was a steady climb to about 530 rigs, and then a decline during the remainder of 1985 to 465 rigs. In early 1986 there was a dramatic drop in oil prices owing to unlimited production by OPEC. The number of MODUs under contract dropped to 275. During 1987 the demand rebounded somewhat to 370 rigs. With rig use less than about 90 percent on a worldwide basis, the day rates dropped to levels where drilling contractors operated at a loss. This led to staff reductions, consolidation of some drilling contracting firms, retirement of older rigs, and shipyard reductions and closures. During this period, offshore service

Fig. 1 Semisubmersible *Henry Goodrich.*

Fig. 2 Semisubmersible *Sovereign Explorer*, a GVA 4500 class.

Fig. 4 Marathon Gorilla class—walk-in at Vicksburg, Mississippi.

Fig. 3 Marathon Gorilla class *Rowan Gorilla I* under tow.

Fig. 5 The *Rowan Gorilla III*, a massive offshore jackup drilling rig, rises 437 feet in the air after its three steel legs had been extended an additional 100 feet. With its 607-foot tall legs, the *Rowan Gorilla III* is capable of drilling in water 450-feet deep in the Gulf of Mexico. The *Rowan Gorilla III* is one of a new generation of large, hostile environment jackup rigs designed and built by Marathon LeTourneau to work in remote, deepwater locations with demanding operating environments.

Fig. 6 Fulmar FPSO.

Fig. 7 Balmoral FPS.

Fig. 8 Early bottom-founded oil storage tank for Dubai Petroleum Company being submerged for installation in the Arabian Gulf.

Fig. 9 Floating oil storage unit (FSU) *Cinta Natomas* underway to Indonesian waters for installation.

and supply personnel were heard saying, "Dear Lord, please send us another boom and we promise we won't screw it up!"

Despite the poor business climate during this period, there were some new, innovative designs developed or built, or both. These included such examples as:

- Large, deepwater jackups
- Trendsetter semisubmersible with dramatically reduced heave motion response
- Large semisubmersible derrick barges with sophisticated ballast systems to compensate for slewing hook loads.

The large, deepwater jackups are primarily an extrapolation of existing technology, i.e., they look a lot like earlier cantilever jackups but are just bigger. The distance between legs must be increased to cope with the increased overturning moments owing to environmental loads, and the hull must be made deeper and stronger to avoid excessive sagging between the widely spaced legs. The Marathon LeTourneau 116-C jackup is rated for a nominal 300-ft water depth, and their Gorilla design for a nominal 400-ft water depth.

The Trendsetter semisubmersible design was a joint effort among Friede & Goldman, SeaTek, and Reading & Bates Drilling Co. It set new standards for reduced heave response and self-contained deepwater

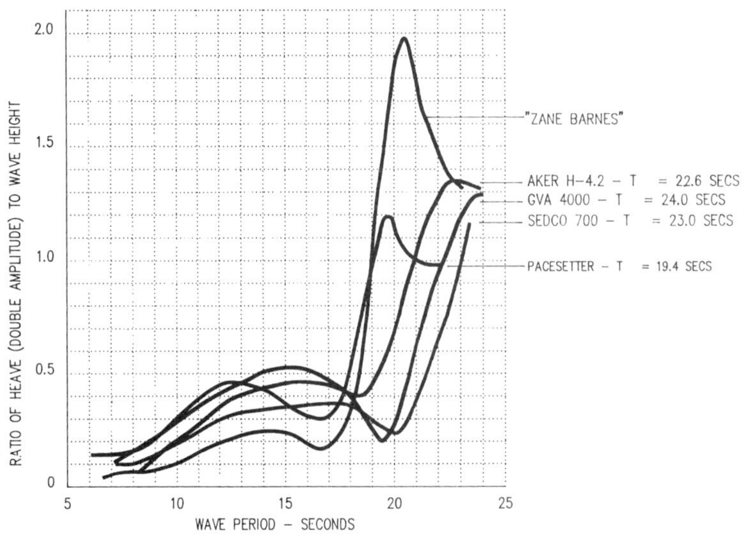

Fig. 10 Heave response curves.

mooring. The design relies on "dog-bone" shaped pontoons and a center caisson to reduce the hydrodynamic heave forcing function. Comparative heave motion response curves for several designs are shown in Fig. 10. The Trendsetter also has gradual changes in waterplane area as it is ballasted down, which results in high variable deck load capacity in all operating modes, including the transition between transit and operating drafts.

Semisubmersible derrick barges of staggering size and capability have been designed and built. One such unit is shown in one of the illustrations. A unique feature of these units is a computer-controlled, high-volume ballast control system that can automatically shift ballast to counterbalance large loads being slewed by the crane. The latest of these rigs has two revolving cranes and a total nonslewing lifting capacity is 13,200 tonnes.

Roger Mowell graduated from the U.S. Coast Guard Academy in 1953. He spent seventeen years working for Exxon, primarily in marine terminal design and floating drilling research. For the past eighteen years he has been in charge of the Engineering Department for the Reading & Bates Drilling Co. He is a member of SNAME, and serves on the Mobil Offshore Drilling Unit advisory groups for the American Bureau of Shipping and Det norske Veritas.

Offshore Industry 1988–1993: Regrowth and Internationalization

NORMAN D. BIRRELL

Dedicated to the honor of Alan McClure, an outstanding engineer.

THE MODEST RECOVERY of oil prices during 1987 was a welcome relief to the industry, which had been hit hard by the precipitous drop from $30 oil to $10 oil over the period from 1983 to 1986. The industry's move to the Arctic, and its deepwater prospects were both slowed somewhat owing to an understandably more cautious attitude about the large front end investments necessary to compete in those areas. Nonetheless, some milestones were noted during the early recovery period. Among these, some were of marked significance to the offshore industry in general and to the profession of naval architecture and marine engineering in particular. Specifically:

1987 Production begins at Endicott, Alaska's first offshore oil field.

1988 Subsea completion installed in 1613 ft of water for Petrobras offshore Brazil.

Shell Oil Co. sets water depth drilling record of 7500 ft in Gulf of Mexico.

Fire destroys Occidental's Piper A platform in the UK North Sea, with loss of 167 lives.

Shell's 1615 ft Bullwinkle platform installed in 1353 ft of water in Green Canyon Block 65.

1989 Placid's Floating Production installed in 1400 ft of water.

Conoco's tension leg well platform (TLWP) is installed in 1760 ft of water at Green Canyon Block 184 in the Gulf of Mexico.

Esso Valdez grounds and pollutes in Alaskan waters

1990 Iraq invasion of Kuwait impacts oil market, raising the price to $36 per bbl.

1991 Desert Storm war ends, bringing downside instability to oil prices.

Heidrun, first concrete tension leg platform, project in 1100-ft water depth approved by Norwegian government.

1992 Offshore industry budgets emphasize international opportunities over U.S. domestic prospects, and oil imports exceed the 50 percent threshold.

Russian and Eastern block countries offer investment prospects to the offshore industry.

Schtockmanskoe tension leg platform (TLP) conceptual design proposed for the world's largest offshore gas field in 1100 ft of water off Russia and north of the Arctic circle.

1993 Auger TLP deck mated to hull and installation in 2700 ft in Gulf of Mexico completed.

Heidrun concrete tension leg platform hull construction begins.

Studies made around 1980 indicated that about 160 billion barrels of oil had been proven at that time. Observers expected, however, that the ultimate offshore recovery would be on the order of 800 billion barrels, of which 80 to 100 billion would be from the deep water slopes. The total of the world's deep water slope area (i.e., water depths between 600 and 10,000 ft) is over 1.5 times the total of the shelf areas. Inasmuch as shallow water oil production during the 1980s accounted for at least a quarter of the world's discovered oil, and could ultimately grow to account for forty percent of future production, the undiscovered reserve potential of the deep water slopes was truly significant.

Clearly the industry was moving into water depths that were beyond the economic limit of the conventional fixed platforms, which were considered the workhorse of the industry for the past fifty years. The development of the compliant platform offered a means of provided the "real estate" needed for economic operations in the deeper offshore fields. The Argyle floating production system, the Lena guyed tower, and the Hutton tension leg platform are prime examples of compliant systems that were prototypes for techniques that evolved during the 1980s and 1990s.

Deepwater petroleum technology has both an exciting history and a challenging future. Just as necessity is the mother of invention, when explorers find a large accumulation of hydrocarbons in deep water, engineers have eventually found an economic means to develop them. Historically, the time lag has been

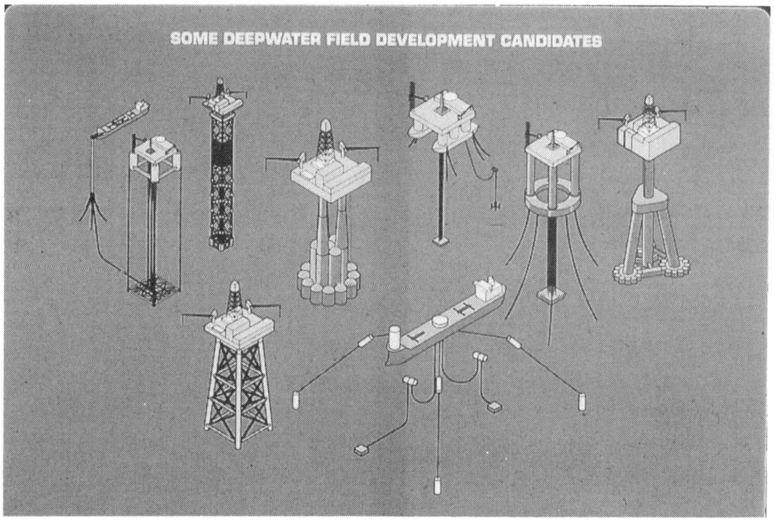

Fig. 1 Some deepwater field development candidates.

about ten to fifteen years for development technology to catch up with exploration technology. This time period factors significantly into the calculation of life cycle costs since it delays the start of income. It has, therefore, challenged the industry to rethink traditional approaches to offshore exploration and development. Thus innovation has become a vital ingredient, since technical concepts and strategic decisions that made sense in shallow and moderate water depths, less than say 600 ft, are inappropriate for extrapolation into deeper water. As illustrated in one of the figures, a wide variety of concepts were being evaluated and tested.

It became apparent during this period that it was no longer possible to rely on larger discoveries, higher oil prices, and lower taxes to offset the increase in drilling and investment costs that accompanied the extension of conventional platform technology into deeper water. One area in particular that continues to receive continual critical review is the topside process facility. The enormous cost of the last few percentage points of recovery efficiency does not always make economic sense. It is important to challenge the logic of multitrain compression, multistage separation, sophisticated gas processing, and a myriad of other weight and space consuming operations on offshore platforms in deep water. Some production platforms designed and installed in the North Sea during the 1970s and early 1980s were virtual offshore refineries. The rise in oil prices at the time did nothing to discourage over-sophistication in design and project execution. The leaner design concepts for the production facilities used during the 1960s needed to be re-introduced.

It was generally agreed that the key to future deep water development was to find ways to make drilling, production, and transport systems simpler, smaller, lighter in weight, and less expensive. Moreover, the industry had to accomplish all this without sacrificing safety of life or environmental protection. Leading candidates that seemed to meet these criteria were floating structures for platforms, multiphase pumps capable of moving process facilities ashore, and a combination of extended reach drilling and horizontal completions that could provide an alternative to the high cost subsea wells being considered for draining the remote corners of the reservoirs.

Deepwater Environment

The engineers working in the offshore industry at this time had to expand their understanding of the physical and geotechnical nature of the deepwater environment. This was necessary in order to determine the long term effects of the environment on the production system. Likewise, the production system's potential impact on the environment had to be better understood as well.

As an example, natural gas hydrates can occur in deep water when gas is present at or near saturation in the interstitial water of near bottom sediments. A gas hydrate is a metastable ice-like material that contains gas molecules in a lattice of water molecules referred to as a *clathrate*. The combinations of pressure and temperature that favor the formation of hydrates are such that affected soil thickness can range from less than an inch under the most shallow waters at which they are stable, to over 3000 ft in water depths of 16,500 ft with near 32 degree bottom water temperature. The effect of the hydrate layer on the performance of a production system is similar to that of permafrost on arctic operations. Specifically, the heat flow from the casing of the producing well can cause the hydrates to melt, which in turn can cause failure of the production system pile foundation. Thus in deep water operations it is important to factor this phenomenon into the design of mooring and foundation systems.

Another geotechnical concern related to deep water is slope stability. While average deep water slopes are one degree or less, some slopes may be as great as 12 to 14 degrees in highly faulted continental slopes such as in the Gulf of Mexico. In many cases such areas are characterized by landslides and creep landforms. Thus, slope stability analysis is a major factor in the design of deep water systems.

The metocean (combined meteorological and oceanographic) criteria for application to design, installation, and operation of offshore facilities presents some special problems in water depths beyond the edge of the continental shelf. In the shallow continental shelf regions, design loads are typically dominated by wave induced velocities. Current induced velocities are relatively small and dominated by a few relatively well understood processes (e.g., winds and astronomical tides). In deep water, currents can have a dominating influence. They are often more energetic and but poorly understood. Velocities over a major portion of the water column in deep water have been reported in excess of two knots for processes like eddy currents that are not directly related to wind or astronomical conditions. Because of the uncertainty and site specific nature of the circulation in deep water, current and water property measurements have become a critical part of the early stages in the engineering process.

Deepwater Platforms

The production platform is always a key component with which other functions are enmeshed to form a suitable production system. With the industry's move into deeper waters, new platform concepts, and new methods of fabrication and installation were evolving, as were new techniques for maintenance and inspection.

In the last twenty years tension leg platforms (TLPs) have developed into a widely accepted solution for deep water development. The process has been similar to most innovations, in that the confidence factor takes time to develop. With the increased understanding of TLP technology throughout industry, acceptance is accelerating.

Historically, the growth of tension leg platform technology occurred in competition with sea floor production systems that attracted much of the industry's attention during the early 1970s. The TLP concept remained an appealing alternative, however, since it retained the traditional fixed platform's proven operational characteristics. It also offered cost advantages over competing systems.

In the early 1970s the industry was evaluating sea floor production systems for producing oil and gas in deep water. The technical data gathered through numerous joint industry projects on the subject was enough to convince many oil companies that the sea floor system was not the total answer. By itself, such a system was not complete and needed a surface platform to process and handle well production.

During the late 1980s the industry began to appreciate the flexibility of the TLP concept. It was applicable over a wide range of water depths, and over a wide range of sizes. The consensus was that the generic TLP design was readily adaptable to most site-specific development requirements anticipated within the foreseeable future. Additionally, the technology gap between exploratory capability and production capability had been narrowed significantly by application of TLPs.

It became apparent that most deep water exploration activities were within the reach of TLP application. Nonetheless, it was recognized that site-specific reserve requirements play the major role in system selection, and that alternative technologies merited evaluation.

Drilling technology development programs concentrating on the adaptation of extended reach and horizontal drilling techniques enhanced the scope of deep water TLP applications.

TLP Evolution

The first joint industry project on model testing a TLP was initiated in 1973. A one-third scale model of a triangular configuration was successfully tested in sheltered waters off the coast of California during the following two years. The test was important because it helped the industry assess the validity of the analytical methods used to design the model. The tests showed that the analytical methods were overestimating forces and motions under heavy seas. It

Fig. 2 TLP *Joliet* being transported from Singapore for installation in the Gulf of Mexico.

also proved that oil field separators can function effectively on a TLP deck subjected to test conditions.

During the 1980s and early 1990s a number of TLPs were developed. Some of them are described below.

Hutton. The Hutton TLP (Fig. 24 of Filson) is probably one of the most talked about structures in existence. Installed during 1984 in 485 ft of water in the hostile North Sea environment, this steel platform displaces about 64,000 tons and has a rectangular configuration with a six-column hull. The hull and deck were fabricated separately, then mated in shallow water, hooked up and towed to location for installing over a partially predrilled template. The TLP was moored by sixteen high strength, forged, thick walled tubular tendons connected to the four foundation templates installed on the sea bed.

Once the tendons were properly tensioned, the platform was ready to accept the tie-back risers connecting the ten predrilled wells to the tree deck. Predrilling the wells allowed first production within 22 days. The platform rig continued to tie back and complete wells, achieving planned peak production in less than three months after the platform was installed.

With many years of successful performance history and normal up-time of about 95 percent (about the same as that of a fixed structure such as Murchison), the Hutton TLP must be considered a success. The project has resulted in the development of some unique equipment; it has also provided confidence and new ideas to extend the concept to even deeper waters. Some specific accomplishments of the Hutton TLP project are as follows:

- Proved that a TLP with restrained motions is just as safe and effective a drilling/production platform as a bottom founded structure.
- Demonstrated successful application of vertical tension legs as a novel mooring system for a floating platform.
- Represented the first offshore use of structural steel castings to improve fatigue life and simplify fabrication of complex column–pontoon nodes in the platform.
- Established novel corrosion protection methods such as flame sprayed aluminum coating for tendons and a high pH potassium carbonate solution to protect the cross load bearings.
- Verified TLP design predictions and allowed refinement of design methods.

Jolliet. The Jolliet tension leg well platform (TLWP) was installed during 1989 in 1760 ft of water in Green Canyon Block 184 in the Gulf of Mexico. This set a record as the world's deepest water production platform. The field lies on the steep continental slope. Water depth varies from 900 ft to 2100 ft within about four miles in the Block. The Gulf's oil reserves are characterized by multiple unconsolidated sands and low well productivity. Moreover, wells in the area have a history of workover (i.e., downhole maintenance) needs. Thus topside completions that allow easy access to the wellhead are considered necessary. After all the alternative solutions were considered, a tension leg well platform was chosen as the most economic system for exploiting these particular deposits. By way of explanation, a TLWP differs from a TLP in that the former has only minimal topside facilities, and serves mainly to support the production risers to the well.

A major feature of a TLWP is its low cost sensitivity to differing water depths. This eliminates the need to re-engineer the basic TLWP design for working in some slightly different water depth. The Jolliet field provided the opportunity to check this premise. As the discovery was being delineated, the center of the reservoir kept moving to deeper water, from initially 1400 ft to 1760 ft. Thus the best location for the well kept moving to deeper water. The use of a TLP allowed this in the midst of the structural engineering studies without affecting its fabrication and installation schedule. In the case of a bottom founded structure, a similar situation would have greatly delayed the project.

The Jolliet platform departs from the full drilling/production approach used with Hutton and serves primarily as a well platform with a completion rig, first stage separator, gas compressors, and accommodations. The lighter, simpler topside facilities on a smaller TLP were made possible by installing a central production platform ten miles away in 615 ft of water. Here the wet oil and gas streams from the Jolliet Field, as well as others, would be further processed before transporting to shore.

The platform has a 140 × 180-ft deck supported by four columns spaced 140 ft. It was built in Singapore and dry-towed (i.e., carried on the deck of a purpose built heavy lift vessel) to the Gulf of Mexico. It was initially moored with eight welded single-piece tendons, two externally attached at each corner. The use of only eight tendons was forced by the unfortunate loss of four while towing. Replacement tendons were fabricated and installed before putting the platform on production.

The Jolliet platform represented considerable progress toward improving economic applicability of a TLP to deeper waters. The Jolliet TLWP hull, with its simpler design and less compartmentation, has a weight density of only ten pounds per cubic ft. This contrasts with the Hutton hull's fifteen pounds per cubic ft. The simpler platform also cost much less.

Heidrun. A four-column concrete hull TLP has been proposed for the Heidrun Field in the Norwegian sector of the North Sea. The concrete TLP could be fitted with either an integrated or a modular deck because of its ability to carry larger loads. The thick walled columns will be connected at the base by a square pontoon frame. The hull will be moored by twelve tendons externally attached to the corner columns at the pontoon level. The tethers, consisting of large diameter thin walled tubes, will be anchored to a concrete gravity foundation.

The external configuration of the concrete TLP is similar to a steel TLP, but has a significantly deeper draft consistent with its larger displacement.

Concrete as a material of construction for the Heidrun TLP offers a number of advantages over steel. Reinforced concrete shells can be designed to resist hydrostatic pressures efficiently. A concrete hull is thus much simpler, with fewer structural elements and smooth surfaces without the local stiffening found in steel hulls. With ample reserve strength against external pressures, any increase in loading of a concrete TLP simply results in an increase in draft without significant increase in tendon tension. Concrete is essentially immune to the marine environment, with long fatigue life and little need for maintenance.

The combination of dry dock construction and slipforming the concrete columns in protected inshore fjords cuts building time in half.

All these positive features of the Heidrun concrete TLP make it an attractive means of developing deep water fields.

Polar Star. The Schtockmanskoe gas field north of Murmansk may be developed with a new generation concrete TLP. The field is above the Arctic Circle in about 1100 ft of water, where occasional ice floes may be encountered. This has led to a proposed floating monocolumn configuration (Fig. 3), which has a downward looking conical shape at the waterline to enable efficient ice breaking. This concept is firmly

Fig. 3 Mono-column TLP for Arctic.

based on experience with both TLPs and ice-resisting structures.

This evolution of the TLP concept retains for the offshore project the good characteristics that have been demonstrated for the more conventional TLP (adapted from the multiple-column stabilized semi-submersible drill rig). It also affords certain prospective benefits in addition to good ice-going features. Fabricating and outfitting can be conveniently completed in construction yards and at a near-shore location before floating to the Schtockmanskoe Field when it is fully ready to produce and process gas. In fact, the monocolumn configuration is expected to be significantly simpler and quicker to build than the multi-column floating structure.

Well templates, foundations, pipelines, and other seabed equipment can be installed separately, off the critical path of the project schedule. (These activities can be done in relatively brief periods and are not sensitive to weather.) Enough wells can be drilled through the sea bed well template by a mobile drilling vessel before the platform is installed. This means that a high rate of gas production can begin soon after the TLP is moored. Access to the wells for maintenance is practically the same as for the large gravity based structures used in the Norwegian sector of the North Sea.

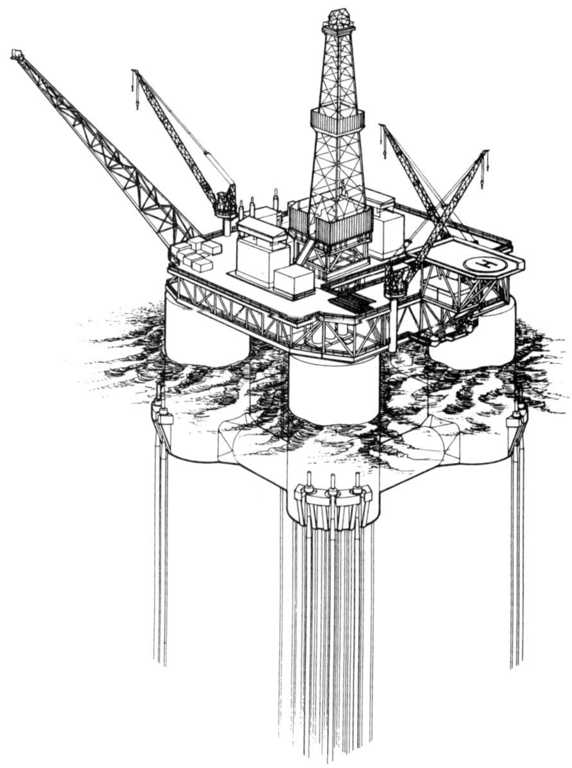

Fig. 4 Concrete TLP concept for Gulf of Mexico (3000 ft depth).

Concept Refinements

The tension leg platform has now become one of the primary options for developing fields in deep water and moderate environments. The great deal of research and development work that is currently underway demonstrates the industry's interest in the concept and leaves little doubt that future TLPs can be successfully installed in water depths of 4000 feet and beyond.

Validation of the finite element model of the TLP system with measured field data has given the naval architect the analytical capability to accurately simulate real time operations. This modelling capability serves as a valuable "laboratory" for testing design refinements and operational conditions in a reliable and efficient manner.

Hull Design, Hull refinements begin with selection of material of construction. Steel hulls use orthogonally stiffened plates and shells supported by steel bulkheads and flats to resist hydrostatic pressures. The design and fabrication of steel hulls thus becomes rather intricate, especially where columns and pontoons intersect. Concrete, as discussed earlier, results in a much simpler design. Owing to the inherently large displacement and deep draft relative to steel, concrete hulls reduce payload sensitivity. Modular decks with a high center of gravity and higher weight than comparable integral decks can be easily accommodated on concrete TLPs. Another advantage of concrete is in its less strict fabrication tolerances. This lowers the fabrication costs as well as simplifying inspection of a platform during its construction and operation.

Design procedures need to be evaluated carefully before starting a design. Development of appropriate strength and buckling formulations in a manner consistent with a reliability based design format is necessary. Using more rational structural design formulations together with improved design procedures, the weight density of steel used in a steel TLP can be reduced significantly.

Mooring Systems. The development of simplified mooring tendons and installation techniques has greatly reduced the cost of these systems. In situ inspection using non destructive techniques is fully developed, as are corrosion protection systems.

Mooring tendons limit the horizontal displacements of the TLP through pretension, and suppress wave induced heave and pitch motions of the buoyant

Fig. 5 Crane ship *Microperi M7000.*

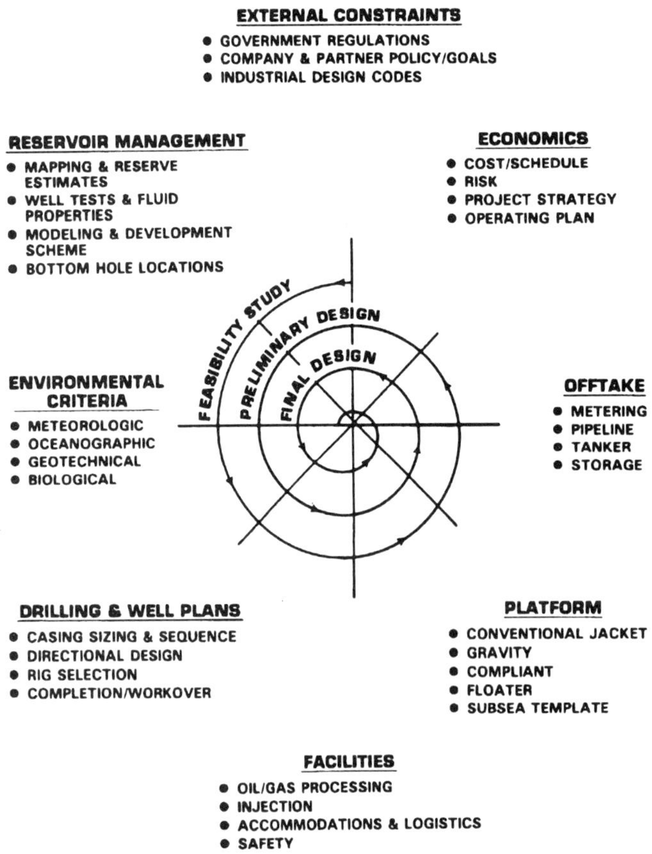

Fig. 6 Offshore development design spiral—heirarchy process (AHP) developed by Prof. Tom Saaty, University of Pittsburgh (Saaty 1990).

hull. The weight of tendons could represent a significant portion of the top tension needed, and increasing weight of tendons in deeper waters, if not reduced by design, would lead to unnecessarily larger TLP structures. Accordingly, various types of tendons are being investigated with accompanying studies of their installation.

The small-bore, forged, thick walled tendon selected for the Hutton TLP was extremely conservative in design. The Jolliet platform uses large diameter thin-walled tubulars with sufficient self-buoyancy to minimize submerged weight and allow easier towing of the complete full length tendon. Normal strength steels could be used with these types of tendons, making them much less expensive than the forged type used for the Hutton TLP.

A single piece welded tendon that is assembled on shore, using pipeline welding and towing techniques, and installed as a unit, offers advantages in ultra-deep waters. These units are simpler, less expensive, and more quickly built and installed than multipiece tendons. Where tendons cannot be installed as a single piece, simple connectors need to be developed. A rapid snap-fit connector would greatly reduce installation time and thus cost, compared to a threaded connector.

TLP tendons today have reliable in-service lives, with full inspection capabilities of fifty years or more. The full cycle economic comparison between tubular TLP tendons and the semisubmersible wire/chain combination favors the former.

Sea Floor Foundations. The Hutton TLP installation involved positioning four foundation templates with accuracy relative to each other and also with respect to the well template. High accuracy acoustic techniques were used. This was time consuming and expensive. In contrast, the Jolliet foundation template was in one piece and was set around the well template, piled, and grouted. The operation was smoother and simpler. Soft bottom conditions and a significant slope of 100 meters per kilometer were experienced. With good engineering and installation techniques all problems were overcome, and confidence in dealing with such concerns has improved greatly. The techniques successfully developed for Jolliet now make elaborate leveling and jacking systems too redundant.

Fabrication and installation of the Jolliet TLWP foundation template, measuring over 60 meters square and weighing 1500 tons, took place while the hull was being fabricated. The crane ship *Microperi M7000* (Fig. 5) was used to set the foundation and drive the piles. An installation procedure using a drilling semisubmersible is also considered feasible and could be the preferred solution in the future.

The Heidrun TLP foundation template may not need piling. It will derive its stability from its own weight and from its skirts penetrating the softer sediment on the sea bed. Other competing foundation concepts may become more appropriate and economical depending upon the soil and bottom topography.

System Selection

The task of synthesizing an offshore development system for deep water is far more complex than for shallow water. For example, when considering the platform, the designer is no longer dealing with the straightforward comparison of a six-pile and an eight-pile jacket, but with structures that are completely different, such as a concrete gravity structure versus a steel floater. The deep water platform contributes a larger portion of the total project cost, and the selection must be made in certain cases with only limited experience. An understanding of the entire design sequence is essential to anyone seeking a rational solution.

The design spiral concept represented in Fig. 6, adapted from the shipbuilding industry, can be used

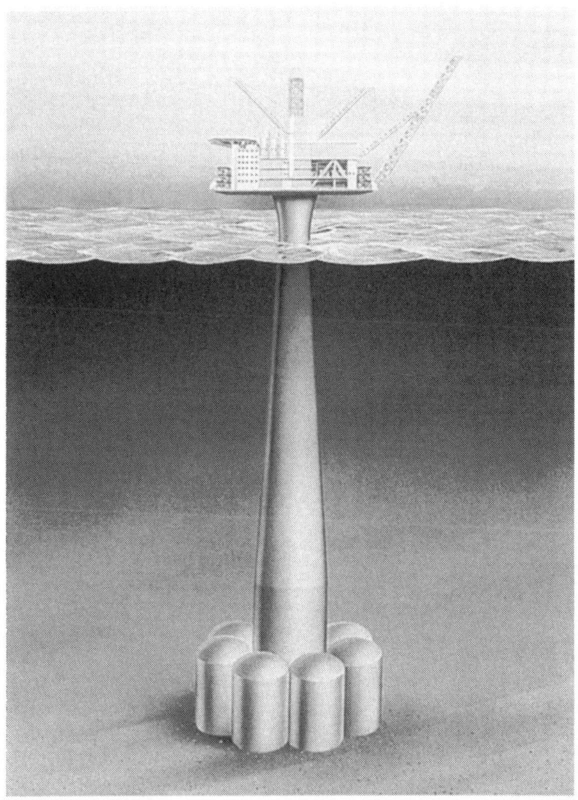

Fig. 7 Artist's impression of Draugen gravity-based structure, GBS, now installed in North Sea in 820 ft of water.

to describe the iterative process required to synthesize an offshore development system. The education of future engineers needs to focus on the interdisciplinary design process, and to recognize the interface interaction required to develop a safe and efficient design.

N. D. (Scotty) Birrell's formal education has won him degrees from the N.Y. Maritime College, Carnegie Tech, and the Polytechnic Institute of Brooklyn. He is, in addition, a graduate of the Management Program for Executives at the University of Pittsburgh. He joined Conoco in 1972, gaining experience in the research, engineering, and construction branches of the offshore marine industry, both domestic and overseas. The author of numerous technical papers, Mr Birrell has been particularly active in professional engineering associations and national committees dealing with advanced technology in the offshore industry. He is a member of SNAME and current chairman of the API Committee on Standardization of Offshore Structures, while also heading a similar international committee. Among his other academic associations, he is an Adjunct Professor at the University of Houston.